KV-025-751

Soil Conductivity Sensing on Claypan Soils: Comparison of Electromagnetic Induction and Direct Methods

K. A. Sudduth
N. R. Kitchen
Cropping Systems and Water Quality Research Unit
USDA– ARS
Columbia, Missouri

S. T. Drummond
Department of Biological and Agricultural Engineering
University of Missouri
Columbia, Missouri

ABSTRACT

Soil electrical conductivity (EC) sensing provides a means for rapidly mapping variations in soil properties such as salinity, moisture, and clay content. On claypan soils, EC measurements are related to topsoil depth above the claypan horizon, an important factor in spatial crop productivity differences on these soils. Soil EC data obtained with a noncontact sensor based on electromagnetic induction principles were compared with data from a direct contact, coulter-based sensor. Differences in EC readings were attributed to differences in sensing depth between the sensors and operating modes. The electromagnetic induction sensor generally provided better estimates of topsoil depth and correlations to crop productivity over the full range of topsoil depths encountered (up to 150 cm). The coulter-based sensor, however, performed well at shallower topsoil depths (up to 90 cm), and also would be useful for investigating soil differences in precision agriculture practice.

INTRODUCTION

Soil electrical conductivity is a function of soil salinity, clay content, and water content (Rhoades et al., 1989). Therefore, soil conductivity measurements have the potential for providing estimates of within-field variations in these properties. Care must be taken, however, to understand the effects of the other, nonestimated properties on the conductivity measurement. In areas with low soil salinity, spatial variation in soil moisture content is often a major factor determining variations in bulk soil EC. For example, Sheets and Hendrickx (1995) measured EC along a 1950 m transect in New Mexico during a 16-mo period, and found a linear relationship between conductivity and profile soil water content. Independent measurements of soil water at several calibration points along the transect were required for each measurement date.

Copyright © 1999 ASA-CSSA-SSSA, 677 South Segoe Road, Madison, WI 53711, USA. *Proceedings of the Fourth International Conference on Precision Agriculture.*

Soil EC measurements can provide information on soil texture, in addition to estimating soil water content. Williams and Hoey (1987) used EM measurements of EC to estimate within-field variations in soil clay content. We (Doolittle et al., 1994) found that EM measurements were highly correlated with the topsoil depth above a subsurface claypan horizon. We then used an automated EM sensing system to map topsoil depth over a number of fields. It was necessary to obtain calibration measurements with a soil probe at a number of locations within a field to remove the effects of temporal variations in soil water content and temperature.

Since soil EC integrates texture and moisture availability, two characteristics that both vary over the landscape and also affect productivity, EC sensing also shows promise in interpreting grain yield variations, at least in certain soils (e.g., Sudduth et al., 1995; Jaynes et al., 1995).

OBJECTIVE

The objective of this project was to compare soil EC measurements from a non-contact, electromagnetic induction-based sensor (Geonics EM38) to those obtained from a coulter-based sensor (Veris 3100). Additionally, the ability of each sensor to estimate topsoil depth above a claypan horizon and to estimate crop productivity was investigated.

EQUIPMENT AND PROCEDURES

Electromagnetic Induction Sensor

The electromagnetic induction (EM) instrument used in this study was the EM38 manufactured by Geonics Limited, Mississauga, Ontario, Canada. The EM38 is a lightweight bar approximately 1 m long, and includes calibration controls and a digital readout of apparent EC in milliSiemens (mS) per meter. An analog output port is provided to allow data to be recorded on a data-logger or computer. In this study, the instrument was operated in the vertical dipole mode (upright orientation) to provide an effective measurement depth of approximately 1.5 m. It was later operated over part of the study area in the horizontal dipole mode, with an effective measurement depth of approximately 0.75 m (McNeill, 1992).

The instrument response to soil conductivity varies as a nonlinear function of depth (Fig. 1). Sensitivity in the vertical mode is highest at about 0.4 m from the instrument, while sensitivity in the horizontal mode is highest at the instrument. The apparent conductivity measured by the instrument is determined by the soil conductivity with depth, as weighted by the instrument response functions (McNeill, 1992). Procedures have been developed to infer the soil conductivity profile with depth by means of multiple readings obtained with the instrument held at varying heights above the ground (e.g., Rhoades & Corwin, 1991). However, since our previous work (Doolittle et al., 1994; Sudduth et al., 1995) showed that depth of topsoil over a claypan could be successfully estimated from single-height EM sensor readings, we applied that same approach here. Eliminating the need to take measurements at multiple heights significantly quickened the data collection process,

ENGINEERING TECHNOLOGY

C631/lnt

AC 0011332 8

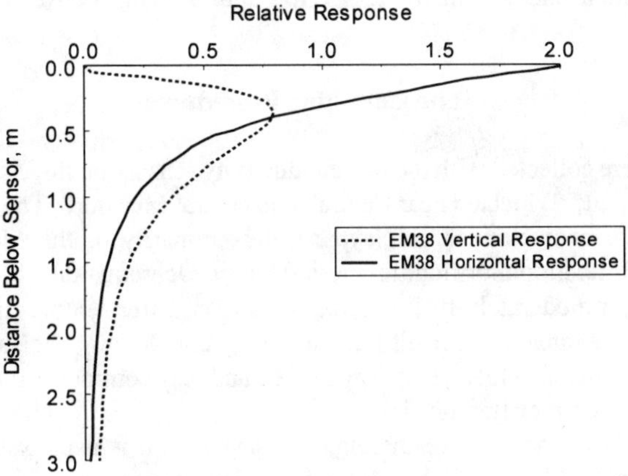

Fig. 1. Relative response of EM38 sensor vs. depth (adapted from McNeill, 1992).

since it was then possible to collect EM38 data on-the-go.

For field-scale mapping, a mobile EM measurement system was used. The EM38 was mounted to a 3 m long cart consisting of a wooden beam supported at the rear by two spoke-wheeled pneumatic tires. The wooden beam was necessary because the EM38 would respond strongly to metallic objects within approximately 1 m. The tongue of this cart was attached to the rear of a second, similar cart, which was in turn attached to the rear hitch of a four-wheel all-terrain vehicle (ATV). The second cart was necessary to increase the distance between the EM38 and ATV, for eliminating the effects of ATV engine noise on the performance of the sensor. With this configuration, the EM38 was suspended approximately 24 cm (horizontal mode) to 28 cm (vertical mode) above the ground surface during data collection.

EM conductivity data were read into a laptop computer mounted in front of the ATV operator through an analog-to-digital converter module. Data obtained from a differential GPS (DGPS) receiver were integrated with the EM38 data to provide the coordinates of each measurement point with an accuracy in the range of 1-2 m.

Colter-Based Conductivity Sensor

The Veris Model 3100 sensor cart (Lund & Christy, 1998) identifies soil variability by directly sensing soil electrical conductivity. As the cart is pulled through the field, a pair of coulter electrodes transmit an electrical current into the soil, while two other pairs of coulter electrodes measure the voltage drop. The measurement electrodes are configured to measure apparent conductivity over an approximate 0- to 30-cm depth (shallow reading) as well as a 0- to 90-cm depth (deep reading). The system georeferences the conductivity measurements using an external DGPS receiver, and stores the resulting data in digital form. The Veris sensor cart records data on a 1 s interval, and data density can be modified by the operator through changes in travel speed and/or spacing between measurement transects.

Data Collection Procedures

Data were collected with the two conductivity sensors on three fields (denoted as Fields 1, 2, and 3) located near Centralia, in central Missouri. The soils found at these sites are characterized as claypan soils, primarily of the Mexico-Putnam association (fine, montmorillonitic, mesic Udollic Ochraqualfs). Mexico-Putnam soils formed in moderately-fine textured loess over a fine textured pedisediment. Surface textures range from a silt loam to a silty clay loam. The subsoil claypan horizon(s) are silty clay loam, silty clay or clay, and may commonly contain as much as 50 to 60% montmorillonitic clay.

Because of extensive weathering, the claypan soil is usually low in natural fertility and pH. Plant available water from the claypan is low because a large portion of the stored water is retained with the clay at the wilting point. With these characteristics, variations in the depth of topsoil above the claypan can lead to significant variations in crop productivity. For each of these three fields, topsoil depth ranged from less than 10 cm to greater than 110 cm.

Data were collected on transects spaced approximately 10 m apart on a 1-s interval, corresponding to a measurement every 3 to 5 m along the transects. This procedure resulted in a data density of 200 to 300 points per ha. Calibration points were established to develop field-specific relationships between EC data and soil properties. Between 15 and 22 locations, spanning the range of landscape positions and topsoil depths present in each field, were selected for soil characterization. Topsoil depth (depth to the Bt horizon) was measured in the field by an experienced soil scientist, through a combination of visual and tactile observations.

RESULTS AND DISCUSSION

Soil conductivity data collected in each of the two operating modes with each of the two sensors exhibited similar trends at a field scale (Fig. 2). However, conductivity readings obtained with each sensor (in mS m^{-1}) were considerably different in magnitude. In general, the Veris 3100 exhibited lower readings than did the EM38, and also tended to span a slightly wider range. In this application, where calibration to the physical parameter of topsoil depth was performed, the difference in the magnitude of the EC reading was relatively unimportant. Understanding the reasons behind the discrepancy, however, would be important in situations where EC might not be as readily related to soil physical factors.

Linear correlations obtained between the four EC methods for each of the three fields also showed good agreement (Table 1). Highest correlations were generally found between the EM38 horizontal and vertical readings (r = 0.84 to r = 0.92). High correlations were also found between the Veris deep reading and the EM38 horizontal and vertical readings (r = 0.81 to r = 0.90). Correlations of Veris shallow to Veris deep readings were relatively constant over the three fields (r = 0.79 to r = 81). Correlations of Veris shallow to EM38 readings were somewhat lower (r = 0.65 to r = 0.75).

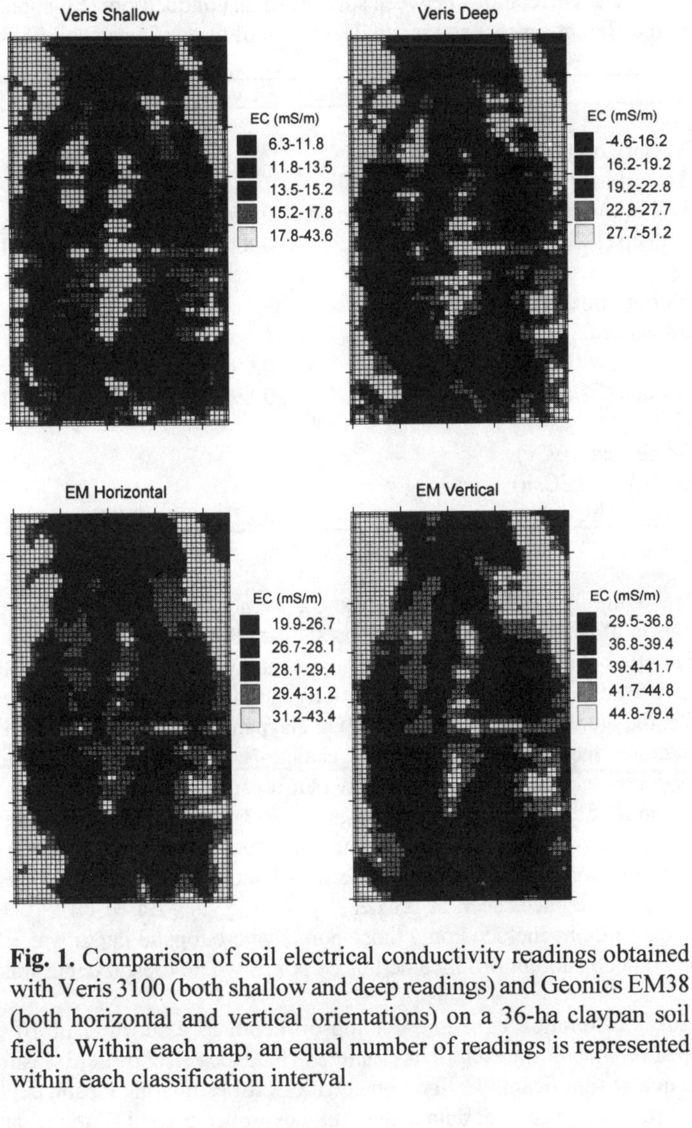

Fig. 1. Comparison of soil electrical conductivity readings obtained with Veris 3100 (both shallow and deep readings) and Geonics EM38 (both horizontal and vertical orientations) on a 36-ha claypan soil field. Within each map, an equal number of readings is represented within each classification interval.

Topsoil Depth Estimation

Topsoil depth data obtained at the calibration points in each field were used to develop linear regression equations for estimation of topsoil depth from the inverse of electrical conductivity (1/EC). This inverse transformation worked well to linearize the relationship between EM-measured EC and topsoil depth (Fig. 3), agreeing with our previous experiences using the EM38 on claypan soils.

However, the relationship between the Veris EC readings and topsoil depth still

Table 1. Linear correlations between soil electrical conductivity (EC) obtained with different sensors and methods on each of three claypan soil fields.

	ECh	ECv	ECsh	ECdp
Field 1				
EM38 horizontal (ECh)	1			
EM38 vertical (ECv)	0.85	1		
Veris shallow (ECsh)	0.75	0.75	1	
Veris deep (ECdp)	0.85	0.82	0.79	1
Field 2				
EM38 horizontal (ECh)	1			
EM38 vertical (ECv)	0.92	1		
Veris shallow (ECsh)	0.78	0.69	1	
Veris deep (ECdp)	0.90	0.89	0.81	1
Field 3				
EM38 vertical (ECv)	–	1		
Veris shallow (ECsh)	–	0.65	1	
Veris deep (ECdp)	–	0.81	0.79	1

exhibited a highly nonlinear trend (Fig. 3), even after transformation. Closer examination of the transformed Veris data showed that it was linear up to a certain topsoil depth, but exhibited what seemed to be random scatter past that depth. This was interpreted to mean that soil conductivity variations below a certain depth did not affect the output of the Veris sensor. On claypan soils, our hypothesis has been that the major factor causing sensor EC changes is the variation in topsoil depth above the high-clay claypan horizon. If that topsoil depth was greater than the sensing depth of the Veris 3100, the sensor would not see the claypan horizon, and variations in Veris EC output would be due to conductivity differences in the surface soil. This situation is in contrast to that seen with the EM38, which responds to soil conductivity differences seen at greater depths (Fig. 1). Based on Fig. 3, Veris shallow calibrations included only those points where topsoil depth was 30 cm or less. Veris deep calibrations included those points where topsoil depth was 90 cm or less.

Linear calibration equations for topsoil depth as a function of 1/EC were developed separately for each field (Table 2), since the effect of field was found to be statistically significant. Calibration equations for estimating topsoil depth from EM38 horizontal or vertical data consistently exhibited a good fit to the data (r^2 = 0.84 to 0.88). Calibrations to Veris data were more variable (r^2 = 0.59 to 0.93). This may have been due in part to the reduced number of calibration points available for use with the Veris data (Table 2).

Field 1 topsoil depth maps were developed from each dataset (Fig. 4). Agreement of overall topsoil depth patterns was good between the EM38 horizontal and vertical maps, although the EM38 vertical data tended to estimate greater topsoil depths in some areas of the field. The Veris data, especially the Veris shallow data, did not accurately reproduce areas of deep topsoil, such as that seen in the drainage channel which extends southward from the north edge of the field.

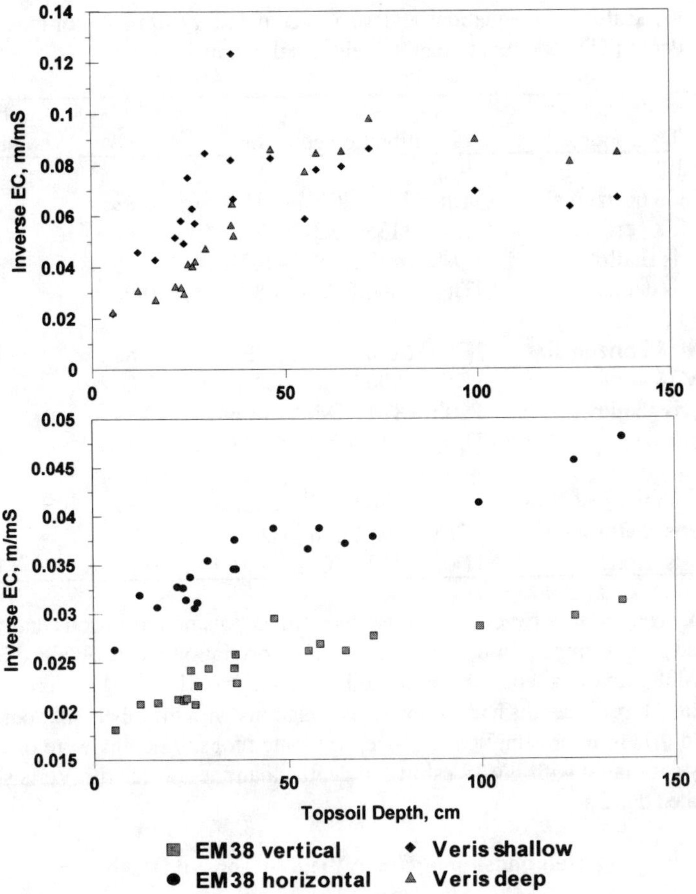

Fiig. 3. Relationship between topsoil depth and the inverse of soil electrical conductivity (1/EC) as measured with Veris (top) and EM38 (bottom) sensors at 21 calibration points in Field 1.

To more closely examine these trends, two adjacent east–west data transects were selected near the north edge of the field. These transects crossed areas of both shallow (<20 cm) and deep (>100 cm) topsoil. Topsoil depth as estimated by each EC dataset was graphed for comparison (Fig. 5). These data showed that topsoil depth estimations were similar when comparing EM38 horizontal and vertical data, across the range from shallow to deep topsoil. Veris deep data followed similar trends, but its upper range was limited to about 90 cm, agreeing with the calibration results described above. Topsoil depths estimated from Veris deep data were generally shallower than those estimated by EM38 data. The upper range of topsoil depth estimated by Veris shallow data was limited to about 30 cm, causing this data to under-predict topsoil depth over the majority of the length of the transects.

Table 2. Calibration equations and associated data for estimation of topsoil depth (TD) as a function of soil electrical conductivity.

Data source	Calibration equation	r^2	Calibration points used
Field 1			
EM38 horizontal	$TDh = 7350\ ECh^{-1} - 215$	0.88	21
EM38 vertical	$TDv = 11350\ ECv^{-1} - 229$	0.85	20
Veris shallow	$TDsh = 490\ ECsh^{-1} - 6.0$	0.76	10
Veris deep	$TDdp = 800\ ECdp^{-1} - 8.1$	0.93	18
Field 2			
EM38 horizontal	$TDh = 6100\ ECh^{-1} - 143$	0.88	12
EM38 vertical	$TDv = 1990\ ECv^{-1} - 28.4$	0.84	19
Veris shallow	$TDsh = 870\ ECsh^{-1} - 11.6$	0.88	7
Veris deep	$TDdp = 2130\ ECdp^{-1} - 34.8$	0.59	12
Field 3			
EM38 vertical	$TDv = 3970\ ECv^{-1} - 84.2$	0.89	18
Veris shallow	$TDsh = 763\ ECsh^{-1} - 9.7$	0.72	8
Veris deep	$TDdp = 1320\ ECdp^{-1} - 13.4$	0.86	14

Correlations between topsoil depths estimated using the various data sources showed similar trends for all 3 fields. Highest correlations were obtained between the EM38 horizontal and vertical topsoil depths ($r = 0.81$ to 0.91). Veris shallow-estimated topsoil depths had the lowest correlations with the other data sources ($r =$ 0.47 to 0.76). Interestingly, Veris deep-estimated topsoil depths were often more highly correlated with EM38-estimated depths than they were with Veris shallow-estimated depths.

Relationship of Grain Yield to Topsoil Depth

In previous work, we have found that topsoil depth, estimated by EM38 vertical data, has been one soil parameter often correlated with spatial yield variations, especially in years where water availability is a limiting factor for grain production (Sudduth et al., 1996). Correlations of grain yield with topsoil depth, as estimated from each of the four EC datasets, were developed for each field (Table 3). In general, correlations of yield to topsoil depth were greater in 1994 and 1997 when moisture stress reduced yields in the low-topsoil portions of Fields 1 and 2. EM38 horizontal data produced slightly better correlations with grain yield than did EM38 vertical data or Veris deep data over the range of years and fields investigated. The correlations of Veris shallow data to grain yield were similar to those of EM38 vertical and Veris deep data for Fields 1 and 2, but markedly different for Field 3, where the Veris shallow data was the most highly correlated to yield. In this field, the deepest topsoil was in a low, poorly drained area, where yields were consistently limited due to early-season ponding; thus the deeper topsoil = higher yield relationship did not hold for the deepest topsoil. Since the Veris shallow reading did not locate this area, but mapped it as shallower topsoil, the correlation for that sensor data to yield was improved compared to the other data sources.

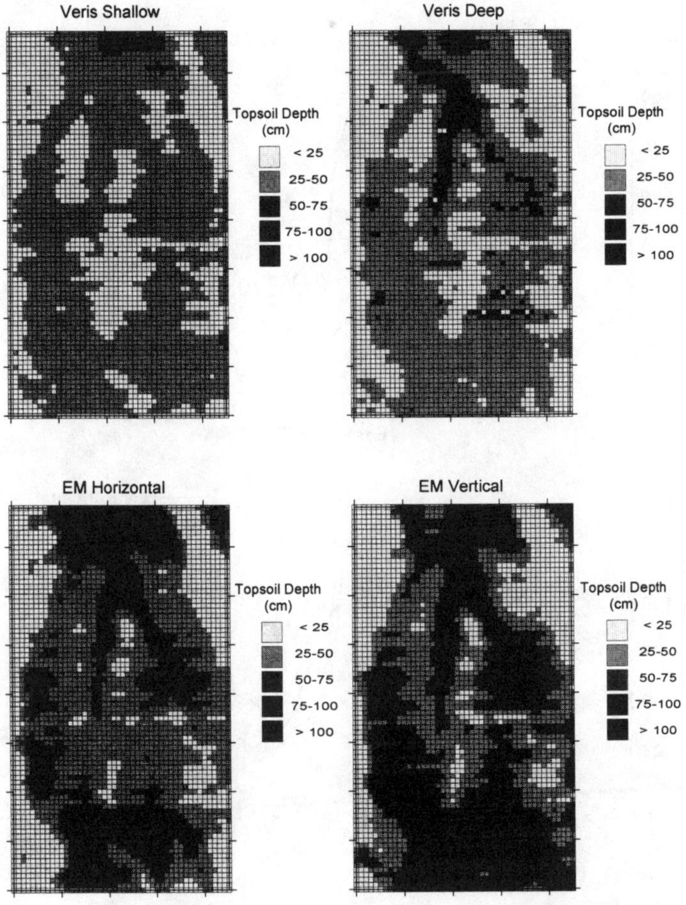

Fig. 4. Topsoil depth estimated from four different soil electrical conductivity datasets obtained on a 36-ha claypan soil field.

CONCLUSIONS

Coulter-type and EM sensors were both able to measure EC on claypan soils, and maps generated from the two sensors exhibited similar patterns at the field scale. Differences between maps were attributed to the differences in sensing depth between the different sensors (Geonics EM38 and Veris 3100) and data collection modes (vertical vs. horizontal or deep vs. shallow, respectively.)

Soil EC data from all four sensing modes successfully estimated measured topsoil depth (depth to Bt horizon) for shallow topsoils. Veris shallow readings,

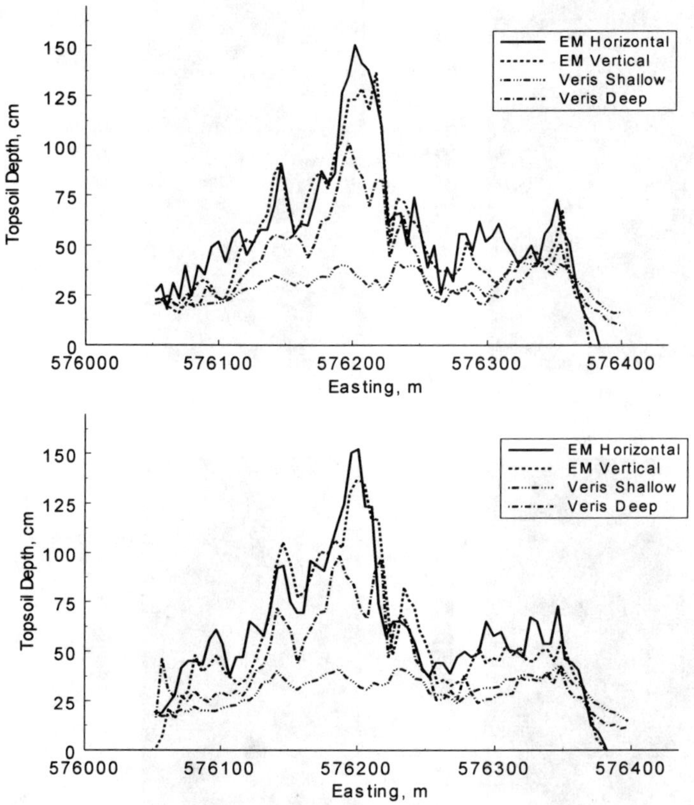

Fig. 5. Comparison of topsoil depth estimated from each of four soil conductivity datasets on two adjacent transects in Field 1.

however, were not able to estimate topsoil depths greater than about 30 cm and Veris deep readings were not able to estimate topsoil depths greater than about 90 cm; these are the approximate sensing depths for this device, as stated by the manufacturer. This limitation of the Veris sensor might not be of concern in practice, since the majority of crop yield variation and associated management modifications would occur at topsoil depths <90 cm.

Correlations of EC-estimated topsoil depth and grain yields were highest for corn crops in those years that water stress limited yields in the low-topsoil portions of fields. EM38 horizontal data were most highly correlated to yields for two fields, while Veris shallow data were most highly correlated to yields for the third field, where the deepest topsoil was located in a poorly drained area that consistently exhibited poor crop stands.

Both EM38 and Veris sensors provided information that would be useful for investigating soil variability and potentially for controlling variable management on

Table 3. Linear correlation of topsoil depth (TD) calculated from soil electrical conductivity readings to grain yield for each study field.

	Topsoil depth estimated from:			
	EM38 horizontal	EM38 vertical	Veris shallow	Veris deep
Field 1				
1993 corn	-0.19	-0.13	-0.13	-0.14
1994 soybean	0.40	0.23	0.13	0.35
1995 grain sorghum	0.05	0.19	0.22	0.02
1996 soybean	-0.05	-0.01	-0.11	-0.10
1997 corn	0.41	0.42	0.27	0.36
Field 2				
1996 soybean	-0.18	-0.20	-0.32	-0.23
1997 corn	0.68	0.63	0.65	0.64
Field 3				
1995 soybean	–	0.02	0.49	0.00
1996 corn	–	0.19	0.37	0.16
1997 soybean	–	0.13	0.35	0.05

claypan soils. Although the EC data from the two sensing approaches was generally similar, it is possible that one sensor could be better than the other for a specific application, due to the measurement differences that were observed. It might also be that integration of data from multiple sensor readings at different depths could provide enhanced information. Additional research comparing Veris and EM38 measurements on claypan soils as well as on other soil types is warranted.

ACKNOWLEDGMENT AND DISCLAIMER

Appreciation is extended to Eric Lund and Veris Technologies of Salina, Kansas for the loan of the Veris 3100 sensor cart used in this study.

Mention of trade names or commercial products is solely for the purpose of providing specific information and does not imply recommendation or endorsement by the U.S. Department of Agriculture or the University of Missouri.

REFERENCES

Doolittle, J.A., K.A. Sudduth, N.R. Kitchen, and S.J. Indorante. 1994. Estimating depths to claypans using electromagnetic induction methods. J. Soil Water Conserv. 49(6):572–575.

Jaynes, D.B., T.S. Colvin, and J. Ambuel. 1995. Yield mapping by electromagnetic induction. p. 383–394. *In* P.C. Robert et al., (ed.) Proc. 2nd Int. Conf. on Site-Specific Management for Agricultural Systems. ASA, CSSA, and SSSA, Madison, WI.

Lund, E.D., and C.D. Christy. 1998. Using electrical conductivity to provide answers for precision farming. p. I-327–I-334. *In* Proc. First Int. Conf. on GeospatialInformation in Agriculture and Forestry. ERIM Int., Ann Arbor, MI.

McNeill, J.D. 1992. Rapid, accurate mapping of soil salinity by electromagnetic ground conductivity meters. p. 201–229. *In* Advances in measurements of soil physical properties: Bringing theory into practice. SSSA Spec. Publ. 30. ASA, CSSA, and SSSA, Madison, WI.

Rhoades, J.D., and D.L. Corwin. 1991. Determining soil electrical conductivity-depth relations using an inductive electromagnetic soil conductivity meter. Soil Sci. Soc. Am. J. 45:255–260.

Rhoades, J.D., N.A. Manteghi, P.J. Shouse, and W.J. Alves. 1989. Soil electrical conductivity and soil salinity: New formulations and calibrations. Soil Sci. Soc. Am. J. 53:433–439.

Sheets K.R., and J.M.H. Hendrickx. 1995. Noninvasive soil water content measurement using electromagnetic induction. Water Resourc. Res. 31(10):2401–2409.

Sudduth, K.A., N.R. Kitchen, D.F. Hughes, and S.T. Drummond. 1995. Electromagnetic induction sensing as an indicator of productivity on claypan soils. p. 671–681. *In* P.C. Robert et al., (ed.) Proc. 2nd Int. Conf. on Site-Specific Management for Agricultural Systems. ASA, CSSA, and SSSA, Madison, WI.

Sudduth, K.A., S.T. Drummond, S.J Birrell and N.R. Kitchen. 1996. Analysis of spatial factors influencing crop yield. p. 129–140. *In* P.C. Robert et al., (ed.) Proc. 3rd Int. Conf. on Precision Agriculture. ASA, CSSA, and SSSA, Madison, WI.

Williams, B.G., and D. Hoey. 1987. The use of electromagnetic induction to detect the spatial variability of the salt and clay contents of soils. Aust. J. Soil Res. 25:21–27.

Development of A CAN Bus Based Air Seeder Monitor

Frank Lang
Electronics Department
Flexi Coil LTD.
Saskatoon, Saskatchewan, Canada

ABSTRACT

The development of a CAN (Controller Area Network) based control system for an air seeder system will be presented. The CAN bus is a high-speed serial communications system, which was developed by Bosch and is being used in automotive, truck and bus industries. It is also being adopted as an ISO standard (11783) for use on agricultural equipment. This paper will present the development of a controller–monitor for a variable drive air seeder. The variable drive air seeder allows changes in application rate on the go, making it well suited for precision farming applications. A discussion of the design process involving possible architectures of the system, expandability considerations, and compatibility of the electronics across various systems. The system architecture had to be a balance between cost, cable complexity, ECU complexity, and system flexibility.

INTRODUCTION

New air seeder technologies have presented the need for increasingly advanced electronics monitoring and controls in their operation. The air seeding equipment and its peripheral equipment such as seed treatment units, flow monitoring has brought challenges to monitoring equipment. This paper presents the development of an architecture for a CAN bus based monitoring system for an air seeder.

OVERVIEW OF AN AIR SEEDER

Figure 1 illustrates a cut away view of a Flexi Coil Air Seeder. It consists of a number of tanks in which different products are contained, a metering system to dispense the product in a controlled and accurate manner, an air distribution system, an air source, and a number of sensors. The meters dispense product into the air stream proportional to their rotational speed. The product is then carried by the air to distribution manifolds which divide the air–seed flow into individual runs. Product from different tanks can either be sent together in the same distribution lines (single shoot), or separated into individual flows (double or multi shoot). A variety of openers are available for applying the seed–fertilizer together, or separated by horizontal and vertical spacing.

Copyright © 1999 ASA-CSSA-SSSA, 677 South Segoe Road, Madison, WI 53711, USA. *Proceedings of the Fourth International Conference on Precision Agriculture.*

Depending on the rates and the type of product being applied, the seeder requires different setups. There are coarse, fine, extra fine meter rollers, which are used according to the physical size and amount of product being applied; coarse for large product–high rates (such as peas), extra fine for small product–very low rates (such as canola). When the meter rollers are installed and the product is in the tanks, a displacement calibration is performed on the meter rollers. This is done by turning the meters a number of revolutions and weighing the amount of product metered. This gives a mass–revolution calibration number for the roller.

The amount of airflow must be set according to the amount of product being carried. The fan speed and damper settings are used to set the correct air velocities for the different tanks. More air is required for more products, but too much air can cause seed damage and/or product bouncing out of the ground.

Fig. 1

Sensors

In order to provide the necessary feedback to the operator, many sensors are required on the air seeder system, which become particularly important when used for precision farming (variable rate application). The sensors required on the air cart are as follows:

Ground Speed — a hall effect sensor picks up the rotational speed of one of the cart tires. A calibration is performed using the monitor to produce the correct reading.

Meter RPM — a hall effect sensor senses the rotational speed of the meter rollers and provides a number of pulses per revolution to the monitor system.

Fan Speed — a hall effect sensor similar to the meter sensor provides a number of pulses per revolution to the monitor system.

Air Velocity — a mass air flow sensor determines the air velocity in the individual meters.

Bin Depth — ultrasonic sensors determine the depth of material in the individual tanks.

Work Switch — a switch on the implement which indicates when the implement is in or out of the soil.

Actuators

Actuators are required on the air seeder system as well. These include motor drives for the meters and linear actuators for adjusting the air dampers and fan speed. A control system is required to regulate the speed of the motors based on feedback from the meter RPM sensor, meter calibration values and the desired application rate.

Auxiliary Functions

A number of auxiliary functions can be added to the air cart. These include a seed treatment system, which allows treatment of seeds on the go, using the air delivery system, and a flow monitoring system, which monitors the flow of product and determines if a line becomes plugged or product stops flowing. These systems require information from the air cart to operate, creating a need for

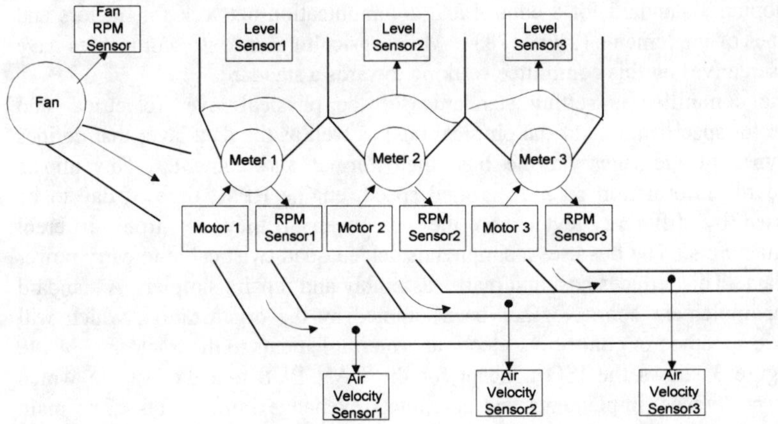

Fig. 2

an interoperable, expandable system. Figure 2 illustrates the types of inputs and outputs required by an air seeder system.

Monitor System

A system is required to tie all of the sensors and actuators together and to provide a user interface for the operator. A console in the cab of the tractor provides an alphanumeric display to provide information, alarms, and diagnostics to the operator, as well as a keypad for selecting information to display and for entering set up and calibration information.

The monitor user interface must be able to present the required information in an understandable and appropriate manner. Visual and audio cues are used to alert the operator of failures, empty bins, incorrect rates or other anomalous conditions. Setting up and calibrating the system must also be easy to do and understand.

The system must provide expandability and versatility. There are options for the air seeder which must be accommodated by the system, such as the on-the-go seed treatment system, product flow monitoring for the distribution of the product, as well as external controllers such as GPS variable rate controllers.

CAN BUS

The CAN (Controller Area Network) bus was developed by Bosch in Germany and has been used on heavy trucks, buses and construction equipment for a number of years. It has been adopted by the International Organization for Standardization (ISO) for the agricultural communications bus standard, 11783. The ISO CAN bus is a 250 kbits s^{-1} (capable of 1 Mbit s^{-1}) serial communication bus used to transfer information between devices on the bus. The bus is designed to allow efficient distribution of power and information, reducing the amount of cables required and reducing Electromagnetic Interference (EMI).

ISO Technical Committee 23/Sub Committee 19/Working Group 1 is developing a standard for a control and communication network for tractors and all types of implements (ISO 11783). Many agricultural industry companies have representatives on this committee working towards a standard.

The committee is setting standards for the physical layer (electrical and connector specifications of the physical bus) as well as the data layer that defines the types of messages on the bus, their format and content. This allows "standard" information such as ground speed, engine RPM, time of day to be accessed by different devices on the bus, even if they are from different manufacturers. The bus uses a simple unshielded quad twist cable to carry power and data. This reduces cost and makes assembly and repairs simpler. A standard tractor-implement connector has been adopted by the organization, which will provide a method of connecting electronics on implements to the tractor.

Figure 3 shows the ISO concept for the CAN BUS and the way in which different devices, implements and computers exchange information. The main bus carries all of the traffic between the different Electronic Control Units (ECUs). There also may be bridge devices, which provide a connection of the main bus to a secondary or proprietary bus. This may be, in the case of a tractor, the engine control system and instrumentation. The engine controller would have proprietary messages and the bridge may only allow certain information to and from the main bus.

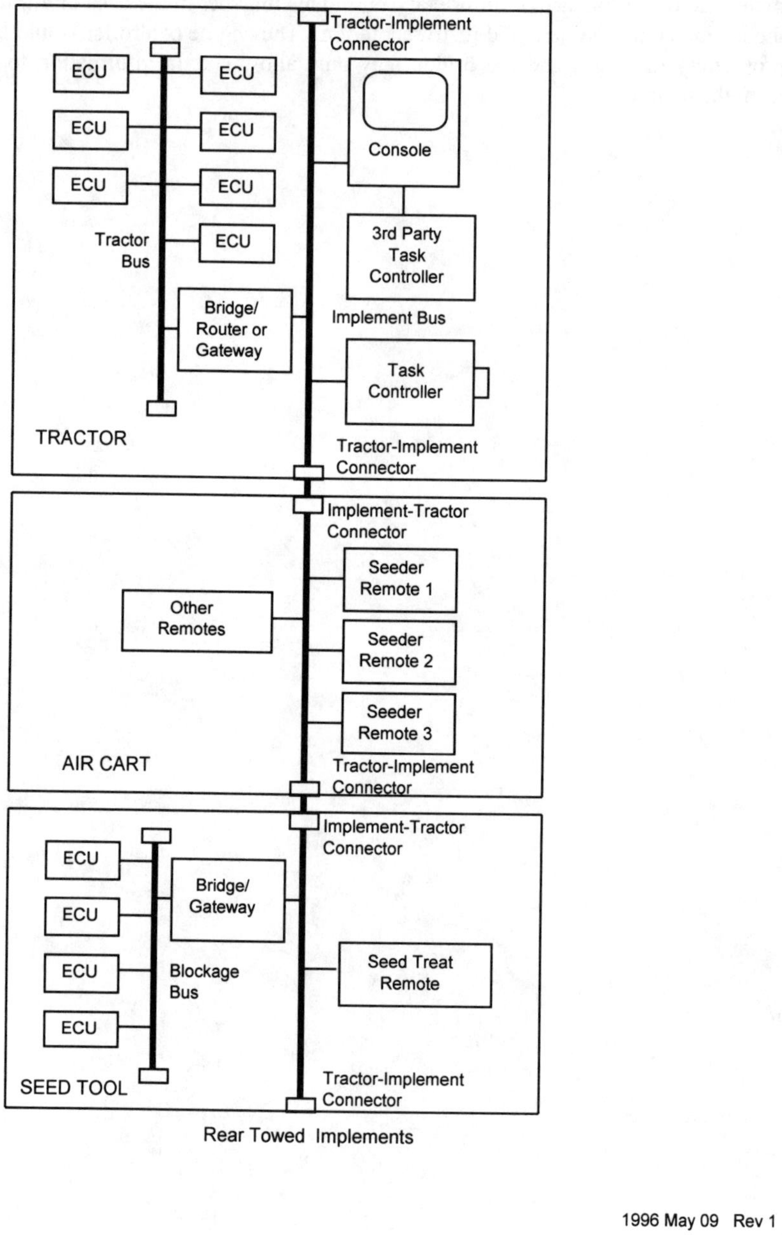
Tractor-Implement Connector
ECU
ECU
ECU
ECU
ECU
ECU
ECU
Console
Tractor Bus
3rd Party Task Controller
Bridge/ Router or Gateway
Implement Bus
Task Controller
TRACTOR
Tractor-Implement Connector
Implement-Tractor Connector
Seeder Remote 1
Other Remotes
Seeder Remote 2
Seeder Remote 3
AIR CART
Tractor-Implement Connector
Implement-Tractor Connector
ECU
Bridge/ Gateway
ECU
ECU
Blockage Bus
Seed Treat Remote
ECU
SEED TOOL
Tractor-Implement Connector
Rear Towed Implements
1996 May 09 Rev 1

Fig. 3

The console or virtual terminal would be a general-purpose monitor unit which any ECU can use to display information on, or request user input from. This would eliminate the need for many different consoles in the cab of a tractor, each performing different function, but probably duplicate functions, such as ground speed.

ARCHITECTURAL OPTIONS

When developing the new monitor system for the air seeder, there were many trade offs to be made in determining the architecture for the electronics system. Three possible architectures are presented in Fig. 4, 5 and 6.

The first architecture presents a number of small special purposes ECUs providing monitoring and control. Each ECU has a dedicated function such as motor control, shaft monitoring, air velocity monitoring. A master ECU would collect and distribute information to and from the display unit. Some of the advantages to this approach are that functions of the units are very defined and easily identified for service and troubleshooting. In the event of a failure, the part to be replaced would be lower cost since it has a small number of components.

A disadvantage of this architecture is the cost of implementing the large number of units. The cases, power condition circuitry and connectors are duplicated several times, increasing overall cost. Configuration of the system becomes more complicated since there are a large number of ECUs that need to be

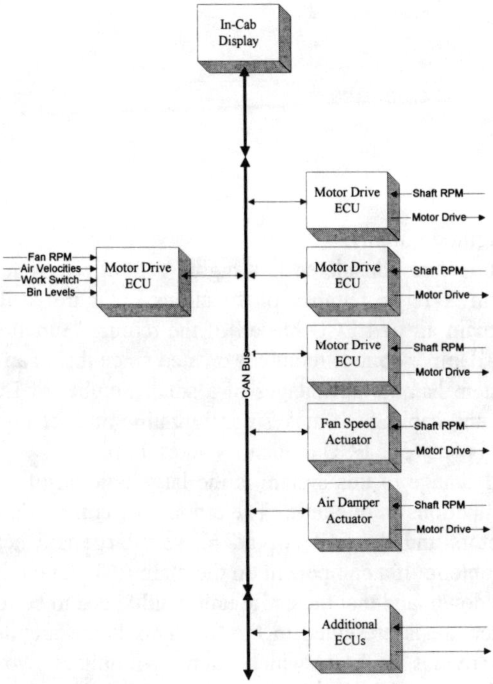

Fig. 4

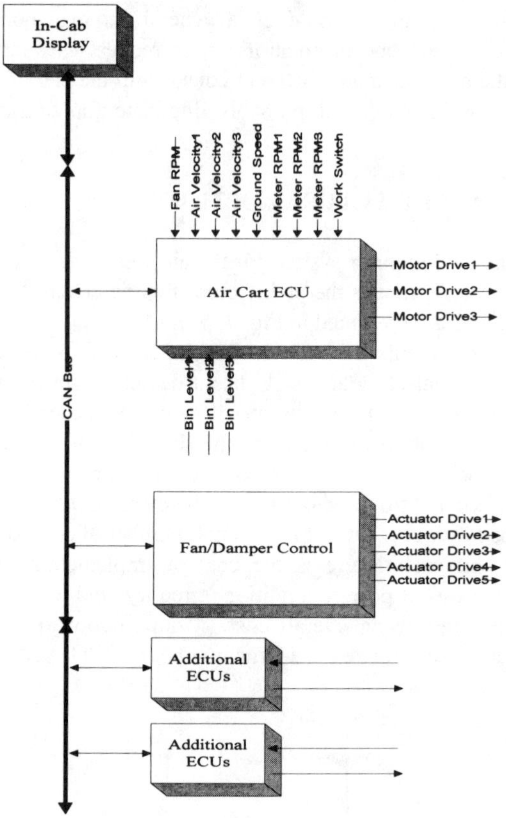

Fig. 5

identified and configured initially.

The second architecture considered is a highly centralized system, where the ECUs each perform a large number of functions. Figure 5 illustrates this architecture. The main air cart ECU has all of the required functions for the air cart. A second fan–damper control remote is created since this is an option on the machine. This system has the advantages of a small number of ECUs reducing costs in packaging and cabling. It makes configuration and set up more straight forward since there is only one ECU to identify and set up.

The major disadvantage to this system is the large amount of I/O required to achieve all of the functions in one unit. The amount of current that is needed to drive the three motors and the ECU would be very large and heat dissipation would become a problem. If a component on the main ECU failed, it would mean the entire air cart is down, and that the entire unit would need to be replaced.

The third architecture is presented in Fig. 6. This is the architecture, which was adopted. It provides an ECU which controls–monitors two tanks. This provides expandability as pairs of tanks. It strikes a balance between the amount of I/O and current required for an ECU as well as cost of the ECU. The current air cart is a three-tank system, therefore two-air cart ECUs of this type allow for a spare tank controller that can be used for a 4th tank option in the future. The ECUs were designed with a 60-pin connector which has two 30-pin halves. Each half is labeled and has all of the cables required to control and monitor a tank. Configuration is simply done by assigning a cable to a tank.

The CAN bus system requires ECUs on the bus to be named as an identification method. When new ECUs are added to the bus, they must be assigned to their appropriate function. This is a one-time process, done either at the factory, or when a replacement unit is installed. The display unit displays the ECUs serial number and asks the operator to name the ECU and to assign its cables to tanks or other appropriate functions. This configuration is stored in non-volatile memory and on subsequent power on events, the configuration is automatic. This process allows optional ECUs such as seed treatment, flow monitoring or sprayers to be added at any time, giving excellent expandability.

The ECUs and sensors were designed such that all connected sensors are detected at power on, and based on the attached sensors the system can determine if all functions on the ECU are being used. Since only one workswitch or fan sensor is required on each air cart, only the ECU which detects their presence

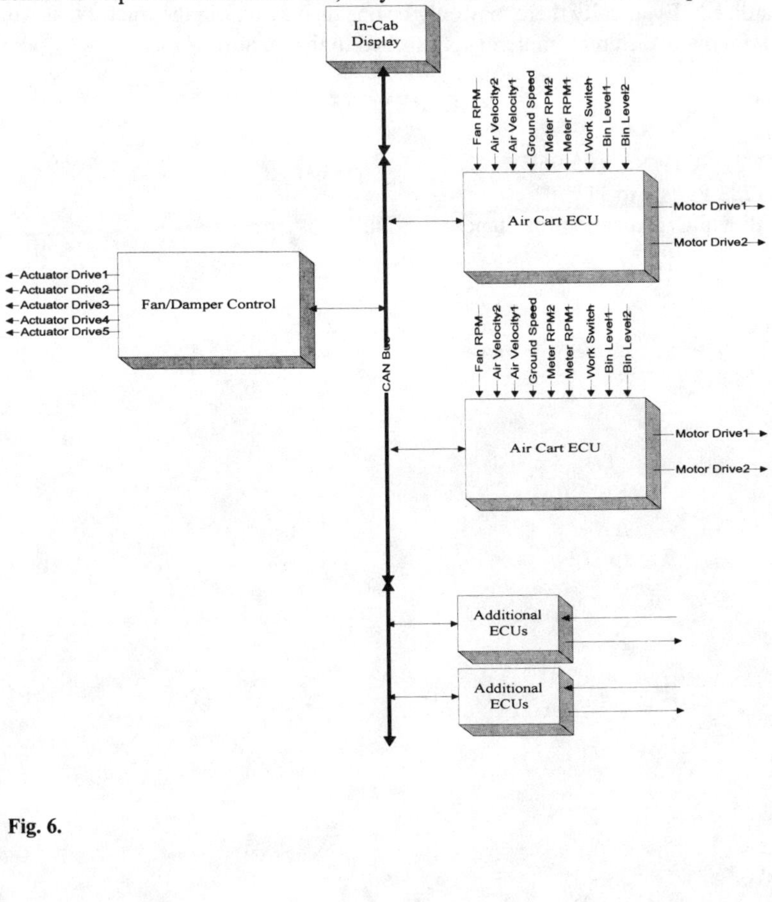

Fig. 6.

would report their existence and their information. The same is true for air cart meters. If both halves of an air cart ECU are not used, it only reports the half that has sensors connected.

The ISO standard also provides for a diagnostic connector on the tractor. This allows diagnostic tools to connect to the CAN bus and query the ECUs in the system for status and error information.

CONCLUSIONS

The CAN bus system is a communications network which provides a means of designing a robust, expandable, and versatile system of electronics for agricultural implements. Choosing an architecture suited to the application is an important part of the design process which impacts the flexibility of the system in the future as new devices or options are developed.

As the ISO standards are finalized for the agricultural communications bus, the long-term benefits of the CAN bus will be realized through the standardization. It will provide farmers with a standard implement interconnect system for electronics similar to the way hydraulic and PTO connections have been standardized. Eventually there may only be one display unit in the tractor that will work with many different implements attached to the tractor.

REFERENCES

ISO/TC23/SC19/No. 105/96E
ISO 11783 Parts 1 to 11
International Organization for Standardization

Spatial Variability of Water Application from Center Pivot Irrigation and Precipitation

Robert W. Jordan
Harold R. Duke
Dale F. Heermann

USDA-ARS Water Management Research Unit
Fort Collins, Colorado

ABSTRACT

The water application patterns of two center pivot irrigation systems in eastern Colorado were simulated over their entire area of operation and mapped for the 1997 growing season, using GIS, as part of a multidisciplinary precision fanning project. The system performance was evaluated over the entire field including locations where the operating characteristics were altered due to end gun operation and topography. Precipitation was measured at the center and around the perimeter of each field. Precipitation surfaces were created using four different interpolation methods and combined with the irrigation application depths to calculate a total water application uniformity coefficient. System mean application depth and uniformity coefficient varied with the topography and end gun settings. Total water application uniformity increased with the addition of the precipitation values and was insensitive to the interpolation method used.

INTRODUCTION

Center pivot irrigation systems are capable of highly uniform water application when adequately designed and operated over level terrain. The depth of application over a center pivot irrigated field is often assumed to be uniform; however, significant changes in the water distribution can occur with location and time.

Topography will affect the pressure distribution along the center pivot lateral, which in turn will affect the discharge and the pattern radius of an individual sprinkler head. The two basic conditions that can occur are an increase or a decrease in elevation between towers. An increase in elevation creates lower pressures at the outer tower, resulting in smaller discharges from the affected sprinkler heads. A decrease in elevation creates higher pressures at the outer towers, resulting in higher discharges from the affected sprinkler heads. The effects of topography on center pivot water distributions have been investigated by James (1982) and Beccard and Heermann (1981). These two researchers used simulation models based on the equations developed by Heermann and Hein (1968) and investigated the effects of topography on center pivot irrigation system performance in areas where the topography of the field would influence it. Irrigation uniformity over the entire field was investigated by Evans et al. (1995) using computer simulation and a polar coordinate interpolation scheme.

Copyright © 1999 ASA-CSSA-SSSA, 677 South Segoe Road, Madison, WI 53711, USA. *Proceedings of the Fourth International Conference on Precision Agriculture.*

The operation of an end gun may also alter the performance characteristics of a center pivot system. When the end gun is turned on, the discharge and head loss increase and force the pump to a different equilibrium operating point than when the end gun is turned off. This results in the pump supplying a larger discharge at a lower head when the end gun is on and a smaller discharge at a higher head when the end gun is off. The end result of these two conditions is different pressure distributions along the lateral, which can result in different mean application depths and different application uniformities if the sprinkler heads are not equipped with pressure regulators or flow control devices.

The objective of this work was to create maps showing the spatial variability of irrigation application and precipitation for use in a multidisciplinary precision farming project in which yield variability will be multiply correlated with input variability.

METHODS

The water application of two center pivot systems in eastern Colorado was simulated over their entire area of operation (Jordan, 1998). The total application of each system was the sum of simulated irrigation events during the 1997 season based on records of the system operation. The irrigation application was simulated using the model developed by Heermann (1990) (USDA Center Pivot Evaluation and Design [CPED] program) which is based on the equations developed by Bittinger and Longenbaugh (1962) and solved by Heermann and Hein (1968). CPED solves the Darcy-Weisbach equation to determine pressures at each head along the machine, taking into account changes in topographic elevation. Then it simulates water application based on manufacturer specifications of the pump and sprinkler heads. CPED does not include corrections for wind effects on sprinkler application patterns. The systems were simulated with the assumption that each tower is moving continuously. There is some nonuniformity in the direction of travel; but it is assumed that it will be negligible for mapping the water distribution over the entire field. Operation of the center pivot was simulated at 5^0 increments around 360^0.

Two farmer-operated fields in eastern Colorado were selected for the study. Field A is 72 ha and Field B is 52 ha. Surface elevation varies by 2 m and 6 m over fields A and B respectively. Center pivot A has rotator heads mounted on drop tubes, center pivot B has spray heads mounted on the center pivot lateral.

Elevation surfaces were created from detailed topographic surveys of both fields. These surfaces were then used to estimate the tower elevations at specific azimuths using a geographic information system (GIS). Individual polygons of the tower radius plus or minus 6.1 m by five degrees (polar coordinates) were created for each 5^0 increment and overlaid onto the elevation surface. The mean elevation within each polygon was calculated and used as the representative elevation of each tower as it moved through each 5^0 arc. These tower elevations were used for input to CPED. Figures 1 and 2 show elevation contour maps created from the survey data.

Can tests, pump tests, pressure measurements and sprinkler discharge measurements were conducted to verify the simulation program results.

The simulation program outputs application depths at a user specified interval. Each simulation was associated with Cartesian coordinates to allow mapping. These points were then overlaid onto a polar grid because a center pivot irrigation system moves in a polar reference system. A polar grid cell can then characterize a specific area of movement while reflecting the operating characteristics of the system (outer nozzles move faster and cover a larger area than the inner nozzles).

Analysis of the water distribution was done using an area weighted mean application depth and a seasonal uniformity coefficient. The seasonal uniformity coefficient is the Heermann-Hein uniformity coefficient for center pivot irrigation machines (ASAE, 1992) calculated using the sum of the simulated irrigation depths from each irrigation event during the season.

A Tipping bucket rain gage network was also installed at each of the fields. Each network consists of six tipping bucket rain gages around the perimeter and one mounted on top of the riser pipe at the pivot point. The rain gages were monitored throughout the season to determine the total seasonal precipitation depth distribution at each location. The environment in which the rain gages were installed lead to some of the gages malfunctioning during specific precipitation events (e.g., spider webs, bird droppings, etc.). During precipitation events in which a rain gage malfunctioned, it was assigned a precipitation depth equal to the mean of the surrounding rain gages.

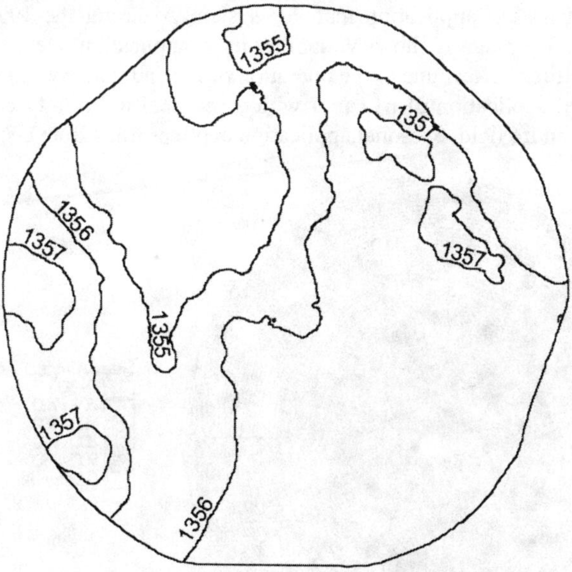

Fig. 1. Field A elevation contours (elevation in m).

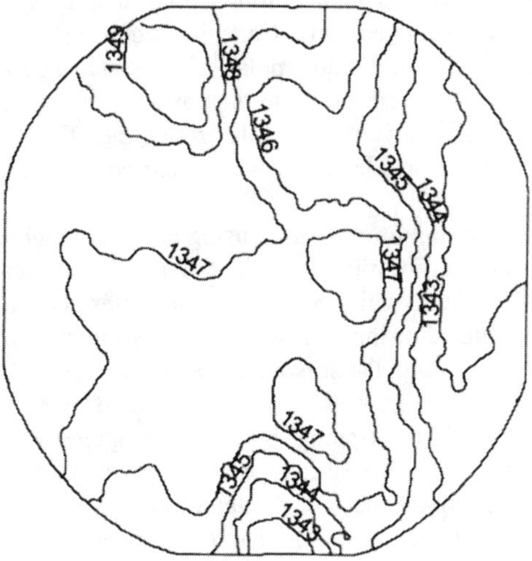

Fig. 2. Field B elevation contours (elevation in m).

RESULTS AND DISCUSSION

The irrigation application map for system A during the 1997 season is shown in Fig. 3. System A shows variability in its application over the field, with the most significant effect due to the operation of the end gun. System A applied a mean seasonal application of 612 mm with a seasonal uniformity coefficient of 0.86 over the entire field. Seasonal application depth profiles from center pivot A

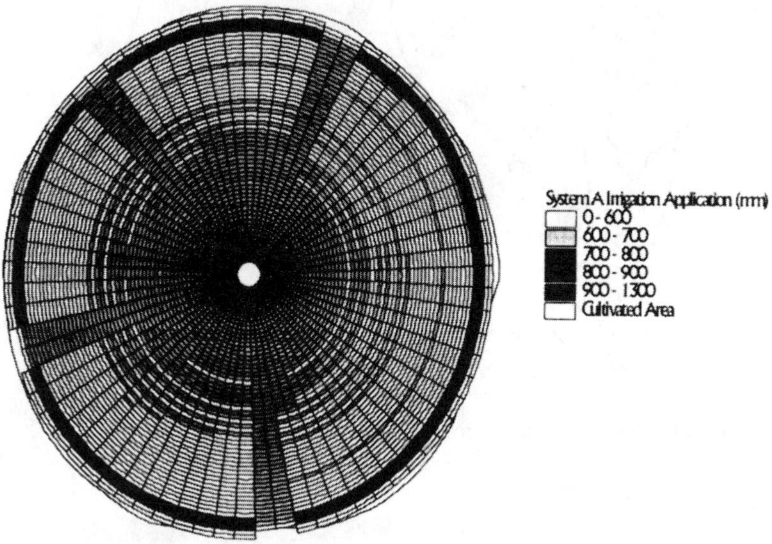

Fig. 3. 1997 Irrigation application for system A.

at 20° clockwise (cw) from north (end gun off) and 15° cw from north (end gun on) (areas with relatively level topography) are shown in Fig. 4. The profile with the end gun off has a mean seasonal application depth of 625 mm and a seasonal uniformity coefficient of 0.92. The profile with the end gun on has a mean seasonal application depth of 608 mm and a seasonal uniformity coefficient of 0.85. Topography had little effect on the overall water distribution of this system because relief was small.

The irrigation application map for System B is shown in Fig. 5. System B shows more interesting variability than System A due to the larger variations in topography and wider end gun settings. System B applied a mean seasonal depth of 608 mm with a uniformity coefficient of 0.86 over the entire field. The low application area in the northwest is created by the high elevation area (approximately 3 m higher than the pivot point). The low elevation area in the east side of the field shows variability due to both the topography and operation of the end gun.

Figure 6 shows the profiles from System B where there is an increase in elevation (335°), a decrease in elevation (90°) and over relatively level topography where the end gun is on (245°) and the end gun is off (250°).

The increasing elevation profile has a mean seasonal application depth of 564 mm and a uniformity coefficient of 0.82. The decreasing elevation profile has a mean application depth of 633 mm and a uniformity coefficient of 0.90. The profile with the end gun on has a mean application depth of 573 mm and a uniformity coefficient of 0.83. The profile with the end gun off has a mean application depth of 604 mm and a uniformity coefficient of 0.89. The profiles and performance calculations show that there can be significant differences in performance due to the different operating conditions of a center pivot around the field.

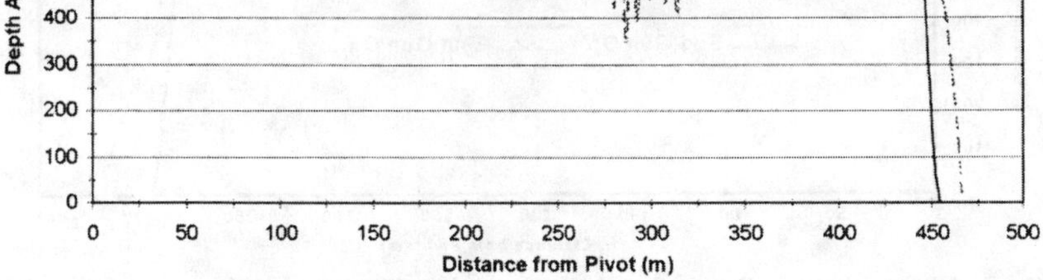

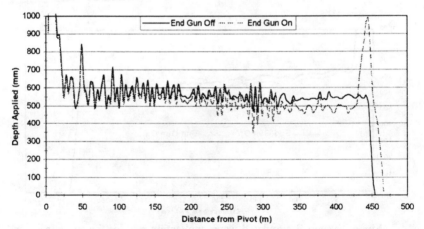

Fig. 4. System A simulated water application profiles for end gun on and off.

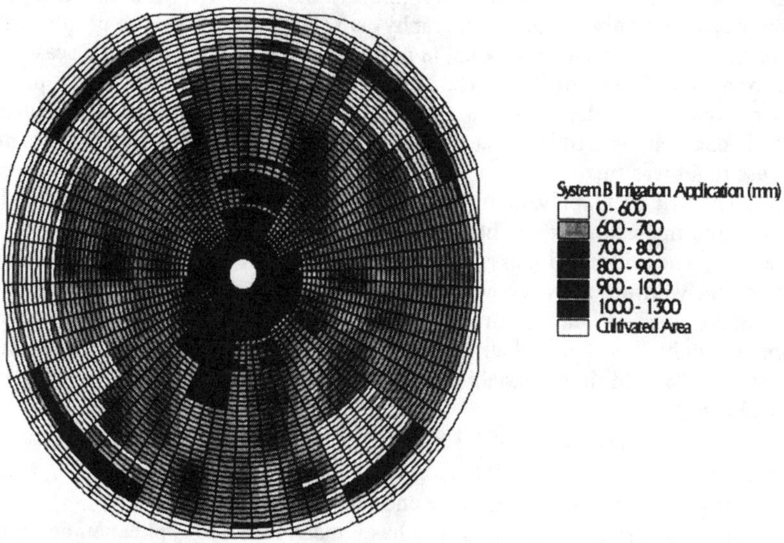

Fig. 5. 1997 Irrigation application for System B.

The often used assumption of an equal depth applied across the field is not an accurate representation of the water application from a center pivot system operating over varying terrain or using an intermittent end gun. These differences are particularly apparent when the system does not use pressure regulators or other flow control devices.

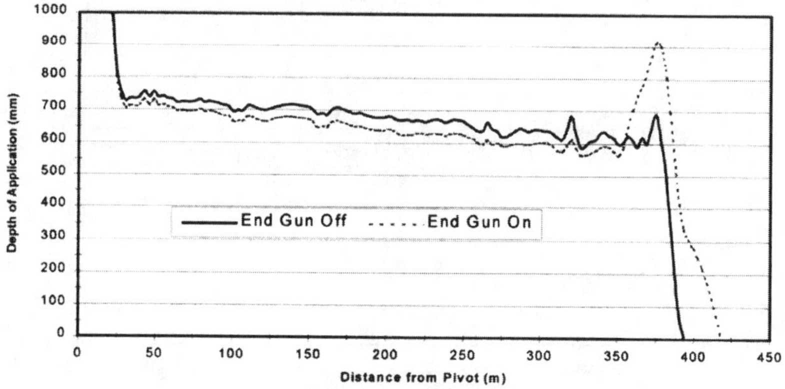

Fig. 6. System B simulated water application profiles for end gun on and off.

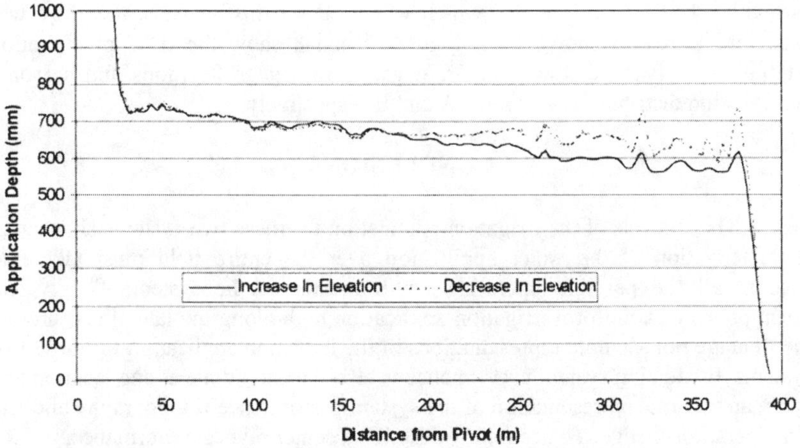

Fig. 7. System B simulated water application profiles for elevation increase and decrease.

The uniformity coefficients over the entire field are similar for both fields; however, the spatial application patterns are quite different. System A has a very small area of the field irrigated with the end gun off which is the high uniformity operating condition of the system. System has a larger area with the end gun off but has greater variation in the spatial application pattern due to the topography present. While a uniformity coefficient of 0.85 is generally considered good, this analysis shows that significantly different uniformities can occur throughout the field when topography and/or end operation changes.

To analyze the effect of precipitation on the uniformity of total water application (irrigation and precipitation), estimates of precipitation distribution within the field were required. Mean seasonal precipitation measured at Fields A and B was 25 and 28% of total water application, respectively. The precipitation measurements from the rain gages were interpolated using four different methods, a least squares fit plane, a bivariate interpolation, a spline fit and an inverse distance weighted (IDW) interpolation.

It is thought that higher wind velocities at greater height around the center-mounted rain gage may have affected the amount of precipitation it received. Interpolations were done with each method on data sets with and without the center rain gage.

These interpolated surfaces were then overlaid onto the polar grid used to represent the irrigation application. The estimated seasonal precipitation depth was added to the seasonal irrigation depth to achieve a total application depth within each of the polar grid cells used to characterize the irrigation application. Uniformity coefficients and mean application depths were then recalculated to quantify the effects of precipitation on the uniformity of total water application.

The calculated uniformity coefficients for the total water application show little effect with interpolation technique except for the case of the bivariate interpolation, which results in uniformity coefficient values two percent higher than the other techniques used. The IDW interpolation is the most convenient to use within ESRI ArcView GIS which was used for this analysis, therefore was used for the results shown here. Figures 7 and 8 show the IDW precipitation interpolations (without the center rain gage), rain gage locations and seasonal precipitation distribution for Fields A and B respectively.

CONCLUSIONS

The analysis of the irrigation application for these two systems shows that characterization of the water application over the entire field must take into account all the possible operating environments of the system. The typical assumptions of a uniform irrigation application both along the lateral and around the field are not accurate representations of the irrigation application for these two systems. Evaluating system performance is also typically done at one location and is not an accurate representation of the system performance if topography and end gun operation change. To accurately describe a center pivot's performance, it must be evaluated at the locations where conditions change such as the angle at which the end gun comes on or off as well as any topographic differences which may significantly alter the pressure distribution in the lateral. Similar results are expected if pumping rate and operating pressure change during the season as often occur when ground water levels decline.

Management of a particular center pivot can significantly influence the spatial variability of the water application as well. The simulated water application from these two systems show that the size of the angles in which the end gun comes on or off can create a considerably different looking spatial pattern. Changes in the speed of the system du-ring an irrigation event or an incomplete irrigation event (one in which the center pivot does not make a complete revolution) can also have a significant impact on the spatial water application pattern.

Particularly where water supplies are limited, the spatial variability of water application demonstrated in this study is likely to significantly affect variability of crop yield. Where water supplies are adequate, this variability will undoubtedly result in variable leaching of agricultural chemicals. In either case, the implications of water application variability to input management in a precision farming scenario are significant.

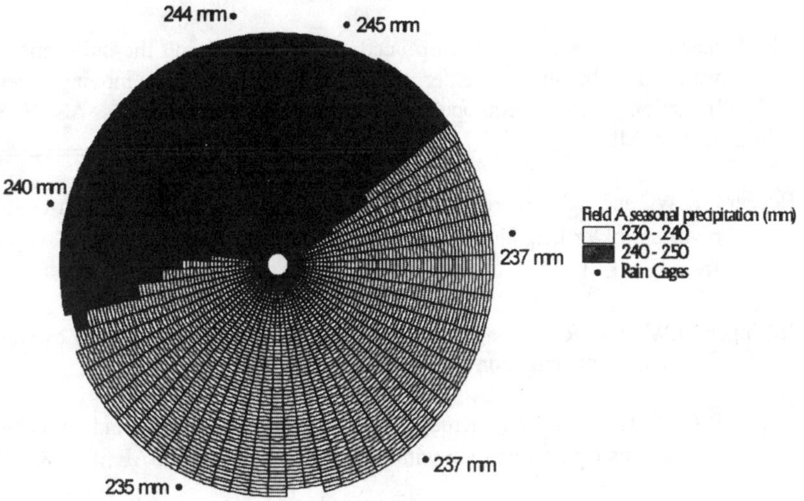

Fig. 8. Field A IDW interpolated precipitation and seasonal measurements overlaid with polar grid.

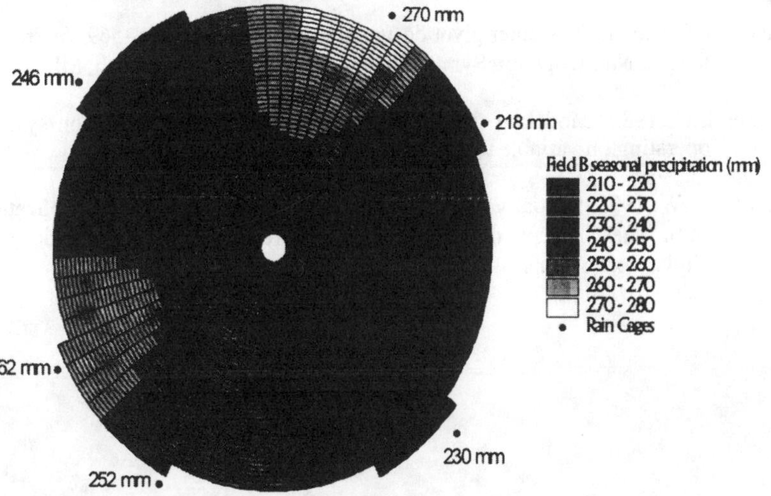

Fig. 9. Field B IDW interpolated precipitation and seasonal measurements overlaid with polar grid.

REFERENCES

ASAE Standards. 1992. S436. Test procedure for determining the uniformity of water distribution of center pivot, comer pivot, and moving lateral irrigation machines equipped with spray or sprinkler nozzles. ASAE, St. Joseph, MI.

Beccard, R.W., and D.F. Heermann. 1981. Performance of pumping plant-center-pivot sprinkler irrigation systems. ASAE Paper No. 8 1-2548. ASAE, St. Joseph, MI.

Bittinger M.W., and R.A. Longenbaugh. 1962. Theoretical distribution of water from a moving irrigation sprinkler. Trans. ASAE 5(1):26–30.

Evans, R.G., S. Han, and M.W. Kroeger. 1995. Spatial distribution and uniformity evaluations for chemigation with center pivots. Trans. ASAE 38(1):85–92.

Heermann, D.F., and P.R. Hein. 1968. Performance characteristics of self propelled center pivot sprinkler irrigation systems. Trans. ASAE 11(1):11–15.

Heermann, D.F. 1990. Center pivot design and evaluation. p. 565–569. *In* Proc. of the 3rd Nat. Irrigation Symp., Phoenix, AZ. ASAE, St. Joseph, MI.

James L.G. 1982. Modeling the performance of center pivot irrigation systems operating on variable topography. Trans. ASAE 25(l):143–149

Jordan, R.W. 1998. Spatial variability of center pivot irrigation water application. M.S. Thesis. Dept. of Chemical and Bioresource Eng., Colorado State Univ., Fort Collins.

Soil Doctor Multi-Parameter, Real-Time Soil Sensor and Concurrent Input Control System

J. W. Colburn, Jr.
Crop Technology
Spring, Texas

ABSTRACT

Conceived in 1982, the Soil Doctor system is one of the earliest fundamental tools envisioned for Precision Agriculture. It senses soil and/or crop conditions *on-the-go* and *immediately responds* to the assays to provide real-time corrective chemicals and inputs to farm fields. The technology's suite of real-time sensing and controls produce high fidelity field treatments that optimize economic and environmental benefits. Alternatively, for other embodiments, the overall patent and product base also provide for a step-wise, GIS-based approach to field assessment and treatment. The system features both contacting and non-invasive sensor techniques embodying scientific principles drawn from electrochemistry, *soil complex resistivity*, and soil conductivity technical disciplines. Systems include equipment and methods for transient fertilizer and seeding rate control as well as data management, visualization, and interpretation.

INTRODUCTION

Crop Technology (CTI) broke new ground in 1982, proposing the revolutionary concept that soil chemistry could be assessed on-the-go with treatments concurrently prescribed and delivered through farm tractor-based equipment.

Well documented for many decades, soil physics, soil chemistry, and crop yields are highly variable, typically far greater than many precision agriculture proponents are prepared to address. Like our conventional farming predecessors, many precision agriculture approaches and methods are designed to fit within the capabilities of the majority of the tools at hand, not to specifically address, with fidelity, documented, highly variable farm field conditions nor to prescribe the highly variable treatments to which the crops within these fields will favorably respond.

Many proponents of GIS and GPS technologies are genuinely unaware of the extent of variability present within farm fields and logically approach variability on the scale of existing practices rather than through documented, measured properties. Consequently, data collection frequency using today's GPS–DGPS and farm level GIS analytical tools are not yet well suited to most farm fields and the true scale of field variability that exists therein.

Today, despite the existence of >50 yrs of published, peer-reviewed research, many academics and federal researchers even scoff at the existence of high variability within farm fields.

Copyright © 1999 ASA-CSSA-SSSA, 677 South Segoe Road, Madison, WI 53711, USA. *Proceedings of the Fourth International Conference on Precision Agriculture.*

Many have gone so far as to rationalize that since field variability ostensibly is low, soil sensing technologies conceptually are not required, and therefore any soil sensor technology can be dismissed to favor more conventional practices.

Others admit to the prevalence of variability, but only at the scale that their own technology can address. They subsequently reason that more detailed technology is just overkill. Still, others argue that the accuracy of laboratory testing is the unbeatable standard by which all other technologies must conform or be judged, so the goal of their precision practice becomes taking as many laboratory tests as one can afford, and striving no further. The fact is that precision agriculture should embrace all technologies, from the conventional to the highly technical, if it is to address the economic needs of growers, as it addresses the environmental needs of this country. As government administrators now concur, it is the final analysis of the economic benefits derived in real cropping systems by real growers that ultimately determines technology efficacy, not the many theoretical heated debates on what may or may not result in one theoretical field or another.

Contrary to common belief, CTI is not confined to corn N fertilization technology and soil chemistry assay as its first and only contribution to precision agriculture. Some 15 yrs after initial brain-storming, Houston-Texas based CTI is the premiere company bridging technologies from the petroleum exploration, production, and petrochemical industries to meet the economics, food and fiber, and environmental challenges of precision agriculture.

This paper describes the Soil Doctor technology perspective and approach, including its initial conceptualization focus on N management, background analytical theory, practice, and product performance. Today, Soil Doctor system features enable not only multi-sensor-based variable rate technology (VRT) of numerous inputs, but also optional yield monitoring based on existing measurement circuitry in concert with the system's multi-processor computer design.

CTI sensor technology combined with one-step, integrated VRT today enables:

- VR Anhydrous ammonia responding to soil type and topsoil depth
- Planting population variation responding to soil CEC and topsoil depth.
- VR herbicides responding to soil organic matter and texture variations
- Starter fertilizer tailored to soil CEC variations, and
- Sidedress N fertilizer metered in response to soil CEC, topsoil depth, and soil nitrate levels.

The original Soil Doctor system technique offers the conceptual advantage of not depending on a field location or positioning system, but the overall patent and product base provide the ability to undertake a step-wise, GIS approach to field treatment as well.

Within Soil Doctor technology arts, the user may:

- Obtain and archive soil surveys of physical and chemical properties

(Assaying soil type, CEC, Organic Matter, Soil Moisture, Salinity, Complex Resistivity Values, N, Topsoil Depth, and Soil Fertility.)

- Review data, integrate it within crop models, and determine prescriptive treatments, and subsequently,

- Return to the field and apply treatments based on either or both of the above steps.

Although CTI provides with its system a money-back performance guarantee, the guarantee does not apply when the user elects to write his prescription himself and implement it, whether with or without the Soil Doctor Applicator technology.

"Real-time sensors offer some benefits over map-based techniques for VRT. Real-time sensing is a direct and continuous measure of the attribute of interest thus allowing the user to reduce the amount of unsampled area in a given application. In map-based applications, maps are based on a limited number of samples thus creating the potential for errors in estimating conditions between sample points" (Sonka, 1997)

THEORY

Precision agriculture technical interests in the literature embrace a variety of disciplines. For the purposes of this brief paper, we confine our focus to management of humid regions (nonirrigated) production, where fertilizers are routinely applied annually. We also confine our attention to the manageable factors most often cited by growers in these regions in their personal analysis of nitrogen management for crop production. Under these circumstances, first order effects on cultivar productivity (yield, **Y**) can be related to local soil yield potential (**C**) and soil nutrient level(s) (**N**) by the functional relationship:

$$\mathbf{Y} = f(\mathbf{C}, \mathbf{N})$$

The production variability gradient ($d\mathbf{Y}/dx$) of the field in any direction (x), say for example, along a crop row or transect is given by:

$$d\mathbf{Y}/dx = P\mathbf{Y}_C\, d\mathbf{C}/dx + P\mathbf{Y}_N\, d\mathbf{N}/dx$$

where,

$P\mathbf{Y}_C$ = The first partial derivative of yield with respect to soil yield potential

$\mathrm{P}\mathbf{Y}_N$ = The first partial derivative of yield with respect to soil nutrient status

Despite the fact that many in precision agriculture have set out to derive all of these quantities from field data acquired with instrumentation, experience has taught CTI scientists that some of these values can be best defined based on existing field experience. For example, $P\mathbf{Y}_C$ is the change in cultivar yield with a variation in soil type (e.g., as measured by CEC/topsoil depth). Soil yield potential varies directly with the ability of the soil to provide water during the growing season.

In a given rainfall environment, these amounts are proportional to soil water storage capacity, which is in turn directly related to available soil pore volume. In general, a sandy soil has approximately 2/3 the pore volume of a silt-clay-loam soil.

So, $P\mathbf{Y}_C$ approximates a 100% yield potential increase for each 20 point (meq L^{-1}) increase in CEC. Thus $P\mathbf{Y}_C$ can be considered as an approximate constant in a given mineralogy. $P\mathbf{Y}_N$ is a more complex function to derive if one is operating near the limit in a very high fertility environment. In fact, this partial derivative is often masked by many agronomic practices where heavy N applications or heavy P and K applications are the rule. It is far easier to estimate this function accurately if fertility is maintained below the yield response plateau. PSNT data provided for corn (Magdoff, 1984) and normalized (relative yield) data provides a suitable reference for estimates of $P\mathbf{Y}_N$. This partial derivative is a near-constant positive value in the linear response range of a particular cultivar, but, of course, drops to zero (or becomes negative) when nutrient levels reach and exceed the critical value. Relative yield reaches 100% as NO_3–N approaches 40 ppm.

Critical soil test values are also known for the other major nutrients such as P and K. Therefore, it should come as no surprise that in many landscapes and cropping conditions, which many farm magazines have observed and described that there is generally no correlation between fertility products applied and crop response. This conclusion was reached from field studies where growers are investigating these relationships in fields with conventional fertility management histories and fertility build-up.

In summary, both of the partial derivatives ($P\mathbf{Y}_C$ and $P\mathbf{Y}_N$), to a first approximation, are positive constants. Therefore, the yield gradient within a field can be considered to be directly proportional to a linear combination of the spatial variations of yield potential and soil nutrients within the field. These also are the two factors most often noted by those who practice the arts of Precision Agriculture and the ones most often addressed in field trials assessing Precision Agriculture efficacy.

The derivatives $d\mathbf{C}/dx$ and $d\mathbf{N}/dx$ are spatial variations in soil physical and chemical properties that can be acquired with Soil Doctor technology. Maps of these derived properties can be used to predict changes in yields and yield variability. With excessive use of nutrients, however, such that $P\mathbf{Y}_N = 0$. (i.e., operating on the yield response plateau of a particular cultivar), yield variation will be reduced to a function of only one main variable. *Under these circumstances, yield becomes directly related to soil types alone and soil CEC maps and yield maps are easily, visually correlated.* The independent variable, C, and the dependent variable V are covariant under these conditions.

$$d\mathbf{Y}/dx = K\,(d\mathbf{C}/dx)$$

Throughout the history of development and operation of Soil Doctor technology, CTI has determined that the sensor signals observed and mapped have been directly related to yield potential and yield results.

It is encouraging that USDA-ARS research (Jaynes, 1995) has now confirmed that yield potential and yield maps themselves are well depicted by EM conductivity measurements. Indeed, the world of crops and agricultural practices is not so chaotic that it must be analyzed and modeled by extraordinarily complex algorithms.

Prior production practices (before precision agriculture), at best, sought uniformity in generous fertility levels, rather than fertility tailored to local potential or fertility accounting for pre-existing levels. However, operating at the point of excess nutrient availability (i.e., $\mathbf{\mathit{P}Y}_N = 0$) does not optimize the use of fertility resources over a landscape. For the full range of soil fertility, where $\mathbf{\mathit{P}Y}_N$ = a Constant >0, resource application can be optimized. *Repeatedly demonstrated in typical Midwest crop production since 1987, yields can be increased, while overall fertilizer use is decreased.*

For this discussion, we will use N studies as our example. The large database of relative yield response and side-dress soil NO_3 values (e.g., Blackmer, 1989) demonstrates that relative yield is correlated with PSNT (pre-sidedress nitrate test) level (Magdoff, 1984) at $R^2 = 70\%$. Although the critical value of the PSNT (the partitioning decision for yield response probability) may be judged by some to be approximately 21 ppm NO_3–N (12 in sampling depth), the relative yield response formulae, if differentiated to determine the maximum yield value, demonstrate that relative yield continues to increase until a soil test value of about 40 ppm NO_3–N is achieved. Thus, a soil test value that exceeds the conventional PSNT criteria by about 100% can reasonably be expected to result in higher yields than those available at 21 ppm.

Implicit in the statistical determination of critical PSNT levels is a decision that a relative yield (in some studies) of 84% (compared with 100%) is a reasonable crop production trade-off. Although a pure environmental protection perspective may be consistent with this 16% yield loss, farmers are not receptive to this trade-off and generally have not been informed that with the adoption of the conventional ISU PSNT they should expect a yield decrease.

Because the Soil Doctor system is not based solely on the ISU PSNT, the 16% yield loss is neither inevitable nor expected with its use. CTI customers are advised of this material issue of yield loss with the conventional PSNT to differentiate the Soil Doctor system from some conventional approaches to the PSNT at low soil test values such as 21 ppm NO_3–N. If the Soil Doctor system were strictly based on the conventional perspective of the ISU PSNT, the Soil Doctor system would have decreased economic, and therefore market, value.

Since the Soil Doctor system addresses available N on a spatial basis, it can permit locally higher soil NO_3 levels than 21 ppm, while still reducing total N consumption, and produce an increase in crop yields, rather than a reduction. While not readily understood by many proponents or practitioners of precision agriculture, spatial variation in nitrogen application rates are to be expected and have been predicted (since 1984) by the following analysis to increase yields.

Many of those practicing precision agriculture techniques are still unfamiliar with the concept that redistribution of fertility products can result in both fertilizer savings and simultaneously increased yields.

At Successful Farming's Precision Agriculture Chat Line, a prominent purveyor of GIS software, summed up his lack of knowledge by drawing the analogy: "Breathe less oxygen and live longer." That analogy, of course, was poorly conceived. The availability of soil-supplied N and the nonlinearity of crop yield response to applied nutrients indeed makes it easy to reduce overall N use and at the same time increase crop yield as described herein.

As reported (Meisinger, 1984) soil NO_3 levels are highly variable. A measure of this variability is the coefficient of variation (CV) of the distribution of soil NO_3 levels. The average CV of data reviewed in this reference is about 45% with ranges from near 30% to well over 100%. (Note: CTI's field research from 1984 through 1992 revealed Midwest corn CVs ranging from 33 to 65% in plots, subplots, and whole fields.)

The significance of the CV is doubly important when it is realized that an intensive soil sampling effort is necessary if one is to hope to have scientific confidence in any mean soil value determined for a field. ISU extension publication Pm-1381 for PSNT sampling recommends that only two dozen cores are to be taken in about 10 acres. Meisinger indicates that when the sampling density is this low, the confidence interval widens drastically. A 90% confidence interval at only 24 samples means that, if a field had a mean of 20 ppm NO_3-N, there would be a one in 10 chance that values of 0 and 30 ppm would also occur within that sampled field.

In the use of the conventional PSNT, only a mean value of field soil NO_3 level is sought and a single prescription is recommended for whole field application. *Since only a mean is documented*, the CV of the field is masked, field variability is unknown, dismissed, and often falsely regarded and relied upon as nonexistent.

The significance of NO^3 availability to the crop can be assessed both by analysis and through on-farm field tests of the Soil Doctor applicator. To analytically assess the effect of NO_3 variability on corn yield, we present below the analysis procedure CTI first employed in 1984.

The Mitscherlich equation and the normal distribution function were incorporated into a TK!Solver model. The yield response to the PSNT level is dictated by a Mitscherlich model as:

$$Y = Ym\,(1 - e^{-C\,(NO_3\text{–}N)})$$

Here, *Y*m is the maximum yield attained (100%) and *C* represents a yield efficiency index after Bock (Bock, 1984). The value of *C* is obtained after defining the relative yield (*Y*/*Y*m) at a specified critical mean soil test level. For this analysis, the distribution of soil NO_3–N values is assumed to be normally distributed. The fraction, *F*, of the field producing yield at any given soil test N (i.e., NO_3–N) level in the distribution is given as:

$$F(N) = (\mu\, CV*2\pi)-1\; e\; -.5\; ((N - \mu) / \mu\, CV)2$$

The total yield produced by the distribution fraction is the integral of the individual contributions of the N values in the field, i.e.,

$$Y = Ym * Fdx,$$

where the integral is performed between 0 and the upper limit of N defined by the mean and coefficient of variation for the soil test NO_3–N distribution.

As an example, a typical 160 bushel maximum yield model was assumed with a yield decrement to 95% of maximum (152 bushels acre^{-1}) at 21 ppm NO_3–N. Note, however, that the relative yield percentage (95%) for the 21 ppm NO_3–N criterion differs significantly from the 84% relative yield later reported (Blackmer, 1989). The referenced equations provide relative yield response correlation.

$$\text{Relative Yield (RY)} = 27.28+3.71*NO_3\text{–N}-.05*(NO_3\text{–N})^2, \text{ and}$$

$$RY = 33.43+3.24*NO_3\text{–N}-.04*(NO_3\text{–N})^2$$

Analysis of these equations demonstrates that the relative yield increases to a maximum (100%) when soil NO_3 levels reach about 37 to 40 ppm NO_3–N. Beyond 40 ppm NO_3–N, toxicity may be indicated because relative yield decreases as determined by these relationships. The preceding two equations were used in this paper to define the upper limit for NO_3–N fertility and the relative yield at a selected critical PSNT level. The selected critical point (e.g., 84% RY and 21 ppm NO_3–N) was incorporated into the Mitscherlich equation (in lieu of the quadratic fit above) for response analysis. No toxicity penalty has been included in our Mitscherlich equation analyses since yield response data normally indicate a broad plateau or asymptote rather than toxic effects.

Depending on how soil N is applied on a variable basis, corn crops can be produced on soils producing levels of 40 ppm NO_3–N that are economic, even though interpretation of the conventional PSNT recommends that N not be applied to such a field. Alternatively, with Soil Doctor variable rate management, at *mean soil levels of about 25 ppm NO_3–N* (but with small CVs), yields comparable to or exceeding those at 40 ppm are not only possible but demonstrable through field trials.

A parametric study of yield as a function of (0-12 in) soil mean NO_3–N (from 10 to 40 ppm) and CV (from 0 to 50%) was conducted. The results are graphically illustrated in Fig. 1 and 2. At relatively common CVs of about 30%, improving the uniformity of N availability will result in about 3.5 bushels acre^{-1} yield increase. If, however, the field is of average variability (CV = 45%), a yield increase of about 15 bushels acre^{-1} can be expected. These results are for the analysis case of assuming that there is only a 5% change in field mean yield (95% RY) as the soil mean NO_3–N ranges from 21 to 40 ppm NO_3–N.

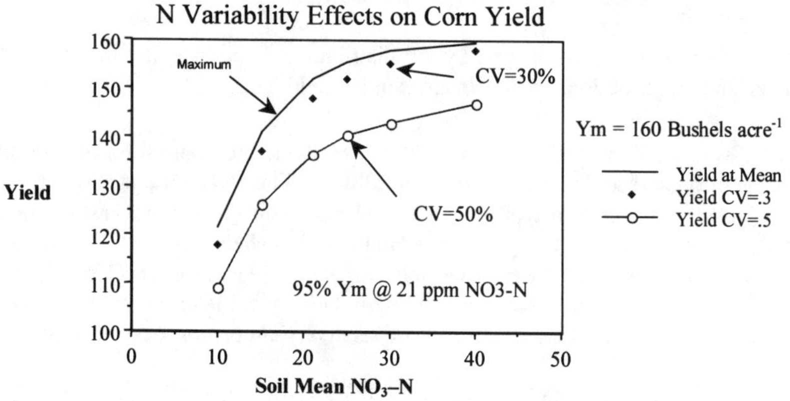

Fig. 1. Nitrogen variability effects on corn yield (95% *Ym* @ 21 ppm NO_3–N).

When an alternate yield response curve (Blackmer, 1989) model is included (i.e., 84% RY @ 21 ppm NO_3–N), the analysis of the sensitivity of yield to both soil mean NO_3–N level and CV indicates a comparable range of yield benefits. Results of this modeling analysis are summarized in Fig. 2. From both of these analyses, it is readily seen that the higher the soil mean NO_3–N, the higher the corn yield, and that the lower the CV of the distribution, the higher the yield. Thus, the same yield can be obtained at a lower available N, if the distribution has a low CV.

Further, at any soil mean NO_3 level, *the potential for a yield increase* (in a nominal 160 bushel acre^{-1} maximum environment), ranges from *an analytical minimum of about 3.5 bushels acre^{-1}* to a *maximum of about 15 bushels acre^{-1}*, depending upon the soil variability (the actual CV). If the CV of a field is >50%, greater yield benefit and N conservation is possible.

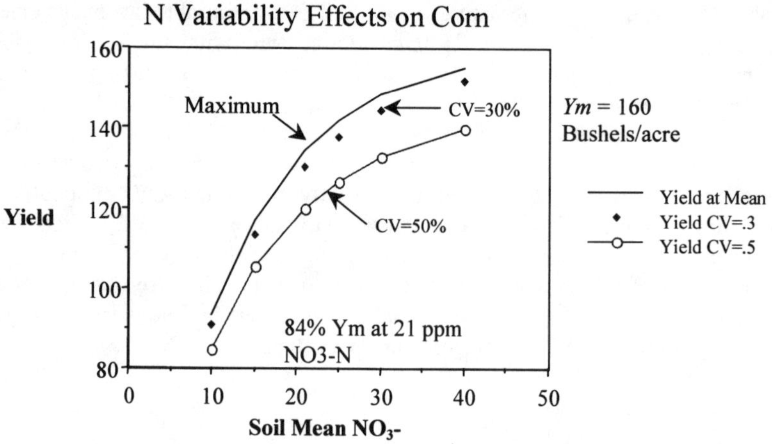

Fig. 2. Nitrogen variability effects on corn yield (84% *Ym* @ 21 ppm NO_3–N).

Those who have intensively studied the PSNT theory agree that 40 ppm NO_3–N is clearly sufficient for a 160 bushel acre^{-1} corn crop. There is, however, no agreement on a single, precise critical level of NO_3–N above which there is low potential for a positive yield response to applied N. The 21 ppm level first advanced by ISU, although favorable to water quality standards, was an arbitrary choice that did not fit all cropping conditions. University of Connecticut and New Hampshire research indicate 35 and 30 ppm to be their choices for critical levels. In the absence of a regulatory mandated level, a critical range from 25 to 35 ppm is judged to be reasonable, particularly when combined with the concerns of the customer for yield. Detailed N management with Soil Doctor variable rate technology enables yield optimization within the context of any regulatory constraint. The appropriate critical value, like the recommendations of many states, will vary with crop yield goal, as will the application rate to be applied.

As a first example from these data using the ISU 84% RY yield model at 21 ppm NO_3–N, a corn yield of 139.7 bushels acre^{-1} is projected at a mean PSNT of 40 ppm and a representative CV of 50%. Increasing the uniformity of availability to the crop, but restraining the mean NO_3–N to 25 ppm, a yield of 140.14 bushels acre^{-1} is indicated at the resulting smaller CV = 20%. Reducing variability to CV = 0%, would result in a yield of 141.94 bushels acre^{-1}, at 25 ppm NO_3–N soil mean.

Thus, a 2.2 bushel acre^{-1} advantage could be expected at a Soil Doctor equivalent On-the-Go PSNT of 25 ppm, along with a total N savings of 60 lbs acre^{-1}. If the Soil Doctor variable applicator were operated to provide a tight (CV = 0%) distribution at 30 ppm NO_3–N, and the comparison is again made to the PSNT of 40 ppm at CV = 50%, an 8.6 bushel acre^{-1} advantage could be expected along with a lower N savings of only 40 lb acre^{-1}.

If we turn our attention to a second example, i.e., the relatively flat yield response curve (*RY* = 95% at 21 ppm NO_3–N), a yield increase of 8.5 bushels acre^{-1} would be expected by comparing 40 ppm, CV = 50% with 25 ppm, CV = 0%. At 30 ppm, CV = 0%, the yield increase would rise to 10.8 bushels acre^{-1}.

These examples indicate that field variability, yield response curve characteristics, and mean soil NO_3–N all influence the potential amount of N savings and yield increase possible. The critical PSNT level is not yet precisely defined, and the precise yield efficiency of any seed variety on the market in its growing environment is largely unknown. Since the acceptance of any product in the agricultural marketplace is tied to its economic benefits to the customer, the operating strategy of CTI's Soil Doctor system has been and continues to favor higher soil NO_3–N limits and delivering yield advantages with contemporaneous placement. The more uniform the resulting distribution of total N becomes, the greater the yield increase that is possible.

By inspection of the analysis of the ISU example (*RY* = 84% @ 21 ppm NO_3–N), adhering to a precise, uniform value of 21 ppm would result in a yield loss of 5.3 bushels acre^{-1}.

This projected *loss is consistent with* a conventional management assessment of the impact of *adopting any manual version of the PSNT* in comparison to a precision approach. In comparison, then, on yield alone, variable rate Soil Doctor technology provides an economic advantage of 8 to 15 bushels acre^{-1} over the conventionally interpreted PSNT.

Irrespective of the growing interest in the environmental consequences of N use in agriculture, most growers have yet to come to grips with the concept of nitrate in their groundwater and the deleterious effects of hypoxia in our coastal waters. Growers have, however, the clear perspective that production economics drives their business.

Thus the Soil Doctor system advantage is not in its ability to adhere to an industry standard of a precise soil test value, but in its ability to optimize crop production while conserving expensive, added nutrients. Responding to in-field variability at any soil test level (even at 40 ppm mean NO_3–N with a non-zero CV) will provide an economic advantage to the grower. The more samples processed and spatially responsive treatments made, the greater the economic return from VRT.

FIELD PERFORMANCE AND CONCLUSIONS

The time-honored, wide-spread tradition of asking the plant what it thinks is the only true, nonspeculative measure of Soil Doctor performance or any variable rate technology performance. Any other measure of accuracy, albeit fascinating or supportive to engineering development by others in research, is, at best, incidental since crop production efficiency alone represents the acid test of efficacy.

Landscape and cropping conditions set the stage upon which any precision technology will operate and are primary factors which influence the magnitude of benefits any technological system or procedure may produce in a test. Despite this caveat, Crop Technology guarantees that its Soil Doctor System, when "properly installed, maintained, and operated, will provide a measurable economic advantage" OVER the grower's current practice, the first year on the farm – or the purchase price will be cheerfully refunded. The economic guarantee is measured by side-by-side tests of fertilizer consumption and yield production to provide objective assessment of economic benefits. No matter where one farms, if CTI sells that person a Soil Doctor applicator, it comes with that guarantee.

The earliest production model Soil Doctor sensor-knife systems (1990 and 1991) in the Sensors On mode, achieved a yield increase in the range of 2 to 4 bushels acre^{-1} while concurrently reducing N consumption by roughly 50% of that which would normally be side-dressed by a grower.

Since CTI's performance guarantee was initiated with its 1992 RE (Rolling Electrode) model, a number of independent tests under a variety of cropping conditions have been completed.

Test data provided by researchers and Soil Doctor system customers to CTI indicate typical economic benefits, in production corn, ranging from a low of $17 acre^{-1} to over $40 acre^{-1}, with an average of $25 acre^{-1}. *Typical N savings are 35 lbs N acre^{-1} compared with a typical customer's best conventional practice, with a concurrent yield increase of 6 bushels/acre.*

These economic benefits are tangible. The learning experience operating behind the tractor wheel however is intangible. It provides a grower insight into soil science in a manner similar to, with lessons far different from, the yield monitor learning experience. Additionally, Soil Doctor soil properties and application record maps provide a robust data set which visually and statistically solve many of the mysteries which most yield maps have collaterally generated with their information.

CTI holds two issued domestic patents which describe further details of its technology base. We invite you to view these publicly accessible documents: U.S. 5 033 397 and 5 673 367 to address the technical scope embodied in these arts. For further information please contact us directly or visit our web site at http://www.soildoctor.com.

REFERENCES

Bock, B.R. 1984. Efficient use of nitrogen in cropping systems, p. 274–277. *In* nitrogen in crop production. ASA, CSSA, and SSSA, Madison, WI.

Blackmer, A.M. et. al., 1989. Correlations between soil nitrate concentrations in late spring and corn yields in Iowa. J. Prod. Agric., 2:103–109.

Jaynes, D.B., et. al., 1995. Yield mapping by electromagnetic induction. p. 383–394. Site-specific management for agricultural systems. ASA, CSSA, and SSSA, Madison, WI.

Magdoff, F.R., et. al., 1984. A soil test for nitrogen availability to corn. Soil Sci. Soc. Am. J. 48:1301–1304.

Meisinger, J.J. 1984. Evaluating plant-available nitrogen in soil-crop systems. p. 400–403. *In* nitrogen in crop production. ASA, CSSA, and SSSA, Madison, WI.

Sonka, S.T., et. al., 1997. Dimensions of precision agriculture. p. 32–33. *In* precision agriculture in the 21st Century. Nat. Academy Press, Washington, DC.

Centimeter-Precision Guidance of Agricultural Implements in the Open Field by Means of Real Time Kinematic DGPS

R.P. van Zuydam
Agrosystems Department
Institute of Agricultural and Environmental Engineering IMAG-DLO
Wageningen, the Netherlands

ABSTRACT

Multi-year precision guidance of agricultural implements in the open field enables the exact placing of inputs, and to grow crops on predetermined spots. The on-line system proposed here, uses a digital map that contains all co-ordinates to describe the intended path of the implement in the field, a sensor to measure the actual position of the implement, a comparator to calculate the position error, a controller to generate a correction signal and an actuator, mounted between the tractor and the implement to side-shift the implement onto the intended path. On a test track a tractor was driven at a speed of 5.2 km h^{-1}, what resulted in a repeatable sideways sway of the tractor plus and minus ±10 cm. The implement, mounted on the backside of the actuator, was side-shifted to a straight path programmed in the digital map. The true path of the implement was recorded and showed a deviation from the straight line of less than ±2 cm.

INTRODUCTION

Decrease of farming inputs demands, among other things, an accurate placing of the inputs. Prevention of the application of inputs on spots where they do not serve the crop, saves inputs and helps to relieve the environmental burden. This is even truer at increasing row distances, in row cultures.

For precision placing of inputs, a very precise steering of agricultural implements is a key technology. To successively fertilize, prepare seedbed, sow or transplant, and spray uniquely on the same spot or band in the field demands high accuracy steering of the implements. Also the placing of adjacent runs to prevent overlap demands fine steering.

The design of the automatic steering system, presented in this study, is based on the assumption that the tractor drives more or less accurate along the desired path. How the tractor itself is steered, by man or automatic, is not relevant here. Only the presence of the tractor in the vicinity of the exactly defined path—within the capture range of the correction mechanism—is assumed. It also should be understood, that the exact position *on the path* is not relevant in this approach either. The only target is to have the implement covering the predetermined trajectory with high precision, at any speed, time after time.

Copyright © 1999 ASA-CSSA-SSSA, 677 South Segoe Road, Madison, WI 53711, USA. *Proceedings of the Fourth International Conference on Precision Agriculture.*

Van Zuydam et al. (1997) described a format to identify a predetermined path on an agricultural field. This format is a very compact way to lay down all information needed to construct the ideal path the implement has to follow on a field. The same paper described the set-up of a correction system to keep the implement on this path, when the tractor position differs slightly from the predetermined path. De Wit (1953) indicated that band fertilizing of sugar beet could save N-inputs. Van Zuydam *et al.* (1993) proved in experiments with band fertilizing of sugar beet by laser-controlled equipment these savings to be about 23% at the same sugar yield on medium clay soils at the experimental farm Oostwaardhoeve. Mechanical weed control could be done very effectively, with a wide working width between the rows and at high speed (14 km h^{-1}) due to the laser steering of the equipment (Van Zuydam et al., 1993). In these experiments the vehicle, a 12 m wide Dowler gantry, was steered by man while the implement was controlled by a stationary laser transmitter on the headland. The average steering accuracy was ±2.6 mm over the full 220 m long field. In this set-up, 88% of the surface of the field was hoed, at least in the first two sessions when the root development of the crop allowed the use of the wide hoeing knives. The saving on the use of herbicide, by spraying only the narrow crop rows 10 cm wide, was 80% compared with full field application.

The results induced further research on the development of a universal guidance system, as described hereafter, to make it possible to guide any implement with centimeter-accuracy along a predetermined path, applicable with normal tractors and normal implements, and with no field-bound investments like beacons or guide wires.

METHOD AND MATERIALS

To measure the position of the implement Real Time Kinematics DGPS was chosen. A set of two Trimble 7400 MSi receivers was used, one as a base station and the second as the rover. Via a radio link, also from Trimble, the differential position information was transmitted by the base to the rover once a second. The rover was programmed to measure its position five times a second, and to transfer a very short message to the computer including time, position coordinates in X- and Y-direction, and an indicator for the quality of the data.

The use of satellite navigation had the great advantage that no field-bound investments, like buried cables or corner reflectors, had to be used, so that the system is universal in use and virtually any field can be used without great preparations.

In 1996 and 1997 tests were done to measure and check the accuracy of the position sensor. Therefore, measurements were done continuously on one spot, and also when the antenna was mounted on a revolving table, so that the antenna covered a circle-shaped path with a diameter of 78 and 39 cm at 11 rpm.

Another problem to be solved was the creation of an electronic map of the trajectory to be followed. Especially the accessibility of the measured momentary position in the map for the computer within the limited time interval between two position measurements caused a lot of trouble. Therefore, it was decided to describe the trajectory as a collection of adjacent straight lines, each one identified by the coordinates of its beginning and its end. These lines were short when they formed a curve together, and (very) long when the trajectory in between was a straight line. The coordinates of each measured position were transformed to co-ordinates in another grid, the Rotating Grid, as described by Van Zuydam (1998). This transformation made a very fast comparison possible of the measured position with the desired line to follow, and so to calculate the positional error square to the direction of travel.

Correction of the position of the implement was done by side-shifting the implement in relation to the tractor. Therefore, a side-shift device was mounted to the three-point linkage of the tractor, being the side-shift of a forklift. The forklift itself represented the agricultural implement. The hydraulic cylinder used to side-shift the fork lift was controlled by a proportional hydraulic valve, that in turn was controlled by the computer. The feedback of this system was obtained by mounting the satellite receiver antenna on the forklift, i.e., the moving part of the system.

To measure the performance of the system, experiments were done on the experimental farm Oostwaardhoeve in Slootdorp, the Netherlands. The experiments included the driving of a tractor over a test rail, where it was forced to cover a well-known, repeatable swaying movement, at different speeds. This test trajectory also was described by Van Zuydam et al. (1997). The mounted implement, connected to the tractor by means of a side-shift device, should not follow the swaying movement of the tractor, but cover a straight line instead. This intended line to follow was stored in a computer memory. The RTK DGPS sensor measured the actual position of the implement, and this position was compared with the desired position from the computer memory. An error was calculated and a correction signal generated which caused the side shift to move and to compensate for the tractors deviation from the straight line. The implement was locked in this way on the straight line from the map in the computer memory.

Navigation and Datum

All positional coordinates refer to a coordinate system, a *datum*. A lot of co-ordinate systems are described in the past, as a result of various geodetic surveys. The latest, and most precise datum is identified as WGS 84 (World Geodetic System 1984).

Positional data, obtained by GPS (Global Positioning System; satellite navigation), are measured and expressed in co-ordinates referring to WGS 84. This is important only when using geodetic information (maps) as a reference, because they refer to a datum as well. The accuracy of coordinates is always related to a datum, as datum differ from each other in position. When navigating on cm-level, the position of the datum is of importance.

Using coordinates related to a local coordinate system, or grid, can have characteristics that suit the specific purpose better. The origin of such a local grid can be placed anywhere, e.g., at the position of the base station of the DGPS navigation system. It can be of advantage to maintain the orientation of the local grid similar to that of the WGS 84 datum, i.e., the *X* and *Y*-axis of the local grid representing the directions N and E in WGS 84. Also, the dimensions of the co-ordinates can be meters in the local *X*–*Y* grid, where they have to be in degrees longitude and latitude in WGS 84. Within the limited dimensions of an agricultural field, earth rounding can be neglected. The errors, made due to undulations in the field, are only errors in an absolute sense. As *repeatability* of the trajectory is the most important, constant errors that are made every time in the same way, do hardly affect the overall result. This means, so far, the existence of two coordinate systems. The first is the WGS 84, located fixed to earth, with zero longitude corresponding with Greenwich meridian and zero latitude corresponding with the equator. Within this system a second grid exists, with *X–Y* coordinates, of which the origin is at a known point in WGS 84 and orientation is parallel to WGS 84. On the field, a third coordinate system is of advantage, the so-called Rotating Grid. This grid, which axes are labeled *R–S*, is trajectory oriented, and will be described hereafter.

Position Sensor

To obtain high-quality measurements of a moving object in general the sensor needs to have a sufficiently high output rate as well as sufficient accuracy. Also the delay, or latency, of the information may not be too great. The information about the coordinates of a measured position must not reflect on a position too far in history. The longer the data processing in the sensor proper takes to calculate the coordinates of the position before transfer, the further the implement has moved already.

In sugarbeets, as an example, the row distance is 0.5 m. Auernhammer (1990) indicates a steering accuracy for implements working on or between plant rows of ±1 cm. This performance demands a very high accuracy of the position sensor. When maximum working speed is limited to 2 m s^{-1} (7.2 km h^{-1}) the frequency of the successive measurements has to be high as well. A repetition frequency of 5 Hz for instance will mean only one position fix per 40 cm of travel. The same calculation counts for the delay: if it takes 200 ms to calculate position coordinates, the implement has moved another 40 cm in this time span.

In this research, after ample discussions, RTK DGPS has been chosen as positioning system. This type of Differential Global Positioning System uses Real Time Kinematics to improve the accuracy of a position fix by phase comparison of the carrier signal. This technique claims sub-centimeter accuracy at a 1 s interval, or ±3 cm accuracy at a 200-ms interval. Latency will not exceed this 200 ms period, and the exact time of positioning is retrievable in the output message. Processing and filtering the data will potentially make the system even more accurate.

RTK DGPS uses a stationary base station as a reference point. The corrections calculated by this base station are transmitted to the moving sensor or rover, to improve accuracy. The *X* and *Y* coordinates (a vector) of the rover in the local grid are calculated in a format MMMMM.MMM meter, so the resolution is at mm level. The origin of this vector is on the position of the base station. In this experiment the base station was positioned at about 80 m from the test site, in the free field on a known geodetic point. The coordinates of this reference point were 52° 49' 27.3879" N and 4° 54' 41.7984" E in the WGS 84 datum. Antenna Ellipsoidal Height (not antenna altitude) was 42.290 m. This reference point is the origin (0,0) in the local grid in which the position data *X* and *Y* of the rover are expressed.

Electronic Map

The data-file that contains the information of the trajectory to be covered was called the electronic map. The map contained all information reflecting on a certain field and on a certain implement to allow the steering system to guide the implement over the field. It is clear that this map has to follow a certain format, and that the access to this format has to be very fast. The demands for operation were:

1. The system must automatically detect if the implement is within reach of the correction system on any place of the field
2. The system must automatically detect the direction of travel, it must be possible to follow the trajectory in reverse direction or to only partly drive or repeat the trajectory
3. The system must automatically tune in when only a part of the field is cultivated, or when operation is abandoned.

The trajectory that describes the predetermined path over the field, that the implement had to follow, was made as a list of numbered coordinates. The trajectory was supposed to be an array of adjacent, straight lines, which built up the total trajectory to drive (polygon). On straight parts of the path these coordinates are far apart, in curves they are close together. Each coordinate had also a code attached that defined if the line to follow (in numerical order) was a crop row or an idle move. Idle moves were defined to be the passes going from one crop row to another, e.g., on the headlands. Here, the correction system could be disabled, while the implement was lifted out of working position by the driver. The coordinates were expressed in the local grid, as *X–Y* co-ordinates (in m) with respect to the origin, the position of the base station. So each line of the map had a number, an *X*-coordinate, a *Y*-coordinate, and a code.

In the experiments, the correction of the position of the implement was done by means of a side-shift device that was mounted between the tractor and the implement. When the tractor was approximately in the right position, the side-shift displaced the implement to the exact position, precisely on the centerline of the trajectory. Alongside the centerline of the described path was an imaginary band that corresponded with the capture range of the side shift. Only when a measured position was within this band, correction of the position of the implement was potentially possible. In other words this band represented the deviation of the tractor that the driver was allowed to make when the implement was still on its course (Fig. 1).

Calculation Algorithm for the Position Error – Rotating Grid

One of the features to make the calculation of the error very simple (and fast) was the introduction of the Rotating Grid. This third coordinate system, with axes *R–S*, was at any moment in time crop row oriented. That means, that the grid had its origin at the same point of the electronic map as the origin of the *X–Y* grid (so the position of the base), but its *R*-axis was oriented parallel with the momentary (part of the) trajectory to cover, so parallel to the straight line connecting the IPP (Immediate Passed Point) and the NPTP (Next Point To Pass, Fig. 2). When the NPTP was

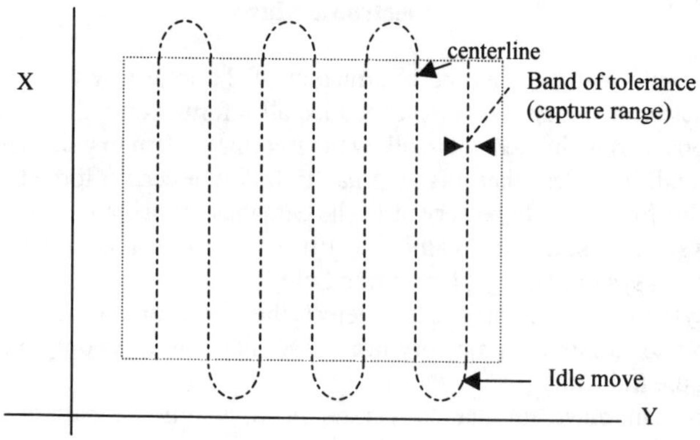

Fig. 1. Field plot, trajectory of the implement and band of tolerance for the tractor to drive.

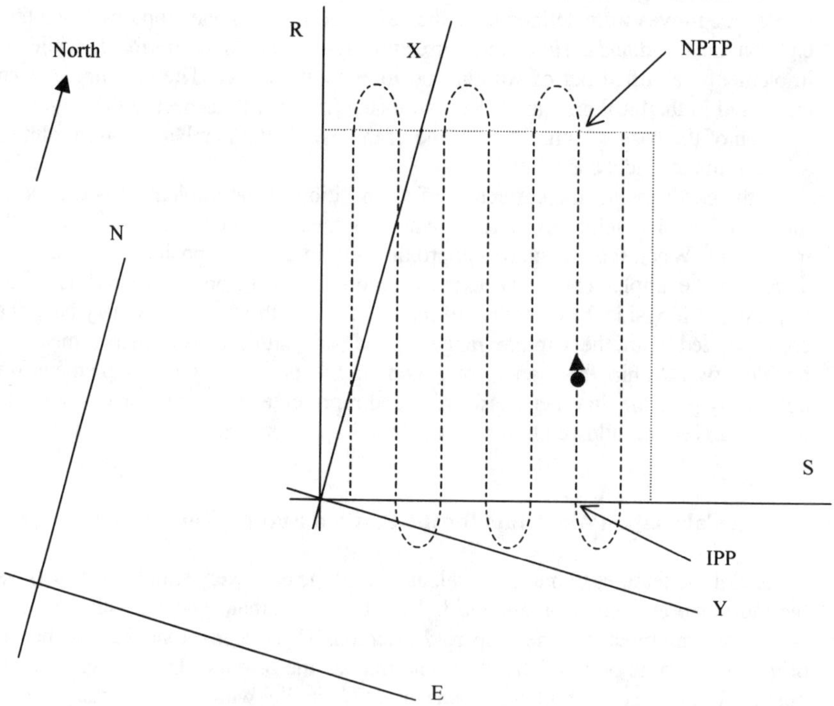

Fig. 2. The rotating grid *R–S*, the local grid *X–Y* and the global grid *N–E* (WGS 84). IPP: Immediate Past Point; NPTP: Next Point To Pass.

reached, the Rotating Grid rotated to its next orientation, its origin still at the position of the base station, but its R-axis parallel with the next part (line) of the trajectory. The line to travel is then easily described as $s = s_{row}$.

Transformation of any measured position, expressed in the *X–Y* grid to the *R–S* position immediately by subtraction of the constant value s_{row} from the value of the s-coordinate of the measured position. This procedure results in fast calculations. During the time that the tractor traveled from the IPP to the NPTP, the *R–S* grid was stable and semi-permanent.

Side-Shift Device

As indicated above the steering system will control the position of the implement. It will have to fine-tune the position of the implement to exactly the predetermined path while the tractor is steered already approximately along this path. (Whether the tractor is steered manually or automatically is not relevant) Thus the implement will be side-shifted to the correct position, on the path, with respect to the tractor that is driving more or less accurately along this path. As a matter of fact, the deviation of the tractor from the path shall not exceed the capture range of the side-shift mechanism, e.g., 20 cm. The implement is locked to the path and will follow the path with high accuracy.

The total system can be imagined as an accessory between the tractor and the implement, constructed and built as short as possible and adaptable to every tractor-implement combination. Necessary electronics can be housed on top of the side-shift or be mounted at a remote position on the tractor, from which battery and hydraulics it is powered.

The mechanical part of the system can be of a simple design: a hydraulic cylinder moving the two parts of the side-shift device with respect to each other in lateral (with respect to the direction of travel) direction, and ample free space to let the PTO shaft - through which some implements are driven by the tractor - pass in every position. The implement, which is mounted to the backside of the side-shift is in this way sideways positioned onto the correct path.

Simulation

In practice, the following factors affect the tracking error,

- RTK DGPS measurement errors.
- side shift measurement errors.
- deviations between the calculated and actual shift caused by the control system.
- the connection between the tractor, actuator and implement is not stiff.

In the simulation only the first and third factor were taken into account. The errors caused by the second factor may be neglected if accurate sensors to measure the side shift are used (e.g., optical encoders) and if the actuator is sufficiently stiff and rigidly connected to the tractor. Then the side shift can be measured up to about 1-mm accuracy. The fourth component might in practice cause errors of the same order of magnitude as the first. With respect to the RTK DGPS sensor errors, it is important to note that they consist of position measurement errors, and, if the sensor (tractor) is moving, of errors due to *latency,* i.e., the fact that the position changes during its computation. Now the errors caused by latency are not simulated because the measurement error signal was obtained from true RTK DGPS measurements with the sensor at a *fixed* position. Also the construction of the measurement error signals used in the simulation presumed that there are no offsets. In practice there will be offset-errors, which add to the simulation errors. The MAE (Mean Averaged Error) turned out to be $1.17 * 10^{-2}$ m (Dijksterhuis et al., 1998).

The errors caused by the PID control system relates to the settling-time, the overshoot and the varying setpoint and were found through simulation with undisturbed sensor data. The tracking error which resulted from this simulation had a MAE of $2.62 * 10^{-3}$ m. After the PID controller had settled the errors were <0.5 cm (Dijksterhuis et al., 1998).

Applying a Chebyshev filter to try to improve the measurements in the simulation resulted in an increase of the MAE tracking error ($1.29 * 10^{-2}$ m). Most probably this is caused by the delay introduced by the filter that affects the tracking error, again because the sensor (tractor) is moving.

RESULTS

Accuracy of the Sensor

As the quality of the position determination will be of key importance for the system, several tests have been done to establish the accuracy of the RTK DGPS receivers.

The first test was to establish if the signal is consistent in time. Therefore, the rover was placed in a stationary position about 1 m aboveground level, on a round table with a top of wood and a metal support. The test was executed at midday, and the signal was monitored for 2.5 min at 5 Hz resulting in 771 fixes. The table was positioned in the free field, and the nearest obstacles (20 m high trees) were at 150 m distance but only to the south. The rover had a free sight to the sky over 360°, in any case at heights exceeding 15° aperture angle over the horizon. The

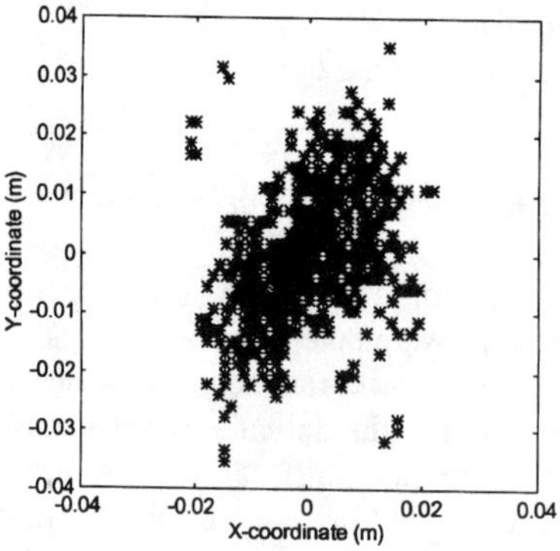

Fig. 3. Plot of the raw position data of a stationary rover at 5 Hz. The true position is assumed to be (0,0).

sensor used eight satellites and indicated an RTK-fix quality of the position, which means that it worked in the highest quality mode. All individual fixes (raw data) were within a rectangle of 4.26 $\ast$ 7.03 cm. Statistic interpretation of the individual position fixes gave var (X) = 7.38 $\ast$ 10^{-5} m^2 and var (Y) = 1.22 $\ast$ 10^{-4} m^2 (Fig. 3).

A second trial was with the rover moving along a circular path. This was done by mounting the rover on the same table as in the first test, but revolving the table at constant 11 rpm. The rover aerial was mounted in two positions, resulting in circles of 78 and 39 cm diam.. The test was done at the same place as the first test, so the sight was identical, at 14.00 h UTC. The aim of the test was to see if the measured path was a circle, if the diameter of this circle was correct, and if repetitive revolutions described the same circle. The table was rotated for approx. 1.3 min, resulting in 14 monitored circles as an array of successive position fixes. In total there were 400 position fixes. Plotting these data gave images that coincided well with the true trajectory. Covariance of the X and Y data were 6.97 $\ast$ 10^{-4} m^2 and 1.18 $\ast$ 10^{-3} m^2, respectively. The true scale images are given in Fig 4. The standard deviations of the position fix were 2.64 $\ast$ 10^{-2} m and 3.43 $\ast$ 10^{-2} m, respectively.

The third test reflected on the subject of repeatability of the measurement of determined motions as well. After all, it does not matter too much, whether the measured position data, e.g. degrees longitude and latitude are correct and accurate in the geodetic meaning of the word, but most important is it that a certain position gives the same readings every time it is measured. These relative coordinate readings or bearings can have the origin of their coordinate system at any point, e.g. at the position of the base station. When successive machine operations, like fertilizing, sowing and hoeing are executed with the use of the same navigation system, all operations will be on the same spot.

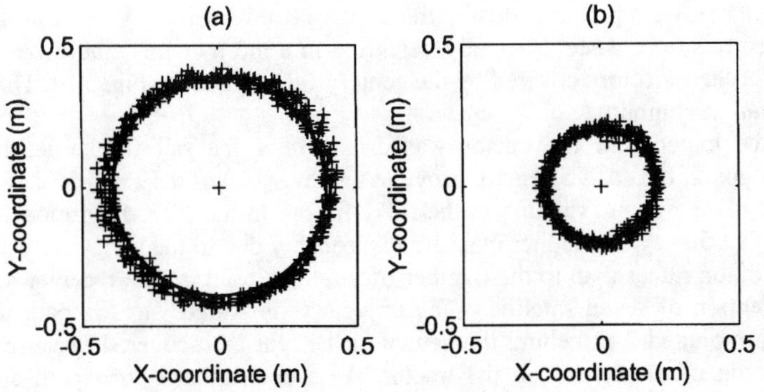

Fig. 4. Plot of position fixes (raw data) of an RTK DGPS receiver in circular motion. (a) true diameter 78 cm, (b) true diameter 39 cm, both at 11 rpm. The drawn circles indicate the true path.

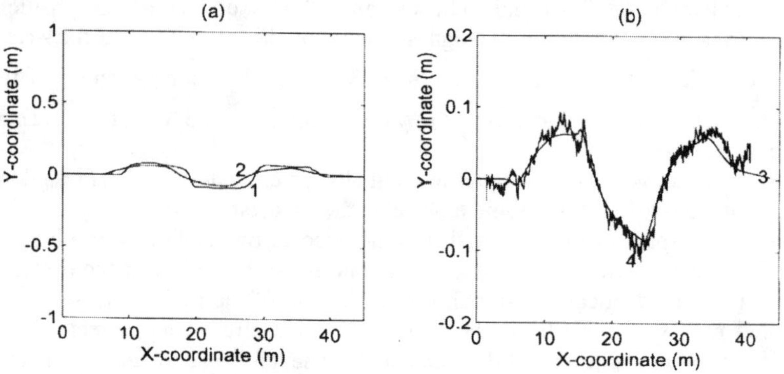

Fig. 5. Graph 1 represents the shape of the rail over which the nose wheel of the tractor was driven. Graph 2 indicates the trajectory of the center of the rear axle for the given tractor dimensions. Graph 3 is the modeled path of the rovers antenna (2 m behind the rear axle), Graph 4 indicates the measured raw position data of the rovers antenna.

A special construction was installed to make it possible to drive a tractor repeatedly along a predetermined path, which deviated in a known way from the straight line, which the implement would have to follow. A tractor, IHC 523, was modified to drive over a steel track that was bolted to a concrete field road.

The steel track was made of an inverted T-bar that was bent (undulated) in horizontal direction, to create the undulating path for the tractor. The tractor was equipped with a retractable nose wheel that was lowered onto the rail so the tractor front wheels were floating. In this way the now three-wheeled tractor could drive the 50 m trajectory with a known, repeatable deviation from the straight line. The exact shape of the rail is recorded in a data file containing the exact coordinates of the rail with an interval of 15 cm (Fig. 5-a). This curve represents only the path traveled by the tractor nose wheel: the vehicle as a whole followed this nose wheel gradually, resulting in a more or less sine wave-shaped curve for the trajectory covered by the center of the rear axle (Fig. 5-b). The curve of the rail has amplitude of 7.5 cm.

In the experiment the tractor was driven over the rail five times logging NMEA position data. The tractor drove at a slow speed of 0.19 m s-1 (0.68 km h-1). The rover was not running in the RTK-fix but in the RTK-float mode, which means that there was a higher inaccuracy, probably due to the momentary satellite constellation rather than to the number of satellites used, as the receiver reported the reception of seven satellites. The trajectory measured was the path that the antenna mounted 2 m behind the center of the rear axle covered because of the positioning of the sensor on the tractor. As a consequence, this path does not correspond with the shape of the rail. The graph of one of the five repetitions is shown in Fig. 5-4.

Graph 3 in Fig. 5b indicates the calculated line the antenna followed using a fire truck calculation model (Bushnell et al., 1993), and using line 1 from Fig. 5a as input.

Accuracy of the simulation

RTK DGPS gives position fixes with a few centimeters inaccuracy with current equipment, even in mobile applications. Given these errors, simulation revealed that interrow guidance is possible using RTK DGPS.

The control of the implement position, as described in this paper, has the advantage that the control system does not need any information concerning the tractor dynamics and is more or less independent of the way in which the tractor is driven, given the limitations (capture range) of the correction mechanism, the side-shift (here ±10 cm). On the other hand, given the independence from the way in which the tractor is driven, the way in which the desired path is traveled *in time* is not controlled by the system

Accuracy of the System on the Test Track

When the tractor drove the rail trajectory a swaying movement of the GPS antenna resulted (Fig. 6 line disabled). When the control system was enabled, and the predetermined trajectory defined as a straight line, the recorded line of the true path of the GPS antenna showed a big decrease of the amplitude of the swaying

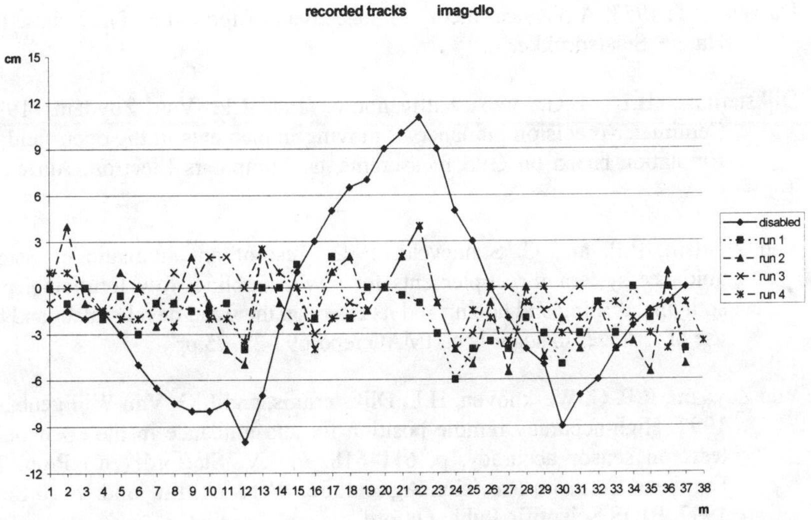

Fig. 6. Graph of the recorded tracks. *X*-axis: direction of travel, Y-axis: deviation of the programmed straight line at $Y = 0$.

movement (Fig. 6 lines run 1,2,3,4). The standard deviation of the measurements of Run 2 (the deviations from the straight line at $Y = 0$) was 1.84 cm.

CONCLUSIONS

RTK DGPS has the potential to be used as an on-line positioning device to keep an agricultural implement on its predetermined course within a tolerance of ±2 cm. Further development of the system could possibly halve this tolerance. Nowadays computers can run fast enough to control the steering of the implement.

Further tests on real soil are needed to prove the universal applicability of the system.

Research on the subject of also guiding the tractor, and not just the implement, can add to the value of the system.

REFERENCES

Auernhammer, H. 1990. Landtechnische Entwicklungen für eine umwelt- und ertragsorientierte Düngung. *Landtechnik,* 45:272–278.

Bushnell, L.G., D.M. Tilbury, and S.S. Sastry, 1993. Steering a three-input chained from nonholonomic systems using sinusoides: The fire truck example. p. 1432–1437. *In* Proc. of the 2nd European Control Conf., ECC '93.

De Wit, C.T. 1953. A physical theory on placement of fertilizers. Ph.D. diss. The Hague, Staatsdrukkerij,.

Dijksterhuis, H.L., L.G. Van Willigenburg, and R.P. Van Zuydam, 1998. Centimeter-precision guidance of moving implements in the open field: A simulation based on GPS measurements. Computers Electron. Agric., *in press.*

Van Zuydam, R.P. and C. Sonneveld. 1993. Test of an automatic precision-guidance system for implements for row crumbling, row fertilizing, row spraying, drilling and hoeing and its effect on the weed development and the use of fertilizer in sugar beet. IMAG report 93-29, 25 pp.

Van Zuydam, R.P. C. Werkhoven, H.L. Dijksterhuis, and L.G. Van Willigenburg. 1997. High-accuracy remote position fix and guidance in the open field: tests on sensor accuracy. p. 611–618. *In* J.V. Stafford (ed.) Proc. 1st European Conf. on Precision Agriculture, Warwick, England. 7–10 Sept. 1997. BIOS Scientific Publ., Oxford.

Van Zuydam, R.P., 1998. Navigation and position fix of an agricultural implement in the open field. Computers Electron. Agric., *in press.*

Comparing GPS Guidance with Foam Marker Guidance

R. Buick
Precision Agricultural Systems, Precise Positioning
Trimble Navigation New Zealand
Christchurch, New Zealand

E. White
Precision Agricultural Systems, Precise Positioning
Trimble Navigation
Sunnyvale, California

ABSTRACT

This paper reports the results of two studies comparing the efficiency of foam marker and submeter differential GPS (Trimble AgGPS132 with Parallel Swathing Option) guidance methods. Efficiency was determined by measuring overlap or skip, and by analyzing operational variables such as ground speed and offline distance. Both tests were conducted in ideal foam marking conditions. In the first study, an experienced foam marker and GPS guidance operator, found each method equivalent in terms of overall swathing accuracy. However, the operator drove 13% faster with GPS guidance than when using foam. In the second study, an inexperienced foam marker and GPS guidance operator proved more efficient when using submeter GPS guidance than when using foam. Additionally, average ground speed was 20% faster with GPS guidance than with foam.

INTRODUCTION

Precision agriculture aims to improve management of farm resources and their variability using technologies that enhance site-specific practices. The overall goal is to improve farm efficiencies by lowering costs or increasing returns. Application efficiency (i.e., optimal application of agri-chemical across the field) is often assumed to be at acceptable levels, based on traditional methods such as foam marker systems, tramlines, the use of white bags placed at swath width spacing along the edge of the field or simply by eye and dead reckoning. The development of GPS and GPS-based guidance systems enables the opportunity to assess exactly how well a field application is made and to compare GPS-based guidance with traditional methods.

This paper reports the results of two studies designed to compare the application efficiency of foam marker and Sub-Meter GPS (Trimble AgGPS 132 with Parallel Swathing Option) guidance systems. Swathing efficiency is the ability to minimize skips and overlaps in an application as well as a consideration to other operational variables that can improve farm efficiency and profitability

Copyright © 1999 ASA-CSSA-SSSA, 677 South Segoe Road, Madison, WI 53711, USA. *Proceedings of the Fourth International Conference on Precision Agriculture.*

(e.g., ground speed, time in the field, distance offline from swath center). The first study was performed in New Zealand using a spray vehicle mounted with a foam marker system. The second study was completed in Tulare, CA, using an AgChem Terra-Gator Soilection machine.

METHODS

Table 1. Summarizes the two swathing studies in terms of the tested variables, driver experience, guidance treatments used and general test day conditions.

	Swathing Study #1	Swathing Study #2
Location	Canterbury, New Zealand	Tulare, CA
Driver	Experienced 10 yrs foam marker 9 mo GPS guidance	Inexperienced 1 d foam marker 1 d GPS guidance
Vehicle	Spray truck with 78.7ft boom *AirTech* spray system with *RDS* (Clean Acres, England) spray controller in the cab and an installed foam marker system	AgChem Terra-Gator Soilection unit with a 70 ft boom width and a foam marker system
Field conditions	Flat field, minimal crop residue, smooth surface	Flat field, cultivated crop field, rough surface
Weather	Sunny, light wind	Overcast, light wind.
Variables measured	❑ Overlap % and skip % for field ❑ Distance offline from swath center ❑ Vehicle ground speed	❑ Overlap % and skip % for field ❑ Distance offline from swath center ❑ Vehicle ground speed
Guidance treatments compared	❑ Foam marker system ❑ Submeter GPS (AgGPS132 Parallel Swathing Option) with lightbar inner LED = 6.3" (0.16m)	❑ Foam marker system ❑ Submeter GPS (AgGPS132 Parallel Swathing Option) with lightbar inner LED = 9.8" (0.25m) ❑ Submeter GPS (AgGPS132 Parallel Swathing Option) with lightbar inner LED = 6.3" (0.16m)

Foam guidance was achieved by dropping foam blobs at the ends of the sprayer truck or the Terra-Gator boom. Drivers follow the previous trail of foam blobs by lining up the boom end on a new swath so the new foam blobs are dropped on top of the previous swath's foam trail.

GPS-based guidance was achieved using Trimble's AgGPS 132 with Parallel Swathing Option. The operator marks the beginning of and end of the first swath. A straight line is created from the two points (A and B). At a user-specified swath width spacing, parallel swaths are generated extending away from the A-B line. The generated swaths are used to guide the operator.

Using a centimeter accurate Real Time Kinematic (RTK) Trimble 7400Msi receiver, each application was recorded. The receiver recorded GGK NMEA messages to a rugged computer and to a TSC1 handheld data logger.

Using Microsoft Excel, the vehicle track data was filtered to remove end of swath turns and low quality RTK positions (i.e., data only includes fixed RTK positions). Distance offline measurements along the actual A-B line (Swath no.1) also were removed because operators don't use guidance along the first swath.

The data was summarized and analyses of variance (ANOVA) were performed (Snedecor & Cochran 1980). The analysis goals were to search for significance in the following swathing efficiency variables:

- Difference in ground speed, distance offline, overlap %, skip %, and driver skill.
- Variability in swath accuracy within a single application.

Overlapped and skipped areas were calculated between adjacent swaths. The values were summed and divided by the total field area to derive field overlap and skip percentages Three different analysis methods were used; Trimble's survey office software *TrimMap* (Trimble, 1997), a simulation modeling software *MatLab* and a Geographic Information System (GIS) *ArcView* (ESRI 1997a,b). Results were comparable across the different methods. Details of these analyses and comparisons between the analysis methods are not included in this report, but further details are described in Buick and Lange (1998).

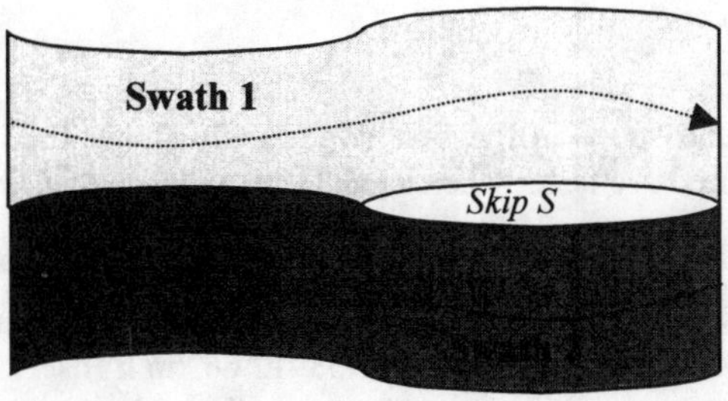

Fig. 1. Conceptual overlap and skip between adjacent swaths.

$$\text{\% Overlap for field} = \frac{\Sigma\ (\text{overlap areas b/t all swath pairs})}{\Sigma\ (\tfrac{1}{2}\ \text{Area Sw1} + \text{AreaSw2} + \ldots + \tfrac{1}{2}\ \text{Area Sw}_N)} * 100\%$$

$$= \{(\Sigma\ O) / [(0.5 \times A1) + A2 + A3 + \ldots A_{N-2} + (0.5 \times A_N)]\} * 100\%$$

$$\text{\% Skip for field} = \frac{\Sigma\ (\text{overlap areas b/t all swath pairs})}{\Sigma\ (\tfrac{1}{2}\ \text{Area Sw1} + \text{AreaSw2} + \ldots + \tfrac{1}{2}\ \text{Area Sw}_N)} * 100\%$$

$$= \{(\Sigma\ O) / [(0.5 \times A1) + A2 + A3 + \ldots A_{N-1} + (0.5 \times A_N)]\} * 100\%$$

Fig. 2. Equations used to derive percent of overlap and skip.

RESULTS AND DISCUSSION

General

The weather conditions during the New Zealand field exercise were sunny, no rain with light wind. Differential correction accuracy for the submeter GPS treatment, using a satellite differential service, was 1.0 to 1.5 m accurate RMS.

The weather conditions during the Tulare, CA field exercise were overcast with light wind. Differential correction accuracy for the submeter GPS treatment, using a satellite differential service, was 1.0 m RMS.

Conditions during both studies were deemed suitable for ground spraying and spreading.

Ground Speed

This section presents the results of Ground Speed measurements calculated in both studies.

Study No.1

Average ground speed in the first swath study shows that the experienced operator was able to drive faster when using GPS guidance than with foam.

Observations:

- Submeter GPS guidance average ground speed across all swaths was 7.8 mph (12.6 km h^{-1}, 95% confidence range 12.53–12.66 km h^{-1})
- Foam marker average ground speed across all swaths was 6.9 mph (11.2km h^{-1}, 95% confidence range 11.12–11.24 km h^{-1}).
- Average GPS guidance ground speed was 13% faster than with the foam.
- Foam application ground speed increased as more swaths were completed. By swath 8, foam and GPS guidance ground speeds were essentially identical.

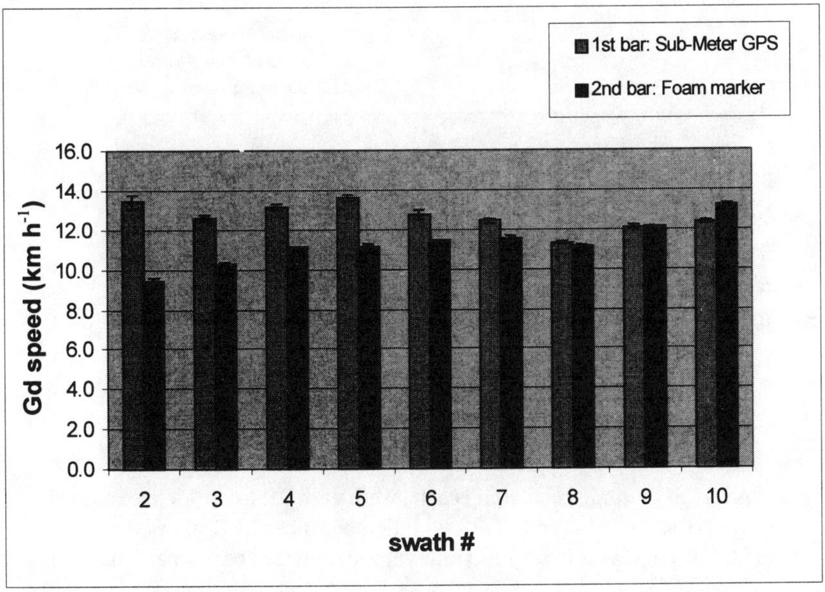

Fig. 3. Average ground speed (km h^{-1}) along each swath with 95% confidence interval error bars shown (1st bar = submeter GPS with inner LED at 6.3 in or 0.16m, 2nd bar = foam marker). Initial A-B line values not included.

Study No.2

As in the first swathing study, the operator drove faster with GPS guidance than with foam.

Observations:

- Submeter GPS guidance, with 9.8 in (0.25 m) lightbar LED spacing, ground speeds averaged 11.6 mph (18.6 km h^{-1}, 95% confidence range 18.41–18.74 km h^{-1}).
- Submeter GPS guidance, with 6.3 in (0.16 m) lightbar LED spacing, ground speeds averaged 12.2 mph (19.6 km h^{-1}, 95% confidence range = 19.0–20.2 km h^{-1}).
- Foam marker guidance ground speeds averaged 9.9 mph (16.0 km h^{-1}, 95% confidence range = 15.74–16.18 km h^{-1})
- Ground speed during the two submeter GPS guidance methods averaged 20% faster than with foam.

- During all treatments, the beginning swaths were slower than later swaths. This suggests that some time was required for the inexperienced operator to get accustomed with each guidance method.

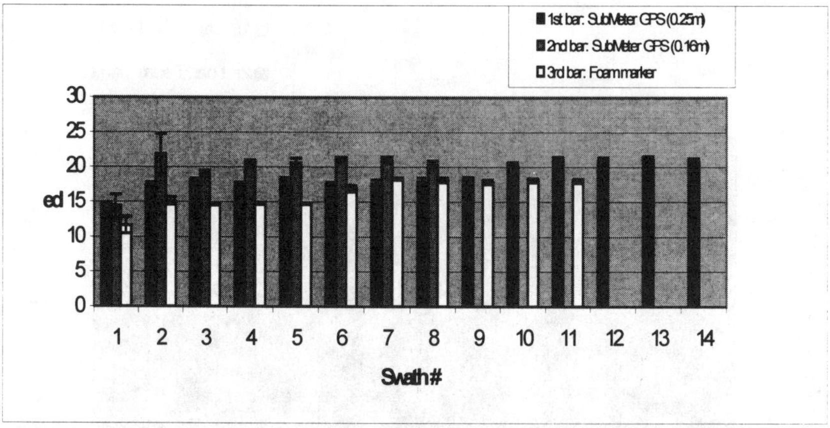

Fig. 4. Average ground speed along each swath with 95% confidence interval error bars shown (1st bar = GPS with lightbar inner LED interval = 0.25m, 2nd bar = GPS with inner LED interval = 0.16m, 3rd bar = foam marker).

Ground speed results from both studies suggest that operators are more productive, (i.e., Able to cover more ground in less time) when using GPS guidance than when using a foam marker system.

Distance Offline

Average distance offline (cross-track error) was recorded from the set of swaths pregenerated from the original A-B line. Distance offline values were analyzed by determining the absolute value of the distance offline values (i.e., absolute value implies the actual distance offline independent of which side of swath center the offline occurred).

An accumulative effect was observed with foam in both studies. In the first study, foam offline distance from pregenerated swaths continued to increase due to high overlap. Given the high overlap occurring with foam in this first study (see Application Overlap and Skip), a better way to measure cross-track error on each swath would be from the previous swath driven. Offline distance on the initial foam swath can be used to estimate actual foam offline distance.

In both studies, foam first swath offline distance was similar to sub-meter GPS offline distance across the field (Study no.1: 1.5 ft or 0.46 m for foam compared with 1.3 ft or 0.4 m for submeter GPS. Study no.2: 3.1 ft or 0.94 m for foam compared with 3.7 ft or 1.13 m for submeter GPS with inner LED = 9.8 in and 2.0 ft or 0.61 m for submeter GPS with inner LED = 6.3 in). The LED setting of 6.3 in on one of the submeter GPS treatments reduced average distance offline below that of foam offline distance along many of the swaths driven.

In the first study, foam offline distance increased from 1.5 ft (0.46 m) at swath 2, to 10.5 ft (3.20 m) at swath 10. This apparent increase in offline is due to the high overlap that occurred with the experienced operator. In the second study, foam offline distance increased from 3.1 ft (0.94 m) at swath 2, to 21.0 ft (6.42 m) at swath 11. This apparent increase in offline distance was more a result of high skip that occurred with the inexperienced operator (see Application Overlap and Skip).

Application Overlap and Skip

This section shows overlap and skip results calculated in both studies.

Observations:

- Within each of the GPS guidance methods, the levels of skip and overlap were equivalent (Fig. 5 and 6), with values all being <2.5%.
- Foam tended to have either a very high overlap (2.0%) and low skip (0.4%) as seen with the experienced operator (Fig. 6) or a high skip (2.3%) and low overlap (1.1%) as seen with the inexperienced operator (Fig. 5).

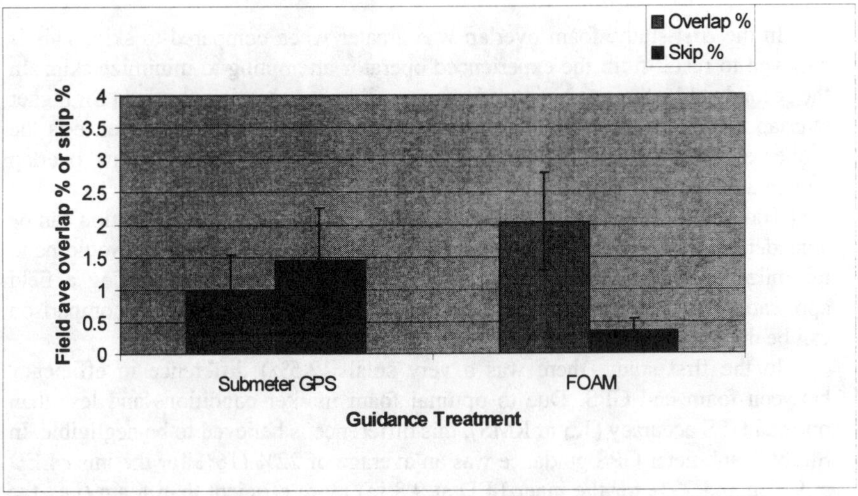

Fig. 5. Average overlap percentage and skip percentage for each guidance treatment in swathing Study No.1, with 95% confidence interval bars.

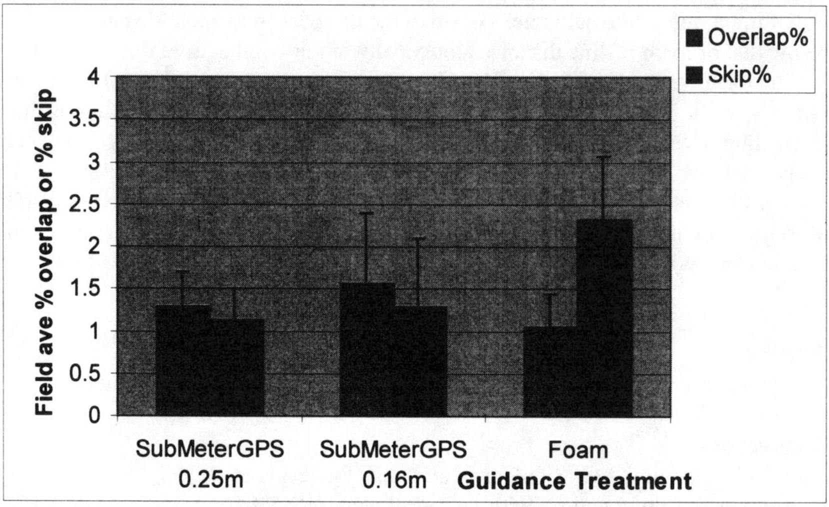

Fig. 6. Average overlap percentage and skip percentage for each guidance treatment in swathing Study No.2, with 95% confidence interval bars.

In the first study, foam overlap was greater when compared to skip. This is believed to result from the experienced operator attempting to minimize skip. In the second study, the inexperienced operator did not appreciate the need for higher overlap to obtain less skip when using foam (Fig. 6). Lack of experience is the probable cause for the higher skip percentage and consequently low overlap percentage.

Too much skip and too much overlap occurring in any one application can be considered an inefficient way to apply agri-chemical. The ideal goal would be to minimize the occurrence of both overlap and skip percentages during a field application. Adding overlap and skip together for each treatment, a comparison can be made (Table 2).

In the first study, there was a very small (2.5%) difference in efficiency between foam and GPS. Due to optimal foam marker conditions and less than optimal GPS accuracy (1.5 m RMS), this difference is believed to be negligible. In Study 2, submeter GPS guidance was an average of 22% (16% for the inner LED at 9.8 in and 29% for the inner LED at 9.8 in) more efficient than foam (i.e., the river minimized both overlap and skip percentages more across the field with submeter GPS than with foam).

Table 2. Difference in efficiency, where maximum application efficiency occurs with minimum field (overlap + skip percentages). Positive % difference is when GPS guidance is more efficient than foam (i.e., lower overlap + skip).

	Experienced operator	**Experienced operator**	**Inexperienced operator**	**Inexperienced operator**	
	Foam	**Sub-M GPS**	**Foam**	**Sub-M GPS**	
				9.8" LED	*6.3" LED.*
Overlap %	2.04	1.00	1.06	1.28	1.55
Skip %	0.35	1.45	2.31	1.12	1.29
Overlap % + Skip %	2.39	2.45	3.37	2.40	2.84
Difference between Treatments as % of foam skip+ overlap	**0 %**	**-2.5%**	**0 %**	**28.8%**	**15.7%**
Efficiency	Only a 2.5% difference in efficiency between Foam and GPS i.e., no difference		GPS is an average 22% more efficient than Foam		

SUMMARY AND CONCLUSIONS

This research showed that foam markers have an accumulative effect. Any errors in application will be followed and propagated across the rest of the field application. Experienced operators tend to perform more overlap than skip when using foam, ensuring that skip and striping effects do not occur. In the first study, by the time nine swaths were applied with the foam marker, the vehicle was applying at 10.5 ft (3.2 m) offline from the ninth pregenerated swath. In a large field using a 78 ft (24 m) spray boom and a foam marker, an entire extra swath would be made due to this overlap after each 67th swath (i.e., 78 ft ∗ 9 swaths, divided by 10.5 ft = 67 swaths) assuming a constant driver performance. After many hours of driving with a foam marker, the driver performance is unlikely to improve.

Minimizing skip can also be important when double application of agri-chemicals do not cause crop damage. GPS guidance allows the operator to easily adjust the relationship of skip to overlap in the field application. The operator can reduce the swath width setting in the GPS guidance system to minimize skip throughout the field application. For example, if the spray boom is 78 ft, the swath width setting in the GPS guidance system can be set to 77ft to minimum skip. Adjusting the application in this way and managing this consistently across an entire field application is not typically dome with foam markers. Similarly, adjustments can be made when using spraying equipment where the outermost spray nozzles on the boom distribute one-half application rates and the operator must encourage some nominal amount of overlap between adjacent swaths.

Overall, it is conclusive that with an experienced foam marker driver and under ideal conditions for foam, the overall swathing efficiency of submeter GPS guidance and a foam marker are at minimum, equivalent. However, operators drove faster when using GPS guidance than they did when using foam. The experienced operator drove an average of 13% faster when using GPS than with foam. It must be noted that after several swaths the operator achieved equivalent driving speeds with foam and GPS guidance. The inexperienced operator drove an average of 20% faster with GPS than with foam. Unlike the first study, velocity differences persisted for the entire application.

Conceptually, optimum swathing efficiency would strive for the minimal occurrence of both overlap and skip within a single application. With an experienced operator, there did not appear to be guidance treatment differences in terms of the ability to simultaneously minimize both overlap and skip. The inexperienced operator was unable to minimize both overlap and skip with foam compared to submeter GPS (an average of 22% more efficient with GPS than with foam).

Although, these studies suggest the importance of driver experience in application swathing efficiency, there were only two drivers used here. For more conclusive results on the impact of driver experience on swathing efficiency and how easy each method is to learn, more extensive research is required where multiple drivers, both inexperienced and experienced, are tested.

In this study both foam and GPS guidance methods proved extremely accurate. The ground speed measurements indicate an improvement in productivity when using GPS based guidance over foam markers. Apart from this, there are many reasons besides swathing efficiency that makes GPS guidance an attractive option over foam and other traditional guidance methods. Many of the benefits GPS guidance provides are listed below:

- GPS allows operation in fog, at night or other conditions that make foam visibility poor (e.g., tall crops, dusk, night time). Winds are generally calmer at night reducing drift potential. In addition, many agri-chemicals perform better when applied at night (e.g., avoid spraying beneficial crop pollinating insects, higher humidity conditions).
- Foam visibility is reduced in hot dry conditions. Foam can be lost in tall crops.
- No on-going maintenance costs of foam marker chemicals and additives.
- GPS guidance eliminates foam marker set-up time.
- No frozen foam lines.
- GPS guidance is ideal for spinner spreaders that often can not use foam markers. The operator must eyeball or dead reckon where to drive or use a white bag technique. Additionally, GPS guidance does not restrict the spreading width in order for the operator to still see and estimate from the previous swath's vehicle tire marks. For example, an increase from a 49 ft to 65 ft (15 to 20 m) spreading width is possible with GPS guidance (assuming an even spreading rate across the swath width is achievable with a specific agri-chemical product). This results in at least 30% less time in the field, translating to lower vehicle operation and maintenance costs.
- GPS receivers can be used in other farm activities

- Driver fatigue is reduced as the operator can look strait ahead when applying material. The operator does not need to look left and right lining boom ends over a previous foam trail.

ACKNOWLEDGMENTS

Many thanks to the Trimble staff who assisted with the testing and analysis, particularly, Joan Hollerich and Chris Dietsch, Phil Jackson, Jim Veneziano, Art Lange, Ian Viney, and Greg Price. Thanks also to those who assisted with arrangement and provision of equipment to conduct the studies including INSAT, AgChem (second study) and David West (first study).

REFERENCES

Buick, R.D., and A. Lange, 1998. Assessing efficiency of agricultural chemical application with differential GPS, ArcView and Spatial analyst. *In* ESRI Users Conf. Proc., San Diego. 27–30 July 1998.

ESRI. 1997a. ArcView version 3.0a software and on-line help.

ESRI. 1997b. Spatial analyst version 1.1 (pre-release) extension software for ArcView and on-line help.

Snedecor, G.W., and W.G. Cochran, 1980. Statistical Methods. 7th ed.. Iowa State University Press, Ames.

Trimble. 1997.TrimMap. Software version 6.50.

Omnistar Virtual Base Station System

Lee Ott
Omnistar
Houston, Texas

ABSTRACT

The OmniSTAR Virtual Base Station system was originally developed in 1990 to provide users reliable differential GPS corrections for high accuracy real time applications. At the present time there are OmniSTAR systems covering all the World's major land masses (excluding the Poles and extreme Northern and Southern latitudes). The system delivers its corrections via geostationary satellite, thus providing seamless coverage over the entire service areas. The OmniSTAR system is unique in that corrections from the entire network are used in the user system solution to provide RTCM corrections for GPS that are at least as good as a dedicated base station within one kilometer of the users position. This paper will describe the OmniSTAR network and user system and address recent advances resulting in enhanced accuracy. It will show results accumulated over the last 4 yrs demonstrating how final position results are dependent on the quality of the GPS engine.

INTRODUCTION

The OmniSTAR Virtual Base Station system was originally developed in 1990 to provide users reliable differential GPS corrections for high accuracy real time applications. Its concept and inception goes back to 1983. It was patterned after the first commercial satellite navigation system STARFIX also developed by the OmniSTAR group. It was not OmniSTAR at that time but was under the name John E. Chance and Associates. OmniSTAR was formed in 1995, for the sole purpose of provided positioning data to land users. Prior to that time the core business of John E. Chance was the offshore Oil and Gas Industry and still is today. OmniSTAR still provides positioning services to John E. Chance and uses common GPS reference stations in many of its networks. This paper will briefly give a historical review of the development of OmniSTAR followed by a description of the OmniSTAR networks and user systems followed by a brief explanation of the Virtual Base Station system. It will then show results accumulated during the last 4 yrs demonstrating how final position results are dependent on the quality of the GPS engine. The last section will address the results and future advances.

Copyright © 1999 ASA-CSSA-SSSA, 677 South Segoe Road, Madison, WI 53711, USA. *Proceedings of the Fourth International Conference on Precision Agriculture.*

HISTORICAL REVIEW

The predecessor of OmniSTAR was the STARFIX system. The STARFIX system was the first commercial satellite navigation system and was developed by John E. Chance and associates. The STARFIX system uplinked a spread spectrum signal to four geostationary C-Band satellites. The signals from the satellites were received by four demodulators and the pseudo range from each satellite was measured. From this pseudo range measurements a position on the earth's surface could be calculated. Data were modulated on each of the four satellites. The information consisted of the position of the satellite in space and measurements from the satellites to receivers at known fixed points in the area of coverage. One of conditions of the STARFIX system was that it needed to be calibrated at a known point before it could be used. Then to determine that nothing happened to the calibration it was checked at the completion of the job. If it checked then there was not a problem. If it did not check then the question was then asked when it changed. Since the normal length of jobs could be as long as forty days, the accuracy of the whole job was suspect. In order for the industry to accept this system they required an independent means of verification of accuracy. This was accomplished by sending Differential GPS data over the C-Band satellites as well. In 1986 when the system was introduced to the industry GPS satellites were only visible 2 h a day; however, this was sufficient to provide a quality check to the STARFIX system. As more GPS satellites were placed into service over the years the longer the coverage per day. In 1989 the GPS coverage was 24 h per day and after a year of continuous operation of two completely independent navigation systems the STARFIX system was shut down in 1990. At this time only a Differential GPS service was provided that became known as the OmniSTAR service.

Not surprisingly, the OmniSTAR service was patterned after the STARFIX service. Not because it was convenient, but because it provided the most reliable and accurate system and the best ability for growth. While the transition from STARFIX to OmniSTAR was taking place John E. Chance was acquired by Fugro. The technology that was developed from STARFIX was transferred to the other operation centers of Fugro throughout the world. There now exists OmniSTAR B.V responsible for Europe, Middle East and North Africa and OmniSTAR Pty who is responsible for Australia, South Africa, and SouthEast Asia. Like OmniSTAR in the USA, which is responsible for the Americas, they operate and manage the network for the Marine divisions as well.

OMNISTAR NETWORKS

There are three Network Control Centers (NCC) that comprise the world wide OmniSTAR Virtual Base Station System. These centers are located in Perth, Australia, Oslo, Norway, and Houston, TX. The Perth NCC uplinks data from Asia, Australia, New Zealand over two satellites. The Oslo NCC uplinks data from Europe, Middle East and South Africa over two satellites. The Houston NCC uplinks data from North and South America over three satellites. A coverage diagram of the OmniSTAR system is shown in Fig. 1. The shaded areas of the map represent where there is coverage into an omni-directional antenna receiver system. Each of the NCC's also uplinks data to additional satellites for the marine users. These satellites are low power satellites requiring directional antennas. For shipboard installations this is of no concern. The OmniSTAR NCC's are uplinking to a total of five low powered systems along with the high powered satellites. The high powered beams allows the OmniSTAR Virtual Base Station system to be a small and portable user system which will be described in more detail in the next section.

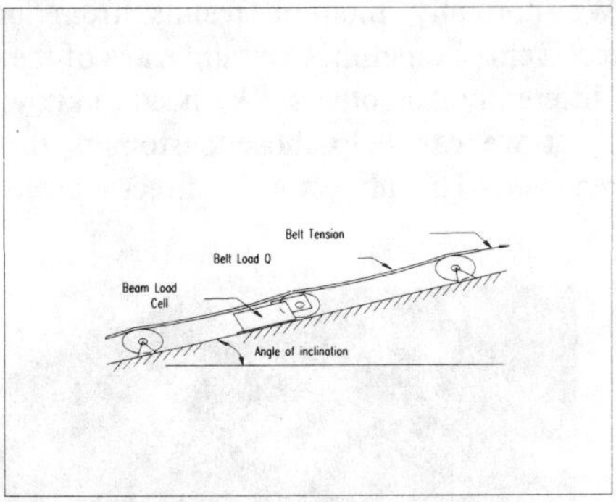

Fig. 1 Shaded areas show OmniSTAR coverage.

The worldwide OmniSTAR network has >70 reference stations that are connected to the NCCs. The GPS receivers at the reference stations are precisely surveyed. The GPS system tells us what the positions are for the GPS satellites and by knowing the precise coordinates of a GPS receiver one can calculate the exact value of a satellite range. The GPS receiver measures the ranges from the satellites and the calculated ranges are subtracted. What is left over are the errors in the GPS system. These errors for each reference site are then sent to the OmniSTAR NCCs. The NCCs then perform integrity checks on the data and eliminate any bad measurements. Each NCC directs the appropriate reference station data to an uplink processor. The uplink processors purpose is to compress the data and send it to a satellite uplink for transmission. Thus the data that is received by a OmniSTAR demodulator is in effect the correction data from each reference station assigned to that uplink.

By the time the correction data is received by the user it represents data that was collected at a reference station 2 to 4 s ago. This is often times referred to as latency. Since there is a certain amount of data that needs to be sent and the bandwidth of the satellite is limited it will take time to send a new packet of information. In the OmniSTAR systems this takes about 2 to 4 s. By the time the fresh information is received the previous data can be as old as 8 s. Deterioration of position accuracy does not begin until the age of data reaches 12 s.

The primary purpose of the NCC is to monitor the results and reliability of the entire system. For example, the NCC monitors the received data from each of the satellites it sends information to and compares it to what was believed to be transmitted. Should the comparison not check then procedures are in place to switch in alternate data paths to the uplinks in a very short time to alleviate the problem. Another function of the NCC is to continuously monitor the position results of the system. Should the position error deviate procedures are executed to determine what is causing the position error. Often times the OmniSTAR NCCs detect and eliminate satellites from the solutions before the U.S. Air Force has set the satellite unusable. Since there are a number of GPS receivers that are used by our clients, we monitor position results using our corrections into numerous receivers. We normally monitor results from one-half dozen different manufacturers receivers. Sometimes certain states of the GPS satellites will effect some receivers differently than others. We need to know these conditions as soon as possible so that we can help those customers that are using a particular manufacturers receiver. This allows us to direct a customer to his solution more quickly.

Fig. 2 OmniSTAR 7000 C-Band receiver.

OMNISTAR USER

The OmniSTAR users consists of a satellite demodulator and a GPS receiver or GPS engine. A GPS receiver is a self-contained unit that has its own operator interface. An engine is a P.C. board that is embedded into a special product such as the OmniSTAR 7000 receiver as shown in Fig. 2. This receiver was especially developed for the agriculture industry. It was jointly developed by John Deere and company and John E. Chance and Associates. The OmniSTAR Company was formed close to the time this project was nearing completion. The receiver receives its signals from a C-Band satellite located in a geostationary orbit located at 87 degrees West Longitude. The signal structure of the satellite is a spread spectrum signal with a chipping rate of approximately 10 Mhz. This generates a signal that has a bandwidth of 20 MHz. This spread spectrum signal is phase modulated at a rate of 1200 bits per second.

The OmniSTAR 7000 receiver is a completely self-contained unit with an omni-directional C-Band antenna and a GPS antenna. Inside the box is a Motorola GPS engine, which provides the positioning. The C-Band signal is received and the data is then processed to form a RTCM data string that very closely represents a RTCM data string from a GPS receiver as if were setting right next to the user. This is how the name Virtual Base Station (VBS) was coined. A description of the VBS will be presented in the next section.

Prior to the development of the 7000 receiver OmniSTAR produced a C-Band demodulator called the OmniSTAR 6300. This unit provided VBS RTCM corrections to GPS receivers. However a requirement of the 6300 VBS unit was that a position had to be supplied to the unit. This could either be done manually or connecting a NMEA string from the GPS receiver. Numerous problems were

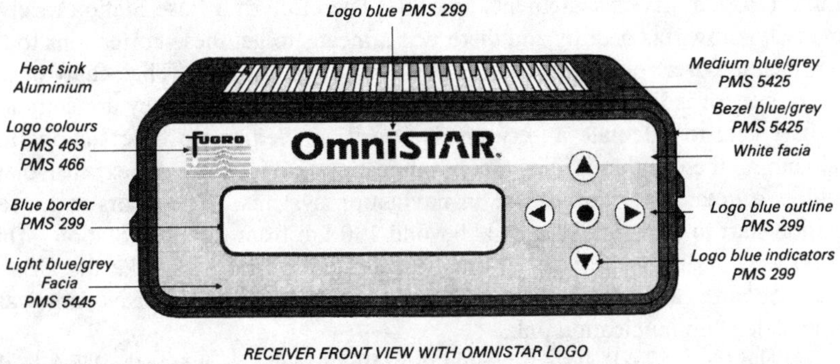

Fig. 3 OmniSTAR 3000LR receiver series.

occurring. Clients were incorrectly entering the position or interface problems to the 6300 developed that meant that the 6300 VBS was supplying RTCM corrections for the wrong location. If the user system had been moved a large distance, it could result in a large position error. This is why an inexpensive GPS engine was built into the 6300 and the 7000 receivers.

In order to expand the OmniSTAR service to the remainder of the world and since there were not many C-Band satellites available a new demodulator was developed that operated on L-Band satellites. This signal structure is a BPSK modulation system. It is phase modulated with a rate 1/2 Viterbi encoding scheme. This means that for a 1200-baud channel the bandwidth is 5 KHz vs. the 20 MHz in the C-Band system. The latest model of the L-Band receiver developed by the OmniSTAR companies is the 3000LR as shown in Fig. 3. This receiver can be acquired either with or without a GPS engine. This receiver can be ordered with several types of GPS engines. At the present time it has been integrated with Motorola Oncore engines, Trimble SK8 engines, Rockwell Jupiter engines, Trimble DSM engines and the Ashtech G12 engines. These engines vary significantly in price and the results of these engines will be shown in the results section. Also if a client elects, he can buy a 3000LR receiver without an engine that he can interface to his own GPS receiver. Again as was discussed previously, care must be taken to assure that the correct GPS position is being fed into the VBS demodulator so it can generate the best RTCM string.

VIRTUAL BASE STATION PROCESSING

The concept of the Virtual Base Station System is quite simple. But first lets back up and discuss Differential GPS. As mentioned earlier in the paper if a GPS receiver is placed at a known point and it measures the ranges from the GPS satellites and subtracts the calculated distances, then one has a set of corrections. If one knew the location of the GPS satellites exactly and the position of the receiver on the ground then in theory one would have a measure of the errors of the system. Now a user receiver who is trying to determine where he is at has similar errors in his measurements as well. Therefore, if a Base Station receiver (precisely known) is near by and there was a means to get these corrections to the user, then the user could calculate his position very accurately. The beauty of this type of system is that it does not matter what the errors are, if they are common, the user gets to calculate a very precise position. Setting up base stations and maintaining them and providing a communication link is, however, expensive and requires a great deal of expertise in navigation systems. The errors in a users position start to grow until a user is beyond 100 km from the Base Station. Thus if one were trying to provide coverage over a large land mass like the USA it would require a large number of stations and a tremendously large and complicated communication link.

The OmniSTAR system is able to provide accuracy across the USA in the order of 20 to 30 cm with only a dozen base stations across the USA. This is accomplished by breaking down the error sources and compensating for them. The error sources are numerous, but fortunately due to the differential concept a large number of them cancel. The remaining error sources in a differential system are basically, satellite errors, atmospheric errors and clock errors.

The satellite errors are a result of several factors. One of them has to do with the fact that when the GPS constellation was originally conceived there were to be two types of service. A precise service was to be used by the military and was to provide 10 m accuracy. A civil service was to be used by the civil community and was to provide 100 m accuracy. When the system was placed in service, however, the manufacturers and designers of the system out did themselves and made a civil service that had an accuracy of 10 m, which was intended to be used only by the military. Hence, to protect the military use of the system the U.S. government went to great pains to degrade the civilian service. They did this by inducing clock errors that changed often enough and random enough so that no one could predict what the clock errors were going to be. They also created the ability to lie to us about the position of the satellites; however, the satellite positions are relative small errors since they have since guaranteed to the civil community that the civilian service would always be less than a 100-m system.

The last error source is the atmospheric errors. The radio signals from space have to travel to the user through the earth's atmosphere. There are two components to the atmospheric errors. These are the ionosphere and the troposphere. The ionospheric errors are caused by the electron count in the upper atmosphere. The electron counts are effected by the sun and hence there is a 24 h time variation of this effect. A fixed point on the earth will rotate through sunshine and nighttime. Sun spot activity will adversely effect the ionospheric errors. The effect on the user is that the satellite range measurement will appear to be longer due to the delay by the ionosphere.

The tropospheric errors are cause by the lower part of the atmosphere. This is also another delay to the satellite signal and is proportional to humidity, temperature and pressure.

With a network of receivers then one can make enough measurements to model the errors and eliminate them. These types of measurements are made at the NCCs and parameters to these models are calculated and transmitted to the users so they can apply them. In the OmniSTAR virtual base station system, we send all of the base station information to the users so that the user can always achieve an optimal solution depending where he is located. The VBS software function is to first determine the elevation and azimuth angles for all the satellites in view for all of the base stations. One set of parameters that is transmitted to the user periodically is the location of all the base stations in the network. Once all of the elevations and azimuths of the satellites are obtained along with the parameters for the ionospheric and tropospheric model then these errors are removed from the individual base stations correction data. Also after adjusting for the satellites position errors then the only errors that remain are the high frequency clock errors. The clock errors are calculated by performing a weighed least squares solution. The weights are determined by the distance to the respective base stations and the age of data.

Once an estimate of the clock errors is determined then the RTCM correction set is made up; however, since the VBS processor knows the position of the GPS receiver, then it also knows the errors that the receiver will be seeing. It will be the clock errors, plus the satellite position errors and atmospheric errors. These errors are calculated for that position and added to the clock error terms, before formatting into a RTCM message. Now when the user receiver adds these corrections to his range readings he is removing all of the known error sources thus providing the most accurate position fix from the GPS receiver.

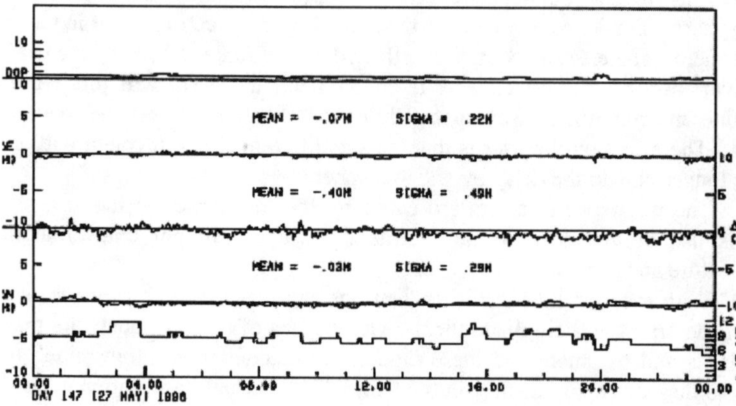

Fig. 4. Position of a survey grade receiver with 15 km baseline RTCM data only.

RESULTS

The price of GPS receivers can vary from \$100 to \$200 for a GPS engine to \$10,000 for a GPS type survey receiver. Figure 4 is a good example of what one can achieve with a survey grade receiver. All of the result plots represent 24-h of data. Starting at the top of the graphs, the first plot is the DOP (dilution of precision). This is a figure of merit used by the GPS industry as to the quality of the position solution at any given point in time. It is assuming that if each GPS satellite range had a 1 m random error then the position solution would have an uncertainty of the DOP value in meters. The second plot shows the error from a known point for the Longitude (E). Printed above the graph is the mean error and the one-sigma error in meters. In this case the mean is 0.07 m and the one-sigma value is 0.23 m. The middle graph is the height error (H) and again with its mean and one-sigma value. The next graph represents the Latitude error (N) with the same means and sigma's displayed. The bottom graph shows the number of satellites that were used in the position calculation. This graph is the results of a 15-km baseline feeding RTCM corrections from a base station receiver directly into a receiver without VBS processing. This gives us a baseline for comparing VBS solutions and different types of engines.

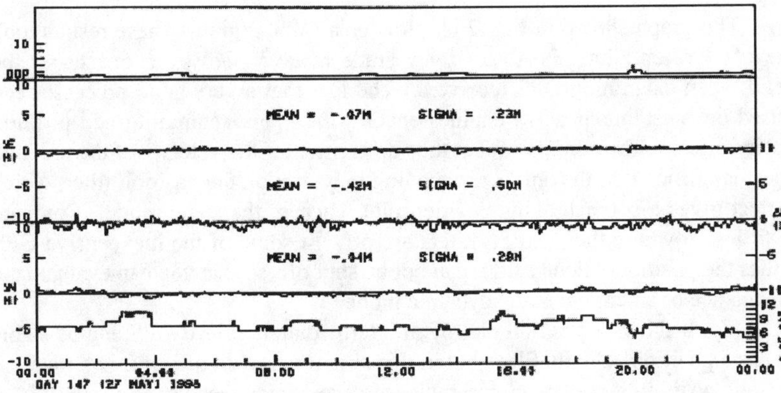

Fig. 5. Position plot of VBS RTCM at 400 kilometer minimum baseline.

Fig. 5 shows the results of a VBS solution into the same type of survey grade receiver where the minimum baseline distance was 400 km. If you will note the one-sigma errors between Fig. 4 and 5 they are practically the same. The only difference is 0.01 m in Longitude and height. The advantage of using a VBS solution is that if a base station goes down for any reason the position accuracy is not effected since it is getting its information from the remaining stations in the network.

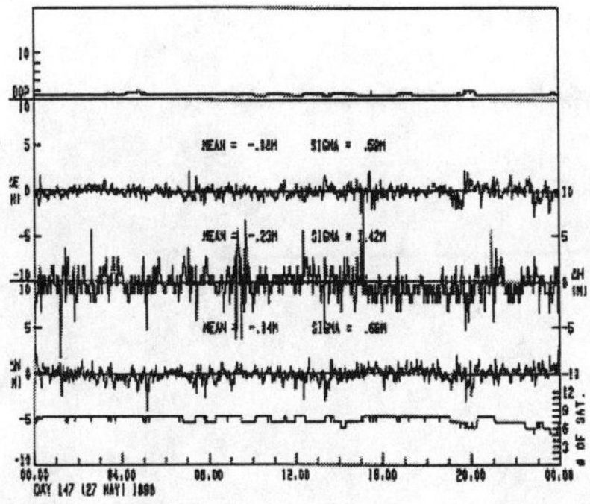

Fig. 6. Position plot of inexpensive GPS engine with filters off.

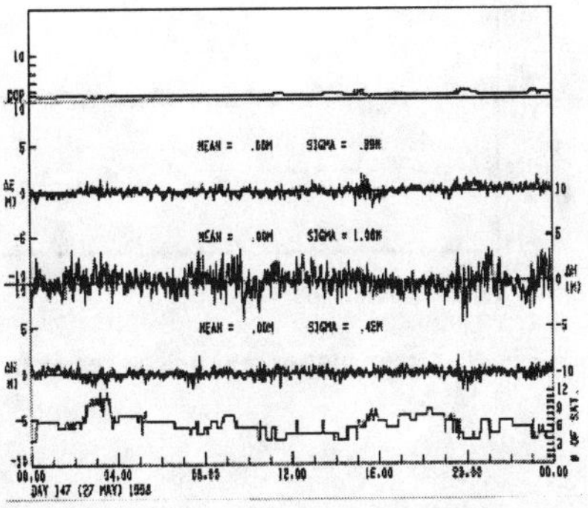

Fig. 7. Position of inexpensive GPS engine with position filters on.

The graph shown in Fig. 7 is a low cost GPS engine. These results look quite good when compared to a survey grade receiver; however, one has to be careful when examining GPS receivers. The fact that a very good processor can be used on these inexpensive engines enables the manufacturers to add position velocity filters to the outputs. What they do is monitor the velocity of the unit and if they determine that the unit is not moving they turn on the position filter, which in effect gives a better looking position plot when in the static mode. Compare this to Fig. 6 where the position filters are off. In some of the inexpensive GPS engines the position-velocity filter can not be shut off, so that you can evaluate the performance of an engine in the dynamic mode.

Fig. 8 shows a position plot of an intermediately priced GPS engine. This unit has a position velocity filter, but in this case it was shut off. If it is turned on, at a static position it very much resembles a survey grade receiver as seen in Fig. 4 and 5. What should also be pointed out in these plots is that the same antenna was

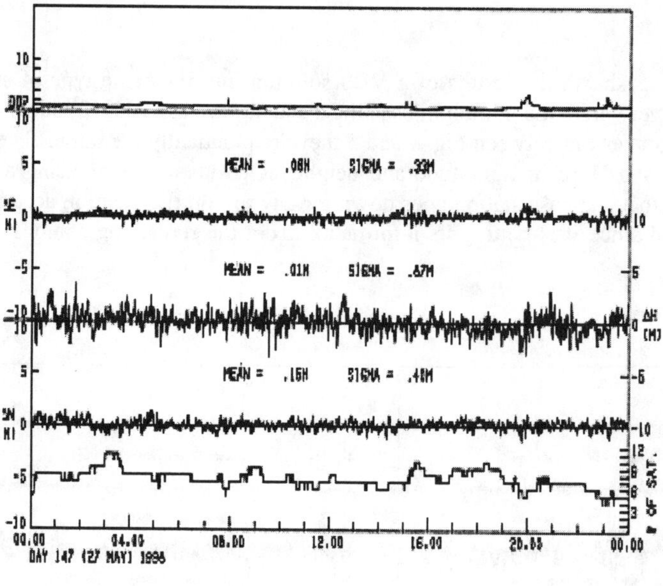

Fig. 8. Error plot of mid priced receiver.

used on each receiver. This was accomplished by having an active 8-way splitter. In this way eight different types of receivers can be monitored and compared at the same time without introducing any uncertainties. Different types of antennas can change the outcome of the results.

CONCLUSIONS

A brief history of the OmniSTAR Virtual Base Station System was presented showing that the OmniSTAR companies were formed to provide position services to the land based communities. Some individuals might think that the criticality of land based positioning is not that important, because there are no high dollar services going on and that the cost of the service is very inexpensive. However, the OmniSTAR companies were born out of operations in the offshore oil and gas industry, where downtime in service could cost a client millions of dollars and hours of lost time. Since the navigation for the marine users is the responsibility of the OmniSTAR companies the land based users reap the benefits of the reliability and quick response times to problems that are demanded by the oil and gas industry.

The OmniSTAR companies are continuing to spend money on improvements to the NCC's and the networks. They continually are improving on the level of service and looking for means to provide an even higher level of accuracy to the land users. One such project that is now in progress is a decimeter wide area positioning system. We are also in the process of designing an OEM card of our L-Band demodulator.

We have demonstrated that the OmniSTAR solution is not the limiting factor in the positioning accuracy. We are continuously monitoring and looking at new GPS receivers hardware so that we might be able to obtain better data from our base station receivers so that we can improve our level of service.

This paper showed comparisons between different types of GPS receivers and it was mentioned that all the test were done with a common antenna. Another paper could be written on the effect of GPS antennas. Not only does one have to have a good understanding of GPS receivers and how they operate, one also needs to understand the effect of GPS antennas. For example a good antenna is not going to make a bad GPS receiver good. But a bad antenna can definitely make a good receiver look bad.

System for Precise Dosage and Documentation of P, K and Mg application

R.E. Lütticken

Hans Hölzl
Precision Fertilizing
Rauhöd 2
83137 Schonstett, Germany

ABSTRACT

The site specific application of P, K, and Mg requires a highly accurate dosage system and specific features for the transport of fertilizer from the spreader tank to the outlets of the boom. The pneumatic spreader developed has the facility to regulate separately the application rates of its four 6 m sections via four different electrohydraulic systems. The unique design of the pneumatic system, dosage cells and outlets reduces variability within each section to 2%. A weighing system, which senses the weight of the tank, serves as a permanent control for the amount of fertilizer applied. This enables a continual recalibration of the system, reducing the negative effects of varying fertilizer qualities. The data of the working path and width, application rates and weighing sensors are stored by the on-board computer, providing a fully automated system with accurate documentation of the whole fertilizing process.

INTRODUCTION

Site specific fertilizer application mainly concentrates on nitrogen application neglecting the application of the base nutrients P, K and Mg. One of the prime objectives of precision fertilizing is yield improvement and therefore the correction of base nutrients, as potential limiting factors, has to be considered. Consequently site specific application of nitrogen might only appear worthwhile when major limiting factors are eliminated. Although numerous studies have been carried on the precise recording of spatial temporal variable data and its interpretation, there is still a lack of technology and suitable methods to put site specific techniques into practice.

The precision fertilizing system presented hereafter has been developed in Germany and is provided for the farmer as a complete site specific fertilizing service. The concept is based on soil nutrient sampling of fields in a grid system and mapping of soil analysis for P, K, Mg, pH and soil texture. A nutrient balance is then determined, considering the measured soil nutrient availability and the expected nutrient uptake of the crop rotation. For the subsequent fertilizer application, the development of a new spreading system was regarded as indispensable, as the application of base nutrients has specific technical requirements.

Copyright © 1999 ASA-CSSA-SSSA, 677 South Segoe Road, Madison, WI 53711, USA. *Proceedings of the Fourth International Conference on Precision Agriculture.*

SYSTEM CONSIDERATIONS

Unlike nitrogen, the application of base nutrients often has to deal with 3 to 5 times higher application rates. This necessitates a different spreading technology to efficiently apply the fertilizer. For site specific application an instantaneously reacting system to change mass flow rates, as well as a part switch, is required. Considering existing spreading systems, such as spinning disc spreaders and pneumatic spreaders, the former often show a change in distribution pattern when flow rate or speed changes. Furthermore spinning discs do not have the capability to properly part switch the application width. Therefore the objectives for site specific application of base nutrients are to develop a system with:

- Continuos mass flow change
- Precise dosage
- Adjustable part switch options
- Automatic regulation of application rates
- Support of navigation within the field
- Complete documentation of application rates and locations

Part switch is considered as one of the central elements of site specific operating systems, but often has only been realized in an on/off mode, such as is available for site specific sprayers. A fully adjustable part switch is indispensable in site specific application, not only to allow precise application on very uneven fields and avoid overlapping, but also to better differentiate between the mapped areas of higher and lower nutrient requirement. In this context an efficient DGPS guiding system is needed to provide the spreader operator with precise navigation information, taking into account that base nutrients are often applied out of the cropping period and consequently not within tramlines. Using a simple arrow directional indicator would not allow the operator to accurately follow up the previous working path. The mealtime display on the computer screen is therefore indispensable. The full documentation of the spreading process finally ensures a permanent control of the quality of application.

Soil nutrient pattern often varies in different ways, which requires the application of single nutrient fertilizers. This means either mixing of fertilizers and application at the same time or application of straight P, K and Mg at different times. Splitting the application has advantages with regard to reaction times of the system and accuracy of rates, as mixing of fertilizers with different grades is always a source of error. A precise fertilizing system run by a service provider also needs efficient logistics, which is difficult to realize if a customer has to be served at the same time with largely differing amounts of fertilizer. From experience most farmers prefer to split the cost of fertilizing, which means, for example, application of phosphorus in one year and potassium in the following year. This also allows flexibility to place the fertilizer within the rotational application to the crop with the highest nutrient requirement.

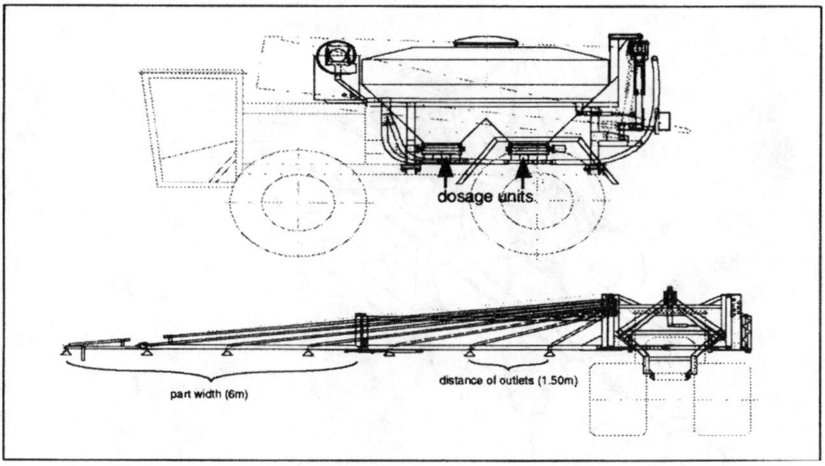

Fig.1. Pneumatic spreader with 24 m working width and a part switch every 6 m.

SYSTEM DESIGN

The spreader developed has a single compartment tank with a capacity of 6 tons (Fig. 1). Under European farming conditions a 24 m boom and a part switch every 6 m seems favorable, as this allows travel in the most commonly used tramline widths of 18, 24 and 36 m. The decision for a 24 m boom was also taken after technical considerations of boom weight, fold up and transport. In context with an even distribution pattern, the question raises 'how to guarantee under varying field conditions an even application pattern and rate'. Technically, only a total variable drive can ensure constant revolutions of the different dosage units without any impact of speed, because there is no gear changing and power output can be stabilized. Consequently travelling over undulating fields (which are common in western Europe) or fields with varying soil conditions will not influence the performance of the dosage system and therefore the application rates, whereas even a power shift drive will still cause differences in revolution rates. Topographical changes within the field also require a boom, which can be adjusted in its vertical position to the ground. Thus a sensor has been installed at the suspension point of the boom, reacting to changes in horizontal position.

The part switches of the spreader can be regarded as a system consisting of four independent working widths of 6 m. Four dosage units have been implemented each feeding one 6 m section of the boom. The dosage units are driven electrohydraulically by one common hydraulic drive. The reacting time for changing application rates at the outlets is only about 2 seconds.

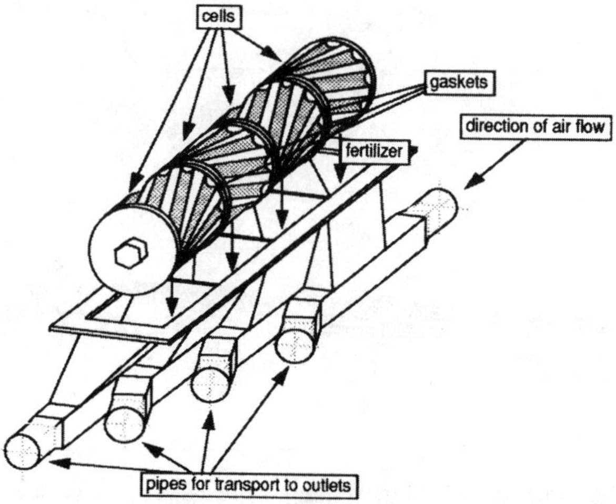

Fig. 2. Dosage unit consisting of 4 segments with sicke-shaped cells and receiving unit underneath the cells.

The short reaction time becomes extremely important as the spreader is working at relatively high speeds averaging 17 - 20 km/h.

The four dosage units are completely independent of each other and therefore allow totally autonomous regulation of each 6 in section. The units themselves consist of four segments sitting on a drum, separated by gaskets (Fig. 2). The segments have sickle-shaped cells, which serves the objectives of a high degree of cell filling and an easy clearance. The cell design also allows a large range of application rates (70-900 kg/ha) and fertilizer types without changing the cell type. The cells are made of VA steel to reduce abrasion and are sitting on the drum with a spin of 40°. The latter considerably reduces the risk of blockage and ensures an even turning of the drum. At the entrance point of the fertilizer from the tank to the dosage unit, a soft plastic lip ensures a uniform filling of the cells, but also prevents any filling of cells when the spreader is standing still.

A pneumatic system is imperative for a quick transport of fertilizer from the dosage units to the outlets. The transport mechanism itself, to the cells and from the cells to the outlets, strongly influences the accuracy of application rates. From the central tank the fertilizer is fed into the cells of the four dosage units (Fig. 3). As the tank is under atmospheric pressure, the same as in the pneumatic transport system, the fertilizer is gravity fed into the pneumatic system. Using the traditional injector principle, the capacity of fans would have been excessive, working with four different dosage units. The principle of the atmospheric balance between tank, cells and pipes has also the advantage that the airstream will not influence the filling of the cells from the tank as well as the complete emptying of the cells. The fertilizer is transported from the cells into the pipes, which lead to the outlets of the boom. This means that each outlet is directly fed by one cell (cf. Fig. 2), allowing by direct transport and short distances, a considerable reduction in variation of application rates within the boom. The distance between the four outlets in each 6 m section is 1.50 m. The application rates are varied via speed of cell rotation.

For control and calibration purposes a weighing system has been installed to ensure that application rates will be maintained, even if fertilizer quality changes. Four load cells are mounted between the tank and independent frame recording the weight of the tank content. A calibration of the dosage can be started at any time for an individual length of time. The weight of the fertilizer in the tank is measured respectively at the start and end of the calibration process, when the spreader is standing still. The signals from the weighing cells are processed via a separate processor unit and then transmitted to the on-board computer terminal. Further control of application rates is realized by speed limitation. As the application rate is a function of speed, radar is used as a ground speed sensor, indicating to the operator the maximum driving speed with respect to the actual application rate.

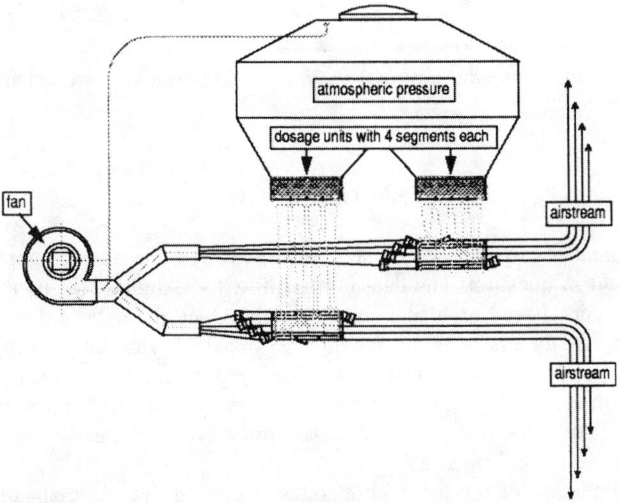

Fig. 3. Principle of fertilizer transport within the pneumatic spreader.

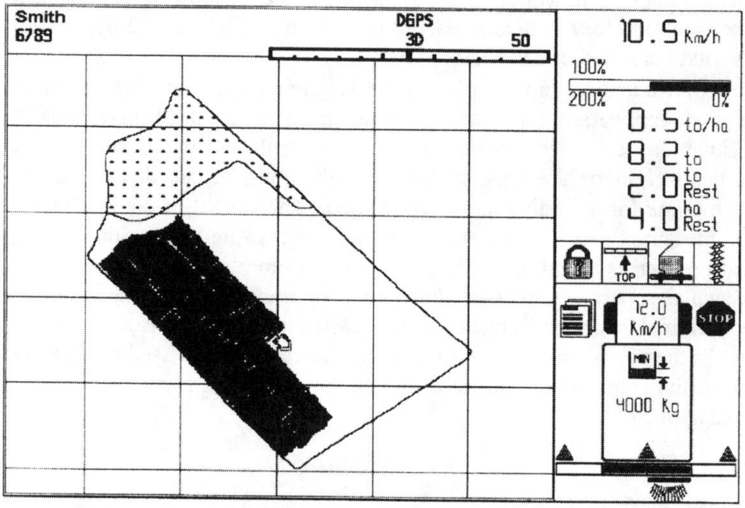

Fig. 4. Display of the spreader terminal showing field borders, restricted areas and spreading path.

SYSTEM OPERATION

The spreader can only be operated in context with the relevant application data of the field in question. The data is transmitted via digital telephone to and from the spreaders, based on a mailbox principle. Once the data is downloaded into the computer, the operator can navigate to the field with the help of DGPS and a compass, as position and heading are displayed in realtime on the computer screen. Furthermore field borders and restricted areas as well as application rates are shown on the screen (Fig. 4). In this way the driver can clearly identify the field and the areas to be fertilized.

The regulation of the application rates is carried out automatically once the driver enters a field. Changing from one application zone horizontally or vertically results automatically in an adjustment of the rates by interpolation of application rates in both directions. Fertilized areas are stored with fun record of application width and rate. This enables the total control of the whole fertilizing process with an automatic on/off switch of the relevant boom sections, when the spreader approaches the field boundaries or any previously fertilized areas. Also it is no longer feasible that the regulation of part switches is operated manually. Therefore the spreader operator only has to concentrate on steering and controlling general machine functions. Changes in speed automatically result in an adjustment of dosage cell revolution to keep application rates independent of driving speed, whereas exceeding the speed limit provokes a warning for the operator to slow down.

The calibration of the dosage system is an interactive process in which the operator has to enter the start and stop points, so that the respective weights are recorded. The measured weights are then compared with the values calculated by the on-board computer. Any difference between measured and calculated values results in an adjustment of the calibration factor.

DOCUMENTATION

The application data, which is stored onto hard disk of the on-board computer, includes positions in north and east coordinates recorded at an average interval of 3 seconds, application rates for each 6 m section and calibration factors. The operator can add for each field, individual remarks concerning the field conditions or spreading, selecting from a prepared list on the computer terminal. The data is then transferred via digital telephone to the office.

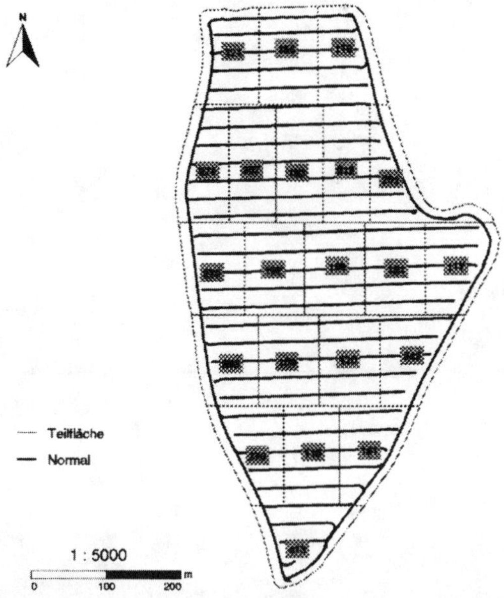

Fig. 5. Documentation of magnesium application; working path and application rates (kg/ha).

The data processing embraces checks of the working paths and differences between default and real application values. In order to draw maps from the spreading data, the same grid, which has been previously generated for the soil nutrient mapping, is used. For each grid cell the data is averaged and can then be displayed showing the real values for application rate. Using the working path as a second layer, a full documentation of the fertilizing process can be realized (Fig. 6). This enables the farmer to easily check the spreading work and the applied amount of fertilizer. In addition to documentation purposes, the application map provides a sound data basis for fertilizer calculation of the following years and also can serve the farmer for any farm assurance schemes.

CONCLUSIONS

The spreading system developed by Hoelzl for application of P, K and Mg offers a substantial improvement of application accuracy by a novel design for the dosage units, with special cell shapes and orientation on the drum. The implemented principle of fertilizer transport from the tank to the outlets considerably reduces errors in applications rates, caused by uneven filling or emptying of cells. As part switch of the boom has been realized for each 6 m section, the automatic control of the spreading process as wen as the documentation are essential components of the system. Further development win concentrate on automatic steering as well as implementation of additional sensors for application rate control at the outlets.

Surveying Biomass in Plant Populations

Detlef Ehlert
Institute of Agricultural Engineering
Max-Eyth-Allee 100
Potsdam, Germany

ABSTRACT

In addition to yield mapping in combines and in other harvesters, biomass distribution in growing plant populations is a possible parameter for surveying heterogeneity inside of fields. To develop the technique of acquisition of biomass data by mechanical scanning of plants with stems, it has been shown that definite, simple relationships between dynamic effect and biomass arise through strip scanning of plant populations. When a cylindrical body and a pendulous pivoted cylindrical body are moved horizontally through a plant population, the forces acting on the cylindrical body and also the angle of deviation are determined by the parameters of plant growth, mainly biomass. By measuring the forces on the cylindrical body and the angle of pendulum, conclusions can be drawn about the biomass of crops. With measurements are determined differences in still growing crops in order to optimise crop management and yields in Precision Agriculture.

INTRODUCTION

The local grown biomass inside a field is an integrating indicator for existing soil conditions. While soil fertility cannot be estimated by a measurable quantity, biomass can be measured as an area-specific quantity (e.g., in kg m^{-2}). The following fundamental methods can be used to survey biomass or its components:

- sequences of aerial photographs to determine the vegetation indexes in the growth phase,
- vehicle based recording,
- manual surveying

Sensor-supported and vehicle-based biomass surveying and yield-mapping is an applicable method that can be used especially in practice to survey heterogeneity for Precision Agriculture.

Yield-mapping has progressed furthest for threshable crops harvested by combines. For yield mapping in combines, a large number of measurement devices has been developed (Borgelt, 1990) and some of them are already available on the market. The technical literature shows, furthermore, that yield mapping with forage harvesters has reached the stage of the research and development (Vansichen & De Baerdemaeker, 1993; Ehlert & Schmidt, 1995; Auernhammer et al., 1995). Technical solutions for yield-mapping with self-loading forage boxes and round bale presses show the same stage of development (Auernhammer et al., 1994). Literature also includes references to solutions for

Copyright © 1999 ASA-CSSA-SSSA, 677 South Segoe Road, Madison, WI 53711, USA. *Proceedings of the Fourth International Conference on Precision Agriculture.*

yield-mapping of sugar beets (Hien & Kromer, 1995) and potatoes (Baganz, 1991; Champbell et al., 1994; Ehlert, 1996). The first manufacturers are already offering yield-mapping in harvesters for sugar beets and potatoes on the market.

All the methods described have one disadvantage. The yield as an essential component of the biomass is estimated at the time of harvest and so this information cannot be used for crop management in the same season. In practice a further problem encountered with yield-mapping lies in the exact measuring and locating of the specific harvested site because of the different working widths of headers and offsetting of the position as the plant material passes through the machines. In addition to yield-mapping in harvesting machines, the technical literature describes sensor-supported manual surveying methods for pastures. (Gonzalez, 1990; Hutchings, 1990).

Vehicle-based biomass surveying is conceivable not only in harvesting machines, but also in standing plant formations. The advantages of this include the following:

- During measuring only minimal changes appear in the plants due to the use of the tram lines.
- No varying working widths need to be taken into account.
- By using various basic machines (mineral fertiliser spreaders, field sprayers, special-purpose vehicles) measurements are possible during passage along the tram lines.
- Data on the biomass formation can be used to direct following crop management activities (fertilisation, plant protection).

MATERIAL AND METHODS

Investigated Crops, Measuring Principles, and Parameters

Since stalk crops (corn, maize, rice, legume, feeding grasses and others) are cultivated predominantly throughout the world and on most of the land used for agricultural production purposes, it is expedient to concentrate work on these. This paper reports on investigations conducted regarding winter rye and winter barley. Since first tests with contact-free ultrasonic scanning of stalk crops did not produce accurate results, investigations were oriented to mechanical scanning. To develop the technique of acquisition of biomass data by mechanical scanning of plants with stems, it has generally been assumed by the author that definite simple relationships between dynamic effect and biomass arise through strip scanning of plant populations.

For the development of the principle of plant mass determination by mechanical screening, the following general theses were defined:

- A force that is determined by friction, the mass moment of inertia, and moments of resistance of the stems acts on a cylindrical body moved horizontally through plant stocks.
- Single plants of a species produce higher mass inertial forces and bending

resistance moments with increased mass.

- Scanning several plants simultaneously produces both a measurement value amplification and signal averaging (smoothing).
- With strip-type scanning clear, simple relations arise between dynamic effect and biomass of plant formations.

The hypotheses put forward were examined for the scanning principles rigid cylinder and pendulous cylinder. Passing a rigid cylindrical body in a horizontal movement direction (Fig. 1) through in a plant formation produces a measured force F_M corresponding to the horizontal component F_{Zh} of the resulting dynamic effect of F_Z on the cylinder. The size of F_{Zh} depends on the parameters friction, scanning height h_A , travel speed v_F , mass of the stems (including mass moments of inertia) m_{Hi} , bending resistance moments of the stalks W_{bi} and number of stalks n_H . Since as a rule inside a field the parameters scanning height h_A and driving speed v_F are constant, F_{Zh} depends on only the plant parameters, which can be used for surveying the grown biomass m_B .

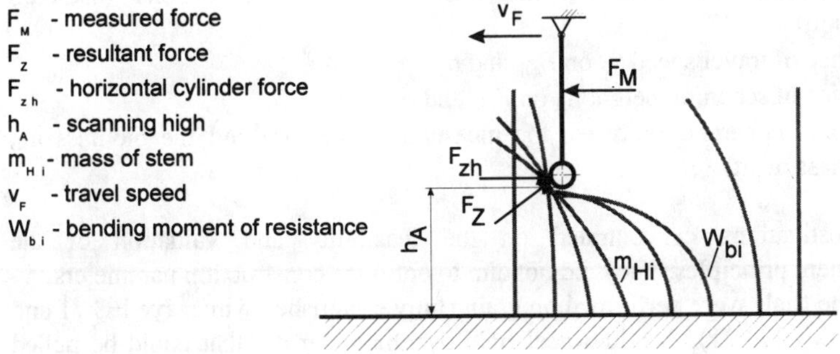

Fig. 1. Biomass measurement by moving cylinder.

When a pendulous pivoted cylindrical body is moved horizontally through a plant population (Fig. 2), the angle of pendulum deviation α is determined by the following parameters: height of the pivot point h_D , length of the pendulum l_P , mass of the pendulum m_P , travel speed v_F , mass of the stems (including moments of inertia) m_{Hi} , moments of resistance of the stems W_{bi} , and the number of the stems n_H . Since it is possible to make h_D ; l_P ; m_P , and v_F within a field and a crop roughly constant, the angle of deviation α only is still dependent of the stock parameters m_{Hi} , W_{bi} and n_H . This measurement is to be considered to describe the biomass formed relatively.

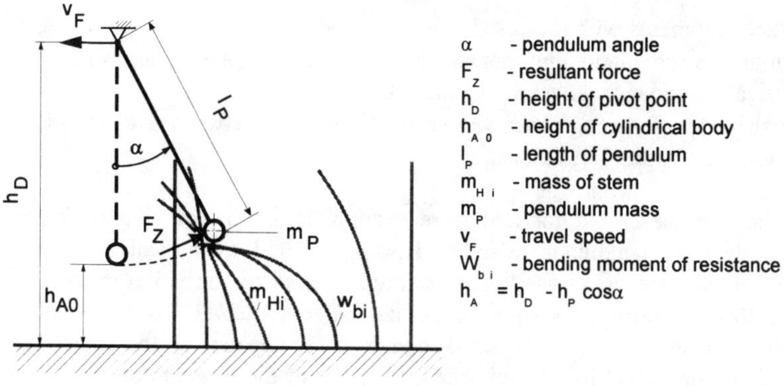

Fig. 2. Biomass measurement by moving pendulum.

According to the assumptions, experimental investigations were focused on the:

- reproducibility of measurements under the same conditions (standard deviation)
- influence of travel speed v_F on F_{Zh} and α
- influence of scanning height h_A on F_{Zh} and α
- functional coherence between the measurands (F_{Zh}; α) and the biomass m_B (goodness of fit)

The investigations concentrated on the feasibility and valuation of the measurement principles. They did not aim to optimize construction parameters.

The trials were performed on grain (fully established winter rye EC 71 and winter barley EC 71). A four-wheel field dynamometer car that could be pulled along the tram lines was developed. The physical parameters of its frame helped to avoid disturbing the crop stems. For both principles the scanning width was 1370 mm. The dynamometer served to carry both the sensors and the measuring system. By using one of the four wheels in connection with an incremental angle sensor, path measurement and speed indication were also possible.

In order to examine measurement deviations occurring under the same conditions, a measured distance of 50 m was driven through 25 times while taking measurements simultaneously of the horizontal cylinder force, F_{Zh} and the angle of deviation, α. Further measurements with both principles should supply information about the fundamental influences of the travel speed and the scanning height. The plant mass was determined by cutting the stems directly above the ground over plot lengths of 5 m and weighing the harvest.

RESULTS AND DISCUSSION

Both measurement methods under the same conditions showed a slight scattering of results which leads to the conclusion that there is only a small influence by random factors (Fig. 3).

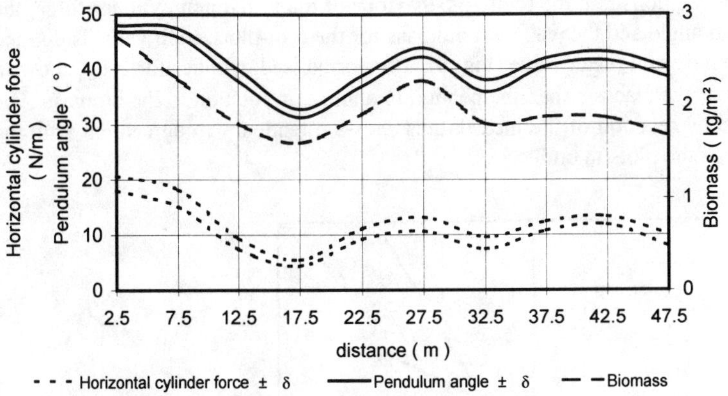

Fig. 3. Standard deviation of horizontal cylinder force F_{Zh} (h_A = 700 mm; v_F = 1 m s^{-1}) and pendulum angle α (h_D = 1250 mm; l_P = 800 mm; m_P = 2,1 kg m^{-1}; v_F = 1 m s^{-1}) in relation to biomass.

The pendulum-width-related horizontal cylinder forces F_{Zh} vary between about 5 and 20 N m^{-1} and over a course of 50 m show similarities with that of the assigned plant mass. The pendulum angle α varies from 31 to 47° under the same conditions. It is apparent that extreme values of F_{Zh} ; α and the biomass are recorded at same places over the 50 m distance.

Since the plant stock strip was traversed by dynamometer 25 times for estimation of a standard deviation under same conditions, slight damaging could not be excluded. A trend-calculation was conducted to check the possible, unavoidable damaging due to passage of the equipment. The mean horizontal cylinder force decreased by 2 N m^{-1} (20%) and the pendulum angle decreased gently by 1.3 ° (4%) (Fig. 4). These tendencies should be considered to obtain the true standard deviation.

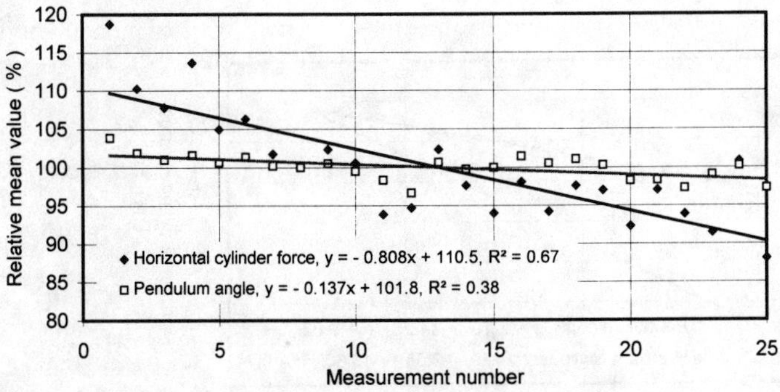

Fig. 4. Relative mean value of measurands depending on measurement number (same measurement conditions like in Fig. 3).

To demonstrate the relative behaviour of the horizontal cylinder force, the pendulum angle and the weighted biomass for the 5 m-plots, each value is divided by its own overall mean value (Fig. 5). The normalised cylinder force F_{Zh} shows a more intensive reaction than the pendulum angle α in relation to the biomass. The fundamental reaction of the measurands shows a tendency to agreement with the biomass of the plots in both cases.

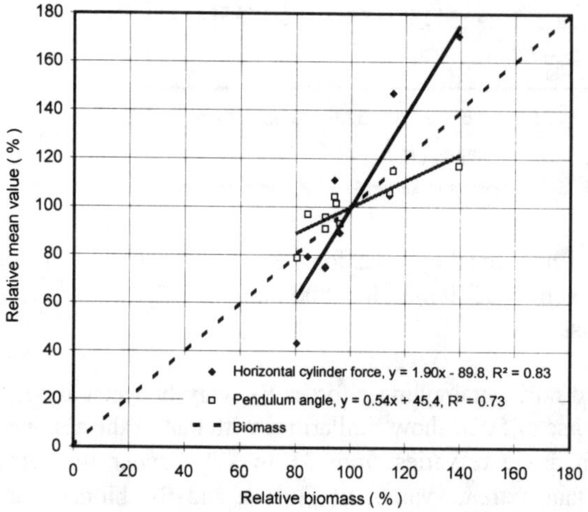

Fig. 5. Relative behaviour of the horizontal cylinder force F_{Zh}, the pendulum angle α and the weighted biomass for the 5 m-plots.

With regard to the mass inertial forces, the question posed is what influence the speed has on the two measurands. The speed range examined was selected in relation to the usual speeds in crop management. For given conditions, increasing speed from 0.5 to 3 m s^{-1} means that the horizontal cylinder force F_{Zh} rises from 15 to 22 N m^{-1} and the pendulum angle from 45.4 to 50.8° with linear characteristic (Fig. 6).

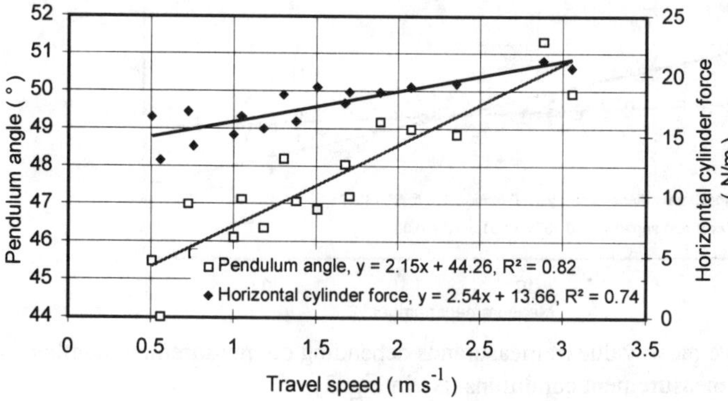

Fig. 6. Horizontal cylinder force F_{Zh} (h_A = 600 mm) and the pendulum angle α (h_D = 1150 mm; l_P = 800 mm; m_P = 2,1 kg m^{-1}) depending on travel speed.

The influence of the scanning height h_A on F_{Zh} and α is more intensive (Fig. 7). For the 50 m-distance at h_A = 900 mm the mean value of F_{Zh} is only about 7 N m^{-1}. It rises at 500 mm progressively to more than three times this value. This can be explained in theory also. The same reduction in scanning height from 650 to 250 mm also causes a linear increase in the mean pendulum angle from 30 to 50 °.

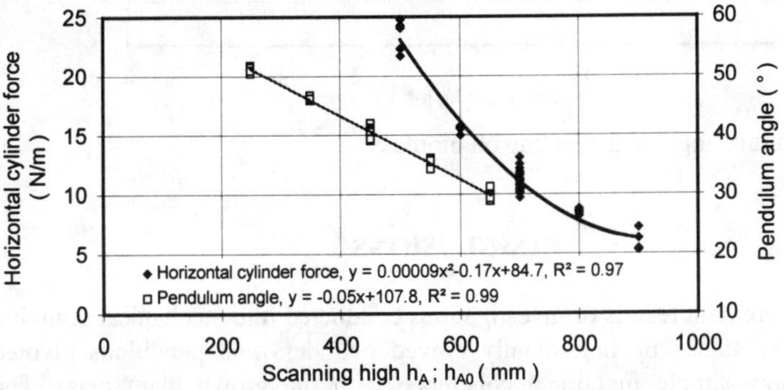

Fig. 7. Horizontal cylinder force F_{Zh} ($v_F = 1$ m s^{-1}) and the pendulum angle α (l_P = 800 mm; m_P = 2,1 kg m^{-1}; $v_F = 1$ m s^{-1}) depending on scanning height.

For the assessment of both surveying methods, the correlation and deviations between measured parameters F_M; $\propto$ and the biomass m_B are of decisive importance. The horizontal cylinder force F_{Zh} and pendulum angle $\propto$ ascertained show a clear functional dependence on the estimated biomass (Fig. 8; Fig. 9). Deviations in the single test points of the regression curve are, however, considerable. In spite of strongly differing stock relationships, the goodness of fit reached R^2 = 0.75 ... 0.91.

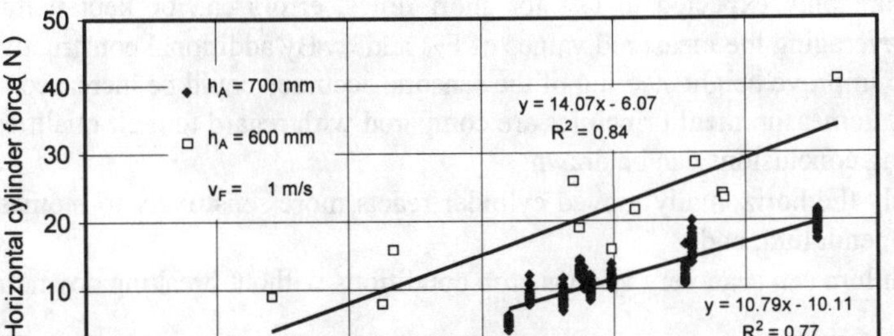

Fig. 8. Horizontal cylinder force F_{Zh} depending on biomass.

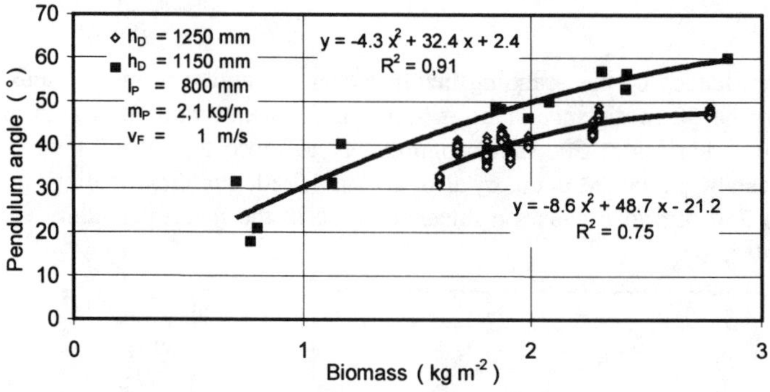

Fig. 9. Pendulum angle ∝ depending on biomass.

CONCLUSIONS

Measurement results of investigations conducted into mechanical scanning of grain plant stands by horizontally moved cylinders and pendulous pivoted cylinders show simple functional coherence with the grown plant mass. The measurements can be reproduced very well with low coefficient of variance arising under the same conditions.

Various speeds affect the measurement values only insignificantly in the examined range of 0.5 ... 3 m s^{-1}. It is assumed that for speeds that differ for example by not >0.5 m s^{-1}, the horizontal cylinder force changes by about 5% and the pendulum angle changes by about 2.5%. It can be concluded from this, that an arithmetic compensation is not necessarily required at low speed differences.

If a vehicle is moved in tram lines, only an insignificant change of the constructive parameters h_A and h_D in the range of centimetres is expected. If it is assumed that these quantities vary by 50 mm, the cylinder horizontal forces would vary by approx. 13% and the pendulum angle by approximately 6%. Because such deviations are only expected to last for short times, errors can be kept within bounds by averaging the measured values of F_{Zh} and ∝. By additional constructive measures to improve height steering of the sensors, accuracy could be increased.

If both measurement principles are compared with regard to their qualities, the following conclusions can be drawn:

- Obviously the horizontally moved cylinder reacts more sensitively to biomass than the pendulum, and
- the pendulum can scan very various crop conditions without breaking down the plants

The results achieved justify further examination of the measuring principles with the aim of optimising the construction parameters h_A; h_D; l_P; m_P as a function of the crop plant species for yield mapping and precision agriculture.

REFERENCES

Auernhammer, H., M. Demmel, Th. Muhr, J. Rottmeier, and K. Wild. 1994. Rechnergestützte Ertragsermittlung für eine umweltschonende Düngung. Landtechnik Weihenstephan, Landtechnik-Schrift Nr. 4:111–134

Auernhammer, H., M. Demmel, and P.J.M. Pirro. 1995. Yield measurement on self propelled forage harvesters. ASAE-Paper no. 95–1757. ASAE, St. Joseph, MI.

Baganz, K. 1991. Yield estimation on potatoe harvesters. *In* Int. Symp. on Locating Systems for Agricultural Machines , Gödöllö.

Borgelt, S.C. 1990. Sensing and measurement technologies for site specific management. p. 141–157. *In* Soil Specific Crop Management. ASA, CSSA, and SSSA, Madison, WI.

Champbell, R.H., S.L. Rawlins, and H. Shufeng. 1994. Monitoring methods for potatoe yield mapping. ASAE-Paper no. 94–1584. ASAE, St. Joseph, MI.

Ehlert, D., and H. Schmidt. 1995. Ertragskartierung mit Feldhäckslern. Landtechnik 4:204–205.

Ehlert, D. 1996. Massestrommessung bei Kartoffeln. Landtechnik 1:20–21.

Gonzalez, M.A., M.A. Hussey, and B.E. Conrad. 1990. Plant height, disk, and capacitance meters used to estimate bermudagrass herbage mass. Agron. J., 82:861–864.

Hien, P., and K.H. Kromer. 1995. Sensortechnologie zur Ertragsbestimmung und Ertragskartierung von Zuckerrüben. VDI-Berichte 1211, Landtechnik:187–190.

Hutchings, N.J., A.H. Phillips, and R.C. Dobson. 1990. An ultrasonic rangefinder for measuring the undisturbed surface height of continuosly grazed grass swards. Grass Forage Sci. 45:119–127.

Vansichen, R., and J. De Baerdemaeker. 1993. A measurement technique for yield mapping of corn silage. J. Agric. Eng. Res. 55:1–10.

Increasing Productivity of Equipment for Variable Rate Technology

Andrey V. Skotnikov
Case Technology Center
Burr Ridge, Illinois

D. E. McGrath
Tyler Industries
Benson, Minnesota

ABSTRACT

Site-Specific Crop Management refers to a rapidly developing new direction in agriculture, promoting variable agricultural management practices within a field according to site conditions. Existing agriculture equipment for variable rate technology (VRT) for multiple input has a significant disadvantage - a constant size of compartments or bins for applying components. When one of the compartments is empty, the whole unit has to stop for a refill. Plus, it requires the presence of a tender's fleet on a field. This reduces productivity and increases the cost of operation significantly. To improve the performance and productivity of equipment for applying agriculture inputs it is necessary to have changeable size of compartments on VRT equipment, matching tenders, and automated systems for management.

Possible technical solutions suitable for fertilizer spreaders, liquid chemical sprayers, seed drills, and tenders are proposed.

INTRODUCTION

Precision Agriculture refers to a rapidly developing new direction in agriculture that promotes a site-specific agricultural technology management within a field. This is a new multidisciplinary concept based on a systems approach for an optimization of management.

Some farm equipment necessary for Precision Agriculture has been developed. Including are: variable rate seed planters, pesticide and fertilizer applicators, irrigation systems, yield monitors, and mapping software packages.

Equipment for variable rate technology working with one variable practically does not have any problems. But equipment for applying of several components simultaneously has a problem mostly due to constant size of bin compartments. When one of the compartments is empty, the whole unit has to stop for a refill. Every additional stop reduces the productivity and increases the cost of operation significantly (Skotnikov & Robert, 1996). This concerns seed drills (seed variety and fertilizer applied), sprayers (combination of herbicides),

Copyright © 1999 ASA-CSSA-SSSA, 677 South Segoe Road, Madison, WI 53711, USA. *Proceedings of the Fourth International Conference on Precision Agriculture.*

spreaders (different combinations of fertilizer) and potentially — harvesters for separating of harvested grain on varieties or quality.

In order to increase productivity of existing equipment we are suggesting several technical solutions for dry and liquid products handling.

DISCUSSION

The suggested design for existing disk and pneumatic spreaders is presented in Fig. 1.

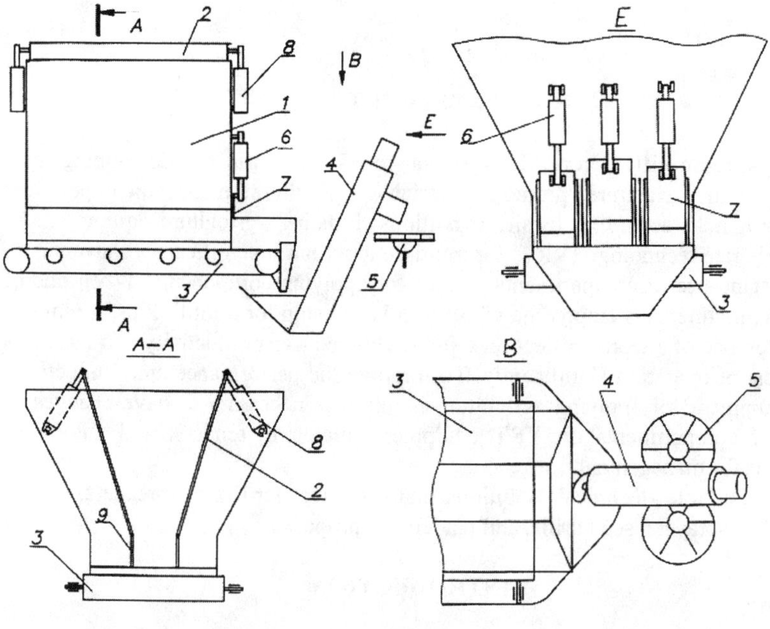

Fig. 1. A kinematics scheme of a spreader with changeable bin compartments.

The spreader has bin 1 with two inside moveable walls 2, under which is mounted conveyer 3 (Skotnikov A.V., and McGrath, 1995.). At the end of conveyer 3 in direction of movement of the spreader is installed mixing auger 4, delivering materials to spreading discs 5 or airboom (not shown). On the rear wall of bin 1 are installed hydraulic cylinders 6 moving up and down metering gates 7, and hydraulic cylinders 8 moving walls 2 inside of bin 1. Movable walls 2 are jointly connected to firmly fixed conveyer dividing boards 9.

Thus bin 1 may have three adjustable compartments for nitrogen, potassium, and phosphorus fertilizers. Depending on requirements and proportion of applying fertilizers, it is possible to change the volume of each compartment and empty all of them simultaneously. For sake of simplicity it is assumed that a speed of conveyer 3 is mechanically synchronized with a ground speed of a spreader.

The block-diagram of an automated system for a spreader with variable bin compartments is presented in Fig. 2.

The spreader works as follows. An automated regime of spreader's work is considered. It also is assumed that the programs of fertilizer application and a spreader

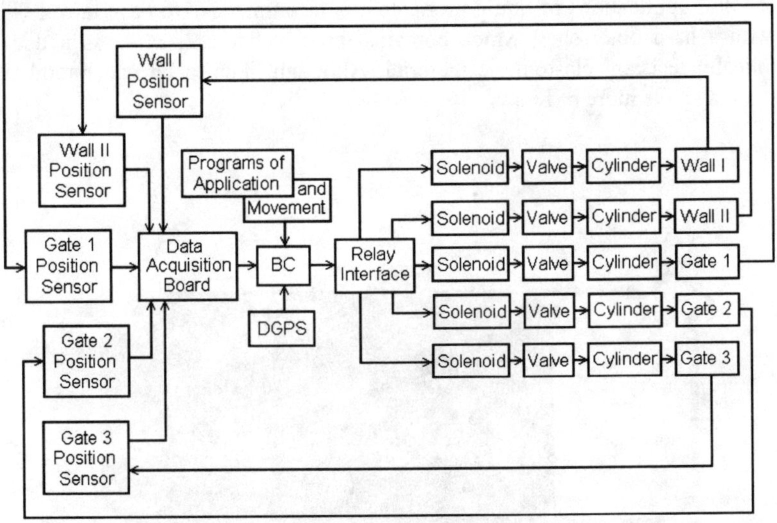

Fig. 2. Block-diagram of automated system for fertilizer applicator with changeable bin compartments

movement are known, and the volume of each applying component for a load capacity of a spreader is calculated. For example, the load capacity of a spreader is six ton, and proportion of applying fertilizers is 2.5 ton of N, 1.6 ton of K and 1.9 ton of P. Such calculations are made along a spreader route for each load. The results of these calculations are added to a program of fertilizer application and fertilizer delivering. The program of fertilizer delivering consists of fertilizer combination for each load and location of its unload. Then the diskette with the program of applying is installed into a board computer BC (Fig. 2). The board computer compares signals from wall position sensors with numbers in programs corresponding to a geographic position of spreader and discrepancy signals go to a relay interface. A relay interface engages corresponding solenoids moving hydraulic valves pressurizing respective chambers of cylinders and moving inside walls to the necessary positions. Next, a tender of a similar design discharges its identical compartments to a spreader. With a beginning of an application, the board computer compare signals from all gate position sensors coming through a data acquisition board with numbers in programs of application corresponding to a spreader's geographic position. The discrepancy signals go to a relay interface. A relay interface engages corresponding solenoids moving hydraulic valves pressurizing respective chambers of cylinders moving gates up and down and manage in this way the rate of applied fertilizer. The fertilizers then are mixed in auger 4 (Fig. 1) and supplied to spreading discs 5 or to the airboom. After the first load is applied, board computer again rearranges position of inside walls for the next run. Tender, using fertilizer delivering program, rearranges compartment sizes, take calculated fertilizer combination for each run and delivers it to a specified program location, unloads it, and a spreader continues to apply fertilizer in accordance with the provided program.

For applications of liquid products it is recommended to have tank 1 (Fig. 3) with a hard outer shell which contains three bladders 2, 3, 4, formed of a chemically resistant elastomeric material. Although, it is to be understood that tank 1 may have more or less

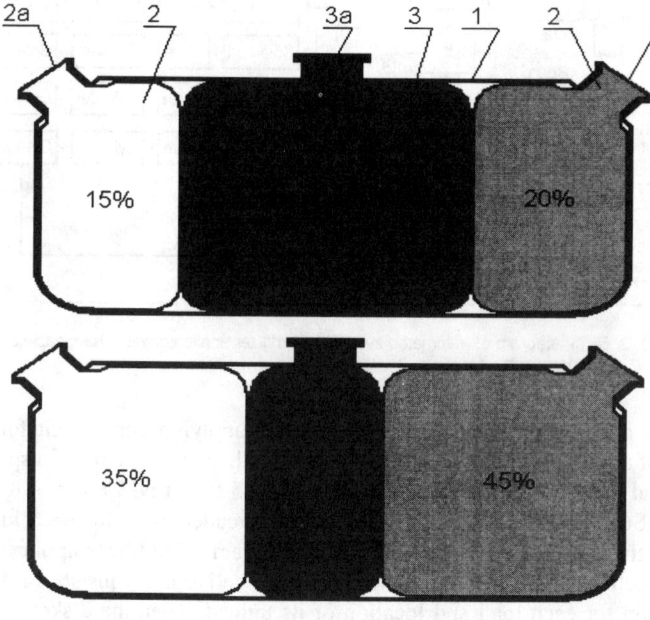

Fig. 3. Sprayers tank with variable size compartments.

numbers of bladders. Each bladder has a respective fill port 2a, 3a, and 4a. Each bladder can expand (or contract) to allow the maximum volume of the tank shell 1 to be used, while simultaneously adjusting the relative capacities of the bladders 2, 3, and 4. For example, bladders 2, 3, 4 may fill the volume of shell 1 with a 15/65/20% ratio or may be expanded and contracted to a 35/20/45% ratio between the bladders while keeping the overall capacity of tank 1 substantially constant.

The sprayer can apply chemicals in accordance with the program of spraying (Fig. 4).

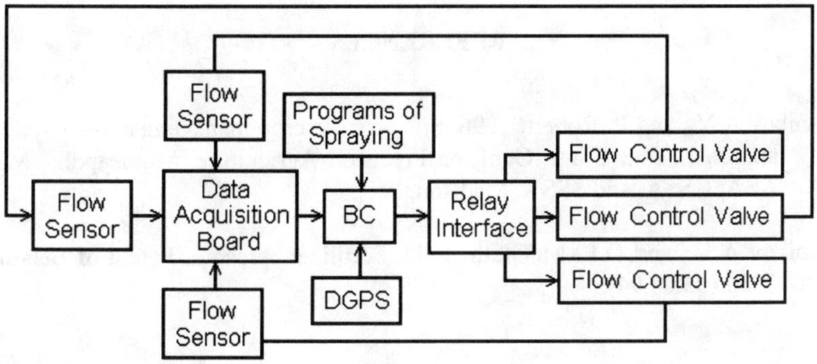

Fig. 4. Block-diagram of an automated system for multiple liquid product application.

The program of spraying (calculated rate for determined area) is prepared on a diskette in advance and is based on results of soil analysis or yield monitoring. Simultaneously, the program of chemical delivery (combination of chemicals and location of each load delivery) is developed. The automated system of programmed application of chemicals works is as follows. The diskette with the program of spraying is inserted into the board computer BC. After beginning of spraying the signals from the DGPS (Fig. 4) and flow sensors go to the BC. The signals of discordance between the flow sensor readings and program requirements from BC through the relay interface RI go to the electric valve of the flow control to increase or decrease the flow of chemicals to the boom. The signal from DGPS also serves to compute the transport delay of the system and the necessary flow to provide the required rate of spraying in accordance with the program. After the sprayer has applied the first load, the next can be delivered by a tender of similar design to the known location in accordance with the program of chemical delivery.

The chemical tender works analogously to the fertilizer tender described above.

CONCLUSIONS

Suggested technical solutions permit

- using full load capacity of equipment for application of agriculture inputs,
- increasing its productivity,
- eliminating preliminary blending of fertilizer,
- reducing amount of tenders involved in support of applicators.

REFERENCES

Skotnikov A.V., and P. Robert, 1996. Site-specific crop management — a system approach. *In* 3rd Intl. Conf. on Precision Agriculture, Minneapolis, MN. ASA, CSSA, and SSSA, Madison, WI.

Skotnikov A.V., and D.E. McGrath. 1995. Fertilizer spreader. Patent of Belarus, 1956 (03.06.94).

Precision Agriculture Requires Precise Tuning

Thomas S. Colvin
Erick W. Kerkman

National Soil Tilth Laboratory
USDA–ARS
Ames, Iowa

ABSTRACT

In the rush to adopt the new technology that many people assume is the essence of precision agriculture, we often overlook the simple requirements of good system management. We have measured the poor distribution of ammonia across application tool bars. We have seen the results of using planters without checking calibration. Lawsuits can be based on sprayer drift or poor technique in loading or operating. The goal of this paper is to remind all of us of the importance of every step in the management of crops. Checking, calibration, understanding instructions (in other words basic good technique) are necessary to achieve the benefits that should be available through the use of precision agriculture.

INTRODUCTION

Let us agree for the discussion in this paper that a reasonable definition of precision agriculture is doing the right thing, at the right place, at the right time. The second part of that statement can be thought of as bringing in the high tech tools of GPS, GIS, and assorted electronics and other paraphernalia. The last part relates to things that we can sometimes control such as choosing an anticipated starting date. Sometimes we don't know till the season is over what the right time was. The first part, the right thing, is the part that we want to focus on in this discussion.

Knowing the right thing to do may involve all kinds of high tech equipment and fancy statistics or other analysis. Doing the right thing however starts with good managers and good operators doing a good job of using common tools such as planters, fertilizer applicators, harvesters, and whatever else might be needed.

Picture this. Your wife or husband goes out to do just a bit of mowing within 25 m of your house in the country. An airplane comes roaring in over your house to begin applying insecticide to the field across the road. The first pass goes fine but as the airplane returns and clears the electric wires at the edge of the field,

Copyright © 1999 ASA-CSSA-SSSA, 677 South Segoe Road, Madison, WI 53711, USA. *Proceedings of the Fourth International Conference on Precision Agriculture.*

either there is a valve malfunction or the pilot is late at hitting the control and insecticide is applied to your yard, house, animals, and the person mowing the yard. That is not precision agriculture.

Two days later you drive to scout a farm for weeds and other pests. It is several miles from your home and you notice the same airplane applying material one-half way between your house and your destination. As you leave your car that you have parked in some trees in the middle of the farm you hear an airplane approaching. You hear what sounds like rain and then you see the airplane that dusted your house and feel the granules hit you. The airplane was on a direct path to base. That is definitely not precision agriculture.

The charitable view of the two incidents is poor equipment or poor operation. The less charitable view is criminal intent. In either case, the right thing is not application outside an intended field, which one would have to assume included the right place. This is common sense and simple courtesy.

These examples could continue to fill the space allowed but let us look at one more example. Assume that you hire a local chemical dealer to apply a mixture of fertilizer and herbicide because you do not have the equipment to handle the materials to be applied. You go to the field and observe the application and see what appears to be part of the boom not working and tell the operator what you have seen. The operator replies that there should not have been a problem because the spray monitor did not indicate any problem. Later you check the field and find 3 m wide strips through the field with green weeds between 20 m wide strips of dead weeds. You call the supervisor and are told that there could not have been a problem because—first there was no herbicide applied that would have killed the weeds (which you had requested) and second—that the spray monitor did not suggest any problem. Three days later the strips are sprayed after considerable argument and foot dragging. This is not precision agriculture.

We need monitors and good equipment but even more critical are good operators and good managers. The best system enhances the effectiveness of each part but as systems become more complicated we must not be lulled into poor habits by high tech assistance. These limited examples are cited to remind us that basic good practices are necessary with or without the high tech assistants.

Proper use of equipment is important. Another issue however is the equipment itself. There can be problems in the design of common equipment. The remainder of this paper will discuss problems with the distribution of anhydrous ammonia across tool bars that have been measured during the past 15 yrs.

MATERIALS AND METHODS

Anhydrous ammonia is used as a common N fertilizer in commercial production systems in the Midwest. We wanted to use it on experimental plots but we wanted to be sure that we had uniform application. In 1982, 1995, 1996, and 1997 we measured the output of individual knives on several applicators by capturing the ammonia in individual containers containing water, which were weighed before and after the tests. This was similar to the method later described by Kranz et al (1994). We had equipment including weigh bars to weigh the tank in the field similar to Weber et al (1993) but that did not answer the question of distribution across the bar.

RESULTS

The research plot applicator that was measured in 1982 showed differences as large as 400% between the knife with the low relative rate for the bar and the knife with the high relative rate. The pattern almost seemed random and was not necessarily repeatable. We chose not to use that applicator. The problem seemed to be in the design of the distribution manifold.

In 1995, we were working with a cooperator on his field and we wanted to put on three rates of N for an experiment on the loss of N to tile drainage water. We used the water-can method on his 16 outlet applicator by working on each half of the applicator independently (Fig. 1 and 2) and established that the flow for each half and between the halves was quite uniform (each knife was no more than about 25% from the mean) at a high flow rate. The individual manifolds were connected by a common plumbing T to the main supply. After the field application it became apparent that the T or some other part of the system was not

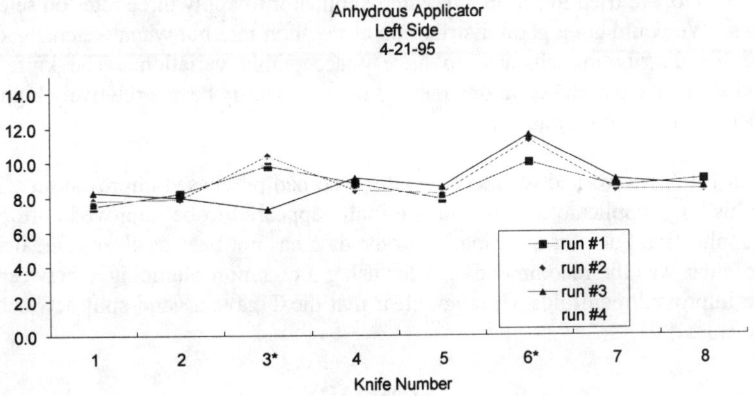

Fig. 1. Measured flow through individual knives on one-half of an applicator

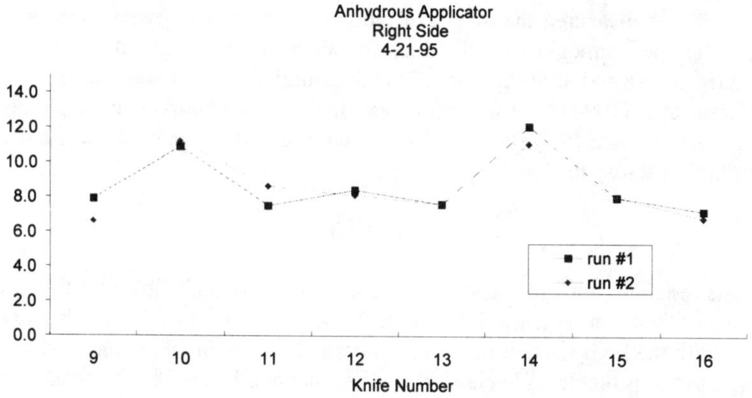

Fig. 2. Measured flow through individual knives on one-half of an applicator

splitting the flow accurately between the halves. The problem was worse with low and medium flow rates. Plant growth was affected by the uneven application but the rate per unit area appeared to be correct and the measured losses responded as expected. Knives 3, 6, 10, and 14 had the heat exchanger cooling flow vented through them, which explains the spikes that stand out from the rest of the knives

In 1996, we tried to set up a research applicator to apply three rates on selected plots. We could get a good distribution at the high rate but when we changed the rate the distribution changed to have unacceptable variation. The knife that tended to have a high rate at one rate did not necessarily have a relatively high rate when the application rate was changed.

In 1997, we worked with a cooperator who had purchased improved manifolds for his large applicator. The results initially appeared to be improved uniformity of application but later examination of the data has not been as clear. One item in particular was the recommendation for using a common plumbing T between the two improved manifolds. It is not clear that the T gave a good split between the two sides.

SUMMARY

Precision agriculture holds out the promise of improved management. That promise will not be fulfilled unless good equipment is used by excellent operators backed by enlightened managers. High technology tools will not compensate for poor equipment or sloppy procedures.

It is unfortunate, but most computers and machines do what the operator commands. The operator must be the master of the system not a slave to the system or any of its parts.

REFERENCES

Kranz, W., C. Shapiro, and R. Grisso. 1994. Calibrating anhydrous ammonia applicators. Nebraska Coop. Ext. Publ. EC 94-737-D. Nebraska Coop. Ext., Lincoln, NE.

Weber, R.W., R.D. Grisso, W.L. Kranz, C.A. Shapiro, and J.L. Shinstock. 1993. Instrumenting a nurse tank for monitoring anhydrous ammonia application. Computers Electron. Agric. 9:133–142.

Applying Soil Electrical Conductivity Technology to Precision Agriculture

Eric D. Lund
P.E. Colin
D. Christy
Paul E. Drummond
Veris Technologies
Salina, Kansas

ABSTRACT

The site-specific application of inputs such as seed, fertilizer and crop protection chemicals has the potential to reduce input costs, maximize yields, and benefit the environment. The economic returns currently received by the early adopters of precision farming methods need to be improved before wide-scale acceptance of this practice will occur. These improvements include cost-effective identification and management of the spatial variability of soil and nutrients, applying inputs based on each site's productive capacity, and correct decision-making using the available layers of information. Electrical conductivity (EC) measurements of soil have long been used to identify contrasting soil properties in the geological and environmental fields. The purpose of this paper is to discuss the applications where EC maps are proving useful in improving economic returns to precision farming.

WHY MAPPING OF SOIL PHYSICAL PROPERTIES IS NEEDED

Precision farmers have a wide assortment of data layers to assist them in their decision-making: yield maps, fertility–grid sampling data, USDA soil surveys, a grower's historical knowledge, and visible changes in soil appearance and topography. Of more limited availability are aerial or satellite imagery, elevation maps, and soil conductivity maps.

While each layer has value, the challenge is to use each one properly, not making assumptions by extending the layer's information beyond its utility. Yield maps, for example, are valuable tools in describing the results from a specific field in a specific year. Yield maps help quantify the effect-they do little to help understand the cause of the variability. (Kitchen et al., 1995) Due to the number of factors that affect yield data-e.g., weed pressures, planter problems, herbicide drift, in-field weather phenomena, yield monitor calibration and anomalies, yield data alone can be misleading in establishing a site's long-term productivity. As a result, if yield maps are used to establish site-specific yield goals, they need to be accompanied by other layers that confirm the productivity of a given area.

Soil surveys and the accompanying soil databases published by the USDA provide excellent background information on soils present in a field, but they are frequently too coarse to guide precision applications. GPS technology was not available when most states were surveyed, resulting in the actual location of soil unit transitions frequently being a considerable distance from the mapped lines. Most counties were mapped according to Order 2 survey guidelines, which allow

Copyright © 1999 ASA-CSSA-SSSA, 677 South Segoe Road, Madison, WI 53711, USA. *Proceedings of the Fourth International Conference on Precision Agriculture.*

for 2 $^{1}/_{2}$ acre inclusions within map units. Layering a soil survey with a detailed map of geo-referenced soil information can be a powerful tool for precision farming.

A layer of information that requires a sizable financial investment from precision farmers is intensive soil sampling. While the results from soil sampling provide extremely accurate information at each sample point, it can be erroneous to make assumptions from a few pounds of soil that will be extended to hundreds of tons of soil. Interpolating between sample points, as is common in grid sampling, assumes there is a spatial relationship between the sample points and therefore the values at the sample points can be interpolated to estimate the values for areas not sampled. If the spatial variability of the field is greater than the sampling density, the actual value between the sample points may be significantly different than that predicted by geo-statistical interpolation techniques. Figure 1 illustrates how ¾ acre and 3-acre grid samples from the same field provide a very different view of its lime requirements. In order to improve the accuracy of the interpolated sample data, a grid size of one acre may be needed. Yet a recent survey of precision input suppliers indicates <5% of grid sampling is performed that intensively (Farm Chemicals. 1997). Sampling efforts that are guided by an additional layer of information, provided that the layer chosen correlates to the nutrient availability in the field, can reduce sampling costs while improving the accuracy of nutrient recommendations for the field.

Fig. 1. ¾ acre and 3 acre grid values from same field.

As soil serves as the growth medium for crops, it is a major factor affecting a crop's potential. For precision agriculture to be successful, efficient and accurate methods of assessing the spatial variability of soil must be used. (McBratney et al., 1997). Interpreting any of the above information layers using a geo-referenced, detailed, accurate map of soil information provides an excellent foundation for precision farming.

PRINCIPLES OF SOIL ELECTRICAL CONDUCTIVITY

Conductivity is a measure of the ability of a material to transmit (conduct) an electrical charge. It is an intrinsic property of the material just like other material properties such as density or porosity. The standard units of measure of bulk soil conductivity are milliSiemens per meter (mS m^{-1}). Siemens are the inverse of Ohms and are the measurement of conductance. The usefulness of soil conductivity stems from the fact that sands have a low conductivity, silts have a medium conductivity and clays have a high conductivity (Fig. 2). Consequently, conductivity (measured at low frequencies) correlates strongly to soil grain size and texture (Williams & Hoey, 1987) (Fig. 3). The value of a soil conductivity map is in its ability to quantitatively delineate similar and contrasting regions of a field.

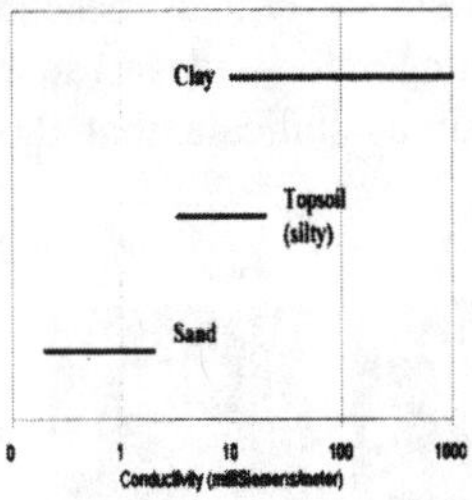

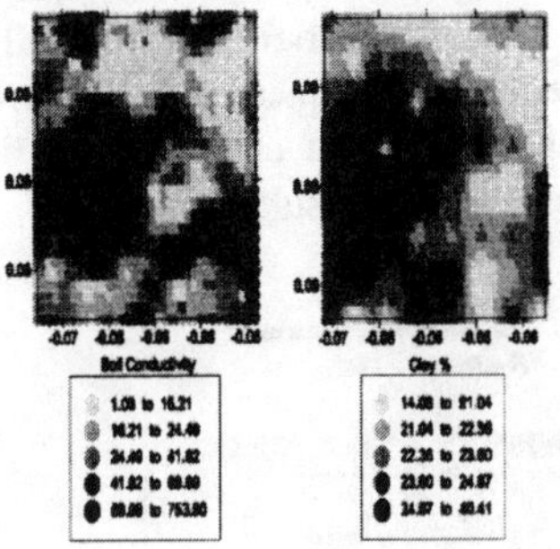

Fig. 2 and 3. Conductivity depends on soil grain size and texture. Correlation is evident in this EC map and clay content map.

METHODS OF MEASURING ELECTRICAL CONDUCTIVITY

The Veris model 3100 sensor cart (patent pending) identifies soil variability by directly sensing soil electrical conductivity. As the cart is pulled through the field, a pair of coulter-electrodes injects electrical current into the soil, while two other pairs of coulter-electrodes measure the voltage drop. (Fig. 4) The system records the conductivity measurements and georeferences them using a DGPS receiver. When used on 40 to 60 swaths at speeds up to 10 miles per hour, the system produces between 80 and 120 samples per acre. The data file is then displayed on standard mapping software as shown in Fig. 5.

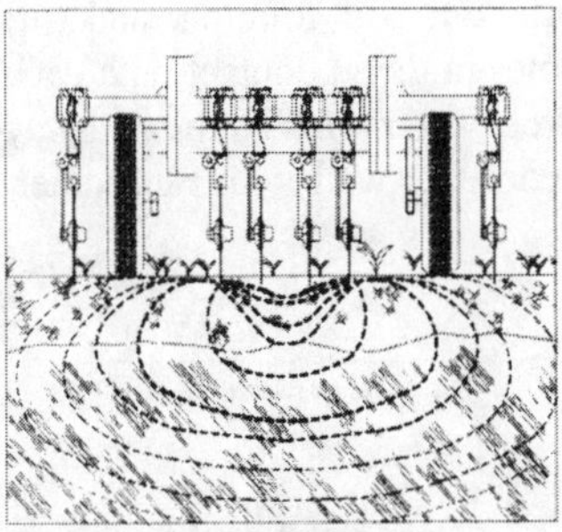

Figure 4. The Veris® 3100 employs two arrays to investigate soil at two depths

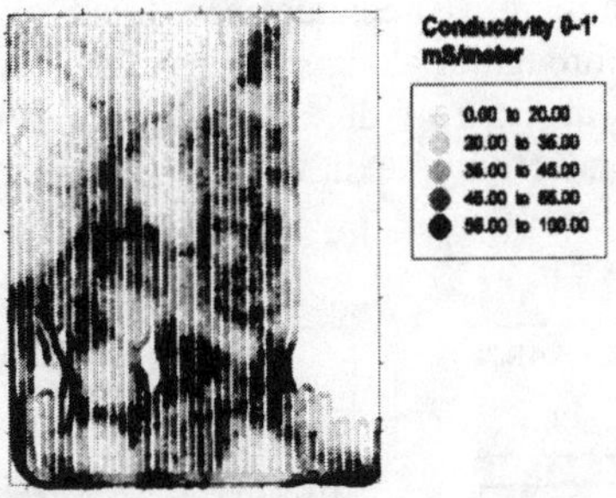

Figure 5. EC map produced by Veris®

The two primary methods of measuring soil conductivity are by electromagnetic induction (EM) or by means of direct contact. Contact methods, like that used by the Veris model 3100, use at least four electrodes that are in physical contact with the soil to inject a current and measure the voltage that results. On the other hand, EM does not make contact but instead uses a transmitter coil to induce a field into the soil and a receiver coil to measure the response. While the contact method has several advantages for widespread use in agriculture, data from fields mapped with both methods indicate that the two methods give very similar results. (Fig. 6 and 7)

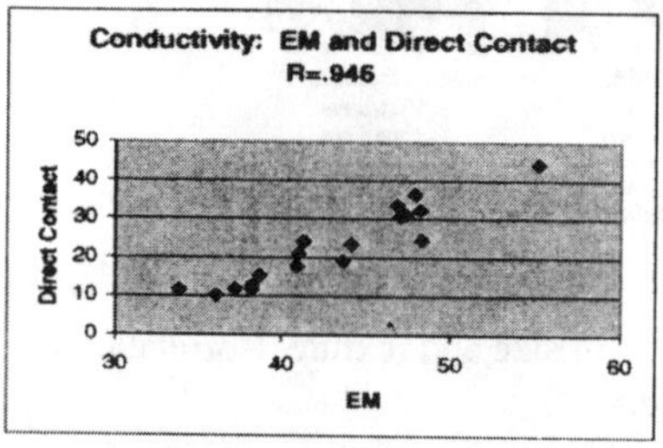

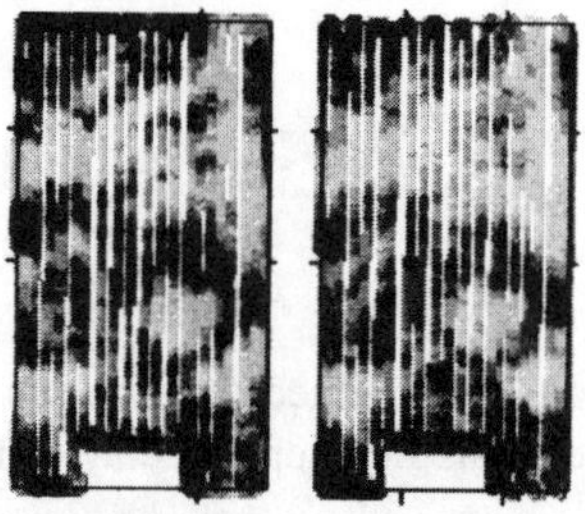

Figures 6 and 7. Correlation between EM and direct contact method of measuring EC. The map on the left was created using EM measurements and the map on the right used direct contact.

EFFECTS OF VARIOUS SOIL PROPERTIES AND CONDITIONS ON ELECTRICAL CONDUCTIVITY MAPPING

Scientific literature on soil electrical conductivity commonly refers to EC being affected by soil texture, moisture, and salinity. Rhoades and Corwin (1990) present an EC model that describes conductance along three pathways acting in parallel: (i) conductance through alternating layers of soil particles and their bound soil solution, (ii) conductance through continuous soil solution, and (iii) conductance through or along surfaces of soil particles in direct contact with each other. In the absence of dissolved salts in the free water present in pathway number 2, conductivity, texture, and moisture all correlate well with each other. See Fig. 8, 9, and 10.

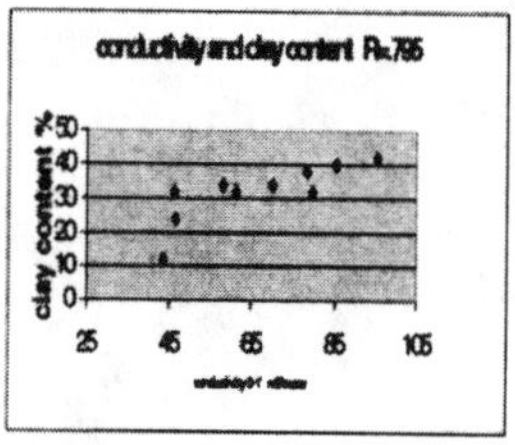

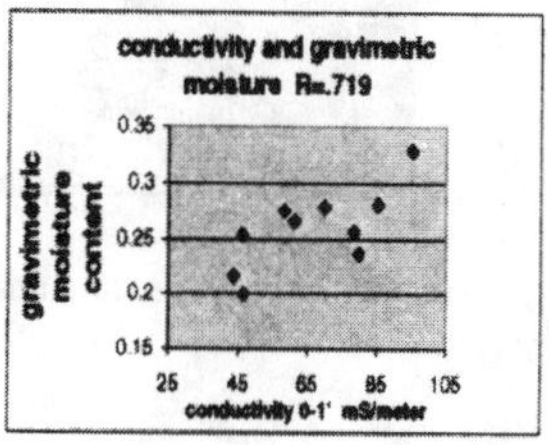

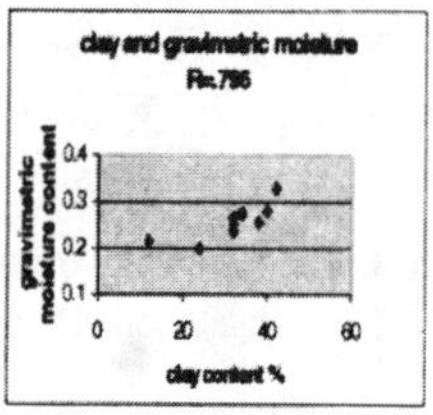

Figures 8, 9, and 10. Relationship between texture, conductivity, and soil moisture

To date, there has been little effort to field calibrate bulk soil EC measurements directly into a soil property value, other than to salinity. As will be covered later in this paper, soil conductivity measurements have been used in bi-variate regression modeling of depth to claypan and cation exchange capacity. In these applications and in zone delineation, ground truthing is a planned component of the methodology, and as such it is the relative value of soil conductivity that is the object of interest. The repeatability of the map zones and the consistency of its relative values are of greatest importance. The question regarding moisture that seems central to evaluating the usefulness of EC mapping in precision agriculture is: *Under different field moisture conditions, do conductivity measurements create different management zones?*

In order to address this question, 10 samples were taken from a Kansas field in April and analyzed for gravimetric water content and texture. EC measurements were also taken from each sample location. One month later, gravimetric water content analysis and EC measurements were again taken at the same points (see Table 1; Fig. 11). As is evident from the data, gravimetric moisture was reduced by 5.8%, while conductivity values increased by 19.7% probably due to an increase in soil temperature. The correlation to clay remained consistent, and the correlation between the two conductivity measurements was 0.94.

Sample #	EC April	EC May	moisture-April	moisture-May	clay %
	mS/meter				
1	38.7	46.2	0.25	0.25	32
2	50	58.2	0.30	0.28	34
3	51.5	61.37	0.27	0.27	32
4	60.6	70	0.30	0.28	34
5	71.6	85.3	0.29	0.28	40
6	40.2	43.6	0.24	0.22	12
7	79.4	95.2	0.30	0.33	42
8	39.8	46.5	0.25	0.20	24
9	70.1	78.3	0.29	0.28	38
10	53	79.7	0.26	0.24	32
average	55.49	66.437	0.28	0.26	

Table 1. Data from replicated sampling

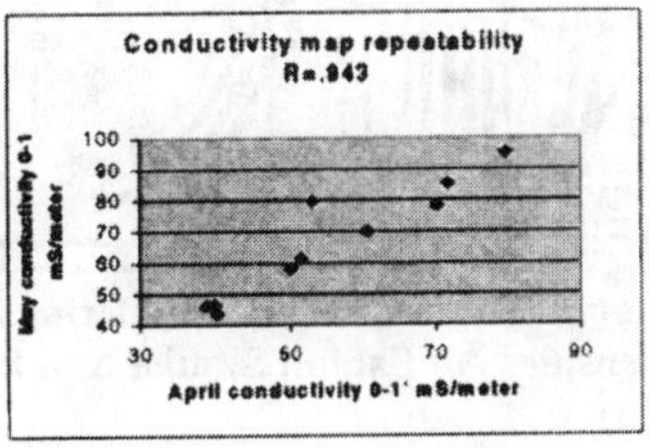

Figure 12. EC repeatabilty

A larger data set with a longer interval between mapping is shown in Figure 12. This plot represents the repeatability of EC measurements from an Illinois field that was mapped in April 1997 and again in April 1998. Correlation between the two data sets is 0.886.

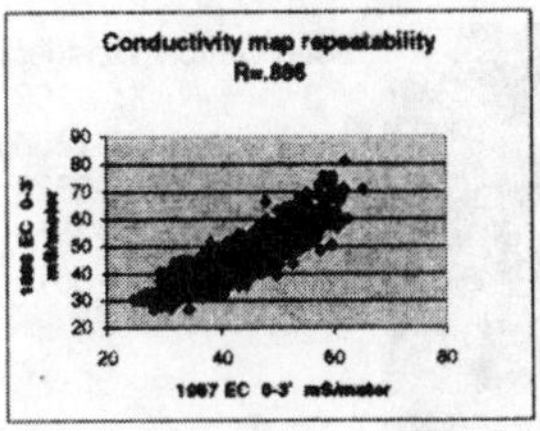

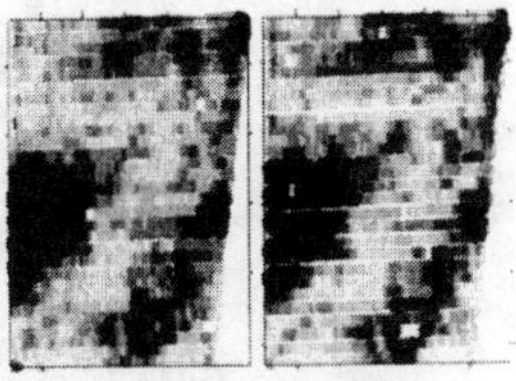

Figure 12. Repeatability of EC mapping of an Illinois field

A soil condition that appears to have a significant effect on EC measurements is soil temperature. EC mapping, whether done with EM or the direct contact method, does not provide reliable data if a frost layer is present in the soil (McKenzie et al., 1989). Soil density appears to have an effect on EC mapping in some conditions. This has been most evident in fields recently tilled prior to mapping. The lowered conductivity in these fields is likely due to the porosity of the tilled soil affecting the electrical current pathways. If two adjacent fields are mapped under significantly different temperatures or densities, they can be effectively normalized and combined.

While calibration curves have been devised for soil temperature and moisture (McKenzie et al., 1989) the complex interaction between temperature, density, and moisture makes it difficult to devise calibrations for all three conditions. Again, it is the relative measurement and the consistency of zones that are important, and these have proven to be repeatable even as temperature, density, and moisture all change. Figure 13 is a silt loam Kansas wheat field that was mapped under four different conditions in 1997. The field was mapped in March when the winter wheat was coming out of dormancy, in July immediately after harvest, in early October after multiple tillage passes, and in November just prior to dormancy. As is evident from the maps, although the conductivity values change, the zones delineated do not.

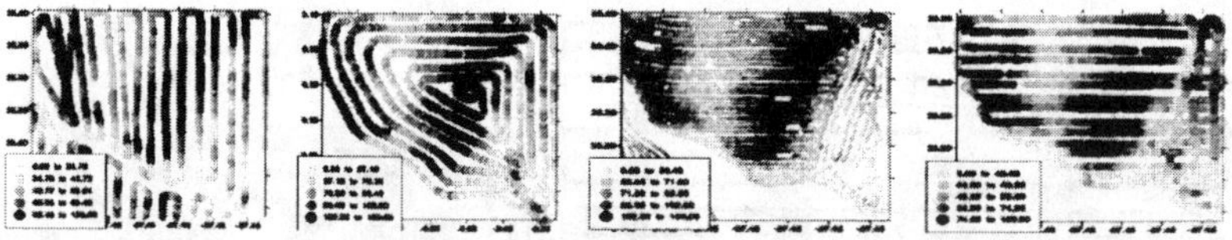

Figure 13. Four Maps At Differing Soil Moisture, Soil Temperature, and Surface Densities All Exhibit Similar Soil Patterns (from left-March, July, October, November)

The responsiveness of EC measurements to the presence of dissolved salts in the water occupying the soil pores has been well documented by work done at the U.S. Salinity Laboratory. (Rhoades & Halverson, 1975) Electrical Conductivity values >150 mS m^{-1} typically represent areas where salinity has a negative effect on field crops, and those areas with EC values in that range on Fig. 14 correspond to the lower cotton production areas on Fig. 15.

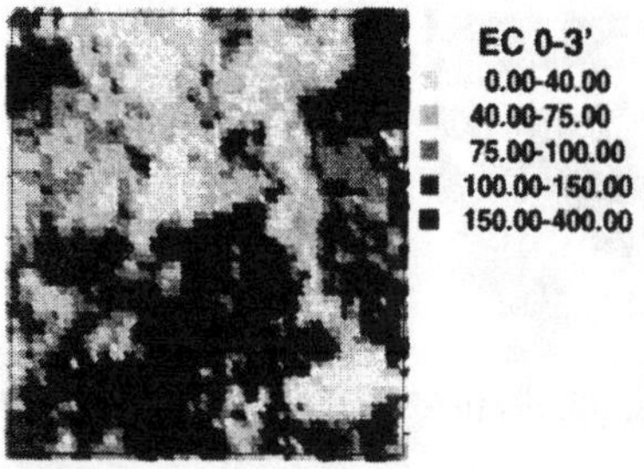

Figure 14. EC map from California field

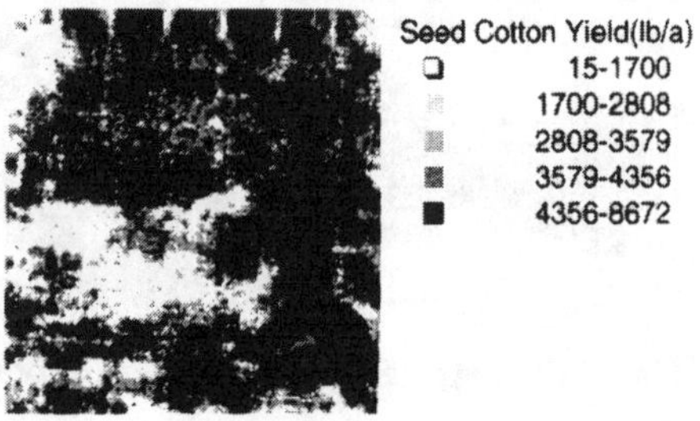

Figure 15. Cotton yield map

VARIABLE RATES FROM ELECTRICAL CONDUCTIVITY MAPS

Creating Management Zones for Directed Sampling

While EC mapping devices do not directly measure soil fertility, conductivity maps frequently relate to nutrients on fields where management-induced variability is not significant. This is because mobile nutrients often follow textural patterns, and also due to the effect soil physical properties have on productivity and subsequently on crop removal of nutrients. Directed sampling using an EC map ideally follow this methodology: a field is EC mapped, zones and sampling points are identified. After sampling, the correlation between conductivity and the given nutrients is evaluated and a precision plan can be created. Depending on the variability, the economics on the field, and the correlation between conductivity and the nutrients, a plan may call for additional samples to be taken, or recipes may be generated using the directed samples.

The field described in Fig. 16 to 19 is part of a Kansas State University precision program. It has been EC mapped and sampled on ¾ acre grids. With such an intense data set, it provides an excellent field to compare various sampling approaches. The field has pH's ranging from 5.4 to 8.1, so it was decided to compare lime recommendations under an EC-directed approach against using 3-acre grid-samples. The procedure using EC maps and directed sampling followed the methodology outlined above. First, sample zones were devised using the EC map. Second, these zones were sampled by selecting 21 sample points from the ¾ acre sampling data set that represented the zones (Fig. 16). Third, the sample data was correlated with the EC value for each point to evaluate its correlation to pH. On this field the correlation between EC and pH was 0.78. Finally the lime recommendations from these points were plotted using an inverse distance interpolation technique. The 3-acre grid-sample method used 21 systematized samples, that was accomplished by using one sample out of every cluster of four ¾ acre samples. The lime recommendation for these points was then plotted using the same procedure as the directed samples. (Fig. 17). Also, a uniform rate for the field was calculated by composting all 84 ¾ acre samples.

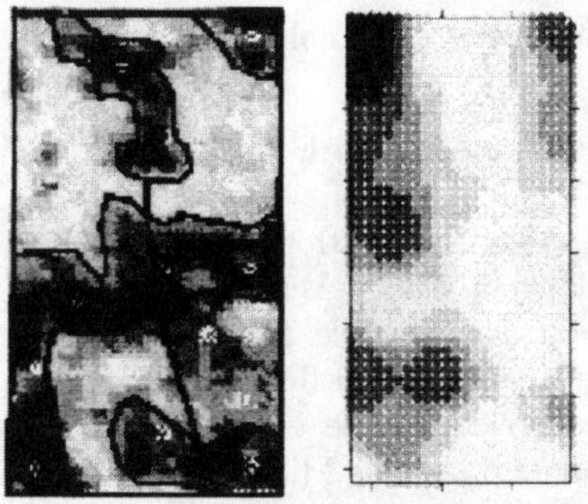

Figures 16. Conductivity/sampling map and lime recipe based on 21 directed samples

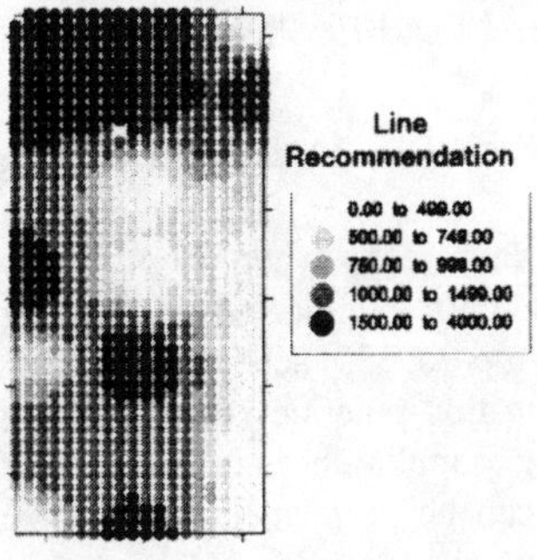

Figure 17. Lime from 3-acre grids

The resulting lime recipes - from directed, 3-ac grid, and uniform rate was compared to the ¾ acre sample point data and evaluated for (i) total error: amount applied using each recipe versus that called for at the ¾ acre points, and (ii) amount applied on soils with pH >7.5. The results show a 23.8% increase in overall accuracy for the directed sampling versus a uniform rate, and an 8.8% increase for directed vs. 3 ac grids. Perhaps more significantly was the improvement in reducing the amount over-applied in already high pH soils (Table 2). It appears from this data that the areas of high pH were identified more accurately using the EC map to ensure that samples were taken that represented each contrasting soil area. (Fig. 18). Both intense sampling approaches improved the overall accuracy of lime application versus a uniform rate, however the 3 ac grid sampling would have erroneously applied over a ton of lime on a 7.9 pH area, more than double the error at that point from a uniform rate. Considering the potential of chemical carryover on high pH soils, minimizing the over-applied lime on already high pH areas could be critical.

Sample ID	pH	Uniform rate	3 acre grid	Directed sampling
		lbs.of lime/acre		
i14	7.9	958	334.26	1.65
l1	7.9	958	2368.97	1491.77
j9	7.6	958	1138.99	107.30
m13	7.9	958	5.26	383.52
n13	7.9	958	206.99	500.40
n14	7.9	958	88.20	32.83
n8	7.9	958	220.66	53.49
Lime recommended at >7.5 pH sample points		**6706**	**4363.31**	**2570.96**

Table 2. Lime over-applied on areas with pH >7.5

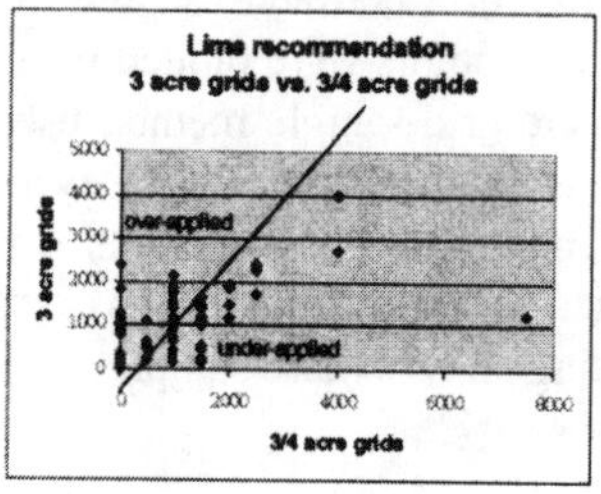

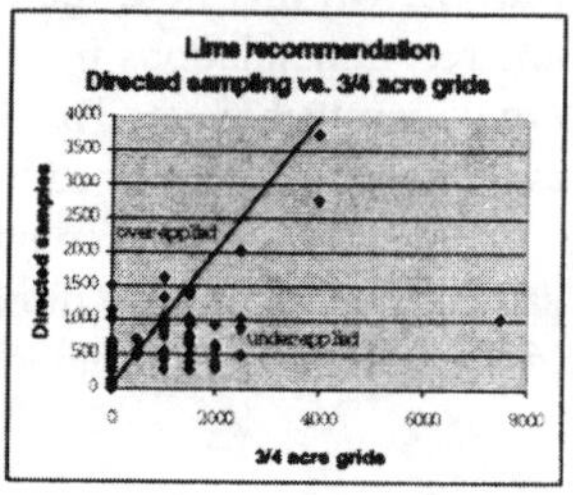

Figures 18 and 19. Plots comparing lime errors between sampling methods

Varying Crop Inputs Based on Soil Depth

In regions where depth of topsoil is a key factor affecting productivity, varying corn populations according to soil depth has proven to increase net returns up to \$33 acre^{-1} (Bitzer et al., 1996). The obstacle facing growers in applying this practice is measuring soil depth throughout a field and creating a planting population map accurately and economically. The example described below can be performed using consumer software such as Microsoft Excel and Farm Works Site Pro. The data is from a Missouri claypan field, but the methodology could be applied to other areas where topsoil is limited.

Steps to vary corn population by soil depth:

1. Create a conductivity map of the field.
2. Select 10 to 20 sample locations based on the conductivity values; select points that will adequately represent the variability and are dispersed throughout the field.
3. Measure the depth to claypan at these locations.
4. Create a regression equation that models topsoil depth at each conductivity data point based on the mathematical relationship between conductivity and topsoil depth on the samples (Fig. 19).
5. Create a recipe for varying population according to soil depth and apply to the model (Fig. 20). In this equation: yield goal = 97(bu $acre^{-1}$) + 2.24 (topsoil depth in inches). Population = 36351 - 1.608445 $*$ 10^8 (yield $goal)^{-2}$. (Note: This equation is from personal communication in May 1998 with Newell Kitchen and Ken Sudduth with USDA-ARS, and is based on unpublished research and experience. It is the subject of on-going research and is included here for purposes of illustration)

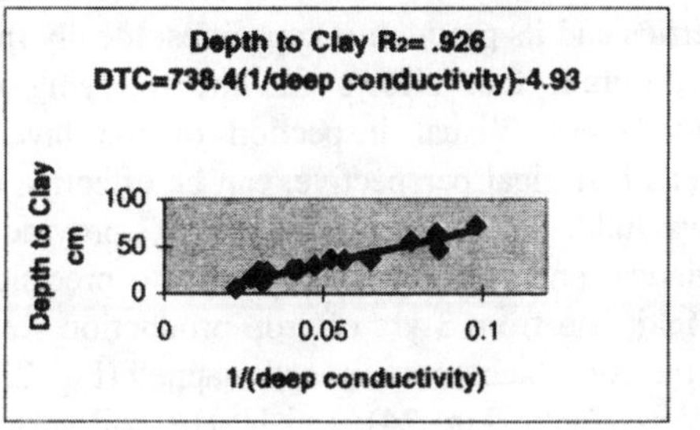

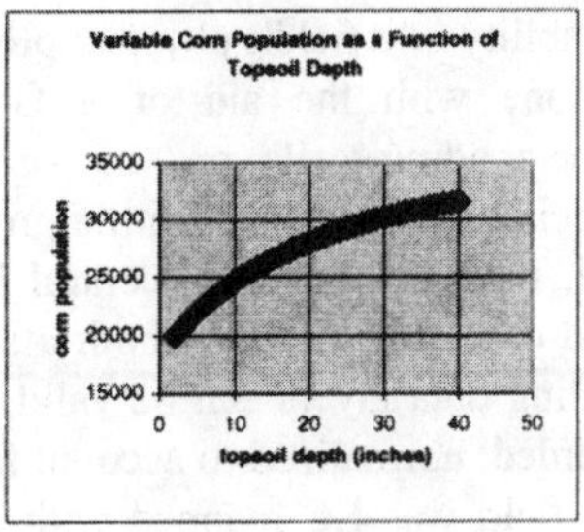

Figure 19. Regression equation for EC and DTC

Figure 20. Sample recipe

Varying Crop Inputs Based on Cation-Exchange Capacity

The methodology used for this application is similar to varying corn populations described above, except the samples are analyzed for CEC and the recipe is created by fitting conductivity to the CEC soil test results. Figure 21 shows the regression equation for conductivity and CEC data from a non-claypan Indiana field. From this information a yield goal and a nitrogen recipe can be generated (Fig. 22). (Note: this recipe is for illustration purposes only and does not reflect any recommendation by the authors for application). It is important to view the conductivity map of the field along with yield maps, or other layers in order to derive a recipe that uses all available knowledge of the field.

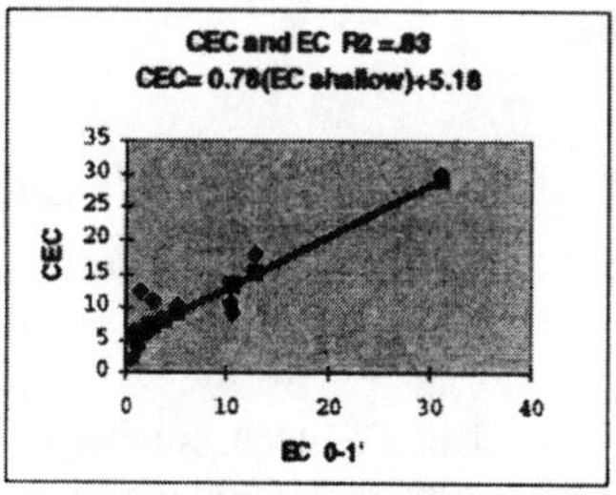

Figure 21. Regression equation for CEC and EC

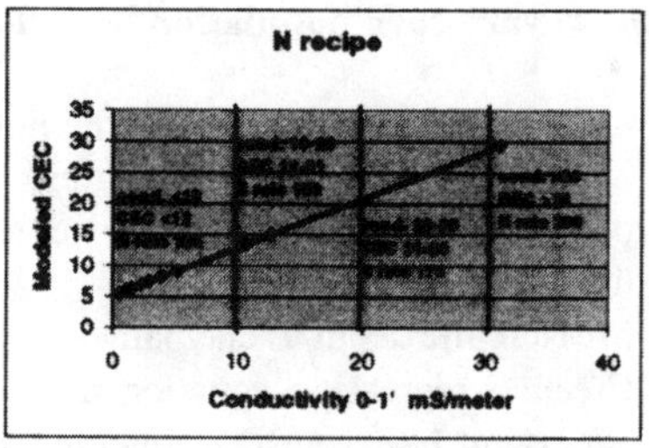

Figure 22. Sample Recipe

Using Electrical Conductivity Maps with Yield and Soil Surveys to Create Recipes

An EC map can be used in conjunction with other layers of information to provide a more detailed understanding of a field. Yield information, soil surveys, and an EC map viewed together can help identify and confirm the spatial variability of a field's physical properties and its productive capacities. Ideally this is done with the aid of a GIS program that allows for the querying of corresponding cells on the various layers. Visual inspection of the layers, especially if combined with a grower's historical perspective, can be effective as well. Data from a north central Iowa field, depicted in Fig. 23 to 27 provide a good case study of how both a statistical and an intuitive approach to precision farming data layers can be valid. Yield data from 3 yrs of crop production were recorded, normalized to account for the corn–bean rotation, and mapped (Fig. 23). The field was EC mapped with a Veris 3100 (Fig. 24), and USDA soil survey maps were layered over both the yield and EC maps. Visual inspection of these two maps reveals that low conductivity areas tend to be low yield, or inconsistent in yield, and the highest conductivity areas yield better, but are not necessarily the highest yielding.

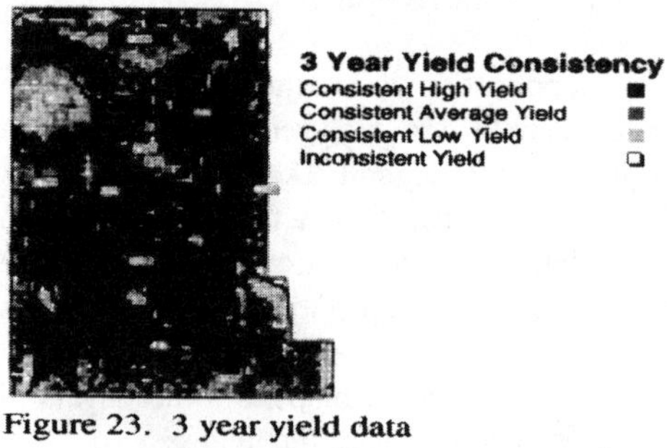

Figure 23. 3 year yield data

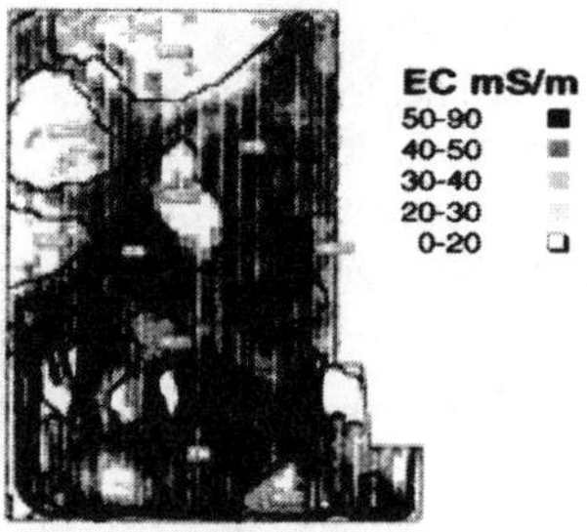

Figure24. EC map

The data was then analyzed by separating the conductivity into divisions of 5 mS m^{-1} and compared with the historical normalized yield for each conductivity class. (Table 3) As is evident from the table and even more striking when plotted (Fig. 25), there appears to be a nonlinear relationship between yield and conductivity in this field. Kitchen et al., (1995) reported the same phenomena in Missouri.

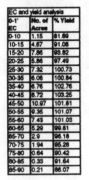

EC and yield analysis		
0-1' EC	No. of Acres	% Yield
0-10	1.15	81.69
10-15	4.67	91.06
15-20	7.86	93.82
20-25	5.86	97.49
25-30	7.32	100.73
30-35	6.06	100.84
35-40	6.76	102.76
40-45	8.72	103.25
45-50	10.97	101.81
50-55	9.35	101.07
55-60	7.43	101.03
60-65	5.29	99.81
65-70	2.9	96.18
70-75	1.94	95.28
75-80	0.64	90.42
80-85	0.33	91.64
85-90	0.21	86.07

Table 3. Yield and EC

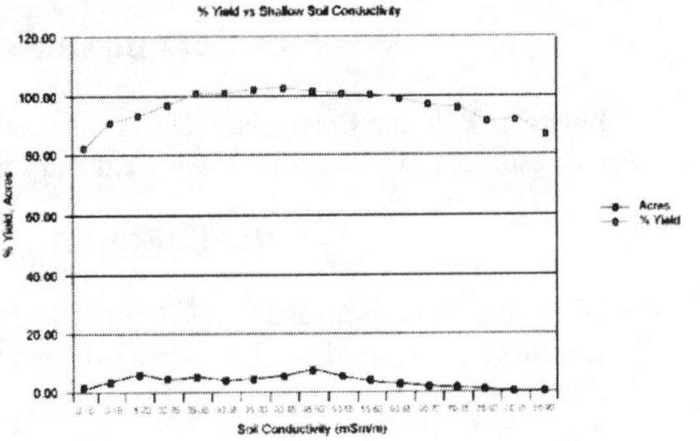

Figure 25. Yield-EC relationship

A recipe for corn population was created using the data from these maps, and another from soil type only (Fig. 26 and 27). Both the yield map and the EC map identify more variability in soils than the three soil types mapped in the soil survey. As a result, the population recipe is more responsive to the field variability.

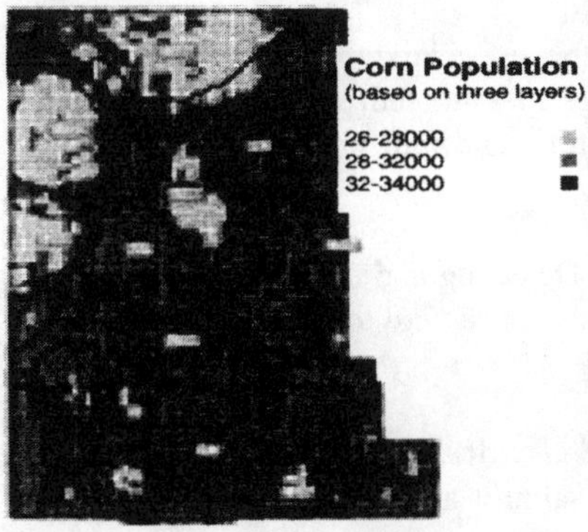

Figure 26. Recipe using all layers

Figure 27. Recipe using soil survey

CONCLUSION

When guided by the proper layer of information, directed sampling can improve the accuracy of variable rate application. By directing samples into each area of the field where soils exhibit unique properties, an EC map has proven to reduce the amount of lime applied on high pH soils. Using EC maps, recipes for variable rate applications can be created that reduce inputs on shallow or less productive soils, and challenge the deeper and highly productive soils to produce even more. Finally, EC maps can be layered under other data from a field to effectively derive a recipe from several sources of information.

ACKNOWLEDGMENTS

Precision Farming Enterprises, Davis CA--cotton yield map.
Geo-Agra Resources LC, Ames IA-Iowa yield and soils data.

REFERENCES

Bitzer, M.J., R.I. Barnhisel, and J. H. Grove. 1996. Varying corn populations according to depth of top soil. *Proceedings 1996 Information Agriculture Conference.*

Farm Chemicals 1997. p.21

Kitchen, N.R., K.A. Sudduth, S.T. Drummond, and S.J. Birrell. 1995. Spatial Prediction of Crop Productivity using Electromagnetic Induction. p. 83–87. *In Missouri soil fertility and fertilizers research update.*

McBratney, A.B., and , M.J. Pringle. 1997. Spatial variability in soil-implications for precision agriculture. p. 3–31. *In Precision Agriculture.*

McKenzie, R.C., W. Chomistek, and N. F. Clark. 1989. Conversion of electromagnetic inductance readings to saturated paste extract values in soils for different temperature, texture, and moisture conditions: *Can. J. Soil Sci.* 69:25–32.

Rhoades, J.D., and A.D. Halverson. 1975. Detecting and delineating saline seeps with soil resistance measurements. *Proceedings of Saline Seep Control Symposium.*

Rhoades, J.D., and D.L. Corwin. 1990. Soil electrical conductivity: Effects of soil properties and application to soil salinity appraisal. *Commun. Soil Sci. Plant Anal.* 21:837–860.

Williams, B., and D. Hoey. 1987. The use of electromagnetic induction to detect the spatial variability of the salt and clay content of soils. *Aust. J. Soil Res.* 25:21–27.

Recent Advances in Yield Monitoring of Conveyor Harvested Crops

Ronald H. Campbell
HarvestMaster
Logan, Utah

ABSTRACT

Yield monitoring instrumentation may be divided into a measurement part and a user interface part. This paper addresses observations and discoveries relating mostly to the user interface of yield monitoring for conveyor harvested crops. Discoveries regarding the use of real time yield monitor information for operational decision making are discussed. Next, operator interface for ease of installation and system performance checking is addressed. Brief attention is given to additional sensors to augment the weight transducer technology to better quantify flow rates. The paper concludes with comments on the value of real time data transfer for management decision support.

INTRODUCTION

Yield mapping of conveyor harvested crops such as potatoes and sugarbeets was the subject of experiment in 1993 and 1994 (e.g., Campbell et al. 1994; Hall et al. 1997). Commercial versions of potato and sugarbeet yield monitor instrumentation were introduced in 1995. While the number of actual field installations in production agriculture was perhaps no greater than a hundred units by the end of 1997, a majority of early users perceive an economic benefit from the data obtained and seem committed to the continued usage of the technology. Relative to growth of demand in this market, however, it seems that more requests are coming from equipment and agri-chemical dealers for this technology than from actual growers themselves. The conclusion might be that although the vast majority of potato and sugarbeet growers have not implemented this technology, many of their direct suppliers are betting they will.

Several variables in the market place indicate that the application of yield monitor technology for potatoes, sugarbeets and other conveyor harvested crops lags that of grain yield monitors significantly both in time and number. Some of our rough estimates have suggested that the combined potential market for conveyor harvested crop yield monitor applications might total 10 to 15% of the potential grain yield monitor applications. In terms of number of units being placed into use, grain combine yield monitors might total somewhere between 4,000 and 8,000 in 1998, whereas those for conveyor harvested crops might be 2 to 4% of those figures.

The apparent small market for yield monitors for conveyor harvested crops and the significantly slower adoption rate of this measurement technology by the farming community has not dampened the enthusiasm of people associated with this area. There appears to be at least three new suppliers of yield monitors for conveyor harvested crops on the market: a significant rate of inquiries

Copyright © 1999 ASA-CSSA-SSSA, 677 South Segoe Road, Madison, WI 53711, USA. *Proceedings of the Fourth International Conference on Precision Agriculture.*

continues to arrive from both dealers and potential users of these systems. A significant portion of inquiries seem to be coming from the fruit and viticulture industries for this type of technology, as opposed to potatoes and sugarbeets.

In addition to the advances of this technology (or lack thereof) in the market place mentioned above, here are some areas of progress in the technology itself which comprise much of the remaining information content of this presentation:

- Use of real time yield monitor information for operational decision making
- Operator interfaces for ease of installation and system performance checking
- Use of additional sensors in addition to weighing technology to better quantify flow rates
- Wireless data communications in real time to facilitate operational decisions

YIELD MONITOR INFORMATION TO SUPPORT OPERATIONS

Up to the present, nearly all of the emphasis on recording of geo-referenced yield data has been for the purpose of mapping. In our field work over these past three years, we have discovered another significant user critical success factor–use of data for controlling ongoing operations.

We repeatedly encounter one school of thought out there that says, Yield monitoring instrumentation should be invisible for the operator–just a black box. On the contrary, as one of the advances in yield monitoring of conveyor harvested crops, we are finding that operator interaction with the yield monitor substantially improves efficiency of operations. Consider the following diagram of the yield monitor screen:

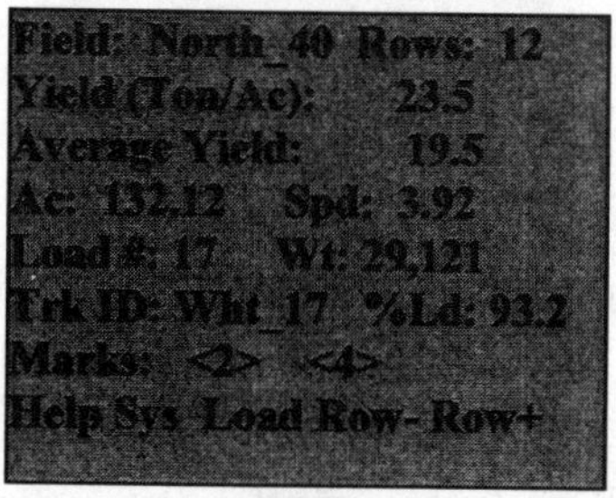

Load information:

For loading truck and managing truck weights
Important for avoiding over-loading of trucks
Optimizing trucking investments

Operator interactive information:

Field Name – to insure collection of data to the right repository
Working Width – row count adjustment
Field Marks

Yield Information and acres harvested:

Instantaneous Yield and Average Yield – useful to ascertain quality of incoming data
Acres covered for the day

There are three main areas of interest on the mobile computer screen which support field operations.

- Load Information

With conveyor harvested crops, management of trucks in the field is typically more intense than in the situation with grain harvesters, since there is usually no, or very limited on-harvester storage for the crop. It must move pretty directly to the truck or wagon alongside as the harvester moves through the field. Therefore, information from a yield monitor becomes very useful in managing this process.

With information from the yield monitor, truck loading managed by the harvester operator provides for trucks being optimally loaded. In this fashion overload fines by state Department of Transportation enforcers are avoided. On the other hand, legal capacity does not go unused due to guessing since the harvester operator has a second by second update of net weight on the truck.

The harvester operator's mobile field computer (yield monitor console) provides for the entry of a truck data base with a set point of net weight capacity for each truck in a fleet. When a new truck drives under the harvester, the harvester operator advances the load count with a single key press. At some point during loading, the harvester operator toggles to the "load" screen to scroll and select the identification of the truck being loaded from a list of truck in the fleet. In this fashion, personalized loading of trucks is operationally feasible. Records are made on the harvester computer of the time and weight of each truck leaving the harvester, thereby providing a backup accounting of trucks and loaded produce leaving the field.

- Operator Interaction for Recorded Data

On-screen yield monitor information and user interaction does not represent a new advance in potato and sugarbeet yield monitoring, but it does firmly establish its utility. Here are some of the reasons why.

The operator adjustable working width (row count) is mostly only used for potatoes, in the crops with which we have dealt. Sugarbeets and tomatoes always have a constant harvest width. Grapes sometimes have varied widths per row within the same field, necessitating on-the-fly working width adjustments for proper area coverage measurement.

The field name registered on the operator's console verifies that the appropriate data repository (field data file) for the information being logged.

The marking of traits (weed patches, rocks) using the mark feature on the yield monitor requires interaction from the harvester operator. This is another information input available from the yield monitor, which provides a basis for optimizing later crop land treatment expenditures. These field marks are truly one of the beneficial by-products of yield mapping instrumentation.

- Yield Information for Quality Control

The third category of information on the operator's main yield monitor screen, the yield data, is perhaps not so useful for any real time operational decision making. Viewed by an individual familiar with the farm, its operations, the farm tillage and seeding practices of the season, etc., real time yield readouts are really helpful in giving a farmer early feedback on the effectiveness of this year's operations—even before the information is mapped.

Additionally, total product harvested during the day and acres covered are a big help in planning crop storage, and storage equipment moves.

One last item is worthy of mention when discussing recent advances in this type of yield monitoring technology: data quality control. Once an operator is familiar with the yield monitor on his harvester, he is in a much better position to insure that recorded data is of high quality by watching the display of instantaneous yield, average yield, ground speed and the other parameters. If the operator notices yield values going abnormally high, or low, it gives early warning of a problem with the yield sensing transducers—perhaps potato vines are tangled in the weight sensors, for instance. In other words, it allows the operator to address a faulty data situation before reams and reams of bad data are collected.

OPERATOR INTERFACES

One of the newer advances in yield monitors for conveyor harvested crops is the display screen for sensors. The screen below shows an example.

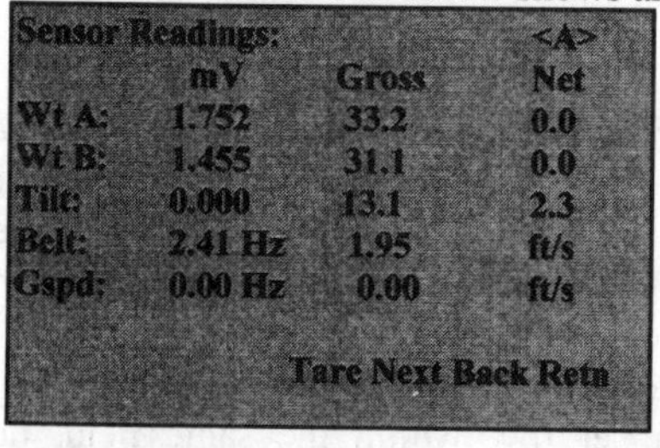
Sensor Readings: <A>
mV Gross Net
Wt A: 1.752 33.2 0.0
Wt B: 1.455 31.1 0.0
Tilt: 0.000 13.1 2.3
Belt: 2.41 Hz 1.95 ft/s
Gspd: 0.00 Hz 0.00 ft/s
Tare Next Back Retn

Some people would say that a regular harvester operator should never have to see this type of information. While this statement might arguably be true, in real life this feature is fairly useful. Sensors on potato, sugarbeet, tomato and grape harvesters are, many times, mounted on adjustable loading booms, subjecting the sensor lead wires to potential mechanical damage. Being able to record readings of individual sensors in the setup and field troubleshooting process is a valuable time saver.

In terms of modern systems, the yield monitor for conveyor harvested crops is a fairly simple device; however, it may become complicated with the combination of analog sensors, digital sensors, the GPS receiver, and mobile field computer. This is especially true where these are all installed on an after-market conveyor type crop harvester and operated in a hostile environment. Having the proper screens available to provide the visibility, which lets the user know that everything is Okay, or that a specific component is malfunctioning is essential to reliable field operation.

ADDITIONAL SENSORS TO SUPPLEMENT WEIGHT MEASUREMENTS

Some of the more recent advances that we have been part of have involved the use of additional sensors to better measure product weight on a harvester mounted harvester system. While many of these trials are as of yet too experimental to present here, the general public should be aware that significant efforts continue on the part of various manufacturers and research organizations to further refine flow measurement of conveyor type crop harvesters.

Whereas the main mode of conveyor weighing has been by rollers directly placed under the conveyor belt, we have experimented with measure of the weight of the entire conveyor in a few different configurations. In two different configurations, we weighed the entire conveyor of a grape harvester. The results were mixed, with one installation coming up with fairly decent yield maps, and the other two with an obvious level of noise in the data. One of the problems identified was stiction in the supporting member joints, which caused the tare values of the load cells to drift. One might imagine that the problem of making weight measurements to better than a pound accuracy is a lot tougher when the tare weight of the measurement system is 200 or 300 lb, and is moving up and down with the harvesting machine as it moves through the field.

Experiments are proceeding with other types of sensors to measure the tare (dirt, clods, and rocks) that are many times a part of the material being loaded from potato harvesters to field trucks. Accurate recording of yields in some types of soils is impossible without this type of a measurement. Again, it is yet too early to disclose the results of this research and the types of sensors being used, with results yet forthcoming. It is interesting to note, however, that some of technologies being tried lend themselves to the measurement of shape factor on potatoes, which could enable the mapping of not only yield, but quality as well.

WIRELESS DATA COMMUNICATIONS

One other experiment underway this season is that of real time, or near real time data transfer from the harvester to a mobile PC in the farm manager's pickup truck, as well as back to the farm's home base – and even to the mapping service agency's computing and mapping center.

This technology is not new, inasmuch as it was reported as having been applied in the field by Schneider et al. (1996) using spread spectrum modems. As of yet, however, we have not seen such a system commercially marketed. In a lot of ways, it might be too early for such a system in production agriculture. It is certainly not needed for the making of maps, since that is generally a post harvest operation anyway. When one considers, however, the real time operational decision support benefits of such a system, it becomes more attractive. It was when we began to see the on-the-fly information utility of yield monitor information from conveyor type harvesters, that we felt it worthwhile to revisit real time data communications from the harvester to the home base, and to the farm manager's pickup truck mobile computer.

CONCLUSION

Many of the advances in yield monitoring for conveyor harvested crops are in the area of user display and user interaction with the system. Current systems allow the operator to more optimally run the harvesting operation, and field managers to more effectively apply their resources.

Further sensing research in making conveyor harvester measurements more accurate will soon enable the mapping of crop quality information as well.

REFERENCES

Campbell, R.H., S.L. Rawlins, and S. Han. 1994. Monitoring methods for potato yield mapping. *In* ASAE Annual Int. Meeting, Atlanta, GA. December, 1994.

Hall, T.L. 1997. Monitoring sugarbeet yield on a harvester. ASAE Annual Int. Meeting, Minneapolis, MN. August, 1997.

Rawlins, S.L., G.S. Campbell, R.H. Campbell, and J.R. Hess. 1993. Yield mapping of potato. *In* Site Specific Management for Agricultural Systems Conf., University of Minnesota, Minneapolis, MN. March, 1994.

Schneider, S.M., S. Han, R.H. Campbell, R.G. Evans, and S.L. Rawlins. 1996. Precision agriculture for potatoes in the Pacific Northwest. Minneapolis, MN. June, 1996.

Evaluation of Sugarbeet Yield Sensing Systems Operating Concurrently on a Harvester

T. L. Hall
L. F. Backer
V. L. Hofman

Department of Agricultural and Biosystems Engineering
North Dakota State University
Fargo, North Dakota

L. J. Smith

NW Experiment Station
University of Minnesota
Crookston, Minnesota

ABSTRACT

Three independent yield-sensing systems were operated concurrently on a 1997 WIC Mini-Tank Sugarbeet Harvester. Data were collected on about 80 ha (200 acres) during the 1997 harvest season. One system was a set of load cells mounted near the end of the outlet conveyor. Another system was a second pair of load cells mounted on the scrub chain discharge conveyor. A third system was a torque sensor mounted in the scrub chain driveline. A HarvestMaster HM-500 yield monitor was used for the data collection. Data from each system were collected simultaneously to produce three parallel sets of yield data for each field. The systems were evaluated by comparing truckload error and standard deviation associated with each system and by comparing yield maps generated from the data produced by each system.

INTRODUCTION

Some research effort has been directed toward bulk crop (i.e., potato, sugarbeet, etc) yield monitoring over the last several years (Hofman et al., 1995; Rawlins et al., 1995; Campbell et al., 1994; and Walter et al., 1996). A result of some of this research has been the development of the HarvestMaster HM-500 yield monitor. During the fall of 1996, research was conducted to determine the effect of different chain support systems on yield monitor accuracy and precision. When the product flow sensors were mounted on a short conveyor section, sugarbeets did not ride smoothly on the conveyor before being weighed. This resulted in lower accuracy and precision; however, reducing the length of the chain support system minimized the effects of the bouncing sugarbeets (Hall et al., 1997).

Copyright © 1999 ASA-CSSA-SSSA, 677 South Segoe Road, Madison, WI 53711, USA. *Proceedings of the Fourth International Conference on Precision Agriculture.*

Another issue that became apparent during the 1996 research was that the hopper available on many harvesters could not be used (Hall et al., 1997). All position data are lost if sugarbeets are stored in the hopper. To overcome this problem, sugarbeets could be weighed before being loaded into the hopper tank. By weighing the sugarbeets earlier, however, some potential cleaning of the sugarbeets is lost. Therefore, there is a concern about how tare dirt affects the data obtained from such a product flow sensor.

The Scrub Chain

Many sugarbeet harvesters use a scrub chain to elevate sugarbeets from ground level to a sufficient height for loading into a truck. A scrub chain is essentially a double vertical conveyor chain mechanism. Sugarbeets are squeezed between the two chains and carried vertically. The two chains usually operate at slightly different speeds to cause tare dirt to be scrubbed from the sugarbeets. A scrub chain typically contains a near-horizontal discharge conveyor section to carry sugarbeets from the vertical section to the hopper. On some harvesters, sugarbeets pass through the scrub chain before entering the hopper. Therefore, if sugarbeet weight is determined in the scrub chain, the problem of not being able to use the hopper is eliminated.

Entering an area of higher yield in a field causes an increased mass of sugarbeets to be elevated by the scrub chain, and a higher torque is required to drive the scrub chain. Therefore, the magnitude of the torque transmitted through the scrub chain driveline is an indicator of the mass of sugarbeets in the scrub chain. Once the mass of sugarbeets in the scrub chain, the ground speed, the speed of the scrub chain, and the harvester width are known, an instantaneous yield can be calculated.

The effect of mud (or tare dirt) is the primary concern of measuring sugarbeet yield by sensing the torque required to drive the scrub chain. The scrub chain is active in the process of cleaning the dirt from the sugarbeets. Dirt removed from a sugarbeet while it is in the scrub chain will affect the torque measurement and consequently the yield measurement. The extent of its effect, however, cannot be predetermined.

The horizontal conveyor section of the scrub chain is a logical place to mount a load cell based product flow sensor to weigh the sugarbeets, since the sugarbeets flow over this section before entering the hopper. Most of the tare dirt has been cleaned by the time the sugarbeets reach this point in the harvester. Therefore, tare dirt is not as much of a concern. The primary concern is the scrub chain discharge conveyor section length. Research conducted during the 1996 harvest season determined that weighing sugarbeets on a relatively short length conveyor section resulted in lower accuracy and precision (Hall et al., 1997). If a small chain support system is used, however, the accuracy and precision of the yield monitor with the product flow sensor mounted in this position may be acceptable.

The standard installation of the yield monitor has the load cells located at the end of the outlet conveyor, just before sugarbeets are loaded into the truck. The advantage to this installation is that the maximum amount of tare dirt is removed from the sugarbeets; however, the load cells are located after the hopper. Therefore, if the hopper is used, position data are lost.

The Yield Calculation

The yield calculation requires the knowledge of the product flow sensor output, instrument calibration factor, system calibration factor, conveyor speed, ground speed, and harvester width. The following equation is used to determine the sugarbeet flow rate over the conveyor.

$$FR = (O - T) \times IC \times CS \times SCF$$

where:

FR = Sugarbeet flow rate, kg s^{-1} (lb s^{-1})
O = Torque sensor or load cells output, mV
T = Tare value, mV
IC = Instrument calibration factor, kg m^{-1}V (lb m^{-1}V) for the load cells or N-m m^{-1}V (ft-lb m^{-1}V) for the torque sensor
CS = Conveyor speed, m s^{-1} (ft s^{-1})
SCF = System calibration factor, m^{-1} (ft^{-1}) for the load cells or m^{-2} (ft^{-2}) for the torque sensor

The tare value is determined by executing a retare function. During this operation, the yield monitor determines an average output from the respective product flow sensor with the harvester operating empty. This value is subtracted from the respective product flow sensor total output to obtain a net output. The tare value is necessary to eliminate the weight of the conveyor chain or the torque required to operate the empty scrub chain.

The instantaneous yield is calculated by the following equation:

$$YLD = \frac{36FR}{GS \times W} = \left(\frac{14.85FR}{GS \times W}\right)$$

where:

YLD = Sugarbeet yield, t ha^{-1} (tons $acre^{-1}$)
FR = Sugarbeet flow rate, kg s^{-1} (lb s^{-1})
GS = Ground speed, km h^{-1} (MPH)
W = Harvester width, m (ft)

HarvestMaster HM-500 Yield Monitor

The HarvestMaster HM-500 (HarvestMaster, Logan, UT) yield monitor consists of four basic components—product flow sensors, speed sensors, a signal conditioner and conversion unit, and the Pro-2000 handheld computer. A differentially corrected Global Positioning System (DGPS) provided position data. The HM-500 is an off-the-shelf yield monitor modified for this project. A torque sensor and a second set of load cells were used as additional inputs to the yield monitor. Essentially, the yield monitor functioned as a signal conditioning and data collection device.

Product Flow Sensors

A set of two 227 kg (500 lb) load cells functioned as a product flow sensor and was mounted under the outlet conveyor chain near the discharge (Fig. 1). A 127 mm (5 in) diameter idler was mounted on each of the load cells to support the conveyor chain. The output was an analog voltage signal proportional to the mass of sugarbeets on the conveyor.

Speed Sensors

Three speed sensors were used—one for ground speed, one for outlet conveyor speed, and one for scrub chain speed. Magnets were mounted on rotating parts and the speed sensor gave a voltage pulse every time the magnet passed the sensor. Single magnets were used on both the outlet conveyor drive and the scrub chain driveline. The ground speed was sensed using four magnets mounted on a ground wheel hub. Multiple magnets were used to prevent a large delay in detecting changes in ground speed.

A speed sensor also was used for a hold–run sensor. A hold–run sensor stops the data logging when the harvester is raised out of the ground, and starts it when the harvester is lowered into the ground.

Signal Conditioning and Conversion Unit (SCCU)

The SCCU is the heart of the yield monitor. It functions to gather signals from all other components of the yield monitor and the DGPS. A low-pass filter eliminates components product flow sensor signals that exceed 3 Hz. It samples and digitizes these filtered signals at 25 Hz and applies a 25-pt moving average to them to obtain one data point per second. The SCCU also converts the output from the speed sensors into frequency values so actual speeds in m s^{-1} (ft s^{-1}) or km h^{-1} (MPH) can be calculated.

Fig. 1. Load cell based product flow sensor.

Pro-2000 Handheld Computer

The Pro-2000 is a 286 DOS-based computer that was mounted in the tractor cab. It uses the frequency values from the speed sensors and the digitized signals from the load cells and torque sensor to calculate yield values. In addition to calculating the yield values, it stores the date, time, position, yield, ground speed, flow rate, and accumulated mass data.

Concord BR6-183 DGPS Receiver

A Concord BR6-183 DGPS receiver was used to give latitude and longitude coordinates accurate to approximately 3 m (10 ft). An FM signal from Differential Corrections (Cupertino, CA) provided the differential correction.

MATERIALS AND METHODS

Three product flow sensors were operated concurrently on a 1997 WIC Mini-tank Harvester as shown in Fig. 2. A HarvestMaster HM-500 yield monitor was modified by HarvestMaster to accommodate input signals from the three product flow sensors. The yield monitor had an independent truckload accumulation feature as well as having independent yield values written to a data file for each sensor.

Two of the product flow sensors were load cell based systems (Fig. 1) mounted on the outlet conveyor and discharge of the scrub chain. The third product flow sensor was a torque sensor mounted in the scrub chain driveline.

Raw Data Collection

A CR-10 data logger (Campbell Scientific, Logan, UT) was used to gather raw digitized signals from each set of load cells and from the torque sensor. Wires were spliced into the output signal wires from each weight sensing system

Fig. 2. 1997 WIC mini-tank harvester.

and input into the data logger. Signals from the three product flow sensors were sampled and recorded at 16 Hz.

A tare value was determined by gathering raw data from the weighing systems at the same time as the yield monitor ran a retare. The tare data were then imported into a spreadsheet and a 16-point moving average was applied to obtain a value for each second. An overall average was then calculated over the entire retare; approximately 25 s in duration.

Raw data were gathered for approximately 5 min over two different harvester passes. It was used to calculate flow rates and yields using the tare value previously found. The conveyor and ground speeds were noted from the output of the yield monitor. These speeds were assumed constant throughout the data collection period. The flow rates and yields were graphed and compared to the respective flow rates and yields recorded in the yield monitor data file.

Raw data were collected for approximately 1.5 min with the harvester operating, but stationary. The data were calculated into accumulations and graphed.

Truckload Error Evaluation

The truckload weight recorded by the yield monitor for each weight sensing system was compared to the actual truckload weight obtained from piling station scale tickets. The following equation was used to calculate truckload error:

$$Truckload \quad Error = \frac{MW - AW}{AW} \times 100\%$$

Where:

MW = Truckload net weight as determined by the yield monitor, t (tons)

AW = Truckload net weight obtained from the piling station scale tickets, t (tons)

The mean and standard deviation of the truckload errors were calculated for each field and for the entire harvest and were graphed. The usefulness of the mean is limited, since it is directly related to the system calibration factor. A mean greater than zero indicates that the system calibration factor was too high. The standard deviation, however, is a measure of consistency or precision. As long as the calibration factor was not changed extensively, it indicates how much precision can be expected from the yield monitor.

An ANOVA was performed on the means of the standard deviations to determine significant differences. Data from the three product flow sensors were regarded as treatments. Each treatment was divided into groups of 5, 10, 15, and harvest day groups. Standard deviations were calculated for each group to give replicated values for the standard deviation of each treatment.

Yield Map Generation

Since high truckload errors were experienced, the yield data were corrected before yield maps were developed. Since each truckload was identified in the yield data file, the area covered when loading a truck could be calculated. A yield correction value was determined for each truckload by dividing the difference between the measured and actual truckload weight by the number of hectares (acres) required to load that truck. This correction value was then added to the recorded yields for that truckload to obtain corrected yield values. This process was repeated for each truckload in that field. The corrected yield values were then imported into AgLink for Windows to generate a yield map.

Yield Map Comparison

The corresponding yield values for each weight sensing system were subtracted from each other to generate yield difference maps. The torque sensor map data were subtracted from the outlet conveyor system map data, the scrub chain system map data were subtracted from the outlet conveyor system map data, and the torque sensor system map data were subtracted from the scrub chain system map data. These maps show how the systems respond relative to each other in different areas of the field. For example, these maps could answer the question about whether the torque sensor gives higher yield values than the outlet load cells in muddy areas of the field

RESULTS AND DISCUSSION

Raw Data

Figure 3 is a graph of the flow rates calculated from the raw data gathered with the CR-10. A 16-pt moving average was applied to the raw data. The 16-pt moving average was then averaged again during the 4-s logging interval to obtain a single value for each 4-s interval.

Figure 3 shows a strong relationship between the sugarbeet flow rates from each weight sensing system. The first 15 s show the flow rates increasing from zero to approximately 23 to 35 kg s^{-1} (50 to 70 lb s^{-1}). Since the harvester was empty at time zero, the increase in flow rates shows the harvester filling to an approximately steady-state condition. The flow variation from approximately 20 to 170 s is assumed to be yield variation in the field, since the ground and conveyor speeds remained constant.

Figure 3 clearly shows the lag times between each of the weight sensing systems measuring the same sugarbeets. A valley in the plot of the flow rate measurement is shown at the 50 s mark for the torque sensor, at the 53 s mark for the scrub chain load cells, and at the 59 s mark for the outlet conveyor load cells. The same lag time appears repeatedly through the entire graph.

Figure 4 is a graph of the accumulations recorded with the harvester sitting stationary and operating empty. The negative accumulations in Fig. 4 show that mud falls from the conveyor chains as the harvester is operating empty and sitting stationary. As mud falls from the conveyor chains and is not replaced by new mud, the output of the product flow sensors decrease. Since the output of each product flow sensor is subtracted from its respective tare value, the net output is negative. Therefore, a negative flow rate is generated and a negative accumulation occurs. The torque sensor was the most affected.

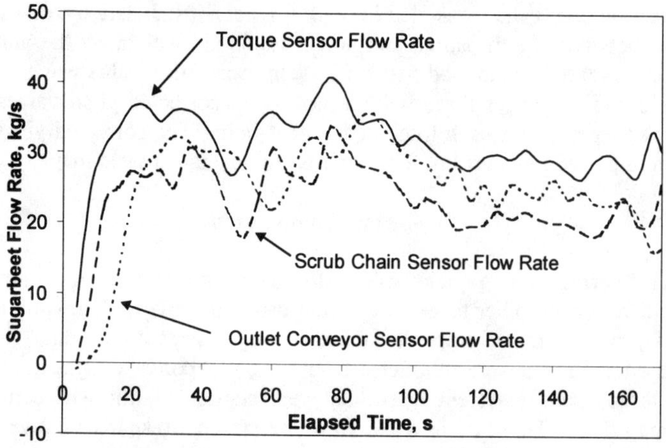

Fig. 3. Raw data flow rates.

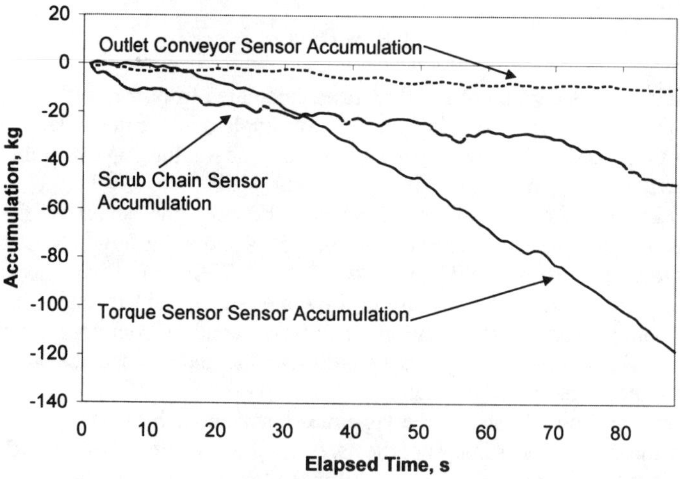

Fig. 4. Stationary harvester accumulations.

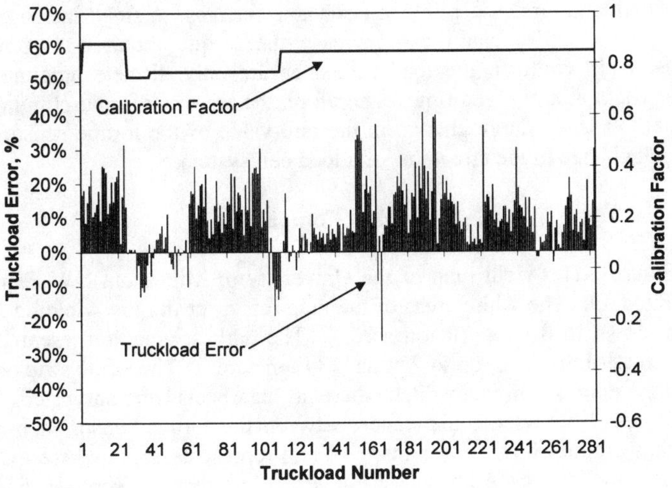

Fig. 5. Outlet conveyor product flow sensor truckload error graph.

Truckload Error

Figure 5 is a graph of the outlet conveyor product flow sensor truckload error for the entire harvest season. The mean and standard deviation were found to be 10.24 and 10.28%, respectively. A total of approximately 280 truckloads were recorded. Similar graphs were developed for the scrub chain and torque sensors. The scrub chain sensor mean and standard deviation were found to be 5.25 and 10.58%, respectively. The torque sensor mean and standard deviation were found to be 12.91 and 16.97%, respectively. The calibration factors varied only slightly over the entire harvest season.

The outlet conveyor product flow sensor had the lowest standard deviation (10.28%) of the three systems, but the scrub chain product flow sensor was close (10.58%). Statistically, the precision of these two systems is not significantly different. Since the scrub chain system allows the use of the hopper, it would be the recommended system.

High standard deviations were encountered for all three product flow sensors. The muddy conditions experienced throughout the 1997 harvest season are suspected as the cause. Muddy conditions cause the tare value for each weight sensing system to vary more than in dry conditions. The yield monitor assumes a constant tare value between retare functions. If the actual tare value varies during that time, high truckload errors and high standard deviations result. It becomes imperative that retares are performed often in muddy conditions.

The torque sensor weight sensing system performed poorly with a 16.97% standard deviation. Muddy conditions are most likely the cause of the high standard deviation. The scrub chain is located where more mud passes through it. Therefore, the probability of the tare value changing often is higher. The physical size of the mechanism and the amount of material powered by the scrub chain driveline make it more susceptible to tare value changes. A small increase in mud clinging to the entire chain can increase the torque required to power the mechanism. In contrast, the load cells measure only discrete amounts of the conveyor chain at any given time. Therefore, the amount of mud clinging to the scrub chain causes a large change in the tare value of the torque sensor system and a small change in the tare value of a load cell system.

Yield Map Comparison

Figure 6 is a yield map of the University of Minnesota NW Experiment Station Field 13. The white areas of the map represent the low yielding areas of the field, down to 0 t ha^{-1} (0 ton acre^{-1}). The darkest areas represent the high yielding areas of the field, up to 77 t ha^{-1} (34 ton acre^{-1}). The white strip down the center of the map is a drainage ditch where no sugarbeets were harvested.

Figure 7 shows the difference between the torque sensor map and the scrub chain product flow sensor map. White represents areas where the torque sensor measured at least 4.5 t ha^{-1} (2 ton acre^{-1}) less than the scrub chain sensor. Black represents areas where the torque sensor measured at least 4.5 t ha^{-1} (2 ton acre^{-1}) more than the scrub chain sensor.

Fig. 6. UMC Field 13 yield map.

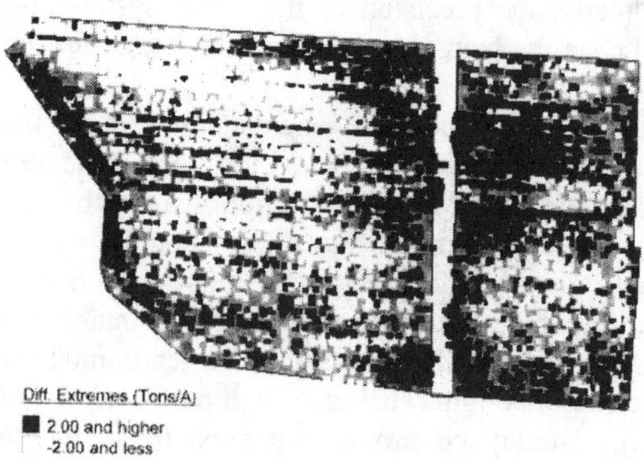

Fig. 7. Map of the difference between the torque sensor and scrub chain sensor yield data.

Figure 7 shows that the torque sensor system measured more yield than the scrub chain system in low yield areas of a field. The same relationship exists between the torque sensor system and the outlet conveyor system; however, comparing the outlet conveyor and scrub chain sensors shows that no distinct relationship existed.

The result that the torque sensor measured more yield in low yielding areas of the field is a secondary relationship. The gap in the center of the field running north and south is a drainage ditch, where the water flows north. The area of low yield along the north end is a drown-out area. Immediately prior to harvesting this field, approximately 76 mm (3 in) of rain was received. Therefore, this area of the field was extremely wet and muddy. The soil conditions during harvest of the high yielding areas were much drier than in the low yielding areas. Therefore, the primary relationship is that the torque sensor measured more yield in muddy areas than in dry areas.

Figure 7 also shows that the torque sensor measured less yield in the high yielding areas than the scrub chain sensor. Once again, this is a secondary relationship. Each system was calibrated over entire truckloads. During the loading of each truck, the harvester passed through both muddy and dry areas of the field. If the torque sensor measured more yield than the other sensors in the muddy areas of the field, it had to measure less yield than the other sensors in the dry areas of the field to obtain a correct measurement for the entire truckload. Therefore, the measurement of less yield by the torque sensor in the dry areas of the field was a direct result of the measurement of more yield in the muddy areas of the field.

The yield data should be corrected for the truckload error if it is 5% or greater. For a 45 t ha^{-1} (20 ton $acre^{-1}$) yield, 5% truckload error on a 13.6 t (15 ton) truckload equates to 2.24 t ha^{-1} (1 ton $acre^{-1}$) error on each individual yield data point.

ACKNOWLEDGMENTS

A special thanks to Amity Technology for providing a harvester on which to mount the instrumentation, Larry Smith and Dale Kopecky of the University of Minnesota NW Experiment Station in Crookston for their patience dealing with the inconveniences associated with this research, Ron Campbell of HarvestMaster for providing the yield monitor software and the technical support, and to GSE for providing the torque sensor. Funding for this project was provided by the North Dakota Power Use Council.

REFERENCES

Campbell, R.H., S.L. Rawlins, and S. Han. 1994. Monitoring methods for potato yield mapping. ASAE Paper no. 941584. ASAE, St. Joseph, MI.

Hall, T.L., V.L. Hofman, L.F. Backer, and L.J. Smith. 1997. Monitoring sugarbeet yield on a harvester. ASAE Paper no. 973139. ASAE, St. Joseph, MI.

Hofman, V.L., S. Panigrahi, B.L. Gregor, and J.D. Walter, 1995. In field yield monitoring of sugarbeets. SAE Paper no. 952114. SAE, Warrendale, PA.

Rawlins, S.L., G.S. Campbell, R.H. Campbell, J.R. Hess. 1995. Yield mapping of potato. p. 59–68. *In* Proc. of Site-Specific Management for Agricultural Systems. Minneapolis, MN. 27–30 Mar., 1994. ASA, CSA, and SSSA, Madison, WI.

Walter, J.D., L.F. Backer, and V.L. Hofman. 1996. Sugarbeet yield monitoring for site-specific farming. ASAE Paper no. 961022. ASAE, St. Joseph, MI.

Development of A Tomato Yield Monitor

M. G. Pelletier, and S. K. Upadhyaya
Biological and Agricultural Engineering Department
University of California
Davis, California

ABSTRACT

A continuous mass flow type yield–load monitor was developed for mapping spatial yield variability in processing tomatoes. The load monitoring system consisted of a load cell to measure weight of tomatoes over a section of a boom elevator and an angle transducer to measure the inclination of the boom elevator. A differential global positioning system [DGPS] was added to the load monitoring system to provide a yield monitoring system. The yield monitoring system was calibrated and validated using a GT weigh wagon during the early part of the 1997 harvesting season. It was found to be accurate within 2.5%. This yield monitor was used to map the variability in tomato yield. There were significant spatial variations in tomato yield. Typically lowest 20% of the yield was less than half the highest 20% of the yield.

INTRODUCTION

A yield monitor is one of the key elements in a precision farming system. It helps to quantify the variability in yield within a field and raise questions regarding management practices. Yield monitors have become commercially available for grain, cotton, and potatoes, but most of the efforts in precision farming have focused on grain crops due to the availability of proven combine grain yield monitors (Sudduth et al., 1997). Reliable yield monitors, however, are currently not available for many of the specialty crops, such as processing tomatoes.

The objectives of this paper are (i) to develop a yield monitor for processing tomatoes, (ii) to varify the yield monitor for accuracy and repeatability, and (iii) to develop yield maps for processing tomatoes.

BACKGROUND

Several methods have been developed for the measurement of mass flow for yield monitoring. Vanischen and De Baerdemaeker (1993) developed a technique for measuring corn silage yield using torque transducers on the silage blower shaft and cutter-head drive shaft. They found the system to perform well when integrated over a 2500 kg (5500 lb) load of silage. Auernhammer et al. (1995) developed a radiometric yield measurement system for a forage harvester using a radioactive source and detector (not currently legal for use in the USA for food). Wild et al. (1994) reported a hay yield monitoring system for round balers with

Copyright © 1999 ASA-CSSA-SSSA, 677 South Segoe Road, Madison, WI 53711, USA. *Proceedings of the Fourth International Conference on Precision Agriculture.*

strain gages on the tongue and axle of the vehicle, providing a measure of the weight of the baler and the bale. Additionally, they added accelerometers to measure vertical accelerations during operation. When stationary, they were able to determine the load within 2% of the actual weight. Measurements under dynamic conditions were still under investigation at the time of this writting. An optical sensor has been developed by Wilkerson et al. (1994) for yield monitoring in cotton. The system used an array of lights and photo-detectors. Laboratory tests provided an integrated measurement correlation over an entire load with a coefficient of determination $r^2 = 0.93$. Commercial grain monitors produced by AgLeader, Deere & Co., and Micro-Trak use an impact plate for yield monitoring, which sensed the force produced from a change in the momentum in a small quantity of grain as it hit a pressure plate.

Campbell et al. (1994) described a system for potato yield monitoring. The system consisted of a single idler wheel (steel hub–rubber tire) instrumented with a load cell, one on each side of the conveyor chain. The load cells measured the weight, and the integration required the addition of a speed sensor to measure the conveyor speed. The conveyor in this case was set at a fixed angle. A similar system was described by Hofman et al. (1995) for measuring sugarbeet yield. Another system for potatoes was described by Rawlins et al. (1995), which used a pivoted table supporting the load on the conveyor, where the load was measured using a load cell at the opposite end of the pivoted table. Calibrations of the systems were performed by correlating the output of the yield monitor to the truck weights (approximately 23 000 kg or 25 ton). The weights from the yield monitor were associated with a ground position based upon DGPS measurements. A correction for the harvester transport lag from digging to weighing was performed by a simple time delay.

Nolan et al. (1996) used a simple technique for correcting the harvester lag by using both a simple time delay model and performing nearest-neighbor filtering on yield maps. Boydell et al. (1996) reported much better results for correcting the yield data by deconvoluting the dynamics of the peanut harvester, and then passing the data through a low pass filter to reduce the noise amplified through the deconvolution filter (high pass filter). Pierce et al. (1997) reported that a simple delay model with smoothing has provided better yield maps than the more complicated maps that model the harvester dynamics as a first order system, and then performs a deconvolution of the harvester.

Belt weighing in nonmobile applications is a very mature science. Colijn (1983) reported that belt weighing could provide very accurate (on the order of ¼ to ½% error) continuous weighing for automated processes, provided that a number of fundamental criteria are followed in the use and installation of belt weighers. Unfortunately, most of the recommendations could not be adhered to due to the retrofit nature of this research. The most serious deviation included the operation of the scale on a dynamic, off-road vehicle with a changing conveyor angle. All of the recommendations were for stable, nonmobile platforms of long belt runs (over 30 m) with a maximum of 12° inclination.

The coupling of DGPS to a load monitor provides the necessary information for producing a yield map (a map depicting spatial location versus crop yield for a desired field). For tomato harvesting, this yield is the marketable red fruit before weight deductions on quality are taken. There are, however, several issues that pertain to this regarding data-acquisition, recording, filtering, correction of harvester dynamics, and aggregation of the final data to produce a useful and accurate yield map.

TOMATO YIELD MONITOR DEVELOPMENT

Impact plates, optical volumetric measurements, radiometric techniques, and continuous weighing methods are some of the more commonly used methods for monitoring crop yield. For tomatoes, the coefficient of restitution is highly variable depending upon maturity. This would preclude the use of an impact type of measurements. Similarly, due to the variability in the tomato density among various varieties, an optical volumetric measurement is also an unlikely method for providing the required accuracy. As stated before, the use of a radiometric or a x-ray technique is not feasible in the USA for health and environmental reasons. This leaves only a few options: (i) bulk weighing–drop bucket, (ii) instrumenting the entire harvester for total weight, or (iii) the use of a belt weighing technique similar to Campbell et al. (1994), and Hofman et al. (1995) used in potatoes and sweet potatoes. As the belt weighing option provides the easiest retrofit and has been successfully used by the industry, we decided to pursue this technique.

The tomato harvesting operation is unique in that right up until the point where fruit is deposited into the truck, material is separated from the fruit. Even on the boom elevator, extremely mature (broken) fruit and mud can slip through the slats in the elevator chain. This made the ideal location for measurement the highest point right before the fruit is delivered into the gondola. Unfortunately, this also is the worst place to position a belt weighing system since this section of the elevator boom is a short section of about 2 m, which is rarely level and averages in inclination about 22° from the horizontal, fluctuating throughout a load between ±30°. This angle must be measured accurately and properly incorporated into the calibration. Moreover, the change in angle should not affect the calibration of the belt system due to varying tension and belt effects. The developemnt of the yield monitor took place over a three year period. Our preliminary studies indicated that a three-roller system that consists of a weigh roller in the middle and a support idler on each side of this weigh idler is the preferred system compared with either a table type design or a single weigh roller design (Fig. 1).

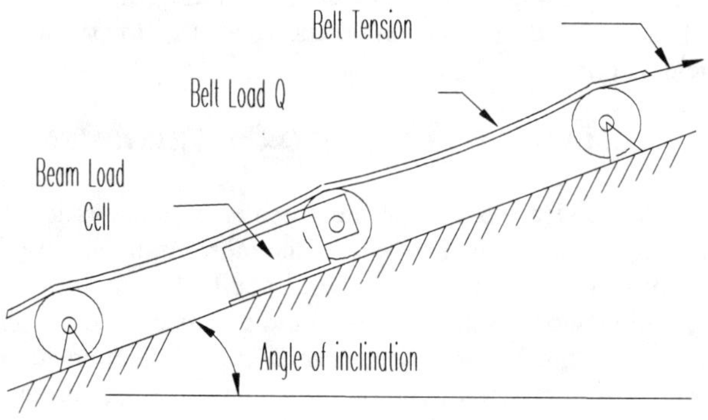

Fig. 1. Continuous weigh type load monitoring system for tomatoes.

Colijn (1983) reported that although a conveyor belt is a flexible tension member, under tension the belt behaves like a beam. Analyzing the belt scale as a beam leads to a statically indeterminant system in which the loading on the load cell is dependant not only on the weight over the active weigh region of the belt, but is also dependant upon the loading taking place in bays adjacent to the active weigh region (bay). Pelletier (1998) analyzed this statically indeterminate system and showed that its frequency response can be represented by Fig. 2.

This figure shows the nonminimum phase characteristic of the impulse response (negative track prior to positive loading). This dramatically shows the effect of how weight in the adjacent bays affects the readings on the load cell. A weight in an adjacent bay reduces the reaction of the load cell from weight located in the primary bay. This would not be a problem if it is consistent; however, if there is a misalignment between adjacent idlers then this unloading will not be consistent (Colijn, 1983). The ratio of the positive loading area to the negative loading area is a dramatic 10%. From a design standpoint, this demonstrates the need for precise alignment between the weigh idler and the adjacent idlers and also the idlers in adjacent bays.

The tomato harvester imparts a series of delays simply due to the transportation of the fruit and also due to the system dynamics. Observation of the harvester system dynamics points out two separate sections on the harvester where a certain mass of fruits needs to build up before any fruit can be conveyed beyond these points. The first location occurs immediately at the entry point to the harvester, just above the cutting bar that frees the plant and sends it up the conveyor toward the shaker. The plants must build up at this point to force

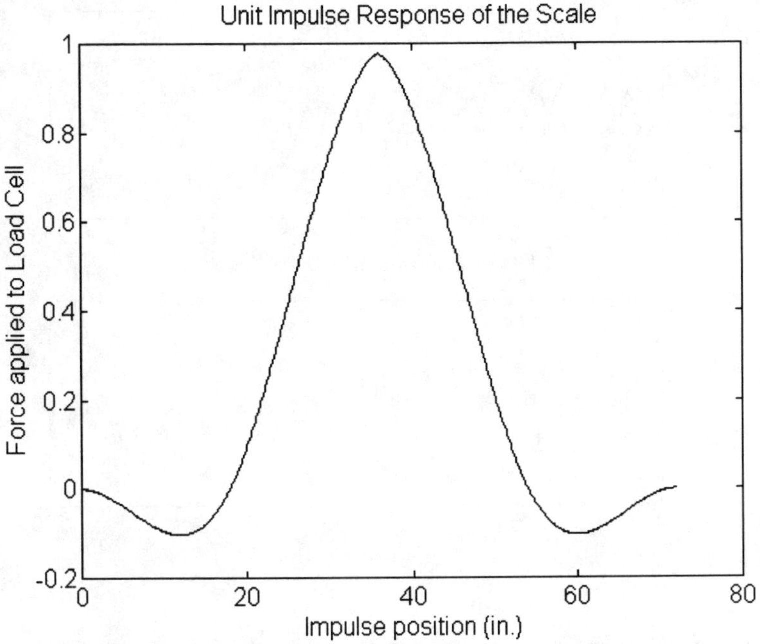

Fig. 2. Unit impulse response of the belt scale as measured by the load cell.

movement up the conveyor in a manner similar to a trash compactor. It takes a steady stream of plants feeding into this point to push the plants on to the conveyor. This is very noticeable when the harvester encounters long sections of 15 m or more of bed without any plants. The plants harvested in the last 15 m are still sitting just above the cutter. Once the harvester reaches the end of this bare area, it takes only about 2 s for the plants to build up to the point where the system begins to flow again. This build-up can be modeled as a simple first order transfer function in which the system is exhibiting pure compliance (no loss) with a time constant of 2 s. The second point where a build-up of material is necessary occurs just before the color sorter. At this point, the fruit is conveyed across a chain that runs up a short 30° incline. The fruit will not climb up this incline until there is more fruit pushing behind on the level section. Again, it takes approximately 2 s to build up enough fruit to push the mass up on to this conveyor. The last point of concern is mixing of the fruit. Mixing occurs by splitting the fruit into two separate paths as it travels through the harvester and rejoins the groups later on; however, one-half of the fruit travels across an extra distance equal to the length of the rear cross conveyor. This

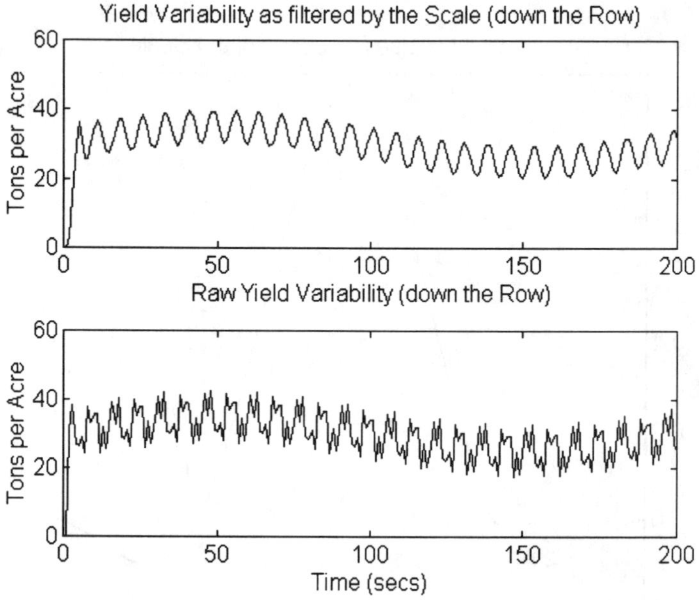

Fig. 3. Yield response as measured by the data acquisition system for the given field variability signal, harvested at a constant velocity of 1.2 m s^{-1}.

path difference causes approximately[1] a 3 s overlap between the two different routes. In 3 s, the harvester has covered from 2 to 3 m of bed row, or 5 to 10 plants. As there is no clean way to mathematically remove this effect short of adding two scales to these areas, this can be regarded as a limiting bandwidth for the system. Simulated studies were conducted by combining the frequency response of the scale with the three time delays described above to determine the abilty of this scale to measure and map yield variability. The results of the simulation study are shown in Fig. 3. This figure clearly indicates that the high frequency component of the signal has been removed. Therefore, this system is capable of measuring the low frequency field variability only. If finer measurements are needed to observe field variability below 9 m, the harvester–scale dynamics will have to be removed through a deconvolution process. This is required as the harvester and scale dynamics severely attenuate the high frequency components of the signal.

In this study a three-idler weigh bridge as shown in Fig. 1 was developed and retrofitted on to a tomato harvester owned and operated by Button and Turkovich Ranch, Winters, CA. An angle transducer and a submeter accuracy GPS system was added to this system. Figure 4 shows the schematic of the data acquistion and signal processing system.

[1] The time is dependent upon the conveyor belt speed, which is variable.

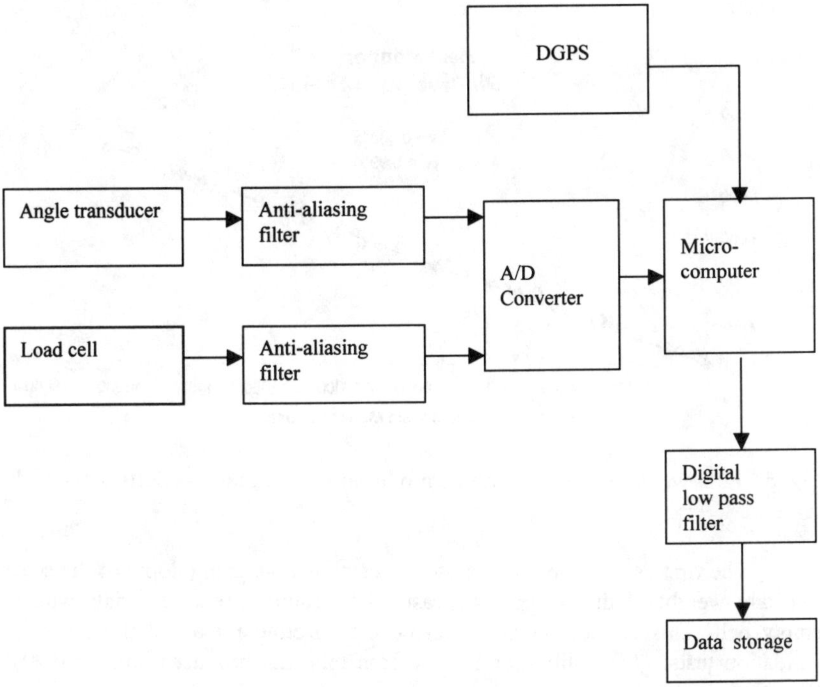

Fig. 4. Schematic of the data acquistion system.

The yield monitoring system was field tested after two weeks of continued use and one more time 3 wks later. The first test consisted of a short calibration trial followed by a validation trial. The validation trial was repeated again 3 wks later to verify that the system was stable and still performing to specifications.

The device was tested using the University of California, Davis, instrumented bulk weighwagon (GT cart). The field trial was performed by loading the weigh wagon from the tomato harvester. The tomatoes were passed across the yield monitor before being loaded into the weigh wagon. A digital read-out display on the weigh wagon provided the bulk fruit weight as tomatoes accumulated in the weigh wagon. The elevator loading belt was stopped periodically to read the weigh wagon bulk weight, and a corresponding mark was made on the data set for the tomato yield monitor. This was done to allow a comparison of the two measuring systems later. After the test, the two data sets were compiled and a regression was performed to measure the accuracy of the system.

RESULTS AND DISCUSSION

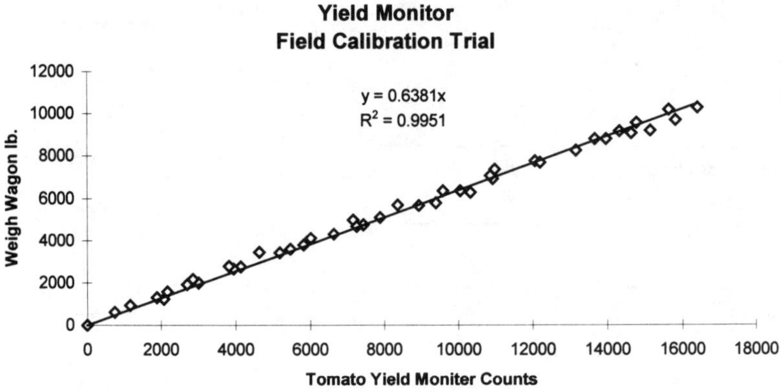

Fig. 5. First weigh wagon test results produced the calibration coefficients of the scale.

The first calibration set was used to establish the gain (slope) of the scale. The tare weight of the scale was measured by running repeated trials with an empty belt. This calibration then became the baseline for all of the following validation tests. The calibration results from this trial produced an $r^2 = 0.9948$ (Fig. 5).

After the calibration had established the gain at 0.6381, a validation test was conducted the following day (Fig. 6). The validation trial resulted in an $r^2 =$ 0.9948 (Fig. 6) with a standard error of 107 kg (235 lb). Another validation trial was performed three weeks later near end of the season (Fig. 7). The system had remained stable with no change in the gain of the system. The accuracy achieved

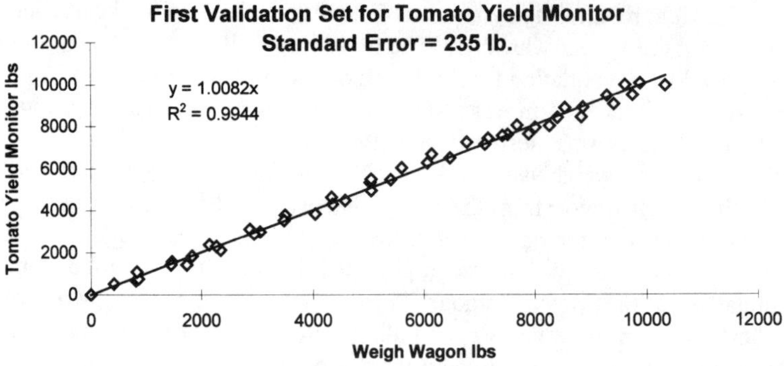

Fig. 6. First validation trial of the tomato yield monitor.

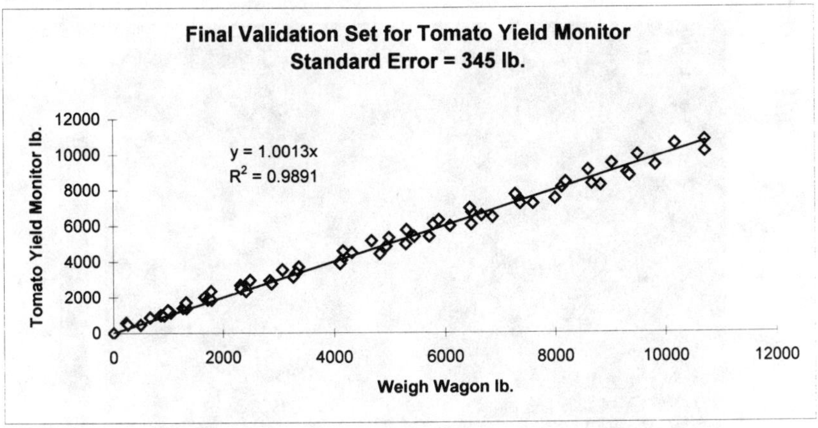

Fig. 7. Final validation trial of the tomato yield monitor.

by the device was based upon the 4500 kg (10 0000 lb) capacity of the weigh wagon. Over many trials, the system was found to provide a standard error of 227 kg (500 lb), 95% of the time.

One problem noticed over the course of the season was that the front set of rollers (on the driver side of the elevator) had a tendency to build up a layer of mud. This was significant enough to reduce the accuracy of the front set of rollers. All previously stated calibration and accuracy were reported without the use of the front set of rollers. This was beneficial from a economic point of view, but it did tend to slightly decrease the overall accuracy of the system. Fortunately, the rear set of rollers were unaffected by this and never needed cleaning throughout the entire season; however, it was advisable to check them from time to time.

The validated calibration equation was used to estimate truck loads using an extrapolation technique. This extrapolation technique led to an unexplainable 15% systematic error in computing truck loads. This aspect needs to be investigated further. Pelletier (1998) developed a correction technique and employed it to generate yield maps. This yield monitor was used to map the yield variability of tomato fields during the 1997 harvesting season. Typical field variability (within a single field) ranged from 25 to 45 tons acre^{-1} with an average yield of about 35 tons (Fig. 8). Analysis of this map indicates that the lowest 20% of yield was less than one-half of the highest 20% of the yield.

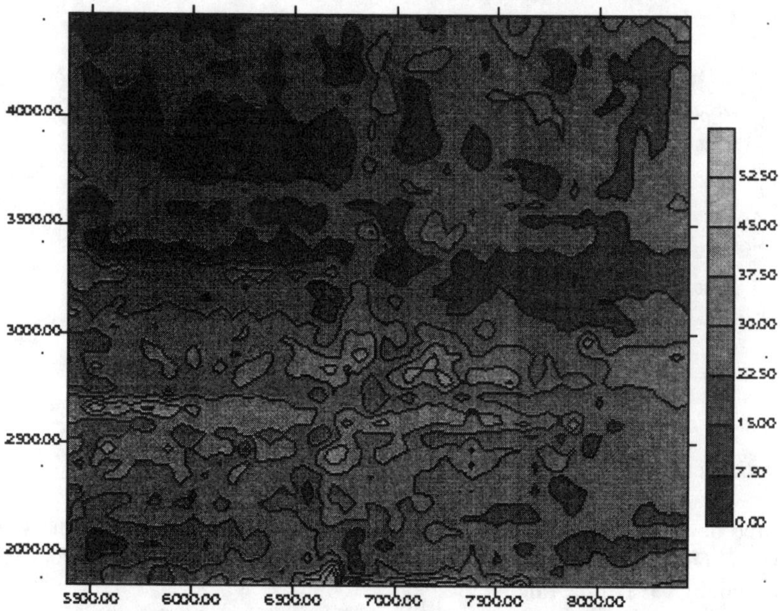

Fig. 8. Yield map from farm operated by a cooperative grower (yield tons per hectare).

CONCLUSIONS

On the basis of this study, we reached the following conclusions:

1. A continuous weigh type yield monitor was successfully developed to measure and map tomato yield.

2. The yield monitor calibration was validated at two separate times during the season. The first after 2 wks of use during the harvest season and once more after three additional weeks of use. The system was found to produce repeatable results to within 2.5% accuracy.

3. A typical yield map indicated considerable yield variability within a tomato field–lowest 20% of yield was found to be less than one-half the highest 20% of the yield.

REFERENCES

Auernhammer, H., M. Demmel, and P.J.M. Pirro. 1995. Yield measurement on a self propelled forage harvesters. ASAE Paper no. 95-1757. ASAE, St. Joseph, MI.

Boydell, B., G. Vellidis, and C. Perry. 1996. Dynamics of peanut flow through a peanut combine. p. 805–814. *In* Precision agriculture. ASA, CSSA, and SSSA, Madison, WI.

Colijn, H. 1983. Weighing and proportioning of bulk solids. 2nd ed. Trans Tech Publications. ISBN 0-87849-047-7.

Hofman, V., S. Panigrahi, B. Gregor, and J. Walker. 1995. In field monitoring of sugarbeets. ASAE Paper no. 95–2114. ASAE, St. Joseph, MI.

Nolan, S.C., G.W. Haverland, T.W. Goddard, M. Green, and D.C. Penney. 1996. Building a yield map from geo-referenced harvest measurements. p. 885–892. *In* Precision agriculture. ASA, CSSA, and SSSA, Madison, WI.

Pelletier, M.G. 1998. Development of a tomato yield/load monitor. Ph. D. diss. Biol. And Agric. Eng. Dept., Univ. of California, Davis.

Pierce F.J., N.W. Anderson, T.S. Colvin, J.K. Schueller, D.S. Humburg, and N.B. McLaughlin. 1997. Yield mapping. p. 211–243. *In* The site-specific management for agricultural systems. ASA, CSSA, and SSSA, Madison, WI.

Rawlins, S.L., G.S. Campbell, R.H. campbell, and J.R. Hess. 1995. Yield mapping of potato. p. 59–68. *In* P.C. Robert et al., (ed.) Site-Specific Management for Agricultural Systems. ASA Misc. Publ. ASA, CSSA, and SSSA, Madison, WI.

Sudduth, K.A., J.W. Hummel, and S.J. Birrell. 1997. Sensors for site-specific management. p. 183–210. *In* F.J. Pierce and E.J. Sadler (ed.) The state of site-specific management. ASA Misc. Publ. ASA, CSAA, and SSSA.

Vanischen, R., and J. De Baerdemaeker. 1993. Continuous wheat yield measurement on a combine. Automated agriculture for the 21st Century. ASAE Publ. no. 1191 ASAE, St. Joseph, MI.

Wild, K., H. Auernhammer, and J. Rottmeier. 1994. Automatic data acquisition on round balers. ASAE Paper no. 94–1582. ASAE, St. Joseph, MI.

Wilkerson, J.B., J.S. Kirby, W.E. Hart, and A.R. Womac. 1994. Real-time cotton flow sensor. ASAE Paper no. 94–1054. ASAE, St. Joseph, MI.

Mapping Peanut Yield Variability with an Experimental Load Cell Yield Monitoring System

J. S. Durrence
C. D. Perry
G. Vellidis
D. L. Thomas
Biological and Agricultural Engineering Department
University of Georgia
Tifton, Georgia

C. K. Kvien
National Environmentally Sound Production Agricultural Laboratory
Tifton, Georgia

ABSTRACT

During the 1997 harvest season, a peanut combine equipped with a load cell yield monitoring system was used to harvest 37.6 ha (119 ac) of irrigated land in Georgia. Yield data from two fields are presented to demonstrate the yield monitor's ability to characterize spatial yield variation as well as quantify cumulative yield. Data recorded from individual wagon loads averaged less than 5% error, and whole field errors were <1%. Analog and digital signal conditioning methods used in the combine instrumentation are discussed. Practical techniques for correcting erroneous data are also described. Interpolation methods used for map creation are identified and justified according to the spatial characteristics of the collected data. Improvements in the yield monitoring system over previous systems are discussed along with the system's limitations and suitability for commercial use.

INTRODUCTION

Peanut represents one of the crops that seems to remain on the potential list for precision agriculture applications. Peanut is a major crop in the southeastern USA, and pressure to maximize yield and/or efficiency has risen in response to threats against the existing government support program. High spatial variability in yields is typical for southeastern states, likely due to the number of soil types within individual fields as well as nematode, disease, and weed stressors. On paper, peanut makes an excellent candidate for the benefits boasted by precision agriculturalists. Efforts to implement such practices, however, have been stymied by the lack of a reliable peanut yield monitoring system.

Understanding the peanut harvest procedure is helpful in realizing the difficulty in obtaining site-specific yield measurements. Peanut is considered a root crop, maturing beneath the soil. Harvest begins when the peanut plants are mechanically

Copyright © 1999 ASA-CSSA-SSSA, 677 South Segoe Road, Madison, WI 53711, USA. *Proceedings of the Fourth International Conference on Precision Agriculture.*

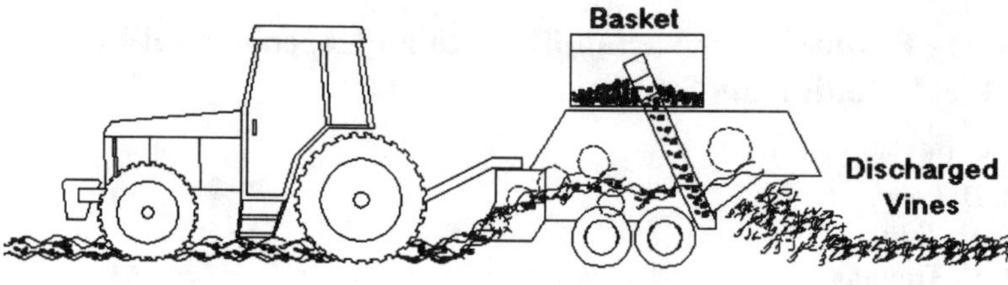

Fig. 1. General representation of peanut combine operation.

dug and inverted to expose the pods for drying. The digging machine places a pair of inverted rows into a single windrow, and when the crop has air dried to a suitable moisture content (usually around 18%), it is ready to be gathered with a peanut combine. The peanut combine functions include lifting the peanut plants from the ground, separating the pods from the vines, and collecting the pods.

While a variety of peanut combines are used, the principle harvesting mechanisms are essentially the same. Figure 1 provides a basic representation of a tractor and combine used to harvest peanut. A pick up reel feeds the vines into a series of threshing and separating sections that remove the vines from the pods. Circular saws then cut the stems away from the pods, and a large centrifugal fan blows the vines and trash out the end of the machine. The remaining portion is ideally the wanted crop. In reality, however, the remainder is a mixture of peanuts and foreign material including small rocks, dirt and leaves. This mixture is pneumatically conveyed from the machine belly to a collecting basket located on top of the machine.

Real-time crop yield measurement requires a method of measuring mass flow of the crop through the combine. For peanut, only pod mass is of concern; hence, for all practical purposes, the flow measurement must be made after the pods have been separated from vines and foreign material. This narrows the potential measurement locations to some point along the pneumatic duct or inside the collecting basket. Past research has demonstrated that reasonable yield measurements can be made by weighing the crop collected in the basket (Thomas et al., 1997). Durrence et al. (1997) showed that these yield measurements are significantly improved using a combination of analog and digital signal processing techniques.

Weighing the collected crop has definite disadvantages for peanut yield monitoring. For each measurement taken, the material weighed includes both the peanut and foreign material that reaches the basket. Without some method of removing the foreign material or measuring its mass separately, there is no obvious solution to this problem. Perhaps a more important disadvantage to the weighing approach pertains to the time delay involved rather than the sensing technique used. Before reaching the basket, the crop must endure the vine separation procedures, during which it is constantly mixed. Boydell et al (1996) performed experiments to characterize the inflow–outflow relationship of a peanut combine, and found that this relationship was considerably more complex than a simple time delay. Attempts to model this relationship have been unsuccessful, however, and the time delay has been used as an approximation in subsequent studies. This *convolution* problem is common among most commercial yield monitors, and the effect of the time delay simplification has not been thoroughly evaluated. Finally, the moisture sensors used in grain yield monitoring systems to correct wet yield measurements are not directly applicable to peanut harvesters. Since no real-time moisture sensor is available for peanut, the yield measurements collected from a peanut yield monitor must be post processed to obtain dry yield.

MATERIALS AND METHODS

The peanut yield monitoring system used in this research was developed at the University of Georgia's Tifton Campus in Tifton, GA (Thomas et al., 1997). In this system, four load cells are used to weigh the crop collected in the combine basket. The system was installed on a Kelley Manufacturing Company[1] (KMC, Tifton, GA) Model 3355 4 Row Wide Body peanut combine and a John Deere (Moline, IL) Model 7700 tractor. An Omnistar (Houston, TX) OS7000 DGPS receiver was used to provide position data, and a computer controlled data acquisition system was used to collect the data from the load cells, ground speed radar, and the DGPS receiver. The computer also served as the user interface for the driver, providing a graphical display of sensor outputs and facilitating control of data logging. The computer and data acquisition system are shown mounted in the tractor cab in Fig. 2. Perry et al. (1997) provide complete descriptions of the sensors, data acquisition system, and installation procedures of the peanut yield monitor.

The two fields discussed in this paper were 11.3 ha (28 ac) and 26.3 ha (65 ac), respectively. Located in Calhoun County, Georgia, these fields have been the subject of precision farming research for the past 4 yrs. Commercial grain yield monitors have been used to map corn, canola, and soybean yield. Had a peanut yield monitor been available, complete yield information for all crops grown would be available for use in an intensive precision farming operation. Figure 3 shows enhanced bare soil images of the two fields harvested. Center pivot irrigation was used in both fields, but the pivot coverages did not include all of the planted area.

The tractor–combine combination equipped with the yield monitoring system was the only machine used to harvest the two fields described; hence, yield data was

Fig. 2. PC Data acquisition system used with the peanut yield monitor.

---

[1] The use of trade names in this publication does not imply endorsement by the University of Georgia of the product named nor criticism of similar products not mentioned.

Fig. 3. Enhanced soil images of two fields used for peanut yield monitoring.

obtained for the entire area of each field. The beds (two planted rows inverted and windrowed) were numbered and flagged to facilitate field notes detailing distinguishing features or management practices used in specific areas. Each time the basket was filled, the crop weight indicated by the yield monitor and the actual crop weight were recorded in a field notebook. Truck scales were used to obtain the tare and loaded weights for each peanut wagon as described in Perry et al. (1997). Each wagon identification number was recorded for comparison to buyer data sheets. About every ninety minutes, a new data files was created after emptying the combine basket. This was done to minimize data loss in the event of system failure. The majority of data processing was done after the harvest; however, a number of improvements of the previous year's system were implemented in real time. An analog anti-aliasing filter was added to the input of the 16-bit A/D converter used in the data acquisition system. Data was collected at 256 Hz to further protect against aliasing, but these points were averaged and stored at 1 Hz to match the GPS data throughput. The output of the signal conditioning circuit used for the load cells was trimmed to remove the dead load of the basket from the yield data. As with previous versions of the system, a manual harvest indicator (pick flag) was used to tag data collected while harvesting peanut.

The data files from each field were divided so that each wagon load harvested could be compared to its respective yield data. The pick flag data was used to remove data collected when the combine was not actually harvesting (e.g., during a dump or turning around at the end of a windrow). A median filter was used to remove spikes from the data possibly caused by rough terrain or electrical noise. During harvest of field A10, the operator noticed that using radio communication in the cab caused the yield monitor output to drop drastically. Examination of the yield data collected prior to this observation revealed severe drops in the cumulative yield curves for short periods of time. After this problem was identified, radio communication was used only when the pick flag was turned off.

Data sheets for the wagon loads were collected from the grower's buying point to obtain data for correcting the recorded yield measurements. This data included the average moisture content and foreign material percentages as measured at the buying point. These percentages were combined to calculate a correction factor for each wagon load, and these factors were applied to their respective data sets.

To create contour yield maps, experimental semivariograms were computed for each of the fields using programs written in the MATLAB programming language

(The MathWorks, 1996). These programs allowed the user to select a rectangular area graphically from a plot of the data. The data from this area was then extracted, and the semivariance was calculated for lags of every 2 m up to ten meters and then every 5 m from 10 to 60 m. For each field, three 1.44 ha regions were extracted, and the resulting semivariograms were averaged to provide a representation for the entire field. The averaged semivariograms were compared with several theoretical models until an acceptable model was found. The parameters from the selected model were then entered into the SURFER software package (Golden Software, Golden, CO) to perform Kriging and create the contour maps.

RESULTS AND DISCUSSION

The yield data collected for each wagon load of field A10 is presented in Table 1. Moisture and foreign material measurements taken from the buyer records are included to demonstrate the difference in the yield data collected and that used to create yield maps. The corrected weight represents the data to which the combined correction factor has been applied; hence, this weight indicates the actual crop weight for each wagon load. The area harvested was calculated by integrating the ground speed data over each sampling interval, and the yield monitor weights represent the sums of the individual yield points collected for each wagon load.

The average yield (corrected for moisture and foreign material) for field A10 was 3845 kg ha^{-1}, and the average percent difference between the yield monitor and wagon scales was 3.1%. The maximum 8.9% error experienced for Load 2 could be attributed to the aforementioned radio interference. Also, the first wagon loads of harvest were characterized by much stop and go action to make combine adjustments and check system performance. The total field error of <1% is not indicative of the site-specific accuracy of the system, but it does emphasize the system's ability to quantify yield accurately.

Table 1. Field A010 peanut yield data.

Wagon Load	Area	Wagon scales	Yield monitor	Percent difference	Foreign material	Moisture content	Corrected yield
#	ha	kg	kg	%	%	%	kg
1	0.724	3529	3416	3.19	4	15.40	2754
2	0.947	4831	4401	8.90	3	13.90	3657
3	0.696	3325	3340	-0.45	4	15.50	2688
4	0.834	3960	4046	-2.18	3	17.80	3205
5	0.753	3731	3679	1.40	4	14.20	3009
6	0.498	2227	2095	5.95	2	14.10	1758
7	0.955	4014	4090	-1.89	4	13.26	3384
8	0.991	4688	4833	-3.10	3	12.19	4099
9	1.008	4463	4594	-2.93	3	11.00	3951
10	0.51	2862	3028	-5.78	2	11.00	2634
11	0.971	4472	4573	-2.25	3	12.30	3874
12	0.854	3699	3706	-0.20	2	11.00	3225
13	0.987	4121	4113	0.19	2	14.20	3447
14	0.409	1404	1333	5.07	2	14.20	1117
Totals	11.13	51326	51246	0.16	n/a	n/a	42801

Table 2 provides the yield information for Field A071, which averaged 3434 kg ha^{-1}. The average wagon load error for this field was 2.06 %, and again the total field error was <1%. The maximum wagon load error was 5.70 %, significantly less than that of Field A10.

Table 2. Field A071 peanut yield data.

Wagon load	Area	Wagon scales	Yield monitor	Difference	Foreign material	Moisture content	Corrected weight
#	ha	kg	kg	%	%	%	kg
1	1.23	5039	5089	-0.98	1	20.3	4005
2	1.18	4901	5020	-2.42	1	20.3	3951
3	1.07	4742	4472	5.70	3	19.4	3470
4	0.88	3885	3938	-1.37	2	18.8	3119
5	0.94	4382	4272	2.51	3	18.8	3341
6	1.01	4574	4755	-3.96	3	16.8	3814
7	1.08	4915	4975	-1.24	2	19.5	3906
8	0.93	4198	4372	-4.15	2	18.3	3484
9	0.36	1567	1538	1.88	1	18.3	1241
10	1.08	4969	5002	-0.67	4	16.3	3987
11	1.1	4965	4947	0.35	2	16.3	4042
12	1.11	4587	4587	0.00	1	17.7	3729
13	1.09	4656	4639	0.36	3	17.7	3679
14	1.11	4851	4965	-2.36	2	17.7	3988
15	1.06	4990	5097	-2.15	3	20	3924
16	1.06	4633	4614	0.42	3	20	3553
17	0.35	1669	1687	-1.06	2	19.2	1329
18	1.04	4910	5006	-1.95	4	22.5	3680
19	1.03	4701	4537	3.50	3	18	3584
20	1.16	4894	4953	-1.20	3	18	3913
21	1.27	5280	5127	2.90	3	17.3	4086
22	0.55	2059	2096	-1.78	3	15.8	1702
23	0.96	3973	3761	5.35	3	19.5	2914
24	1.23	5364	5187	3.30	4	19.5	3968
25	1.13	5114	5022	1.80	4	18.5	3892
26	1.33	5420	5415	0.11	1	18.5	4359
Totals	26.4	115241	115073	0.15	n/a	n/a	90658

In Fig. 4, the experimental semivariogram data for each field are shown with the selected models. In both cases, a spherical semivariogram model was selected. The model for field A10 has a range of ten meters, while that of Field A071has a range of 12 m.

Figure 5 shows the contour yield map for field A10, with the field border shown with a dotted line. The contour map highlights the problems observed for the southwest corner of the field. During the growing season, this area of the field

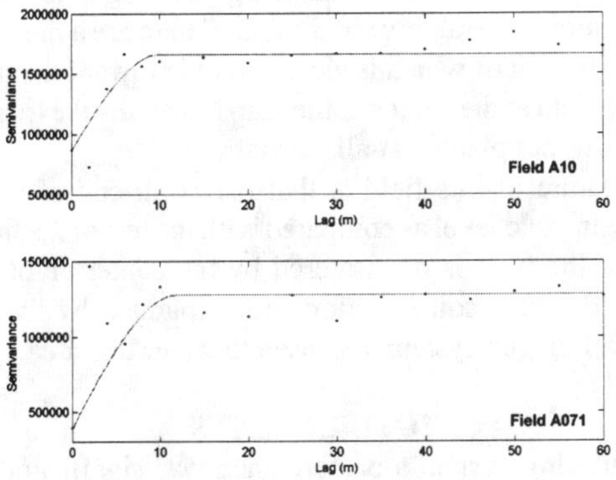

Fig. 4. Semivariogram data and selected models for the two fields harvested.

became saturated with water, and the resulting crop was very poor. A sandy region on the eastern edge of the field also is clearly visible in the yield map. In this area, the plant population was extremely low. The streak of low yield area that begins at the center of the north edge of the field and extends in a slight southwest direction was caused by an unplanted strip, so this feature may be misleading.

Figure 6 shows the contour yield map for field A071. The most obvious low yielding region was in the center of the south end of the field. During the season, this area of the field was inspected and a large degree of disease incidence was observed.

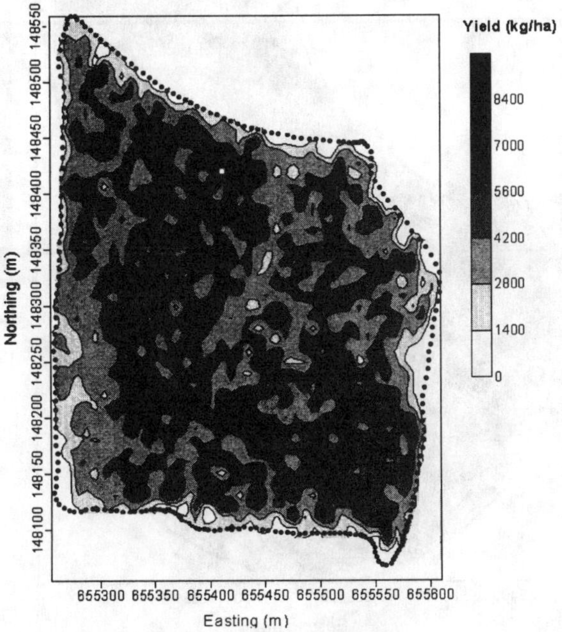

Fig. 5. Contour yield map of Field A10.

Many of the plants were killed by these diseases, resulting in low yield. The eastern edge of the map features several low yield spots. There are a number of sandy areas along this edge, the largest of which is clearly visible toward the northern end. The circular area of low yield at the center of the map is actually the location of the center pivot, so this area was not planted at all.

A surprising feature of this field is that the north end of the field shows no apparent difference in yield level as compared with the rest of the field. This is ironic because this end of the field is not watered by the center pivot irrigation system installed in the field. This contradiction was explained by the grower when he revealed that a traveling gun system was used to water this area.

FUTURE PLANS

The yield monitoring system's performance was significantly improved over previous seasons and comparable to grain yield monitors. Although this was encouraging to those growers who observed the system, the amount of instrumentation used in the system gave it an impractical appearance. Also, the amount of post processing required for the collected data makes peanut yield mapping an intricate process.

To encourage area growers to experiment with precision farming and invite industry to examine the peanut yield monitor, a black box system is under preparation for the coming season. This system will incorporate the majority of the data processing techniques formerly used in post process, and the instrumentation will be reduced to small user interface box in the tractor cab and a stand alone box located

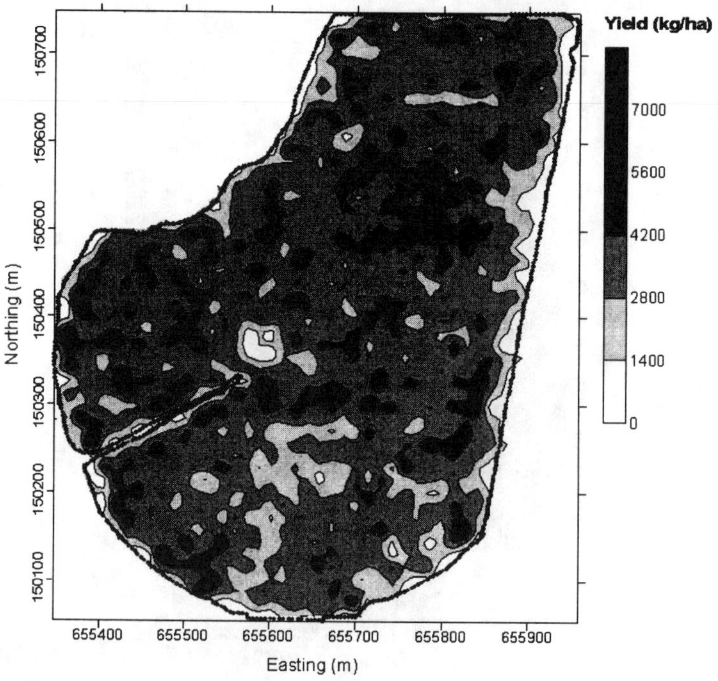

Fig. 6. Contour yield map of Field A071.

on the combine. The prototype systems will be installed on the research combine as well as four grower combines at different locations in south Georgia. The systems will be operated by growers with assistance from the research team.

CONCLUSION

The peanut yield monitoring system tested during the 1997 season showed significant improvements over versions tested during previous seasons. With per-wagon-load errors averaging around 3% and whole field errors of <1%, the yield monitor accuracy is quite comparable to commercially available grain yield monitors. Despite the yield monitor's ability to measure the quantity of harvested material, it is not capable of measuring the moisture content or percentage of foreign material harvested along with the crop. In lieu of real-time sensors for these parameters, the collected data was adjusted after the harvest using measurements taken at the buying point.

After correcting the yield data for moisture and foreign material content, contour yield maps were created to observe the spatial yield variation in the two fields harvested. Interpolation by Kriging was performed with commercial software after proper parameter estimates were obtained with basic geostatistic procedures. The resulting contour maps showed areas problem areas related to soil moisture, soil type and disease incidence.

Considering the favorable system performance, the newest version of the system is now being prepared for the coming season. This system will implement most of the signal processing in real time, and it will be streamlined for easy installation on several grower combines. These efforts will likely draw the attention of many Georgia peanut growers and possibly attract an industry to commercialize the system.

ACKNOWLEDGMENTS

The authors gratefully acknowledge the contributions of the following UGA-BAE personnel toward the success of this research: Dewayne Dales, Mike Gibbs, Rodney Hill, Andy Knowlton, Charles Welsh, Jerry White, and Terrell Whitley. The grower-cooperators provided the peanut fields and valuable assistance during harvest (providing fuel, obtaining buyer data). The growers were W.P. and Tony Smith.

REFERENCES

Boydell, B., G. Vellidis, C. Perry, D.L. Thomas, and R.W. Vervoort. 1996. Deconvolution of site-specific yield measurements to address peanut combine dynamics. *In* P.C. Robert et al., (ed.) Proc. of the 3rd Int. Conf. on Precision Agriculture, Minneapolis, MN. 23–26 June 1996. ASA, CSSA, and SSSA, Madison,WI.

Durrence, J.S., T.K. Hamrita, G. Vellidis, C.D. Perry, D.L. Thomas, and C.K. Kvien. 1997. Digital signal processing techniques for optimizing a load cell peanut yield monitor. ASAE Paper no. 97-3009. ASAE, St. Joseph, MI.

Perry, C.D., D.L. Thomas, G. Vellidis, J.S. Durrence, L.J. Kutz, and C.K. Kvien. 1997. Integration and coordination of multiple sensor and GPS data acquisition for precision farming systems. ASAE Paper no. 97-3143. ASAE, St. Joseph, MI.

The MathWorks, 1996. MATLAB version 5 user's guide. The MathWorks, Natick, MA.

Thomas, D.L., C.D. Perry, G. Vellidis, J.S. Durrence, L.J. Kutz, C.K. Kvien, G. Boydell, and T.K. Hamrita. 1997. Development and implementation of a load cell yield monitor for peanut. ASAE Paper no. 97-1059. ASAE, St. Joseph, MI.

Vector Method for Determining Harvest Area Using Combine Position Data

Scott T. Drummond
Department of Biological and Agricultural Engineering
University of Missouri
Columbia, Missouri

Clyde W. Fraisse
Kenneth A. Sudduth
Cropping Systems and Water Quality Research Unit
USDA-ARS
Columbia, Missouri

ABSTRACT

The measurement of actual harvested area per unit time is an important component in the creation of accurate crop yield maps. For row crops, such as corn (*zea mays* L.), these measurements can be made manually on most conventional yield monitors. However, in drilled or broadcast crops a more accurate and automated method is required. In this study, a vector method is developed to determine actual combine harvest area at each time step of the harvest process from a global positioning system (GPS) trajectory. The algorithm was coded into a geographic information system (GIS) and modifications were made to increase computational efficiency. The method was compared with a previously reported raster method of swath width determination on data collected during the 1997 drilled soybean harvest. The vector method greatly improved yield accuracies over the assumption of constant swath width, and provided a number of distinct advantages over the raster method.

BACKGROUND

The equipment and techniques necessary for crop yield measurement have progressed significantly in the last decade. Pierce et al. (1997) reviewed a variety of grain flow sensors, several of which are commercially available today. Through the inclusion of position determination techniques, most common of which is the differential global positioning system (DGPS), the ability to create yield maps has become fairly commonplace, both among researchers and producers.

The accuracy and quality of these yield maps are of extreme importance to the development of effective precision management strategies. However, the creation of accurate yield maps is a process made difficult by a number of possible error sources. Blackmore and Marshall (1996) identified six main groups of errors affecting yield map accuracy, and intuitively ranked them with regard to their effects on yield maps. They ranked the error sources in this way:

Copyright © 1999 ASA-CSSA-SSSA, 677 South Segoe Road, Madison, WI 53711, USA. *Proceedings of the Fourth International Conference on Precision Agriculture.*

1. Unknown crop width entering the header during harvest.
2. Time lag of grain through the threshing mechanism.
3. The inherent wandering error from the GPS.
4. Surging grain through the combine transport system.
5. Grain losses from the combine.
6. Sensor accuracy and calibration.

It is interesting to note that these authors felt that swath width determination was the largest error source affecting yield map accuracy. Although varying degrees of effort have been applied to all of these potential error sources, the problem of swath width determination has been widely studied from a variety of approaches. What follows is a survey of the techniques which have been used (or might possibly be used) to measure swath width or harvested area, or to minimize its effect on yield map accuracy.

Operator Estimation

Some commercial systems allow the combine operator to manually adjust on-the-go the recorded cutting width value on their yield monitor. This approach is an improvement over the assumption of constant swath width; however, it is extremely difficult to implement in the field for at least two reasons. First, the combine operator seldom has the time and concentration available to make the right corrections precisely at the right times. Secondly, even when the combine operator can make the adjustments properly, or has a second person available to do so, the effective resolution (and resulting accuracy) of these systems would likely be measured in feet.

Effect Minimization

Methods have been suggested that could minimize the effect of varying swath width on yield maps. One such method was suggested by Birrell et al. (1996). They proposed a raster accumulation method by which the geographically harvested area corresponding to the header travel during each time interval was determined from the GPS position information. The grain mass harvested during that time interval was then distributed to each grid cell traversed, according to the ratio of the harvested area in any particular grid cell to the total area harvested for the time interval. After all sample points had been processed, each cell would contain the total grain mass harvested in that cell. A very similar technique was suggested by Blackmore and Marshall (1996) and given the name Potential Mapping.

As long as the cell size is large in relation to the combine header width and the GPS positioning error, this method could be quite effective in removing the effects of combine swath width on yield maps. However, this method is obviously quite susceptible to missing data, or even data that has been shifted improperly (i.e., improper time lag, position errors). Additionally, these methods by definition provide a raster format for the final yield data, and this may or may not be desired for a particular application.

Ultrasonic Sensors

Several groups have investigated the use of ultrasonic sensors for automatic swath width measurement (Vansichen & De Baerdamaker, 1991; Reitz & Kutzbach, 1996; Stafford et al., 1997; Wang et al., 1997). In general, the results from these studies have been quite promising, particularly when the distance between crop edge and sensor is relatively small. In practice, however, these systems may be difficult to implement, for a variety of reasons. First, when the crop edge is dense and crisp, the sensor measurements are usually quite accurate (various authors have reported errors in the range from 2 to 10 cm). However, under certain relatively common conditions (i.e., sparse crops, crops bent over from a previous pass, crops that have fallen or matted, extremely windy conditions) ultrasonic swath width sensing may be problematical or even impossible.

Machine Vision

Hoffman (1996) described the Demeter project, an automated vehicle guidance system for a forage harvester based upon high speed (5–30 Hz) crop edge detection from camera images. While the current system does not employ any yield measurement device, the machine vision approach might be useful in either minimizing the effect of swath width variation through vehicle guidance, or as a means to measure cutting width outright.

While it may be possible to achieve quite acceptable accuracies through this approach, the implementation of such a system for swath width detection could be quite difficult. Without a doubt the implementation of a machine vision system for swath width measurement would require a significant investment in terms of additional hardware costs.

Combine Position

Han et al. (1997) proposed a method for determining effective combine harvest area from the GPS positions of the combine in the field. The method consists of initializing a high resolution bitmap that represents the pre-harvest crop conditions of the field (zero = no crop, one = crop). The bitmap is then progressively updated at each step of the harvest process, by turning all of the cells whose centers fall within the combine header area to zeros. The cells that changed from ones to zeros during that time interval represent the area harvested for that time interval.

The major advantages of this approach are that it requires no additional equipment, beyond that required for yield mapping, and no additional user input. Another important consideration is that the accuracy of methods based upon combine position should significantly improve as the accuracy of the positioning equipment improves. Additionally, though the method might require significant processing power, it could likely be implemented in real-time. As long as cell size is very small in relation to header width, this method provides a simple, efficient means of estimating harvest area; however, the memory necessary to store a high resolution bitmap for a large field is considerable. For example for a 50 ha square field with a

5 cm cell size, about 25 MB of memory is required. As the accuracy of GPS equipment steadily and significantly improves, cell sizes this small, or possibly even smaller, may be required to minimize harvest area errors.

PROPOSED VECTOR METHOD

In an attempt to build upon the advantages of the raster method suggested by Han et al. (1997), we developed a simple vector method that similarly determines the effective harvest area from combine position data. As in the raster method, the polygon representing the area covered by the header (based upon position and trajectory information) during each time interval is determined. Each polygon is then processed in chronological harvest order by subtracting, in the boolean sense, all previously processed polygons from the current one. In a more formal sense, let the area traversed by the combine header for time interval i of the harvest process be represented by a_i. The actual harvested area for each time interval, A_i, is given by:

$$A_i = a_i - (a_1 \cup a_2 \cup \ldots \cup a_{i-1})$$

The resulting polygons represent the actual harvested area during each time step. Figure 1 shows a graphical representation of the concept for three time intervals for which the combine header areas overlap. Intuitively, it is obvious that this technique should be able to produce the actual harvested areas to the level of positioning accuracy for each time interval of the harvest process.

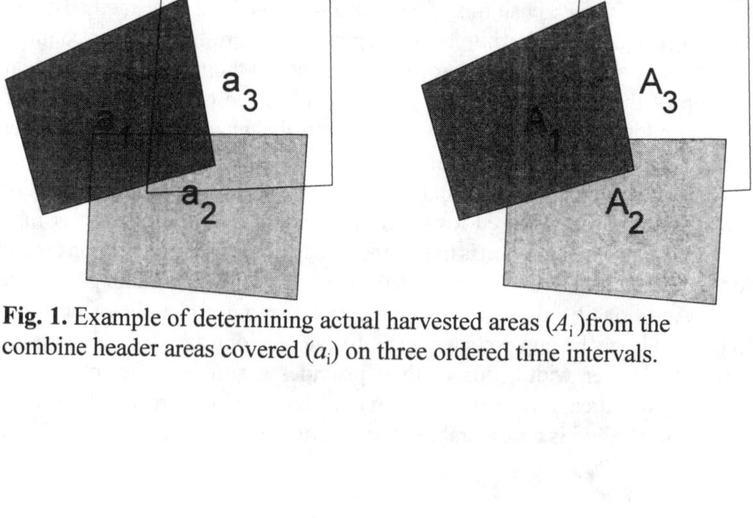

Fig. 1. Example of determining actual harvested areas (A_i) from the combine header areas covered (a_i) on three ordered time intervals.

Reducing Computational Complexity

This approach, however, is not without drawbacks. Boundary conditions at the edges of fields, or at islands within the fields, whether real or due to missing data, may cause overestimation of actual harvested areas along these boundaries. If the true locations of these boundary conditions are known, judicious use of boundary polygons could be used to correct these problems.

A more serious drawback of this simplistic approach is the computational complexity of the algorithm. Given that we have N discrete time intervals and associated polygons that require processing, and we assume the polygon clipping routine is at most order N (a very reasonable assumption), then the complexity of the algorithm is order N^2. In other words, as we double the size of our dataset, we quadruple the amount of processing required to find the actual harvested areas.

There are techniques we can use to reduce the processing time of this problem. The most obvious of these is a bounding window check of the two polygons upon that we are about to perform the polygon clipping algorithm. If the bounding windows of the two polygons do not overlap, the clipping algorithm need not be performed for that particular pair of polygons. While this modification to the algorithm significantly reduces processing time, it does not reduce the computational complexity of the problem below order N^2.

A technique that could be used to reduce the complexity of the algorithm, would be to group adjacent, consecutive time interval polygons together into individual passes or coverages. Coverages might be distinguished from one another by a number of possible techniques, i.e., header raised–lowered, combine heading or speed changed, break in time sequence. Once an individual coverage has been processed, and the resulting individual harvest area polygons (A_i) have been output, the entire coverage could be represented by a single boundary polygon (Fig. 2.).

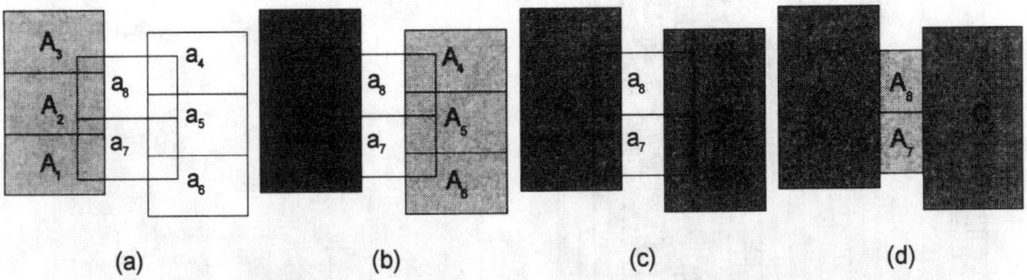

Fig. 2. Example of algorithm with coverage implementation: (a) actual areas $A_{1,2,3}$ in coverage C_1 are processed and recorded; (b) actual areas in coverage C_2 are processed and recorded; (c) header areas $a_{7,8}$ are processed against C_1 and C_2 and (d) all actual harvest areas have been processed and output.

This technique should significantly affect the computational complexity of the problem. For example, consider the best case of a square field with $\sqrt{N}$ parallel swaths with each coverage containing $\sqrt{N}$ individual polygons. Now, each new polygon a_i must be clipped with less than $\sqrt{N}$ other coverages and less than $\sqrt{N}$ individual polygons within its own coverage. This yields an overall computational complexity of order $N^{3/2}$ - a considerable improvement over N^2. It is also clear that the algorithm is inherently parallel in nature, and could easily be implemented in a multiprocessor environment, further reducing the total amount of time necessary to achieve a solution.

PROCEDURES

The algorithm outlined in the previous two sections was implemented using ARC/INFO Version 7 Arc Macro Language (AML) routines. While it was understood that this approach would incur significant overhead and thus affect processing speed, we felt that the reduction in development time and the portability of the finished product were acceptable trade-offs for these initial investigations.

Our trial dataset consisted of an approximately 10 ha region of a soybean field located in central Missouri. The crop was harvested with an R42 Gleaner combine with a 4.6 m width grain platform, and yield data were collected with an AgLeader 2000 yield monitoring system. Position information was collected using a pair of Ashtech Z-Surveyor GPS receivers operating in real time kinematic (RTK) mode. The nominal accuracy of these receivers was generally on the order of 3 to 10 cm.

A significant fraction of the area under study required replanting due to water ponding problems in the early season, and the replanted soybeans were harvested approximately two weeks after the soybeans from the first planting. As a result, the harvest pattern for this field was quite complex, and it was difficult for the driver to keep a constant swath width entering the header at all times. Position and heading data were used, along with knowledge about the geometry of the header and its relationship to the GPS antenna, to create a map of the areas covered by the header during each second of the harvest process (Fig. 3).

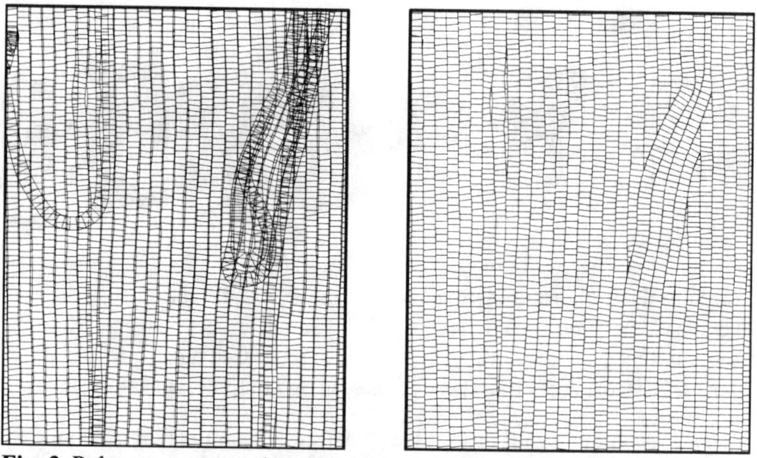

Fig. 3. Polygons representing the areas covered by the header, and the actual harvested areas determined by the proposed vector method, for a small region of the field.

Each of the polygons was identified by a time tag, such that grain flow information collected by the yield monitor could be recombined with the actual harvested areas after our processing was complete. New coverages were initiated whenever: (i) a time-break occurred in the polygon list; (ii) the heading of the combine changed rapidly and significantly; and (iii) shortly after the combine header sensor indicated a raising of the combine head. These steps were taken to insure that our coverages would consist of contiguous but likely non-overlapping polygons. Next, our algorithm was applied to the polygon dataset, and the actual harvested polygons and their associated areas were determined (Fig. 3).

The bitmap method suggested by Han et al. (1997) was implemented on this same initial polygon set, with both 10 cm and 30 cm^2 cell sizes. This was done in order to compare the proposed vector method with the previously reported raster method. Additionally, to approximate the conventional assumption of a constantly full swath width, the total area covered by the header during each time interval was calculated and the results were compared to both the raster and vector methods.

RESULTS AND DISCUSSION

A surprising amount of variability in actual harvested area versus total area covered by the header was observed (Fig. 4). It is interesting to note that two distinct peaks occur in this distribution. The smaller one occurs at 100% of header width, and represents those areas where we were harvesting lanes within the field with crop on both sides of the header. The second occurs at about 93% of total header area, and represents the more common harvest situation where the operator is trying to keep a full header while making certain to leave no crop standing. Point rows and

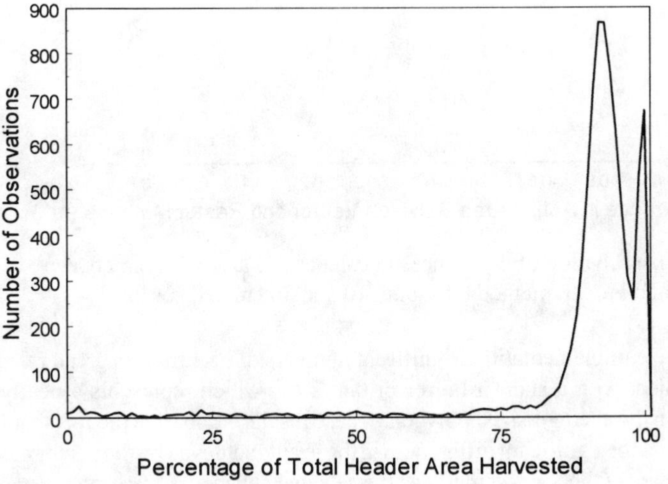

Fig. 4. Distribution of the fraction of actual area harvested to the total area covered by the header during each time interval.

relatively small cleanup strips would account for the observations <80%, and the distribution in this region appears to be quite random. Over all time intervals, the average harvested area was approximately 89% of the total header area covered. In other words, for our dataset we could expect an average of 11% error in yield estimation on a point-by-point basis, just from the assumption of a constant, full header width of crop entering the combine. If all observations less than half full were removed from the dataset, this was reduced to approximately a 7% yield error. This still represents a very significant error term in the calculation of instantaneous crop yield. These results were supported by the results of the bitmap method, with yield errors for both the 10- and 30-cm cell sizes of approximately 11 to 12%.

A comparison of the results from the proposed vector method to those from the bitmap method proved interesting. Figure 5 shows the distribution of the differences in calculated harvest areas for each yield observation between the bitmap method and the vector method, for both cell sizes. While the average area of the actual harvest polygons was 8.24 m^2, for the 10 cm cell size the difference between methods was seldom more than ± 0.2 m^2, with a standard error of 0.157 m^2. This would correspond to less than a 2% difference on the average-sized polygon. For the

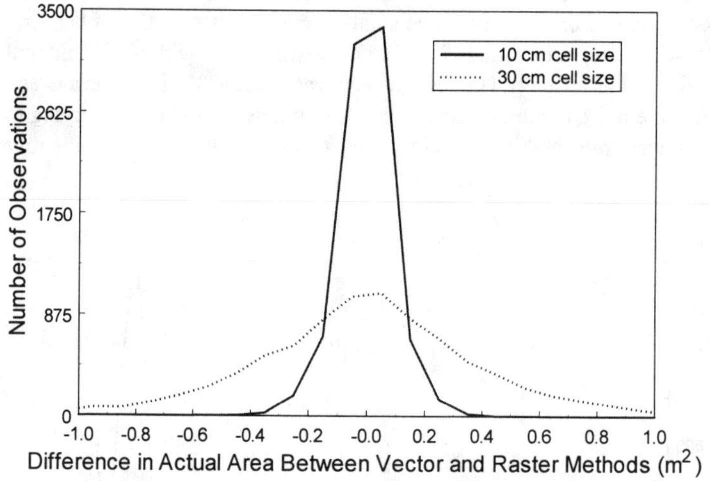

Fig. 5. Distribution of differences in calculated harvest areas between vector and bitmap methods, for both 10 and 30 cm cell sizes.

30 cm cell size implementation, significant numbers of occurrences greater than ± 1 m^2 were noted, with a standard error of 0.402 m^2, which represents a nearly 5% difference on the average-sized polygon. These results emphasize the fact that there is the potential for significant differences in the results achieved by the vector method and the bitmap method, particularly as the selected cell size increases.

In addition to the potential for improved accuracy, a useful byproduct of the proposed vector method is that it provides us with the actual polygons representing the harvested areas. With these polygons, it is possible to create a yield map of classed polygons, as opposed to the more conventional classed point maps. Figure 6 shows a small area of a yield map created both with constant swath width (a), and by using the vector method to estimate actual harvest area (b,c). The data is displayed both as classed point maps (a,b) and as a classed polygon map using precisely the same class definitions (c).

Classed polygons provide several advantages in the interpretation phase of yield

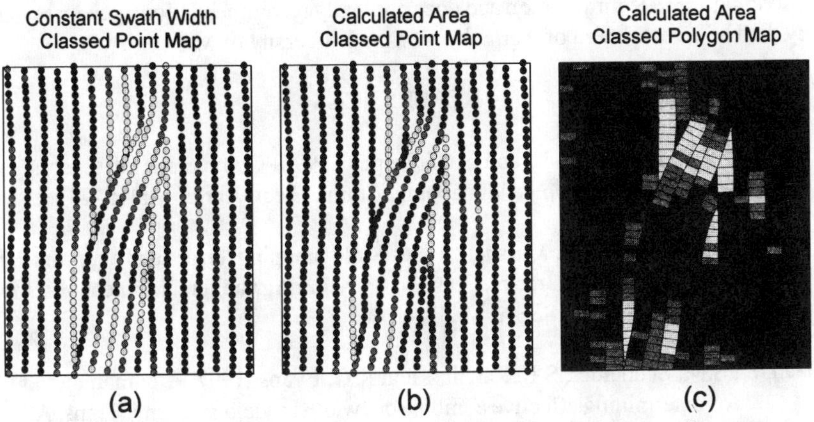

Fig. 6. Comparison of classed points and classed polygons for mapping crop yields with varying combine cutting width.

mapping. First, they allow the user to intuitively see how and in what order the crop was harvested, helping the user to locate areas of the yield map that may provide unreliable results. Due to the effects of combine dynamics, (ramping, starting and stopping in the crop), knowledge of the exact harvest pattern can dramatically affect yield map interpretation. Secondly, with classed point maps, points which represent extremely small areas are given precisely the same symbol size as points which represent larger, more representative areas. This can mislead the user in his evaluation of yield maps, particularly where harvest patterns are unusual. Classed polygon maps allow the user to more easily ignore small areas in the interpretation, since these areas will be represented by a proportionally small area.

SUMMARY

The measurement of actual harvested area is an important component in the creation of accurate crop yield maps. In this study, a vector method was developed and implemented to determine the actual combine harvest area at each time step of the harvest process using combine position information. The method provides several advantages over a previously reported raster method. It can provide significant error reduction over the raster method, particularly as cell size increases in relationship to positioning accuracy. Although the vector algorithm is computationally more complex, it does not require any additional computational resources as positioning technologies become more accurate. The raster method would require steadily smaller cell sizes and larger memory requirements to achieve comparable accuracies. The vector method also allows for the creation of classed polygon maps, which provide distinct advantages over the more conventional classed point maps in the area of yield map interpretation.

DISCLAIMER

Mention of trade names or commercial products is solely for the purpose of providing specific information and does not imply recommendation or endorsement by the U.S. Department of Agriculture or the University of Missouri.

REFERENCES

Birrell, S.J., K.A. Sudduth, and S.C. Borgelt. 1996. Comparison of sensors and techniques for crop yield mapping. Comp. Elect. Agric. 14:215–233.

Blackmore, B.S., and C.J. Marshall. 1996. Yield mapping; errors and algorithms. p. 403–415. *In* Proc. 3rd Int. Conf. on Precision Agriculture, Minneapolis, MN. ASA, CSSA, and SSSA, Madison, WI.

Han, S., S.M. Schneider, S.L. Rawlins, and R.G. Evans. 1997. A bitmap method for determining effective combine cut width in yield mapping. Trans. ASAE 40(2):485–490.

Hoffman, R. 1996. A perception architecture for autonomous harvesting. ASAE Paper no. 963068. ASAE, St. Joseph, MI.

Pierce, F.J., N.W. Anderson, T.S. Colvin, J.K. Schueller, D.S. Humburg, and N.B. McLaughlin. 1997. Yield mapping. p. 211–243. *In* The State of Site-Specific Management for Agriculture. ASA, CSSA, and SSSA, Madison, WI.

Reitz, P., and H.D. Kutzbach. 1996. Investigations on a particular yield mapping system for combine harvesters. Comp. Elect. Agric.14:137–150.

Stafford, J.V., B. Ambler, and H.C. Bolam. 1997. Cut width sensors to improve the accuracy of yield mapping systems. p. 519–527. *In:* Proc. of Precision Agriculture '97. Vol. 2. SCI, London, UK.

Vansichen, R., and J. De Baerdemaker. 1991. Continuous wheat yield measurement on a combine. p. 346–355. *In* Proc. ASAE Symp. on Automated Agriculture for the 21st Century, Chicago, IL. ASAE, St. Joseph, MI.

Wang, W., K.A. Sudduth, S.J. Birrell, S.T. Drummond, and M.J. Krumpelman. 1997. Combine cutting width measurement for yield monitoring. ASAE Paper no. MC97-104. ASAE, St. Joseph, MO.

Evaluation of Weighing and Flow-Based Cotton Yield Mapping Techniques

Stephen W. Searcy
Agricultural Engineering Department
Texas A&M University
College Station, Texas

ABSTRACT

In 1997, commercial yield mapping was available for cotton pickers on a limited basis. The Zycom AgriPlan 600 system was installed and evaluated in Texas and Arizona. An experimental system based on weighing the harvested cotton mass was also tested in 1997 on a cotton stripper operated in the southern High Plains of Texas. Both systems were evaluated for their accuracy in predicting yield at individual points within a field. Two yield samples were manually harvested at each evaluation point. The sample variance was used to determine the confidence interval of the manual yield estimates, as a basis for judging the accuracy of the yield mapping systems. The relatively large sample variance resulted in a 95% confidence interval of approximately 0.8 bales of lint per hectare (1/3 bale ac^{-1}). The weighing system confidence interval was approximately equal to that of the manual samples, while the Zycom confidence interval was approximately 50% greater.

INTRODUCTION

Yield mapping will play an important role in the precision management of cotton, as in all other crops. While research efforts have been underway for several years, 1997 was the first year that commercial cotton yield mapping systems (Zycom Corp. and Micro-Trak Systems) were offered for use on cotton pickers. In Texas, the majority of the cotton produced is harvested with cotton strippers. Texas A&M has had efforts underway to develop a cotton yield mapping system based on weighing the mass of cotton harvested (Searcy et al., 1997). This system has been developed for and tested on both cotton pickers and strippers. Both the Zycom AgriPlan 600 system and the experimental design were evaluated in 1997. The objective was to determine the accuracy of each system in estimating yield at specific points within a field. The point accuracy is of greater importance than the accuracy of predicting an accumulated mass, and represents a more difficult standard of performance than most yield mapping systems are held to today.

MATERIALS AND METHODS

Zycom Corporation agreed to cooperate with Texas Agricultural Experiment Station (TAES) personnel to evaluate the AgriPlan 600 system. Zycom made two system available. The first unit was installed on a Deere 9965 cotton picker at the

Copyright © 1999 ASA-CSSA-SSSA, 677 South Segoe Road, Madison, WI 53711, USA. *Proceedings of the Fourth International Conference on Precision Agriculture.*

King Ranch, Kingsville, Texas with the assistance of Zycom factory personnel. This picker was used for recording yield data during harvest in Texas during August and again in Arizona in November. A second unit was made available for installation by TAES personnel on a cooperating producer's Deere 9965 cotton picker near Temple, TX.

The Zycom AgriPlan 600 cotton yield mapping system consists of an integral GPS receiver, operator display and data entry boxes, and two optical sensors mounted on the air ducts of the picker. The differential correction signal was supplied by mounting an OmniStar 7000 DGPS receiver on the cab, and using the available RTCM 104 output. The optical sensors use a through beam arrangement, so that cotton that passed through the duct would break the light beams. The cotton yield was estimated from the changes in light beam signals. Since the Deere 9965 pickers pass all cotton harvested by a row unit through a single duct, each sensor measured the entire amount of cotton harvested on that row. For the King Ranch machine, two ducts were measured. This system was operated in the Kingsville, TX, area and in the Colorado River Valley of Arizona. The system at Temple was installed with only one row sensor, due to a shortage of sensors at the time. A Zycom representative indicated that this would result in a decrease in accuracy of approximately 1.5%, compared to a two row system. This system was operated in the central blacklands area, near Temple, TX. Both of the systems were set up to record yield data and position on a 2-s interval.

The system installed at the King Ranch was not calibrated by weighing a basket, but the Zycom representative indicated the factory calibration was reasonable. In Temple, the system was calibrated by weighing a basket load. Subsequent loads had an error of approximately 3% of total mass.

The TAES yield mapping system was installed on a cotton stripper owned by the USDA-ARS Cotton Mechanization Laboratory in Lubbock, TX. Load cells were installed in modified basket supports so the basket and the cotton within could be weighed continuously during harvest. Yield per area was determined from the change in mass over time and the travel distance of the stripper. A calibration procedure for the system was not performed, as each load cell was calibrated prior to installation.

The TAES system recorded the total weight of the basket and cotton harvested along with the position information every second. The yield estimates were calculated by determining weight differences over an area harvested. The size of that area was specified by the user when processing the recorded data. For all of the data reported here, a 25 m travel distance was used. This represents an area of 0.001 ha (0.024 acre). A yield estimate was determined for each GPS position by subtracting the total basket and cotton weight 12.5 m before the point from the weight 12.5 m after the point. As a result, yield estimates were created for every 1 s of travel, but each represented the yield >25 m of travel. Yield estimates were calculated for every GPS point so the points closest to the center of the samples could be used for comparison.

The evaluation of both systems was done by manually sampling within the cotton fields, and comparing the yield per area with that indicated by the mapping system. In Kingsville, the Zycom system was used to harvest an experimental field. Within that field, 12 manual samples of 0.0008 ha (0.002 acre) were taken. In

Temple, Lamesa, and Arizona, the manual samples taken consisted of two areas of 0.0004 ha (0.001 acre) adjacent to the path of the picker. Sample points were selected by walking along a row and flagging areas that visually appeared to be uniform over a width of three machine passes. The intention was to avoid situations where there were clearly visible differences between three adjacent passes of the harvester. The seed cotton was either picked (lint and seeds pulled from the burrs) or stripped (entire boll pulled from the plant), depending on the harvest system being used in the field. The hand harvested seed cotton was weighed as two individual samples (A and B). The two samples were handled separately so that a correlation between those samples could be examined and compared to the other yield estimates. For all fields, the two samples were combined for a yield estimate from a 0.0008 ha (0.002 acre) area. This yield estimate from the larger area was used for comparison to the mapped yield estimates. For the TAES system testing in the Southern High Plains, a weighing boll buggy was available. Most of the basket dumps were weighed and compared with the accumulated weight indicated by the experimental system.

The mapped and sampled yield estimates were compared using correlation analysis. Both values were estimates of the cotton yield in nearby, but not identical areas. Both estimates were subject to unknown variances. The availability of two manual samples at each point allowed the calculation of a variance for the manual estimates. Comparisons were made between the two manual samples and the manual sample A and the combined yield estimate. If sufficient sampling were done to describe the population of yield estimates for a given area, a regression between the samples and the actual yield would give a slope coefficient of 1.0. Using that slope coefficient, the 95% confidence interval on the predicted yield was calculated (Steel & Torrie, 1960). These values gave a reference against which the variances between the mapped and sampled yields could be evaluated.

An accuracy criteria for each data point of ±0.6 bales of lint per hectare (1/4 bale/ac) was arbitrarily selected. This accuracy level was considered to be reasonable for management purposes. Since the yield mapping systems were measuring seed cotton, that accuracy criteria was converted to mass of seed cotton per area by dividing the yield data by a typical gin out ratio (0.33 for pickers and 0.25 for strippers).

While the Zycom and TAES systems were analyzed using the same procedures, the results are not directly comparable. The two systems were not operated in the same fields, and the yield estimates do not represent similar areas.

RESULTS

Manual Yield Estimates

The manual yield data clearly illustrates the problems associated with sampling to estimate yield. Although the samples were taken in areas selected to be relatively uniform and separated by no more than four rows, there were considerable differences between the yield estimates. At some sites, the differences were 50% or greater. Figure 1 shows the scatter of the paired sample yield estimates. Since these samples were to be used as yield estimates, excessive deviation between the samples caused

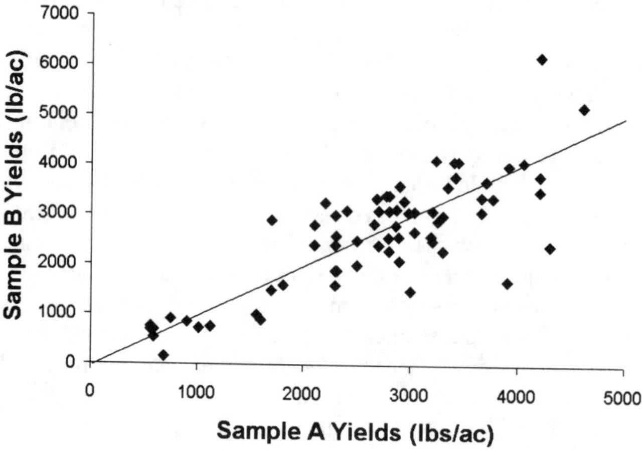

Fig. 1. Paired estimates of seed cotton yield from manual samples of 0.0004 ha (0.001 acre). The line would indicate a perfect correlation between the samples.

both of the estimates to be in question. For this reason, any sample pairs that differed by >25% from their mean were eliminated. Table 1 shows the correlation coefficients between the remaining sample pairs for four fields where split sampling was done. The correlation for all sites is better than the individual fields because of the extended range of yield. Any single field had a smaller range of yield estimates.

For comparison with the yield map data, both samples were combined to give a sample yield estimate for a 0.0008 ha (0.002 acre) area. For the operating conditions in the test fields, each mapped yield data point would represent between 0.0005 and 0.001 ha (0.0013–0.0025 acres), and the combined yield was the best estimate for comparison. With the combined sample, three yield estimates were available at each point; however, only the two small area samples were independent. In order to have a reference for the range of uncertainty relative to the actual yield, 95% confidence

Table 1. Correlation Between Paired Yield Samples

Field name	Correlation coefficient	Sample points no.
Temple	0.68	9
KR AZ 1	0.54	14
KR AZ 2	0.84	15
Lamesa	0.72	26
All sites	0.88	64

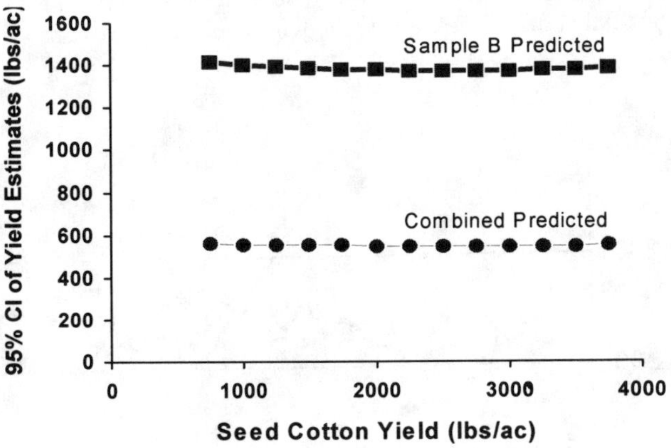

Fig. 2. Confidence intervals for prediction of yield based on the known yield for Sample A.

intervals were calculated for the prediction of the Sample B and the combined area yields from Sample A. Figure 2 illustrates the magnitude of uncertainty at each sample area. The predicted yield could be the measured yield plus or minus that value. It is not surprising that the uncertainty of predicting the combined yield is less than for the second sample, because the combined estimate is not completely independent of sample A. The larger sample area, however, also would be expected to have a lower variance about the actual yield. The confidence interval of ±616 kg ha^{-1} (550 lb ac^{-1}) of seed cotton represents approximately 180 kg ha^{-1} of lint (~ ⅓ bale ac^{-1}). The sample yield estimate from the larger area was used to compare the accuracy of the yield mapping systems.

It should be noted that the uncertainty of the manually sampled yield estimates is greater than the desired level of accuracy (± ¼ bale ac^{-1}). This means that no conclusions can be made regarding the ability of either the Zycom or the TAES system to meet the accuracy goal. This points out the limitations of verifying yield map accuracy. It is desirable to validate the accuracy of a yield mapping system over small areas, but the verifying yield estimates require destructive sampling, making a direct comparison difficult. Sampling areas of >0.0008 ha (0.002 ac) will be necessary to evaluate the performance of the yield mapping systems at the desired level of accuracy.

Zycom Evaluation

Data analyzed from the Zycom system consisted of the individual data points recorded for three adjacent passes of the cotton picker in the areas where the manual samples were taken. Figure 3 is an example of raw yield data from the Zycom system for a field yielding 340 to 450 kg ha^{-1} (300–400 lb ac^{-1}) of lint. At each data point, the yield was integrated over the distance traveled in 2 s. Figure 3 represents a difficult situation for the optical system, as the yield was low and the flow of cotton less continuous. The large changes in the yield estimates from point to point seemed unrealistic for that field. The curve is a moving average of five consecutive points (10

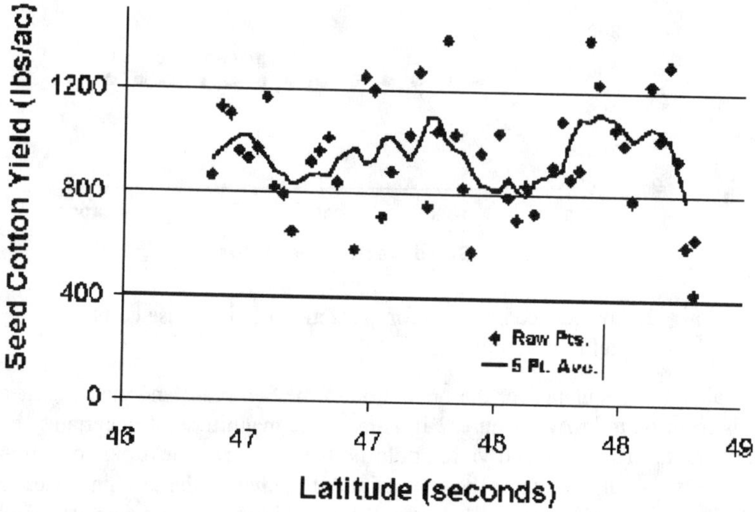

Fig. 3. Single point yield estimates and a five point moving average for a selected harvest pass in a field near Temple, Texas.

s period). At typical operating speeds, the picker would harvest 0.005 ha (0.013 ac) in 10 s. This smoothing removes some of the variability and appears to give a more realistic estimate of yield.

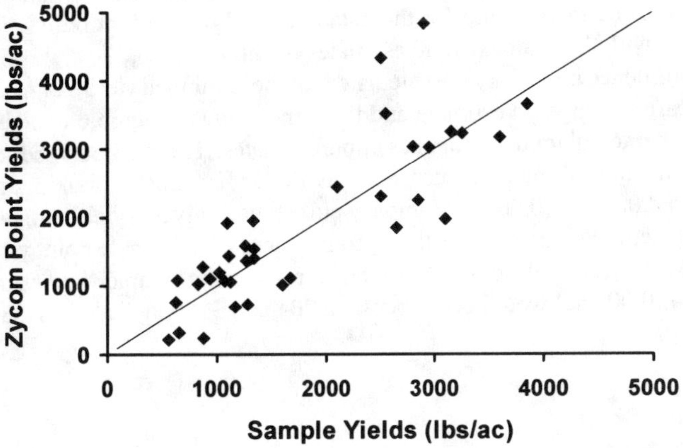

Fig. 4. Comparison of the sample yield estimates and the nearest recorded yield data point from the Zycom system.

For point comparisons of the mapped yield, four different values were generated. These included the single mapped value closest to the manually sampled area and the average of mapped yield estimates for three different radii (5, 7.5, and 10 m). Table 2 gives the correlation coefficients (*r*) between the mapped and sampled yield estimates for each field. The *r* values for the individual fields are relatively low, in part, due to the narrow magnitude range of the manual yield estimates. The comparisons are worse for the King Ranch and Temple fields where yields were low, 340-560 kg/ha (300–500 lb ac^{-1}) of lint. The correlation coefficient for all sites is much higher because of the greatly expanded range of yields. The 5 m radius average provided the best agreement with the manual samples.

Figure 4 shows the relationship of the nearest mapped yield data point and the sampled yield estimates. Most of the samples were taken in the lower yielding range. Although the comparisons in the lower yield range seems to better fit the perfect correlation line than those in the upper range, the deviations are large relative to the

Table 2. Correlation between sample yield estimates and Zycom data.

Field	Sample point	5 m radius	7.5 m radius	10 m radius
King Ranch	0.08	0.36	0.24	0.09
Temple	0.24	0.47	0.06	-0.02
KR AZ 1	0.46	0.48	0.42	0.42
All sites	0.84	0.90	0.90	0.90

yield. This is the reason for the low correlation coefficients in Table 2. Figure 5 shows the same relationship for the data averaged over a 5 m radius. Clearly the agreement with the sampled yield estimates is improved.

Confidence intervals were calculated for the combined yield estimate compared to the nearest map yield estimate and the 5 m average. Figure 6 contains the 95% confidence interval for the different sample estimates. The curves for the nearest point and 5 m average should be interpreted as the 95% probable maximum deviation between a 0.0008 ha (0.002 ac) sample yield estimate, given a yield estimate from the Zycom system. For example, if the Zycom system gave a single point yield estimate of 2242 kg ha^{-1} (2000 lb ac^{-1}), 95% of the corresponding sample yield estimates from 0.0008 ha (0.002 ac) would occur between 704 and 3780 kg ha^{-1} (628–3372 lb ac^{-1}).

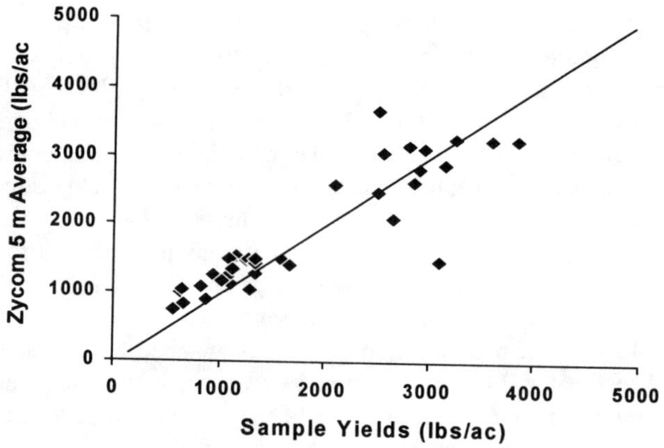

Fig. 5. Comparison of the sample yield estimates and the 5 m radius average of recorded yield data points from the Zycom system.

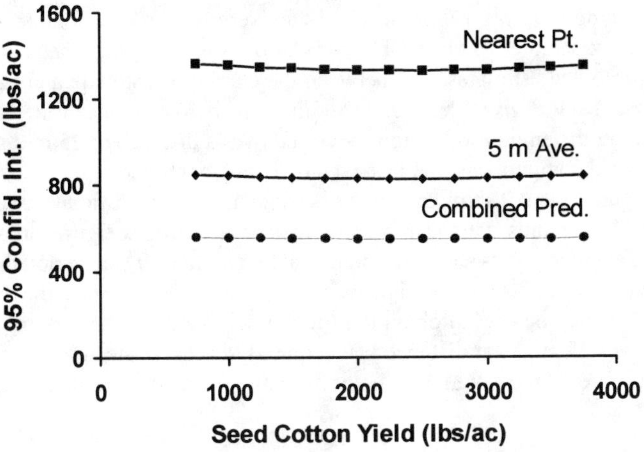

Fig. 6. Confidence intervals for the yield estimates from the single data point nearest the sample estimate and for the average of all points in a 5 m radius around the sample point. The confidence interval for the predicted sampled yield is shown for comparison.

The confidence interval for the combined prediction is interpreted as the range of deviations that might occur in a 0.0008 ha (0.002 ac) sample, given a yield estimate for a 0.0004 ha (0.001 ac) sample. This represents an estimate of the variability occurring in the sample yield estimates. Although not directly comparable, it does give a measure against which to evaluate the performance of the yield mapping system. The yield estimates generated by the Zycom system do have a greater variance than would be expected from the variance in the sample estimates. The confidence interval for the single point yield estimates is approximately 150% larger than that of the sample yield estimates. Averaging several data points within a region greatly reduces the variance. The use of the 5 m radius averaging reduce the confidence interval to 50% of the sample estimates. Data smoothing would be a recommended practice for the Zycom system as configured in 1997.

TAES System Evaluation

The yield mapping cotton stripper was used to harvest several experimental fields in the Southern High Plains region. During most of the harvesting, the basket was dumped into a weighing boll buggy. Although the evaluation of accumulated mass estimates was not the focus of this study, the availability of that data provides an additional means of evaluation. The coefficient of determination for the regression between the boll buggy and mapping system weights was 0.99 (Fig. 7).

At Lamesa, Texas, 17 sample yield sites were available for comparison (mapped data for some samples was lost). The correlations were calculated between the sample A and combined estimates, and between the combined estimate and the yield map estimates. Table 3 gives those values. The map yield estimates were calculated by determining the amount of cotton harvested over a distance of 25 m down the row. For the yield comparisons, the nearest single data point and regional averages were used. Figures 8 and 9 show the scatter between the sample estimates and the nearest point and 10 m radius average, respectively. The data in these figures are similar, with the 10 m average points showing somewhat less scatter. It can be noted that most of the map yield estimates for the higher yielding areas underestimated the sample yields. This effect may be due to obtaining the mass value from the difference in cotton weight over 25 m. That distance was required to achieve sufficient accuracy in the weight difference measurements, but if the entire 25 m distance did not have the same

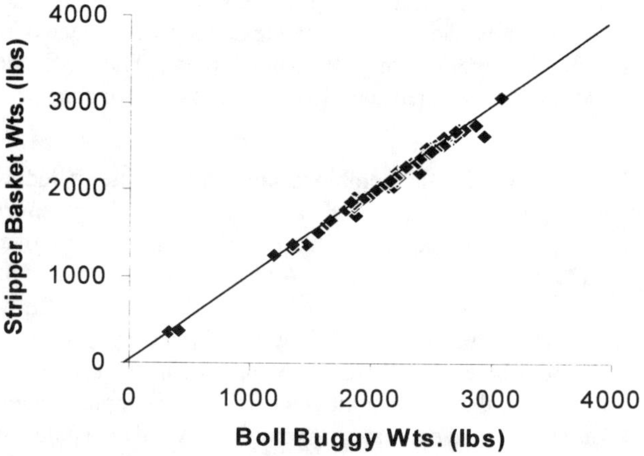

Fig. 7. Comparisons of boll buggy weights with the harvested cotton weights indicated by the yield mapping system.

Table 3. Correlation coefficients for seed cotton yield estimates.

	Combined yield estimates
Sample A	0.92
25 m - nearest point	0.75
25 m - all points in 5 m radius	0.80
25 m - all points in 10 m radius	0.83

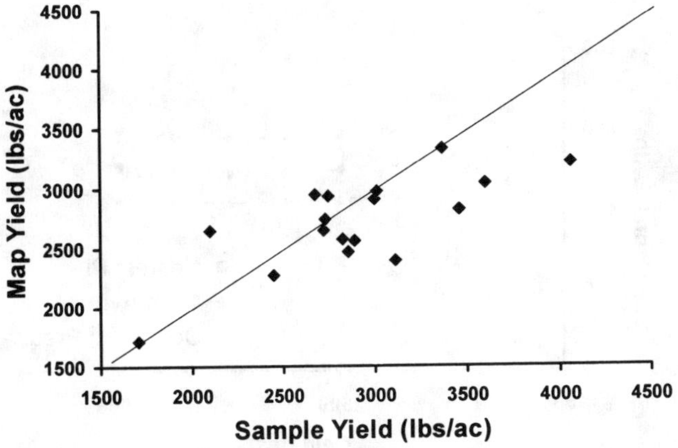

Fig. 8. Comparison of the sample yield estimates and the nearest recorded yield data points from the TAES system.

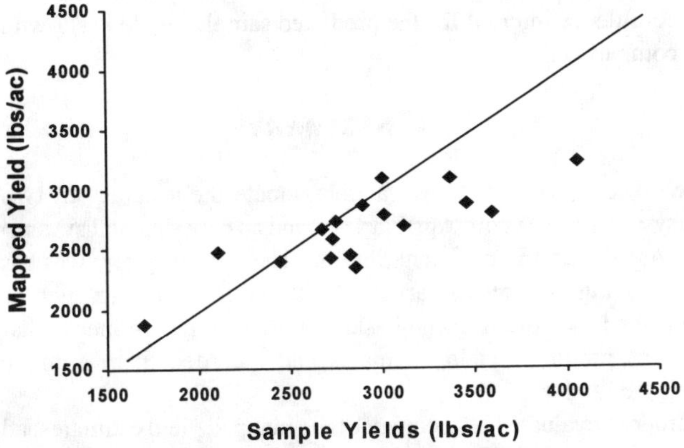

Fig. 9. Comparison of the sample yield estimates and the 10 m radius average of recorded yield data points from the TAES system.

high yield, then underestimation may have resulted.

Figure 10 contains the confidence intervals for the prediction of combined sample yield values, nearest point and 10 m average estimates. The points nearest to the samples had about 20% wider confidence interval than the combined samples, while the 10 m average had approximately 15% smaller interval. The majority of the variability in the yield estimates from the TAES yield mapping system could be attributed to variability in the sample yield estimates used for comparison. Based on this limited data set, it appears that the TAES system could be as reliable for obtaining yield estimates as manually sampling, especially if smoothing is performed on the raw data.

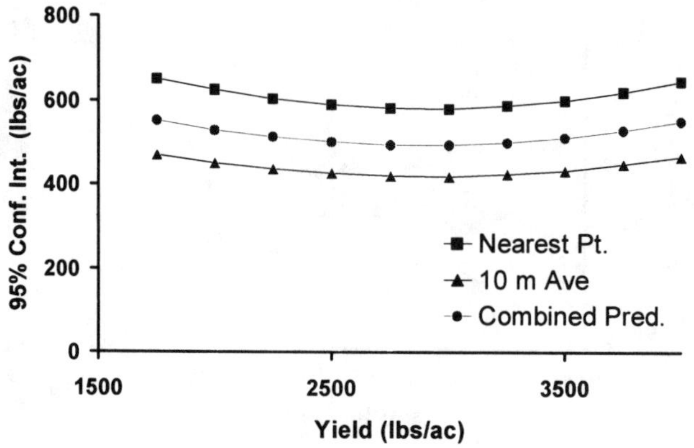

Fig. 10. Confidence intervals for the yield estimates from the single data point nearest the sample estimate and for the average of all points in a 10 m radius around the sample point. The confidence interval for the predicted sampled yield is shown for comparison.

SUMMARY

The objective of the study was to evaluate the accuracy of two cotton yield mapping systems, the Zycom AgriPlan 600 and an experimental system developed by the Texas Agricultural Experiment Station. The Zycom system was based on optical sensors measuring the flow of harvested cotton through the air ducts, and the TAES system weighed the cotton in the basket in order to determine the harvested mass. Both systems produced yield estimates and recorded them along with machine position in the field.

In order to evaluate the accuracy of the mapped yield estimates, additional yield information was needed. This was obtained by selecting relatively uniform locations in the fields and hand harvesting two yield samples (A and B). The two samples were combined into a single value that was used for comparison with the mapped data. The two samples were harvested and weighed separately so that a measure of variability could be determined for the sampled yield estimates. The confidence interval for the prediction of the combined sample yield from the sample A yield indicated that the yield at a point could be ± 0.8 bales of lint ha^{-1} (± ⅓ bale ac^{-1}). This level of uncertainty was used in comparison to the variability of the mapped yield estimates. Unfortunately, this level was higher than the ± 0.6 bale ha^{-1} (± ¼ bale ac^{-1}) target accuracy, resulting in the inability to evaluate the yield mapping systems at the desired level.

The Zycom system yield estimates were highly variable, particularly for lower yielding fields. The use of averaging across a 5 m radius improved the agreement of the data with the sample estimates. With the 5 m averaging, the yield estimates were approximately 50% more variable than the sample estimates. The single nearest points were 150% more variable than the sample estimates. Data smoothing was necessary for the Zycom generated data to be useful; however, this evaluation was for the system that was installed at the beginning of the 1997 harvest season. None of the improvements made by the manufacturer during the season were incorporated here.

The TAES system yield estimates were slightly more variable than the sample yield estimates. The points nearest to the sample areas had a 95% confidence interval that was approximately 20% wider than the sample estimates. When averaged over a 10 m radius, the variability was 15% less than that of the estimates.

The inability to evaluate the yield mapping systems at the desired accuracy demonstrates the difficulty of verifying yield map accuracy. Paired samples of 0.0004 ha (0.002 ac) or greater will be necessary to improve the reliability of the sample yield estimates. This requires considerable labor and time, and would be justifiable only for research purposes.

ACKNOWLEDGMENT

The author would like to acknowledge the support and cooperation of the Texas Agricultural Experiment Station, USDA–ARS, the King Ranch, Coufal Farms, and Zycom.

REFERENCES

Searcy, S.W., D.S. Motz, and A. Inayatullah. 1997. Evaluation of a cotton yield mapping system. ASAE Paper no. 971058. ASAE, St. Joseph, MI.

Steel, R.G.D., and J.H. Torrie. 1960. Principles and procedures of statistics. McGraw-Hill Book Co., New York.

Measurement of Grain Yields in Japanese Paddy Field

Michihisa Iida
Toshikazu Kaho
Choung Keun Lee
Mikio Umeda
Masahiko Suguri

Laboratory of Farm Machinery
Division of Environmental Science and Technology
Graduate School of Agriculture
Kyoto University
Sakyo-ku, Kyoto, Japan

ABSTRACT

This paper describes two methods of yield mapping in Japanese paddy fields. One is the yield mapping of grain and straw by hand at the Takatsuki experimental farm of Kyoto University. The variety of rice was *MINAMI-HIKARI.* The field was divided into three kinds of the rectangular cells. Yield maps showed large variability in relation to position. The variability of grain and straw yield were 5.4 to 9.7t ha^{-1} and 6.2 to 18.0t ha^{-1}, respectively. The moisture content of grain and straw varied 9.5 to 23.1% wb and 37.8 to 73.0% wb, respectively. The second method was conducted as a trial to measure the yield of grain and straw using a head-feeding combine. In this case, measurement of grain yield through straw yield also was investigated. We estimated the relation of yield between grain and straw and produced the grain yield map by this method.

INTRODUCTION

Precision agriculture gains worldwide attention. It also is the focus of attention in Japanese agriculture. As rice in Japan is traditionally farmed intensively, the grain yield in paddy fields is thought as unvarying. It is not clear how grain yield varies in relation to position. This knowledge is vital in determining how efficient the application of precision agriculture would be for rice farming in Japan.

As grain yield is the primary measure for how well a product or management practice works on a farm, we produced a grain yield map to determine the variability of grain yield in a paddy field as first step for Japanese precision agriculture. A field test was conducted at the Takatsuki experimental farm of Kyoto University on 27–31 Oct., 1997. The area of the test field was 0.5 ha. Since it was necessary to generate a yield map that precisely indicated variability, the data were gathered by hand. As a result, yield maps showed a large variation with position.

As investigating yield variability by hand, however, is time-consuming, developing an automatic mapping system for the combines is advisable. Several grain flow sensors have been developed for conventional combines in the USA and Europe and are commercially available. A grain flow sensor has been developed using a permanent magnet motor to drive the clean grain system (Peterson et al., 1989). Wagner and Schrock (1987) developed a pivoted auger grain flow sensor using a load cell. Pang and Zoerb (1990) have developed a grain

Copyright © 1999 ASA-CSSA-SSSA, 677 South Segoe Road, Madison, WI 53711, USA. *Proceedings of the Fourth International Conference on Precision Agriculture.*

flow sensor made from piezo film strips for use under combine sieves. Grain flow sensors commonly used today measure the impact of grain on a sensor mounted atop a grain elevator, and then convert it to a mass of grain using a calibration equation (e.g., Pierce, 1997). Although there has been much research and development of grain flow sensor for conventional combines, there is no grain flow sensor commercially available for attachment to head-feeding combines. As a lot of head-feeding combines are used for rice harvesting in Japan, we gathered relevant data of grain yield using a head-feeding combine. In head-feeding combines, a feed chain holds and conveys the rice straw during threshing. The straw yield input into the thresher can be recorded by detecting the velocity of the feed chain and the thickness of the straw layer. Assuming that grain yield is proportional to the straw yield, this method would allow us to measure grain yields accurately. This end forms the focus of this paper and the resulting grain yield map is presented here.

TEST FIELD

Figure 1 shows the test field used for this experiment. The test field was 100 m long in an east–west direction and 50 m wide in a north–south direction. As the interval and direction of line and row of rice in the shaded part of the test field were not constant, it was reaped beforehand and excluded from the investigation of yield. The squares A-R were the yield mapped by hand. Areas S-U were the yield mapped by combine using the straw yield into the grain yield relationship. The area V could not be investigated due to time constraints. Division into the rectangular cells was performed using a surveying instrument with an optical range finder (SOKKIA total station SET-4100). Each cell A-R was 12 m in width and 10 m in length. Cell A was, further, divided into one hundred cells which were 1.2 m in width and 1 m in length. Each cell B-D and J-M was divided into four smaller cells which were 6 m by 5 m. The areas S-U were 0.6 m in width and 90 m in length. The minimum width of S-U cut was determined to be 0.6 m by multiplying 0.3 m by 2 rows, because the interval distance between each row of rice was 0.3 m and the combine used for this research is able to harvest 2 rows simultaneously.

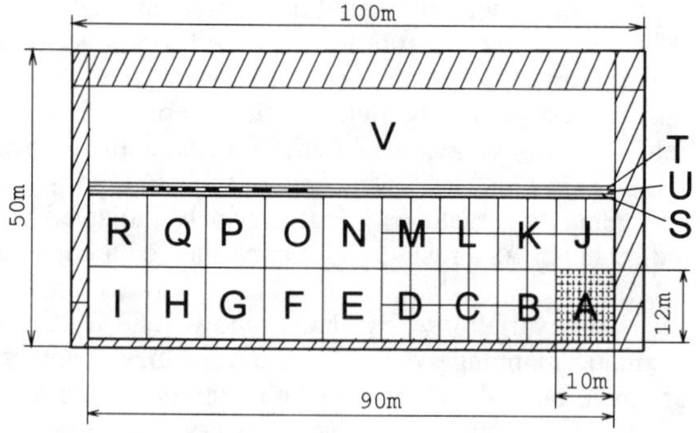

Fig. 1. Test field divided into rectangular cells. The shaded areas were excluded in the yield mapping.

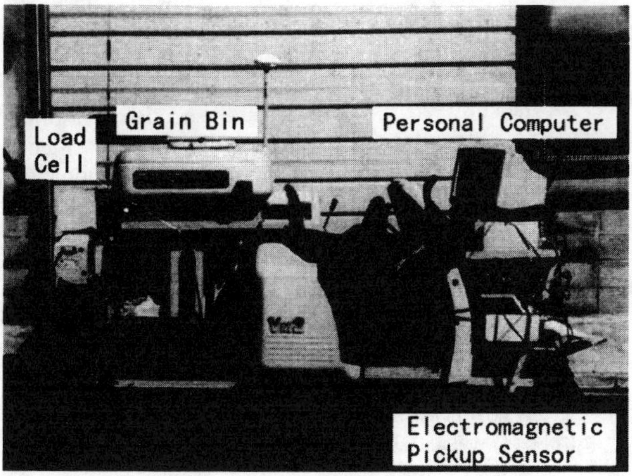

Fig. 2. A head-feeding combine with installed sensors for measuring the yield during harvesting. This figure shows the right side view of the combine.

EXPERIMENTAL EQUIPMENT

Figure 2 shows head-feeding combine used for the field test. This combine can harvest two rows of rice simultaneously. This combine had a load cell installed under the grain bin and determined the total mass of grain by measuring the force applied to the grain bin. Two electromagnetic pickups were used to record the travel of the combine. They were installed on both of left and right crawlers in order to detect the rotation of the sprocket. A personal computer was used for data

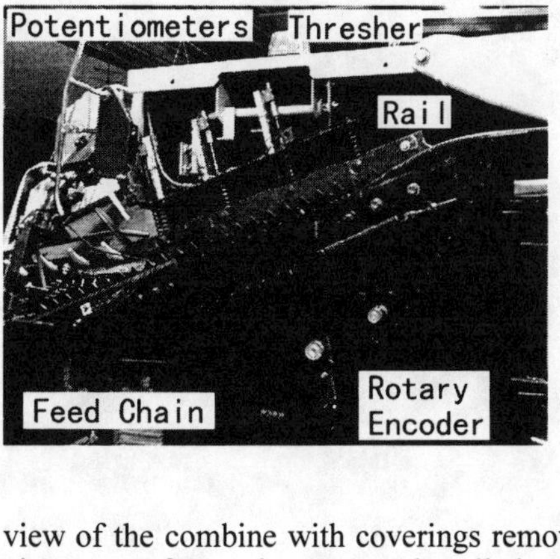

Fig. 3. Left side view of the combine with coverings removed. Sensors for measuring the section areas of straw layers were installed on the combine.

acquisition. The computer was installed on a table in front of operator's seat. It had a 12-bit A/D converter board and a 24-bit counter board. The counter board was used for counting pulses from the electromagnetic pickups. Data acquisition was done at sampling rate of 10Hz.

Figure 3 shows the sensors for measuring straw yield. As it was difficult to measure straw yield directly during harvesting, we substituted the section area of the straw layer for the straw yield. We recorded the section area of the straw layer by installing three potentiometers on a rail of the feed chain which measured the straw layer thickness. Only thickness h of the middle potentiometer was used in this experiment. The rotary encoder detected the rotation N of the sprocket driving the feed chain and converted N into the velocity v of the feed chain to convey straw. The relation between v and N was simply expressed as follows, where k_1 was a constant coefficient.

$$v = k_1 N \qquad [1]$$

The section area A_s of the straw layer held between the feed chain and the rail is shown in Figure 4. A_s was calculated as follows.

$$A_s = \int vhdt \qquad [2]$$

The grain yield W_d for each unit cell was represented by the mass of dry matter inside the area. Here, assuming that the grain yield W_d is proportional to the section area A_s of the straw layer, W_d was calculated as

$$W_d = k_2 A_s \qquad [3]$$

with k_2 as a constant coefficient.

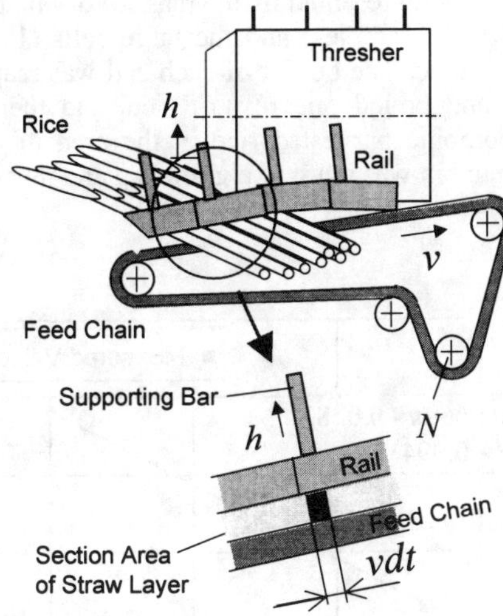

Fig. 4. Method for measuring the section area of the straw layer. Rice was held with a feed chain and a rail. *h* was measured with a potentiometer and *N* with a rotary encoder.

YIELD MAPS

We investigated the yield of grain and straw by hand so that we could obtain their precise variability. A high-yielding rice variety called *MINAMI-HIKARI* was used. The heading date of rice was about August 31. Yield measurement was done in the 0.22 ha area A-R in Fig. 1. As mentioned before, the dimensions of A-R were 12 m by 10 m (large cells), and the cells A-D and M-J were divided into 6 m by 5 m cells (medium cells). Cell A was further divided into small cells measuring 1.2 m by 1 m. Here, as the interval between each row of rice was 0.3m, all swath cuts were multiples of 0.3 m. Division of the field was accomplished by a surveying instrument with an optical range finder. As rice existed on the border of each cell, however, the width and the length of the cells were reconfirmed by tape measure.

In the small cells (100 samples), all rice in Cell A was reaped by hand, and then threshed with the combine. The grain accumulated in the grain bin and the straw discharged from the combine were taken out thirty seconds after threshing. Thirty seconds was the time required to completely discharge grain into the bin. This interval was determined by the following experiment. Different masses of grain (input) were thrown into the thresher. After different time intervals, accumulated grain in the bin (output) was taken out and measured. The relation between the input and the output is illustrated as Fig. 5. This figure shows that the relation between input and output is linear. Therefore, a time interval of 30 s was determined for full grain discharge. The masses of grain and straw taken out from the combine were measured with an electric balance. The moisture content of grain and straw was measured by the oven drying method. The yield for each unit cell of grain and straw was represented by the mass of dry matter inside the area.

In the medium cells (28 samples) and the large cells (10 samples), rice was harvested using the combine. The border of each cell was reaped by hand or with a binder that reaped and bound one row of rice, and then threshed with the combine. After the combine harvested rice in the area of each cell, the grain accumulated in the grain bin was put into a grain bag and measured using a load

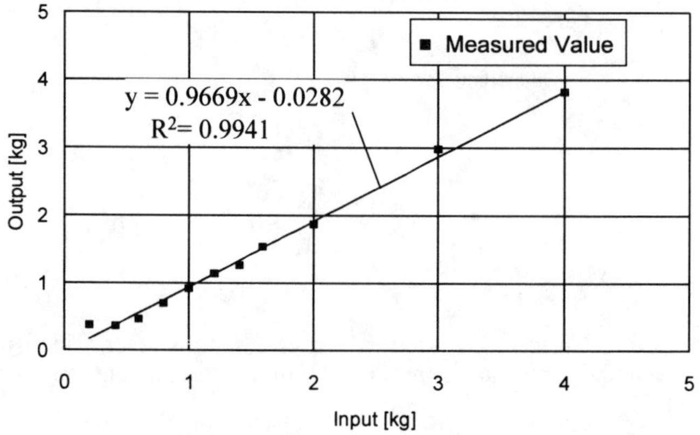

Fig. 5. Relation between the input and output of grain in the thresher. Input means the mass of grain thrown into the thresher. Output means the mass of grain accumulated into the grain bin. A line was calculated by the least square method.

cell. The quantity of straw was too huge to measure by hand in the medium cells and the large cells. The moisture content of grain was measured by the oven drying method. In the case of the medium cells, two samples were taken from each cell in order to measure the moisture content. Similarly, in the case of the large cells eight samples were taken from each cell.

The moisture content maps and the yield maps of grain and straw were generated from the gathered data. Figure 6 shows a moisture content map of grain in the small cells. Similarly, Fig. 7 and 8 shows those in the medium cells and the large cells, respectively. The darker cells indicate higher moisture content. The moisture content of grain in the small cells had the largest variability, from 9.5 to 23.1%wb. Figure 9 shows a map of the moisture content of straw in the small cells. That of straw in the small cells varied from 37.8 to 73.0%wb. The relation of the moisture content between grain and straw had no correlation.

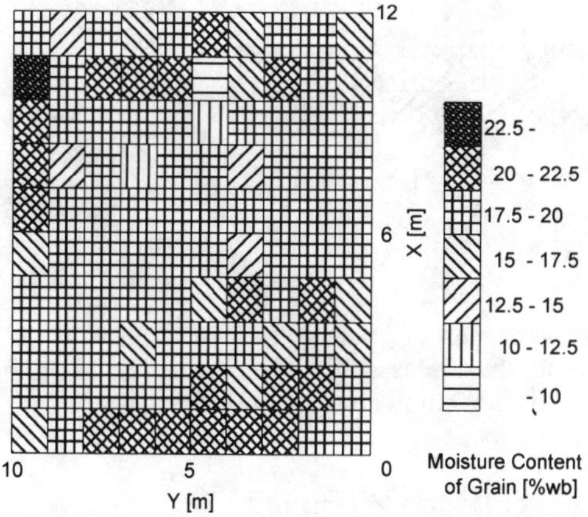

Fig. 6. Moisture content map of grain in the small cells. The cells shown in this figure corresponds to Area A. The *X* axis is 12 m in a north–south direction and the *Y* axis is 10 m in an east–west direction.

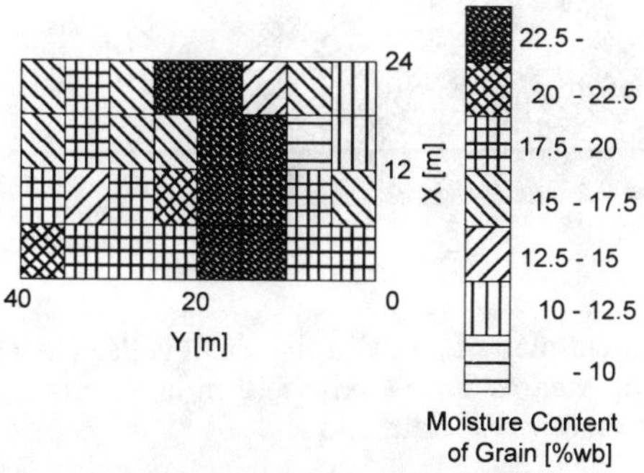

Fig. 7. Moisture content map of grain in the medium cells. The cells shown in this figure corresponds to the Areas A-D and J-M. The *X* axis is 24 m in a north–south direction and the *Y* axis is 40 m in an east–west direction.

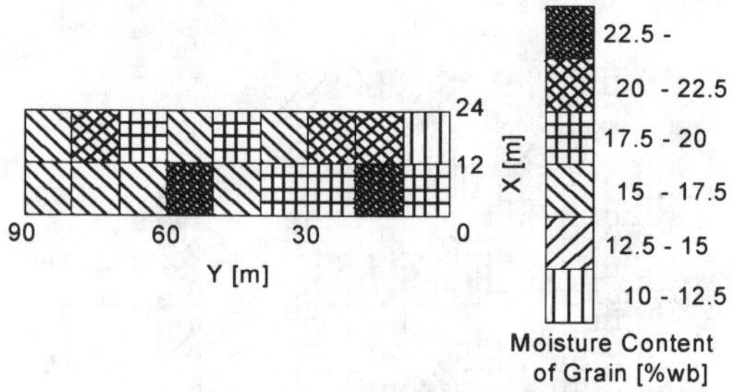

Fig. 8. Moisture content map of grain in the large cells. The cells shown in this figure corresponds to the Areas A-R. The *X* axis is 24 m in a north–south direction and the *Y* axis is 90 m in an east–west direction.

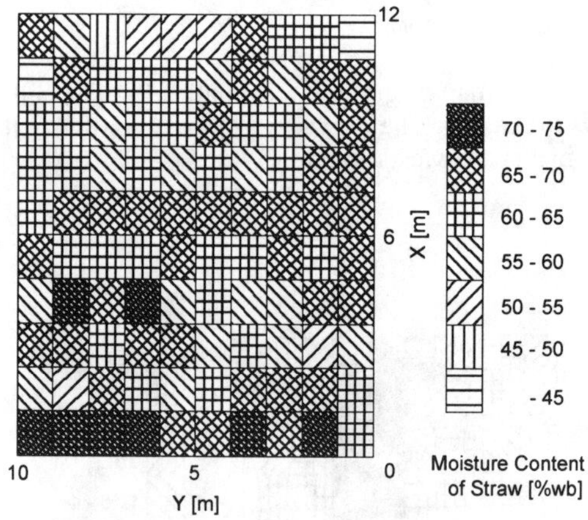

Fig. 9. Moisture content map of straw in the small cells. The cells shown in this figure corresponds to Area A. The *X* axis is 12 m in a north–south direction and the *Y* axis is 10 m in an east–west direction.

Figure 10 shows a yield map of grain in the small cells. Similarly, Fig. 11 and 12 shows those of grain in the medium and large cells, respectively. The darker cells indicate higher yields. The grain yield in the small cells had the largest variability, from 5.4 to 9.7 t ha^{-1}. Figure 13 shows a yield map of straw in the small cells. The straw yield here varied from 6.2 to 18.0 t ha^{-1}.

The yield maps show that Japanese paddy field have a large variability of grain and straw yield. The variability of grain yields in the small cells was the largest, although this is likely due to the inaccuracies of measuring the cell area.

We estimated the relation of yield between grain and straw. The relation between the mass of dry matter of grain and the mass of straw was highly correlative as shown in Fig. 14.

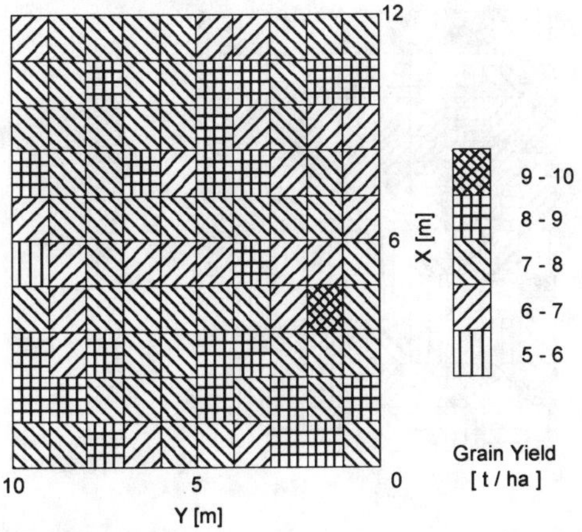

Fig. 10. Grain yield map in the small cells. The cells shown in this figure corresponds to Area A. The *X* axis is 12 m in north–south direction and the *Y* axis is 10 m in an east–west direction.

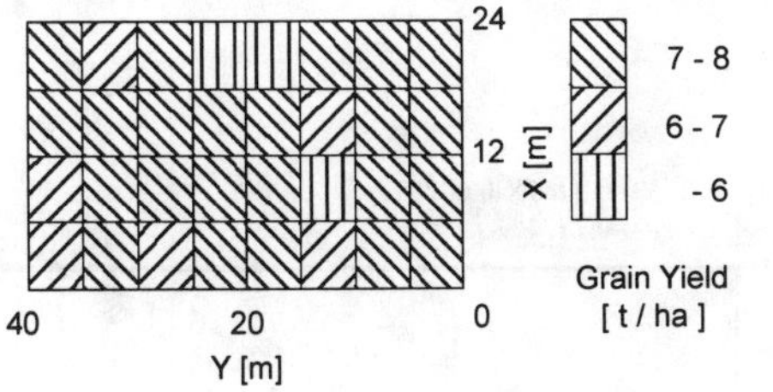

Fig. 11. Grain yield map in the medium cells. The cells shown in this figure corresponds to the Areas A-D and J-M. The *X* axis is 24 m in a north–south direction and the *Y* axis is 40 m in an east–west direction.

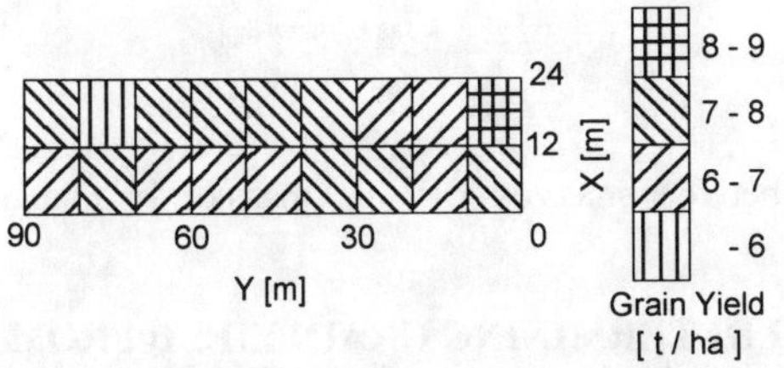

Fig. 12. Grain yield map in the large cells. The cells shown in this figure corresponds to the Areas A-R. The *X* axis is 24 m in a north–south direction and the *Y* axis is 90 m in an east–west direction.

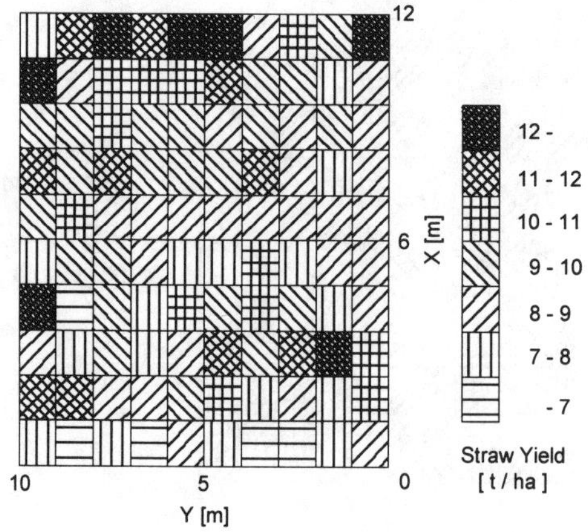

Fig. 13. Straw yield map in the small cells. The cells shown in this figure corresponds to Area A. The *X* axis is 12 m in a north–south direction and the *Y* axis is 10 m in an east–west direction.

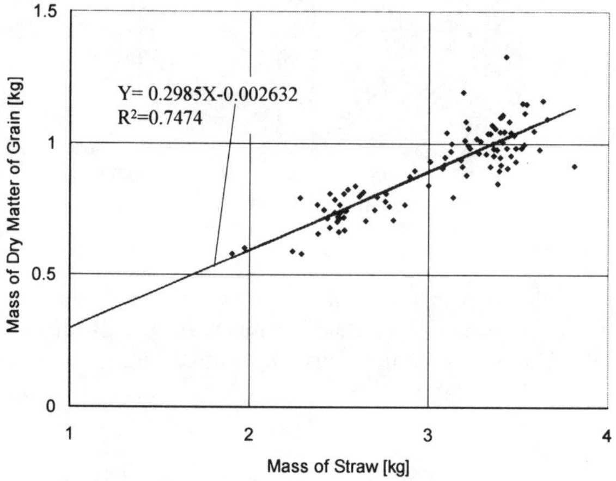

Fig. 14. Relation between mass of straw and mass of dry matter of grain.

METHOD FOR DETERMINING GRAIN YIELD FROM STRAW YIELD

The yield maps produced by hand disclosed that the grain yield in Japanese paddy fields could vary in relation to position; however, as it was time-consuming to investigate the variability of yield by hand, we developed the mapping system using combines as described in the introduction. Figure 14 shows that the relation between the mass of dry matter of grain and the mass of straw was highly correlated. However, as it was difficult to measure the mass of straw directly

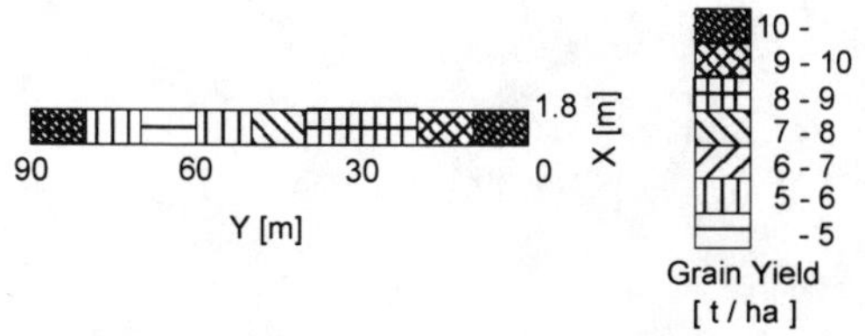

Fig. 15. Grain yield map generated by determining the grain yield from the section area of straw layers. The cells shown in this figure corresponds to the areas S-U. The X axis is 1.8 m in a north-south direction and the Y axis is 90 m in an east-west direction. After the combine harvested 2 rows 3 times, nine cells of 1.8 m by 10 m were divided.

during harvesting, we substituted the section area A_s of the straw layer for the mass of straw. A_s was obtained by Eq. [2]. A_s calculated with this method has some inaccuracies as the straw widths and lengths vary among individual plants. Therefore, the relation Eq. [3] between A_s and W_d was calibrated by experiment.

In order to produce the grain map by this method, three rows of S-U in Fig. 1 were harvested using a combine with measuring instruments. The grain yield map produced by this method is illustrated as Fig. 15.

As a result, the method for determining grain yield from section areas of the straw layer into was successfully accomplished.

CONCLUSIONS

In this paper the main results from our study stem from two method of yield mapping in Japanese paddy fields. One was the yield map of grain and straw investigated by hand. Yield maps with this method showed large variability in relation to position. The variability of grain and straw yield were 5.4-9.7t ha^{-1} and 6.2-18.0t ha^{-1}, respectively. The moisture content of grain and straw varied 9.5–23.1%wb and 37.8–73.0%wb, respectively. The second method was a study to measure the grain yield using a head-feeding combine. We related the straw yield measured during threshing to the grain yield and produced the grain yield map by this method. As the result, the method for determining grain yield from straw yield was successfully demonstrated.

ACKNOWLEDGMENTS

The authors would like to thank several people who made this research possible. At the experimental farm, Division of Paddy Field has been extremely helpful. Funding for this research was provided by the BRAIN.

REFERENCES

Pang, S.N., and G.C. Zoerb. 1990. A grain flow sensor for yield mapping. ASAE no. 90-1633. ASAE. St. Joseph, MI.

Pierce, F.J. 1997. Yield mapping. RESOURCE. February: 9–10.

Peterson, C.L., J.C. Whitcraft, K.N. Hawley, and E.A. Dowding. 1989. Yield mapping winter wheat for improved crop management. ASAE Paper no. 89-7034. ASAE. St. Joseph, MI.

Wagner, L.E., and M.D. Schrock. 1987.Grain flow measurement with a pivoted auger. Trans. ASAE. 30(6):1583–1586.

Yield Deconvolution – A Wetted Grain Pulse to Estimate The Grain Flow Transfer Function?

M. J. Pringle

Australian Centre for Precision Agriculture
University of Sydney
Sydney, Australia
and
CSIRO Division of Land and Water
Floreat Park Laboratories
Wembley, Australia

B. M. Whelan

Australian Centre for Precision Agriculture
University of Sydney
Sydney, Australia

M. L. Adams, and
S. E. Cook

CSIRO Division of Land and Water
Floreat Park Laboratories
Wembley, Australia

G. Riethmuller

Agriculture Western Australia
Merredin Dryland Research Institute
Merredin, Australia

ABSTRACT

The point-to-point accuracy of real-time grain yield mapping is decreased by a process of flow convolution as grain moves through a harvester. Yield data can be deconvolved with a transfer function. This paper outlines the possibility of using a strip of wetted grain to estimate the parameters of the transfer function. Some results are shown for a preliminary trial performed on a barley crop. The conclusion from the work is that the grain pulse was too wet to adequately represent the normal state of the harvester's threshing dynamics.

Copyright © 1999 ASA-CSSA-SSSA, 677 South Segoe Road, Madison, WI 53711, USA. *Proceedings of the Fourth International Conference on Precision Agriculture.*

INTRODUCTION

The point-to-point accuracy of real-time grain yield monitoring is decreased by a process of flow convolution as the grain is processed by a moving harvester. Grain flow convolution is described in detail by Lark et al. (1997).

To gain more accurate site-specific yield data, deconvolution of grain flow has been successfully performed in the frequency-domain with a Fast Fourier Transform of a non-parametric grain flow transfer function (Whelan & McBratney, 1997). Further, Whelan (1998) and Whelan and McBratney (1998) have provided a parametric model for the transfer function that includes a dispersion (or mixing) coefficient and a characteristic pathway length. To estimate the parameters of the transfer function is not easy. In the aforementioned studies, Whelan and McBratney had to physically mark a pulse of sorghum, the width of the harvester's swath and 1.5 m long, with paint and weigh amounts of colored grain in the output. This approach, though effective, is considered too labor-intensive and technically difficult for ready adoption by practitioners; therefore there is a need for the development of an easier method to estimate the transfer function.

One possibility that would be easy, fast and convenient to apply is to wet a pulse of grain to an above-average moisture content. By driving over the pulse with a harvester equipped with a mass flow sensor and nearby moisture sensor, it may be possible to infer the grain flow of the pulse from the observed change in moisture content. This paper describes and discusses results from an experiment aimed at investigating the potential of the wetted grain pulse for estimation of the grain flow transfer function.

MATERIALS AND METHODS

A field of barley, located at Wyalkatchem, Western Australia, was chosen for an experiment with a wetted pulse to estimate the grain flow transfer function. The dimensions of the pulse used for this experiment were 9 m wide × 1 m long. The width of the pulse corresponded to the width of the harvester's cutting bar.

About 15 min before it was due to be harvested, the 9 m^2 area of weed-free barley plants was sprayed with water applied from a hose. The quantity of water used was not measured but the plants were made wet enough so that free water was dripping off them.

The location of the centre of the pulse (t = 0) was recorded by a hand-held Global Positioning System with differential correction. The wetted area was driven over at an average harvest speed (~2.1 m s^{-1}) and the mass flow and moisture sensors' measurements logged to the data card at one second intervals by an AgLeader '2000' yield monitor (Ames, IA). The logged yield data is herein referred to as measured yield.

An empirically modelled grain flow transfer function can be described by a nonlinear convection–dispersion equation (Whelan & McBratney, 1998 – after Jury & Sposito, 1985):

$$C(L,t) = \frac{C_0 L}{2\sqrt{(\pi D t^3)}} \exp(-\frac{(L - Vt)^2}{4Dt}) \quad (1)$$

where,

- $C(L,t)$ is the grain flow transfer function (g s^{-1});
- C_0 is the integrated initial pulse yield (g);
- D is a dispersion coefficient;
- L is a coefficient of the characteristic length the grain travels (m) between intake by the harvester and measurement by the mass flow sensor;
- V is the mean velocity of the harvester (m s^{-1});
- t is time (s).

To obtain deconvolved yield data, measured yield data can be transformed from the time domain to the frequency domain with a Fast Fourier Transform (FFT) (Press et al., 1989). The FFT of the measured yield must be divided by the FFT of the transfer function, or:

$$\text{FFT}[\textit{deconvolved yield}]_i = \text{FFT}[\textit{measured yield}]_i \,/\, \text{FFT}[C(L,t)]_i. \quad [2]$$

where i is the ith reading number of the data set.

Both the measured yield and transfer function must sum to one. The result of the division in the frequency domain (*i.e.*, FFT [*deconvolved yield*]) must then be back-transformed to the time-domain by an inverse FFT, divided by its sum and multiplied by the mean of the original observed data, *i.e.*,:

$$\textit{Deconvolved yield}_i = \overline{\textit{measured yield}} \times \left(\frac{\text{inv.FFT}[\textit{deconvolved yield}]_i}{\sum_{i=1}^{n} (\text{inv.FFT}[\textit{deconvolved yield}]_i)} \right) \quad [3]$$

RESULTS AND DISCUSSION

Estimating the Parameters of the Transfer Function

Figure 1 shows a plot of moisture content against time from t = 0 until 60 seconds after the pulse has been harvested. The moisture content for the part of the field harvested averaged 12.5% (±0.1%). The moisture content rose sharply as a result of the wet grain reaching the moisture sensor approximately 5 to 6 s after harvesting, and reached a maximum of 19.9% at 12 s after harvesting. The moisture content of the grain passing the sensor then slowly decreased during the next 30 s until it returned to the level of 12.5% about 43 s after harvesting.

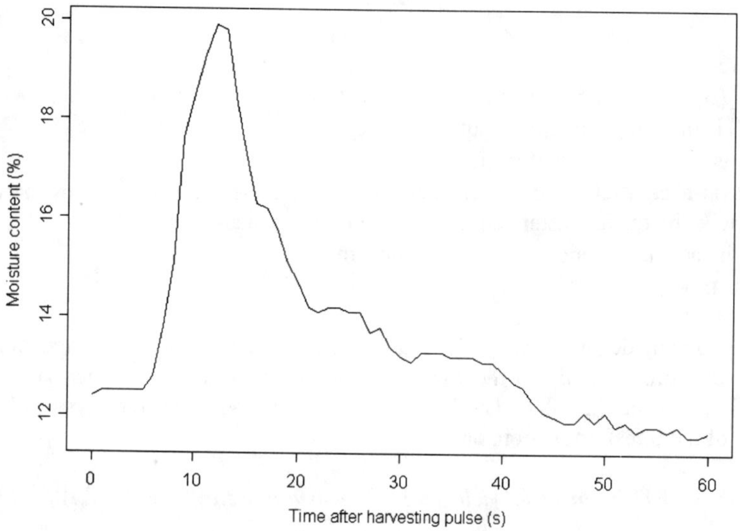

Fig. 1. Plot of moisture content vs. time after harvesting wetted pulse.

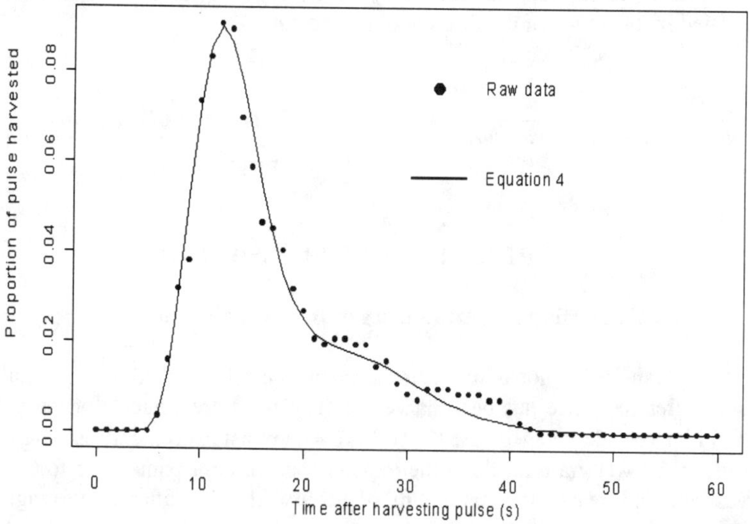

Fig. 2. Plot of the proportion of grain harvested vs. time after harvesting of the wetted pulse.

The difference between the moisture content of the non-wetted grain and the wetted grain was found for each reading and then divided by the sum of the data to make the area under the curve equal to one. The results are shown as 'Raw data' on Fig. 2, where each point represents the proportion of the wetted pulse harvested after $t = 0$. The greatest amount of grain from the pulse that passed the mass flow and moisture sensor at any time was about 9%.

The data are unusual and unexpected in that there are recurring peaks at about 25 and 33 s. The two minor peaks contain approximately 20% of the total grain harvested – a significant amount – so fitting Eq. [1] to the empirical data posed a problem because it can only describe one peak. The data are much better described by treating the two minor peaks as one and summing the transfer functions for the major and minor peaks. This is shown by Eq. [4]:

$$C(L,t) = \left(\frac{C_0 L}{2\sqrt{(\pi D t^3)}} \exp(-\frac{(L-Vt)^2}{4Dt}) \right)_1 + \left(\frac{C_0 L}{2\sqrt{(\pi D t^3)}} \exp(-\frac{(L-Vt)^2}{4Dt}) \right)_2 \quad [4]$$

where the subscripts 1 and 2 respectively refer to the major and minor peaks.

The result of fitting Eq. [4] is shown as the line in Fig. 2. The best combination of parameters for the transfer function was determined by the lowest value of the Akaike Information Criterion (Webster & McBratney, 1989) and are shown in Table 1.

The parameter C_0 is dimensionless because proportions have been used. The major peak accounted for 77% of the pulse. The grain that constituted the minor peak travelled almost twice the distance within the harvester as that in the major peak and, as such, dispersed to greater degree.

Deconvolving Yield Data with the Transfer Function

The barley data were deconvolved in the frequency domain using the parameters in Table 1 and Eq. [2] and [3]. Sixty readings have been extracted from a part of the field and shown in Fig. 3 as a comparison of the measured and deconvolved data.

The deconvolution process has produced a peakiness in the output data, which may be attributed to two sources:

1. an artefact of the FFT's sensitivity to the input data (Press et al., 1989);
2. inaccuracy in the transfer function's representation of point-to-point grain flow.

Table 1. Parameters of Eq. [4] for the grain flow transfer function

Parameter	Major peak	Minor peak
C_0	0.77	0.23
L (m)	31	61
D	2.2	3.3
V (m s^{-1})	2.16	

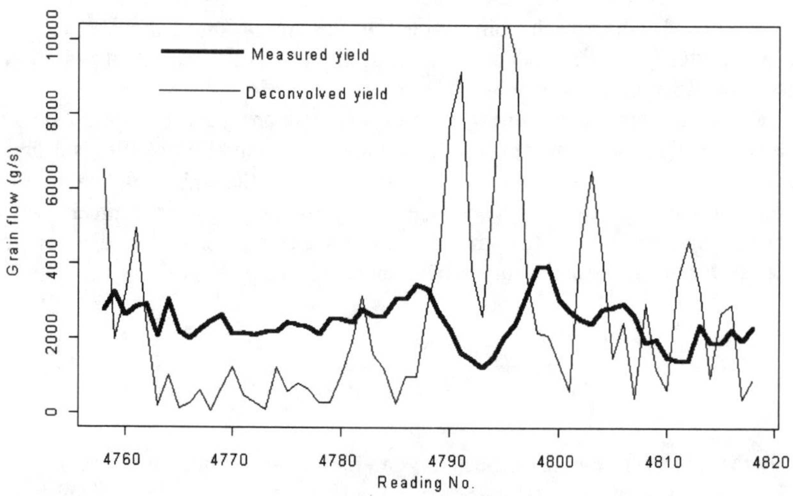

Fig. 3. Plot of measured and deconvolved yields vs. reading number during harvesting of the field using the wetted pulse transfer function.

It must be emphasised that the certainty with which the deconvolved data are shown is speculative as the site-specific truth of the grain yield is not known, however, we infer that the wetted pulse transfer function has not done a very good job of the deconvolution for the second reason given above. It is likely that the time grain took to move through the harvester was changed relative to normal harvest conditions due to increased cohesion (grain sticking to itself) and adhesion (grain sticking to the insides of the machine) caused by the elevated moisture content. If this was the case, it would have contributed greatly to changes in the length (L) and dispersion (D) parameters of Eq. [4] and therefore the wetted pulse transfer function would not be representative of normal grain flow.

To strengthen this inference, consider the difference in threshing characteristics between barley and wheat grain. Intuitively, one would expect a grain such as barley to possess a characteristically shorter threshing time than wheat as wheat grain is enclosed within the lemma and palea of its spikelet, whereas barley grain is less protected. A wheat grain flow transfer function was estimated from another experiment using the method of Whelan and McBratney (1998) on the same harvester as this experiment used (manuscript in preparation). When comparing wheat grain's transfer function with the parameters in Table 1, it was found that the L parameters were decreased to 21.9 and 50.8 m for the major and minor peaks respectively, and the D parameters were reduced to 1.91 and 1.78, respectively, i.e., the wetted barley has taken longer to thresh than the wheat. If we expect barley to thresh more quickly than wheat under normal conditions, these results further consolidate the inference that the wetted barley grain threshed differently than under normal harvest conditions.

CONCLUSION

The nonlinearity of the grain flow system during harvesting, as found by Whelan and McBratney (1998), has been confirmed by this experiment. The essential difference between the data from that study and this is the presence of secondary peaks in our data. This may be due to the different grain being used (barley as opposed to sorghum), but is more likely attributed to a fundamental difference in the mechanics of each harvester confounded with the different threshing properties of wet and dry grain.

It is likely that the grain was too wet in this experiment for the results to be generalised to the normal harvest operation of barley. Future work using this method would have to be more careful with the amount of water applied–perhaps a grain moisture content that is unambiguously measurable that does not influence threshing dynamics can be obtained.

The wetted pulse method for estimating the grain flow transfer function, while easy to implement, did not succeed as we infer the wetted grain threshed differently to that of grain in the normal harvest condition. The wet grain may have distorted the flow dynamics within the harvester and as such the transfer function obtained was possibly unrepresentative. Further work will be required to refine the method before it can be recommended.

ACKNOWLEDGMENTS

Special thanks to Doug Maitland for letting us do this work on his farm. This work forms part of a larger project entitled "A system to apply site-specific management within paddocks," which is funded by the Australian Grains Research and Development Corporation.

REFERENCES

Jury, W.A., and G. Sposito. 1985. Field calibration and validation of solute transport models for the unsaturated zone. Soil Sci. Soc. Am. J. 49:1331–1341.

Lark, R.M., J.V. Stafford, and H.C. Bolam. 1997. Limitations on the spatial resolution of yield mapping for combinable crops. J. Agric. Eng. Res. 66:183–193.

Press, W.H., B.P. Flannery, S.A. Teukolsky, and W.T. Vetterling. 1989. Numerical recipes: The art of scientific computing (Fortran version). Cambridge Univ. Press, New York.

Webster., R., and A.B. McBratney. 1989. On the Akaike information criterion for choosing models for variograms of soil properties. J. Soil Sci. 40:493– 496.

Whelan, B.M. 1998. Reconciling continuous soil variation and crop growth: A study of some implications of field variability for site-specific management. Ph.D. diss., Univ. of Sydney, Australia.

Whelan, B.M., and A.B. McBratney. 1997. Sorghum grain flow convolution within a conventional combine harvester. p. 759–766. *In* J.V. Stafford (ed.) Precision Agriculture '97 1st European Conf. on Precision Agriculture, Warwick Univ. Conf. Ctr., England. 7–10 Sept. 1997. Bios Scientific Publishers, Oxford, England.

Whelan, B.M., and A.B. McBratney. 1998. A parametric transfer function for grain flow within a conventional harvester. Prec. Ag. (submitted).

PREDICTION UNCERTAINTY AND IMPLICATIONS FOR DIGITAL MAP RESOLUTION.

B.M. Whelan
Australian Centre for Precision Agriculture, University of Sydney, Australia.

A.B. McBratney
Australian Centre for Precision Agriculture, University of Sydney, Australia

ABSTRACT

Digital maps are used to visually present the observed spatial variability in many soil and crop attributes. Real-time yield monitor data in particular provides observations which possess inherent measurement errors. These operational or measurement errors are additive and will impart a level of uncertainty to the yield data. The local spatial variation and the method chosen for predicting onto a regular grid for map construction will also contribute to the total uncertainty. This paper proposes that the uncertainty in any estimates should be quantified and that block kriging is an ideal prediction technique for this purpose. It is shown that the optimum block size (minimum map resolution) may be determined from the response of uncertainty to increasing block size. The information content of digital maps is also quantified, and when coupled with the optimum block size, can be used to determine the optimal spectral resolution for a map.

INTRODUCTION

Uncertainty can be generally described as a probability-based doubt in the accuracy of an estimate. Uncertainty is inherent in the maps produced by interpolation (prediction) using data obtained from crop yield and other real-time sensors. Uncertainty is predominantly contributed by measurement errors and prediction techniques.

The current crop yield monitors use sensed data to calculate a yield quantity per unit area for a given harvest duration. The flow sensors in these monitoring systems are generally regarded as accurate to within a maximum 5% on a mass balance basis when assessed over a reasonably large area or grain tonnage (Pringle et al., 1993; Murphy et al., 1995, Missotten et al.,1996; Pierce et al., 1997). Contributing to this error may be variations in grain density, inclusion of foreign material, ground slope, machine vibration and electrical noise, and ambient dust and humidity levels.

Far less study has been directed at the accuracy of the yield value (mass/unit area) calculated from each mass-flow observation. Each yield estimate requires a distance traveled and swath width measurement (to calculate area harvested) in conjunction with the mass flow observation. Measurement error is introduced during speed and cutting width determination (Missotten et al.,1996; Vansichen and De Baerdemaeker, 1991; Stafford et al., 1997), estimation of the time delay between cutting the crop and the relevant grain mass reaching the sensor, and

Copyright © 1999 ASA-CSSA-SSSA, 677 South Segoe Road, Madison, WI 53711, USA. *Proceedings of the Fourth International Conference on Precision Agriculture.*

inaccurate modeling of the flow-path dynamics within a combine harvester (Whelan & McBratney, 1997). A further source of error is introduced in the standardisation of the yield estimate to a constant moisture content. Error in the moisture observations will be propagated through the yield calculation along with the other observation errors. These sensors, along with all electrical and electronic sensors are also subject to electrical and operational interference which adds noise to the output signal.

All the errors identified here will combine to impart an uncertainty in the individual observations made from these sensors. The interpolation or prediction technique used to produce a continuous surface map from these non-aligned data points may add further uncertainty to the resulting map. It is therefore important that these errors be minimised and quantified to provide integrity to the final result – the attribute map.

This paper will discuss the use of block kriging to reduce and quantify the uncertainty propagated through to the final digital maps. It will also propose a method for quantifying the information content within digital maps that includes the inherent uncertainty and provides a quideline for the optimum classification resolution.

MATERIALS AND METHODS

Digital Maps

All digital maps are based on some form of map model (Figure 1) whereby values are represented as a set of blocks (*B*) the centres of which are located on a grid (*G*). According to Goodchild (1992) the blocks may have sides equal to the grid spacing (a raster model), the blocks may be points on a regular grid (a grid model) or they may be points and the grid irregular or infinitely fine with missing values or values equal to zero (a point model).

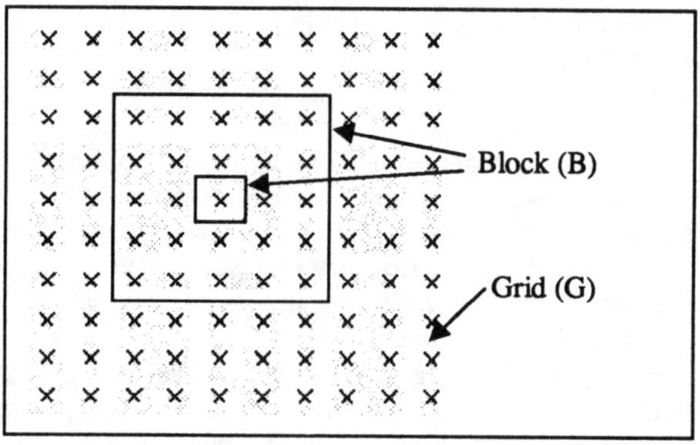

Figure 1. Generalised map model.

Block kriging for prediction

When Burgess & Webster (1980) first introduced geostatistical spatial prediction techniques into soil science, they talked about two techniques – point (or punctual) kriging and block kriging. Since then almost all of the attention has been placed on point kriging, i.e. to a method that interpolates at any given location a variable with a point support. Block kriging has rarely been used and software for doing it is rather scarce. Block kriging attempts to predict the average of a variable over some block of length *(dx)* and width *(dy)* centred about some prediction point *(x0, y0)*. It should be noted that the locations *(x0, y0* - the prediction grid or raster) can be closer together than the block length or width. This in fact gives an aesthetically pleasing smooth map. The major advantage of using block kriging is that the estimate of the block mean, not surprisingly, improves as the block dimensions increase. This is quantified by the kriging variance of the block mean, which according to Burgess and Webster (1980) is given by Equation 1.

$$\sigma^2{}_A = s^T \begin{bmatrix} \lambda \\ \mu \end{bmatrix} - \iint \gamma(x,y) p(x) p(y) dx dy \qquad (1)$$

Here the term on the left estimates the point prediction variance and the right hand term is the within-block variance, which increases with block size provided that the variogram hasn't reached its sill value. One can think about block kriging in another way, it filters out variation less than the block size. So if there is significant nugget variance, blocks as small as 1m by 1m will be significantly better estimated than points.

Block kriging yield data

Block kriging requires of course that the variable to be estimated is additive - clearly crop yield is additive- its total at many points divided by the area over which those are measured is the accepted measure of yield no matter how large the area. The same could not be said for pH or hydraulic conductivity which may not be additive.

Data that has been deconvoluted to account for combine grain dynamics (Whelan & McBratney, 1997) have a considerable short-range variation and point estimates at unmeasured locations are quite uncertain. Figure 2 shows a comparison of uncertainty calculated for original yield sensor data and the deconvoluted equivalent. It shows that the uncertainty in yield predictions is underestimated when data is not deconvoluted. It also shows that as we increase from point predictions to block kriging the uncertainty decreases with block size. The value at 20m being approximately 0.35t/ha in this instance which is probably acceptable. There is no point in increasing the block size further as the uncertainty plateaus and we do not wish to miss the site-specific management opportunities provided by smaller blocks.

(b)

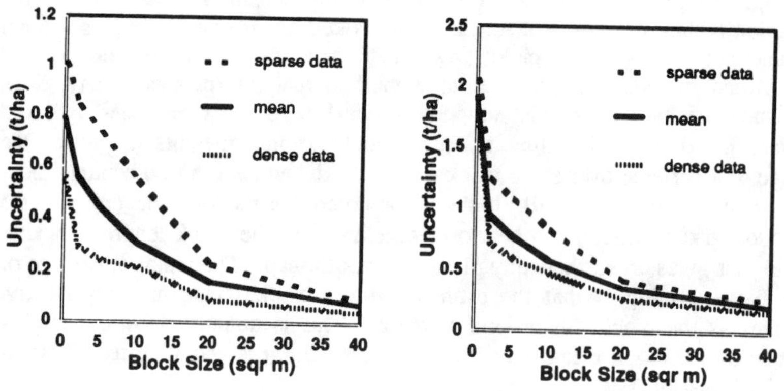

Figure 2. Uncertainty associated with increasing block size. (a) original sensor data. (b) deconvoluted equivalent data.

Figure 3 displays the spatial distribution of the yield monitor data used in Figure 2. The points A and B refer to regions where data for use in prediction are relatively dense and sparse respectively. The effect of this data distribution on the resulting uncertainty is apparent in Figure 1. Less data, greater uncertainty at all block sizes.and vice versa. It is apparent then that the uncertainty may vary within a mapped area depending on the data density.

Figure 4 shows maps produced on a 2m prediction grid while increasing the representative prediction area from points to 40m blocks. As expected, the maps depict an increasing smoothness with increasing block size. This is the trade-off. Blocks of 20m seem optimal – but is larger than we might have expected.

Block kriging can also employ local variogram models for neighbourhood kriging as with punctual kriging (Whelan et al., 1996). We are suggesting here that kriging with a local variogram onto a fine raster of 2m to 5m with 20m by 20m blocks may be a optimal mapping technique.

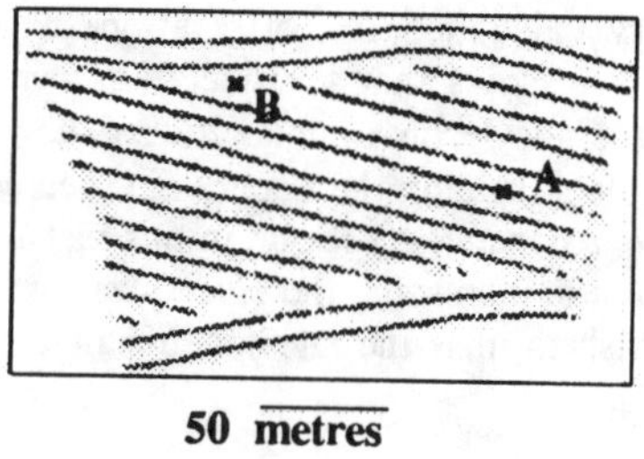

Figure 3. Spatial location of deconvoluted yield sensor data.

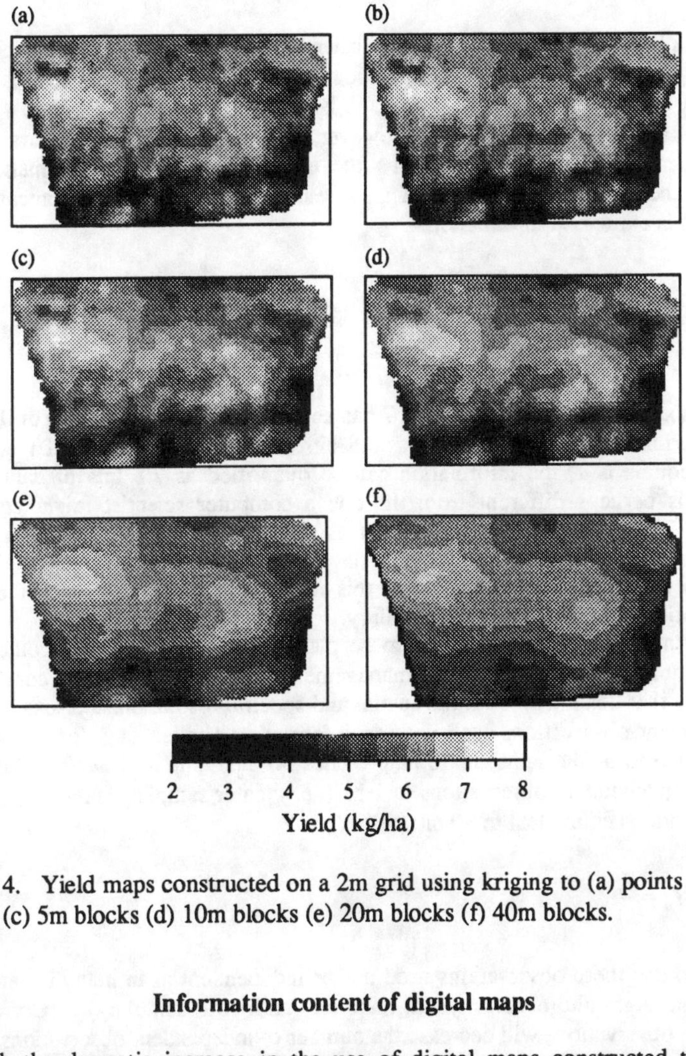

Figure 4. Yield maps constructed on a 2m grid using kriging to (a) points (b) 2m blocks (c) 5m blocks (d) 10m blocks (e) 20m blocks (f) 40m blocks.

Information content of digital maps

With the dramatic increase in the use of digital maps constructed through interpolation (prediction), the question of information content arises. This is because scale is essentially meaningless and it is necessary to deal with the notions of spatial and spectral resolution. Quantifying information content may eventually provide a useful method for assessing the quality of a map and improving its value in decision making.

Information content has been previously mathematically quantified by Shannon (1948) using. the number of alternative choices conveyed (C_a) to the logarithm base 2 ($\log_2$). This is the concept of bits: two choices equals one bit of information (Equation 2).

$$I = \log_2(C_a) \tag{2}$$

For all the map models described earlier it may be generalised that each block (B) has an attribute value (v) and an associated attribute uncertainty (u). A simple analogy of Equation 2 for digital maps is given in Equation (3). Here I_c is the information content per block. However, we suggest that the uncertainty (u) associated with an estimate lessens the effective amount of information and should therefore be considered in an assessment of information content. The concept is expressed in Equation 4.

$$I_c = \log_2(v) \tag{3}$$

$$I_c = \log_2(v) - \log_2(u) \tag{4}$$

For example if a pH value (v) of 7 has an associated uncertainty (u) of 0.5 then the information can be quantified as 3.8 bits of data. Alternatively if v equals 7 and u equals is 1, the information can be quantified as 2.8 bits of data. This notion is perhaps different from the one a computer scientist might consider. They might judge information content as the minimum amount of information requited to store the data (i.e. it would have to include storage of both the data (v) and the uncertainty (u)). However, this may result in an apparent increase in information with increases in uncertainty.

The information content needs to be placed into a per unit area context (I_a) using Equation 5 to be useful for management. It is the information content per unit area that deals with both the spatial and spectral resolution. For example, 10 independent observations each with an information content of 3 (of course they don't have to be the same) in 10 ha provides 30 bits/10ha or 3 bits/ha. Consider now 10 independent observations in 1 ha (i.e. greater sampling intensity) and the information is equivalent to 30 bits/ha.

$$I_a = \frac{I_c}{A} \tag{5}$$

Of course these observations need not be independent as in many instances in Precision Agriculture (e.g. yield mapping and soil sampling). Correlation between observations will decrease the number of independent observations. This should be reflected in the information content of a map. The lower bounds for the number of independent observations (N_{low}) can be estimated by summing the inverse of the observation correlation matrix for the observations (Barnes, 1988).

$$N_{low} = 1'C^{-1}1 \tag{6}$$

Where C^{-1} is the inverse of the observation correlation matrix, 1 is a vector of ones and 1^t is the transpose of 1.

The upper limit is bounded by the absolute number of observations (N). An empirical estimate of the true number of independent observations (N_{ind}) can be found through Equation 7 (Barnes, 1988)

$$N_{ind} = N_{low} \times \exp(1 - \frac{N_{low}}{N}) \quad (7)$$

It is hypothesised here that the information content (I) of a digital map may then be measured by the information content per unit area (I_a) multiplied by the number of independent observations (N_{ind}) as per (Equation 8).

$$I = I_a \times N_{ind} \quad (8)$$

Given the previous scenario, the number of independent observations in 1 ha may now equal 7 and the information content represented in a point map equal to 21bits/ha. This whole procedure assumes that the mapped pattern is complex and cannot be described by a low parametric function of space co-ordinates (e.g. a planar trend).

Extending this concept, now consider that when predictions are made onto a fine grid it often results in an increase in values per unit area compared with the density of initial observations. Does this increase in the density of values result in more information, is the information content merely spread out, or does the increased correlation of prediction point values decrease the information per unit area?

Figure 5 shows a 1 ha region of crop yield data extracted from the data set shown in Figure 2.

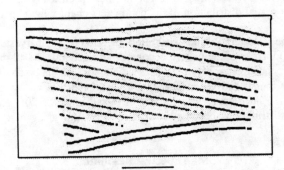

Figure 5. 1 ha subset of yield monitor data

The information content for the observations in this area has been calculated as 43.91 bits/ha using Equation 10. From Table1 it is interesting to note that the number of independent observations is approximately 4% of the original data points. Prediction based on punctual kriging is shown to reduce the information content by 60% mainly due to the increase in correlation between the prediction points and the corresponding decrease in the number of independent observations.

As the block size (*B*) is increased the uncertainty (*u*) decreases but the correlation between the predictions will increase. From the results in Table 1 it appears that the fall in the uncertainty is proportionately larger than the increase in correlation and the information content begins to increase. This increase continues up to a block size of 20m in this example. Thereafter the correlation overrides any gains in the certainty of predictions.

Table 1. Parameter values for calculating the information content within the 1ha area as the size of the prediction block increases.

Data/Estimates	N	N_{low}	N_{ind}	mean *v* (t/ha)	mean *u* (t/ha)	mean Ic (bits)	Ia (bits/ha)
Observations	645	9.48	25.40	5.66	1.67	1.73	43.91
Point kriging	1134	3.87	10.48	5.66	1.80	1.64	17.22
5m block kriging	1134	3.80	10.30	5.66	0.54	3.39	34.90
10m block kriging	1134	3.62	9.81	5.65	0.44	3.68	36.16
20m block kriging	1134	3.23	8.76	5.65	0.32	4.20	36.81
40m block kriging	1134	2.27	6.15	5.64	0.22	4.96	30.51

To illustrate these points, Figure 6 shows the spatial distribution of the *v*, *u* and I_c parameters for 3 block sizes. While the requirement to produce black and white figures has restricted the ability to display a finer spectral resolution, the figures show that the spatial distribution of the parameters changes with increasing block size.

Initially the uncertainty follows a pattern dictated by the harvester runs but as the block size increases, the pattern becomes more general. At 40 meters there is a more pronounced edge effect as data points for use in prediction are more unevenly distributed across the block. This increases the uncertainty and may explain why the information content begins to fall after the 20m block size. In general, the distributions become more even with decreasing block sizes up to 20m.

Map Spectral Resolution

The 20m block size is noted as providing the most information in Table 1 and, not unrelatedly, is the point where the uncertainty plateaus in Figure 1. 20 metres therefore provides an estimate for the optimal block size for this data. By quantifying the optimal block size (spatial resolution) it is conceivable that this

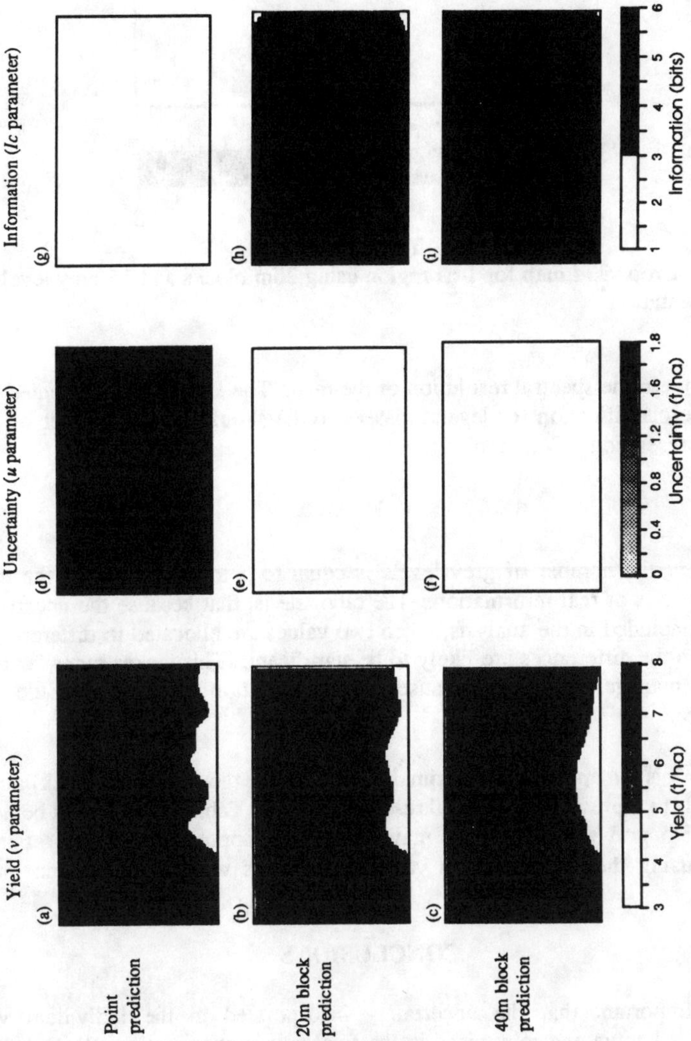

Figure 6. Spatial distribution of the yield, uncertainty and information parameters per block for increasing block sizes

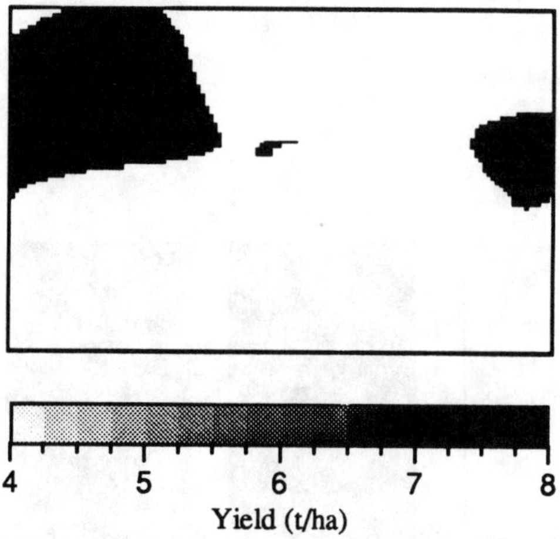

Figure 7. Crop yield map for 1 ha region using 20m blocks and 16 grey levels for map presentation

should govern the spectral resolution of the map. The simple idea presented here is that the classification (or legend classes) (GL) should be the number of grey levels represented by Equation 8.

$$GL = 2^{\overline{Ic}} \qquad (8)$$

In effect the number of grey levels is equal to 2 to the power of the mean number of bits of real information. The rationale is, that because the uncertainty has been included in the analysis, when two values are allocated to different grey levels then the differences are likely to be significant. This makes maps far more powerful management tools because the classification of yield is made with confidence.

Figure 7 is a map using the optimal block size for the data (20m block) and 2^4 grey levels to represent the spectral resolution (as per Table 1). It should be noted that the 3m grid size used here may not be the optimal dimension. Further analysis using these methods on various grid sizes will be undertaken in the future.

CONCLUSIONS

It is important that the uncertainty incorporated in the individual yield estimates is known and minimised in the final map representation. Block kriging provides a very useful technique for achieving these requirements. Further it is desirable that the inherent uncertainty be quoted along with any digital map as a matter of best practice.

The proposed method for determining the information content of a digital map appears useful for incorporating the uncertainty into a map. When used in conjunction with block kriging to determine the optimal block size for a data set, the usefulness of yield maps for management should be enhanced. This work also suggests that in the absence of block kriging software, PA practitioners would be advised to produce maps using a 10m radius moving average than the support-limiting inverse square distance method. A simple estimate of the variance within those blocks would also be also useful.

REFERENCES

Burgess, T.M. & Webster, R. (1980). Optimal interpolation and isarithmic mapping of soil properties: II block kriging. *Journal of Soil Science* **31**: 333-341.

Goodchild, M.F. (1992). Geographical data modelling. *Computers and Geosciences* **18**: 401-408.

Missotten, B., Strubbe, G. & De Baerdemaeker, J. (1996). Accuracy of grain and straw yield mapping. In P.C. Robert, R.H. Rust & W.E. Larson (ed.) *Precision Agriculture: Proceedings of the 3rd International Conference on Precision Agriculture*, ASA, Madison. pp 713-722.

Murphy, D.P., Schnug, E. & Haneklaus, S. (1995). Yield mapping - a guide to improved techniques and strategies. In P.C. Robert, R.H. Rust & W.E. Larson (ed.) *Proceedings of the 2nd International Conference on Site-Specific Management for Agricultural Systems*, ASA, Madison. pp 32-47.

Pierce, F.J., Anderson, N.W., Colvin, T.S., Schueller, J.K., Humburg, D.S. & McLaughlin, N.B. (1997). Yield mapping. In F.J. Pierce & E.J.Sadler (ed.) *The State of Site-Specific Management for Agriculture,* A.S.A, Madison, pp 211-243.

Pringle, J.L., Schrock, M.D., Hinnen, R.T., Howard, K.D. & Oard, D.L. (1993). Yield variation in grain crops. *Paper No, 93-1505, 1993 International Winter Meeting* ASAE, Illinois.

Shannon, C.E & Weaver, W. (1949). *Mathematical Theory of Communication,* Univ. of Illinois Press, Urbana, 117p.

Stafford, J.V., Ambler, B. & Bolam, H.C. (1997). Cut width sensors to improve the accuracy of yield mapping systems. In J.V. Stafford (ed.) *Precision Agriculture '97: Proceedings of the 1st European Conference on Precision Agriculture,* Bios, London, pp 519-527.

Vansichen, R. & De Baerdemaeker, J. (1991). Continuous wheat yield measurement on a combine. In *Automated Agriculture for the 21st Century* . ASAE, Michigan. pp 346-355.

Whelan, B.M. & McBratney, A.B. (1997). Sorghum grain flow convolution within a conventional combine harvester. In J.V. Stafford (ed.) *Precision Agriculture '97: Proceedings of the 1st European Conference on Precision Agriculture,* Bios, London, pp 759-766.

Whelan, B.M., McBratney, A.B. & Viscarra Rossel, R.A. (1996). Spatial prediction for Precision Agriculture. In P.C. Robert, R.H. Rust & W.E. Larson (ed.) *Precision Agriculture: Proceedings of the 3rd International Conference on Precision Agriculture*, ASA, Madison. pp 331-342.

On-Farm-Research Protocols: Determining Vertical Accuracy of Differentially Corrected Carrier and Code Phase Global Positioning Satellite Systems

D. P. Johansen
ASGROW
Atlantic, Iowa

D. E. Clay
C. G. Carlson
K. W. Stange
S. A. Clay
J. A. Schumacher

Plant Science Department
South Dakota State University
Brookings, South Dakota

M. M. Ellsbury
Northern Grain Insect Research Laboratory
USDA-ARS

ABSTRACT

The objective of this study was to demonstrate an approach for determining the vertical accuracy of different global position satellite systems. Two hundred points were selected for comparison in a field with a gently rolling topography. The highest point was approximately 30 m higher than the lowest point. Elevation information from the differentially corrected GPS systems were compared with surveyed elevations. The regression equation between the local differentially corrected carrier phase receiver (y) and surveyed elevations (x) was: $y = 0.01 + 0.99x$; $r = 0.99^{**}$. Based on this analysis, the carrier phase receiver was not biased and the vertical errors were <2 cm. The regression equation between the Satellite differentially corrected code phase receiver and surveyed elevations (x) on different days were: $y = -1.35 + 1.28\,x$ ($r = 0.82^{**}$), and $y = 1.79 + 0.70x$ ($r = 0.78^{**}$). These results showed that the differentially corrected code phase receiver produced inconsistent biased vertical measurements.

INTRODUCTION

Plant nutrient availability, drainage, erosion, pH, and yields are just a few of the parameters influenced by topography. Yield maps superimposed on topography maps improve the ability of a producer or consultant to identify anomalous areas of fields. Producers or consultants using a differential global positioning system have the ability to record elevation information; however, the accuracy of elevation

Copyright © 1999 ASA-CSSA-SSSA, 677 South Segoe Road, Madison, WI 53711, USA. *Proceedings of the Fourth International Conference on Precision Agriculture.*

information received from different differential global positioning systems may not be accurate enough for mapping elevation. Elevation differences between global positioning systems is due in large part to the type of receiver used. Sub-centimeter accuracy achieved with survey grade systems use carrier phase receivers which are substantially more accurate than the code phase receivers. Other sources of global positioning error which influence accuracy include: selective availability, satellite clocks, satellite orbits, atmospheric disturbances, and multipath errors (Morgan & Ess, 1997). Much of the error associated with global positioning systems may be resolved using differential correction; however, simply employing differential correction does not guarantee absolute accuracy. Survey grade receivers typically utilize a base station and transmitter for differential correction, which is placed in close proximity to the users location (<20 km). Code phase receivers, used for precision farming, typically utilize differential correction signals obtained from the U.S. Coast Guard, local FM signals, or satellite-based differential correction. Each of these differential correction techniques offer-varying levels of positional accuracy depending upon the source and type of receiver used. The objective of this study was to demonstrate an approach for determining the vertical accuracy of different global position satellite systems.

MATERIALS AND METHODS

A 65 ha no-tillage field, in a corn (Zea mays L.) – soybean [Glycine max (L.) Merr.] rotation, located in eastern South Dakota with latitude and longitude values of 44.22 °N and -96.65°W, respectively, was used for this study. At the study site, 200 points were surveyed to determine relative elevation using a Sokkia total station (Sokkia, Overland Park, KS). The Sokkia system used a Sokkia Set-3 total station, Sokkia SDR33 electronic field book, Sokkia System A single prism, and Sokkia prism pole. Elevations at these points, were measured by the Leica single frequency (Leica, Norcross, GA) and the Omnistar (Omnistar, Houston, TX) Global Positioning Systems (GPS). The Leica GPS receivers were single frequency 12 Channel L1 continuous tracking sensors capable of tracking C/A narrow code and carrier phases.

Data acquisition approaches for the Leica system were real time stop and go and real time moving (Leica, 1996). For the moving and stop and go approaches, 1 and 3 s of information, respectively were used to determine positional locations.

For the Leica measurements the correction signal was obtained from a base station located in the northeast corner of the field. The base station consisted of a Leica SR 9400 sensor (SR9400B), a Leica AT201 antenna (AT201B), a Leica CR344 controller (CR344B), and a Omni Optical 1700QR tripod. A Pacific Crest Model RFM96W 35 watt transmitter was used to transmit differential corrections. Elevations at each point were measured by a rover that consisted of a Leica SR 9400 sensor (SR9400R), a Leica AT201 antenna (AT201R), and a Leica CR344 (CR344R) controller. For real time differential correction, a Pacific Crest Model RFM96W 2 watt radio was used to receive corrections at the rover.

The Leica base station was located at a point were the latitude, longitude, and elevation values were known. This fixed position was then broadcast to the rover unit to be used for differential correction. Prior to collecting elevation information, static initialization was performed on the rover. Initialization involved placing the rover at an unknown point for 15 to 20 minutes during which the location was of the rover, which was in communication with the base station, was determined. After static initialization was conducted, the CR344R controller indicated that ambiguities were fixed and a position was available. Ambiguity, as defined by Leica (1996), is the unknown integer number of cycles of the reconstructed carrier phase contained in an unbroken set of measurements from a single satellite pass at a single receiver. The ability to resolve the ambiguities on the carrier phase portion of the GPS carrier signal is the critical issue which determines if the positional results meet the stated precision (Clark, 1996). The CR344R controller also calculates the coordinate quality (CQ) value, which is a statistical measure of accuracy of the current position in meters. The maximum CQ was set at 0.05 m. Once the rover position was fixed, this information was recorded and stored in the CR344R. For stop and go measurements, the rover antennae mounted on a pole were placed at a location and 3 seconds of information was averaged to determine a positional location.

For the moving measurements, the rover unit was mounted to a All Terrain Vehicle and driven over the field at a speed of approximately 10 km h^{-1}. For moving measurements a single measurement was used to determine positional locations.

The Omnistar system used in this study used an Omnistar 6300A receiver, an Omnistar 6380 antennae, and a Motorola patch antennae (Omnistar, 1996). The Motorola patch antennae was mounted on a Corvallis Microtechnology Inc. adjustable height pole. The Omnistar system uses a nationwide real-time differential GPS (DGPS) data broadcast system to deliver corrections from an array of ten GPS base stations positioned from coast to coast in the continental USA. The base stations used 12-channel GPS receivers that are set up to provide RTCM SC-104 version 2 corrections every 0.6 s. These base stations provide industry standard formatted corrections to a Network Control Center near Houston, TX, where corrections were decoded, checked, and transmitted to a C-band communications satellite. Corrections were received and processed at the users location by the Omnistar GPS receiver. The patch antenna was mounted on a pole and at the predetermined points the positional locations were determined.

The lowest point in the field as determined by surveying with the Sokkia Total Station was defined as 0. The elevation at this point as measured by the Leica and Omnistar was subtracted from all the values within their respective data set. Simple linear regression was used to compare the relative elevations of the Leica and Omnistar to the surveyed points. Bias in the elevation measurements was evaluated by subtracting relative elevations as measured by the Leica and Omnistar from the surveyed point. Simple linear regression was used to compare these differences with surveyed elevations. The system was biased if the residuals had a significant r-value. The study was conducted between 8 July 1997 and 16 July 1997, and repeated approximately 7 d later.

RESULTS AND DISCUSSION

The Leica moving and stop and go sampling approaches were highly correlated to the relative field elevations and were not biased (Fig. 1 and 2). Both approaches had *y* intercepts and slopes that were not different from 0 and 1, respectively. When the experiment was repeated the results were almost identical (data not shown). The observation that both techniques produced similar results was interesting, in that a much longer period of time was required to collect the information for the stop-and-go than the moving approach, and therefore the stop-and-go approach, which used 3 s of information, did not significantly reduce vertical errors.

The relative elevations as measured by the Omnistar were correlated to the actual relative elevation (Fig. 3); however, the correlation coefficients were much lower than those observed for the Leica and the actual regression equations were different for the two sampling dates. For sampling dates 1 and 2, the regression equations were:

$$y = -1.35 + 1.28\,(x) \qquad r = 0.82^{**} \qquad [1]$$

$$y = 1.79 + 0.70\,(x) \qquad r = 0.78^{**} \qquad [2]$$

respectively, where y was the Omnistar measurement and x was the surveyed elevation. The fact that the slopes and y intercepts changed, indicated that on Date 1 the slopes were steeper and valleys deeper than actually measured, while on Date 2 the Omnistar flattened the landscape.

The magnitude of the Omnistar bias was influenced by averaging the relative surface elevations of a point taken 1 d, with the same point taken on another day (Fig. 4). The regression equations were influenced by the amount of information used to estimate the relative elevation at the points selected for repeat measurements. Generally, the correlation coefficients increased by increasing the amount of information used to estimate each elevation; however, slopes and y intercepts of the regression equations did not follow a systematic pattern. For example, when 1 and 2 points were used the slopes were similar to 0. It should be pointed out that even after six measurements, the averaged Omnistar measurements were still biased.

The differences in accuracy between the Leica and Omnistar were due to the type of receiver used by each system and the location of the differential correction. There are two broad classes of receivers available; code phase (C/A-code) and carrier phase (Lange, 1996). The Omnistar system used a C/A-code receiver that uses the speed of light and the time interval that it takes for the signal to travel from the satellite to the receiver to compute the distance to the satellites. The time interval is determined by comparing the time at which a specific part of the coded signal (chip) left the satellite with the time it arrived at the antennae. The time interval is translated to a range by multiplying the interval by the speed of light constant (c = 298,000 km sec^{-1}) (Lange, 1996). The comparison between sending and receiving a chip can only be made with a single chip, which lasts only about one microsecond. At the speed of radio waves, one microsecond translates into a distance of 300 m (French, 1996). Signal processing techniques are able to refine the observation resolution to approximately one-percent of a signals wavelength, which translates to about 3 m. This represents the theoretical maximum resolution by design of the C/A-code; however, due to the effects of selective availability and other errors associated with GPS resolution, actual resolution can be considerably higher. In most cases, position fixes from C/A-code receivers would be accurate to within 25 m. However, since the U.S. Department of Defense imposes selective availability, position fix accuracy is limited to within 100 m without Differential Global Positioning Systems (DGPS) (Lange, 1996). DGPS enables the user to improve and remove the effects of selective availability and other sources of GPS error such as satellite and receiver clock errors, satellite orbital error, atmospheric error, and multipath errors (Morgan & Ess, 1997). With DGPS correction techniques C/A- code receivers are generally accurate to within a few meters.

The Leica system used carrier phase receivers that are capable of sub-centimeter accuracy. Accuracy is achieved by measuring the length of the carrier wave and the number of wavelengths between the satellites and the receiver antennae. Once the number of wavelengths is known, a pseudo-range (estimated range) may be calculated by multiplying the number of whole wavelengths by the wavelength of the carrier wave plus the partial

Fig. 1. The relationships between the Leica moving (LM) and the relative elevation (RE).

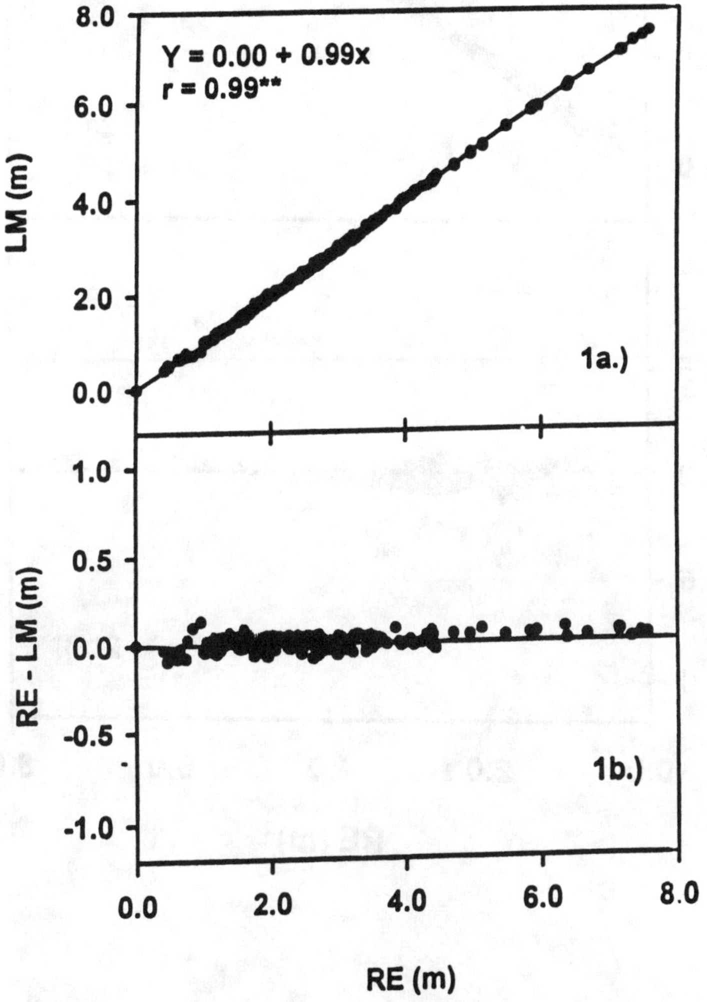

Fig. 2. The relationships between the Leica stop and go (LSG) and the relative elevation (RE).

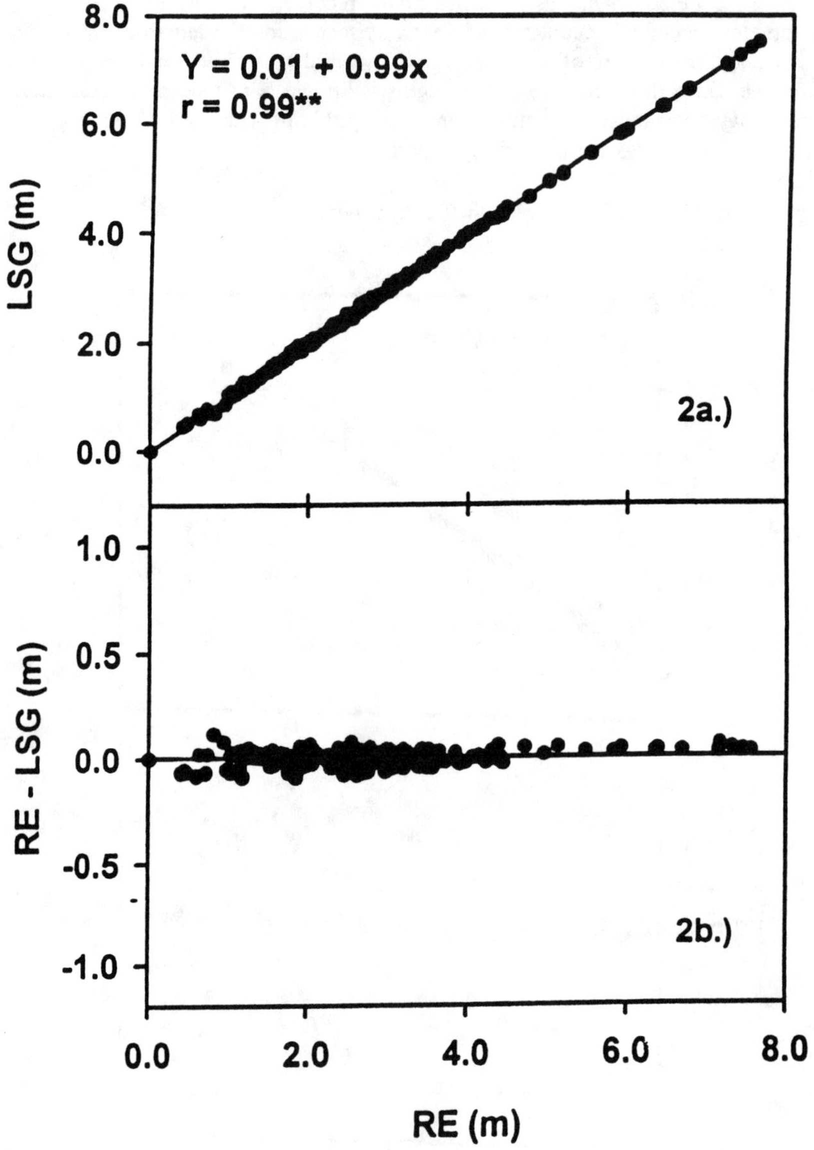

Fig. 3. The relationship between the Omnistar (Os) and the relative elevation (RE).

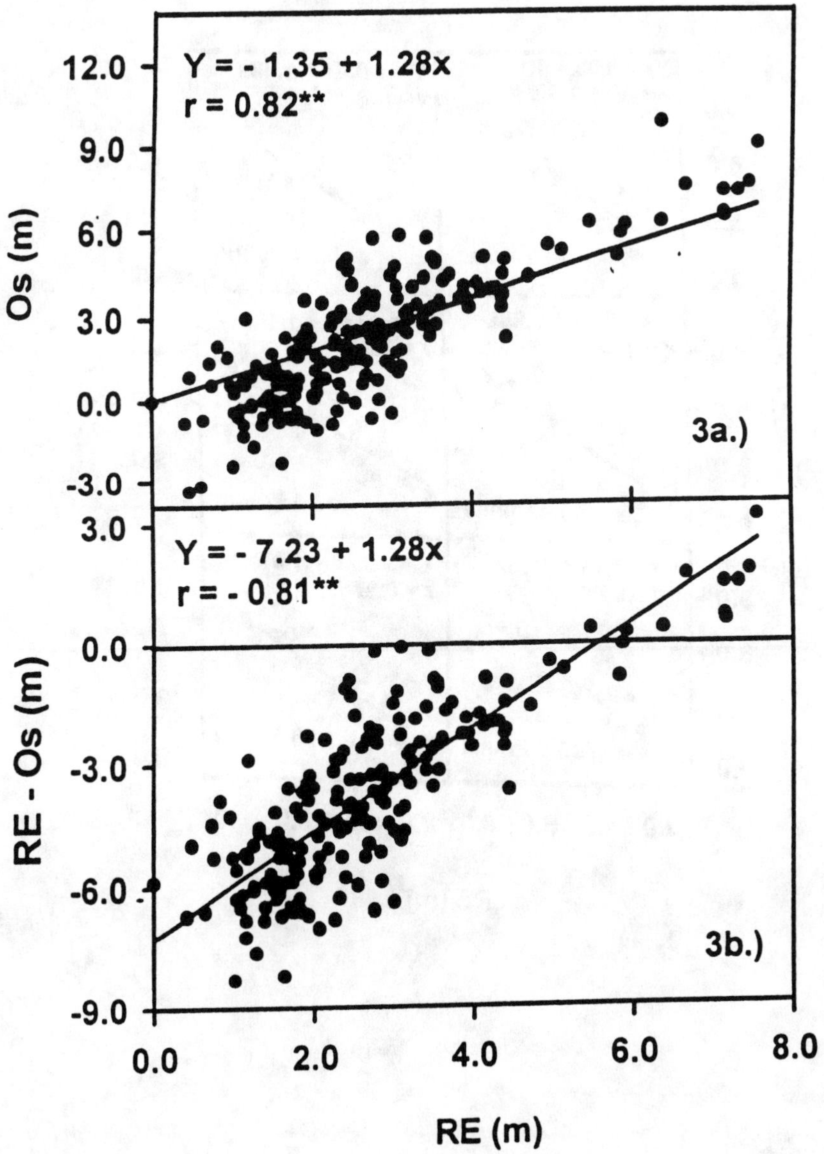

Fig. 4. The relationships between the Omnistar (Os) and the relative elevations (RE) with six repeated measurements.

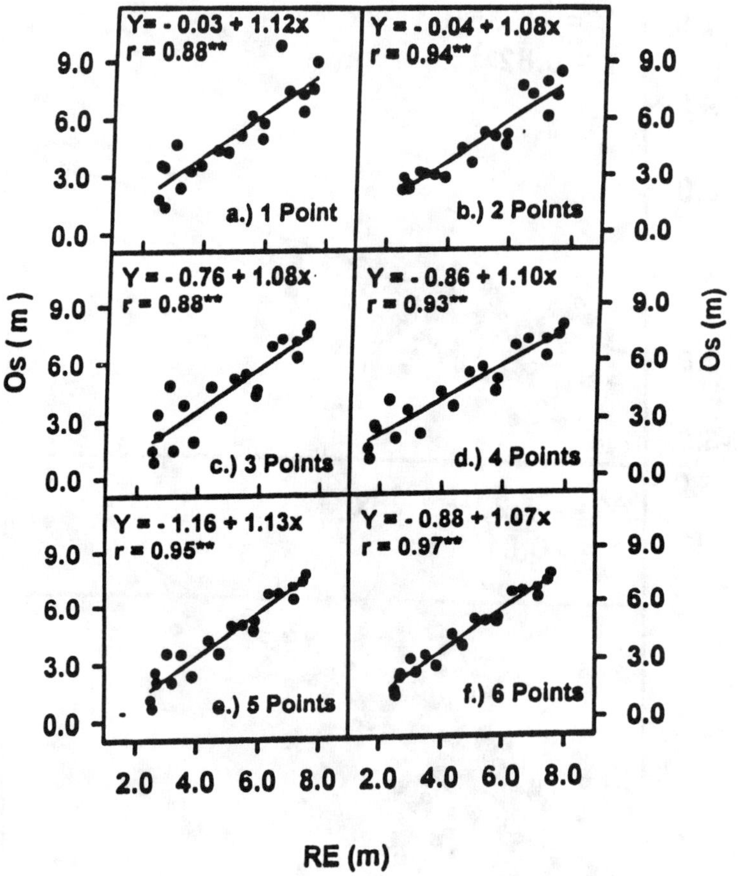

wavelength. With this information a baseline distance and azimuth between any pair of receivers operating simultaneously can be computed. If one receiver is placed at a known point the positional location of the unknown point can be determined (Lange, 1996). Like C/A-code receivers, accuracy is directly related to wavelength size. The wavelength of the L1 carrier is 19 cm, or about 1/1579th the 300 m length of the C/A - code chip (French, 1996). This results in significantly higher accuracy than that observed with C/A-code receivers. Like the code based receivers, signal processing techniques refine the observation resolution to approximately one percent of the signals wavelength, resulting in a theoretical maximum resolution potential for a carrier resolved observation to 1.9 millimeters (French, 1996).

In summery, the advantages of the Leica single frequency carrier phase system were: that these systems provide extremely accurate on-the-go elevation information; that repeated measurements were not required; and that errors associated with the vertical measurements using the moving approach were similar to those using the stop and go approach. It is important to point out, that an elevation map needs only to be developed once, and that once completed, this information can be used for many years. Disadvantages with the Leica single frequency carrier phase system were that it is expensive and that at least an hour was required to set the system up prior to collecting any information. Advantages of the code phase Omnistar system were that it was relatively inexpensive, set up time was <5 min, and horizonal accuracy is acceptable for many precision farming applications (yield monitors and soil sample locations). Disadvantages were that the elevational measurements were biased and that the relationship between actual and measured elevations were not consistent from day to day. This research shows that both code and carrier phase systems have roles in precision agriculture and that due to the relative speed, ease of use, and superb accuracy of information received from the Leica system, a farmer or consultant that wishes to obtain topographic information may consider custom hiring or renting a DGPS system that is capable of sub-centimeter accuracy.

REFERENCES

Clark, R.L., 1996. A comparison of rapid GPS techniques for topographic mapping. p. 651–662. *In* P.C. Robert et al., (ed.) Precision agriculture. Proc. of the 3rd Int. Conf. Minneapolis, MN. 23–26 June 1996. ASA, CSSA, and SSSA, Madison, WI.

French, G.T., 1996. Understanding the GPS: An introduction to the global positioning system , what it is and how it works. 1st ed. GeoResearch, Bethesda, MD.

Lange, A.F., 1996. Centimeter accuracy differential GPS for precision agriculture applications. p. 675-680. *In* P.C. Robert et al., (ed.) Precision Agriculture. Proc. of the 3rd Int. Conf. Minneapolis, MN. 23–26 June 1996. ASA, CSSA, and SSSA, Madison, WI.

Leica. 1996. GPS equipment users manual. Leica AG., CH-9435, Heerbrugg, Switzerland.

Morgan, M., and D. Ess. 1997. The precision farming guide for agriculturists. 1st ed. John Deere Publ., Moline, IL.

Omnistar. 1996. Omnistar technical reference manual. Omnistar, Houston, TX.

Use of Aerospace Structural Sensor Technology for Soil Physical Characterization

John M. Svoboda
J. Richard Hess
Reed L. Hoskinson
Lockheed Martin Idaho Technologies Company
Idaho National Engineering and Environmental Laboratory
Idaho Falls, Idaho

J. Wayne Sawyer
Langley Research Center, Structures Division
National Aeronautics and Space Administration
Hampton, Virginia

ABSTRACT

Tractor hitch pins, instrumented by NASA Langley Research Center, are being used between tractors and cultivating equipment to log the spatial variability in the mechanical resistance or drag force required for cultivation. The INEEL Site-Specific Technologies for Agriculture (SST4Ag) research project, in collaboration with university, other federal agency and industry partners, is correlating soil physical spatial variability, measured as tillage tool drag force, with crop productivity.

INTRODUCTION

Precision farming requires an understanding of the spatial variability of the growing environment. Many growing parameters, such as soil fertility, are measured at sampling locations and estimated among the sampling points with statistical procedures. Other parameters are measured continually, or at a high frequency, with sensors such as yield monitors. Physical soil properties that correlate to many of the characterization parameters may be estimated by measuring soil physical resistance to conventional farming equipment. Load sensors developed to investigate structural joints in advanced aerospace structures may be used to continuously evaluate soil physical characteristics that are spatially variable over large areas. These characterization data would be useful both as input to site-selective decision support systems, and for real-time feedback to equipment controls to improve operational efficiencies. The objective of this study is to develop a calibrated draft measurement system that is adaptable to conventional farming equipment and plot size research equipment. This system has been integrated into existing field tillage operations to site-specifically characterize and define unique soil management zones.

Copyright © 1999 ASA-CSSA-SSSA, 677 South Segoe Road, Madison, WI 53711, USA. *Proceedings of the Fourth International Conference on Precision Agriculture.*

INSTRUMENTED FASTENERS DEVELOPMENT BACKGROUND

Advanced aerospace vehicles, such as the proposed VentureStar reusable launch vehicle, require the development of advanced materials and structural concepts and new measurement techniques to evaluate their performance (Baumgartner, 1997). Ceramic matrix composite materials can be used for structural components that operate at high temperature without the need for active cooling or a thermal protection system. This high temperature use capability results in reduced structural weight and lowers maintenance requirements. Ceramic matrix composites are brittle and require special joining techniques to develop useful structural components. Instrumented fasteners have been developed by NASA Langley Research Center to use in evaluating and developing joining techniques for assembling ceramic matrix composite structural components.

It is this concept of instrumented fasteners on conventional farming equipment that is finding useful applications in precision agriculture.

INSTRUMENTED FASTENER (HITCH-PIN) DESCRIPTION

A standard tractor pin was instrumented with eight precision strain gauges to form a load cell. Two strain gauges were installed on each of four machined flats on the pin and the strain gauge leads were then routed through a hole drilled in the center of the pin. The strain gauges and the leads are held in place and protected by an epoxy potting compound. A constant voltage power supply and a data recording system is required to obtain and record the data from the pin. The instrumented pin, integrated with a Global Positioning System (GPS), can then be used to measure the spatially variable force required to perform various farming operations.

OPERATIONAL CONCEPT OF THE INSTRUMENTED FASTENER (HITCH-PIN) DRAFT SENSING SYSTEM

Soil and landscape physical properties (e.g., soil type, soil moisture, organic matter, topography, etc.) contribute to the tillage tool draft (other confounding factors such as tillage tool type, depth, etc. are also recognized). The integral of these factors can be integrated with GPS and an instrumented hitch-pin.

The instrumented hitch-pin is being developed as part of a robust soil–landscape characterization system. Draft measurements, coupled with data from grid soil sampling and/or other sensors, is expected to allow definition of the draft integral and/or predictable patterns of unique, and possibly somewhat static, management zones for variable rate water application.

The draft measurement system collects data as a component of normal farming operations. Standard equipment bolts–pins are replaced with instrumented bolts–pins at the equipment fastener points for draw-bars, three-point hitches, tillage shanks, etc. to measure draft of any tillage tool–shank being pulled through the soil. Tillage tool type and depth are recorded to determine fluctuations caused by the tillage tool. Topography is determined from the GPS data. Grid soil sampling is used to provide data on other potentially confounding parameters (e.g., moisture, organic matter, etc.).

The instrumented bolts–pins (and other sensors) are connected to the onboard data system complete with GPS input and RF link to a base station. Information from the draft measurement system (including soil sample data) is transferred to an analysis system (computer algorithms, consultant data system, etc.) that identifies, or assists in the identification, of unique soil–landscape management zones. Identified management zones can then be used by the scientist–consultant–farmer to develop water application recommendations for input into a variable-rate irrigation controller.

DRAFT SENSING SYSTEM IMPLEMENTATION

The hitch-pin draft measurement system, as installed on conventional farm equipment, was able to measure spatial variations in tillage tool draft. These measured variations reveal distinct spatial patterns across the field (Fig. 1). Most notable was an area that ran from north to south near the center of the field (Fig. 1). Tractor speed, as recorded with the GPS, was highly sensitive to tillage tool draft, however topography significantly confounded speed measurements. The hitch pin was nearly unaffected by nonextreme topography. Tractor fuel consumption also was monitored for correlation to tillage tool draft. Fuel consumption was not useful in this application, since the tractor remained fully loaded under minimum drafts and on negative slopes.

An apparent correlation between areas of greater draft and wheat/biomass yield was observed along the north–south strip of greater draft and in the corners of the field beyond the reach of the pivot irrigation system (Fig. 1, and 2). Speculation suggests that these suspected correlations may be related to soil

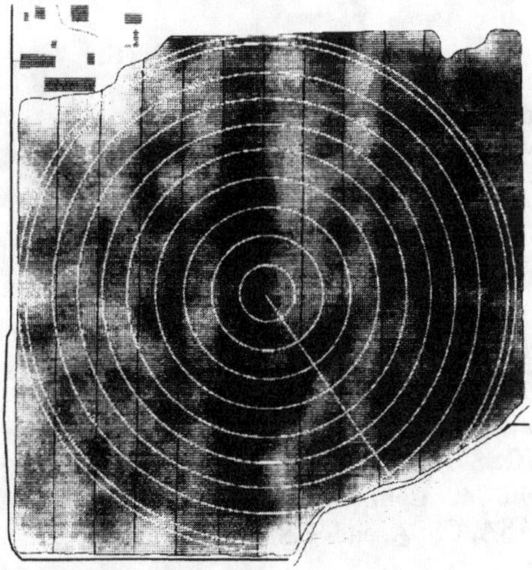

Fig. 1. Spatial map of tillage tool draft. Darker gray = greatest draft.

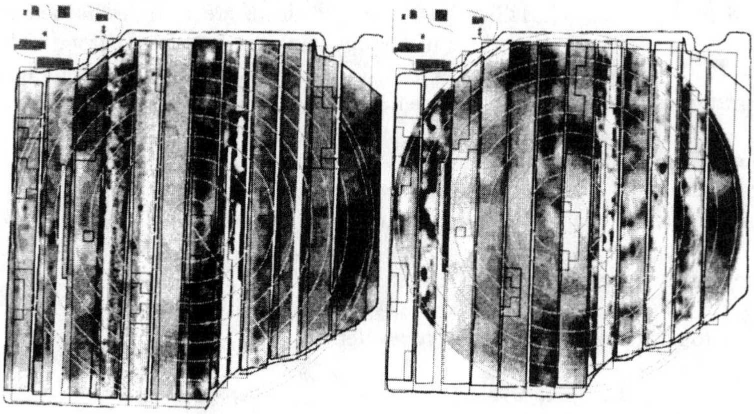

Fig. 2. Spatial map of total wheat biomass yield (left map) and wheat grain yield (right map). Darker gray = areas of highest yield.

moisture patterns.

CONCLUSIONS

The correlations observed between draft and wheat/biomass yield require further analysis, including consideration of possible confounding factors, to validate. To accomplish this, draft measurement systems are being implemented on four sites throughout Idaho and Washington in 1998. These sites will be extensively characterized for moisture, soil physical proprieties and yield. Three of the sites will have variable--rate irrigation systems installed. Artificial Intelligence algorithms included in a decision support system will be employed to identify predictable relationships between draft and crop production (Hoskinson & Hess, 1999, this publication).

ACKNOWLEDGMENTS

A portion of this work was funded by the INEEL Laboratory Directed Research and Development (LDRD) program under DOE Idaho Operations Office Contract DE-AC07-94ID13223.

REFERENCES

Baumgartner, R.L. 1997. VentureStar single stage to orbit reusable launch vehicle program overview. p. 1033–1039. *In* Space Technology and Applications Int. Forum, 2nd Conf. on Next Generation Launch Systems.

Hoskinson, R.L., and J.R. Hess. 1998. Using the decision support system for agriculture (DSS4Ag) for wheat fertilization. *In* P.C. Robert et al., (ed.) 1998 Proc. 4th Int.. Conf. on Precision Agriculture, St. Paul, MN. 19–22 July, 1998. ASA, CSSA, and SSSA, Madison, WI.

Field Comparison of Two Soil Electrical Conductivity Measurement Systems

R. M. Fritz
D. D. Malo
T. E. Schumacher
D. E. Clay
C. G. Carlson

Plant Science Department
South Dakota State University
Brookings, South Dakota

M. M. Ellsbury

Northern Grain Insect Research Laboratory
USDA-ARS
Brookings, South Dakota

K. J. Dalsted

Engineering and Environmental Research Center
South Dakota State University
Brookings, South Dakota

ABSTRACT

Bulk soil electrical conductivity (EC) is influenced by soil water content, salts, and parent material. This study evaluated the ability of the EM-38 or Veris 3100 soil mapping system to describe soil parameters. The Geonics EM-38 is a widely used noninvasive soil electromagnetic induction meter that measures soil (EC) for the top 120 cm of soil (vertical configuration 30 cm from soil surface). Veris 3100 sensor cart is a direct contact soil EC meter that measures soil EC for the surface 33 and 100 cm of soil. A comparison was made between both systems at matched GPS location points. Additional information at these points were gravimetric water content and NO_3–N concentration. The two measuring systems were highly correlated to each other and water content. The EM-38 and Veris deep reading were similar while the Veris shallow reading was lower and higher than the EM-38 in the dry and wet areas, respectively.

Copyright © 1999 ASA-CSSA-SSSA, 677 South Segoe Road, Madison, WI 53711, USA. *Proceedings of the Fourth International Conference on Precision Agriculture.*

INTRODUCTION

On-the-go measurement of bulk soil electrical conductivity (EC) is an inexpensive approach that can be used to conduct an initial survey of a field (Cannon et al., 1994). Soil EC can be measured with instruments like the EM-38 and Veris 3100 sensor cart (Rhoades & Corwin, 1984). These systems measure soil EC directly by contact with the soil surface or indirectly through electromagnetic induction (EMI). The Geonics EM-38 is a EMI meter that uses two coils placed one meter apart to create an electrical field (Rhoades et al., 1989). The EM-38 can be placed in a horizontal or vertical configuration and read to a depth of 75 cm and 150 cm respectively (Corwin & Rhoades, 1990). The Veris Technologies 3100 Sensor Cart is a direct contact soil EC measurement meter that uses a system of coulters to pass current through the soil. The Veris system uses different spacing to create measurements of the top 33 cm and the top 100 cm simultaneously (Christy & Lund, 1997). A field study was conducted to compare the two different soil EC meters using GPS in a real-time application.

METHODS

Two soil EC systems were compared in a 10 ha field (-96.80819 N, 44.34077 W) in October 1997. The field has been extensively manured over many years with a continuous corn rotation. Soil samples from the 0- to 15-cm depth were collected from a 30 by 50m grid. At each sampling point 15 cores were collected and composited. Samples were analyzed for NO_3–N and gravimetric water contents.

The field was marked into 15-m transects and each unit followed the same transects so direct comparisons can be made. Both the EM-38 and the Veris systems used Omnistar (DGPS) differential correction.

The EM-38 was fitted 2 m behind a vehicle and mounted in the vertical position to a wooden trailer. The instrument was positioned 30 cm from the soil surface and pulled over the field at a rate of 11 km h^{-1}. Data was collected every second or 3.05 m and recorded the EM-38 output and the positional location by using Omnistar differential global positioning (DGPS). Over the same transects the electrical conductivities were measured by the Veris 3100 Sensor Cart and recorded every second. The Veris system was pulled behind a vehicle at a speed of 11 km h^{-1} and created output containing GPS coordinates of the shallow (0–33 cm) and deep (0–100 cm) soil EC measurements.

Each data set was used to create kridged maps that were placed over topographical surface maps of the field for visual comparisons. Elevation data was collected on a 15 by 15m grid and kridged to create the surface maps. The two data sets were then compared by GPS output for locations that were <3 m apart. This combined data sct was used to create a linear regression of the Veris and EM-38 data sets. The combined data set was then spatially matched to the NO_3–N and gravimetric water content grid points. The data set was then analyzed statistically by correlation using the SYSTAT version 7.0 statistics package (Wilkinson, 1997).

RESULTS

The soil EC of the field was highly associated with topography (Fig.1–3). The values of the two instruments ranged from 70 dS m^{-1} in the lower (footslope) topographical locations to 20 dS m^{-1} in the higher (shoulder, summit) topographical locations. The change in values and its relation to topography can be attributed to the combined effects of water and salt accumulation in the lower elevations.

The EM-38 and two Veris readings were highly associated. The regression equation between the EM-38 and Veris shallow reading was

$$y = 1.816x - 36.46 \qquad r = 0.910^{**} \qquad n = 697$$

where x was the EM-38 reading and **y** was the Veris shallow reading (Fig. 4). The regression equation between the EM-38 and Veris deep reading was

$$y = 1.297x - 13.36 \qquad r = 0.887^{**} \qquad n = 697$$

where x was the EM-38 reading and y was the Veris deep reading (Fig. 5). Both regression equations had slopes that were > 1 and y-intercepts <0.

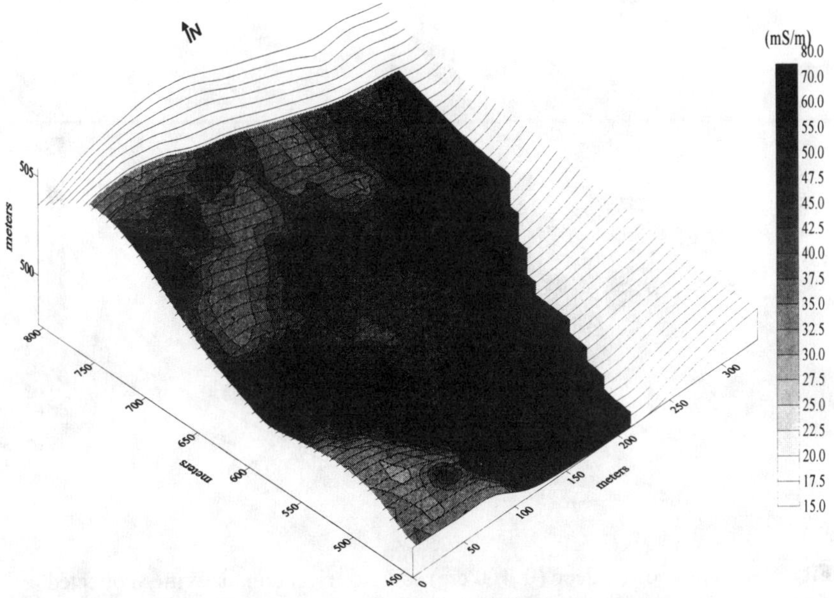

Fig. 1. Actual reported soil electrical conductivities collected every second with DGPS in a 10 ha field by the Geonics EM-38.

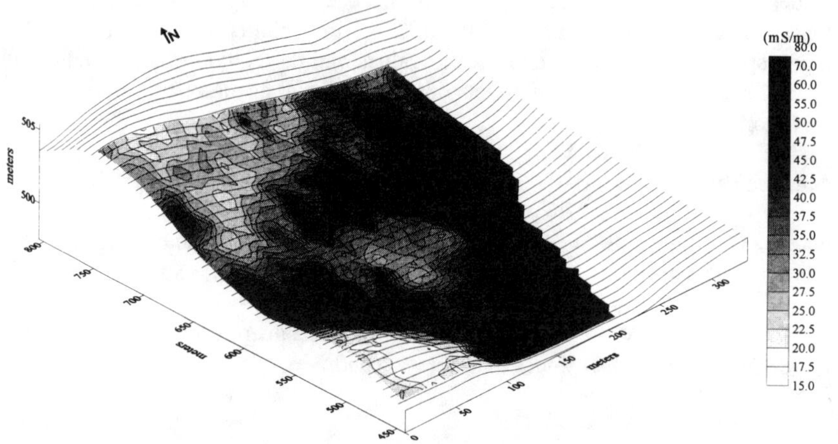

Fig. 2. Actual reported shallow (0–33 cm) soil electrical conductivities collected every second with DGPS in a 10 ha field by the Veris 3100.

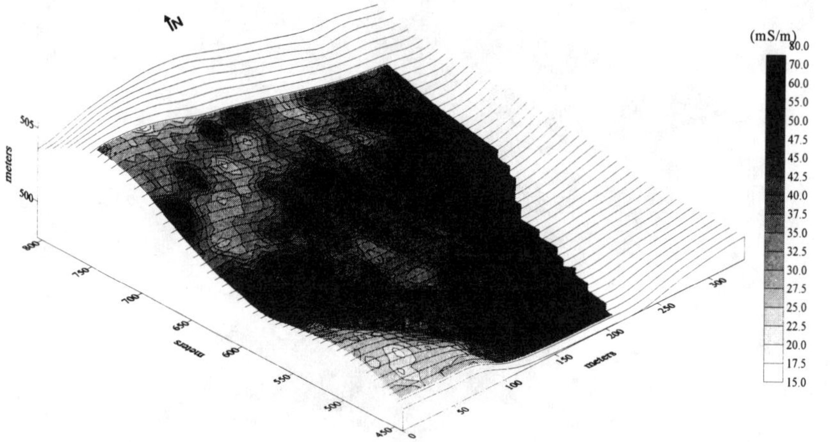

Fig. 3. Actual reported deep (0–100 cm) soil electrical conductivities collected every second with DGPS in a 10 ha field by the Veris 3100.

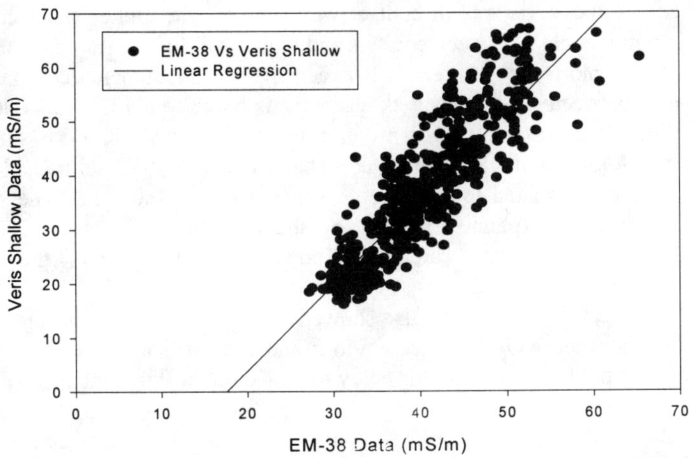

Fig. 4. A comparison between the EM-38 and Veris shallow measurements over the entire field.

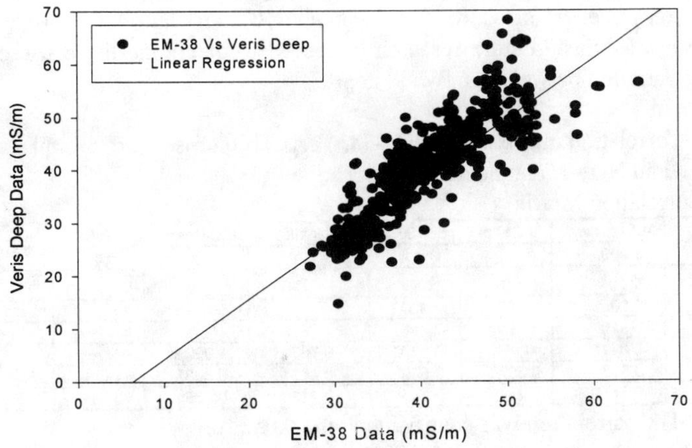

Fig. 5. A comparison between the EM-38 and Veris deep measurements over the entire field.

The EM-38 had higher values in the areas of the field that were drier and lower values in the wetter areas than the Veris shallow (Fig. 4, and 5). This would indicate that the Veris shallow has a measurably different response to water content than the EM-38. These differences may be due to: where the salts have accumulated in the soil relative to the volume of soil measured by the instrument; or that because the Veris was in contact with the soil, the higher water content resulted in better contact between coulters and higher soil EC measurements.

To further explore the differences between the two instruments correlation's were made between the measured values of bulk soil EC, soil NO_3^--N concentrations, and gravimetric soil water contents. The EM-38, Veris shallow, Veris deep, EM-38 subtracted from Veris shallow, and EM-38 subtracted from Veris deep were correlated to gravimetric water content (Table 1). The strong correlations may be explained in part by the combined effects of the high concentration of water and calcium carbonate accumulation in the lower geographical positions.

The data collected in this study also showed that there was what appeared to be a correlation between NO_3–N concentrations and the two soil EC meters. This correlation was not significant in this study but looks interesting and needs further study.

This data set showed that the EM-38 and the Veris 3100 sensor cart provide similar field-based data. In the future we will further examine the EM-38 and/or Veris readings in comparison to field lab data. Maps based on these comparisons may be used to decide where to soil sample for areas of interest to create possible salinity and water content maps for whole fields. These findings could be useful in large-scale mapping of fields for both water content and salinity. Since the two systems appear to systematically differ in there measurements of EC (Fig. 4,5) there is a possibility of subtracting the readings to remove background noise. This technique may better measure field parameters that cause the systematic difference in readings. Future research is needed to determine the cause of the systematic variation between the two systems.

Table 1. Correlation matrix of the EM-38, Veris 3100 sensor cart, % soil water, and NO_3–N at matched GPS locations.

Pearson Correlation Matrix

	V.Shallow	V.Deep	EM-38	EM-shallow	EM-Deep	% water	NO3
V. Shallow	1.000						
V. Deep	0.891*	1.000					
EM-38	0.805*	0.918*	1.000				
EM-Shallow	-0.915*	-0.680*	-0.498*	1.000			
EM-Deep	-0.814*	-0.898*	-0.649*	0.750*	1.000		
% water	0.558*	0.567*	0.520*	-0.463*	-0.508*	1.000	
NO_3	0.347	0.284	0.143	-0.411	-0.385	0.049	1.000

* The correlation was significant at the 0.05 level.

REFERENCES

Cannon, M.E., R.C. McKenzie, and G. LaChapelle. 1994. Soil salinity mapping with electromagnetic induction and satellite - based navigation methods. Can. J. Soil Sci. 74:335–343.

Christy, C.D., and E.D. Lund. 1997. Using electrical conductivity to provide answers for precision farming. Veris Technologies, Salina, KS.

Corwin, D.L., and J.D. Rhoades. 1990. Establishing soil electrical conductivity depth relations from electrical conductivity measurements. Commun. Soil Sci. Plant Anal. 21:861–901.

Rhoades, J.D., and D.L. Corwin. 1984. Monitoring soil salinity. J. Soil and Water. Conserv. 39:172–175.

Rhoades, J.D., S.M. Lesch, P.J. Shouse, and W.J. Alves. 1989. New calibrations for determining soil electrical conductivity depth relations from electromagnetic induction measurements. Soil Sci. Soc. Am. J. 53:74–79.

Wilkinson, L. 1997. SPSS, Inc. Chicago, IL.

Integrated Weighing Platform for Manure Spreader Regulation

F. Thirion
P. Zwaenepoel
F. Chabot

Agricultural and Food Engineering Department
Cemagref
Montoldre, France

ABSTRACT

Livestock producers want to control manure application with more accuracy, for crop management and environment protection. The lack of appropriate equipment for solid manure has lead to the development of an integrated weighing platform, designed for box spreaders with bottom conveyor. After calibration, this sensing device allows measurements of the manure flow during spreading with an acceptable accuracy. In the near future, it could be incorporated into a regulation system with control on the conveyor velocity, to apply a constant rate or variable rates within a field, based on application maps.

INTRODUCTION

Livestock producers have at their disposal large quantities of manure that allow man-made chemical fertilizers savings. The problem faced with manure is to apply the desired quantities, for avoiding lack of nutrients to the crop or contamination of the environment due to nitrate leaching. It must be emphasised that regulations in Europe will limit in the near future the amount of organic N applied annually to 170 kg ha^{-1}, in many regions where density of livestock is high or where NO_3 exceed 50 mg L^{-1} of water resource. Livestock producers have many difficulties to manage this problem with solid manure, due to its variability and also, the incidence of loading techniques: field tests have shown big errors in the estimation of the application rate. For example, it is quite usual to obtain 50 tons of manure applied for 30 tons desired. Moreover, many types of spreaders show an uneven spreading pattern transversally and/or longitudinally, i.e., in the driving direction of the tractor.

Different methods have been studied or are under development to accurately measure and regulate solid manure application rates. An indirect method is based on torque measurements on the spreading device, as there is a direct relation between the p.t.o. power requirement and the quantity of manure spread; however, the comminution resistance of the manure acts also upon the power requirement (Malgeryd & Wetterberg, 1996). A direct method recently developed by the Department of Biological Systems Engineering (University of Wisconsin), consists in equipping the manure spreader with load cells to measure the emptying as the spreader moves through fields ; field irregularities lead to great dynamic

Copyright © 1999 ASA-CSSA-SSSA, 677 South Segoe Road, Madison, WI 53711, USA. *Proceedings of the Fourth International Conference on Precision Agriculture.*

variations in spreader weight which must be treated by an appropriate calculating method based on a high frequency sampling rate. With this method, the total quantity of manure applied can be known with a very good accuracy.

The method presented in this paper, recently patented, has been designed to control the flow of manure of box spreaders with bottom conveyor and spreading device, which are the most common in Europe for the spreading of solid manures. It is well known that box spreaders have an uneven spreading pattern of manure in the longitudinal direction with a proportionally greater amount of manure spread during the first part of the driving distance and a lower amount in the remaining part. This phenomenon is caused by the crumbling of the manure at the end of the discharge; moreover, an uneven load is always transferred into an uneven profile of distribution. So, the method developed should match to adjust the discharge rate at the level desired by the driver or transmitted by the board computer.

MATERIALS AND METHODS

Description of the sensing device

The sensing device has been installed into a box spreader Miro Heywang SH 45/80 with box dimensions of 4.5 m length, 1.43 m width, 1 m height. The bottom conveyor, with transversal U-shape slats every 37 cm, is driven by an hydraulic motor via two bed chains. The spreading device consists of two vertical beaters of 700 mm diam with a frontward angle to get a 6 m width spreading pattern. A rear hydraulic door permits the loading and spreading of thick liquid manures (fig.1).

Fig. 1. Box spreader instrumented to control the flow manure.

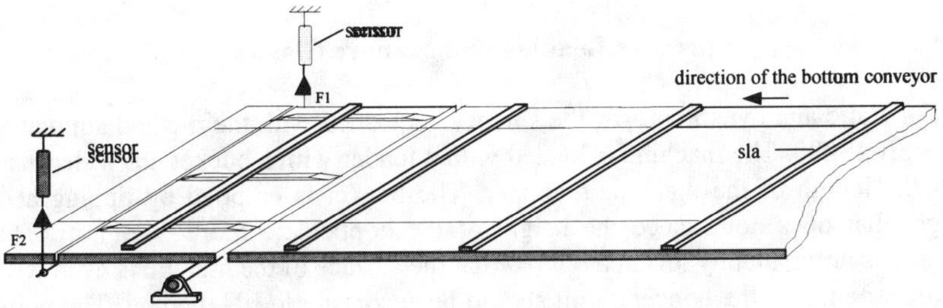

Fig. 2. A schematic diagram of the weighing platform inside the box spreader.

The sensing device has been inserted transversally inside the spreader: the metallic floor has been cut between two struts and replaced at the same level as the floor by a weighing platform of 0.425 m $*$ 1.2 m; the front side of the platform is articulated around an axle, and the rear side is hung to the box through two vertical weight sensors (1000 daN range each with an accuracy of 2 daN). This platform is located on the rear part of the spreader just before ejection of the manure (Fig. 2).

The two weight sensors measure the vertical forces acting on the platform, due to manure and slats moving on its surface. As the axle of the platform is at the same level as the floor, the friction forces due to the sliding of the manure and the slats on the platform surface act on the axle and not on the sensors. On a rough estimate, these vertical forces correspond to the apparent weight of the manure and the slats on the platform; to get the actual weight of the manure, specific corrections should take into account : (i) the influence of the manure cohesion coefficient, as the manure above the platform is not physically independant of the remaining load (ii) the friction forces of the manure along the lateral sides of the box, and (iii) the influence of the slats.

Because of manure variability and heterogeneity of the manure load, these specific corrections are difficult to estimate with accuracy. For this reason, calibration tests have been conducted to determine the ratio between actual load and apparent weight measured with the weighing platform, in static and dynamic conditions, and for a large variety of manures.

Calibration Tests

Static and dynamic calibration tests have been realized. For these tests, elements of the European standard prEN 13080 « Manure spreaders - specifications for environmental preservation - Requirements and tests methods » have been used. This European standard, under CEN enquiry (European Committee for Standardization), specifies requirements for environmental safety and their verification for design and construction of spreaders for spreading manure in agriculture and horticulture; it specifies requirements for transversal and longitudinal spreading characteristics such as acceptable working width, steady flow, stretch with tolerance zone and coefficient of variation for longitudinal spreading.

Spreader Loading and Manure Classes

For static and dynamic tests, the spreader is loaded with the method stipulated by prEN 13080. The machine is loaded with a loader with a bucket not wider than half the length of the machine's hopper. The bucket is emptied by tipping at a height that does not exceed the height of the hopper, no more than 1 m. The manure is not manually levelled out (so that the surface of the manure is even with the upper edge of the hopper), as it should be according to EN method. The point was to observe whether variation of load was observed with the sensor.

The bulk density of the manure is calculated on the basis of mass and rough volume (6.4 m^3) of manure inside the box spreader. Different manure classes have been used, with bulk densities between 480 kg m^{-3} and 850 kg m^{-3}, to get a wide variety of conditions as found on farms.

Static Tests

The static tests are carried out according to the method proposed for the measurement of longitudinal distribution of manure spreaders in prEN 13080.The longitudinal distribution, corresponding to the flow of manure, is measured stationary by registration of the machine's weight and the time elapsed since the beginning of the test. The whole spreader–tractor is weighed continuously during the unloading of the manure with six electronic scales ALCO R/R10, each one under a wheel, connected to an electronic box CAPTEL VPR HS 8. The total weight is registered every 0.734 s with a resolution of 10 kg. The data are filtered by mean of the mobile average technique on 10 values, prior to the calculation of the unloading flow. The signals of the two weight sensors are captured independently at a high sampling rate, to give average data every second during unloading, which are then filtered with the same mobile average technique.

For each test, the calibration coefficient k is calculated, assumed constant for the test, and defined as the ratio between m (mass of the manure on the weighing platform) and m_a (apparent mass of manure on the weighing platform).

$k = m/m_a$

k is calculated in the following way :

$$(F_1 + F_2) \times l = m_a \times g \times \frac{l}{2} \qquad \text{(equilibrium of momentum)}$$

$$m = 2\frac{k}{g}(F_1 + F_2)$$

where F_1 and F_2 are the vertical forces measured by the weight sensors, and l, the width of the platform.

$$M = \int q dt \Rightarrow M = \frac{v}{l}\int m dt$$

$$M = 2\frac{v \times k}{l \times g}\int (F_1 + F_2)dt \Rightarrow k = \frac{M \times l \times g}{2v\int (F_1 + F_2)dt}$$

where q is the flow rate of manure, M the total mass of manure spread and v the velocity of the conveyor.

The calibration coefficient k also can be estimated by means of a mathematical adjustment between the weight curves from the sensing device and the electronic scales, to get the best correlation coefficient between the two sets of data.

Dynamic Tests

The whole spreader–tractor is weighed before and after spreading. The signals of the two weight sensors are registered independently every second during unloading in the field.

For each test, are measured or calculated: (i) the weight of manure spread, the bulk density and the speed of the conveyor, and (ii) the average flow, calculated on the basis of the weight of the manure spread and the unloading time, the calibration coefficient k.

RESULTS AND DISCUSSION

Static Tests

The Fig. 3 gives an example of the static tests carried out, after correction of the weighing platform curve by a calibration coefficient of 2.38. The curves show small peaks due the effect of the slats, which act in compacting the manure in front of them; we may observe the same peaks for both curves: thrust of a compacted block in one case, discharge of a compacted block for the other case. The correlation coefficient 0.94 between the two curves shows that k can be assumed constant for each test and that the weighing platform gives a good estimation of the manure flow.

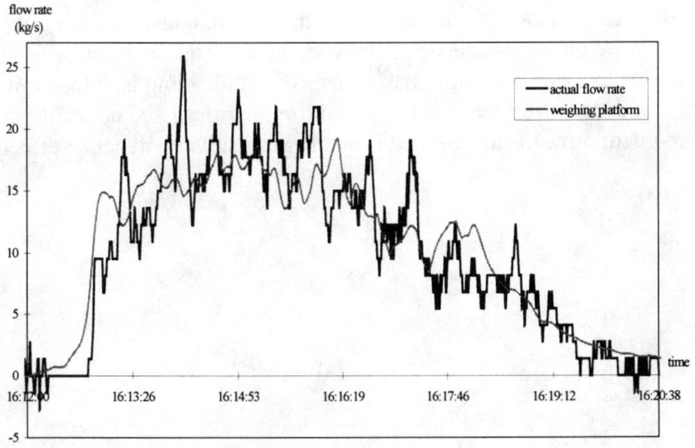

Fig. 3. Comparison between flow rate data from the electronic scales (actual flow rate) and the weighing platform.

All static tests have confirmed the ability of this sensor to measure relative variations of the flow with a good accuracy. As this flow sensor is located before the spreading device, it seems feasible to develop a regulation system based on it, with the advantage of anticipation on the coming flow to get the desired rate.

Dynamic Tests

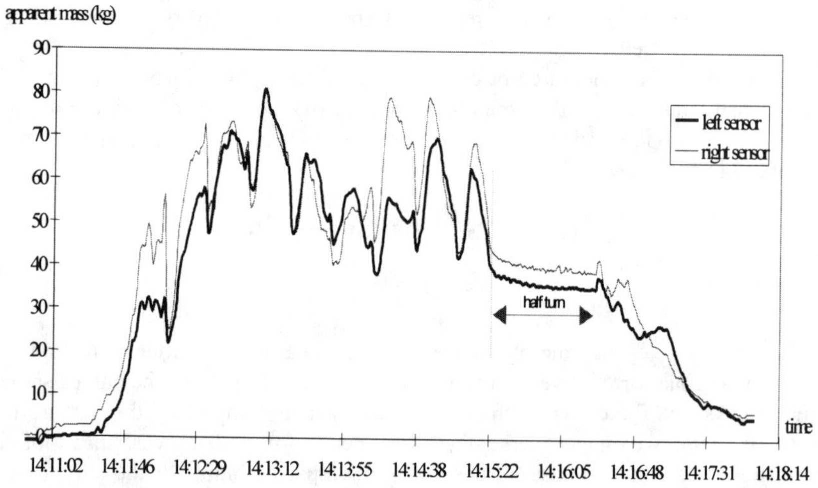

Fig. 4. Apparent mass on the weighing platform during a dynamic test.

The Fig. 4 gives a characteristic example of the dynamic tests carried out. The signal of the two weight sensors is not exactly identical, certainly because of some heterogeneity due to loading. During half turns, when unloading is stopped, it can be observed that dynamic effects due to ground irregularities have no influence on the signal recorded; it means the method is nearly insensitive to dynamic effects.

Table 1. Determination of the calibration coefficient *k* through dynamic tests with different types of manure

Test n°	Mass spread	Bulk density	Conveyor speed	Average flow	*k*
	kg	kg m^{-3}	m mn^{-1}	kg s^{-1}	%
2	5390	840	1.84	32	2.14
3	5460	850	1.65	29	2.2
4	5010	780	1.09	17.6	2.68
6	5150	800	1.08	17.1	2.43
7	5300	830	1.08	17	2.46
25	4500	700	1.04	14.6	2.48
26	4690	730	1.05	17.5	2.12
27	4630	720	1	14.3	2.5
28	4220	660	1	12.8	2.67
29	3070	480	0.98	11.4	2.22
30	3130	490	1.07	13	2.19
Mean	4595				2.37
CV	18 %				8.7

In a wide variety of conditions, with bulk densities from 480 kg m^{-3} to 850 kg m^{-3}, and conveyor speeds varying between 0.98 and 1.84 m mn^{-1}, the average value of *k* is 2.37 with a variation coefficient of 8.7 % (Table 1). These tests permit assessing the capability of this method to estimate the load of manure with <10% error.

CONCLUSION

The static and dynamic tests carried out with different manures show that the weighing platform, designed for box spreaders with bottom conveyor, can be used as a flow sensor for the spreading of solid manures. To get the actual flow value in kg s^{-1}, this method requires determining a calibration coefficient, which depends on spreader–weighing platform dimensions and manure physical characteristics. This coefficient has to be determined once by the spreader manufacturer for each type of spreader, with usual manures; on this basis, the accuracy on the flow measurement should be better than 10%, as shown in the tests with different manures. For particular products, as poultry manures or industrial wastes, this coefficient can be estimated by the user from dynamic calibration tests as described before.

From this flow sensor, a relevant regulation could be developed to control the discharge rate through the adjustment of the bottom conveyor velocity. Some preliminary tests, not described here, have shown the feasibility of this regulation through action on the hydraulic motor driving the bottom conveyor. The next step should concern the incorporation of this box-spreader in a precision farming set up, so that the spreader, combined with positioning, could spread manure at variable rates within a field, according to soil characteristics and crop needs.

REFERENCES

Malgery, J., and C. Wetterberg. 1996. Physical properties of solid and liquid manures and their effects on the performance of spreading machines. J. Agric. Eng. Res. 64:289–298

Evaluation of Commercial Cotton Yield Monitors in Georgia Field Conditions

Calvin D. Perry
Jeffrey S. Durrence
Daniel L. Thomas
George Vellidis
Biological and Agricultural Engineering Department
University of Georgia
Tifton, Georgia

Chris J. Sobolik
John Deere Precision Farming
Moline, Illinois

Alan Dzubak
John Deere Des Moines Works
Ankeny, Iowa

ABSTRACT

Two commercial cotton yield monitoring systems (YMS) were marketed in the Fall of 1997. The systems were purchased and installed on three cotton pickers in south Georgia. Installation of both YMS involved much time, labor, and specialized tools. Field-scale accuracies of the systems were investigated by using the systems to map yields on three fields (113, 19, and 7 ha) representing extremes in cotton production levels in Georgia. Results indicated one system's accuracies were within 3% of actual yield while the other system exhibited a considerably larger error. Instantaneous accuracies of the two systems were researched by harvesting small plots of varying cotton yield levels. The errors for both systems over 33 m (100 ft) plots were large but were not representative of actual system performance. The growers involved in the investigations were impressed that commercial yield monitors were available, but the YMS overall performance was not sufficiently rigorous or accurate to encourage their purchase. Improved versions of both systems are needed before widespread use will begin in Georgia.

INTRODUCTION

In 1997, two companies introduced yield monitoring systems (YMS) for cotton. The Zycom[1] Corporation (Bedford, MA; www.zycomcorp.com) and Micro-Trak Systems (Eagle Lake, MN; www.micro-trak.com) marketed systems to be

[1]The use of trade names in this publication does not imply endorsement by The University of Georgia of products named nor criticism of similar products not mentioned.

Copyright © 1999 ASA-CSSA-SSSA, 677 South Segoe Road, Madison, WI 53711, USA. *Proceedings of the Fourth International Conference on Precision Agriculture.*

mounted on 2- or 4-row cotton pickers from John Deere (Moline, IL) or CaseIH (Racine, WI). The precision farming research team at the University of Georgia Biological and Agricultural Engineering Department at the Tifton Campus (UGA-BAE) began an investigation to evaluate the two commercial cotton YMS.

Three Micro-Trak YMS were purchased and installed on three John Deere 9965 four-row cotton pickers, and two Zycom YMS were purchased and installed on two of the pickers. The 4-row pickers were provided by UGA-BAE, John Deere Precision Farming (JDPF), and a grower in Crisp County, Georgia.

Both the Micro-Trak and the Zycom YMS used light as their sensing medium. The two systems featured sensors that contained light emitting diode (LED) –photodiode pairs that mounted on opposite sides of the picker delivery duct. The emitter/receiver pairs were arranged in arrays and mounted such that cotton bolls passing between the pairs block light and were sensed. The two vendors employed proprietary algorithms for calculating actual mass flow rate from the sensors. Neither system contained a moisture sensor; therefore each system recorded wet yield with no correction for moisture. The yield reported by both YMS was for seed cotton and not lint cotton.

Many Georgia cotton growers are interested in adopting precision farming techniques but are reluctant to make the transition until a reliable YMS is available; therefore, the goal of this investigation was to provide an unbiased evaluation of the two commercial YMS introduced in 1997.

HARVEST SITES

Three field sites and one plot site were selected for use in this study. All three field sites were grower fields and represented different production practices, terrain, soil types, irrigation practices, yield levels, etc. The first field site (Crisp County, 113 ha or 280 ac) had a relatively low average yield[2] (785 kg ha^{-1} or 700 lb ac^{-1}), relatively flat terrain, some weed pressure, and no irrigation. Three 9965 pickers (2 with Micro-Trak and Zycom, 1 with Micro-Trak only) were used in the harvest. The second field site (Worth County #1, 16.2 ha or 40 ac) had some slope and was terraced. This field had low weed pressure and was irrigated with a center-pivot system. Yields for this field were relatively high for Georgia growing conditions (2800 kg ha^{-1} or 2500 lb ac^{-1}). Only one 9965 picker with both sets of sensors was used. The third field site (Worth County #2, 19 ha or 47ac) was characterized by several depressions, gullies, and highly eroded areas. Minimal irrigation was provided by a traveling gun system. Yield variations were visible and were apparently related to drainage, soil type, and relief. Weed pressure was low in the field, and the average yield was relatively low (1790 kg ha^{-1} or 1600 lb ac^{-1}). Again only one 9965 picker with both sensor sets was used to harvest this field.

The plot site was located on a University of Georgia research farm. The 33 m (100 ft) long plots totaled about 2 ha (5ac). The cotton yield could be characterized as better than average (1340–2690 kg ha^{-1} or 1200–2400 lb ac^{-1}). Weed pressure was

---

[2]Approximately 544 kg (1200 lb) of seed cotton produces a 227 kg (500 lb) bale of ginned lint cotton. All values are reported in kg (lb) of seed cotton.

very low, and the terrain was gently sloped. One 9965 picker with both sensor sets was used.

During the 1997 harvest, the El Nino weather phenomenon wreaked havoc on cotton harvesting. Week after week of heavy rain delayed harvest and damaged bolls. The test harvest at Crisp County was hampered by the rain and by the field's consistently low yields. The test also had problems with valid calibration weights. The combination of low yields, rain, and calibration problems led to little acceptable data being collected. The Worth County #1 site had a good cotton crop, but again rain interrupted harvest. Several attempts were made to harvest, but the soil was too wet to support heavy pickers. Finally, when the weather improved slightly, the grower brought in custom pickers to finish the harvest. At that point only 6.9 ha (17 ac) had been harvested with the yield monitors. The Worth County #2 site was interrupted by rain only once. This harvest produced the only valid data set for both Micro-Trak and Zycom sensors on the same picker; however, the other field harvests allowed other parameters (such as operation, calibration, and performance) to be evaluated under somewhat extreme conditions.

YIELD MONITOR EVALUATION RESULTS

Cost

Both Micro-Trak and Zycom offer complete YMS including GPS receivers, sensors, data storage modules, and mapping software. Customers may opt to supply their own GPS receivers with differential correction source or purchase a DGPS from the vendors. Table 1 lists the costs (in 1997 dollars) of a system designed to equip a 4-row picker (4 ducts) with a complete cotton YMS. The pretax prices reflect the costs of the equipment used in the 1997 investigations.

In comparison, both Micro-Trak and AgLeader (Ames, IA; www.agleader.com) grain yield monitoring systems have list prices of about $3200 without DGPS—considerably less than the two cotton YMS. However, the price for a particular cotton YMS should be relative to cotton picker costs—new pickers range from $180,000 to $250,000.

Table 1. System costs for comparable cotton yield monitors.

Manufacturer	*With DGPS*†	*Without DGPS*
Micro-Trak	$10,630.00	$6,130.00
Zycom	$14,383.00	$10,083.00

† Quoted systems include four sensor units, display unit, (2) storage modules, software, and OMNISTAR 7000 DGPS receiver.

Installation

The installation of both systems was quite involved and tedious despite having a fully-equipped research shop and competent technical support. Both systems required mounting sensors on the delivery ducts, installing hardware in the picker cab, installing a GPS antenna on top of the picker cab, and connecting all equipment with proper cabling, including power. Both systems required a clean 12 VDC power source.

The Micro-Trak flow sensor (IRSA100) consisted of two enclosures each housing the 8 emitters and 8 receivers. The emitter enclosure was 2.54 * 25.4 * 2.54 cm (1 * 10 * 1 in) and the receiver enclosure was 4.44 * 25.4 * 2.54 cm (1.75 * 10 * 1 in). As shown in Fig. 1, the two enclosures were mounted on opposite sides of the cotton duct. Sixteen holes must be drilled in the duct, eight in each side. These 13 mm (0.5 in) holes must be spaced to match the sensor spacing on each enclosure, and the holes on opposite sides of the duct must align properly. Care was taken to locate the sensor enclosures away from brackets or other duct features that could interfere with mounting. Micro-Trak provided adhesive templates to facilitate proper hole placement. Framing squares were used to set hole alignment. All metal surrounding holes was smoothed with files to prevent cotton snags. Once the holes were prepared, the sensor enclosures were mounted by placing the emitter and receiver enclosures over the holes and running a metal strap around both sensors and the duct and then tightening the strap to hold the enclosures firmly in place. Note that the recessed sensors did not extend into the air stream.

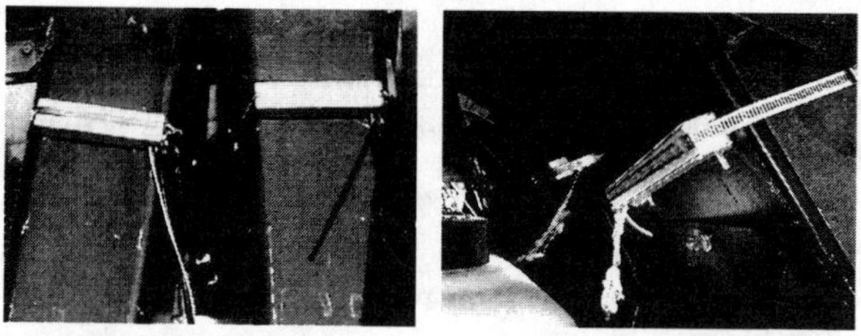

Front view of duct. Rear view of duct.

Fig. 1. Micro-Trak flow sensor mounted on ducts 3 and 4.

The Micro-Trak system also required the installation of a magnetic ground speed sensor. This sensor used a magnetic switch to monitor the rotation of a shaft–wheel in direct proportion to ground speed. The speed sensor was mounted on the drive axle on the left side of the picker. A plastic tie wrap secured the magnet to the axle and the sensor was mounted with a provided bracket.

A harvest indicator, the run–hold switch, was installed next. This magnetic sensor mounted between the picking unit support bar and the cab platform. The sensor was attached to a spring-loaded hinge that opened when the picking units were lowered - indicating harvest action and initiating data collection. When the picking units were raised, the hinge closed and a magnet on the platform closed the magnetic switch, halting data collection.

After all sensors were installed, an interface/control unit, Grain Sensor Input (Tngs100), was mounted underneath the cab platform. The sensors on the ducts were daisy-chained together and connected to the control unit. Then the speed sensor, run–hold switch, as well as display–data collection modules were attached to the control unit.

The display–data collection modules, Grain-Trak (TNy4400) and Data-Trak (TNm1100), were mounted to a bracket that featured a suction cup mount. This mount was attached to the front windshield of the picker for easy viewing and control. An OmniSTAR 7000 DGPS receiver was mounted on the cab of the picker and connected to the Micro-Trak system. Another cable was routed from the display–data collection modules through the cab floor and down to the interface/control unit.

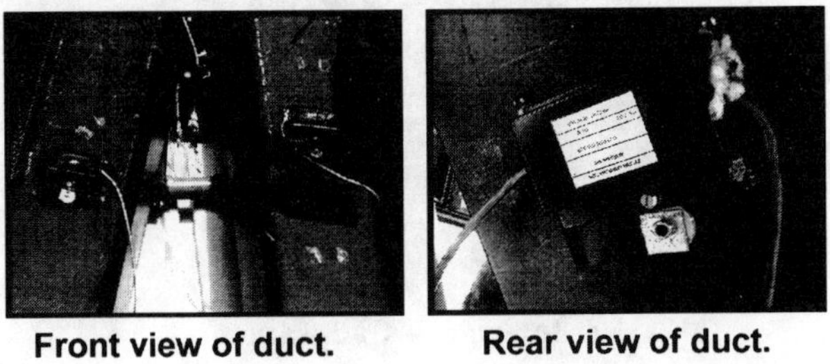

Fig. 2. Zycom flow sensor mounted on ducts 3 and 4.

The Zycom system consisted of pairs of emitter–receiver arrays (AGRIplan 600 Cotton Sensor), one pair on each duct. The emitters and receivers were mounted in similar housings (Fig. 2) on opposite sides of the duct. Two 7.6 mm $*$ 12.7 mm (3 in $*$ 5 in) rectangular holes (one on each side of the duct) were created by first using framing squares to mark center holes, then drilling starter holes, and finishing the cut-outs with a jig-saw. When installed correctly, the Zycom sensor assemblies had a small lip that extended into the air stream by 2.5 mm (0.1 in) but the sensor's LEDs were recessed from the air stream. A small hole was drilled below each rectangular hole for a bolt to secure the housing. Emitter and receiver sensors were connected to the respective mates, and then two pairs were connected to a *y*-cable. Two *y*-cables joined the four sensor pairs to the cab electronics.

The Zycom YMS used velocity data contained in the GPS output string; therefore, no speed sensor was required. Data collection (run–hold) was controlled by logic operation in the system's firmware, and thus data was logged only after cotton began to pass through the ducts.

A switchbox and a display unit (AGRIplan 600 GPS monitor and AGRIplan Interface unit) were mounted in the cab of the picker. The switchbox provided operator controls and sensor connection and the display unit provided visual feedback as well as GPS connection. The same OmniSTAR 7000 DGPS receiver used with the Micro-Trak YMS provided position data for the Zycom system.

The Zycom YMS use of GPS speed and logic run–hold function coupled with the easier duct installation (two rectangular holes vs. 16 drilled holes per duct) made for a relatively easier overall installation. Nevertheless, both systems required several man-hours to install and the installations would be quite difficult for an average cotton grower. Total installation time would be conservatively estimated at 6 h for the Zycom system and 10 h for the Micro-Trak system for a first and only unit installation (typical end-user).

Fig. 3. Micro-Trak interface modules.

User Interface – Calibration

The user interface for the Micro-Trak YMS consisted of two modules, the Grain-Trak (TNy4400) and the Data-Trak (TNm1100), (Fig. 3), that make up the company's grain yield monitoring system. The Grain-Trak featured a backlit LED display of various harvest parameters such as load–field–season counters, current or average yield, number of rows being harvested, setup and calibration parameters, including sensor check, GPS differential confirmation, ground speed, and distance or area harvested. Since the Grain-Trak was being used for cotton sensors, many of the harvest and calibration parameters were not applicable to the cotton YMS. The Grain-Trak provided three toggle switches to navigate through menus, make selections, and change values in such settings as number of rows, manual run–hold, active counter displayed (load–field–season), counter zeroing, and calibration–setup parameters. Navigating menus was fairly straightforward after some practice. The Data-Trak module provided a PCMCIA memory card drive and a DGPS interface to the Grain-Trak module. The Data-Trak used SRAM memory cards to store yield, area, and position data.

Calibration of the two YMS was critical to their successful operation and accuracy. Calibration of the Micro-Trak involved adjusting both the speed sensor and the flow sensor calibration coefficients. The speed sensor was calibrated by first marking off 500 m (1640 ft) of level ground similar to field conditions. The Grain-Trak distance counter was then zeroed, and the picker driven the 500 m. At the end of the 500 m the Grain-Trak distance value was corrected to match the 500 m traveled by adjusting the appropriate calibration parameter. This was repeated to verify the calibration.

Calibration of the Micro-Trak flow sensors was accomplished by first zeroing the load values, picking a basket load of cotton, and unloading the cotton into a buggy or trailer to determine the seed cotton weight and storing the sensor indicated load weight. The picker could then be used to harvest while weighing the cotton. Once the actual weight of the cotton was determined, the flow sensor calibration value could be fine-tuned by adjusting this value until the stored load weight matched

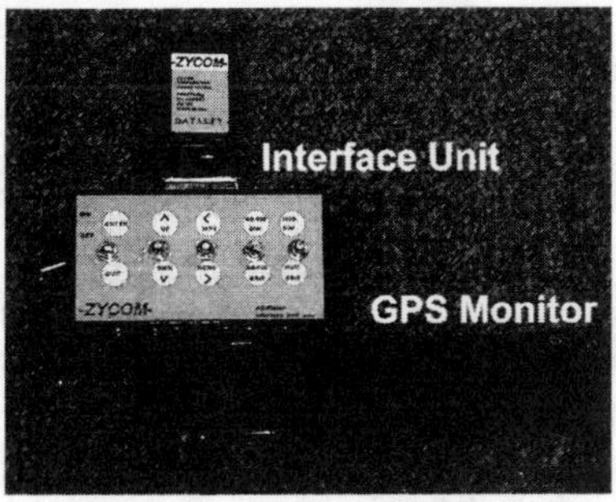

Fig. 4. Zycom user interface.

the actual weight of the cotton. The user manual suggested the calibration procedure be performed twice.

The Zycom YMS user interface consisted of the GPS monitor and the Interface unit (Fig. 4). The GPS monitor display unit was functional but had many limitations. The display provided minimal information and required considerable effort to operate. The separate Interface unit used toggle switches to control the GPS monitor. The display unit indicated acres harvested, current yield (lb/ac), and total weight (lb) during harvest operations. Other menus available included sensor test, setup values, and calibration. Changing or checking parameters required an elaborate combination of switch operations.

One important difference between Zycom and Micro-Trak was the Micro-Trak's ability to maintain individual load, field, and season counters whereas the Zycom only maintained data in fields. If a grower maintained a log of actual weights of individual loads, with a Micro-Trak YMS he could compare those weights to recorded sensor totals for loads–fields–season, whereas he could not make the comparisons with the Zycom system directly unless a new field was designated for every load.

Data storage was provided by a proprietary data module that used a parallel port interface and connected directly to the Interface unit. A special parallel cable was included to connect the module to a computer's parallel port for downloading data to mapping software.

As the Zycom YMS did not require a ground speed sensor, the only calibration required was of the yield coefficients, a procedure similar to the Micro-Trak. The empty picker was operated as usual. A toggle was pressed to start the calibration data collection. A full basket of cotton was harvested then the toggle was pressed again to stop the calibration and to store the yield value. The cotton was dumped into a trailer and weighed. The picker could resume harvesting while the calibration cotton was weighed, however, the first harvest data could not be recalibrated from our limited investigation of the system. When the actual cotton weight was obtained, the weight was entered into the Zycom GPS monitor, thereby fine-tuning the calibration coefficient. Zycom recommended only one calibration per season.

This type of calibration created a dilemma. In low yielding cotton, the area harvested to fill the picker could be considerable. Doing the calibration twice would obviously remove even more area from a potential yield map. Also the researchers had to decide whether to wait for the calibration weight or continue harvesting. It was never clear to the research team whether crop harvested prior to fine tuning the flow sensor calibration could be mapped with the new calibration value. Since this investigation required as many comparisons of yield monitor weights versus actual weights as possible, no more than one additional basket load was harvested while waiting for a calibration weight. Obtaining accurate calibration weights proved to be another challenge. Cotton trailers had to be weighed empty and then with the cotton inside. Weighing stations were often several miles away and had scales of unknown accuracy. Cotton boll buggies with load cells also suffered from accuracy problems.

To attempt to solve the problem with calibration weights, UGA-BAE and JDPF installed load cells on two 9965 pickers. The load cells would be used to measure the calibration loads as well as to compare subsequent loads to the values indicated by the two yield monitors. A portable data acquisition system was installed in the cabs of the pickers. During the harvest at Crisp County, the load cell weights were considered the true weights and all calibrations were performed using the load cell weights. Contrary to the YMS user manuals, calibration was performed after each full load. Later calibration was performed only if unusually large differences occurred between the load cell value and the two yield monitor values or if field conditions changed. This practice led to doing three to four calibrations during a 6-h harvest period.

Results from the Crisp County harvest caused suspicion concerning the load cell performance and accuracy. A re-test of the load cells disclosed problems with hydraulic cylinders, load cell mounts, and ground slope. The errors induced were severe enough to deem Crisp County calibration data unusable. Later calibrations were done with either cotton weighed in trailers, boll buggies, or hand-placed in bags and weighed on laboratory scales.

Operation – Performance

Even though early calibrations were invalid, much knowledge about the operation and performance of the two YMS was gained from the cotton harvest. Since both YMS were installed on two of the 9965 pickers used in the 1997 harvest, direct comparisons were possible. Neither system's sensors failed due to excessive wear. All installations remained intact and required little adjustment. This was despite the fact that the positions of the sensors were vulnerable to damage from foot traffic up and down the ladder, raising and lowering the basket, and extending the ducts. The researchers exercised extreme caution to ensure that no cables were pinched or stepped on during all operations.

Even though care was taken to smooth all drilled and cut edges created when installing the yield sensors, cotton pieces caught on these edges and partially or completely blocked the sensor eyes. After each dump, the researchers checked each sensor for cotton snags and cleaned them if necessary. The Zycom sensor design made checking and cleaning straightforward and simple. The sensor cover was unscrewed and rotated upward, exposing the eyes for inspection and cleaning. If the null frequency displayed by the Grain-Trak indicated a problem with the Micro-Trak flow sensors, a person had to climb onto the picker and peer into the discharge end of each delivery duct. By using a flashlight and a telescoping mirror, inspection and cleaning could be accomplished.

The Micro-Trak had an occasional problem with blocked sensors. The Grain-Trak would indicate that a sensor was blocked but no amount of cleaning would remedy the problem. A work-around was to "null" the system with the sensor blocked and continue harvest. This technique worked unless the blocked sensor suddenly self-cleaned and began functioning during harvest. When this happened, the system collected erroneous yield data until the system was re-nulled. The Micro-Trak sensors also had problems with dust–trash collecting on the sensor eyes after several rounds of harvest. When this happened, a soft cloth was used to clean the eyes. The manufacturer recommended applying clear shipping tape over the sensor eyes to try to prevent this from occurring. One 4-duct set of Micro-Trak sensors was taped but dust buildup continued to cause problems.

Zycom's recessed sensor mounts prevented problematic dust and trash from building on the lenses during harvest; however, the mounting location caused problems for its automatic harvest indicator. If the top end of the lower inner duct section had cotton clinging to it (as happened in this investigation), and the picking units were raised, the cotton clinging to the end of the inner duct caused the Zycom sensor to be blocked and the system to collect data. This problem caused a small addition to the cumulative load with no cotton being added to the basket.

Accuracy

Due to calibration and harvest problems in the Crisp County and Worth County #1 fields, the Worth County #2 field and the small plot study provided the only data sets suitable for accuracy determinations. The UGA-BAE 9965 picker equipped with both Zycom and Micro-Trak YMS was used for both harvests.

The Worth County #2 field (19 ha or 47 ac) harvest consisted of 2 d of picking followed by several days of rain and finally a third day of harvest. Both YMS were calibrated at the beginning of harvest and were not recalibrated during the first 2 d. On the last day of harvest (following rains), the YMS were recalibrated since field conditions had changed. A summary of the harvest is given in Table 2. Each basket of cotton was dumped into a cotton trailer and towed to a nearby agribusiness to be weighed on legal-for-trade scales. The Zycom YMS provided reasonably accurate yield values for the whole field total, while the Micro-Trak system consistently overestimated the actual yield. The majority of this error was attributed to calibration and sensor blockage problems caused by dust–trash accumulation. This field did not have good defoliation; therefore, more leaf material passed by the sensors.

Table 2. Harvest summary for 19 ha (47 ac) Tift County #2 field.

Harvest date	Actual yield kg	Micro-Trak			Zycom		
		Yield kg	% Error	Area ha	Yield kg	% Error	Area ha
25NOV97	17636	20854	18.2	9.3	17295	-1.9	9.7
26NOV97	11839	15130	27.8	6.1	11745	-0.8	6.5
05DEC97	4390	4676	6.5	3.2	4506	2.6	3.3
Total	33865	40660	20.1	18.6	33546	-0.9	19.4

The yield maps shown in Fig. 5 show the data collected by the two YMS. The Micro-Trak output was limited to one sample every three seconds whereas the Zycom stored data once per second. The horizontal bands appearing in the Micro-Trak map are attributed to sensor blockage occurrences. Unfortunately, these sensor problems mask some of the yield variations expected in the field. The Zycom map seems to better represent specific yield level regions created by the field topography and soil types.

Figure 6 indicates the results of the small plot harvest. The plot test was designed to investigate the instantaneous accuracies of both YMS by harvesting nine 33 m (100 ft) long plots (four rows each) of cotton. Three yield levels were created in the plots by manually removing or adding cotton stalks. Several rows of cotton were harvested to generate calibration data prior to harvesting the first plot. After each plot was harvested, the picker was stopped and the cotton was bagged and removed from the basket to be weighed later. Thus the actual weight of all cotton passing through the four ducts could be determined at the end of every plot.

The obtained results did not agree with those seen in whole field testing. Both YMS underestimated the measured plot yields except for one anomaly with the Zycom system. On the average, the Micro-Trak and the Zycom systems underestimated yields by 22.8 and 22.4 percent, respectively. Potential reasons for

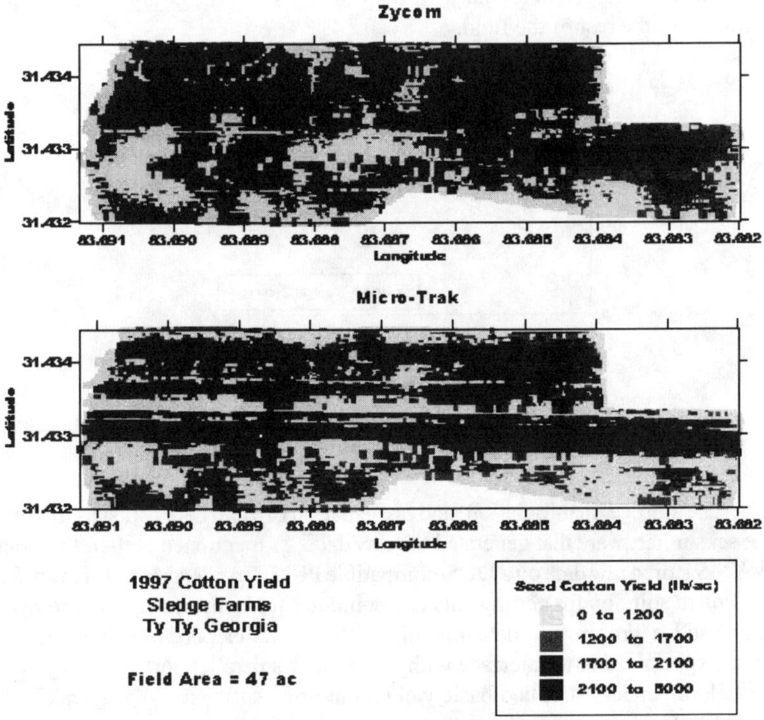

Fig. 5. Yield maps from Worth County #2 field.

the large errors are only a guess since the exact yield algorithms are not known. The most likely cause could be referred to as edge effects. When observing yield maps, most often there are low yields at field boundaries. Sometimes this is actually low yielding areas. With grain combines, the crop is transported from the edge of the field before the sensor senses the yield (i.e., a lag time). However, with a cotton picker there is a very small lag time between the picking unit reaching the bolls and the bolls passing the sensors. The edge effects were likely caused by the yield calculation algorithms used in the two YMS. A simple moving average illustrates this situation. This data analysis technique is employed to remove erroneous yield data from raw samples but requires initial conditions for its calculation. Data collected at the beginning of a row may be improperly weighted since no initial conditions were available. The digital filters likely used in YMS involve some form of moving average and thus depend on initial conditions. Data taken at the beginning of the row will likely be adversely effected by such moving average algorithms. If edge effects were the problem for both YMS, then the large errors found in the small plot study can be explained. One additional reason to suspect edge effects is the high correlation between the Micro-Trak and measured results (99%). Although the Zycom correlation was lower (66%), removing the anomalous data value improved the correlation considerably (95%). To avoid this problem, plots would need to be embedded into a larger field and then both the data and the cotton would require isolation from the rest of the field.

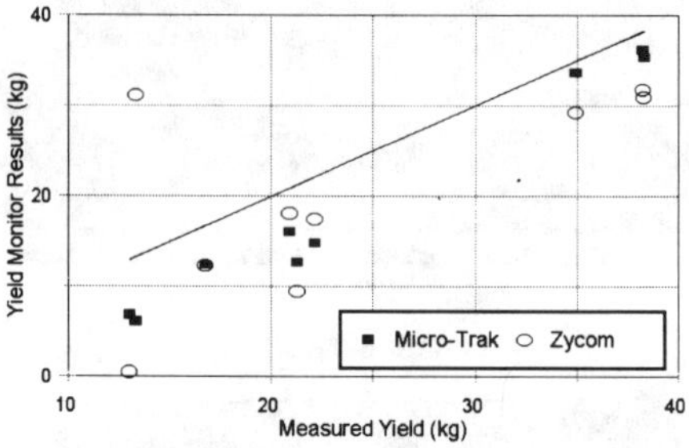

Fig. 6. Results of plot harvest.

Software

The storage, manipulation, and presentation of yield data is just as important as the sensor hardware that generated the raw data. As mentioned earlier, the Micro-Trak YMS stored raw data on a DOS-compatible PCMCIA SRAM card. This offered a convenient solution for getting data to a computer for further analysis. The Zycom system used a proprietary data module with a parallel port interface requiring Zycom's special cable to interface with a computer's parallel port.

Both vendors supplied basic yield monitoring software. Micro-Trak had its own Grain-Trak Utility program that was used to format the data storage cards, download collected data, and display the data in tabular format as well as map form. Just as with the Grain-Trak module, the software was written for grain yield data but was modified to accommodate cotton. Many of the parameters displayed (i.e., dry yield) were meaningless since the system did not have a moisture sensor. The tabular information was useful for verifying field records and eliminating erroneous data files. Basic file utilities included text file export in numerous formats and merging data files.

Zycom included a DOS program to control and upload data from the data storage module and provided a third-party mapping software (FarmWorks) with basic capabilities of displaying raw data, averaged data, and contour maps. This software did not provide tabular summaries of the imported data, but multiple files could be imported to the same yield map. Zycom also supplied an Arcview project file that performed yield and area calculations and created contour maps.

CONCLUSIONS

This evaluation of commercial cotton yield monitors was intended to assess the suitability of the available products for practical use and identify areas of needed improvement.As with any new technology, many modifications and improvements to the original versions will be made; thus, many of the disadvantages identified in this evaluation may be eliminated. Nevertheless, these disadvantages should be noted for the benefit of potential customers. Improvements to both YMS are expected for the coming season.

Installation of both systems was quite involved and may well have been beyond the capabilities of farmers lacking specialized tools. The sensors by design were non-intrusive, allowing free flow of cotton through the duct; however, the ducts must have either 16 small drilled holes or two large rectangular cut holes placed in each duct. These holes would need to be addressed if the sensors were not in place.

As mentioned earlier, the location of the sensors on the ducts made them vulnerable to damage from normal picking operations. The two YMS were after-market items, therefore, a factory installation could not be obtained. Sensor housings, cables, system modules, etc. were installed or routed as securely yet conveniently as possible, but care still had to be taken to prevent damage to expensive components.

Calibration of the two YMS was essentially identical - with similar advantages and disadvantages. A definite disadvantage was the amount of cotton needed for a calibration load. In low yielding areas, harvesting a full basket might represent a considerable area of the total field. The calibration process itself should be shortened and made less frequent (Micro-Trak recommends doing calibration twice). The ability to back calibrate data collected during the calibration load is essential. Without a moisture and/or quality sensor in either YMS, there was a need to recalibrate whenever harvest and crop conditions changed.

The operation of the two YMS was relatively simple once time was spent using the systems. Neither system suffered any breakdowns during the harvest. Harvest data was retrieved and manipulated with little difficulty. Since each system included a different mapping software package, an independent software program was used to display the yield data, thereby providing an unbiased presentation of the data.

System accuracies over small areas were difficult to determine. The data obtained over a small area was apparently subject to the influence of smoothing algorithms. Accuracies were greater over larger harvest areas, such as entire fields, but this data is not practical for analyzing site-specific accuracies of the sensors. Certainly maps of overly-smoothed data would be of little use to a grower trying to pinpoint problem areas and develop detailed application strategies.

Precision farming technology does not come cheap. The two yield monitors evaluated cost \$6130 and \$10083 without a GPS receiver that can cost another \$3000 to \$4500 (all 1997 dollars). An extensive precision farming software package can cost more than \$2000. Other precision farming techniques (soil mapping, variable-rate input application, precision scouting, etc.) add additional expense.

The Micro-Trak and Zycom companies should be applauded for bringing these products to the market; however, the researchers and growers involved with this evaluation are aware of some flaws and needed improvements with each system. With improvements, both in accuracy and operation, more growers should be interested in purchasing a cotton yield monitor.

ACKNOWLEDGMENTS

The authors gratefully acknowledge the contributions of the following UGA-BAE personnel towards the success of this research: Dewayne Dales, Mike Gibbs, Rodney Hill, Andy Knowlton, Charles Welsh, Jerry White, and Terrell Whitley. The farmer-cooperators provided the cotton fields and valuable assistance during harvest (providing fuel, weighing cotton trailers, etc.). The farmers were Phil Adkins, Crisp County; Milton Sledge, Worth County, and Sephus Willis, Worth County. The authors also wish to thank Dr. Gary Herzog for providing the plots of cotton at The University of Georgia Lang Farm near Tifton, GA.

Soybean Moisture Content Measurement Based on Impact Force Parameters

Brian L. Steward
Department of Agricultural Engineering
University of Illinois
Urbana, Illinois

John W. Hummel
Crop Protection Research Unit
USDA-ARS
Urbana, Illinois

ABSTRACT

Previous work has shown that force-time curves for impacting soybeans vary with soybean moisture content. This paper documents an effort to characterize the nature of the relationship between a parameter developed from the force-time curve and soybean moisture content. The underlying objective was to determine if impact force sensing could be used to reliably make real-time moisture content measurements for yield monitoring applications. A parameter, *C*, based on the peak force and duration of the force-time curve was calculated. *C* was regressed onto moisture content using a polynomial model. The model fit the data with $R^2 = 0.84$ and 0.76 for the Conrad and Jack cultivars, respectively. The variability in *C* increased as moisture content decreased, and the strength of the relationship between *C* and moisture content decreased with increasing moisture content.

INTRODUCTION

Accurate measurement of grain moisture content is required for correct yield measurements. Currently, yield monitors use capacitance-type sensors to make real-time moisture content measurements as the grain is being harvested. There is evidence, however, that the calibration of such meters can be disturbed by plant liquids released during threshing (Hummel et al., 1996). Therefore, exploring the potential of alternative real-time grain moisture sensing techniques has value.

It is well established that a relationship exists between moisture content and Young's modulus of a soybean (Misra & Young, 1981; Hoki et al. 1983). In addition, the shape of an impact force-time curve of a soybean impacted against a flat object has been shown to change with moisture content (Stroshine & Hamann, 1996; Bartsch et al., 1980). The maximum impact force and the contact time between a soybean and the plate against which it is impacting can be predicted by Hertz contact theory (Timoshenko & Goodier, 1970; Goldsmith, 1960). Theoretically, maximum force and contact duration of a sphere impacting a plate are functions of impact velocity, Young's modulus, mass, and radius of the sphere. Hoki et al. (1983)

Copyright © 1999 ASA-CSSA-SSSA, 677 South Segoe Road, Madison, WI 53711, USA. *Proceedings of the Fourth International Conference on Precision Agriculture.*

showed that Hertz contact theory adequately predicts maximum force and contact duration for high moisture content soybeans at impact velocities less that 5 m s^{-1}.

Misra et al. (1990) demonstrated that soybeans could be classified as diseased or healthy by using data taken from a impact force-time curve. Soybean mass was also predicted by these data.

Harrenstein and Brusewitz (1986) found that there existed a linear relationship between the sound level of wheat falling on a pile of wheat and the moisture content of the wheat. This relationship was used in the development of an acoustic moisture meter for continuously flowing grain. The sound level of falling wheat, as measured by the acoustic moisture meter, was related to moisture content quadratically with a coefficient of determination of 0.99 (Mexas & Brusewitz, 1987). The relationship between sound level and moisture content for corn, milo, and soybeans has also been shown to be similar with the relationship strongest for soybeans (Brusewitz & Venable, 1987).

Research showing that grain moisture content affects Young's modulus, which in turn results in impact force-time curve differences, suggests that measurement from impact force-time curves could be used to predict grain moisture content. The work on an acoustic moisture meter also supports this contention. Nevertheless, no work has been found in which such relationships have been characterized.

OBJECTIVES

The objectives of this work were to characterize the relationship between a parameter derived from the impact force-time curve of soybeans and soybean moisture content and to determine if impact measurements could be used to measure moisture content of continuously flowing soybeans for real-time yield monitoring applications.

METHODS AND MATERIALS

The impact of soybeans was measured with a PCB Piezoelectronics 200B01 (PCB Piezoelectronics, Depew, NY) impact sensor. Soybeans were dropped from a height of 0.23 m (9.1 in.) above the sensor, accelerated by gravity, and impacted upon the face of the sensor. A small metal tube connected to a vacuum pump was used to hold the soybean over the sensor. When an impact measurement was made, the vacuum holding the soybean was removed, allowing the soybean to fall and impact the sensor (Fig. 1).

The piezoelectric impact sensor developed a charge across its electrodes when compressed by the impact of the soybean. This charge was converted to a voltage by a charge amplifier, and the voltage was sampled by a CIO-DAS16/M1 data acquisition board (Computer Boards, Middleboro, MA). A sampling rate of 400 MHz was used to acquire 1400 samples of the force-time curve. This sampling rate was chosen to get enough resolution of the curve to confidently determine the maximum force peak. This number of samples was adequate to capture all of the

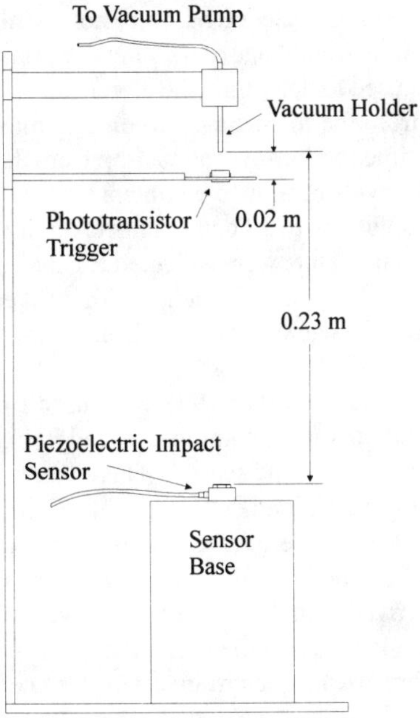

Fig. 1. Experimental setup. Soybean is dropped from the vacuum holder, impacting on the sensor.

impulse part of the force-time curve across all moisture content levels. An infrared (IR) emitter and phototransistor pair was placed 0.02 m below the end of the vacuum holder. When the soybean traveled between the pair, the light was interrupted and the data acquisition board was triggered to start taking data.

Samples of soybeans were selected from 1997 Illinois Foundation Seed seed soybeans. Twelve 100 g samples of two cultivars, Conrad and Jack, were selected from the seed bags. The original moisture content of the soybeans were 11.1% (w.b.) for the Jack cultivar and 12.1% (w.b.) for the Conrad cultivar. To generate an ensemble of samples with varying moisture contents greater than the initial moisture content, nine samples of each cultivar were placed in a germination chamber on trays covered with dry pieces of tissue paper. The temperature of the chamber was initially set at 22 °C, but was increased to 38 °C to cause the soybeans to take on moisture more quickly. The samples were positioned so as to reduce direct contact between wet surfaces and soybeans; however, a few soybeans in some samples did come in contact with water. In two samples, the majority of the soybeans came into contact with wet paper and were discarded. To generate samples with lower moisture content than the initial equilibrium moisture content, two samples of each cultivar were placed in a desiccator until their moisture content arrived at the desired value. One sample of each cultivar received no treatment and thus remained at the initial moisture content. The samples were stored in sealed bags at 5 °C for more than 3 wks to allow for equilibration of the samples. The result of this process was eleven samples of each cultivar at moisture contents in a range from approximately 6%

(w.b.) to 19% (w.b.). The moisture contents were not evenly spaced in some parts of this range, but gave adequate coverage of the range. A randomized complete block experimental design was used to determine the effects of soybean moisture content on a parameter derived from the force-time curve. The mass of each soybean was measured, as well as the time for the soybean to drop from the phototransistor site to the impact sensor. Eight levels of moisture content were selected for each cultivar (from the original 11 samples), with 20 data points taken for each of 5 replications. Therefore, a total of 800 data points were collected for each cultivar. Each soybean was selected randomly without replacement, with broken soybeans excluded. Moisture content was determined gravimetrically in accordance with ASAE Standard S352.2 (ASAE, 1994).

Soybeans harvested in the fall of 1997 were used for validating the model. Samples were taken directly from the combine, placed in plastic bags and stored at 5°C until they were impacted onto the sensor. There were five samples of varying moisture content levels for the Conrad cultivar and four for the Jack cultivar. One hundred soybeans of each sample were selected and impacted on the sensor resulting in 500 and 400 data points for the Conrad and Jack cultivars, respectively. The mean value of the parameter derived from the force-time curve for each level was used in the regression models developed earlier to predict moisture content. This prediction was compared with gravimetrically-determined moisture content measurement.

RESULTS AND DISCUSSION

Initial impact measurements revealed substantial variability in the impact force-time curves of the same soybean for repeated drops as well as those of different soybeans at the same moisture content (Fig. 2 and 3). Since soybeans are composed of different parts with varying elastic moduli, this variability would be

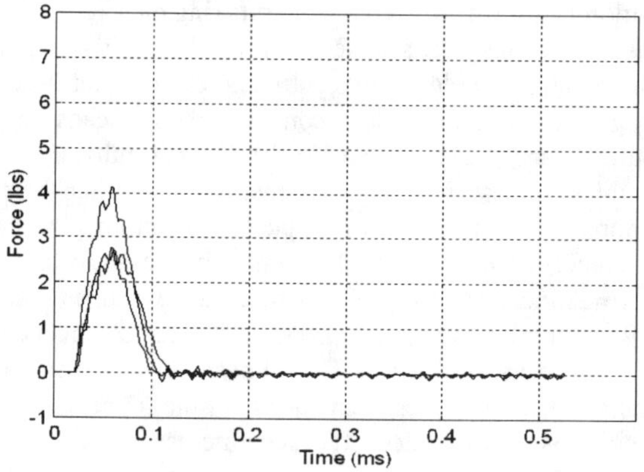

Fig. 2. Impact force curves for three Jack cultivar beans at 17.2 % (w.b.) moisture content.

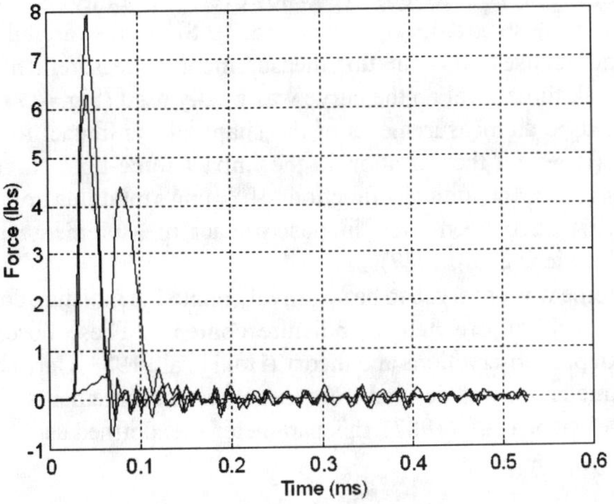

Fig. 3. Impact force curves for three Jack cultivar beans at 9.5 % (w.b.) moisture content.

expected. In addition, soybean are ellipsoids rather than spheres. Because of this shape, it is possible for the center of gravity of the soybean to be rotated from the vertical axis of the soybean on impact. Rebound energy is partitioned into angular and linear kinetic energy depending on the location of the center of gravity relative to the vertical axis (Yang & Schrock, 1994). This partitioning results in different rebound motion which was observed even as the same soybean was dropped starting from the same orientation. It was assumed that this effect due to the soybean's orientation on impact contributes to the variability in the force-time curve. To monitor the effect of controlling the starting orientation of the soybean on orientation at impact, a video camera was used to acquire images of soybeans before and after they contacted the impact sensor. A mark was placed on individual soybeans and drops of these soybeans were repeated with the initial orientation of the soybean in the vacuum holder controlled. Although the initial orientation of a soybean was fixed, repeated drops of the soybean resulted in random orientation at impact. Because of these observations, no attempt was made to control soybean orientation at the vacuum holder in subsequent experiments.

To determine if variation in the velocity of the soybean at impact was a source of variability in the data, the time from when the soybean passed through the phototransistor and IR emitter to the start of the impact force-time curve was measured. This time varied very little, with a c.v. of 0.72% for the Conrad soybeans and 0.485% for the Jack soybeans. Since velocity is linearly related to travel time under a constant acceleration and fixed travel distance, the velocity must be described by similar statistics. It was concluded that velocity variation contributed little to the variability in the impact data.

The peak force and the duration of the curve were measured on each individual curve. In determining how to measure duration, a measurement of the time from the start of the curve to the first zero-crossing was first taken. A randomly occurring gradual slope was observed, however, on some curves and a more commonly occurring steep slope on others (Fig. 3). Since this variability in the slope added substantial noise to this duration measurement, a measurement of duration at the one half peak force level on the curve was finally used (Fig. 4). Peak force and duration were used as measurements of the shape of the impact force-time curve because they (i) reflect the variation in the impact force-time curve as moisture content varies, (ii) are theoretically functions of Young's modulus from Hertz contact theory, and (iii) have been used in past impact measurement applications in agriculture (Delwiche et al., 1987).

Peak force was observed to be negatively related to moisture content whereas duration was positively related to moisture content. These observations were consistent with past observations and theory (Hoki et al., 1983, Bartsch et al., 1980). A single parameter, *C*, was derived from these two measurements. Following the example of Delwiche et al. (1987), this parameter was defined as:

$$C = \frac{Peak\ Force}{Duration} \quad [1]$$

Among other similar parameters tried, *C* had the highest correlation with moisture content (R = -0.807 for Conrad soybeans and R = -0.845 for Jack soybean) and the lowest correlation with mass (R = -0.175 for Conrad soybeans and R = -0.157 for Jack soybean). Thus, *C* — as defined by Eq. [1] — was considered best for the purpose of measuring moisture content.

Stepwise regression was used to develop a model relating moisture content to *C*. Mass and the square and cube of moisture content were included in the model. *C* was considered to be dependent on these factors with the majority of error in the model being error in *C*. Since variance in *C* varied with moisture content, a weighted regression model was used — with the reciprocal of the sample variance in *C* at each moisture content level being the weight for that level.

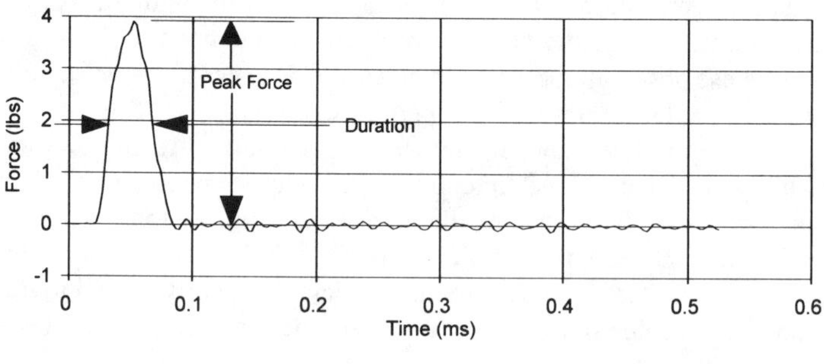

Fig. 4. Measurement of peak force and duration of the impact force-time curve.

For the Conrad data, moisture content, mass and moisture content cubed were all significant factors in predicting *C*. Since mass only had a partial R^2 of 0.0064, it was dropped from the model. This model had a coefficient of determination of 0.8392 (Fig. 5). For the Jack data, moisture content, moisture content squared, moisture content cubed, and mass were all significant factors in predicting *C*. Nevertheless, mass only had a partial R^2 of 0.0020. Thus, mass was excluded from the model that then had a coefficient of determination of 0.7638 (Fig. 6). When both data sets were pooled, moisture content, the cube of moisture content, and mass were all found to be significant. Once again, mass had a low partial R^2 (0.0026) and was therefore excluded from the model. This reduced model had an R^2 of 0.8033 (Fig. 7). Another motivation for excluding mass from the model was that in practical applications, the measurement of mass of individual soybeans would be rather impractical. Nevertheless, it is useful to know that mass is not highly correlated with C.

Variability in the data increased substantially as the moisture content decreased with relatively large variability for moisture contents below 10%. In addition, the slope of the regression curve decreased with increasing moisture content that hinders the measurement of moisture content at the higher levels. Thus, the sensing of moisture content via impact force can be accomplished only in a limited range with a large number of samples.

The data collected from impacting harvested soybeans were as variable as those of the seed soybeans. A coating of residue, apparently deposited on the soybeans during threshing, was observed on several samples of soybeans. This residue could have affected impact measurements. The mean value of C was recorded for each sample of 100 soybean impacts, and then by using the model for

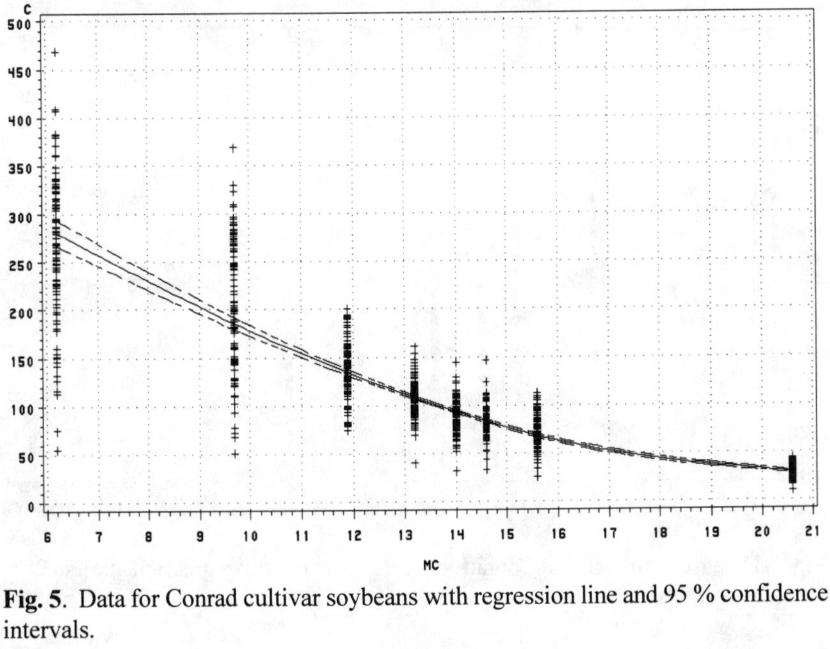

Fig. 5. Data for Conrad cultivar soybeans with regression line and 95 % confidence intervals.

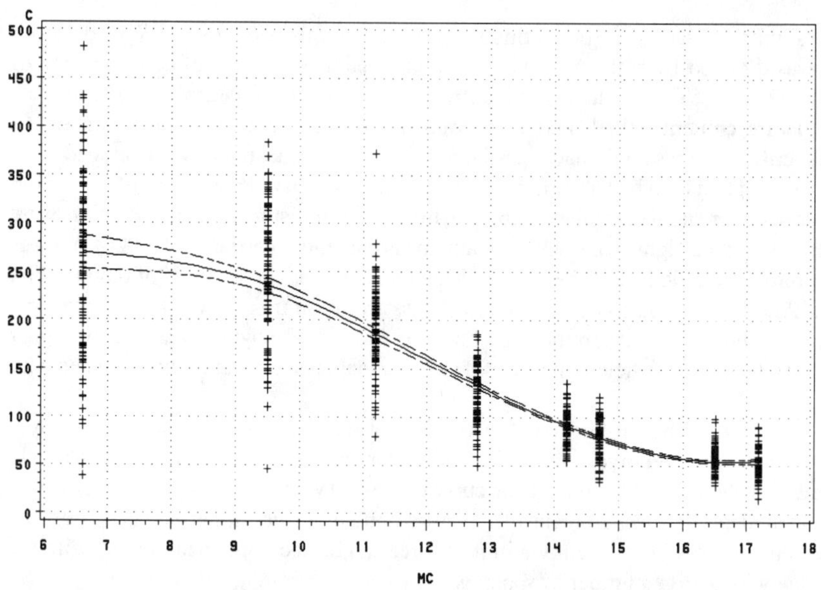

Fig. 6. Data for Jack cultivar soybeans with regression line and 95% confidence intervals.

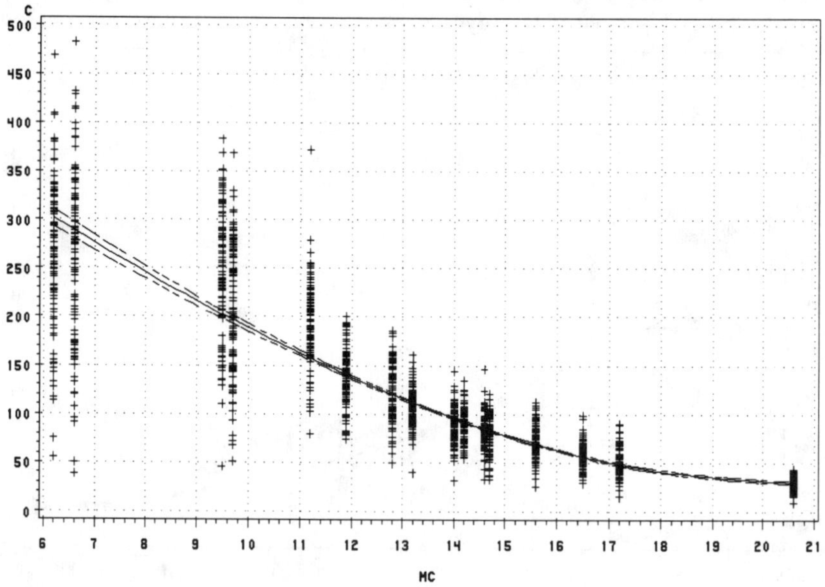

Fig. 7. Data for both Jack and Conrad cultivar soybeans with regression line and 95% confidence intervals.

the corresponding cultivar, the moisture content, which would give us this value of C, was found by solving a quadratic or cubic equation for moisture content. The Conrad model predicted the moisture content of Conrad soybeans to within 4% of gravimetric measurements (Fig. 8). The Jack model predicted moisture content of Jack soybeans to within 14% of gravimetric measurements (Fig. 9). The harvest impact data were pooled, and the combined model was used to predict moisture content. The results were similar to those obtained with the individual models. Moisture content was predicted to within 13% of gravimetric measurements (Fig. 10).

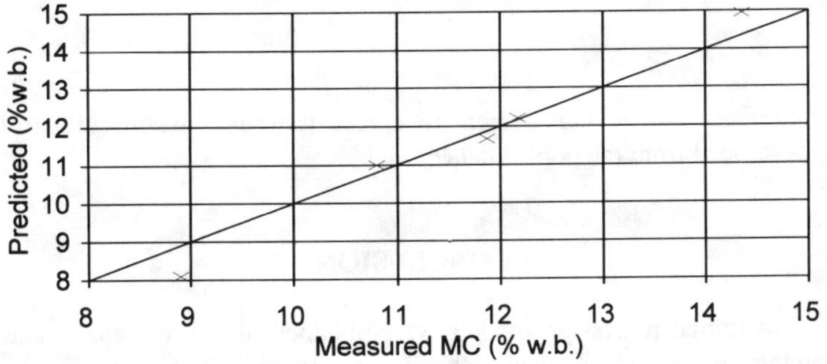

Fig. 8. Prediction of moisture content of harvested Conrad soybeans by the Conrad model.

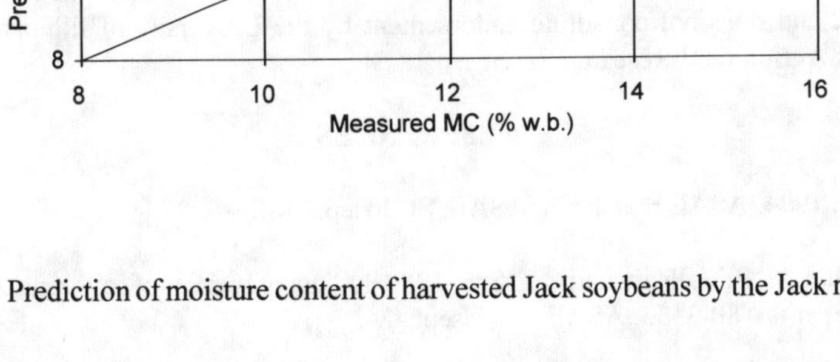

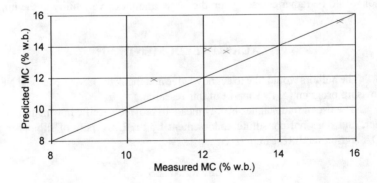

Fig. 9. Prediction of moisture content of harvested Jack soybeans by the Jack model.

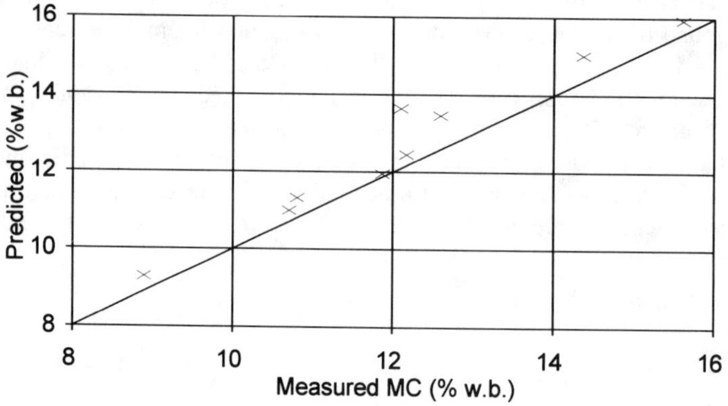

Fig. 10. Prediction of moisture content of harvested soybeans of both cultivars by the model developed from the pooled data.

CONCLUSION

Polynomial regression models were obtained to estimate the relationship between the parameter *C*, defined as the quotient of the peak force and the duration of the impact force-time curve, and the moisture content of two cultivars of soybeans. This parameter was only slightly correlated with mass. The variability of *C* increased with decreasing moisture content. The strength of the relationship between *C* and moisture content decreased for increasing moisture content. Therefore, impact sensing can be a viable sensing technique for only a large sample of soybean impacts and within a limited range of moisture content. In addition, the accuracy of such a sensor would be rather coarse, given the large amount of variability in the impact data.

ACKNOWLEDGMENTS

The authors would like to acknowledge and thank the USDA National Needs Fellowship program for its support of this research.

The use of trade names is only meant to provide specific information to the reader, and does not constitute endorsement by the University of Illinois or the USDA Agricultural Research Service.

REFERENCES

ASAE. 1994. ASAE Standards. ASAE, St. Joseph, MI.

Bartsch, J.A., R.M. Peart, and C.T. Sun. 1980. Impact properties of soybeans. ASAE. Paper no. 80-3535. ASAE, St. Joseph, MI.

Brusewitz, G.H., and P.B. Venable. 1987. Sound level measurements of flowing grain. Trans. ASAE. 30(3):863–864.

Delwiche, M.J., T. McDonald, and S.V. Bowers. 1987. Determination of peach firmness by analysis of impact forces. Trans. ASAE. 30(1):249–254.

Goldsmith, W. 1960. Impact. Edward Arnold Publ., London.

Harrenstein, A., and G. Brusewitz. 1986. Sound level measurements on flowing wheat. Trans. ASAE. 29(4):1114–1117.

Hoki, M., T.H. Burkhardt, M.L. Esmay, and R. Stroshine. 1983. Impact strength of soybeans. ASAE Paper no. 83-3521. ASAE, St. Joseph, MI.

Hummel, J.W., K.A. Sudduth, and S.J. Birrell. 1996. Real-time soil and crop sensors - How well do they work? p. 1–9. *In* Na. Zhang (ed.) New trends in farm machinery development and agriculture, SP-1194. Soc. of Automotive Eng., Warrendale, PA.

Mexas., S., and G.H. Brusewitz. 1987. Acoustic grain moisture meter. Trans. ASAE 30(3):853–857.

Misra, M.K., B. Koerner, A. Pate, and C.P. Burger. 1990. Acoustic properties of soybeans. Trans. ASAE 33(2):671–677.

Misra, R.N. and J.H. Young. 1981. A model for predicting the effect of moisture content on the modulus of elasticity of soybeans. Trans. ASAE 24(5):1338–1341,1347.

Stroshine, R., and D. Hamann. 1996. Physical properties of agricultural materials and food products. Department of Agric. and Biol. Eng., Purdue Univ.. West Lafayette, IN.

Timoshenko, S.P. and J.N. Goodier. 1970. Theory of elasticity. McGraw-Hill. New York.

Yang, Y., and M.D. Schrock. 1994. Analysis of grain kernel rebound motion. Trans. ASAE 37(1):27–31.

The ISO 11783 Standard and Its Use in Precision Agriculture Equipment

C. Strauss
C. E. Cugnasca
A. M. Saraiva
S. M. Paz
Department of Computer Engineering Agricultural Automation Laboratory
Universidade de São Paulo Escola Politécnica
São Paulo, São Paulo, Brazil

ABSTRACT

Nowadays precision agriculture requires new equipment and systems that are mounted in agricultural tractors and implements. These systems usually have distributed architecture and are composed of several devices like sensors, actuators, control elements and supervision and control units, all of them intercommunicating in real time. This application requires robustness, flexibility and expansion possibility, involving devices of different manufactures. In that sense, several standards are being proposed in order to help to achieve these goals. The ISO 11783 standard specifies a serial data network for communication and control in tractors and implements, standardizing the method and format of the data interchange between control elements, actuators, computers, sensors and other intelligent devices connected in a system. It is based on the CAN specification (Controller Area Network), used nowadays in other applications. This paper discusses the main characteristics of the ISO 11783 standard, and presents its adoption in a planter monitor with a GPS receiver, used in precision agriculture in order to generate a planting map. This monitor has distributed architecture, and is composed of a main module in the tractor and a sensor module in the planter. This implementation allows for the use of new intelligent sensors and modules, and is compared with other usual implementations.

THE PLANTER MONITOR

Purpose

The motivation for this paper is the project developed by the Agricultural Automation Laboratory (LAA) team, sponsored by FINEP - Financiadora de Estudos e Projetos (a Brazilian federal funding agency). A Planter Monitor was designed (Saraiva et al., 1993), performing the following functions:

- It shows its operator some useful information, such as seed rate, seed population, tractor speed, and planted field area.
- It warns its operator about fault operational conditions, such as seed rate out of range, planter out of operational position, and lack of communication between its modules.
- It stores operational statistical data, and sends them to a Personal Computer (PC), for further analysis.

Copyright © 1999 ASA-CSSA-SSSA, 677 South Segoe Road, Madison, WI 53711, USA. *Proceedings of the Fourth International Conference on Precision Agriculture.*

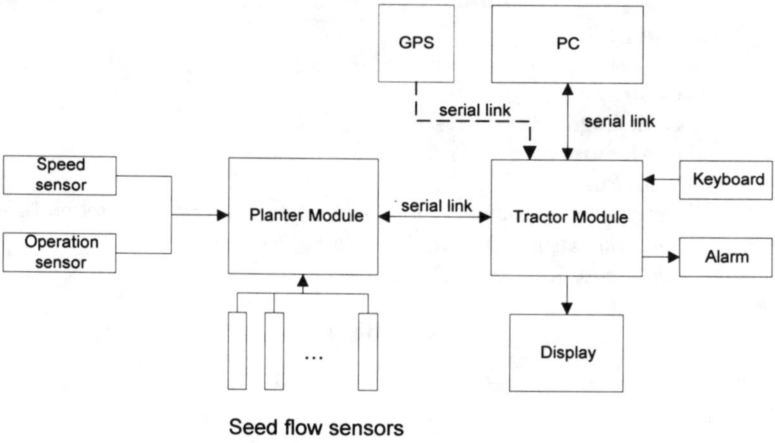

Fig. 1. Architecture of the planter monitor.

Architecture

The system, as shown in Fig. 1, includes sensors of seed flow, speed, and planter position. These sensors are read by the Remote Module, installed on the planter, which sends these data, through a serial link, to the Central Module, installed in the tractor cabin. The Central Module communicates with the operator through the keyboard, the display and the alarm. It also can be linked to an external PC, through a serial interface. The connection of a GPS receiver is being developed, so that the Central Module can also collect positional information (Saraiva et al., 1997).

Protocol in Use

The LAA team created a proprietary protocol that allows for the communication between the Central and the Remote Modules. Furthermore, it is a simple protocol, easy to be implemented, because microcontrollers were being used, with limited processing capacity. Therefore, the RS-232 was chosen, because it allows for independent and simultaneous communication (full-duplex) between two nodes.

After that implementation, a more general protocol was decided on, allowing for several Remote Modules connected to the same Central Module, using an RS-485 bus.

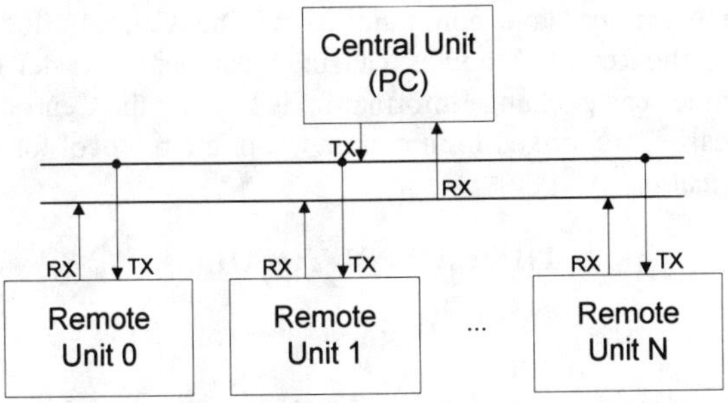

Fig. 2. Physical layer of the RS-485 protocol.

Physical Layer - RS-485 Bus (Fig. 2)

The channel consists of a twisted pair cable, with two buses, TX and RX. Through the TX bus, the Central Module can send data, which are received by all Remote Modules. Through the RX bus, the Remote Modules can send data to the Central Module.

Link Layer - Message Format

The messages have the following fields (ASCII characters):

- address, that identifies the Remote Module
- message type
- data, according to the message type
- cyclic redundancy code (CRC)
- end of message

Application Layer - Types of message

The message types are:

- initialization requisition and response
- configuration
- sensor status
- planter status
- seed counters data

Limitations

As it was presented, this is an LAA's proprietary protocol. Therefore, it is not compatible with different manufacturers' equipment.

In the case of the RS-232 protocol, the expandability is limited, for each new module requires a dedicated I/O port. What is more, each pair of modules requires a separate cable, leading to the problem of having too many cables.

The RS-485 protocol avoids these problems, using a single bus. However, it only allows for a master-slave communication. The Central Module must be constantly polling the Remote Modules, causing a bandwidth waste. The only way the Remote Modules can exchange information is through the Central Module.

For that reason, we looked for a more appropriate protocol for the Precision Agriculture application.

THE CAN PROTOCOL

History

The CAN protocol (Controller Area Network) was developed by the automotive industry, to support the communication of on-board equipment, in trucks and buses. Its good performance caused its adoption as a standard by SAE (Society of Automotive Engineers), and, later, by the ISO (International Organization for Standardization).

Features

- Created specifically for being used in vehicles
- Robustness: the network can operate in a degraded way, even if the cable is broken or in short-circuit
- Simplified cabling: a single cable connects the devices
- Existence of components and cards that implement the protocol
- Existence of international standards

However, the norms only specify how the information is transmitted, with no reference to the data format and type. This allowed for the creation of several protocols based on CAN. Among them, the ISO 11783 standard can be pointed out.

Implementation

The CAN Transceivers are integrated circuits that read and write bits to and from the CAN bus wires. An example is Philips's PCA82C250 - CAN Controller Interface.

The CAN Controllers are devices that format the messages to be transmitted and participate in the decision process of which unit will transmit the next message. Some CAN controllers have an interface to microprocessors, while others are embedded in popular microcontrollers (which simplifies the project). In both cases, the CPU specifies the messages to be transmitted and is notified of the arrival of incoming messages.

Examples of CAN controllers:

- 82527 - Serial Communications Controller (CAN Protocol), from Intel;
- SJA1000 - Stand Alone CAN Controller, from Philips.

Examples of embedded controllers:

- 87C592 - 8-bit microcontroller with on-chip CAN, from Philips
- 80C196CA - Advanced 16-bit CHMOS microcontroller, from Intel

- SAB C505C - 8-bit CMOS microcontroller with CAN, from Siemens
- COP888EB - 8-bit microcontroller with CAN interface, A/D and UART, from National

The slave I/O devices implement remote I/O ports controlled by the CAN bus. For example, instead of having dozens of digital and analog signals transported in separated wires, only a CAN cable and various slave I/O devices are used. A component of this type is Philips's P82C150.

THE ISO 11783 STANDARD

The ISO 11783 standard specifies a serial network for communication and control of agricultural vehicles (tractors) and its implements (ISO, 1994a). Part of the physical layer and of the link layer are based on the CAN protocol.

The Physical Layer (ISO, 1994b)

The physical layer (Fig. 3) consists of a twisted pair cable, with four wires, supporting the rate of 250 kbits s^{-1}. Between two signals, CAN_H and CAN_L, there is a voltage difference that is interpreted as a 0 (dominant) or 1 (recessive), in case it is +1V or -1V respectively. The two other signals are CAN_BAT and CAN_GND, connected to power supply and to ground, respectively. A maximum of 30 electronic devices can be connected to the bus.

At each end of the cable, there is a termination to polarize CAN_H and CAN_L, so that, with no signal, CAN_H = 4V and CAN_L = 5V. If any device issues the dominant state, and another issues the recessive state, the result is the dominant state.

In order to arbitrate conflicts of bus access, after issuing a bit, each device monitors the line, verifying the resulting state. If it is dominant, and the device had issued the recessive bit, it stops sending, losing the bus control, because another device issued the dominant bit.

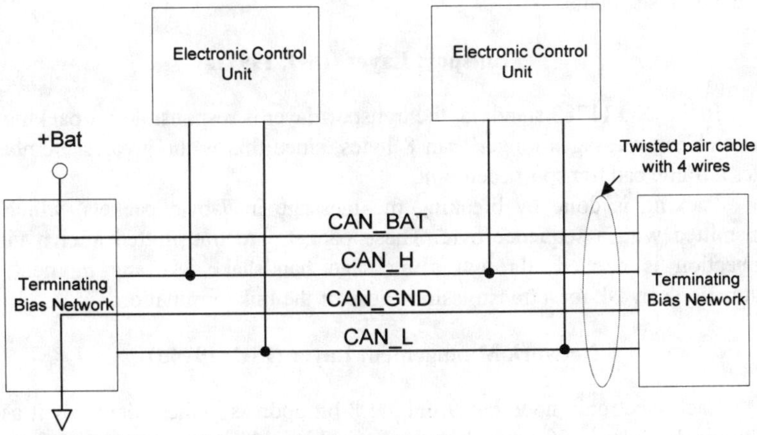

Fig. 3. Physical layer of the ISO 11783 protocol

According to the specification, the network is resistant to several kinds of failure, capable of operating even in case of short-circuits between CAN_H and CAN_GND, between CAN_H and CAN_BAT, between CAN_L and CAN_GND, and between CAN_L and CAN_BAT. It also resists to failure in one of the terminators, to interruptions of CAN_H or CAN_L, and to other occurrences. The communication is not possible when there is a short-circuit between CAN_H and CAN_L, or when a terminator is not powered.

Link Layer (ISO, 1997)

Each message to be transmitted by the upper layers is mounted in one or more PDUs (Protocol Data Unit), with their formats based on CAN standard.

Each PDU has a Start of Frame (SOF) bit, priority, origin address, message type, destination address (or broadcast address), number of data bytes (from 0 to 8), data, CRC, and End of Frame (EOF), besides other control bits.

The message type can be: command (the destination has to perform some action); requisition (the destination has to answer it); broadcast (no determined destination); response; confirmation (ACK or NACK); or a special function (network management, connections management, etc.). Associated to each type of message, there is a number (Program Group Number - PGN), defined by the standard. It is possible to submit a form for registering new PGNs to ISO.

Network Layer (ISO, 1994c)

An ISO 11783 network (Fig. 4) can have several segments, connected by bridges. The tractor segment links devices such as the engine, the transmission, the brakes and the lights. One or more segments in the implement link the implement devices one to each other, and to the tractor segment. A virtual terminal, connected to the implement segment, performs the interface with the operator. The bridges select the messages, and conduct them from one segment to the other, allow for different transmission rates in both segments, and isolate and protect both segments.

Transport Layer (ISO, 1997)

In the ISO 11783 standard, the transport layer is responsible for packing and reassembling messages longer than 8 bytes, since this is the greatest number of bytes a frame can transport each time.

Packing is done by breaking the message in 7-byte packets, which are transmitted with a sequence byte. These packets are transmitted after a virtual connection is opened, through a message handshake. At any moment, the destination can ask for a transmission pause or the link termination.

Network Management Layer (ISO, 1994d)

Each electronic node has a unique 8-bit address, which identifies it as the origin or destination of network messages. This address is independent from the device function, and can even be different each time the device is turned on. On the other hand, the device name is a 32-bit number, defined by ISO for each functionality (e.g., front break actuator, electric system, and planter monitor). The network management layer is responsible for the association between the name and the network address.

When a device is connected to the network, it can send a message to all nodes, demanding their addresses, and then it assumes any address not being used yet. The device can also have a preferential address, and announce it to the network. If any node answers that this address is in use, another address can be chosen. The device is also likely to only accept its preferential address. In this case, if a conflict happens, that device has to stop transmitting, until its address is externally changed, and has to announce that condition to the network.

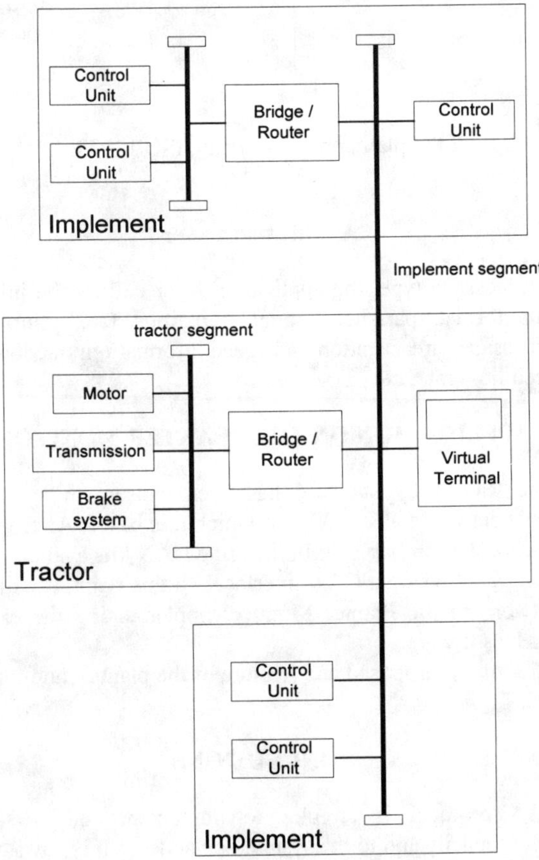

Fig. 4. Architecture of a ISO 11783 network

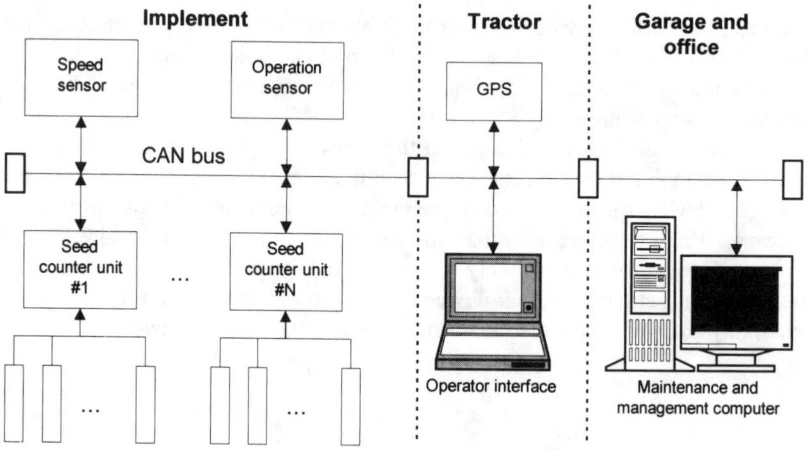

Fig. 5. Architecture of the planter monitor using ISO 11783.

Application Layer

For each message type, the application layer defines the information to be transmitted, its format, parameter range, repetition rate, priority, and others. Examples of messages are: position and speed information, traction applied by the implement, seed flow rate, etc.

APPLICATION ON THE PLANTER MONITOR

Once the ISO 11783 standard has been chosen, the next step will be studying this standard, and also CAN, on which it is based. After that, a prototype of a CAN network will probably be built, using CAN kits available in the market. If necessary, we can develop a CAN interface. At this point, it will be possible to apply that network to the Planter Monitor, implementing the extra ISO 11783 protocol layers by software.

Figure 5 shows a proposed architecture of the planter monitor using the ISO 11783 protocol.

CONCLUSIONS

The LAA protocol is simpler to implement and is suitable to the microcontrollers capacity and to the system demands, as it is now.

The ISO protocol provides good extensibility, favors a modular approach and allows for interoperability between devices of several manufacturers. Non-expensive circuits and cards that implement part of the protocol are available in the market, also minimizing its greater complexity. The reduced cabling allows for easier installation and maintenance, besides lower cost. Furthermore, the standard provides ready solutions, such as message format and contents, avoiding the task of developing them.

Another aspect is easier maintenance. Instead of having to go near each equipment to make tests, it is now possible to plug a test computer in a CAN port of the tractor and to get the equipment status remotely.

The ISO 11783 and CAN standards also provide better reliability (which is important in vehicles), resisting vibration, open and shorted wires and faulty connections.

REFERENCES

ISO. 1997. Tractors and machinery for agriculture and forestry – Serial control and communications data network. Part 3. Data Link Layer. ISO 11783.

ISO. 1994a. Tractors, machinery for agriculture and forestry – Serial control and communications data network. Part 1. General Standard. ISO 11783.

ISO. 1994b. Tractors, machinery for agriculture and forestry – Serial control and communications data network. Part 2. Physical layer. ISO 11783.

ISO. 1994c. Tractors, machinery for agriculture and forestry – Serial control and communications data network. Part 4. Network Layer. ISO 11783.

ISO. 1994d. Tractors, machinery for agriculture and forestry – Serial control and communications data network. Part 5. Network Management Layer. ISO 11783.

Saraiva, A.M., C.E. Cugnasca, A.M.A. Massola, 1993. Planter monitoring: A management approach. 1993. ASAE International Winter Meeting, Paper n.93-1.552, ASAE, St. Joseph, MI.

Saraiva A.M., S.M. Paz, C.E. Cugnasca, 1997. Improving a planter monitor with a GPS receiver. p. 321–327. *In* Proc. 1st European Conf. on Precision Agriculture. Vol.1. Warwick..

Cotton Yield Sensor Produces Yield Map

Mike Gvili

Zycom Corporation
Bedford, MA

ABSTRACT

The design of a cotton Yield Monitoring System, using electro-optical devices, was tested in field and laboratory conditions with better then expected results. As the cotton passes through the chute, it is illuminated by the emitters. The detectors that are positioned across the chute detect the light pulses. The detection circuit then generates pulse train proportional to the length of time the light detected at the detector was blocked. The pulse count is relative to the mass of the cotton passing through the conduit. The sensor is constructed of an array of infra red emitters and detectors assembled in self cleaning brackets. The optical components are kept clean by the application of continuous clean airflow under pressure. Light test is being conducted during operation to verify performance and compensate for inoperative or partially operating light emitter – detector pairs. A sequence of calculations is being performed taking into account the sensor performance, the yield level, moisture of the cotton, the variety, and other parameters. The cotton weight and the yield is displayed on the cab indicator and also stored on a recording media, together with the GPS position of the picker. The data can then be transferred to the office PC, processed and displayed as a yield map.

Copyright © 1999 ASA-CSSA-SSSA, 677 South Segoe Road, Madison, WI 53711, USA. *Proceedings of the Fourth International Conference on Precision Agriculture.*

CROP MODELING

A Methodology to Define Management Units in Support of an Integrated, Model-Based Approach to Precision Agriculture

B. J. van Alphen
J. J. Stoorvogel

Laboratory of Soil Science and Geology
Wageningen Agricultural University
Wageningen, the Netherlands.

ABSTRACT

Over the last decades much effort has been invested in the development of complex simulation models, incorporating and integrating the current understanding of soil–water–plant interactions. The use of such models in precision agriculture has been shown to have great potential. This research presents a methodology to derive basic units for precision agriculture, referred to as management units. Their main purpose is to reduce the theoretically infinite variability of growth conditions in the field to a limited set, which can be evaluated using mechanistic models. A quantitative criterion is applied to ensure that management units accurately represent local variation with respect to growth conditions. Using representative soil profiles for each management unit, real-time simulations can provide insight in crop performance and the nutrient status of the soil. This information can be used to optimise farm management, maintaining crop performance while reducing environmental impacts.

INTRODUCTION

In precision agriculture, fields are managed at a higher spatial and temporal resolution than in traditional agriculture. A major break-through was the development of global positioning systems (GPS), which together with geographical information systems (GIS) and specialised tillage equipment enables farmers to register field variability and vary management on-the-go. However, major problems remain to be solved in the process of translating observed variability (e.g., yield, soil type) into useful information to adapt and improve farm management (e.g., fertilization).

Optimal management, viewed within the framework of precision agriculture, should explicitly incorporate the local variation of growth conditions into farm management operations. Growth conditions are determined by complex interactions between growth determining factors. During a growing season, different factors may become limiting at different locations and for different periods of time. In a spatial context this means that water deficiency may be limiting plant growth at one location, while nutrient availability may be or become limiting elsewhere. The same applies to the time dimension: water deficiency may be limiting at one time, while nutrients may be or become limiting later.

Copyright © 1999 ASA-CSSA-SSSA, 677 South Segoe Road, Madison, WI 53711, USA. *Proceedings of the Fourth International Conference on Precision Agriculture.*

Considering the complexity of soil-water-plant interactions, the main challenge for precision agriculture lies in understanding the fundamental mechanisms, hereby explaining spatial and temporal variation in crop performance and enabling farmers to adjust their management accordingly. Many researchers have tried to establish simple statistical relations between measured yields and one or several growth determining factors (Stafford, 1997). These approaches share a common problem: they hardly contribute to an understanding of observed yield variations. Moreover, derived parameter values may be irrelevant from a physiological point of view, e.g., a negative response to solar radiation (Laird & Cady, 1969). Where statistical relations fail to clarify observed variation, mechanistic models attempt to describe the fundamental processes governing crop production. The primary advantage of mechanistic models is that they can incorporate our collective knowledge of crop, soil and atmospheric processes (Acock & Pachepsky, 1997). As a result, it seems obvious that the application and further development of mechanistic models will play a prominent role in precision agricultural research. So far this has not been the case, disregarding some interesting research projects illustrating the potential value of these models (e.g., Verhagen, 1997).

Probably the greatest obstacle limiting a wider application of mechanistic models, is their data intensive character. An extensive set of parameters is required, characterising soils, crops, climate and nutrient cycling. For these models to become an operational tool, a challenge lies in ensuring efficient data collection and minimising parameter requirements.

The most promising application for models in precision agriculture relates to the real-time simulation of soil-water-plant dynamics. These simulations can provide information on the status of important growth determining factors (e.g., water and nutrient availability), hereby enabling farmers to adjust their management accordingly to meet actual needs. Given the continuous variation of growth conditions in the field, an almost infinite number of simulation runs would be required. In order to solve this problem the number of conditions included in the simulations should be limited.

This research presents a methodology to determine basic units for precision agriculture, referred to as management units. Management units are areas within a field that show relatively little variation in growth conditions. The variation within a management unit is considered to be negligible for management purposes. From a theoretical point of view, this concept is suboptimal; i.e., a certain amount of spatial variation is lost. In practical terms, however, management units may be considered an essential generalisation, needed to enable the incorporation of mechanistic models in precision agriculture. The developed methodology uses simulation models in combination with GIS technology and is applied for two fields on a commercial farm located in the western part of the Netherlands (Fig.1).

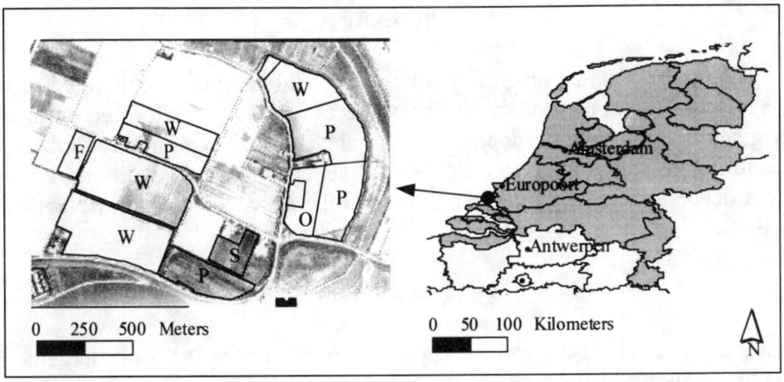

Fig. 1. Location of the study area (van Bergeijk farm) and the farm layout with the crop rotation for the 1997 growing season (W = winter wheat; P = potato; S = sugar beet; O = other; F = fallow).

MATERIALS AND METHODS

The developed methodology to delineate management units is presented schematically in Fig. 2. It is based on a detailed soil inventory, including a quantitative characterisation of soil chemical and soil physical properties. Together with climatic records from an on-farm weather station, these data provide the input for a mechanistic simulation model quantifying crop production and N leaching for individual soil sampling locations. These point observations are classified and then interpolated to derive a limited number of management units.

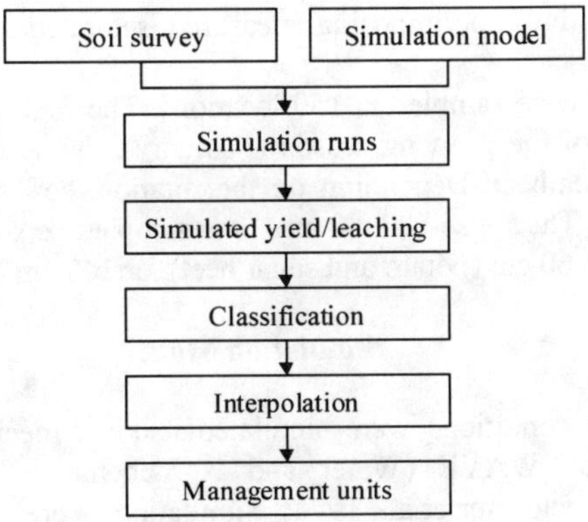

Fig. 2. The methodology to derive management units.

The Study Area

The farm covers an area of approximately 100 ha, mainly cultivated to winter wheat, consumption potatoes and sugar beets (Fig. 1). It is located on an island originating from marine deposits. Soils are generally calcareous with textures ranging from sandy loam to clay. With excellent drainage conditions, controlled by a dense system of pipe-drains, these soils are considered prime agricultural soils.

Soil survey

A detailed 1: 5,000 soil survey was conducted at the van Bergeijk farm, counting approximately six borings per hectare. The survey focused on soil functional properties, therefore neglecting soil morphological characteristics not relevant in this respect (e.g., color). Soil profiles were described in terms of layer sequencing and layer thickness. Layers were coded according to a Dutch classification system (Wösten et al., 1994) and characterized in terms of texture (clay, silt, and sand fractions), organic matter content and lime content.

The soil moisture retention and soil hydraulic conductivity curves are the most important soil functional properties for modelling purposes. These curves can be determined through laborious laboratory measurements, such as the Multistep procedure (van Dam et al., 1994). However, over the last decade pedo transfer functions have been developed, relating basic soil properties such as texture, bulk density and organic matter content, to soil hydraulic characteristics. These empirical relations are derived from extensive measurements for different soil types. A continuous pedo-transfer function for Dutch soils (Wösten, 1997) was used to derive the van Genuchten parameters for individual soil layers (van Genuchten, 1980). Estimated soil hydraulic characteristics were adjusted to fit local conditions by superimposing measured saturated water contents on the estimated values.

Soil N levels were sampled at 150 locations. The first samples were collected before the start of the growing season (February), the second series after potato harvesting (November). Depending on the rotation, soil samples were taken at different depths. The topsoil (0–30 cm) was sampled separately, the subsoil was sampled either to 60 cm (potato and sugar beet), or 100 cm (winter wheat).

Simulation Model

Local growth conditions were simulated using a mechanistic, deterministic model known as WAVE (Water and Agrochemicals in soil and Vadose Environment, Vanclooster et al., 1994). Simulations were executed for individual points, corresponding with the auguring locations from the soil survey. WAVE integrates four existing models: one dimensional soil water flow based on SWATRER (Dierckx et al., 1986), N cycling based on SOILN (Bergström et al., 1991), heat and solute transport based on LEACHN (Hutson & Wagenet, 1992) and crop growth based on SUCROS (Spitters et al., 1988).

WAVE applies a finite difference calculation scheme to solve the differential equations governing water movement and solute transport. For this purpose the soil profiles were divided into one-centimetre compartments. Water movement is described by the Richard's equation, combining the mass balance and Darcian flow equations. Verhagen (1997) made two conceptual changes to the original model:

1. Water uptake by plant roots was originally modelled assuming preferential uptake in the top soil compartments, therefore excluding roots in the deeper layers. After revision the total water uptake is calculated as an integral over the entire root zone.
2. Nitrogen uptake was originally modelled using the nitrogen concentration in the leaves as the driving force controlling total nitrogen uptake. After revision nitrogen uptake is related to biomass production as described by Greenwood and co-workers (Greenwood et al., 1990).

Water stress is calculated according to Feddes et al. (1978). The maximum water uptake is defined by a sink term, which is considered constant with depth. Water uptake is reduced at high and low-pressure head values, according to crop specific thresholds.

Stress resulting from nitrogen deficiencies is calculated using the critical nitrogen concentrations as defined by Greenwood et al. (1990). They describe decreasing N-concentrations with increasing plant mass using a negative exponential function. When the actual uptake is insufficient to sustain the necessary concentration in the produced biomass, production is reduced proportionally to the ratio of actual over required uptake.

Model Performance

Verhagen (1997) successfully validated the WAVE model at a farm in the northern part of the Netherlands. Model performance was tested against measured field data for the 1994-growing season (Fig. 3). Model validation at the van Bergeijk farm was still in progress. The validation focuses on soil moisture contents and mineral nitrogen concentrations.

Simulation Runs

Management units were derived using so called backward-looking simulations. These simulations use historic weather data and provide an excellent means to evaluate the behavior of the soil-water-plant system in space and time. Pedo transfer functions provided the hydraulic characteristics for all soil profiles described in the soil survey. Together with the meteorological records these data were fed into the simulation model, quantifying crop growth, soil water regimes and nutrient fluxes under different weather conditions. Expressed in these terms, spatial and temporal variability within fields and between years was used to delineate the management units, i.e., the functional grouping of soil profiles. Three key indicators were considered:

1. Water induced yield reduction (water stress);
2. Nitrogen leaching;

3. Residual nitrogen content at harvest.

Yield reductions due to water stress were simulated for a dry year, N leaching and residual N contents were simulated for a wet year.

Classification

With respect to the selected indicators, soil profiles were classified using the Jenk's optimisation criterion provided by the ArcView GIS system (ESRI, 1996). Basically, class boundaries are chosen such that the internal variance within the classes is minimised. In practice the class boundaries tend to coincide with clear jumps in the observed parameter values.

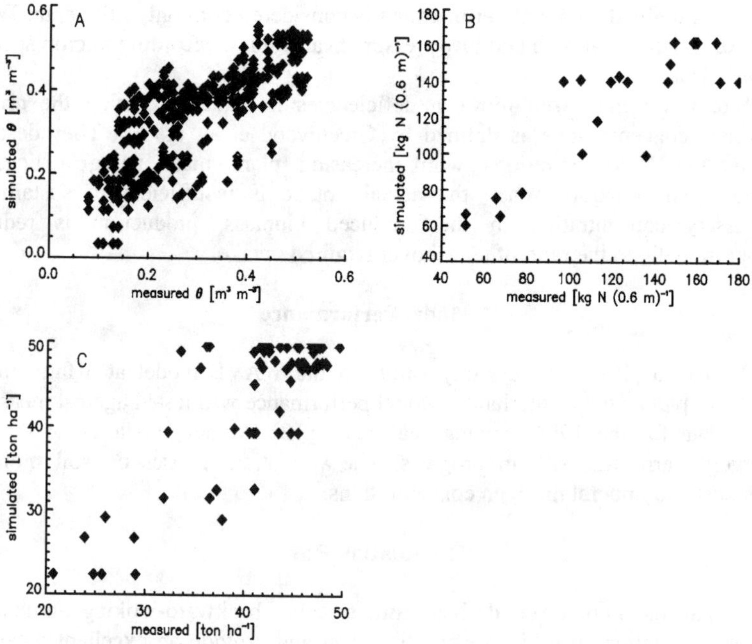

Fig. 3. Validation of the WAVE model. (A) moisture content (R^2=0.65), (B) mineral N [0–60cm] (R^2=0.58), and (C) yield (R^2=0.65). Figure taken from Verhagen, 1997.

Initially two classes were defined for each indicator parameter. Using standard analyses of variance (ANOVA) the resulting degrees of determination were calculated. The degree of determination (R^2 in [0..1]) is defined as (Devore & Peck, 1986):

$$R^2 = 1 - \frac{SSW}{SST}$$

In which R^2 is calculated as a function of *SSW,* the sum of squares within classes, and *SST,* the total sum of squares.

A quantitative criterion is required to accept or reject the initial classification. In this study a minimum degree of determination was introduced. The threshold value was set at 0.75, meaning that at least 75% of the total variation should be explained by the classes. If this was not the case for either of the indicator parameters, the number of classes for this parameter was increased by one and tested again for acceptance.

Interpolation

Once a classification was accepted, extrapolation to spatial units was conducted using simple linear interpolation. Each spatial unit was characterised by at least one, but in general several soil profiles with similar functional characteristics. These units can be used as management units for precision agriculture.

RESULTS AND DISCUSSION

Management units were derived for two fields, indicated with A and B in Fig. 4. Field A covers approximately 10 ha, Field B is larger and covers over 15 ha. Soil variation is described by 90 and 81 borings for fields A and B respectively. Simulations were executed for all profiles described in the soil survey and for two extreme years (Table 1).

Table 1. Summary of weather data for the wet year (1987) and the dry year (1996).

Year	Precipitation (April–July)	Evapotranspiration (April–July)	Precipitation (Year)	Evapotranspiration (Year)
1987	322 mm	321 mm	766 mm	565 mm
1996	124 mm	353 mm	558 mm	599 mm

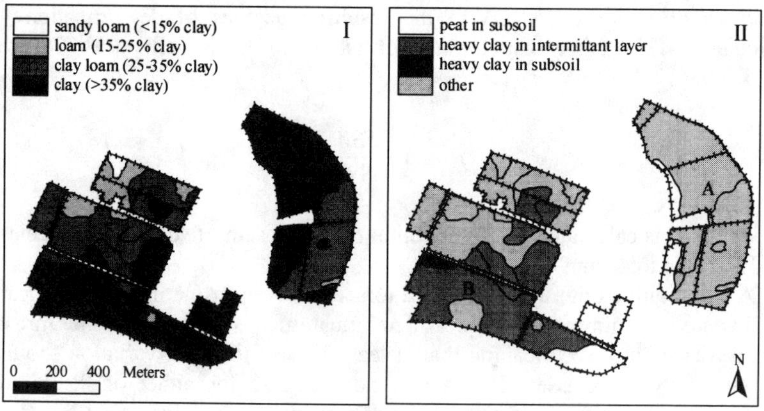

Fig. 4. Soil textural units (I) and soil profile typology (II) based on the 1997 soil survey. Management units were derived for Fields A and B.

Winter wheat was selected as the species for crop growth simulation. Initial nitrogen concentrations were set equal to the average concentration measured in February 1997 (respectively 116 kg N $ha^{-1}m^{-1}$ in Field A and 100 kg N $ha^{-1}m^{-1}$ in Field B). Four homogeneous fertiliser applications were defined in correspondence with common fertiliser dosages applied in Dutch agriculture (Table 2).

Soil profiles were classified based on their functional properties with respect to yield, N leaching and residual N content. For both fields the number of classes explaining at least 75% of the total simulated variation amounted to twelve (2*3*2; Table 3).

Table 2. Selected scenario for split N fertilization.

Application	Date	Dose (kg N ha^{-1})
1	March 14th	80
2	April 20th	80
3	May 5th	50
4	May 31st	40

Table 3: Class boundaries and degrees of determination for the indicator parameters in Fields A and B.

Class	Yield 1996 (ton ha^{-1})		N-residual 1987 (kg N ha^{-1})		N-leaching 1987 (kg N ha^{-1})	
	Field A (R^2= 0.82)	Field B (R^2= 0.75)	Field A (R^2= 0.86)	Field B (R^2= 0.77)	Field A (R^2= 0.91)	Field B (R^2= 0.89)
1	8.9 – 10.5	5.3 – 8.6	-34 – 15	-28 – 14	20 – 43	36 – 50
2	10.5 – 10.9	8.6 – 10.9	-14 – 1	15 – 73	44 – 76	51 – 64
3			0 – 24	74 – 139		

Management units were derived from the classified profiles by simple first order linear interpolation (Fig. 5 and 6). A management unit may consist of several nonconnecting areas originating from the same class.

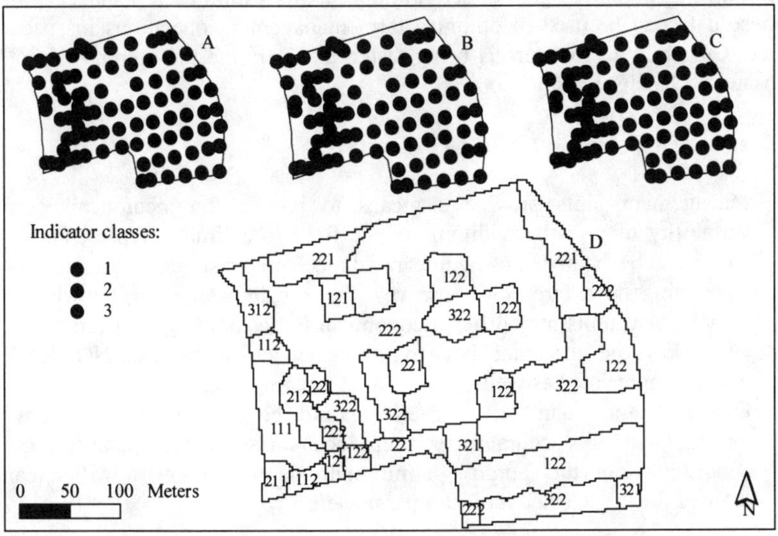

Fig. 5. Soil profile classification and management units for Field A. (A) Yield, (B) residual N, (C) N leaching, and (D) management units. Management units are coded using class numbers for the indicator parameters.

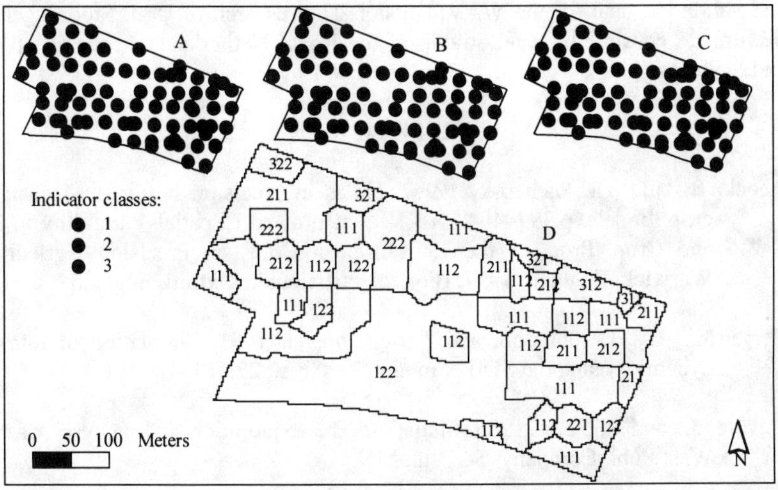

Fig. 6. Soil profile classification and management units for Field B. (A) Yield, (B) residual N, (C) N leaching, and (D) management units. Management units are coded using class numbers for the indicator parameters.

The delineation of management units is not a final objective, but merely a tool to support the incorporation of real-time simulations in precision agriculture. Based on representative soil profiles, these simulations can dynamically quantify the status of important growth determining factors within each management unit. These data can be used to optimise farm management operations in space and time. Currently a procedure is being tested to determine the optimal timing and dosage for fertiliser applications.

CONCLUSIONS

1. Management units provide a means to reduce the theoretically infinite variability of growth conditions in the field to a limited representative set, which can be evaluated using integrated mechanistic simulation models.
2. Depending on a farmer's objectives, the key indicators used to define the management units may differ. Economical indicators (e.g., net returns, yield) will always be important, but environmental indicators (e.g., NO_3 leaching) may be appreciated as well.
3. Objective and quantitative criteria should be formulated to ensure that management units accurately represent local variability with respect to growth conditions. For this purpose a minimum degree of determination can be defined for the selected indicator parameters.

ACKNOWLEDGMENTS

This research is part of a larger study financed by the Dutch Board for Remote Sensing (BCRS). Jan Verhagen is kindly acknowledged for providing calibration and validation data for the WAVE model. The research of Dr J. Stoorvogel has been made possible by a fellowship of the Royal Netherlands Academy of Arts and Sciences.

REFERENCES

Acock, B., and Ya. Pachepsky, 1997. Holes in precision farming: Mechanistic crop models. p.397–404. *In* J.V. Stafford (ed.) Spatial Variability in Soil and Crop. Proc. of the 1st European Conf. on Precision Agriculture, Warwick, England. 1997. Bios scientific publ., Oxford, England.

Bergström, L., H. Johnsson, and Tortensson, G. 1991. Simulation of nitrogen dynamics using the SOILN model. Fert. res., 27:181–188.

Devore, J., and R. Peck, 1986. Statistics: The exploration and analyses of data. West Publ. Company, St. Paul, MN.

Dierckx, J., C. Belmans, and P. Pauwels, 1986. SWATRER, a computer package for modelling the field water balance. Reference manual, Leuven, Belgium.

ESRI. 1996. Using ArcView G.I.S. ESRI, Redlands, CA.

Feddes, R.A., P.J. Kowalik, and H. Zaradny, 1978. Simulation of field water use and crop yield. Simulation Monographs. PUDOC, Wageningen, the Netherlands.

Greenwood, D.J., G. Lemair, G. Gosse, P. Cruz, A. Draycott, and J.J. Neeteson, 1990. Decline in percentage N of C3 and C4 crops with increasing plant mass. Ann. Botany 66:425–436.

Hutson, J., and R.J. Wagenet. 1992. LEACHN, a process based model for water and solute movement, transformations, plant uptake and chemical reactions in the unsaturated zone. Technical Report, Cornell Univ., Ithaca, NY.

Laird, R.J., and F.B. Cady, 1969. Combined analyses of yield data from fertiliser experiments. Agron. J. 61:829–834.

Spitters, C.J.T., H. van Keulen, and D.W.G. Van Kraalingen, 1988. A simple but universal crop growth model, SUCROS87. p. 87–98. *In* R. Rabbinge et al., (ed.) 1988. Simulation and systems management in crop protection. PUDOC, Wageningen, the Netherlands.

Stafford, J.V. (ed.) 1997. Spatial variability in soil and crop. *In* Proc. of the 1st European Conf. on Precision Agriculture, Warwick, England. 1997. Bios Sci. Publ., Oxford, England.

Van Dam, J.C., J.N.M. Stricker, and P. Droogers, 1994. Inverse method to determine soil hydraulic functions from Multistep outflow experiments. Soil Sci. Soc. Am. J. 85:647–652.

Van Genuchten, M.Th. 1980. A closed form equation for predicting the hydraulic conductivity of unsaturated soils. Soil Sci. Soc. Am. J. 892–898.

Vanclooster, M., P. Viane, J. Diels, and K. Christiaens, 1994. WAVE, a mathematical model for simulating water and agrochemicals in the soil and vadose environment. Reference and users manual. Inst. for Land and Water Manage., Leuven, Belgium.

Verhagen, J. 1997. Spatial soil variability as a guiding principle in nitrogen management. Ph.D. diss., Wageningen Agricultural Univ., Wageningen, the Netherlands.

Wösten, J.H.M. 1997. Bodemkundige vertaalfuncties bij SC-DLO: State of the Art. Technical Document. DLO-Staring Ctr., Wageningen, the Netherlands.

Wösten, J.H.M., G.J. Veerman, and J. Stolte, 1994. Waterretentie- en doorlatendheidskarakteristieken van boven- en ondergronden in Nederland: de Staringreeks. Technical Document nr.18. Revised Ed. DLO-Staring Ctr., Wageningen, the Netherlands.

Model-Based Technique to Determine Variable Rate Nitrogen for Corn

J. O. Paz
W. D. Batchelor

Agricultural and Biosystems Engineering Department
Iowa State University
Ames, Iowa

T. S. Colvin
S. D. Logsdon
T. C. Kaspar
D. L. Karlen

USDA-ARS
National Soil Tilth Laboratory
Ames, Iowa

B. A. Babcock
G. R. Pautsch

Economics Department
Iowa State University
Ames, Iowa

ABSTRACT

Past efforts to correlate yield from small field plots to soil type, elevation, fertility, and other factors have been only partially successful for characterizing spatial variability in corn (*Zea mays* L.) yield. Furthermore, methods to determine optimum N rate in grids across fields depend upon the ability to accurately predict yield variability and corn response to N. In this paper, we developed a technique to use the CERES-Maize crop growth model to characterize corn yield variability. The model was calibrated using 3 yrs of data from 224 grids in a 16 ha field near Boone, IA. The model gave excellent predictions of yield trends along transects in the field, explaining approximately 57% of the yield variability. Once the model was calibrated for each grid cell, optimum N rate to maximize net return was computed for each location using 22 yrs of historical weather data. These results were used to evaluate the economic benefit associated with variable rate N prescriptions.

Copyright © 1999 ASA-CSSA-SSSA, 677 South Segoe Road, Madison, WI 53711, USA. *Proceedings of the Fourth International Conference on Precision Agriculture.*

INTRODUCTION

The advent of yield monitors and global positioning systems that can create spatial yield maps has generated much excitement and controversy among farmers and researchers. Site-specific field management promises to maximize field level net return and minimize environmental impact by managing fields using spatially variable management practices. The success of site-specific field management depends upon discovery of relationships between environment, management, and resulting yield variability, and ultimately, how these relationships can be exploited to compute optimum prescriptions. Farmers are faced with trying to determine how to manage variability to improve profits. Researchers are trying to develop methods to analyze causes of yield variability, and determine how to develop prescriptions for fertility, and cultural practices to capitalize on variability across field. While environmental, management, soil, and pest factors have been studied for many years, researchers are just beginning to determine how these factors vary across fields, and contribute to spatial yield variability.

Initial efforts to study yield variability have focused on taking static measurements of soil, management, or plant properties and regressing against grid level yields (Cambardella et al., 1996; Khakural et al., 1996; Sudduth et al., 1996; Jones et al., 1989). However, these efforts have proven to be illusive in determining causes of yield variability. The reason for this is apparent: crop yield is influenced by temporal interactions of management, soil properties, and environment. Traditional analytical techniques, which regress static measurements against yield, do not account for temporal interactions of stress on crop growth and yield. Some successes have been achieved in developing relationships between soil type or elevation and yield variability by using regression approaches. However, these do not directly account for the dynamic interaction of available soil moisture, root water uptake, and water related stresses that can occur and affect plant growth processes. Developing this knowledge is imperative to understanding and quantifying yield variability. Soil moisture stress (drought or excess water) can cause significant variability due to variations in soil moisture holding characteristics, rooting depth and distribution, and drainage patterns across a field. Methods to accurately compute interactions of stress on growth will ultimately lead to the development of optimum site-specific prescriptions.

Assessment of spatial variability within a given field is necessary prior to implementation of variable rate fertilization (VRF). Process-oriented crop growth models are a promising tool to help researchers search for relationships between environment, management, and yield variability. In a recent study, Paz et al. (1997) used a crop growth model and found differences in water availability explained up to 69% of yield variation within transects in a central Iowa soybean [*Glycine max* (L.) Merr.] field.

The objective of this study was to demonstrate the use a corn crop growth model in determining yield variability and variable nitrogen prescriptions in the same field.

PROCEDURES

Site Description

Spatial yield distribution of corn was investigated in a 16 ha field in Boone County, Iowa. The field, which is the southwest (SW) quadrant of the Baker farm, used a conventional farming method consisting of a corn–soybean rotation, conventional tillage, and application of commercial fertilizer and pesticides. Figure 1 shows the arrangement of the eight transects in the field. Each transect consists of 28 corn yield plots or grids. This gave a total of 224 grids with measured yields. Each grid was 12 m wide by 46 m long. Final corn yield was measured from the five center rows in each grid using a plot combine and weigh wagon for 1989, 1991, and 1995.

The site is typical of low-relief swell and swale topography characteristic of broad areas of the Des Moines lobe surface (Steinwand & Fenton, 1995). The field contains nine soil classes that are predominantly from the Clarion–Nicollet–Webster soil association (Steinwand, 1992). A detailed soil map of the field was obtained from the National Soil Tilth Laboratory (NSTL) in Ames, IA. Estimates of soil physical properties were provided by Logsdon (1995, unpublished data) of the NSTL. Thus, estimates were available for lower limit, drained upper limit, saturated moisture content, saturated hydraulic conductivity, bulk density, and organic C at several depths for each soil type. Properties for the predominant soil type were used to represent soil properties in each grid

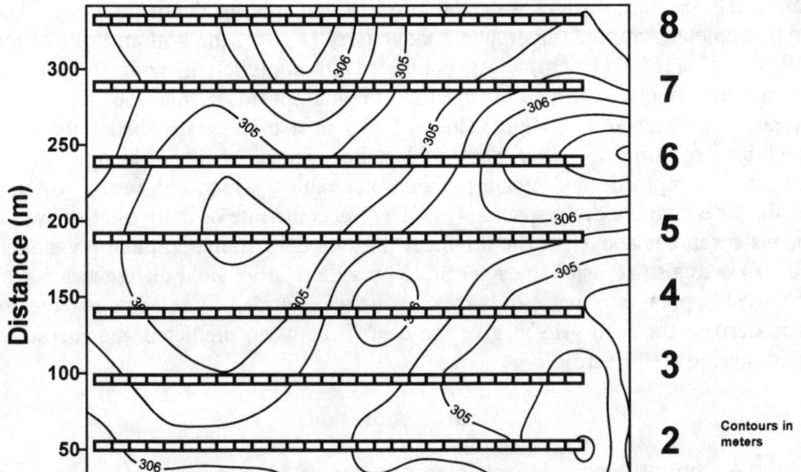

Fig. 1. Contour map and layout of yield transects and grids in Baker farm. Contour intervals are in meters.

Data Collection

In this study, planting date, N application date and rate, and final yield in each grid were collected for 1989, 1991, and 1995. It is important to note that soil water content, initial nutrient levels, and plant population and barrenness were not collected for each grid. In the following analysis, we assumed uniform initial NO_3 and soil water content levels across all grids.

Crop Growth Model

In this study, the CERES-Maize (Jones & Kiniry, 1986) crop growth model was used to characterize yield variability across a corn field. The model was developed to compute growth, development, and yield on homogeneous units (either plot, field, or regional scale), and has been demonstrated to adequately simulate crop growth at a field or research plot scale. CERES-Maize model requires inputs including management practices (variety, row spacing, plant population, fertilizer and irrigation application dates and amounts) and environmental conditions (soil type, daily maximum and minimum temperature, rainfall and solar radiation).

We assumed that two factors dominate spatial and temporal yield variability: water related stress and plant barrenness or population differences among grids. In order to test this hypothesis, we developed a technique to calibrate several input parameters of the corn model to minimize error between predicted and measured yields in each of the 224 grids. Two soil parameters were adjusted to mimic water table and tile flow dynamics in each grid. These parameters primarily affect water table depth and rooting depth progress. The first parameter, saturated hydraulic conductivity (K_{sat}) of the bottom layer of the soil profile (180–200 cm), was calibrated in conjunction with the second parameter, effective tile drain spacing, to attempt to mimic the soil water dynamics in each grid. High values of K_{sat} in a grid creates better drainage conditions resulting in lower water tables. Low values reduce drainage out the bottom of the profile and create higher water tables, which can restrict rooting depth. Effective tile drain spacing (FLDS) affects the rate of daily tile flow when the water table is above the tile drain. A third model parameter, plant population (PPOP), was also adjusted in each grid to provide relative yield differences due to consistently poor emergence or barrenness between grids. Thus, three parameters were derived for each grid to give the best fit between predicted and measured yields during a 3 yr period.

Model Calibration

A control program containing the simulated annealing algorithm was linked with the CERES-Maize model. The program was used to solve for the optimum set of these three parameters for each of the 224 grids in the 16 ha Baker field. Simulated annealing is a very robust algorithm (Goffe et al., 1994) and is used in solving complex combinatorial optimization problems. The algorithm is based on the metaphor of how annealing works: reach a mininum energy state upon cooling a substance, but not too quickly in order to avoid reaching an undesirable state (Greenberg, 1998). This study used simulated annealing routine as described by Corana et al. (1987) and implemented by Goffe et al. (1994).

Model parameters were optimized in each of the 224 grids to minimize the sum of square error between predicted and measured yield for 1989, 1991, and 1995. The objective function established for the model simulations was written as:

$$Min:\ \ SSE = \sum_{i=1}^{i=3} \left(Ym_i - Yp_i\right)^2 \qquad [1]$$

where *SSE* is the sum of square error between *Ym* (measured yield) and *Yp* (predicted yield), and *i* is the *i*th year.

Economic Analysis

After calibrating the model for each grid in the field, we conducted a simple analysis to determine optimum N application rate in each of the 224 grids within the field. Our strategy was to determine the N rate that maximized profit on average during 22 yrs (1975–1996) of historical weather data. A total of 21 different N rates (0–200 lb ac^{-1}) were tested for each of the 22 yrs. The profit ($ ac^{-1}) was computed for each N rate and for each year by:

$$\text{Net Return} = Y * P_c - N * P_n \qquad [2]$$

where *Y* is corn yield (bu ac^{-1}), P_c is the price of corn ($2.50 bu^{-1}), N is nitrogen application rate (lb ac^{-1}), and P_n is the cost of N fertilizer ($0.20 lb^{-1}).

RESULTS AND DISCUSSION

Yield Predictions

The model gave very good results for the average field level corn yields. Yield predictions were within ±14% of measured yields for each of the three corn production year (1989, 1991, and 1995). The percentage of error between field level predicted and measured yields were 6.9, -13.5, and –0.4% for 1989, 1991, and 1995, respectively (Table 1). The 3 yr field level predicted yield of 9027 kg ha^{-1} was only –2.4% the average measured yield of 9248 kg ha^{-1}.

The calibrated model generally gave excellent predictions of grid-level yields over all years, especially for yields in the range of 6000 and 11 000 kg ha^{-1} (Fig. 2). The model over-predicted corn yields in grids with measured yields of 6000 kg ha^{-1} or less for 1991 production year.

The model gave good predictions with regard to yield trends along transects in the field for all production years except 1991. Figure 3 shows an example of yield trends along transect 7. There were instances where the model gave poor agreement between predicted and measured yield on several grids, notably those in low lying areas. However, predicted and measured yield trends

Table 1. Average field-level measured and predicted yield for each corn production year.

Production year	Measured yield, kg ha^{-1}	Predicted yield, kg ha^{-1} and error, %
1989	9303	9946 (6.9)
1991	9343	8080 (-13.5)
1995	9097	9056 (-0.4)
3 years	9248	9027 (-2.4)

matched. A possible explanation is the inability of the model to account for surface run-on or sub-surface flow to a grid coming from several neighboring grids.

Overall, the model explained approximately 57% of the yield variability in all grids during 3 yrs. This indicates that the adjustments of soil parameters, which induced variable water stress across the grids, as well as the adjustment of plant population, which scaled the relative yields in grids, accounted for a significant amount of the spatial and temporal yield variability across the field. While these results are not as good as those found by Paz et al. (1997) for soybean, where the model explained 69% of the yield variability in the same field, they are promising. The interaction of water and NO_3 stresses, as well

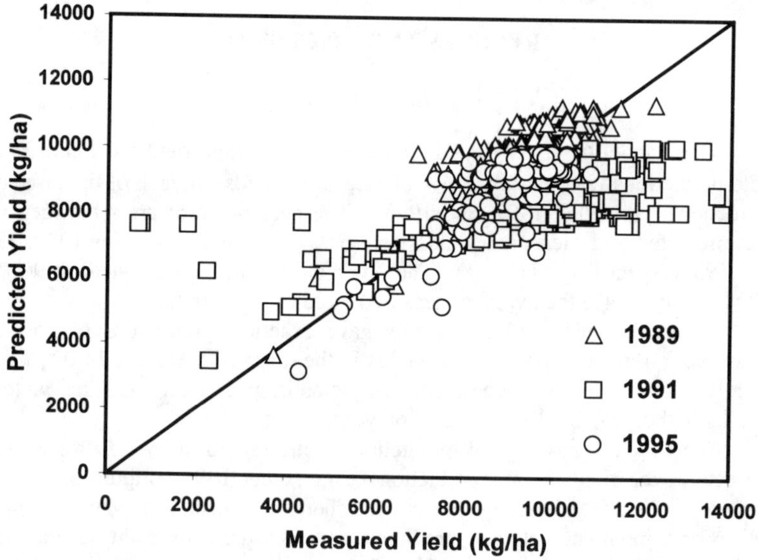

Fig. 2. Predicted vs. measured corn yields for the 224 grids in Baker Farm using 3 yrs of data.

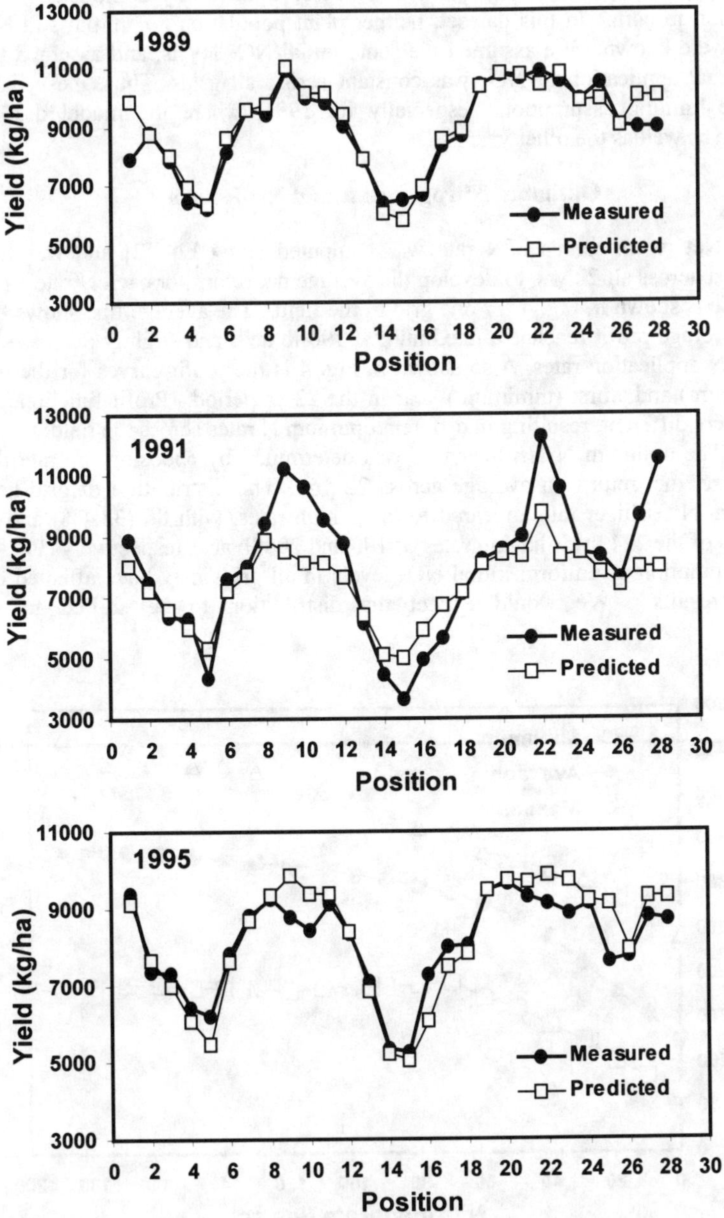

Fig. 3. Measured and predicted corn yields of each grid position along transect 7 for the different production years.

as the difficulty in computing plant barrenness, significantly complicates yield prediction in corn. In this dataset, neither plant population nor initial soil NO_3 levels were known. We assumed the same initial NO_3 levels, and assumed that the annual application of NO_3 was constant across all grids. Thus, these likely became limiting assumptions, especially for 1991, where the model did not perform as well as the other years.

Optimum Nitrogen Rate and Net Return

Net return for each N rate was computed using Eq. [2] and was then averaged across all 22 yrs, to develop the average net return for each N rate. This response is shown in Fig. 4 for one grid in the field. The average line shows that 22-yr average profit reaches a maximum at 190 lb ac^{-1}, and slightly decrease for higher N application rates. Also shown in Fig. 4 is the profit curves for the best (maximum) and worst (minimum) year in the 22 yr period. Profit functions for grids were different, resulting in different optimum N rates across the field.

The optimum N fertilizer rate was determined by choosing the rate that maximized net return on average across 22 yrs. The distribution of grid-level optimum N fertilizer rate appeared to favor high rates, with 86 (38.4%) and 54 (24.1%) of the 224 grids having rates of 190 and 200 lb ac^{-1}, respectively (Fig. 5). The assumption of uniform initial NO_3 levels in all grids may have affected our model results. We would expect the distribution to be shifted to the

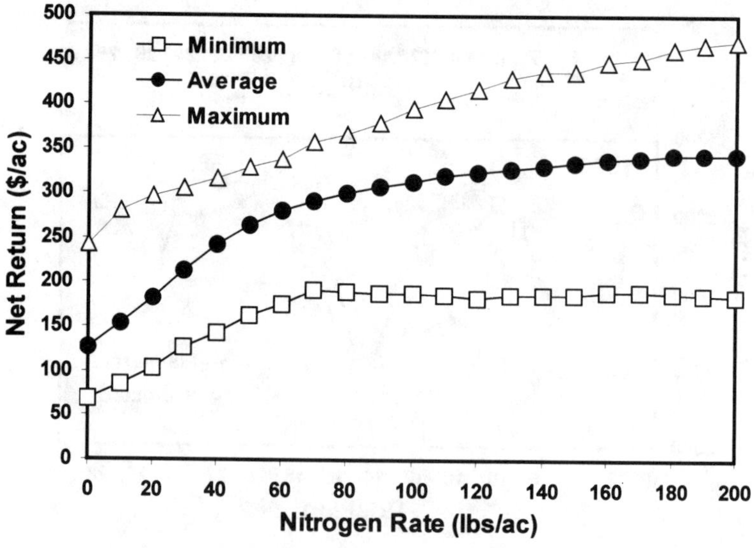

Fig. 4. Diagram showing minimum, average, and maximum net returns for each N fertilization scheme. This scenario is for grid number 2.

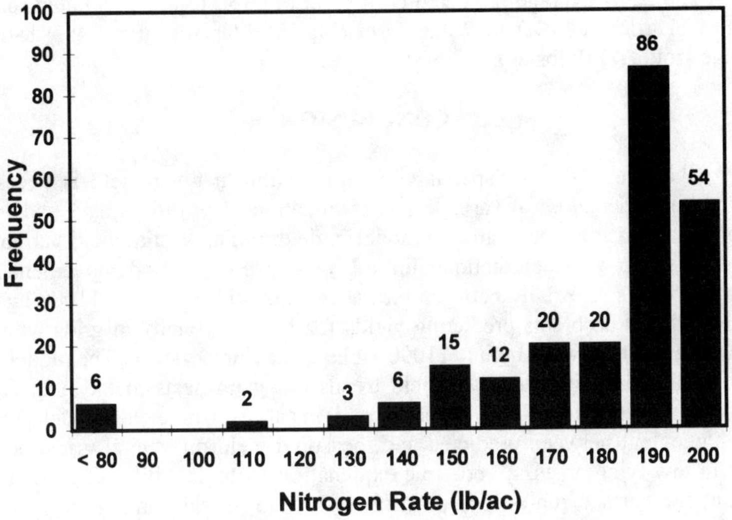

Fig. 5. Distribution of grids with corresponding optimum N fertilizer rates.

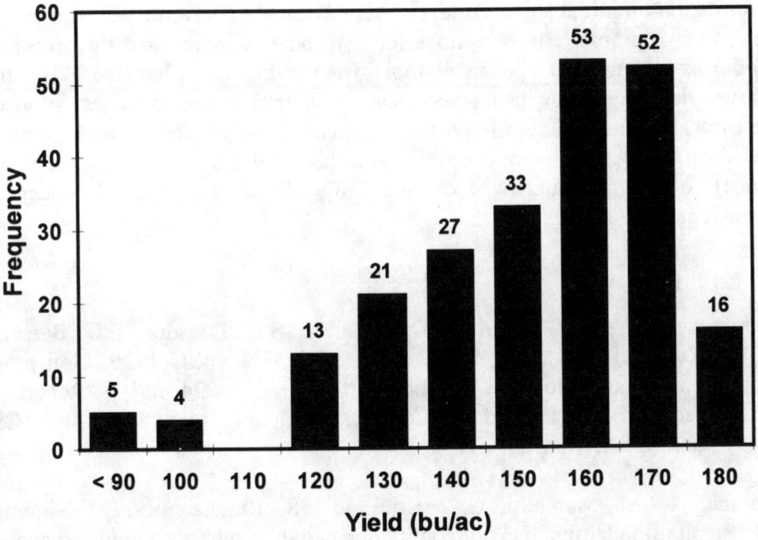

Fig. 6. Distribution of 22-yr average corn yield groups for grids in Baker farm.

left, that is, more grids would have lower optimum fertilizer rates (e.g., 130, 140 lb ac^{-1}) if information on initial soil NO_3 levels were available.

The 22-yr average predicted corn yield of >150 lb ac^{-1} accounted for 154 of the 224 grids (68.8%) in Baker farm (Fig. 6). Only 9 grids (4%) had low average yield (< 110 lbs/ac).

CONCLUSION

Characterization of spatial variability within a given field is necessary prior to implementation of variable rate fertilization. Our efforts have shown the importance of using a crop growth model in determining spatial yield variability. Grid-level corn yield predictions for all years were in good agreement with measured yields especially between the range of 6000 kg ha^{-1} and 11000 kg ha^{-1}. The model had problems predicting yields for 1991 especially in grids with low (<6000 kg ha^{-1}) and very high (>11000 kg ha^{-1}) measured yields. The model gave good predictions with regard to yield trends along transects in the field for all production years except 1991. There were instances where the model showed poor agreement between predicted and measured yield on several grids, notably those in low lying areas. A possible explanation is the inability of the model to account for surface run-on or sub-surface flow to a grid coming from several neighboring grids.

Nitrogen rates of greater than 190 lb ac^{-1} were found to be optimum in 140 of 224 grids (62.5%). Our assumption of uniform NO_3 levels in all grids may have contributed to these rather unusually high fertilizer rates. We would expect that more grids to have lower optimum fertilizer rates (e.g., 130, 140 lb ac^{-1}) if information on initial soil NO_3 levels were available. Nevertheless, our efforts to determine appropriate prescriptions for corn have demonstrated the crop growth model as a viable and powerful tool in achieving this objective. The model allows yield prediction using historical weather data and provides information necessary to make decisions on management strategies that must be employed based on risk and economic benefit. Further analysis will be made on several model output parameters including NO_3 leaching potential under each management strategy.

REFERENCES

Cambardella, C.A., T.S. Colvin, D.L. Karlen, S.D. Logsdon, E.C. Berry, J.K. Radke, T.C. Kaspar, T.B. Parkin, and D.B. Jaynes. 1996. Soil property contributions to yield variation pattern. p. 417–224. *In* P.C. Robert et al., (ed.) Proc. of the 3rd Int. Conf. on Precision Agriculture. ASA, CSSA, and SSSA, Madison, WI.

Corana, A., M. Marchesi, C. Martini, and S. Ridella. 1987. Minimizing multimodal functions of continuous variables with the simulated annealing algorithm. *ACM Trans. Math. Software* 13:262–280.

Goffe, W.L., G.D. Ferrier, and J. Rogers. 1994. Global optimization of statistical functions with simulated annealing. *J. Economet. 60:65–99.*

Greenberg, H.J. 1998. Mathematical programming glossary. World Wide Web, http://www-math.cudenver.edu/~hgreenbe/glossary/glossary.html.

Jones, C.A., and J.R. Kiniry. 1986. CERES-Maize: A simulation model of maize growth and development. Texas A&M Univ. Press, College Station.

Jones, A.J., L.M. Mielke, C.A. Bartles, and C.A. Miller. 1989. Relationship of landscape position and properties to crop production. J. Soil Water Conserv. 44:328–332.

Khakural, B.R., P.C. Robert, and D.J. Mulla. 1996. Relating corn/soybean yield to variability in soil and landscape characteristics. p. 117–128. *In* P.C. Robert et al., (ed.) Proc. of the 3rd Int. Conf. on Precision Agriculture. ASA, CSSA, and SSSA, Madison, WI.

Paz, J.O., W.D. Batchelor, T.S. Colvin, S.D. Logsdon, T.C. Kaspar, and D.L. Karlen. 1997. Characterizing spatial yield variability of soybeans using a crop growth model. Transactions of the ASAE (submitted for publication).

Steinwand, A.L. 1992. Soil geomorphic, hydrologic and sedimentologic relationships and evaluation of soil survey data for a Mollisol catena on the Des Moines Lobe, central Iowa. Ph.D. diss.. Iowa State Univ. Ames.

Steinwand, A.L., and T.E. Fenton. 1995. Landscape evolution and shallow groundwater hydrology of a till landscape in Central Iowa. *Soil Sci. Soc. Am. J.* 59:1370–1377.

Sudduth, K.A., S.T. Drummond, S.J. Birrell, and N.R. Kitchen. 1996. Analysis of spatial factors influencing crop yield. p. 129–140. *In* P.C. Robert et al., (ed.) Proc. of the 3rd Int. Conf. on Precision Agriculture. ASA, CSSA, and SSSA, Madison, WI.

Performance of ICEMM: A Cotton Simulation Model In A Precision Farming Study

M. Y. L. Boone, and J. A. Landivar
Texas A&M University
Research and Extension Center
Corpus Christi, Texas

ABSTRACT

Precision farming revolves around the use of information-based technologies. The biological information link between the crops and these technologies are obtained from scouting reports and plant mapping. However, these activities are time consuming and expensive. Crop simulation model running on real time can provide many of the soil–plant–weather information at a fraction of the cost. Initial performance evaluation of the ICEMM-cotton simulation model to a precision farming study at the King Ranch in Kingsville, TX, shows promising use of the models. A 40-ha field divided into 20 blocks was used for the study. Agronomic practices, daily weather data and soil physical and chemical properties (collected at the start of the planting season) were used as inputs to the model. The date of occurrences of developmental events were predicted within 2 d of actual occurrences. The final plant height and node numbers were close to the observed values in the 20 blocks. Although the overall average yield data approximated the actual yield, individual block yield comparisons were inconsistent partly due to incorrect soil hydrologic parameters.

INTRODUCTION

Precision farming holds tremendous potential for improving the economic and environmental sustainability of agriculture. It combines extensive knowledge about the soil and crop variability within production fields with satellite and computer technologies in an attempt to optimize the amount of production inputs (seed, fertilizer, chemicals, etc.) at all points in the field. Optimizing production inputs is difficult because of the complex interactions of many variables that impact plant production, and the expense and availability of information needed to make the management decision (G. Sabbagh, 1997, personal communication).

With the availability of information-based technologies such as Geographic Information System (GIS), Geographic Positioning System (GPS), and variable rate implements, precision farming is becoming a reality. However, most of the biological data necessary for management decision making are still obtained by plant mappings and scouting reports. Mechanistic and dynamic crop models can provide this information on a daily or weekly basis or as often as the user desires. One such crop model is the ICEMM cotton simulation model.

Copyright © 1999 ASA-CSSA-SSSA, 677 South Segoe Road, Madison, WI 53711, USA. *Proceedings of the Fourth International Conference on Precision Agriculture.*

ICEMM COTTON SIMULATION MODEL

The Integrated Crop Ecosystem Management Model (ICEMM) is based on the 1988 version of GOSSYM (Baker, et al., 1983). It has been improved and extensively validated as well as calibrated by Landivar (1991) for the Coastal Bend areas of Texas. It simulates the physiological and soil processes. As a mechanistic and dynamic model, it requires for its inputs:

1. Daily weather: temperature, rainfall, radiation and wind;
2. Initial soil fertility status: nitrate, ammonia and organic matter content;
3. Soil physical properties: bulk density, hydraulic conductivity, and moisture retention characteristics; and
4. Cultural inputs–row spacing, cultivar, plant density, fertilizer and plant growth regulators applications, etc.

It provides the user with a daily status report on a number of plant parameters, such as: plant height; node, square, green boll and open boll counts, nitrogen , water and carbohydrate stresses, etc. At the end of a full season run, it creates a summary table on date of maturity, plant height, yield, and for each designated developmental events, the number of nodes, squares, green bolls and open bolls.

PRECISION AGRICULTURE STUDY IN SOUTH TEXAS

Water is the major limiting production input for most of the South Texas Coastal Bend cotton farmers. Their dryland cotton production on the average yielded about 400 kg ha^{-1} on dry years and roughly 1000 kg ha^{-1} on good years. Because of their close proximity to the Gulf of Mexico and the heavy shrinking and swelling clay soils of the region, most of these farmers do not irrigate their fields. There are economic and environmental issues that concerns these farmers regarding the viability of their agronomic practices. This 3 yr project was initiated in 1996 to isolate the sources of in-field cotton production variability that can be identified, measured and used profitably in management decisions.

Two adjacent 40 ha area fields under sorghum–cotton rotation have been chosen for the study. Both fields located at the King Ranch, Kingsville, TX, have been georeferenced using a differential GPS. Each field has been subdivided into 20 2-ha area blocks. Figure 1 shows the field layout of the cotton field.

Soil samplings using systematic grid sampling were done in 30 Jan., 1997 and 21 Jan., 1998 for each of the soil horizons of the Mercedes clay (fine, smectitic, hyperthermic Udorthentic Pellusterts), Raymondville clay loam (fine, mixed, hyperthermic Vertic Calciustolls), and Victoria clay (fine, montmorillonitic, hyperthermic Udic Pellusterts). These samples were analyzed for bulk density, hydraulic conductivity, soil texture and moisture retention characteristics. Results of these analyses were used to develop the soil hydrology input files of ICEMM.

Soil samples were also taken at 15-, 46- and 76-cm depths for soil nutritional analyses. These are for organic matter content, N, P, K, Mg, Ca, pH, SO_4, Zn, Mn, Fe, Cu, and B.

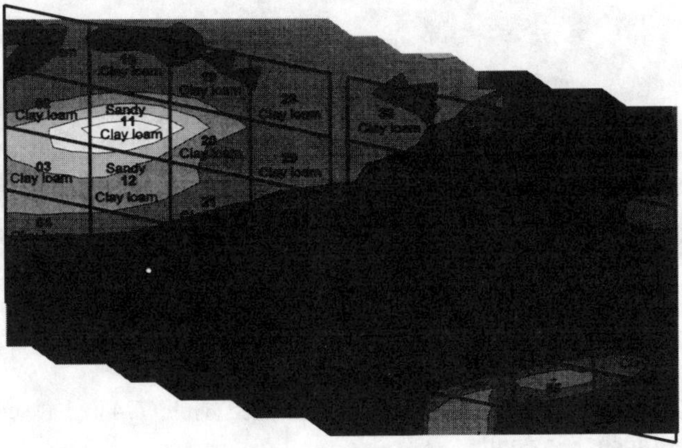

Fig. 1. Field layout of the precision farming study site at the King Ranch, Kingsville, TX, showing the block numbering and the soil texture at 15-cm depth.

Plant mappings were done at 25, 45, 63, 91, and 121 d after planting. These data were used to validate the model. The plant height data were necessary for the estimation of PIX applications that were applied on 3 and 18 June , 1997. At harvest, yield was estimated by hand sampling and yield monitor. Plant parameters data collection were obtained using transect sampling.

DISCUSSION

ICEMM is continuously being improved to include the latest knowledge on cotton growth and development. For example, the soil routines (Boone et al., 1995) of the 95 release of GOSSYM/COMAX have been incorporated into ICEMM. This revised version of ICEMM is used in the 1997 model evaluation .

Figure 2 shows the yield maps of the observed and the simulated yields. Although the maps showed that ICEMM estimation of the yield was fairly close for about 50% of the area, it is interesting to note that the model can predict the yield variation. The actual yield in the 40-ha field ranged from 442 to 844 kg ha^{-1} with an average of 585 kg ha^{-1}. The simulated yield varied from 385 to 1089 kg ha^{-1} with a mean of 619 kg ha^{-1}. The average yield values differ by 6%. which is an acceptable margin.

Typical examples of blocks that ICEMM had performed well and poorly are shown in Fig. 3. The node estimates differed by 1 to 2 nodes for both blocks, which is the same as the average node difference of 1.5 nodes between actual and predicted node count. The plant height predictions were greatly off on Block 54, which also overestimated the yield by almost 65%. On the other hand, height predictions on Block 55 were fairly close. Moreover, the yield estimate differed by -21 kg ha^{-1} on an actual yield of 844 kg ha^{-1}.

The close node count estimation illustrates the robustness of the model in predicting developmental events. The model predicted the occurrences of the

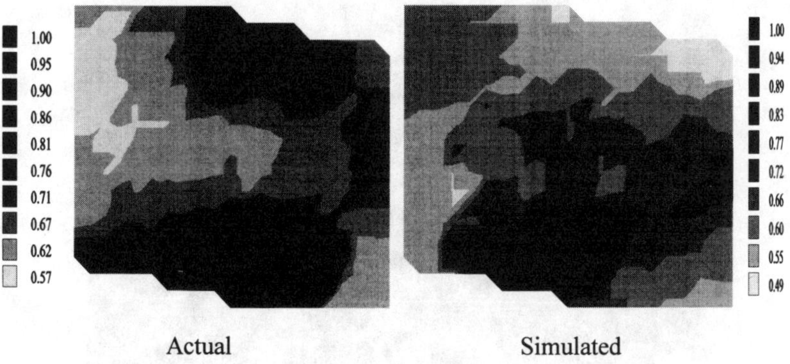

Fig. 2. 1997 Actual (yield/692 kg ha^{-1} and simulated (yield/774 kg ha^{-1}) cotton yield at the King Ranch, Kingsville, TX.

phenological events within 2 d of the actual occurrences. Although plant height estimations were off by 15 cm on the average, this problem is likely because of the lack of accountability of the PIX applications by ICEMM.

A major data input of the model is the soil hydrology files. There were some problems with the laboratory analysis of the hydraulic conductivities, which for Victoria clay typically ranged from 1 to 12 cm d^{-1}. The laboratory results have some values that are in the hundreds of cm/day. Although some of the hydraulic data that were used in the model had been adjusted by using the average of the nearest surrounding sample points, still they were inadequate. These erroneous data files are the likely explanations for the differences in the yield map.

The final plant height, node and boll count did not have an effect on the observed yield variations in the 40-ha study site. The weather patterns and the cultural practices were practically the same with the exception of the variable PIX applications.

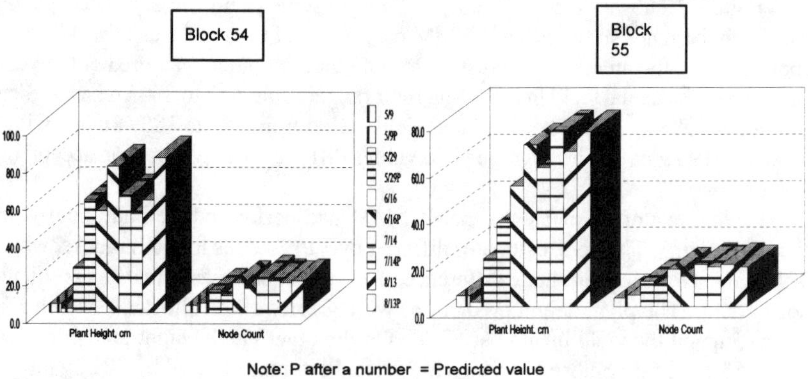

Fig. 3. Actual vs. predicted plant height and node count in Block 54 and Block 55.

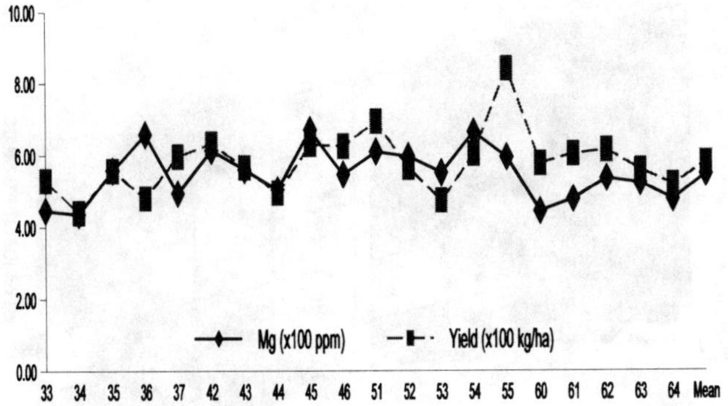

Fig. 4. Plot of Mg content (ppm) vs. yield (*100 lb ac^{-1}).

Considering all of the above factors, what can possibly be the source of yield variations? The soil nutritional status could have a likely influence on them. Figure 4 shows the plot of the Mg content of the field at 15-cm depth and the yield by blocks. Note that the points fall on top of one another. Mg content seems to be related to yield. Moreover, the three soil types in the field would likely affect the amount of plant water available throughout the cropping season. Unfortunately, the soil physical properties data that are so vital in the ICEMM estimation of crop yields were suspect.

LESSONS LEARNED FROM THE FIRST YEAR OF THE STUDY

The greatest hurdle to the success of precision farming is the determination of the optimum values. Precision agriculture and crop models require reliable information to furnish credible information. The failure of the soil physical laboratory analysis (despite the meticulous handling of the samples) led to questionable soil hydrology files that limited their use in the crop model, ICEMM.

Soil surveys provide approximation of the soil type boundaries. Out in the field to do the soil sampling and trying to determine the boundary locations was a frustrating experience because detailed sampling for both the physical and chemical laboratory analysis rely heavily on where the samples are taken. A more refined boundary location estimation is critical because soil sampling is laborious, time consuming and expensive.

Maps of the soil texture at 15-cm depth and yield provided clues to the locations of the soil type boundaries (Fig. 5). The aboveground cultural inputs were practically the same throughout the area and there appeared to be no soil nutritional elements that influenced yield. Thus, it was safe to assume that the observed yield variations were more a result of the soil physical properties. Subsequent soil samplings were based on this redefined soil type boundaries.

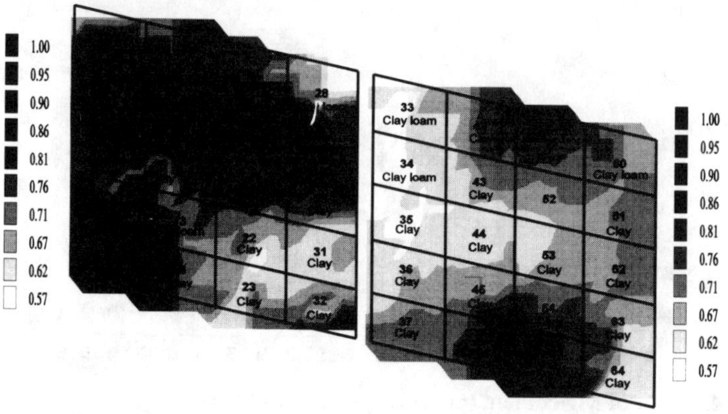

1996 Cotton yield (Yield/692 kg/ha) at the King Ranch. 1997 Cotton yield (Yield/841 kgl/ha) at the King Ranch.

Fig. 5. Map of the 1996 and 1997 cotton yield at the King Ranch, Kingsville, TX.

CONCLUSION

The revised ICEMM model performed favorably in its first year of testing. The predicted node count and yield values are within reasonable range. ICEMM's estimates of the occurrences of phenological events are within 2 days of the actual events. Soil Mg content at 15-cm depth seemed to influence the observed yield variations in the study site.

To fully evaluate the performance of ICEMM , adequate soil samples need to be taken to correct the problems in the soil hydrology input files. Consequently, a better sampling strategy needs to be explored and developed for both the soil physical and chemical properties evaluation.

REFERENCES

Baker, D.N., J.R. Lambert, and J.M. McKinion. 1983. GOSSYM: A simulator of cotton crop growth and yield. S.C. Agric. Exp. Stn. Bull. 1089.

Boone, M.Y.L, D.O. Porter, and J.M. McKinion. 1995. RHIZOS 1991: A simulator of row crop rhizosphere. Natl. Technical Inf. Serv., U.S. Dep. of Commerce, Springfield, VA.

Landivar, J.A., B.R. Eddleman, J.H. Benedict, D.J. Lawlor, D. Ring, and D.T. Gardiner. 1991. ICEMM, An integrated crop ecosystem management model. p. 453–457. *In* Beltwide Cotton Conf., San Antonio, TX. 8–12 Jan. 1991. Natl. Cotton Council of Am., Memphis, TN.

Evaluation of Crop Models to Simulate Site-Specific Crop Development and Yield

C. W. Fraisse
K. A. Sudduth
N. R. Kitchen
USDA - ARS
Cropping Systems and Water Quality Research Unit
Columbia, Missouri

ABSTRACT

Crop simulation models have been used historically to predict average field crop development and yield under alternative management and weather scenarios. The objective of this paper was to evaluate and test a new version of the CERES-Maize model that was modified to improve the simulation of site-specific crop development and yield. Seven sites within a field located in central Missouri were selected based on landscape position, elevation, depth to the claypan horizon, and past yield history. Detailed monitoring of crop development and soil moisture during the 1997 season provided data for calibration and evaluation of the model performance at each site. Mid-season water stress caused a large variation in measured yield with values ranging from 3.0 Mg ha^{-1} in the eroded side-slope areas to 11.7 Mg ha^{-1} in the deeper soils located in the low areas of the field. The results obtained demonstrated that the modifications improved the ability of the model to simulate site-specific crop development. Areas of potential model improvement and further investigation were discussed.

INTRODUCTION

The interest demonstrated in recent years by farmers and researchers alike in the crop management system known as precision agriculture has caused a surge in the collection of such geospatial data as crop yield and soil properties. Although the collection of geospatial data is relatively easy, it is more difficult to know how to most effectively use that data in making crop management decisions (Sudduth et al., 1998). Several researchers have used statistical analysis to better understand the functional relationship of crop yield to other spatial factors (Sudduth et al., 1996; Mallarino, 1996; Pierce et al., 1994). However, crop production is a function not only of spatial factors but also of temporal variability. Climate variability may often be even more important than spatial variability. In fact, the impact of spatial variability on crop yield may be negligible in some years (Mulla & Schepers, 1997). Crop models, when well calibrated and validated, are able to integrate soil and weather conditions and management decisions to predict crop development and yield under alternative scenarios. Bouma (1997) discussed the use of simulation models in a proactive management approach to anticipate occurrences of crop stress, in contrast to the reactive approach where crop scouting, perhaps coupled with remote sensing, is used to identify stress after it has occurred.

Copyright © 1999 ASA-CSSA-SSSA, 677 South Segoe Road, Madison, WI 53711, USA. *Proceedings of the Fourth International Conference on Precision Agriculture.*

The CERES-Maize model (Jones & Kiniry, 1986) was selected for this study because it is a process oriented model capable of simulating water balance, nitrogen balance, and crop growth and development, while maintaining reasonable input requirements that would not prevent it from being used by crop consultants and farmers. The version used was a modification of the CERES-3.1 version with several new features added to improve the model's performance under a site-specific crop management approach (Batchelor & Ritchie, 1998; Garrison, 1998). The model was modified to include the effects of limited soil aeration on crop growth and development. The concept of using a root distribution weighting factor to estimate the relative root growth in all soil layers was replaced by a layer-specific root hospitality factor. In addition, this modified version of the model does not allow roots to extend into saturated layers. It also increases root senescence if a soil layer becomes saturated to account for poor respiration under oxygen depleted conditions. An additional factor, the hardpan factor, has been included in the model to slow down root penetration through a hardpan. Other modifications include a tile drainage routine to better simulate water table dynamics and root interactions under tile drainage conditions and to reduce leaf and stem expansion and photosynthesis under water-logged conditions.

OBJECTIVES

The overall objective of this research project is to evaluate the use of crop models as decision aid tools under a precision agriculture management approach. This first study focuses on the testing and evaluation of a new version of the CERES-Maize model. Model calibration and observation of the model behavior under different growing conditions were the first phase of the study and are reported here. The next phase of the research will focus on spatial and temporal validation of the model for the study area.

MATERIALS AND METHODS

Data were collected during the 1997 cropping season in a 36 ha field located near Centralia, in central Missouri. The field is managed in a high yield goal, high input, minimum till corn (*Zea mays* L.) soybean [*Glycine max.* (L.) Merr.] rotation. The corn hybrid Northup King RX790 was planted in the 1997-cropping season with a target population of 62 000 plants ha^{-1}. Fertilizer and chemical inputs were applied at a single rate over the entire field. The growing season was characterized by a wet spring followed by mid-season drought stress during the pollen shed period (only 0.5 cm of precipitation occurred during the first two weeks of July) that caused yields to be reduced in the areas of the field with shallow, eroded soils.

Site Description

The soils of the field are characterized as claypan soils (fine, montmorillonotic, mesic, Udollic Ochraqualfs and Albaquic Hapludalfs). These soils are poorly drained and have a restrictive, high-clay layer (the claypan) occurring below the topsoil. Figure 1 shows the elevation and aspect maps of the field. Field elevation, determined by a topographic survey, ranges from 261.9 m at the north edge of the field to 265.8 m at the southeast corner. Aspect values were calculated using the TAPES-G (Terrain Analysis Program for the Environmental Sciences - grid version; Gallant & Wilson, 1996) model. Surface and subsurface water flows from the west and east sides of the field to a central natural drainage channel that carries the water to an outlet located along the north side of the field. A detailed first-order soil survey of this field established the presence of three distinct soils - Adco silt loam, Mexico silty clay loam (eroded) and Mexico silt loam (overwashed phase, Fig. 2). Based on previous work (Sudduth et al., 1995), topsoil depth above the claypan was estimated from soil electrical conductivity measured by a commercial electromagnetic induction (EM) sensor (Fig. 2). Yield in the field has been monitored and mapped since 1992. The monitoring program has shown yield variability following landscape patterns during dry or wet years and a more uniform distribution of yield during years with well distributed rainfall.

The selection of the monitoring sites for this study was based on existing information for the study area. The main goal was to select enough sites to characterize the yield variability measured in the field. The site selection was primarily based on topography, topsoil depth, and previous yield patterns. Topsoil depth and topography are primary yield limiting factors that cannot be corrected easily and their effects on yield are of major interest. Although the field was previously sampled on a 30 m grid for soil properties such as phosphorus, potassium, pH, organic matter, calcium and magnesium, this information was not used in the site selection because soil fertility factors other than N are not taken into account by the model and could not be properly simulated. Seven monitoring sites were selected (Table 1; Fig. 1 and 2).

Table 1. Monitoring site characteristics.

Site	Elevation m	Topsoil depth cm	Slope %	Aspect degrees
1	263.1	25	1.25	67
2	262.1	100	0.53	35
3	263.3	15	0.56	87
4	263.6	44	0.37	315
5	263.5	36	0.26	12
6	264.4	30	0.32	12
7	264.9	32	0.10	304

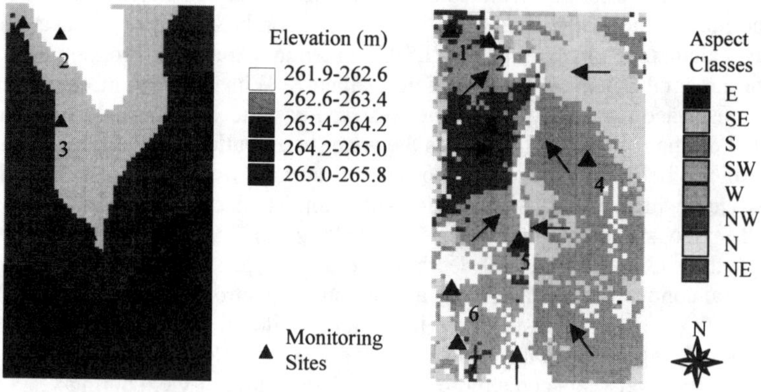

Figure 1. Elevation (m) and aspect classes of the study area.

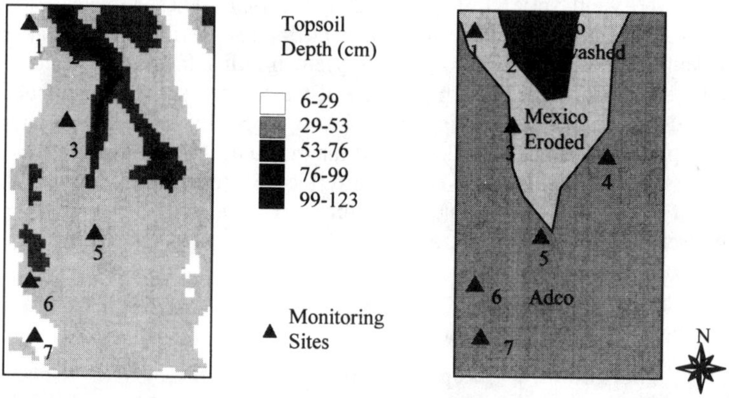

Figure 2. Topsoil depths (cm) and soil types of the study area.

Data Acquisition

Soil samples were obtained before planting at each site to a depth of 120 cm for profile characterization and mineral N analysis. Soil horizons were determined by a combination of visual and hand-feel inspection. The amount of surface residue from the previous crop was estimated based on residue material randomly collected at each monitoring site. Neutron probe tubes were installed for root zone soil moisture monitoring during the growing season. Neutron probe readings were obtained every week at 15-, 30-, 45-, 60-, 80-, 100-, and 120-cm depths. Weather data required by the model, including rainfall, maximum and minimum daily air temperature and incoming solar radiation were collected on site, by an automated weather station located at the west side of the field.

Soil textural composition and other soil physical and chemical properties required by the model were based on the results of a first-order soil survey conducted by the USDA-NRCS (Natural Resources Conservation Service). The lower limit (LL) and the drained upper limit (DUL) of available soil water were assumed to be the reported volumetric water contents corresponding with soil matric potentials of -1500 and -33 kPa, respectively. Saturated hydraulic conductivity (K_{sat}) values were not available from the NRCS survey and were estimated based on the method by Rawls et al. (1982). In this method representative K_{sat} values are given for each soil textural class.

Each monitoring site consisted of a rectangular area of approximately 12 m^2 (3.8 m $*$ 3.0 m) with the neutron probe tube located at the center of the area. A buffer zone of 2 m was established around each site. Destructive biomass sampling for characterization of crop development was carried out beyond the buffer zone at different crop developmental phases. A 1-m stake was used to randomly select a row section for harvesting. The stake was moved down the row at the sampling location until each end of the stake fell midway between two plants. At the beginning of the season, all plants within the sampled row section were transported to the laboratory for measurement of leaf area and dry biomass. Later in the season, only one representative plant (not the largest, not the smallest) was used for leaf area and dry biomass measurements. However, the fresh weight of the entire sample was used to back calculate the total dry biomass.

RESULTS AND DISCUSSION

The CERES-Maize model was calibrated against the measured data for root zone soil moisture content and seed yield. The calibration procedure involved first adjusting the soil water lower and upper limits, K_{sat}, root hospitality, and hardpan factors so that the simulated soil moisture values closely matched the measured values. In soils with a high-clay restrictive layer such as the claypan soils, the root development is an important factor in determining yield, especially in the case of mid-season drought stress. The hardpan factor and the root hospitality factor were key factors in calibrating the model for the measured yield on these soils.

The phenological development of the crop was reported during field trips for biomass sampling or occasionally during neutron probe readings. No attempt was made to record the exact dates of changing growth stages. The corn hybrid planted requires 2580 heat units to reach maturity or approximately 114 to 115 d. The genetic coefficients required by the model include the thermal time (growing degree days) from seedling emergence to the end of juvenile stage (P_1), the photoperiod sensitivity coefficient (P_2), the thermal time from silking to physiological maturity or black layer (P_5), the maximum kernel number per plant (G_2), and the potential kernel growth rate (G_3). The coefficients were set for initial testing according to recommendations provided by one of the researchers involved in the modifications of the model (W. Batchelor, Iowa State Univesity). Table 2 shows the initial values that were used based on the number of growing degree days required by the hybrid to reach maturity (GDD).

Table 2. Recommended initial values for genetic coefficients based on the number of growing degree days (GDD) to reach maturity.

GDD	P1	P2	P5	G2	G3 mg d^{-1}
2500–2600	160	0.75	780.0	750.0	8.5
2600–2650	185	0.75	850.0	800.0	8.5
2650–2700	212	0.75	850.0	800.0	8.5
2700–2750	240	0.75	850.0	800.0	8.5
2750–2800	260	0.75	850.0	800.0	8.5

Figure 3 shows the simulated and measured volumetric soil water contents at three different depths for sites 1, 2, 3, 4, 5, and 7. Results obtained for Site 6 were very similar to Site 7. At most sites simulated and measured soil water content matched well. It was observed that the correspondence was better at the sites where the claypan layer was found within 25 to 40 cm from the soil surface (Sites 1, 5, and 7). The reason for that seems to be that the hardpan factor added to the model to slow down root penetration operates only within the 30- to 45-cm layer.

The high clay layer (B_t or claypan horizon) at Site 1 is located at 25 cm from the soil surface. Minor adjustments in the lower and upper soil water limits and the root hospitality factors were required for calibration of this site. Site 5, with a 36-cm topsoil depth, also was easily calibrated. In the case of Site 7, the simulated soil moisture contents at the 45- to 60-cm layer were, in general, lower than the measured ones. This could be corrected by decreasing the root hospitality factor for that depth; however, this would affect seed yield and obtaining a good match between the measured and simulated seed yield was a higher priority in the calibration procedure than matching soil moisture.

Site 3 has shallow topsoil and the B_t horizon is found 15 cm from the soil surface. The model over-predicted the soil moisture contents in the 45-to 60-cm layer. This was a consequence of the low values set for both the root hospitality and hardpan factors, in order to properly simulate the low yield (in the range of 2.5 Mg ha^{-1}) measured for this site. Site 2 has deep topsoil and the hardpan factor was set to 1.0 in order to facilitate root penetration through the soil layers. The root hospitality factors were also set to 1.0 in order to facilitate root development and water uptake throughout the profile and to increase the simulated yield. The simulated soil water contents were always lower than the measured ones for root zone depths >30 cm. This site is located in a low area of the field and near the field water outlet (Fig. 1). It not only has a deep topsoil, facilitating root development, but it also receives surface and, probably, subsurface water contributions from the upland areas of the field. The water balance in the model does not account for run-on or lateral subsurface water contributions and can not be properly calibrated in this case. The same problem may have occurred at site 4. The simulated soil water contents agreed well with measured ones in the upper layers but were lower in the 45- to 60-cm layer. Site 4 also has deeper topsoil and receives water contribution from adjacent areas.

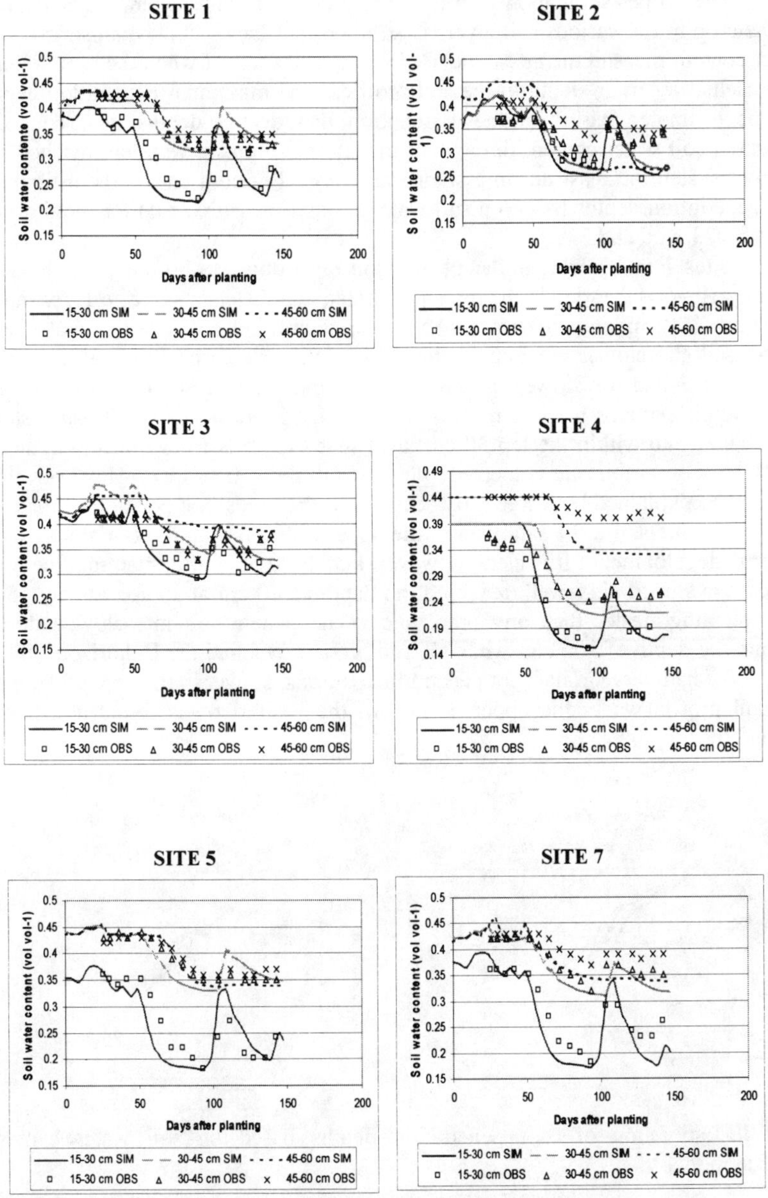

Fig. 3. Simulated and observed volumetric soil water content at 15- to 30-cm, 30- to 45-cm, and 45- to 60-cm depths for six sites within the study field.

The proper simulation of root development and consequent water uptake by the crop in the various soil layers is an important factor in claypan soils. Soil water content measurements can be used as an indication of when the root system has reached a certain depth in the soil profile. The maximum penetration depth can be estimated based on the assumption that deeper depths with constant measured soil water content throughout the cropping season were not reached by the root system, at least not in a significant way. Figure 4 shows the estimated root penetration depths based on soil water content measurements for monitoring Sites 1, 2, 3, 4, 5 and 7.

Sites 1 and 3 had similar penetration rates down to 30 cm from the soil surface. Beyond that depth, the roots penetrated at a faster rate in Site 1, reaching a depth of 100 cm in approximately 85 d after planting. Root penetration at Site 3 progressed at a slower rate and reached a maximum depth of 80 cm at the same time. Site 3 had the slowest penetration rate and shallowest root system of all sites, which explains the low measured yield of 2.5 Mg ha^{-1}. Site 2 had a slow penetration rate within the top 30-cm layer that should not have occurred, since the B_t horizon at this site is located 100 cm from the soil surface. However, this fact can be explained by a wet early season and the landscape position of the site that caused the soil to be saturated in the upper layers during the initial stages of the crop development. It is clear, however, that 40 to 50 d after planting, the root system developed very well, reaching the depth of 80 cm at approximately 65 d after planting, faster than any other site. The penetration rate slowed down beyond the depth of 100 cm, when the root system reached the B_t horizon. Sites 4, 5, and 7 had very similar root penetration patterns, slower in the upper 50 cm of the soil profile where the upper portion of the B_t horizon is found and faster beyond this depth.

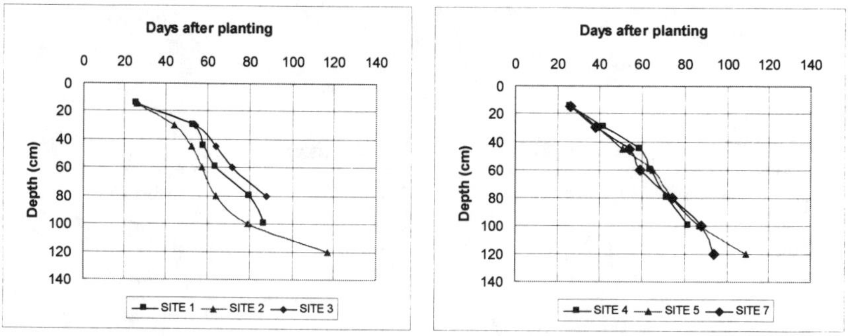

Fig. 4. Estimation of root penetration depths based on soil water content measurements.

Figure 5 illustrates the fact that seed yield predictions for the seven monitoring sites were well calibrated to the observed values. The mid-season water stress that occurred during the cropping season enhanced the importance of proper simulation of root development. In claypan soils, the depth of the high-clay layer is an important factor determining and simulating root density and water uptake at the various depths. Seed yield simulation was very sensitive to the hardpan factor, and proper estimation of this factor based on the soil profile characteristics is required for validating the model.

In the CERES-Maize model potential dry matter production of a plant is a linear function of intercepted photosynthetically active radiation. The actual rate is calculated by multiplying the potential dry matter production by the most limiting of four stress factors due to temperature, limited soil water, excess soil water, and temperature. The actual dry matter production is then partitioned among the different plant organs growing in any phenological stage. Plant leaf area growth is affected by the most limiting of two factors, a turgor factor and a saturation factor. The turgor factor is a function of potential root water uptake and plant transpiration, and the saturation factor accounts for soil water saturation conditions in the root zone. The newly introduced root hospitality and hardpan factors define the ability of roots to penetrate and explore a soil layer, thus affecting the potential root water uptake.

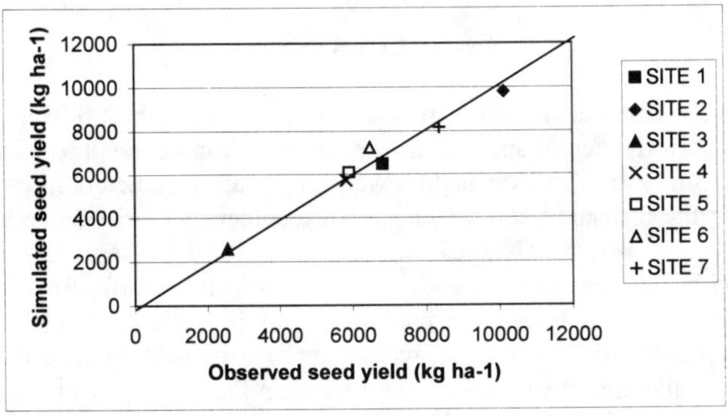

Fig. 5. Comparison of simulated and observed seed yield for the seven monitoring sites.

The leaf area expansion rates simulated by the model for most of the monitoring sites were lower than the measured values. Figure 6 shows a comparison of simulated and measured leaf area index for Sites 1, 2, and 5. The values simulated for Site 2 were reasonably close to the measured ones; however, the simulated values for Sites 1 and 5 were low when compared with the maximum measured values. The modifications introduced in the model are probably enhancing the water stress effects on leaf area expansion due to a reduction in the water uptake from layers with low root hospitality factors.

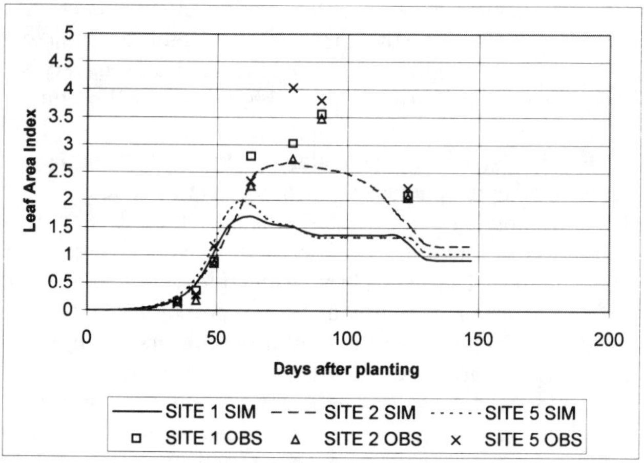

Fig. 6. Comparison of simulated and measured leaf area index for monitoring sites 1, 2 and 5.

CONCLUSIONS

The new features included in this version of the CERES-Maize model improved its performance in simulating site-specific crop development and yield. The root hospitality and hardpan factors were important parameters in calibrating the model for the measured soil water moisture content at various depths of the soil profile. The results obtained for most monitoring sites were in good agreement with the measured values. Further flexibility in adjusting the depth from the soil surface at which the hardpan factor is applied, instead of "hard wiring" it to the 30- to 45-cm layer, would improve its performance in claypan soils. Perhaps, allowing the user to input a hardpan factor (impedance factor) for each layer of the soil profile would result in a better characterization of root penetration rates and improved simulation results.

Areas that receive run-on or subsurface flow contributions from the upland areas of the field are difficult to simulate with the current modeling approach. The coupling of hydrologic models with crop models seems to be the best way to properly simulate the water balance in these areas; however, it might increase the complexity and input requirements of the combined model to a level that would prevent its adoption as a management tool.

The modifications introduced in the model have probably enhanced the water stress effects on leaf area expansion to levels that exceeded the actual effects. This aspect is currently under review and evaluation by the model developers.

Soil cores were extracted from each monitoring site at the end of the season and are currently being used for determination of root density at the various depths. The results obtained will provide additional information for evaluation and calibration of the root hospitality and hardpan factors introduced in the new version of the model.

It can be concluded that modifications recently introduced in the CERES-Maize crop model are improving performance for site-specific crop development simulation. In addition to the CERES-Maize model, the CROPGRO-Soybean model also has been recently enhanced and will also be tested in this research project.

REFERENCES

Batchelor, W.D., and J. Ritchie. 1998. Modifications made to the CERES-Maize to incorporate new water balance and tile drainage features (personal communication).

Bouma, J. 1997. Precision agriculture: Introduction to the spatial and temporal variability of environmental quality. *In* Optimizing Management for Precision Agriculture: A Systems Approach (Course Notes). Agric. and Biol. Eng. Dep., Univ. of Florida, Gainesville.

Gallant, J.C., and J.P.Wilson. 1996. TAPES-G: A grid-based terrain analysis program for the environmental sciences. *Computers Geosci.* 22(7):713–722.

Garrison, M.V. 1998. Validation of the CERES-Maize water and nitrogen balance modules under tile drained conditions. M.S. thesis. Agric. and Biosystems Eng. Dep. Iowa State Univ., Ames.

Jones, C.A., and J.R. Kiniry. 1986. CERES-Maize A Simulation Model of Maize Growth and Development. Texas A&M Univ. Press, College Station.

Mallarino, A.P. 1996. Evaluation of optimum and above-optimum phosphorus supply for corn by analysis of young plants, leaves stalks and grain. Agron. J. 88:377–381.

Mulla, D.J., and J.S. Schepers. 1997. Key processes and properties for site-specific soil and crop management. p.1–18. *In* F.J. Pierce and E. J. Sadler (ed.) The State of Site-Specific Management for Agriculture. ASA, CSSA, and SSSA, Madison, WI.

Pierce, F.J., D.D. Warncke, and M.W. Everett. 1994. Yield and nutrient variability in glacial soils of Michigan. p. 133–150. *In* P.C. Robert et al., (ed.) Proc. of the 2nd Int. Conf. on Precision Agriculture. ASA, CSSA, and SSSA, Madison, WI.

Rawls, W.J., D.L. Brakensiek, and K.E. Saxton. 1982. Estimation of soil water properties. ASAE Paper no. 81–2510. ASAE, St. Joseph, MI.

Sudduth, K.A., N.R. Kitchen, D.F. Hudges, and S.T. Drummond. 1995. Electromagnetic induction sensing as an indicator of productivity on claypan soils. p. 671–681. *In* P.C. Robert et al., (ed.) Site-Specific Management for Agricultural Systems. ASA, CSSA, and SSSA, Madison, WI.

Sudduth, K.A., S.T. Drummond, S.J. Birrel, and N.R. Kitchen. 1996. Analysis of spatial factors influencing crop yield. p. 129-140. *In* P.C. Robert et al., (ed.) Proc. of the 3rd Int. Conf. on Precision Agriculture. ASA, CSSA, and SSSA, Madison, WI.

Sudduth, K.A., C.W. Fraisse, S.T. Drummond, and N.R. Kitchen. 1998. Integrating spatial data collection, modeling and analysis for precision agriculture. *In* Proc. of 1st Int. Conf. on Geospatial Information in Agriculture and Forestry. Lake Buena Vista, FL (in press).

Simulation of within Field Variability of Corn Yield with Ceres-Maize Model

J. E. Corá
Department of Soil Science
São Paulo State University (UNESP)
Jaboticabal, São Paulo, Brazil

F. J. Pierce
B. Basso
J. T. Ritchie
Department of Crop and Soil Sciences
Michigan State University

ABSTRACT

The CERES-Maize model was used to estimate the spatial variability in corn (*Zea mays* L.) yield for 1995 and 1996 using data measured on soil profiles located on a 30.5 m grid within a 3.9 ha field in Michigan. The model was calibrated for one grid profile for the 1995 and then used to simulate corn yield for all grid points for the 2 yrs. For the calibration for 1995, the model predicted corn yield within 2%. For 1995, the model predicted yield variability very well (r^2 = 0.85), producing similar yield maps with differences generally within ± 300 kg ha^{-1}. For 1996, the model predicted low grain yields (1167 kg ha^{-1}) compared with measured (8928 kg ha^{-1}) because the model does not account for horizontal water movement within the landscape or water contributions from a water table. Under nonlimiting water conditions, the model performed well (average of 8717 vs. 8948 kg ha^{-1}) but under-estimated the measured yield variability.

INTRODUCTION

Simulation models are valuable tools for predicting crop yields and examining alternative agricultural management practices for crop production and their potential impacts on the environment. Intuitively, crop models have considerable value in evaluating site-specific management (SSM) systems and associated component precision farming practices but have found limited use (Van Uffelen et al., 1997; Han et al., 1995). Sadler and Russell (1997) suggested that limited application appear to be caused by lack of knowledge about within-field variability of soil properties needed for predicting crop yield.

Copyright © 1999 ASA-CSSA-SSSA, 677 South Segoe Road, Madison, WI 53711, USA. *Proceedings of the Fourth International Conference on Precision Agriculture.*

Traditionally, models were developed assuming soil properties to be homogenous (Han et al., 1995). Several modeling attempts have been reported using soil map delineation as a means to express spatial variability patterns, using representative soil profiles for each delineation (Kiniry et al., 1997; Pang et al., 1997; Gabrielle & Kengni, 1996). However, the assumption that delineated areas on a second order soil map can serve to adequately represent soil spatial variability is flawed because considerable variation occurs within map units (Beckett & Webster, 1971). Nonetheless, since SSM aims to maximize crop production through efficient use of managed inputs according to localized variability in soils, pests, and crop condition (Pierce & Sadler, 1997), model inputs should include details on within field variability. If soils in a field exhibit large spatial variability, many different management zones may be identified for SSM. According to Verhagen and Bouma (1997), modeling results can be very useful for evaluating effects of spatial and temporal soil variability on crop yield because they can cover a wide range of conditions that are relevant to agricultural production and environmental quality concerns. Static soil properties will not always provide a reasonable explanation of factors affecting yield variability when spatially variable water content exists at planting time. Such variability is related to landscape position and may influence plant populations or stand uniformity leading to yield losses.

The ability of models to predict yield based on soil variability on a finer spatial scale still needs to be demonstrated in order to strengthen confidence in their usefulness for SSM (Sadler & Russel, 1997). Therefore, results of simulations must be validated with real data, obtained by field measurements on a finer scale and several growing seasons (Verhagen & Bouma, 1997). Interpolation techniques make it possible to extend results of simulations obtained at point locations to large land areas (Finke, 1993). Therefore, simulations across the range of annual weather conditions expected for a given area could be used to delineate areas displaying consistent patterns over time that can be used for SSM.

Both complex mechanistic models and simple functional ones have been used to estimate crop yields. The Decision Support System for Agrotechnology Transfer (DSSAT) integrates several models, which are in the functional class, with standardized input and output (IBSNAT, 1989). The DSSAT maize model (CERES-Maize) has been tested and used in the USA and around the world with promising results. Kiniry et al. (1997) tested the model at one county in each of the nine states in the USA (Minnesota, New York, Iowa, Illinois, Nebraska, Missouri, Kansas, Louisiana, and Texas) and concluded it was appropriate for predicting yield for most counties. Hodges et al. (1987) found that earlier versions of CERES-Maize accurately simulated maize grain yield in the northern U.S. Corn Belt. Researches have adapted the model in Michigan (Algozin et al., 1988), California (Pang et al., 1997), France (Gabrielle & Kengni, 1996), and China (Wu et al., 1989).

While CERES-Maize model has been used successfully in many situations, like most crop simulation models, it has not been tested under conditions where soil spatial variability within a field was taken into account. A successful application of CERES-Maize to known conditions of spatial variability over time would

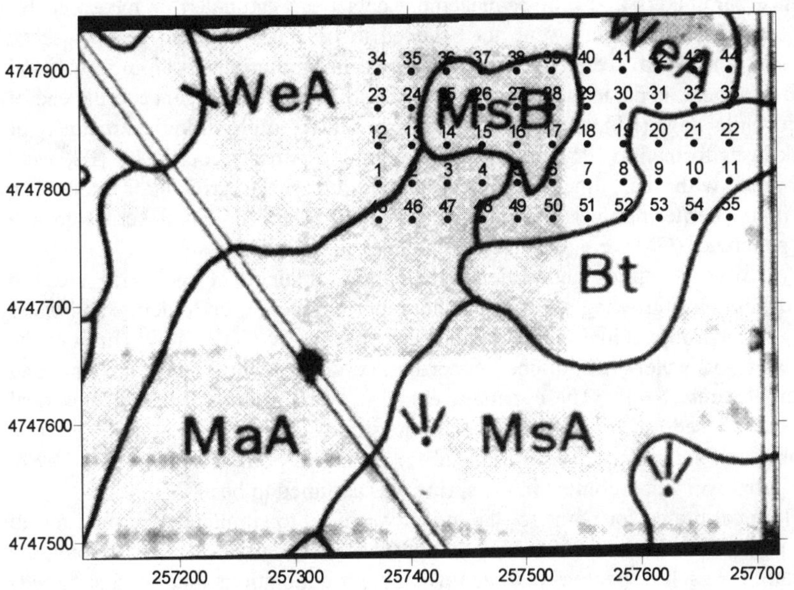

Fig. 1. Experimental site showing the location of the grids with respect to the soil survey map delineations.

encourage the use of crop simulation models in SSM. This study evaluated the capability of the CERES-Maize model to account for known soil spatial variability on simulated corn yield within a field.

MATERIAL AND METHODS

The DSSAT version 3 (1994) of the CERES-Maize model was applied to an extensive data set consisting of soil profile properties and corn performance data obtained on a 30.5 m grid positioned in a 3.9 ha portion of a field near Durand, MI (Fig. 1). The model was first calibrated on one grid site (soil profile #15) chosen because it was representative of the soil mapping unit, was nearly level, a water table not present within 1.5 m, and plant populations were close to the target populations. The model was calibrated for 1995 only in order to establish appropriate crop coefficients for the corn hybrid used both years.

Input to the model included on-site weather data collected both years, soil data (soil profile properties), soil N balance parameters, crop management data, and genetic parameters and crop coefficients for the corn hybrid used. The minimum set of weather data (daily minimum and maximum temperatures, solar radiation, and rainfall) was collected by a weather station established at the site. The soil N balance parameters and crop management data were estimated or measured for each site. Plant nutrients were not believed to be a limiting and pests (insects, diseases, and weed) were controlled and posed no limitations to crop growth and yield. The genetic parameters: the thermal time from seed emergence to the end of the juvenile stage (P1), the thermal time from silking to physiological maturity or black layer formation (P5), and the photoperiod sensitivity coefficient (P2) were calibrated with real time periods observed during the growing season. The maximum kernel number per plant parameter (G2) and potential kernel growth rate parameter (G3) were estimated from the calibration site results.

Soil water contents and water table depths obtained at each grid location throughout each growing season (data not reported) were used to derive soil water related model inputs including: drainage coefficient (SWCON), lower limit of the available soil water (LL), drained upper limit water content (DUL), and saturated water content (SAT). These parameters also were calculated using empirical equations based on the texture and bulk density of each horizon. The SCS curve number and soil albedo were estimated for each grid point (Ritchie et al., 1990). The initial soil water content in the spring was assumed to be at DUL.

Once calibrated for point 15, the model was used to simulate corn yield for all grid points for 1995 and 1996, varying the profile soil data for each grid point, the weather for each respective year, and using plant populations measured at harvest as model input. The performance of the CERES-Maize model was evaluated by regressing actual corn yields measured for each grid location within the experimental site with yields predicted by the model.

RESULTS AND DISCUSSION

Model Calibration

The calibration of the CERES-Maize model on profile 15 resulted in soil inputs of 0.5 for SWCON, 0.13 for albedo, and 67 for SCS curve number using the procedure of Ritchie et al. (1990). The calibrated genetic coefficients were 200 dd for P1, 680 dd for P5, 0.5 h^{-1} for P2, 650 and 9.6 mg $kernel^{-1}$ d^{-1} for G2 and G3, respectively. The calibrated genetic coefficients generally agreed with the values of G2 and G3 were close to those for corn hybrids grown in the northern U.S. Corn Belt. These input values gave a harvest index (ratio of total biomass to grain yield) close to 0.5, which is expected for corn grain yield when neither nutrients nor water are limiting. Plant population for the grid point 15 was 60 000 plants ha^{-1} for 1995. The model predicted corn grain yield well for 1995 differing by only -2% with a standard deviation of about 600 kg ha^{-1}. The CERES-Maize model was considered ready for use in the evaluation of spatial variability.

Simulation of Spatial Variability

Corn grain yields simulated by CERES-Maize corresponded well to measured yields in 1995. The simulated yields generally fall on the 1:1 line and a simple linear

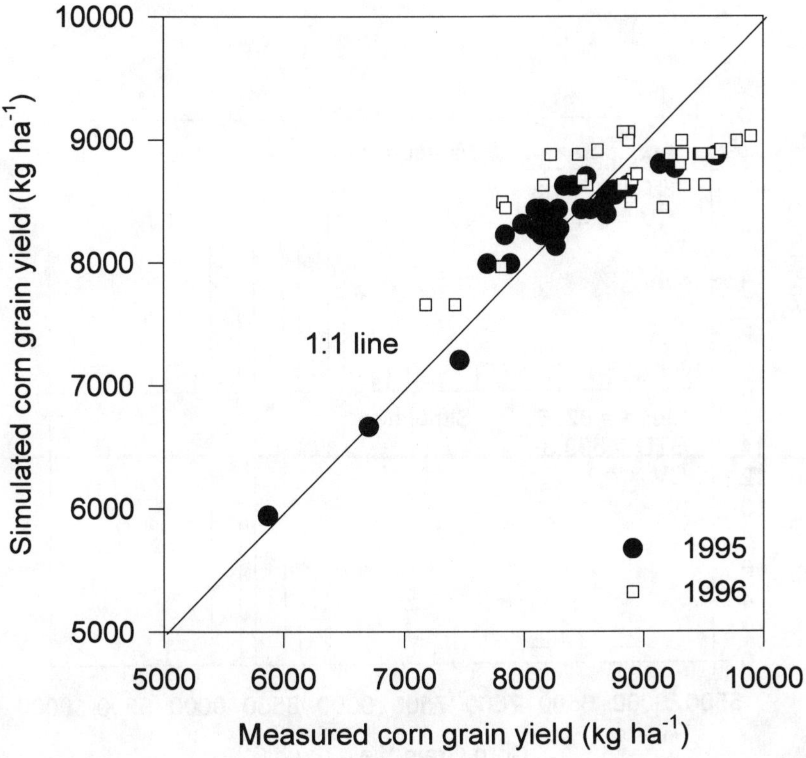

Fig. 2. Plot of simulated vs. measured corn yield for the Durand site for 1995 (●) under natural rainfall and for1996 (□) under a no water stress condition.

regression of simulated on measured yield accounted for 86% variability (Fig. 2). There is a tendency of the model to underpredict at yields above 9000 kg ha^{-1}, as illustrated in the comparison of frequency diagrams of measured and simulation yields in Fig. 3.

The average, standard deviation, and CV for the measured and simulated data for 1995 for the whole area (33 profiles) are similar, with the difference in mean of only 27 kg ha^{-1} (Fig. 3). The yield maps interpolated from measured and simulated yields are comparable (Fig. 4a,b), with consistent patterns within the area, i.e., lower yields in the eastern part of the field and higher yields in the central southern part. Differences between yield maps were generally within ± 300 kg ha^{-1} (Fig. 4c). The performance of CERES-Maize in predicting the corn yield variability within

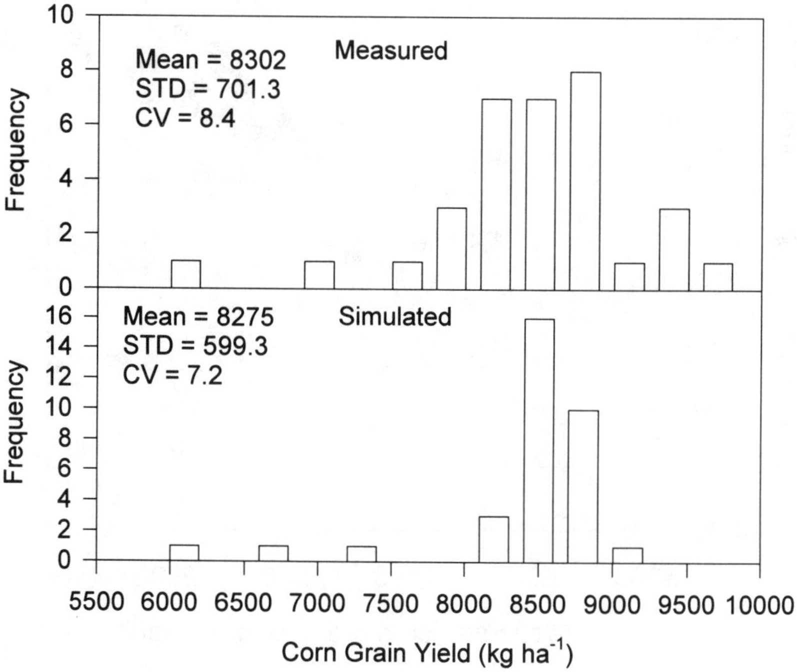

Fig. 3. Frequency distribution of measured and simulation corn yields for the Durand site in 1995.

the experimental area for 1995 is encouraging and is consistent with earlier evaluations of the model (Kiniry et al., 1997).

Simulations for 1996 were performed using the same model coefficients and soil data sets used for 1995 and using1996 weather data and 1996 plant population for each grid point. The CERES-Maize model predicted very low grain yields for 1996, with an average of 1167 kg ha^{-1} and a range of 469 to1946 kg ha^{-1}, while the measured yields averaged 8928 kg ha^{-1} and ranged from 7 180 to 10 159 kg ha^{-1}. Plant populations for 1996 were similar to that for 1995.

The model greatly under predicted grain yields because of a drought during grain filling from 1 to 19 August during which cumulative rainfall was 3 mm. Regular observations made during this period indicated no visible plant water stress symptoms suggesting water was not limiting to the corn during this period.

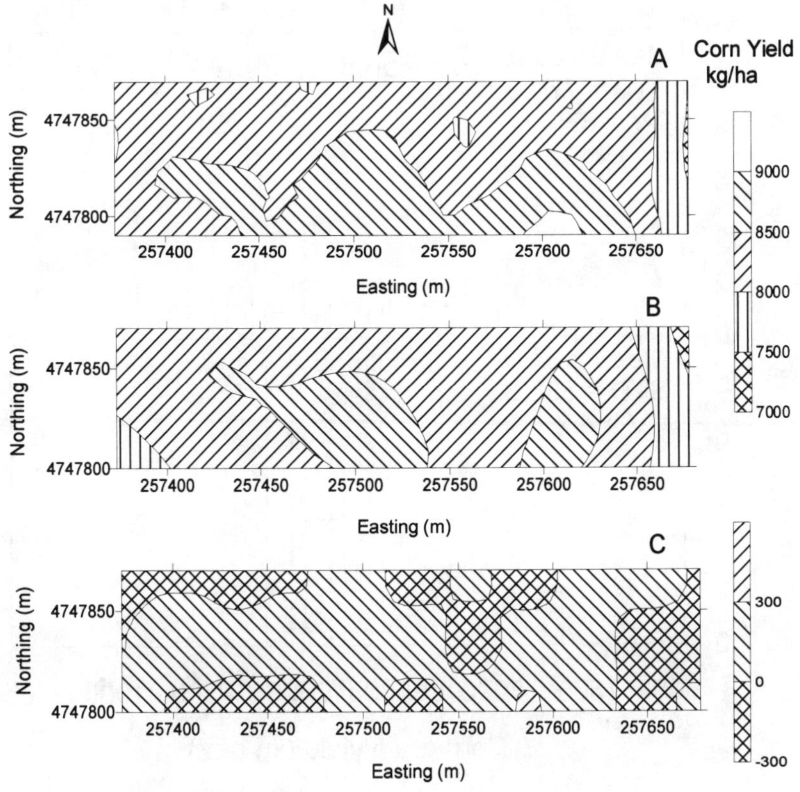

Fig. 4. Interpolated maps of (A) measured, (B) simulated, and (C) difference between measured and simulated corn yields at the Durand site in 1995.

Water was available from the water table even for soil profile 15 for which the water table was not within 1.5 m. Water was available to the plant at site 15 below 75 cm even though the upper profile water content was at LL of water availability during this dry period. The problem for the CERES-Maize model is that it does not account for the presence of a water table and the upward flux of water into the root zone.

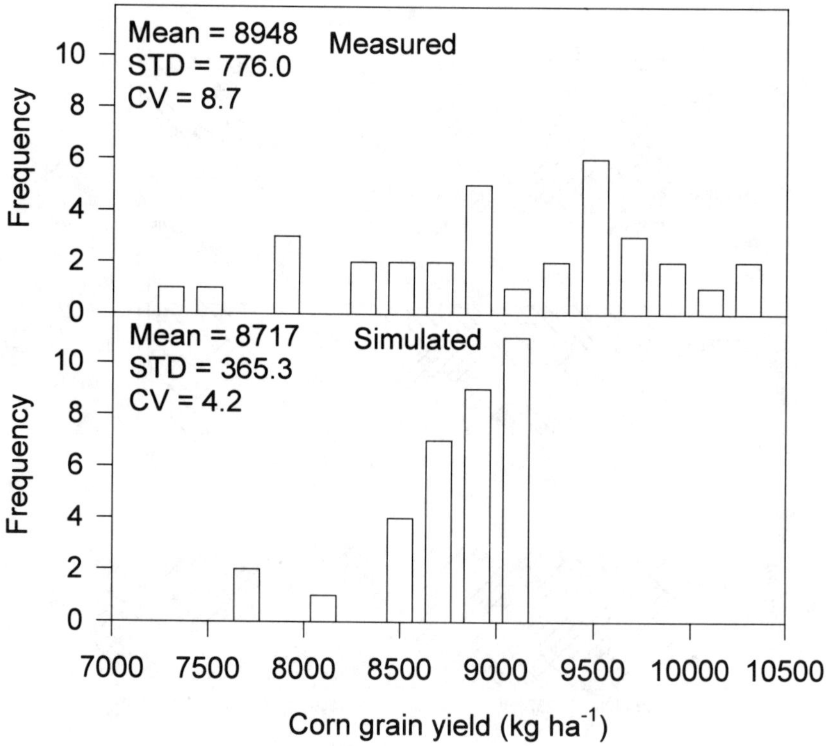

Fig. 5. Frequency distribution of measured and simulation corn yields for the Durand site in 1996 under a no water stress condition.

Since water did not appear to be limiting in this field in 1996, the simulations were rerun with the model set to eliminate water stress during the growing season. Thus, simulated corn grain yields were determined primarily by weather conditions and plant populations at harvest. Under these conditions, the model did predict the average corn yield well (8717 versus 8948 kg ha^{-1}) but under predicted

For 1996, the results indicate that variation in soil properties in combination with differential effects of the water table affected the spatial distribution of corn yield not accounted for in the model. There is little correlation between the yield map interpolated from the measured data and that interpolated from the simulated yields, although yield differences were generally within 600 kg ha^{-1} (Fig. 6).

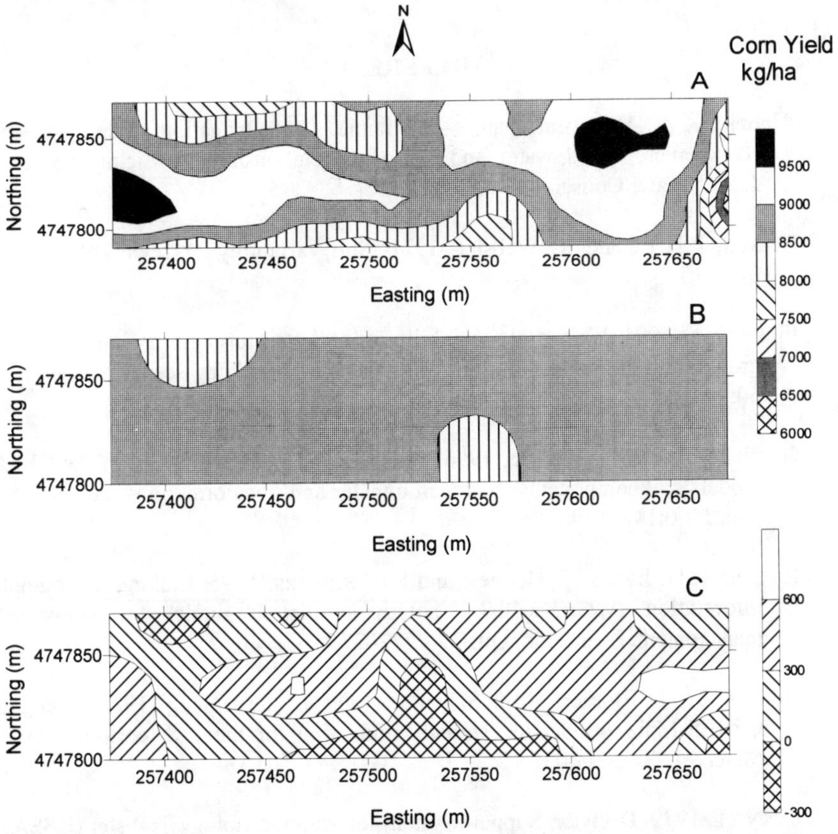

Fig. 6. Interpolated maps of (A) measured, (B) simulated, and (C) difference between measured and simulated corn yields at the Durand site in 1996.

CONCLUSION

The CERES-Maize model simulated corn yields and yield variability well in 1995, accounting for a large portion of the variability in yield and producing a yield map very similar to the measured values. A drought during 1996 showed the importance of water table and horizontal water movement into lower landscape positions in supplying water to corn within this field. The CERES-Maize does not account for water table or horizontal water contributions and will need modification for soils where these are significant to crop yield. These results show the potential for crop models to simulate the spatial and temporal variability in yields and clearly demonstrate the importance of the presence of a water table and runon in understanding variability in crop yield and in the delineation of management zones within fields for site-specific management.

REFERENCES

Algozin, K.A., V.F. Bralts, and J.T. Ritchie. 1988. Irrigation strategy selection based on crop yield, water, and energy use relationships: A Michigan example. J. Soil Water Conserv. 43:428–431.

Beckett, P.H.T., and R. Webster. 1971. Soil variability: A review. Soil Fertil. 34:1–15.

Finke, P.A. 1993. Field scale variability of soil structure and its impact on crop growth and nitrate leaching in the analysis of fertilizing scenarios. Geoderma 60:89–107.

Gabrielle, B., and L. Kengni. 1996. Analysis and field-evaluation of the CERES models'soil components: Nitrogen transfer and transformations. Soil Sci. Soc. Am. J. 60:142–149.

Han, S., R.G. Evans, T. Hodges, and S.L. Rawlins. 1995. Linking a geographic information system with a potato simulation model for site-specific management. J. Environ. Qual. 24:772–777.

Hodges, T., D. Botner, C.M. Sakamoto, and J. Hays Haug. 1987. Using the CERES-Maize model to estimate production for the U.S. corn-belt. Agric. Metereol. 40:293–303.

IBSNAT. 1989. Decision Support System for Agrotechnology Transfer (DSSAT)-User's Guide. IBSNAT Project. Dep. Agronomy and Soil Sci., Univ. of Hawaii, Honolulu.

Kiniry, J.R., J.R.Williams, R.L. Vanderlip, J.D. Atwood, D.C. Reicosky, J. Mulliken, W.J. Cox, H.J. Mascagni Jr., S.E. Hollinger, and W.J. Wiebold. 1997. Evaluation of two maize models for nine U.S. locations. Agron. J. 89:421–426.

Pang, X.P., J. Letey, and L. Wu. 1997. Yield and nitrogen uptake prediction by CERES-Maize model under semiarid conditions. Soil Sci. Soc. Am. J. 61:254–256.

Pierce, F.J., and E.J. Sadler. 1997. The state of site-specific management for agriculture. ASA Misc. Publ. ASA, CSSA, and SSSA, Madison, WI.

Ritchie, J.T., D.C. Godwin, and U. Singh. 1990. Soil and weather inputs for the IBSNAT crop models. p. 31–45. *In* Proc. of the IBSNAT symposium: Decision support system for agrotechnology transfer: Part I. Las Vegas, NV. 16–18 Oct. 1989. Dep. of Agronomy and Soil Science, College of Tropical Agriculture and Human Resources, Univ. of Hawaii, Honolulu, HI.

Sadler, E.J., and G. Russell. 1997. Modeling crop yield for site-specific management. p. 69–79. *In* F.J. Pierce and E.R. Sadler (ed.) The state of site-specific management for agriculture. ASA Misc. Publ., ASA, CSSA, and SSSA, Madison, WI.

Van Uffelen, C.G.R., J. Verhagen, and J. Bouma. 1997. Comparison of simulated crop yield patterns for site-specific management. Agric. Syst. 54:207–222.

Verhagen, J., and J. Bouma. 1997. Modeling soil variability. p. 55–67. *In* F.J. Pierce and E.R. Sadler. (ed.) The state of site-specific management for agriculture. ASA Misc. Publ. ASA, CSSA, and SSSA, Madison, WI.

Wu, Y., C.M. Sakamoto, and D.M. Botner. 1989. On the application of the CERES-Maize model to the North China Plain. Agric. Meteorol. 49:9–22.

Machine Learning Methods in Site-Specific Management Research: An Australian Case Study

M. L. Adams
S. E. Cook

CSIRO Division of Land and Water
Floreat Park Laboratories
Wembley, West Australia, Australia

P. A. Caccetta

Satellite Remote Sensing Services
WA Department of Land Administration
Floreat, West Australia, Australia

M. J. Pringle

Australian Center for Precision Agriculture
University of Sydney
Sydney, New South Wales Australia

CSIRO Division of Land and Water
Floreat Park Laboratories
Wembley, West Australia, Australia

ABSTRACT

Two machine learning methods based on the induction of regression trees and Bayesian networks from data were used to predict wheat yield in 1996 from fourteen soil, remotely sensed, and other variables for a field located near Wyalkatchem, Western Australia. The regression tree model explained 39.7% of the variation in the data, while the Bayesian network model explained 69.1% of the variation. Both models predicted similar spatial trends in yield, and both models predicted yield to within ±0.25 t ha^{-1} of measured yield >50% of the area of the field. In comparing model types, both models suffered from an inability to predict yields <0.33 t ha^{-1} or yields >1.89 t ha^{-1}. We suggest that Bayesian network models are more suitable frameworks for site specific management research than regression trees, or other machine learning methods similar to regression trees.

Copyright © 1999 ASA-CSSA-SSSA, 677 South Segoe Road, Madison, WI 53711, USA. *Proceedings of the Fourth International Conference on Precision Agriculture.*

INTRODUCTION

Several different machine learning model types have been explored in the precision agriculture/site-specific management literature. Neural networks were used by Sudduth et al. (1996) to analyze yield-limiting factors retrospectively from sets of multi-year yield maps, soil and topographic data. Factor analysis was used by Mallarino et al. (1996) to identify underlying yield limitations such as soil fertility, weed control and conditions for early growth. Fuzzy clustering has been used by Lark and Stafford (1996) to identify areas of consistently high or low yield, which were then used to explain consistent trends using soil maps (Stafford et al., 1996). Bayesian networks have been used by several researchers in a decision support capacity (Cook et al., 1996b; Tari, 1996; Audsley et al., 1997).

Little attention has been given to comparing the performance and capabilities of the different machine learning model types. In this paper, we compare the ability of regression trees (Breiman et al., 1984) and Bayesian networks (Pearl, 1988) to predict yield, and briefly discuss the advantages and disadvantages of each model type.

METHODS

The data is taken from a field experiment conducted in 1996. The field, Rowlands A1, located near Wyalkatchem, Western Australia, was in the second year of wheat (*Triticum aestivum* cv. Blade) following lupins (*Lupinus angifolius*, cv. Merritt). A single treatment, top-dressing with N, was applied at 0, 40, 80, and 120 kg ha^{-1} of urea (0, 18.4, 36.8, and 55.2 kg N ha^{-1}.). The treatments were applied according to a checkerboard design, in which the four rates of urea were applied in a regular grid of plots with dimension 30 m by 30m. All other paddock operations were applied uniformly over the field. Further details of the field site and experimental plans can be found in Cook et al. (1998). The data for this analysis consist of measurements from 940 randomly selected points from the data layers given in Table 1. The soil chemical property data layers (e.g., soil K and soil pH) were derived from an inverse distance weighted (IDW, power = 1) interpolation of soil analyses from soils sampled on a 100 m grid (total of 64 samples in the field). The yield data also was IDW interpolated (refer to Table 1). The other data layers did not require interpolation.

Regression Tree Model

A regression tree analysis (Breiman et al., 1984) was performed to predict 1996 grain yield from the 940 observations of the 14 variables in Table 1 with S-Plus version 3.3 (Mathsoft, Seattle, WA). The initial tree derived from the S-Plus algorithm was bushy, and overfit the data (Breiman et al., 1984). The initial tree was pruned to nine terminal nodes based on the results from a cross validation procedure (see Breiman et al., 1984; Venables & Ripley, 1996, for a more detailed description of the pruning process) applied to the initial tree.

Bayesian Network Model

As the induction of Bayesian networks are perhaps less familiar to the general audience, we briefly describe it here. A Bayesian network may be represented by a directed graph and an associated set of conditional probability tables. The nodes

Table 1. Variables used in the analysis

Variable	Abbreviation	Method of determination
Wheat yield in 1996 (t ha^{-1})	yld96	AgLeader 2000 Yield monitor (Ames, IA). Data interpolated with IDW (power = 1) to a 5m grid.
Wheat yield in 1995 (t ha^{-1})	yld95	AgLeader 2000 Yield monitor. Data interpolated with IDW (power = 1) to a 5m grid.
Soil type map	soilmap	refer to Cook et al. (1998)
Topdressed urea (kg ha^{-1})	urea	refer to Cook et al. (1998)
Midseason Normalized Difference Vegetation Index	NDVI	refer to Cook et al. (1998)
Soil Reactive Fe (mg kg^{-1})†	soilFe	Tamm (1922)
Soil K (mg kg^{-1})	soilK	Colwell (1963)
Soil S (mg kg^{-1})	soilS	Anderson et al. (1994)
Soil pH	soilpH	1:5 soil:0.01M $CaCl_2$ extract
Soil organic carbon (%)	soilOC	Walkley and Black method (see Rayment and Higginson, 1992)
Gammaradiometric K	radK	refer to Cook et al. (1996a)
Gammaradiometric Th	radTh	refer to Cook et al. (1996a)
Gammaradiometric U	radU	refer to Cook et al. (1996a)
Gammaradiometric total counts	radtot	refer to Cook et al. (1996a)

† All soil measurements were taken from the top 10 cm of the soil surface.

of the graph represent random variables and the edges of the graph represent conditional independence relationships that exist between the variables. Bayesian model selection was performed using Moby version 1.0 (see Caccetta, 1998), which is an implementation of the K2 procedure described by Cooper and Herskovits (1992), to which the reader is referred for a detailed discussion of the assumptions regarding the algorithm. The K2 algorithm employs a forward stepwise selection procedure in developing a best model. Furthermore, the K2 algorithm requires a user-specified ordering of the input variables. The ordering is typically specified according to a user perception of the causal ordering of the variables. In some cases, it may be natural to do so. In other cases, it may not be so obvious. For example, the application of urea was performed before yield was determined, so it makes sense that urea should come before yield in the ordering. On the other hand, if a number of soil chemical variables (e.g., Colwell-extractable P) are determined from the same soil sample, it is difficult to determine how to order the soil chemical variables. The variable ordering used in this analysis was soilmap, radtot, radU, radK, radTh, soilS, soilpH, soilOC, soilK, soilFe, yld95, urea, NDVI, and yld96.

The basis for comparing two candidates Bayesian networks, B_1 vs. B_2, for example, relies on computing the ratio

$$\frac{P(B_1|D)}{P(B_2|D)} = \frac{P(B_1,D)}{P(B_2,D)} \quad [1]$$

For a candidate model, *P(B,D)*, i.e., the joint probability of Bayesian network, *B*, and the data, *D*, is calculated as

$$P(B,D) = P(B)\prod_{i=1}^{n}\prod_{j=1}^{q_i}\frac{(r_i-1)!}{(N_{ij}+r_i-1)!}\prod_{k=1}^{r_i}N_{ijk}! \quad [2]$$

where n is the number of variables, r_i is the number of states of variable v_i, q_i is the total number of states defined by the cross product of the states of the parents of v_i, and N_{ijk} is the number of observations for each state of v_i given its parents. If $\frac{P(B_1,D)}{P(B_2,D)} > 1$ then model B_1 is considered to be a better fit to the data than B_2.

Predicting Grain Yield

The pruned regression tree and the Bayesian network models were used to predict grain yield from the 940 observations used to build the models. The results from each model were kriged to generate a map of predicted yield. Predicted yield from each model was compared to actual yield (yld96) measured by an AgLeader 2000 yield monitor by a simple map subtraction of actual minus predicted.

The terminal nodes, or leaves, of the pruned tree (i.e., the predicted yield) were sorted back to the 940 original locations of the data set. Various theoretical variogram models (Webster & Oliver, 1990) were fitted to yld96 and the predicted yield (pruned tree) data using the weighted least-squares method of Cressie (1985). The exponential model, given by

$$\gamma(h) = Nugget + (Sill - Nugget) \times \left(1 - \exp^{-\left(\frac{h}{Range}\right)}\right) \quad [3]$$

fitted the yld96 and predicted yield (pruned tree) data best according to the Akaike Information Criterion (AIC) (Webster & McBratney, 1989). The spherical model, given by

$$\begin{cases} \gamma(h) = Nugget + (Sill - Nugget) \times \left(1.5 \times \left(\frac{h}{Range} \right) - 0.5 \times \left(\frac{h}{Range} \right)^3 \right) & \text{for } h \leq Range \\ \gamma(h) = Nugget + (Sill - Nugget) \text{ for } h > Range \end{cases} \quad [4]$$

Table 2. Variogram parameters for yld96 and predicted yield by the pruned regression tree and Bayesian network model.

Parameter	yld96 †	Predicted yield (pruned tree)	Predicted yield (BN model)
Nugget $(\text{t ha}^{-1})^2$	0.096	0.013	0.043
Sill-nugget $(\text{t ha}^{-1})^2$	0.095	0.081	0.109
Range (m)	213.2	317.0	774.0

† Parameters for yld96 and predicted yield (pruned tree) are for an exponential model. Parameters for predicted yield (BN model) are for a spherical model.

fitted the predicted yield (BN model) best according to the AIC. The parameters for each are shown in Table 2. The variogram parameters of the pruned tree and Bayesian network model were point kriged to a 10 m × 10 m grid.

RESULTS AND DISCUSSION

Regression Tree Model

The pruned regression tree is displayed in Fig. 1, which shows that the most important properties for the explanation of yld96 were (in decreasing order of importance) soilpH; soilOC and soilK (tied); soilmap and radTh (tied); radtot, and soilFe. The yield predictions range from 0.53 to 1.46 t ha^{-1}.

The variogram *Nugget* and *Sill-Nugget* parameters for the pruned regression trees are lower than yld96, although the *Range* is longer (refer to Table 2). The significant reduction in the *Nugget* parameter between yld96 and the pruned tree indicated there is much less random variation in the predicted values than in the actual values. This isn't surprising since the regression tree can only predict 9 discrete values, while yld96 varies over a continuous range. The predicted yield from the pruned tree explained 39.7% of the variation of yld96 according to the method of Shatar and McBratney (1998).

Bayesian Network Model

The results of the K2 algorithm are depicted graphically in Fig. 2. For the purposes of this paper, only the yield measured in 1996 is of interest. The results suggest that (in decreasing order of importance) the soilmap, radTh, and soilpH were the most influential variables in determining yield in 1996. We found that topdressed urea, the treatment variable in Cook et al. (1998), and midseason NDVI in 1996 were directly related to each other (indicated by an arrow from urea to NDVI in Fig. 2), but not directly related to final yield. This is similar to the results presented in Cook et al. (1998), and leads to the conclusion that the application of urea in 1996 resulted in a crop which had increased leaf growth due to the urea application (as detected by NDVI), but the increased leaf area was not translated into yield.

The variance of the predictions was compared to the variance of the actual data using the method employed by Shatar and McBratney (1998); the Bayesian network explained 69.1% of the actual variance in yld96. We note here that the selected Bayesian network is described by many more parameters than the selected regression tree and hence a greater explanation of the variability is expected.

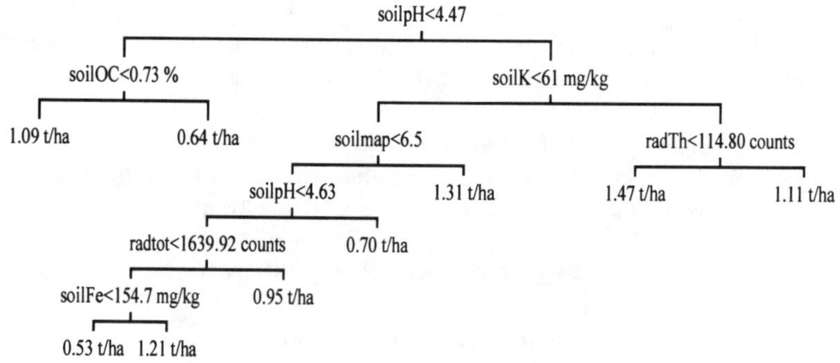

Fig. 1. Pruned regression tree with nine terminal nodes for 1996 yield data (yld96).

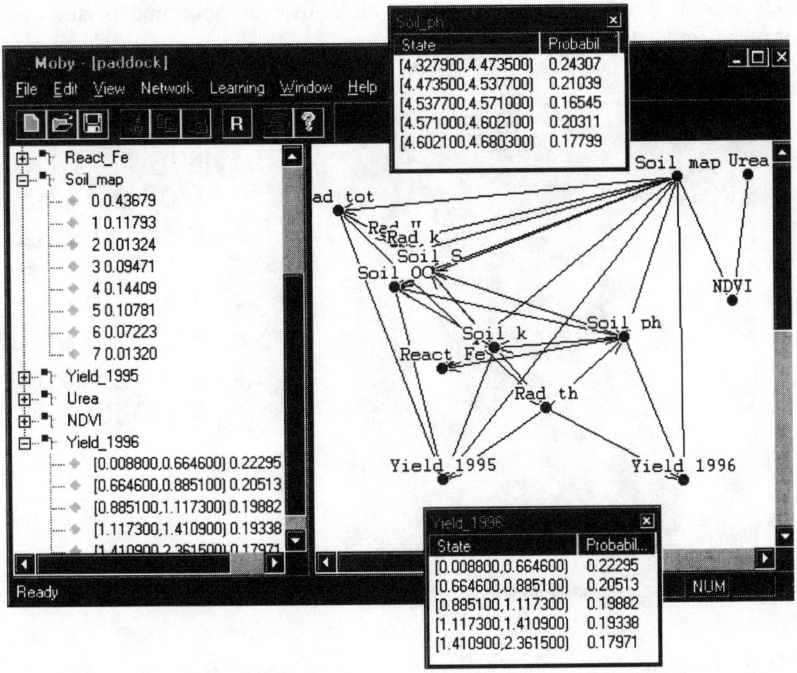

Fig. 2. Graphical depiction of the selected Bayesian Network by the K2 algorithm. The network is shown in its initialized state.

Grain Yield Prediction

For reference purposes the yield map from 1996 is show in Fig. 3. Predicted yield and difference maps (actual minus predicted) for the pruned regression tree and the Bayesian network model are shown in Fig. 4 and 5, respectively.

The kriged predicted yield maps from the regression tree and Bayesian network models are similar (Fig. 4, left map for the regression tree, and Fig. 5, left map for the Bayesian network). Both models predict lower yield in the western, eastern and northeastern parts of the field, and greater yields in the central and southeastern parts of the field. The main difference between the two predicted maps is the spatial extent and patterns of predicted yield. The similarity of predicted yield maps is expected, in part, because soilpH and soilmap are important predictor variables in both models, especially at the scale of the legend.

The differences between actual and predicted yield derived from the two model types also show similar trends. In particular, both models predicted inaccurately yield in the eastern and southeastern portion of the field. This is most likely caused by a weed infestation in that portion of the field, which adversely affected yield (Cook et al., 1998). Because a weed density-related variable was not included in the development of either model, it is reasonable to expect that yield predictions in areas of dense weeds would be poor. Approximately 50% of the field was predicted accurately to ±0.25 t ha^{-1}; however, both models either under or overestimated yld96 by up to 1.0 t ha^{-1}.

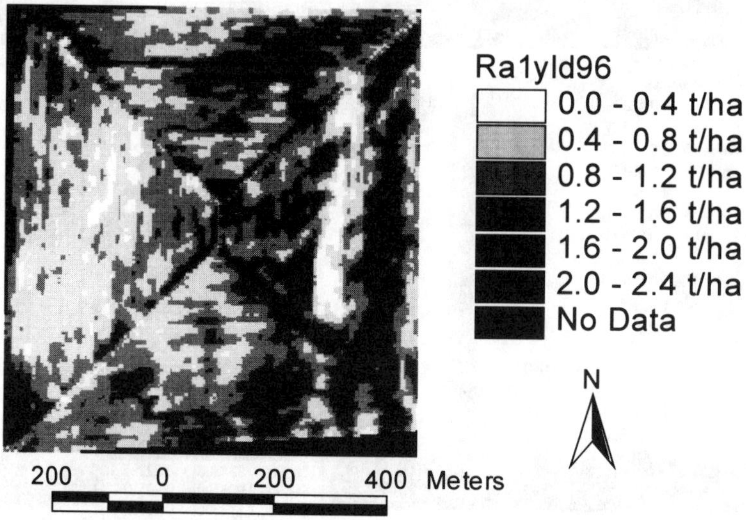

Fig. 3. Wheat yield map for Rowlands A1, Wyalkatchem from 1996 harvest season.

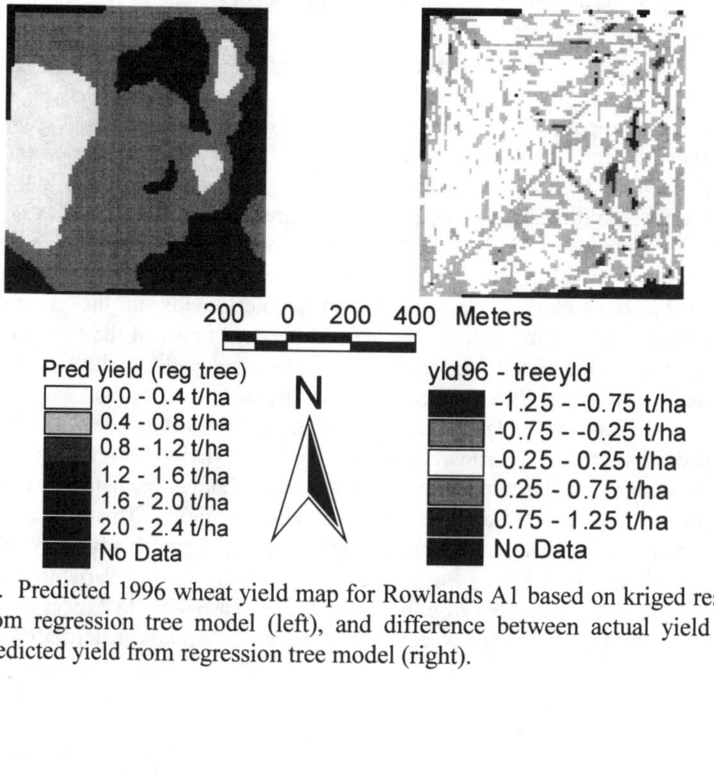

Fig. 4. Predicted 1996 wheat yield map for Rowlands A1 based on kriged results from regression tree model (left), and difference between actual yield and predicted yield from regression tree model (right).

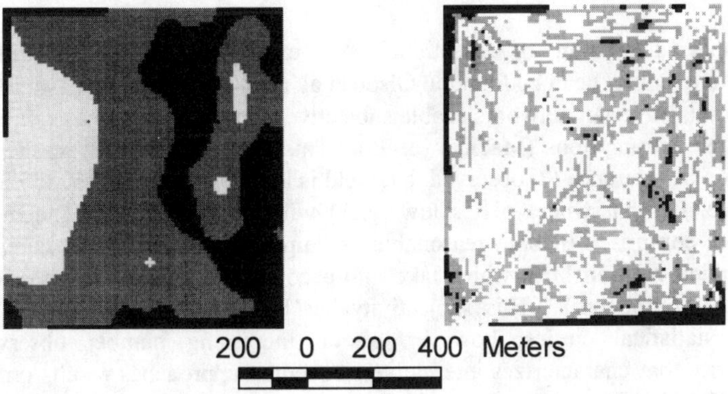

Fig. 5. Predicted 1996 wheat yield map for Rowlands A1 based on Bayesian network model (left), and difference between actual yield and predicted yield from the Bayesian network model (right). Legends are the same as those given in Fig. 4.

A weakness in both models in regards to predicting yield is the inability to predict extremes of yield. The regression tree predicts only nine values of yield ranging from a low of 0.53 t ha^{-1} to a high of 1.47 t ha^{-1}. The Bayesian network predicts yield ranging from a low of 0.33 t ha^{-1} to a high of 1.89 t ha^{-1}.

Comparison of Model Types

Regression trees and Bayesian networks both operate and perform similarly. Both are empirical machine learning models of the learning-from-example type (inductive learning). Both models divide up sample space into hypercubes, although they evaluate the splits in different ways (see Breiman et al., 1984; Venables and Ripley, 1994; MacKay, 1995). Both models predicted similar distributions over the field, and both, being global models, failed to predict the extremes of the statistical distribution. Neither predicted the effects of weeds, which were not included in the model.

Regression trees and Bayesian networks, however, differ in two major characteristics. First, they use very different algorithms. Using the terminology of Pearl (1988), regression tree models are extensional type models, that is they are structured as a series of logical "and" statements, such as

```
if soilph<4.47 and soilOC>0.73%, then yield = 0.64 t ha-1
```

whereas Bayesian networks are intentional type models that can express *degree* of belief such as

```
if soilpH<4.47 and soilOC>0.73% then the probability that yield >
0.64 t ha-1 is 0.75.
```

A further discussion of intentional and extensional modeling within an agricultural context can be found in Olson et al. (1990).

Second, Bayesian networks enable subjective inference on the basis of explicit, reasonable assumptions (MacKay, 1995). This can be illustrated by the classic inversion of inference ("I observed that yield is low where pH is low, so I infer (to a degree, *p*), that where pH is low, yield will also be low"). The model is therefore comprised of both reasonable assumptions (the subjective component) and explicit relationships, which take into account all available information (the objective component). This subjectivity has been the target of heated debate within statistical circles; however, as an increasing number observe, the objectivity that characterizes alternative frequentist approaches results only *after* the problem has been defined by subjective choice and judgement (Savage, 1990). So, for many decision problems Bayesian networks promise a flexible approach for machine-learning. Bayesian networks have been applied by several researchers in decision support systems (e.g., Tari, 1996; Audsley et al., 1997) and we believe that Bayesian networks to be a useful model structure for site-specific management applications.

CONCLUSIONS

Regression trees and Bayesian networks, two machine learning methods, were used to predict wheat yield in 1996 for a field located near Wyalkatchem, Western Australia. Both models identified soilpH and soil type as being important in describing yield variation in 1996, while the treatment variable, topdressed urea, was unimportant. Both models predicted similar spatial trends in yield and predicted actual yield to within 0.25 t ha^{-1} across 50% of the field area. However, both models were incapable of predicting yields less than approximately 0.3 t ha^{-1}, or greater than approximately 1.9 t ha^{-1}, which is a severe limitation for generalizing the models to other situations. We suggest that Bayesian networks provide a more suitable framework for site specific management research than regression trees, because of the more flexible inferential mechanism.

ACKNOWLEDGMENTS

This research was partially funded by the Grains Research and Development Corporation. Thanks are due to Doug and Noela Maitland for allowing us to use their farm for these trials.

REFERENCES

Anderson, G.C., R.D.B. Lefroy, N. Chinoim, and G.J. Blair. 1994. The development of a soil test for sulphur. Norwegian J. Agric. Sci. 15: 83–95.

Audsley, E, B.J. Bailey, S.A. Beaulah, P.J. Maddaford, D.J. Parsons, and R.P. White. 1997. Decision support systems for arable crops: Increasing precision in determining inputs for crop production. p. 843–850. *In* J.V. Stafford (ed.) Precision Agriculture 1997. Proc. of the 1st European Conf. on Precision Agriculture, Warwick Univ., England, 7–10 Sept., 1997. Bios Scientific Publ., Oxford, England.

Breiman, L., J.H. Friedman, R.A. Olshen, and C.J. Stone. 1984. Classification and regression trees. Wadsworth, Belmont, CA.

Caccetta, P.A. 1998. Learning Bayesian networks from data - a case study. Technical Report [on-line]. Availabe WWW: http://www.rss.dola.wa.gov.au/staff/caccetta/publications.html.

Colwell, J.D. 1963. The estimation of the phosphorus fertilizer requirements of wheat in southern New South Wales by soil analysis. Aust. J. Exp. Agric. Anim. Husb. 3:190–198.

Cook, S.E., J.M. Nolan, G. Riethmuller, M.L. Adams, and R.J. Corner. 1998. The checkerboard: A demonstration of on-farm experimentation using variable rate technology. J. Precision Agriculture. In Review

Cook, S.E., R.J. Corner, P.R. Groves, and G.J. Grealish. 1996a. Use of airborne gamma radiometric data for soil mapping. Aust. J. Soil Res. 34:183–194.

Cook, S.E., R.J. Corner, G. Mussell, G. Riethmuller, and M.D. Maitland. 1996b. Precision agriculture and risk analysis: An Australian example. p. 1123–1132. *In* P.C. Robert et al., (ed.) Proc. of the 3rd Int. Conf. on Precision Agriculture, Minneapolis, MN. 23–26 June, 1996. ASA, CSSA, and SSSA, Madison, WI.

Cooper, G.F., and E. Herskovits. 1992. A Bayesian method for the induction of probabilistic networks from data. Machine Learning 9:309–347.

Cressie, N. 1985. Fitting variogram models by weighted least squares. J. Int. Assoc. Math. Geol. 17:563–586.

Lark, R.M., and J.V. Stafford. 1996. Consistency and change in spatial variability of crop yield over successive seasons: Methods of data analysis. p. 141–149. *In* P.C. Robert et al., (ed.) Proc. of the 3rd Int. Conf. on Precision Agriculture, Minneapolis, MN. 23–26 June, 1996. ASA, CSSA, and SSSA, Madison, WI.

MacKay, D.J.C. 1995. Probable networks and plausible predictions: A review of practical Bayesian methods for supervised neural networks. Network: Comput. Neural Syst. 6:469–505.

Mallarino, A.P., P.N. Hinz, and E.S. Oyarzabal. 1996. Multivariate analysis as a tool for interpreting relationships between site variables and crop yields. p. 151–158. *In* P.C. Robert et al., (ed.) Proc. of the 3rd Int. Conf. on Precision Agriculture, Minneapolis, MN. 23–26 June, 1996. ASA, CSSA, and SSSA, Madison, WI.

Olson, R.L., T.L. Wagner, and J.L. Willers. 1990. A framework for modeling uncertain reasoning in ecosystem management: I. Background and theoretical considerations. AI Appl. 4:1–10.

Pearl, J. 1988. Probabilistic reasoning in intelligent systems: Networks of plausible inference. Morgan Kaufmann, San Mateo, CA.

Rayment, G.E., and F.R. Higginson. 1992. Organic carbon. p. 29–37. *In* Australian Laboratory Handbook of Soil and Water Chemical Methods. Inkata Press, Melbourne, Australia.

Savage, L.J. 1990. The foundations of statistics reconsidered. p. 14–20. *In* G. Shafer and J. Pearl (ed.) Readings in uncertain reasoning. Morgan Kaufmann, San Mateo, CA.

Shatar, T.M., and A.B. McBratney. 1998. Empirical modelling of relationships between sorghum yield and soil properties. J. Prec. Agric. (submitted).

Stafford, J.V., B. Ambler, R.M. Lark, and J. Catt. 1996. Mapping and interpreting the yield variation in cereal crops. Computers and Electronics in Agriculture 14:101–119.

Sudduth, K.A., S.T. Drummond, S.T., S.J. Birrell, and N.R. Kitchen. 1996. Analysis of spatial factors influencing crop yield. p. 129–136. *In* P.C. Robert et al., (ed.) Proc. of the 3rd Int. Conf. on Precision Agriculture, Minneapolis, MN. 23–26 June, 1996. ASA, CSSA, and SSSA, Madison, WI.

Tamm, O. 1922. Eine method zur Geotemmung de anorganischen komponente des glekomplexes in Boden. Medd. Skogsforoksanst. 19:1–20.

Tari, F. 1996. A Bayesian network for predicting yield response of winter wheat to fungicide programmes. Computers Electron. Agric. 15:111–121.

Venables, W.N., and B.D. Ripley. 1994. Tree-based methods. p. 329–348. *In* Modern Applied Statistics with S-Plus. Springer, New York.

Webster, R., and A.B. McBratney. 1989. On the Akaike information criterion for choosing models for variograms of soil properties. J. Soil Sci. 40:493–496.

Webster, R., and M.A. Oliver. 1990. Statistical methods in soil and land resource survey. Oxford Univ. Press, New York.

A Landform Segmentation Model for Precision Farming

R. A. MacMillan
LandMapper Environmental Solutions
Edmonton, Alberta, Canada

W. W. Pettapiece, and L. D. Watson
Research Branch, Lethbridge Research Station
Agriculture and Agri-Food Canada
Edmonton, Alberta, Canada

T. W. Goddard
Conservation and Development Branch
Alberta Agriculture, Food and Rural Development
Edmonton, Alberta, Canada

ABSTRACT

A robust new approach for describing and segmenting landforms that is directly applicable to precision farming has been developed in Alberta. The model uses derivatives computed from DEMs and a fuzzy rule base to identify up to 15 morphologically defined landform facets. The procedure adds several measures of relative landform position to the widely used classification of Pennock et al., (1987). The original 15 facets can be grouped to reflect differences in complexity of the area or scale of application. Research testing suggests that a consolidation from 15 to 3–4 units provides practical, relevant separations at a farm field scale. These units are related to movement and accumulation of water in the landscape and are significantly different in terms of soil characteristics and crop yields. The units provide a base for benchmark soil testing, for applying biological models and for developing agronomic prescriptions and management options.

INTRODUCTION

Background

There has been increasing interest in western Canada, and indeed in many other locales, in developing generic procedures for automatically classifying landscapes into functional landform or soil-landform spatial entities. Motivation for such research includes the need to define effective management units for precision farming and requirements for mechanisms for scaling the results of site specific modelling and analysis up to local and regional scales (Groffman, 1991 cited in Pennock et al. 1994; Penney et al., 1996).

Automated procedures for segmenting landforms into landform elements or facets have been described by, among others, Pennock et al. (1987, 1994),

Copyright © 1999 ASA-CSSA-SSSA, 677 South Segoe Road, Madison, WI 53711, USA. *Proceedings of the Fourth International Conference on Precision Agriculture.*

Skidmore et al., (1991), Irwin et al., (1997) and MacMillan and Pettapiece (1997). The landform classification procedures of Pennock et al. (1987) have been among the most widely accepted and applied in western Canadian prairie landscapes, but have some limitations (Pennock et al., 1994; MacMillan & Pettapiece 1997). These are: (i) a fragmented or chaotic spatial pattern, (ii) confusion in differentiating level areas and depressions in upper versus lower landscape positions and (iii) the inflexibility of fixed, rigid Boolean classification rules that require adjustment for different types of landforms.

Objectives

The overall objective of the current project was to develop a practical procedure for partitioning soil-landscapes into landform elements that display significant differences with respect to soil properties and the management requirements for precision farming.

The procedure was also to address limitations identified with the Pennock model and was to be flexible enough to apply to the variety of landforms in western Canada.

MATERIALS AND METHODS

Defining Landform Elements From Digital Elevation Data

The procedure for automatically defining landform elements or facets was based entirely on processing a high-resolution digital elevation model (DEM) formatted as a regular raster grid. No other input data were required. The procedure involved six main steps.

Step 1 involved acquiring appropriate digital elevation data. DEMs may be produced from real time GPS acquisition of *x, y, z* data, from traditional floating dot photogrammetry or from automatic extraction of DEMs from stereo aerial photographs via stereo auto-correlation. Most agricultural landscapes in Alberta are reasonably well represented by DEMs with a horizontal grid spacing of 5 m and a vertical resolution of 0.1 m (10 cm) or better. DEMs often exhibit artificial patterns of high frequency noise, which can be removed with a 3∗3 mean (averaging) filter.

Step 2 involved identifying and computing appropriate terrain derivatives. Ten terrain derivatives were selected as useful for automated classification of DEMs into landform elements. The derivatives were selected based mainly on their ability to reflect how landform shape and slope position affect the spatial distribution and redistribution of water and energy in the landscape. The required 10 derivatives were computed using an integrated suite of custom programs.

Step 3 involved converting the terrain derivatives into fuzzy landform attributes. Twenty landform attributes were identified that expressed fuzzy semantic concepts such as relative slope steepness or relative slope position as integer numbers scaled from 0 to 100. Fuzzy landform attributes were computed from the initial terrain derivatives by selecting appropriate values for standard index (***b***) and dispersion index (***d***) based on expert judgement and applying the appropriate fuzzy Semantic Import (SI) model as per Burrough et al. (1992).

Step 4 was the conversion of fuzzy attributes into fuzzy landform element classifications. Expert judgement (Table 1) was used to establish guidelines for a heuristic rule base that defined 15 fuzzy landform elements in terms of a weighted linear combination of fuzzy landform attributes. The final heuristic rule base identified the number of landform elements to define and described each landform element in terms of a combination of fuzzy semantic constructs such as nearly level or near a divide An overall joint membership function (JMF) ranging from 0 to 100 was computed for every grid cell for each of the 15 defined landform element classes.

Step 5 involved assigning the most likely fuzzy landform classification to each cell. The landform element class with the largest JMF value was identified and nominated as the most likely initial classification for each grid cell.

Step 6 involved simplification and generalization of the initial 15 unit classification. Application of a 3 $*$ 3 or 5 $*$ 5 modal filter to the initial 15 unit classification was used to reduce spatial fragmentation and to produce larger, more continuous, spatial entities. The original 15 landform element classes were reduced to a more limited number of classes (3–4) to simplify use and to maximize contrast.

Testing the Landform Segmentation Model

A series of linear transects were laid out oriented perpendicular to trends in the landscape. Soil profile observations were recorded at regular intervals of 20 m (1/20 of the slope length) along each transect. All site locations were documented using GPS. Soil samples were taken by horizon at every second site and analyzed for organic carbon, pH, bulk density and available nutrients.

The landform element classification was determined for each transect site. The distribution of soil classes (Subgroup phases), soil properties (organic carbon, pH, thickness of topsoil, depth of profile) and yield was determined for each of the 15 original landform classes. Summary statistics were computed to determine the mean and variance for soil properties and yield by landform element class.

These data were used to guide simplification of the original 15 classes into more generalized groupings.

Table 1. Initial guidelines proposed for the landform classification rule base.

Landform Category	Landform Element (also called Landform Facet)				Slope (%)	Slope Curvature (deg 100 m^{-1})	
	no	name	abbr.	comments	slope	Profile	Plan
Upper Slope	1	Level crest	LCR	level area in upper slope	0 – 2	+10 to -10	-
	2	Divergent shoulder	DSH	convex upper, water shedding element	> 2	>+10	-
	3	Upper depression	UDE	depression in upper slope position	0 – 2	< -10	< -10
Mid-slope	4	Backslope	BSL	rectilinear transition mid-slope segment	> 2	+10 to -10	+10 to -10
	5	Divergent backslope	DBS	Sloping ridge	> 2	+10 to -10	> +10
	6	Convergent backslope	CBS	Sloping trough	> 2	+10 to -10	< -10
	7	Terrace	TER	level mid-slope > 2m above base level	0 – 2	+10 to -10	na
	8	Saddle[2]	SAD	special case of a divergent footslope	Na	< -10	> 10
	9	Midslope depression	MDE	depression in midslope position	0 – 2	< -10	< -10
Lower Slope	10	Footslope	FSL	concave, water receiving element	> 2	< -10	na
	11	Toeslope	TSL	rectilinear in lower slope > 20% of low slope	> 2	+10 to -10	+10 to -10
	12	Fan	FAN	special case of a divergent toeslope	> 2	+10 to 10	> +10
	13	Lower slope mound	LSM	crown in lower slope < 2m above base level.	> 2	>+10	>+10 ?
	14	Level lower slope	LLS	level in lower slope, > 20% of low slope	0 – 2	+10 to -10	+10 to -10
	15	Depression	DEP	concave element in lowest landform pos.	0 – 2	< -10	<0

RESULTS AND DISCUSSION

Computing Terrain Derivatives

All of the computed terrain derivatives provide some information on the most likely pattern of distribution and redistribution of water and energy in the landscape and the logical reflection of these patterns in soils and soil properties. For example, profile curvature (Fig. 1a) can be interpreted in terms of rapid runoff, erosion and thin soils in convex areas, decelerating runoff, deposition and thicker soils in concave areas and neutral or modal conditions in areas that are planar in the downslope direction. Relative relief provides an indication of the landform context of each cell in terms of moisture availability (Fig. 1b).

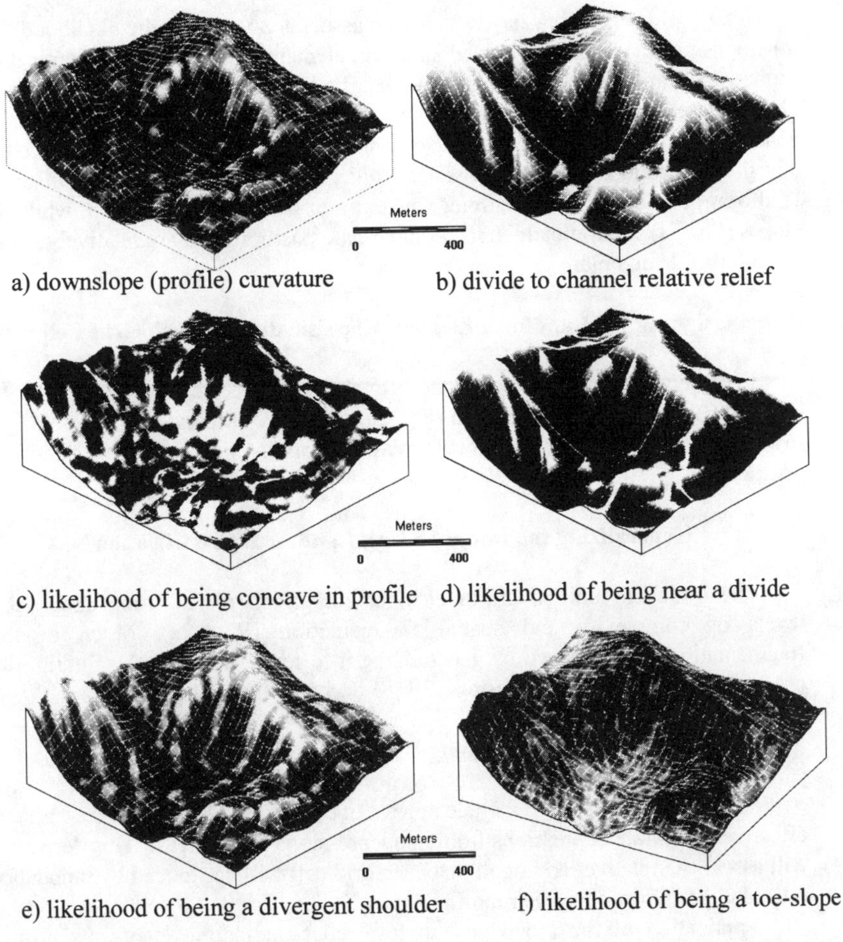

a) downslope (profile) curvature b) divide to channel relative relief

c) likelihood of being concave in profile d) likelihood of being near a divide

e) likelihood of being a divergent shoulder f) likelihood of being a toe-slope

Fig. 1. Three Dimension views illustrating the process of fuzzy landform classification at the Hussar site.

Converting Terrain Derivatives into Fuzzy Landform Attributes

The fuzzy landform attributes computed from each of the initial terrain derivatives express the degree to which a particular value of a given terrain derivative reflects the semantic construct associated with a given fuzzy landform attribute. For example, two of the fuzzy landform attributes considered in the semantic import model are the degree to which each grid cell can be considered to be strongly concave in profile (Fig. 1c) and the degree to which a cell is located relatively close to a divide (Fig. 1d).

Converting Landform Attributes into Fuzzy Landform Element Likelihood Values

Every grid cell in a raster DEM has an associated value for the likelihood that it belongs to each of the 15 defined landform element classes. Visual examination of the two examples provided confirms that the likelihood of belonging to a given class ranges from very high (white) to very low (black) in a manner that is consistent with the shape and relative position in the landscape of a grid cell at any given location. Thus divergent shoulders (Fig. 1e) occur mostly in upper landform positions and have strong convexity in both profile and plan, while toe slopes (Fig. 1f) occur mostly in lower landscape positions and are relatively planar in both profile and plan.

Assigning the Landform Element Classification to Each Grid Cell

Each cell was assigned to the landform element class with the largest JMF for that cell. The 15 unit classification (Fig. 2a) effectively separated upper, mid, and lower slope units as well as level areas in upper, mid, and lower landform positions.

Generalizing the Initial 15 Unit Landform Classification

The initial 15 unit landform element classification had some undesirable levels of complexity and spatial fragmentation (Fig. 2a). Much of this fragmentation was removed by generalizing it to fewer units and by filtering the original 15 unit classification using a 3 by 3 or 5 by 5 modal filter (Fig. 2b).

Ongoing Developments

Efforts are underway to include at least one new derivative in the rule base to effectively separate depressions from nondepressional level areas. This derivative will assess the relative likelihood that each grid cell will be flooded by inundation related to ponding of surface runoff.

Application of the procedures to low relief undulating landforms is now being tested.

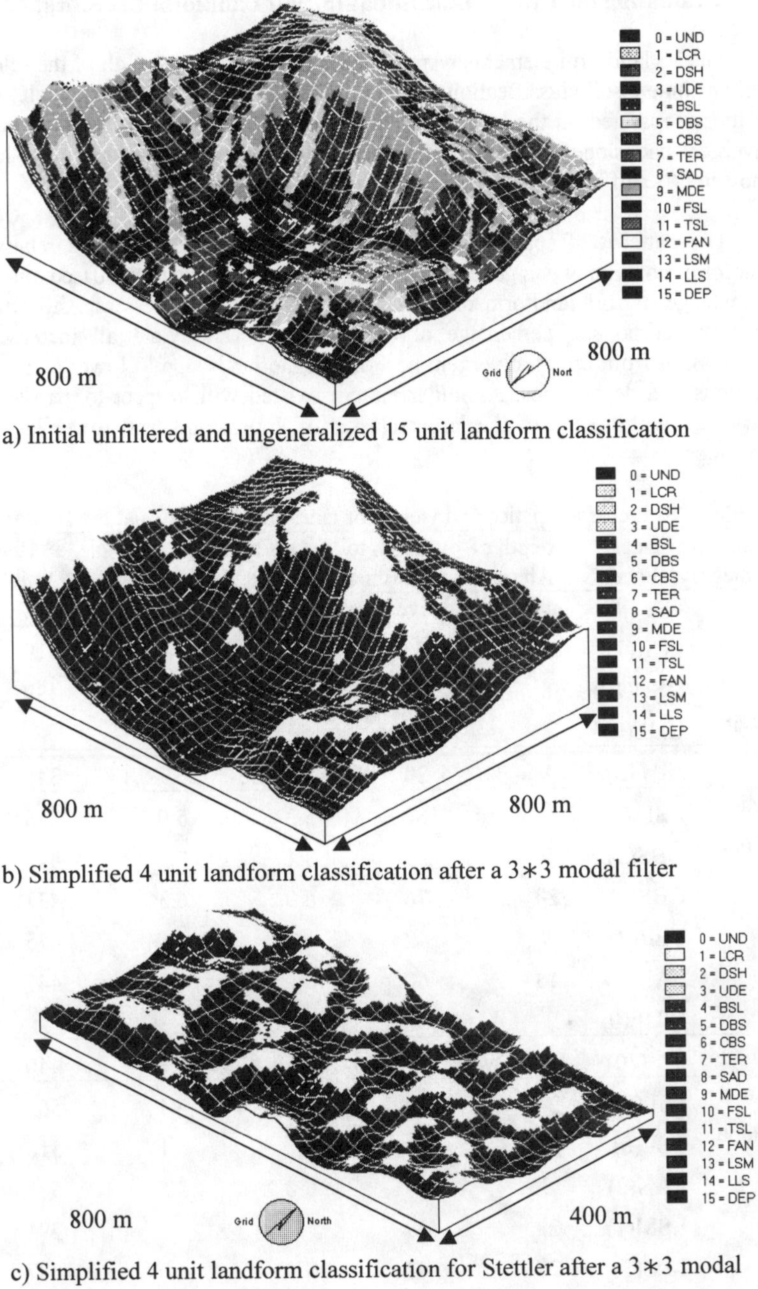

a) Initial unfiltered and ungeneralized 15 unit landform classification

b) Simplified 4 unit landform classification after a 3∗3 modal filter

c) Simplified 4 unit landform classification for Stettler after a 3∗3 modal

Fig. 2. Three Dimension views illustrating the initial and generalized landform classifications at two study sites (a, b = Hussar and c = Stettler).

Evaluating the Utility of the Initial 15 Unit Landform Classification

The 15 landform elements were not all different in terms of all of the selected soil attributes, soil classifications or yields (Tables 2 and 3). For example, some of those separated on the basis of across-slope or plan curvature (convergent vs. divergent backslopes) showed differences while others (toeslopes vs. footslopes) showed little difference at this scale for this site.

The patterns of spatial distribution of soil properties and yield relative to landform position were generally consistent with expectations based on local knowledge of soil–landform relationships (Table 2). Thickness of Ah horizons, depth to carbonate, percentage of organic matter and yield all increased in progressing from upper, divergent portions of the landscape to lower convergent portions and depressions. A similar pattern existed with respect to across slope curvature within each of the larger classes of upper, mid, and lower landform positions.

Table 2. Soil characteristics and yields for landform categories at the Hussar site.

Landform facet Category	Facet	depth of Ah (cm)		depth to Ca (cm)†		Org Mat %†		pH		1996 yield	
		ave	(cv)	ave	(cv)	ave	(cv)	ave	(cv)	ave	(cv)
Upper slope	LCR(1)	16		58		5.3		6.3		419	
	DSH(12)	8		17		4.1		7.1		339	
	UDE(0)										
	all (13)	8	31	20	99	4.2	20	6.7	13	345	27
Mid slope	BSL(8)	9		62		5.3		5.9		435	
	DBS(4)	9		23		3.1		7.2		447	
	CBS(5)	20		76		6.9		6.3		437	
	TER(2)	8		42						455	
	SAD(1)	13		60						447	
	MDE(0)										
	all (20)	12	79	56	63	5.5	42	6.2	9	440	16
Lower slope	FSL(8)	24		79		8.0		6.4		563	
	TSL(8)	24		85		6.1		6.1		518	
	FAN(2)	8		38		6.8		5.5		521	
	LSM(1)	28		79						291	
	LLS(3)	28		92		9.4		5.9		587	
	DEP(2)	41		100						570	
	all (23)	25	55	81	35	7.7	19	6.0	7	538	14

† As not all sites were sampled, there were a total of 22 analysis for OM and pH.

Evaluating the Utility of a Simplified 3 to 4 Unit Landform Classification

There were several groups related to general landform position and downslope or profile curvature that exhibited distinct, quantifiable differences in soil attributes and yields (Tables 2 and 3; Fig. 3).

Upper, generally convex, water shedding slopes had the thinnest soils and lowest yields while lower, generally concave or level, water receiving slopes had thicker soils and higher yields. The water neutral mid-slopes exhibited intermediate or average conditions.

These three generalized classes exhibited most of the significant variation in soil properties and yield. Some systematic variation in soil properties observed within these larger classes could be related to across slope curvature. For example, thicker, more organic rich soils occurred in convergent elements and thinner, less organic rich soils in divergent elements within each of the larger groupings of upper, mid, and lower. These differences in soil properties within the three main classes were not mirrored by significant corresponding differences in yield within the period of study (Fig. 3).

The three generalized groups also differed in a systematic pattern with respect to soil classification (Table 3). This systematic variation can facilitate extrapolation of results to other locations based on soil characteristics.

The consolidation to 3 to 4 classes (depressions sometimes require separate treatment) has additional benefits for the definition and identification of land based management units (Fig. 2 and 3). The procedure worked equally well on the hummocky landscape of the Stettler test site and the rolling landform at Hussar site (Fig. 2b, c). Application of the procedures to a other sites with different landforms indicated a need for modification of the rule bases in areas of low relief.

Table 3. Number of soil profiles of each soil type within the three generalized landform categories at the Hussar site.

Landform element groupings	Canadian System of Soil Classification soil types (phases of Subgroups)[†]						
	REG	RDB	ODB _er	ODB	EDB _tk[‡]	SOL	Totals by class
water shedding -upper slopes [§]	5	3	3	1			12
water neutral -mid-slopes [§]		3	4	11	6		24
water receiving -lower slopes [§]			2	3	13	6	24
Totals by soil type	5	6	9	15	19	6	60

[†] REG = Regosol; RDB = Rego Dark Brown; ODB_er = Orthic Dark Brown (Ah < 10cm) ODB = Orthic Dark Brown (Ah 10–15 cm); EDB_tk = Eluviated Dark Brown (Ah >15cm).

[‡] EDB_tk includes all soils developed on > 50 cm of slope wash (some ODB and Sz DB) and having > 15 cm Ah.

[§] Upper slopes includes DSH; mid-slopes includes BSL, DBS, CBS, TER, SAD, LCR, LSM, UDE; Lower slopes includes FSL, TSL, FAN, LLS, DEP.

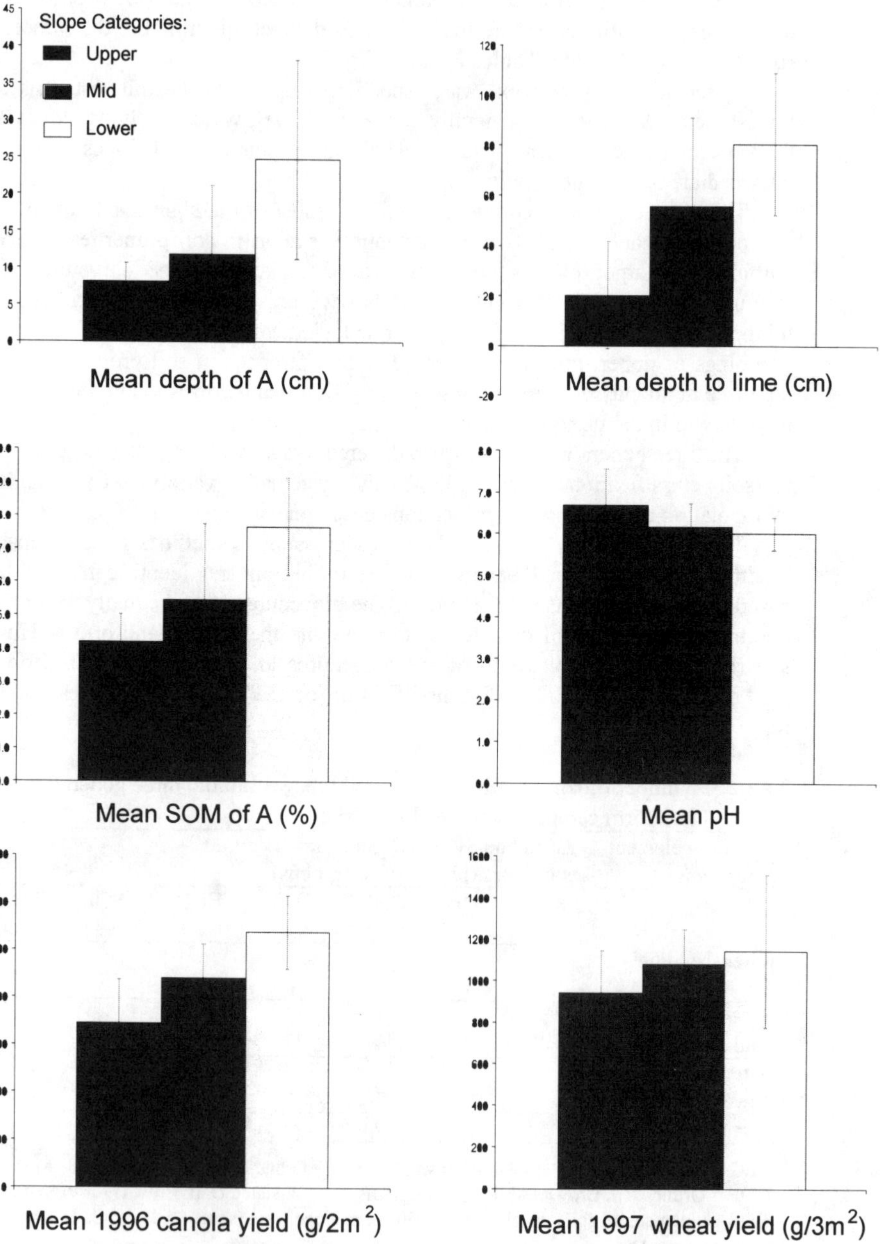

Fig. 3. Relationship of soil attributes and yields to generalized landform categories for a Mh landscape at the Hussar site.

CONCLUSIONS

With Respect to the Procedures Used to Produce the Landform Elements

- A new landform segmentation methodology has been developed that can be applied to a wide variety of agricultural landscapes in western Canada and generalized according to the magnitude and scale of the landscape and data.
- Inclusion of several measures of relative and absolute landform position resulted in improvements in several aspects of the model relative to that of Pennock et al., (1987), specifically:
  - spatial fragmentation was reduced, producing larger more spatially continuous areas.
  - misallocation of cells into classes at variance with the relative slope position was reduced (e.g., cells in upper landform positions were not classed as footslopes based on shape alone).
- The use of fuzzy logic allowed a single set of rules to be applied to a wide variety of landforms.

With Respect to Evaluation of the Effectiveness of the Landform Model

- All 15 landform elements chosen were not individually or exclusively different in terms of the selected soil attributes, soil classification or yields. However, several groups related to general landform position and downslope or profile curvature exhibited distinct, quantifiable differences in soil attributes and yields.
- The recommended 3 to 4 class grouping offers an optimum combination of simplicity, spatial coherence, and meaningful differences in soil properties and crop yields.
- These spatial entities appear particularly well suited for defining management units for precision farming or for locating benchmark soil sampling (test) sites.

ACKNOWLEDGMENTS

This project is based on the guidance and financial support of a number of industry partners including: Western Cooperative Fertilizers (John Harapiak), Norwest Research (Jean Crepin), Agrium (Doug Beever) and NovAtel. Funding was also provided by Agriculture and Agri-Food Canada (Matching Investment Initiative Program). Alberta Agriculture Food and Rural Development (Sheilah Nolan, Germar Lohstraeter, and Annette Svederus) provided the DEM and GPS and field support.

REFERENCES

Burrough, P.A., R.A. MacMillan, and W. Van Deursen. 1992. Fuzzy classification methods for determining land suitability from soil profile observations and topography. J. Soil Science. 43:193–210.

Irwin, B.J., S.J. Ventura, and B.K. Slater. 1997. Fuzzy and isodata classification of landform elements from digital terrain data in Pleasant Valley, Wisconsin. Geoderma. 77:137–154.

MacMillan, R.A., and W.W. Pettapiece. 1997. Soil landscape models: automated landform characterization and generation of soil-landscape models. Tech. Bull. no. 1997-E. Research Branch, Agric. and Agri-Food Canada, Lethbridge, AB.

Penney, D.C., S.C. Nolan, R.C. McKenzie, T.W. Goddard, and L. Kryzanowski. 1996. Yield and nutrient mapping for site specific fertilizer management. Commun. Soil Sci. Plant Anal. 27:1265–1279.

Pennock, D.J., B.J. Zebarth, and E.DeJong. 1987. Landform classification and soil distribution in hummocky terrain, Saskatchewan, Canada. Geoderma. 40: 297–315.

Pennock, D.J., D.W. Anderson, and E. DeJong. 1994. Landscape changes in indicators of soil quality due to cultivation in Saskatchewan, Canada. Geoderma. 64:1–19.

Skidmore, A.K., P.J. Ryan, W. Dawes, D. Short, and E. O'Loughlin. 1991. Use of an expert system to map forest soils from a geographical information system. Int. J. Geograph. Infom. Syst. 5:431–445.

Identifying Agroecozones in the Central Great Plains

R. L. Anderson
R. M. Aiken

USDA-ARS
Central Great Plains Research Station
Akron, Colorado

H. R. Sinclair
S. W. Waltman
W. J. Waltman

USDA-NRCS
National Soil Survey Center
Lincoln, Nebraska

ABSTRACT

Winter wheat–fallow is the conventional cropping system in the semiarid central Great Plains. Fallow weed control relies on tillage, but this practice leads to erosion and loss of organic matter. Cropping systems studies in this region, conducted since the early 1980s, are demonstrating that annualized yield can be doubled and soil quality improved with continuous no-till cropping. However, with only four research sites in the central Great Plains, producers question how far experimental results may apply across this region, given the extreme variation in soil properties and climatic characteristics. Therefore, we identified agroecozones of similar soil and climatic characteristics in relation to the four cropping systems sites. The winter wheat-fallow region can be divided into two agroecozones, with a zone of uncertainty identified. Within an agroecozone, soil differences may have a greater impact on land productivity than climatic factors.

INTRODUCTION

The central Great Plains is a semiarid region where crop production systems are developed around winter wheat (Fig. 1). In the region, there are 8 million hectares of cropland, with approximately half of the land planted to winter wheat. Proso millet also is grown in this region, but occupies < 5% of the cropped hectarage (Hinze & Smika 1983).

The conventional production system is to rotate winter wheat with fallow, because fallow enhances nutrient availability and minimizes the impact of erratic

Copyright © 1999 ASA-CSSA-SSSA, 677 South Segoe Road, Madison, WI 53711, USA. *Proceedings of the Fourth International Conference on Precision Agriculture.*

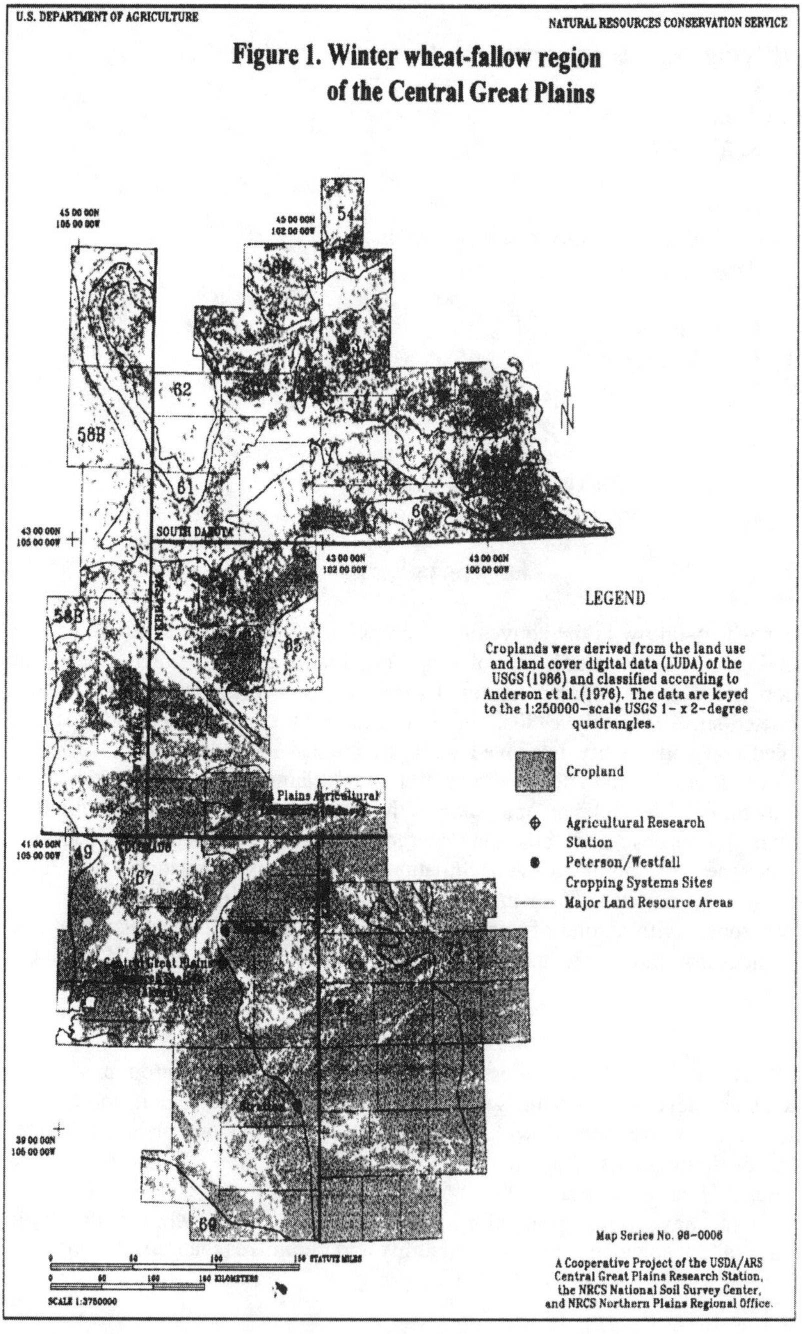

Figure 1. Winter wheat-fallow region of the Central Great Plains

precipitation on grain production (Greb, 1983). However, fallow degrades soil quality by increasing soil erosion and loss of organic matter (Peterson et al., 1993).

Improved technology has led to no-till production systems, which maintain more crop residue on the soil surface, subsequently increasing precipitation storage (Smika & Unger, 1986) and water use efficiency of cropping systems (Peterson et al., 1996). With improved water relations due to no-till systems, more intensive cropping is now possible in this region. Furthermore, no-till systems enable producers to move production practices of humid regions westward, such that corn and sunflowers are now grown successfully in this region. Along with winter wheat and proso millet, producers now have four crops with viable markets that can be included in their rotations.

Cropping systems studies are currently located at four sites in the region: Akron, Sterling, and Stratton CO, and Wall, SD. These cropping systems studies are demonstrating that annualized yield can be almost doubled and soil quality improved with no-till continuous cropping. With only these four research sites, producers question how far research results apply across the region, given the extreme variation in soil properties and climatic characteristics. For example, mean annual precipitation ranges from 300 to 600 mm across the region, whereas precipitation at the four cropping systems sites ranges between 380 and 430 mm per year. Furthermore, soil types are similar at all four sites, in contrast with the drastic differences in soils of the region.

Our goal was to define areas (agroecozones) in the region that have soil and climatic characteristics similar to the cropping systems research sites. An agroecozone is developed by relating soil and climatic characteristics with a spectrum of crops, and can serve as a guideline for producers who are evaluating the cropping potential of their farm.

MATERIALS AND METHODS

We followed a three-step procedure to define possible agroecozones. Initially, a terrain modeling approach was used along with the Newhall simulation model to aggregate land areas into zones of similar climatic and soil factors. Factors analyzed included growing degree day accumulation, root zone available water holding capacity, soil water balance, yearly precipitation, and potential evapotranspiration.

Secondly, we grouped areas based on comparison with the climatic conditions to the four cropping systems sites: Sterling (Logan County) CO; Akron (Washington County) CO; Stratton (Kit Carson County) CO; and Wall (Pennington County) SD.

Boundaries of each zone were then modified by potential yield calculations for corn and sunflower based on long-term precipitation records from local weather stations. Corn and sunflower yields in this region are related to summer precipitation (Lyon et al., 1995), thus potential yield was calculated with following formulas: for corn, Yield (kg ha^{-1}) = 2130 + 18.5*July-August rainfall (mm), r^2 = 0.70, and for sunflower, Yield (kg ha^{-1}) = 462 + 7*August-September rainfall (mm), r^2 = 0.80 (Nielsen et al., 1996). Based on suggestions by producers, we used a yield limit (break-even point) of 2800 kg ha^{-1} for corn and 1000 kg ha^{-1}

for sunflower to guide zone boundaries. To further modify zone boundaries, we evaluated cropping systems used by progressive producers located throughout the region.

With this three-step procedure, we estimated potential production zones for the central Great Plains region in relationship to the four cropping systems sites.

RESULTS AND DISCUSSION

Agroecozone Delineation

Our results suggests that this winter wheat–fallow region can be divided into two agroecozones (Fig. 2).

Zone 1 is a production area where cropping systems more intensive than wheat–fallow are successful. The zone can be subdivided into two subzones by the adaptability of sorghum (1A and 1B), as night temperatures at higher elevations are too cool for successful sorghum production in the western part of the zone (1A).

Rotations used by progressive producers in zone 1 include wheat–corn–corn, wheat–corn–proso millet, wheat–proso millet, or wheat–corn–sunflower–proso millet. At the western edge, producers include fallow in the rotation, such as wheat–corn–proso millet–fallow or wheat–corn–sunflower–fallow.

The four cropping systems studies, located in the western edge of Zone 1, are also demonstrating that this region can support more intensive cropping. At Wall SD, successful rotations include wheat–suntlower–proso millet, wheat–field peas–proso millet, and wheat–corn–proso millet (Stymiest, 1997). At Akron, annualized yield of wheat–corn–proso millet is double that of wheat–fallow. Other successful rotations at Akron are wheat–proso millet, wheat–corn–sunflower–fallow, and wheat–corn–proso millet–fallow (Anderson, 1998). At Sterling and Stratton CO, grain yield of wheat–corn–proso millet–fallow is 70% greater than wheat–fallow (Peterson et al., 1993).

Zone 2 is a production area where either low rainfall, cool temperatures, or shallow soils prohibit the use of corn or sunflower. The prevalent crop rotation is wheat–fallow, however, producers may be able to expand their cropping options in this zone by including low-water-use forages in rotation with winter wheat or proso millet.

Zone of Uncertainty

We also identified a zone of uncertainty, which is to the west of the four cropping systems sites (Fig. 2). This area has low precipitation during July and August or cool temperatures which make successful corn and sunflower production questionable. Scientists are exploring short season corn varieties as a possible option for this area: yields of 70- to 85-day corn varieties are promising, approaching 4000 kg ha^{-1} at various locations in this zone.

Soil Effect on Cropping Systems

Within an agroecozone, soil differences may have a much greater impact on land productivity than climatic factors. For example, in southeastern Wyoming and western Nebraska, there are soils where bedrock is only 0.8 m below the surface, thus

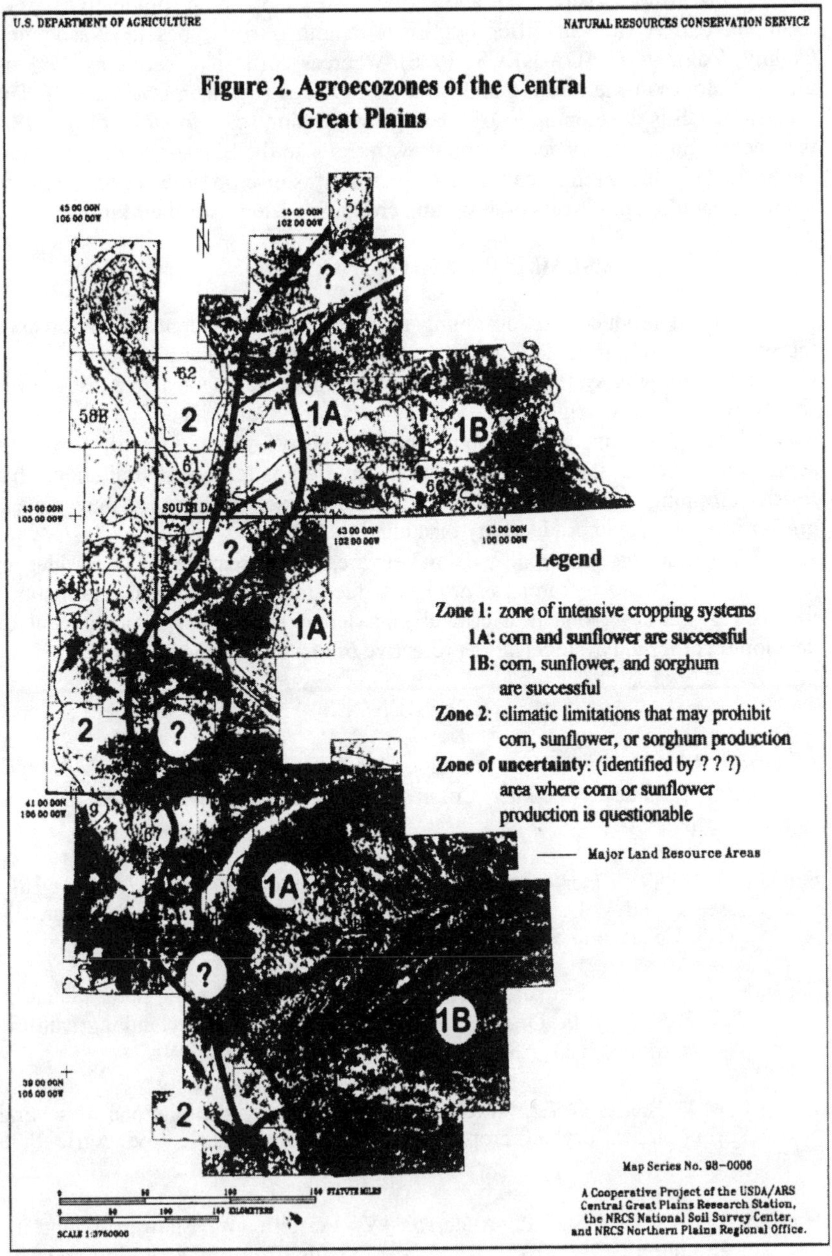

water holding capacity is limited. Currently, producers use a wheat–fallow system on these soils, and cropping options may be restricted due to the shallow soil.

Furthermore, soil type and slope also affect crop productivity. For example, corn yield can differ by two-fold among soil types in Washington County, Colorado (USDA-NRCS, 1986), whereas corn yield varies by 35% on different slopes at the Sterling and Stratton CO sites (Peterson et al., 1997). The USDA-NRCS is developing a soil-plant growth rating scale for soils of the USA that incorporates soil impact on crop growth (refer to the Sinclair et al., 1999, this publication). This rating scale's assessment of soil-crop interactions will be helpful in guiding producers in designing cropping systems for their farms.

SUMMARY AND CONCLUSIONS

To aid producers in designing new rotations, we identified landscapes (agroecozones) of similar soil and climatic characteristics that producers can use to relate to cropping systems developed at the four research sites. We divided the winter wheat-fallow region into two agroecozones, and identified a zone of uncertainty. One zone will support more intensive cropping with corn and sunflowers whereas the second zone may have climatic or soil limitations that restrict cropping options. Within an agroecozone, soil differences may have a greater impact on land productivity than climatic factors.

The zone of uncertainty is a key area for researchers to consider in planning their future research, as producers face more environmental limitations in this region. This zone is a critical knowledge gap in our regional goal of developing cropping systems more protective of our soil resource.

REFERENCES

Anderson, R.L. 1998. Designing rotations for a semiarid region. p. 4–15 *In* Proc. 10th Annual Meeting, Colorado Conservation Tillage Association, Sterling, CO.

Greb, B.W. 1983. Water conservation: Central Great Plains. p. 57–73. *In* H.E. Dregne and W.O. Willis (ed.) Dryland agriculture. Agron. Monogr. 23. ASA, CSSA, and SSSA, Madison, WI.

Hinze, G.O., and D.E. Smika. 1983. Cropping Practices: Central Great Plains. p. 387–395. *In* H.E. Dregne and W.O. Willis (ed.) Dryland agriculture. Agron. Monogr. 23. ASA, CSSA, and SSSA, Madison, WI.

Lyon, D.J., F. Boa, and T.J. Arkebauer. 1995. Water-yield relations of several spring planted dryland crops following winter wheat. J. Prod. Agric. 8:28 1–286.

Nielsen, D., G. Peterson, R. Anderson, V. Ferreira, W. Shawcroft, and K. Remington. 1996. Estimating corn yields from precipitation records. Colorado Conservation Tillage Association Fact Sheet #2-96. Colorado Conserv. Tillage Assoc., Ault, CO.

Peterson, G.A., D.G. Westfall, and C.V. Cole. 1993. Agroecosystem approach to soil and crop management research. Soil Sci. Soc. Am. J. 57:1354–1360.

Peterson, G.A., A.J. Schlegel, D.L. Tanaka, and O.R. Jones. 1996. Precipitation use efficiency as affected by cropping and tillage systems. J, Prod. Agric. 9:180–186.

Peterson, G.A., D.G. Westfall, L. Sherrod. 1997. Sustainable dryland agroecosystem management. Colorado State Univ. Agric. Exp. Stn. Tech. Bull. TB97–3. Colorado State Univ., Ft. Collins, CO.

Smika, D.E., and P.W. Unger. 1986. Effect of surface residues on soil water storage. *In* B.A. Stewart (ed.), Adv. Soil Sci. 5:111-138.

Stymiest, C.E. 1997. Annual progress report. South Dakota State Univ. Plant Science Pamphlet No. 90.

USDA-NRCS. 1986. Soil survey of Washington County, Colorado.

An Object Model for Information Systems for Precision Agriculture

A. M. Saraiva
A. M. A. Massola
C. E.Cugnasca
Department of Computer Engineering, Agricultural Automation Laboratory
Universidade de São Paulo, Escola Politécnica
São Paulo, Brazil

ABSTRACT

Adequate information systems are one of the main needs of precision agriculture. Although many systems have been proposed as research tools and some others have been available in the market, there are many issues concerning their completeness, compatibility, cost and user-friendliness, to mention a few. Aiming at contributing to the understanding and to the solution for the availability problem of more adequate systems a model for such a class of systems was developed. It was based on software engineering fundamentals such as domain analysis concepts and object oriented modeling methods. This paper discusses the need for a broader understanding of the role of information systems in precision agriculture in view of the quick changes in the technology. The modeling method is described and some of the results are presented.

INTRODUCTION

Precision agriculture (PA) is a recent technology, going through quick evolution, and still being consolidated. Research in PA shows a great diversity of methods and techniques being proposed and tested, without a definite consensus and frequently with a disparity of results. The product and service market for PA reflects this changing situation, showing a constant innovation and a certain instability of the related companies and their products.

Users and researchers still deal with tough challenges, until technology can be well established among producers. This difficulty can be understood if we consider the great natural complexity of the agricultural production process, increased by the PA approach, that sees the process from a global and, at the same time, localized point-of-view.

Among the three typical phases of PA—data acquisition, information management and variable rate application —, those related to field work quickly benefited from the specific products that appeared in the market. For these phases (data acquisition and variable rate application), there are several pieces of equipment available, even though many points in them still have to be improved.

Between those phases, the information management phase has been pointed out as critical, for two basic reasons.

First of all, there is the lack of knowledge about what to do with the field data in order to get an effective management. Data processing supports decision

Copyright © 1999 ASA-CSSA-SSSA, 677 South Segoe Road, Madison, WI 53711, USA. *Proceedings of the Fourth International Conference on Precision Agriculture.*

making that leads to variable rate application, but the adequate methods for data analysis and processing are still an issue.

The second reason is that the available information systems (IS) do not support decision making as they ought to, being an additional problem for PA practitioners.

This text is part of a work aiming at contributing to the understanding and to the solution for the availability problem of adequate ISs for PA.

ISSUES ON INFORMATION SYSTEMS FOR PRECISION AGRICULTURE

Literature offers several examples of papers dealing with software development for PA, and a revision of them can be found in Saraiva et al. (1997) and Saraiva (1998). Some of them focus on providing a project or experiment with supporting computational tools, while others focus on the IS problem itself.

Among the former, there are very simple systems concerning some aspects, such as the operating interface and the degree of automation for performing the ordinary tasks required for the information management. This characteristic, however, was acceptable for the early systems, considering that their users were basically researchers, who knew about the tools and processing methods. Of course there were no systems supporting all details of this new research area, and investing in this problem was not advisable. The greatest effort was the study of other problems, such as data acquisition, variable rate application, accurate field positioning and the related field equipment to allow for this technology.

After some time, this situation was altered. Because of research evolution, more data being collected, more field parameters being considered, and more complex problems to be solved, those simple systems were no longer acceptable. Even researchers accustomed to inadequate tools were now demanding more friendly systems for information processing. At the same time, technology dissemination among producers was demanding support systems more adequate to operators not so specialized and not used to computational tools and to analysis methods.

Some other examples in the literature point out to this concern with the IS itself, as a fundamental component for the success and dissemination of technology, as in Jürschik (1997).

The same was happening in the market. Until recently, there were no ISs specific to PA. There were only some small packages for map exhibition and printing, using the data collected by yield monitors for example. These programs have a relative utility, however, because they do not allow data manipulation, or even when they do allow, they do not indicate the methods used. As a consequence, the maps obtained are mostly useless or hard to interpret, usually showing a great field variability. What is more, they show just one field, and do not allow for data concatenation from other areas or from previous years.

In the last few years some specific PA products came out, many of them self-designated agricultural GIS (geographical information system). Their features and comprehensiveness are significantly different, but this paper does not intend to compare them. Although they represent some improvement when compared with the previous situation, there are still some remaining problems, that must be solved so that these systems can be effective decision support tools.

A preliminary evaluation of the available examples in the literature and of several packages from the market shows some points to be improved in ISs for PA:

- The operation interface is not always customized for the application, mainly in systems built from subsystem integration. On the contrary: they maintain the characteristic of the programs they are based on (typically, a commercial GIS is such a package), making the operation more difficult.
- The cost is high. Systems built in a totally proprietary mode may have a high development cost, due to the potential complexity of a good system. A component integration approach can lead to lower costs, but, because commercial systems are mainly based on GISs from the market, their final cost also includes an additional software layer cost. That could be attenuated if function libraries or GIS objects were only incorporated to the systems as they are needed, since several generic GIS functions are not used by the final system.
- Data formats accepted are limited to proprietary formats, or also accept some market standard. They are usually not open, in a strict sense.
- The possibility of integration with other systems is generally reduced, making it difficult to exchange data in an automatic way and to expand the systems.
- They do not allow for customization of the available functions, the methods used, the operation interface and the information presented to the user.
- They do not deal with meta-data that would allow to trace data origin and quality.
- Usually, they do not allow for automation of the processing according to the user profile. In some cases, where that automation already exists, there is no complete information about the methods used for data manipulation.

There are several reasons for those problems. Some of them can be granted to the situation of information technology, in a broader sense. For example, the question of component integration and the open data formats are related to the question of the existence of open standards for software interoperability. This is a problem that extrapolates the domain of PA, and pervades in several levels of IT.

That is the case of software platforms and the mechanisms they make available for interoperability. Today, several platforms, such as OLE/COM (Microsoft), CORBA (OMG), Java (Sunsoft) compete for the market, and we hope that, besides the effective software interoperability within the same platform, there will be some degree of interoperability among the ones that endure.

In a more restrict level, this is also a GIS problem (surely, a fundamental component for PA) and for the lack of spatial data standard. For this level, the proposition of an open GIS, by OpenGIS Consortium, is the main initiative (OpenGIS, 1996).

The problem reaches to the agricultural domain which lacks standard data dictionaries for that specific knowledge area. At this level, the proposals of ISO and AEA could provide the base for interoperability (ISO, 1995; AEA, 1997).

However, not every problem is caused by Information Technology (IT). Several of them are consequence of inadequate system specification and structuring.

Many aspects pointed here as flaws in the systems analyzed are due to the fact that they were developed primarily concerning support for the initial research in PA. The urgency for obtaining them, a lesser concern for handling, and the lack of a clear vision on their potential and importance could have been some reasons for the adopted options.

Clearly, ISs requirements change with time, particularly if the application itself is not well defined and is still evolving, as is the case of PA. It is necessary, then, to provide the IS with capacity of evolution, so that it can follow that dynamics.

With the fast changing scenario of PA, and also of IT, however, new system proposals must incorporate the new domain requirements, and also the new technology trends.

Concerning the trends of IT, we see that the dominant computational paradigm is going from closed to open systems; from isolated to real time interoperating systems; from independent and strongly packed applications to application environments, where interoperable software components provide more flexible capacities to the user - the componentware. At the same time, global computing is quickly coming up, and the paradigm is going to a net-based computing. Internet and other networks provide access to countless data sources and services for millions of users. The advantage of using such technologies are evident for users, who feel the need for integrating data and processing resources both at desktop level and through local and global networks, in order to obtain a complete interoperability (OpenGIS, 1996)

The maturity of computing industry has some parallels with agribusiness, where, in developed countries, a few rural producers supply several consumers. A widely accepted view in the computing industry is that there will eventually be a community with few producers (implementers of software objects) satisfying many consumers (object users). Consumers will be the programmers and power users, who will build applications from a bank of interoperable objects, before writing their own code or before developing their own objects for filling gaps in functionality.

A basic question in this change is the software development process itself, and working methods. The object-orientation paradigm has been widely disseminated, based on the advantages it offers to modeling and the understanding of problems, associated to good solution structuring and to the possibility of contributing to the stability given the changes in requirements.

MODELING FIELD INFORMATION SYSTEMS

Objectives

Once this situation was stated, we detected that, more important than developing another system for PA is studying and formalizing the requirements of this class of systems, given the application needs and IT reality. The method used for this task was modeling of ISs.

The purpose of the model is to allow a clear, deep and stable understanding of PA needs, in terms of ISs for supporting farm management. We hope the model will act as a paradigm for detailing ISs that fulfill the needs for specific applications, in the wide universe of PA.

In order to achieve this goal, here we propose an abstract model, regardless of implementation. It is a meta-system model that reflects the desirable characteristics for a system class. Therefore, it is not a single system model, but a conceptual structure based on the application domain, so that the systems to be implemented can be detailed, and the existing systems can be evaluated.

Method

The method definition was based on the following objectives:

- to cover the preanalysis and analysis phases of the system development cycle, based on the proposed focus, i.e., a meta-system model
- to try and obtain a robust structure concerning requirement changes, aiming at the current domain characteristics
- to incorporate a previous and deep study of the problem domain, to allow for a comprehensive view of the environment where these ISs are inserted
- to use concepts and notations that facilitate communication with users

Based on these objectives, the object paradigm was adopted, because it is suitable for the approach chosen. The method itself results from a composition of concepts and notations of two different object-oriented modeling methods proposed in the literature: the Object Modeling Technique method - OMT (Rumbaugh et al., 1997), and the Object Oriented Software Engineering method - OOSE (Jacobson et al., 1993).

A pre-analysis phase was also incorporated, based on concepts from the Domain Analysis for acquiring knowledge, aiming at the definition of the requirements and characteristics of the PA systems.

Phases, Activities, and Products of the Method

The pre-analysis phase aims at the system class requirements definition. For that we use concepts and resources from both the Domain Analysis (Arango; Prieto-Díaz, 1991) and the OOSE method.

Concerning the Domain Analysis, we used the identification of information sources, their accessibility and their study. Among these sources, there was traditional technical literature, domain experts, system users and existent systems.

From that study, a basic reference text - Problem Enunciation - was created. The text defines an initial system requirements set, mainly from the high level functionality point-of-view. Despite the ambiguities inherent to the textual description, it is a starting point for the Requirement Model description, which goes deeper into system requirement formalization, using the concept and the notation from the OOSE method. The objective, here, is to consider the users' point-of-view on systems functionality. Based on that text, and considering the analysis of other existing and proposed systems, the next step is the elaboration of a Use Cases Model, in a diagram that shows the system delimitation, concerning the environment. The concept of actors is used to express all entities, human or not, that interface with the system, exchanging information with it.

The use cases symbolize the ways these external entities use the system, or are used by it, and can be seen as a macro-abstraction of the system functionalities.

Associated to the method, the Use Case Description is a text that details the sequence of interaction events between the system and the agents, with each sequence representing a use case, detailing its meaning and the information exchanged by the systems and their users. This detailed description allows identification not only of the aspects of the operating interface, but mainly the contents interchanged between actors and system, which will be the base for the further modeling steps, when the object classes integrating the system structure are identified. The use cases shown in the diagram are further divided subcases in the description.

With these three descriptions, the Requirement Model and the pre-analysis phase are concluded.

In the Analysis phase, the objective is to obtain a problem domain model, regardless of implementation aspects. That is done according to an object-oriented approach, identifying those problem domain objects, their relationships and their structure, for a better domain description. The base for this activity is the requirements identified by the use cases from the previously created Requirement Model, and also the domain knowledge and its terminology, both acquired during the domain study, in the pre-analysis phase.

The analysis phase is the one that effectively uses the object-oriented approach. The result of this domain mapping activity, in terms of objects, is an Object Model. This is the most important model in the analysis phase. As a complement to these model diagrams, an Object Class Description is presented as a text.

The Subsystems Model is additionally used to help the structure description, dividing the Object Model into subsystems. This hierarchic structuring of the object classes in subsystems, besides improving the understanding of the several model components, has the objective of identifying the perspective of implementing specific systems from the integration of independent subsystems satisfying the model premises.

RESULTS

The initial results obtained from the domain study were texts analyzing PA in a general manner and its ISs in a particular manner.

After that, the Problem Enunciation was stated, trying to identify the basic IS functionalities for AP to be modeled.

The next product is the Use Case Model, presented in Fig. 1. In this diagram, four actors can be identified: operator, manager, data acquisition equipment and application controller equipment.

Three use cases involve the actors and the system: Insert field data, Generate maps, and Configure the system.

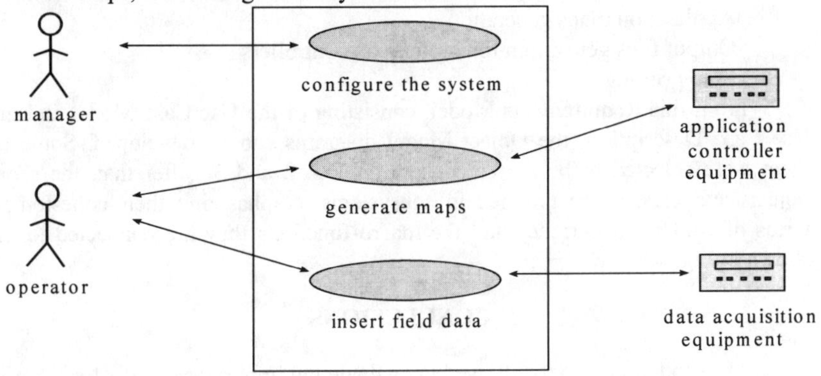

Fig. 1. Use case model.

The use case insert field data involves the input of several data from the field, from samples as well as from the farm. The use case generate maps concerns all data processing as far as the obtainment of application maps. The use case configure the system concerns system programming for a determined user, crop or other particular conditions, including data insertion and knowledge about crops, machines, the market, the user. The diagram is deliberately simple, as to allow a quick view of the system limits and of the application environment.

The next result, the Use Case Description, divides the use cases, and details the actor-system interaction for each of them. Therefore, it emphasizes the contents exchanged in the interactions. The subdivisions of the three primary use cases provided the following use sub-cases, only listed below:

Use case: Insert field data

- New farmer register
- New farm register
- New field register
- Sampled data reading, from external equipment
- Manual input of sampled data

Use case: Configure the system

- Management rules configuration
- Data base configuration of inputs and equipment
- Task recipes configuration
- Task automation shortcuts configuration
- New operator register

Use case: Generate maps

- Sampled data files correction
- Primary data layers creation
- Basic data layers creation
- Data layers manual edition

Secondary data layers creation
Scenario simulation execution
Application maps generation
Output files generation for application controllers
Map printing

From this Requirements Model, consisting of the Use Case Model and the Use Case Description, the Object Model diagrams can be developed. Some of them were selected to be presented here on Fig. 2 and 3. After that, the same objects and classes are grouped in subsystems, emphasizing their cohesion in terms of content, occurrence and the macro-functions they are connected to, as shown in Fig. 4.

CONCLUSIONS

The modeling activity allowed to evaluate the importance of ISs for PA, the existing problems, and the requirements to be fulfilled, so that the technology can be successful

Several activities were performed, based on software engineering fundamentals, aiming at obtaining a set of models and descriptions that will hopefully be a conceptual infra-structure for the understanding of requirements, in order to guide future implementations and to help the evaluation of existing systems. Part of these model results were summarized here.

The Use Case Model obtained, despite its simplicity, allowed stating the limits of the system in its environment, and to visualize the basic functions required. The simplicity is a consequence of a series of interactions and of several identified sub-cases groups, and is a virtue of the model. Completing the Requirements Model and closing the pre-analysis phase, the Use Case Description that followed offers a second level of detail of this functionality, and provides a more detailed view of the complex processing needed to support decision making in PA.

In the resulting Object Model, a positive point is that the identified classes are strongly based on the problem domain, and their structure reflects this domain, and not a possible particular implementation. This characteristic provides the model with robustness, facilitating its understanding because of the direct relationship with real world objects. Operating interface aspects could be easily included in the model, concerning the detail level of some descriptions, such as the use cases. However, the inclusion of computational domain object classes would endanger this fundamenting in the problem domain, a characteristic that guarantees the model robustness.

The model contributes for macro-functions (the use cases) identification, which can guide the system development in an incremental way, based on the proposed infra-structure. It also contributes to the identification of additional subsystems besides the ones usually found in the researched systems, and to clarify their roles in the ISs for PA. It provides a wide view on the needs and activities concerning information management in PA, contributing to proprietary systems analysis and to the evaluation of their adequacy to the domain.

The proposed model allows locating the current initiatives of standardization, and their products, and contributes to their use and development.

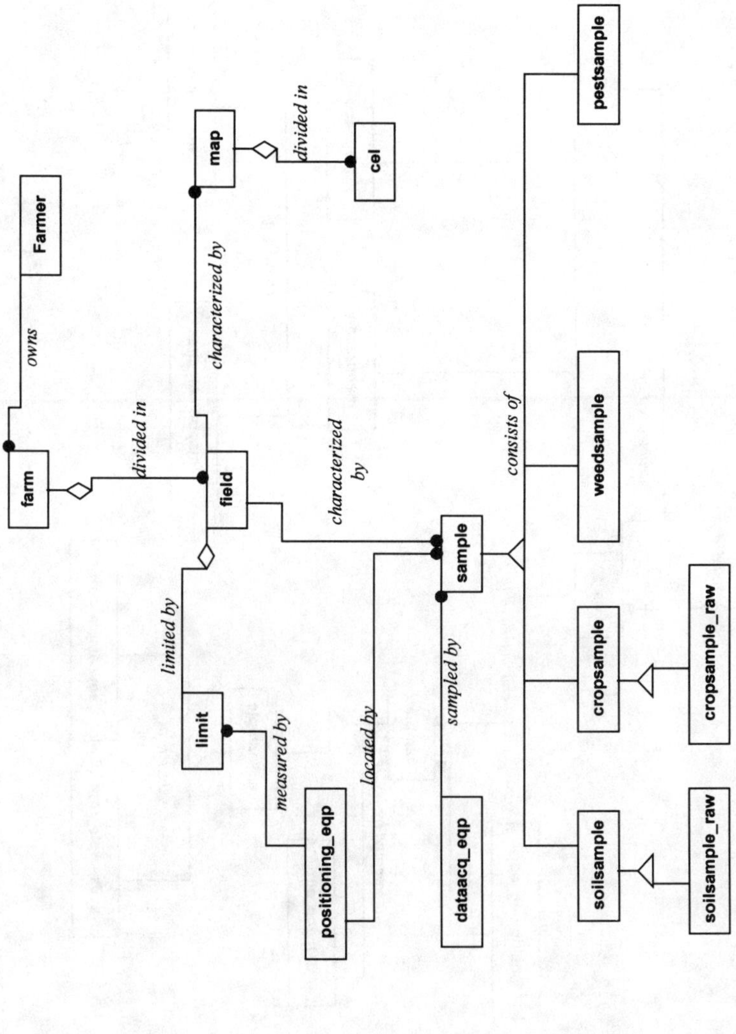

Figure 2 - Class diagram - Use case "Insert field data"

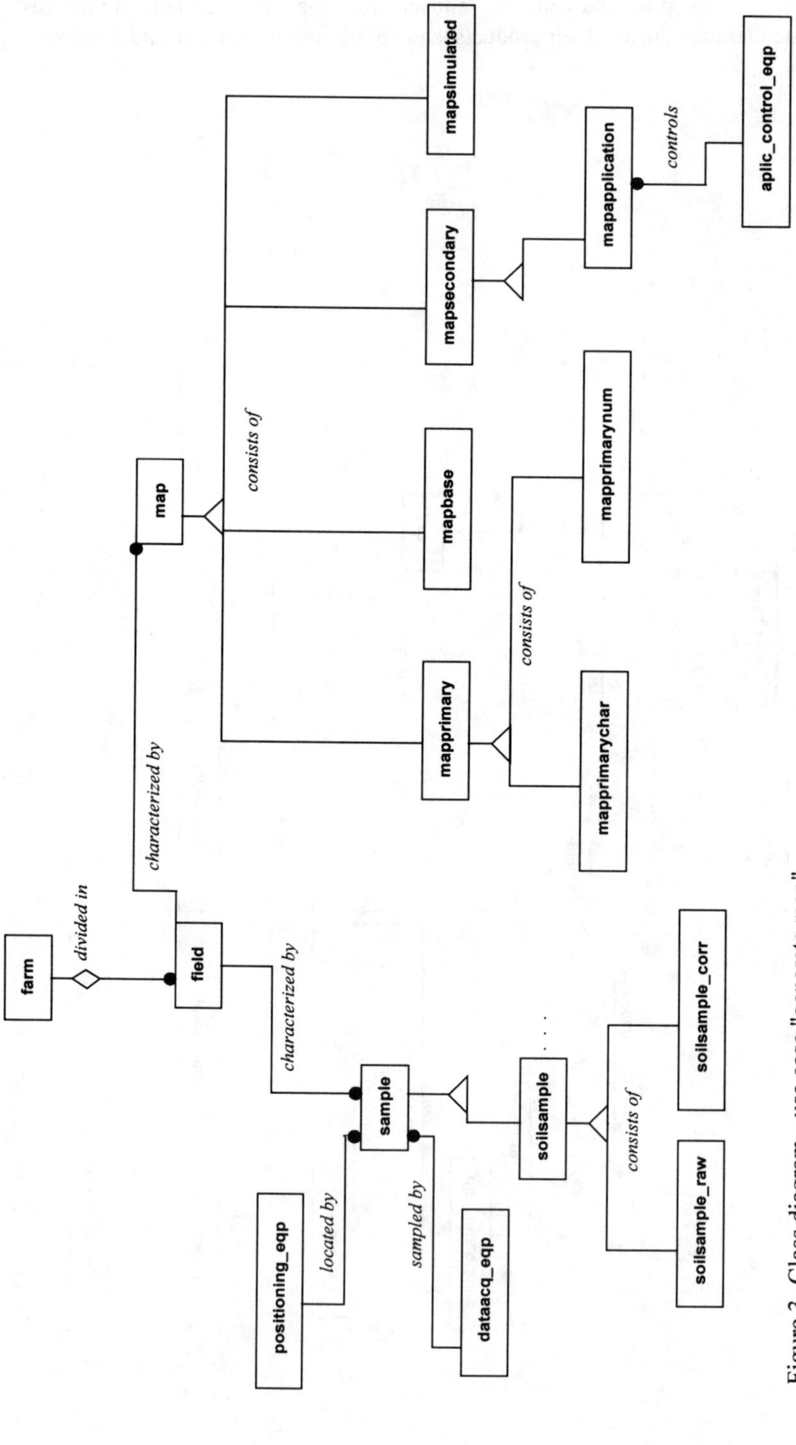

Figure 3- Class diagram - use case "generate map"

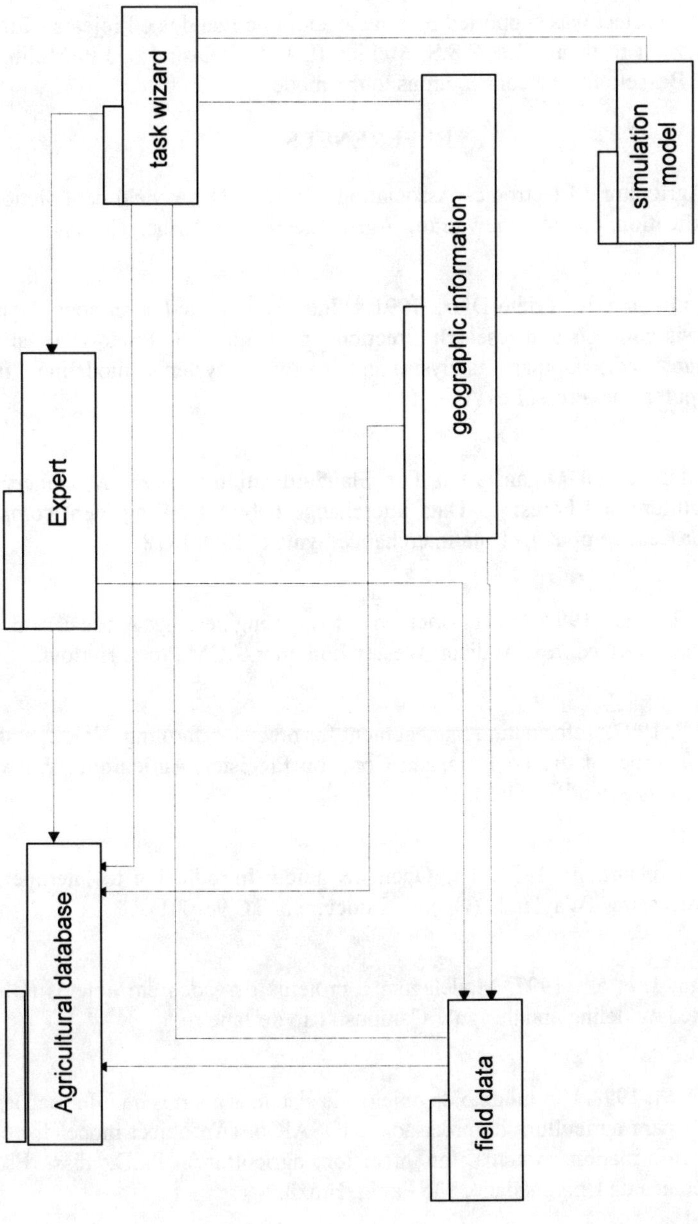

Figure 4 - Subsystems and relationships

ACKNOWLEDGMENTS

This project was supported by Financiadora de Estudos e Projetos - FINEP. We also want to thank Drs. S.S.S. Melnikoff, L.A. Balastreire, J.P. Molin, and Mr. B.A. Basseto for the contributions to the model.

REFERENCES

AEA - Agricultural Electronics Association. 1997. A*E*A yield data dictionary specification: public review draft. Agric. Electronics Assoc., Chicago.

Arango, G., and R. Prieto-Díaz, 1991. Introduction and overview: Domain analysis concepts and research directions. p. 9–32. *In* R. Prieto-Díaz, and G. Arango, (ed.) Domain analysis and software systems modeling. IEEE Computer Soc. Press, Los Alamitos.

ISO - International Organization For Standardization. 1995. Machinery for Agriculture and Forestry - Data interchange between management computer and process computer - Data Interchange Syntax. ISO, 11787.

Jacobson, I. et al. 1993. Object-oriented software engineering: A use case driven approach. 4th ed. rev. Addison Wesley Longman/ACM Press. Harlow.

Jürschik, P. 1997. Information management for precision farming. Vol. 2, p. 477–484. *In* Proc. of the 1st European Conf. on Precision Agriculture, Warwick. 1997. Bios Scientific, Oxford.

OpenGIS Consortium. 1996. The OpenGIS guide: Introduction to interoperable geoprocessing. Wayland. (OpenGIS document TC 96-001).

Rumbaugh, J. et al. 1997. Modelagem e projetos baseados em objetos (Object oriented modeling and design). Campus. Rio de Janeiro.

Saraiva, A.M. 1998. Um modelo de objetos de sistemas abertos de informações de campo para agricultura de precisão - MOSAICo. (An object model for open field information systems for precision agriculture). Ph.D. diss. Escola Politécnica da Universidade, São Paulo, Brazil.

Saraiva, A.M. et al. 1997. Object oriented approach to the development of a field information system. Paper no. 97–3015. ASAE, St. Joseph, MI.

NLEAP Simulation of Soil Type Effects on Residual Nitrate-Nitrogen and Potential Use for Precision Agriculture

Jorge A. Delgado

Soil Plant Nutrient Research Unit
USDA-ARS
Fort Collins, Colorado

ABSTRACT

Well water NO_3^-–N concentrations in the central part of the San Luis Valley had been found as high as 72 mg NO_3^-–N L^{-1}. Precision management practices according to differences in soil type found under a center pivot irrigation sprinkler may have the potential to improve N-use efficiency. No Nitrogen Leaching Economic Analysis Package (NLEAP) model simulation of effects of different soil types and crops on residual soil NO_3^-–N (RSN) has been conducted under a similar irrigation system. Lettuce (*Lactuca sativa* L.), potato (*Solanum tuberosum* L), winter cover rye (*Secale cereale* L.), and winter wheat (*Triticum aestivum* L.) were grown on sandy loam and loamy sand areas of a center-pivot. A new NLEAP 1.2 version simulated soil type and crop effects on RSN ($P < 0.001$). NLEAP is a potential technology transfer tool that can be used to protect water quality by evaluating the effects of crops and soil types on RSN and can potentially be used with precision agriculture management practices.

INTRODUCTION

Stogner (1996) found that NO_3^-–N concentrations in water samples collected from 16 National Water Quality Assessment (NAWQA) sampling wells for the shallow unconfined aquifer in the central region of the San Luis Valley (SLV) of South Central Colorado, ranged from 0.07 to 72 mg NO_3^-–N L^{-1}. Implementation of best irrigation and nutrient management practices are being conducted in the SLV to reduce the movement of NO_3^-–N out of the root zone of planted crops in order to conserve water quality (SLVWQDP, 1994). One of the best management practices being used in the SLV to protect soil and water quality is the inclusion of winter cover crops in crop rotations (Delgado et al., 1998; Delgado, 1998). Winter cover crops can contribute significantly to the conservation of water quality by recovering NO_3^-–N left below the root zone of previous crops that otherwise could be susceptible to leaching (Holderbaum et al., 1990; Brinsfield and Staver, 1991; Decker et al., 1994; McCracken et al., 1994; Delgado et al., 1998).

Computer models are technology transfer tools capable of assessing impacts of agricultural systems on residual soil NO_3^-–N available to leach and movement of NO_3–N out of the root zone. Beckie et al. (1994) found that the Nitrate Leaching and Economic Analysis Package (NLEAP), the Crop Estimation through Resource and Environment Synthesis (CERES), the Erosion–Productivity Impact Calculator

Copyright © 1999 ASA-CSSA-SSSA, 677 South Segoe Road, Madison, WI 53711, USA. *Proceedings of the Fourth International Conference on Precision Agriculture.*

(EPIC), and the Nitrogen Tillage, and Residue Management (NTRM) models performed similarly in estimating NO_3^-–N and water content in the rooting zone of wheat. Khakural and Robert (1993) found that simulated NO_3^-–N leaching values were correlated to measured NO_3^-–N leaching values for NLEAP or LEACHM-N. Other investigators have also found NLEAP to be a useful tool in evaluating the effect of agricultural practices on residual soil NO_3^-–N (Shaffer et al., 1991; Follett et al., 1994; Shaffer et al., 1995). NLEAP version 1.10 simulated residual soil NO_3^-–N in the root zone and NO_3^-–N leaching from the bottom of the root zone (Shaffer et al., 1991; Khakural & Robert, 1993; Beckie et al., 1994; Follett et al., 1994; Shaffer et al., 1995). Delgado (1998) used a new NLEAP version 1.2 (Shaffer et al., 1998) to evaluate N dynamics in a lettuce–winter cover rye–potato rotation.

This manuscript presents initial results of a technology transfer effort conducted by USDA-ARS and USDA-NRCS-SLVWQDP. Results from three of 25 sites where multiple years of data have been collected for calibration–validation of NLEAP are presented in this manuscript. Different crops with significant rooting depth changes were studied. A new version of NLEAP capable of evaluating not only the residual soil NO_3^-–N on the root zone of each crop, but also the residual soil NO_3^-–N based on a base line layer similar for winter wheat, lettuce, winter cover rye, and potato systems was needed.

The new NLEAP 1.2 version can simulate the residual soil NO_3^-–N for three layers, 0 to 0.3 m, 0.3 m to bottom of the root depth, and bottom of the root depth to a base line (Shaffer et al., 1998). This base line can be set from a minimum of 0.3 m to a maximum of 1.5 m, by 0.03-m increments. The base line also can be set to be equal to the bottom of the root depth. The 1.2 version also simulates NO_3^-–N leaching from the bottom of the root depth and from the base line.

Another new feature in the NLEAP 1.2 version is that the maximum root depth can be entered to the nearest 0.03-m, from a minimum root depth of 0.3 m to a maximum of 1.5 m (Shaffer et al., 1998). Previously for the NLEAP 1.10 version, maximum root depth was entered to the nearest 0.3 m increment, (e.g., 0.3, 0.6, 0.9, 1.2, or 1.5 m). The last new feature of the NLEAP 1.2 version is that it can simulate for a single year two crops with different rooting depths.

This new 1.2 version of NLEAP allowed the evaluation of the effects of management practices on residual soil NO_3^-–N of a winter wheat, lettuce-winter cover rye rotation and potato grown under a loamy sand and sandy loam. The objective of this study was to test the potential use of the NLEAP 1.2 model to assess the effect of different soil types on residual soil NO_3^-–N and the potential application of NLEAP as a technology transfer tool for precision agriculture.

MATERIALS AND METHODS

Field Sites

The studies were conducted in three farmer's field located about 10 km north and 20 km south east of Center, CO. At each farmer's field, four plots 20.9 m^2 each were established in a center-pivot irrigation sprinkler on a sandy loam. Four additional plots also were established in the same center-pivot irrigation sprinkler, but on a loamy sand. For site one the soils were McGinty sandy loam (Coarse-loamy, mixed, frigid, Typic Calciorthids) and Gunbarrel loamy sand (Mixed, frigid, Typic Psammaquents) (USDA-SCS, 1973). For site two and three we located plots in a San Luis sandy loam (Fine-loamy over sandy or sandy skeletal, mixed, Aquic Natrargids). The loamy sand used in sites two and three was a Kerber loamy sand (Coarse-loamy, mixed, frigid Aquic Natrargids) (USDA-SCS, 1984). At each site, two transponders were placed permanently in the sandy loam and loamy sand plots so that soils could be sampled during the length of the study. Plots were located under the same sprinkler span to minimize spatial variability due to differences in sprinkler nozzles.

Information about Climate and Management Practices

Climatic data from the nearest weather station in Center, CO (10–20 km from our three sites) were collected. During the growing seasons, rain and/or snow amounts were measured locally at both sites. Potential evapotranspiration (E_{tp}) was entered into the NLEAP model using the modified Jensen-Haise (JH) method (Follett et al., 1973; Jensen et al., 1990). Air temperature, monthly rain and irrigation amounts are shown in Fig. 1, 2, and 3, respectively.

On all three sites the management practices such as irrigation, N fertilizer application, planting, harvesting, cultivation, and other agricultural practices were similar for the sandy loam and loamy sand. Table one shows the crops studied at each site with their growing seasons and selected N inputs. On site two, the scavenger winter cover rye was not fertilized. Center pivot sprinklers were calibrated for accuracy. Irrigation water samples were collected three times during the growing season and analyzed for NO_3^-–N.

Plant and Soil Samples

Prior to harvesting, plant parameters such as biomass yield of crops were collected. Plant samples for winter wheat and winter cover rye were collected by

Table 1.Crop, growing season, and selected N inputs.

Site	Crop	Planted	Harvested	DNF	NF	BN
				--------	kg ha^{-1}†	------
One	Winter Wheat	9/18/1993	8/ 9/1994	0	202	95
Two	Lettuce	5/18/1994	8/ 4/1994	49	104	24
	Winter Cover Rye	8/ 8/1994	4/11/1995	0	0	10
Three	Potato	5/15/1994	9/15/1994	101	131	49

† DNF, Dry N fertilizer applied; NF, N applied in fertigations; BN, background N content in irrigation water.

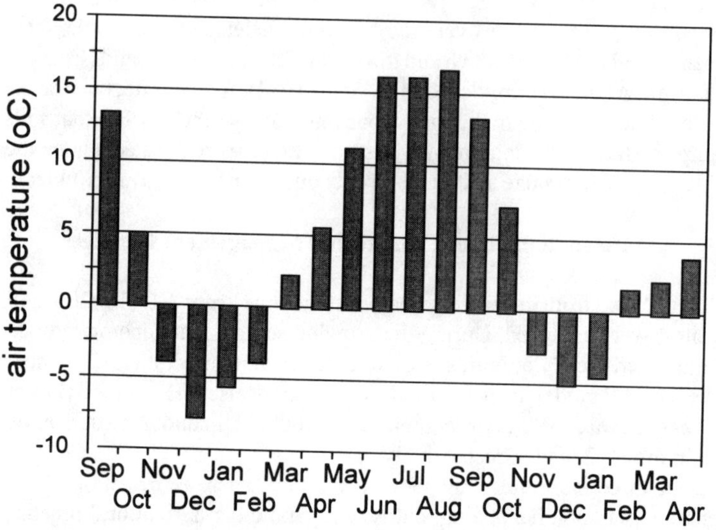

Fig. 1. Monthly air temperature for Sites 1, 2, and 3 starting September (Sep) 1993.

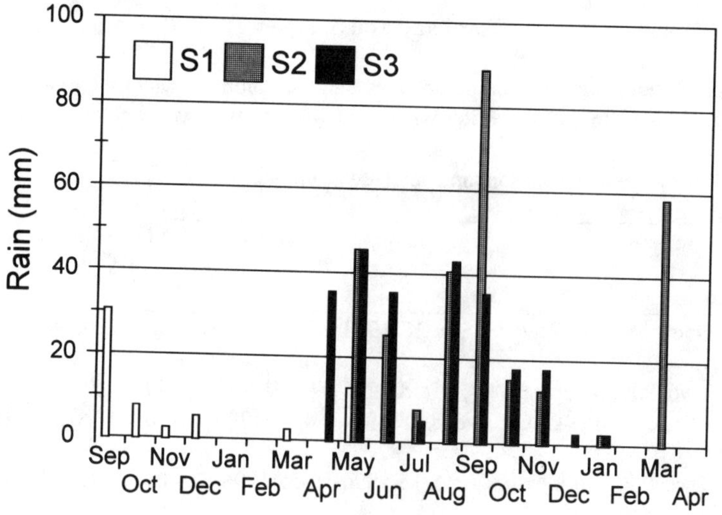

Fig. 2. Monthly rain for Sites 1 (S1), 2 (S2), and 3 (S3) starting September (Sep) 1993.

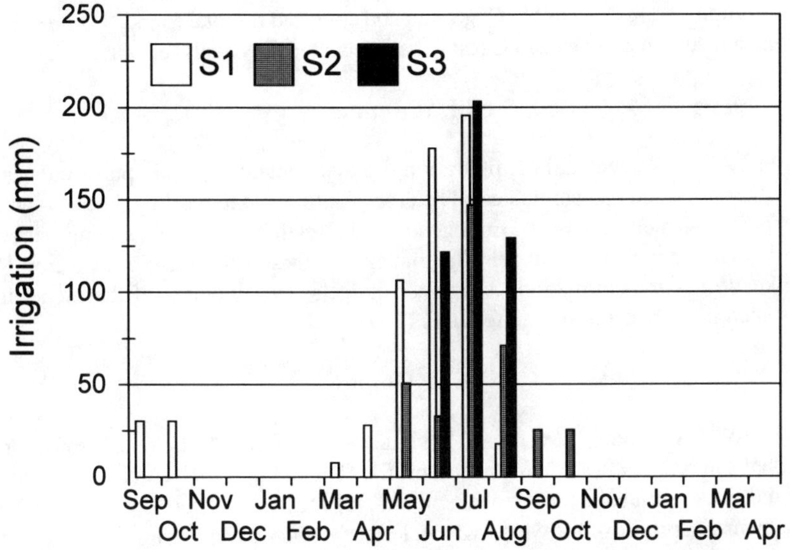

Fig. 3. Monthly irrigation for sites 1 (S1), 2 (S2), and 3 (S3) starting September (Sep) 1993.

harvesting 0.4 m^2 in each plot. Aboveground rye was collected before spring kill in 1995. During the fall of 1994 the farmers harvested the lettuce and potato plot areas prior to sampling, therefore lettuce and potato samples collected outside the plots were used. For potato and lettuce, yield and N concentrations were assumed to be equal for sandy loam and loamy sand plots. The mean bottom of the root depth was 0.4 m for potato, 0.39 m lettuce, and 0.9 m for winter wheat and winter cover rye.

To monitor the effect of management practices on residual soil NO_3^-–N, soil samples were collected in each plot as follows: (i) at site one, in fall of 1993 before winter wheat planting and in fall of 1994 after harvesting; (ii) at site two in spring of 1994, before lettuce planting and summer of 1994, after lettuce harvest; (iii) at site two, in spring of 1995 before winter cover rye kill; (iv) at site three, before potato planting and in fall of 1994 after harvest. At each time, soils were sampled in 0.3-m intervals to 1.5 m. Soil measurements included: % coarse fragments by weight and by volume, % organic matter, pH, cation-exchange capacity, and water content.

Soil samples collected from each 0.3-m depth increment were air dried, and sieved through a 2 mm sieve. The percentage weight of the coarse fragments was used to calculate the percentage coarse fragments by volume (Delgado et al., 1996). Bulk densities were estimated from texture as described by USDA-SCS (1988). Sieved samples were extracted with 2*N* KCl. The NO_3^-–N and NH_4^+–N contents of the extracts were determined colorimetrically by automated flow injection analysis. Plant samples were dried at 55 °C, ground and analyzed for total C and N content by automated combustion using a Carlo Erba automated C/N analyzer.

NLEAP Inputs

NLEAP uses a regional configuration file that contains crop N uptake indices, and other plant and soil parameters. The crop planting and harvesting dates, N- and water- management inputs and timing, soil and climate information, and measured yields were entered in the model. The model uses these values to simulate crop N uptake, soil N transformations and water budgets. Development of the model is presented in more detail in Shaffer et al. (1991).

NLEAP Outputs

The NLEAP model 1.2 was used to simulate the effects of crop management on residual soil NO_3^-–N in a 0 to 0.9 m soil profile. The initial soil NO_3^-–N was entered and the model simulated the residual soil NO_3^-–N for the three soil layers within the 0 to 0.9 m soil profile of the winter wheat, lettuce–winter cover rye, and potato crops grown in the loamy sand and sandy loam; 0 to 0.3 m, 0.3 m to bottom of the root depth, and bottom of the root depth – to 0.9 m.

Statistical Analyses

Statistical analyses were performed using the SAS analysis of variance GLM procedure (SAS Institute, 1988). Differences in dry matter and N uptake between the winter wheat and winter cover rye grown in the sandy loam and loamy sand were conducted with the PROC GLM. Differences between the sandy loam and loamy sand in residual soil NO_3^-–N at planting and after each cropping sequence were also tested with PROC GLM.

NLEAP simulated values after winter wheat, lettuce, winter cover rye, and potato harvests, for the sandy loam and loamy sand were correlated to observed values. Correlations between predicted *vs* observed residual soil NO_3^-–N were conducted using SAS REG procedure. The intercept (b_0) and slope (b_1) were tested with SAS REG for differences from 0 and 1, respectively. The following three regression tests were conducted between predicted *vs* observed residual soil NO_3^-–N data set: (i) that include only the means of the plots where we have the aboveground plant production (Site one, winter wheat and site two winter cover rye); (ii) that includes only the means of the plots where we don't have the aboveground plant production and we used plant samples from outside the plots (site two lettuce and site three potato); and (iii) that includes all the means of sites one, two, and three.

RESULTS AND DISCUSSION

Plant and Soil Samples

Nitrogen uptake by grain (winter wheat) of 169 kg N ha^{-1} was significantly higher in the sandy loam than the 139 kg N ha^{-1} in the loamy sand ($P < 0.01$)(Table 2). Total N uptake by aboveground biomass in the sandy loam (227 kg N ha^{-1}) was higher than

Table 2. Nitrogen uptake and yield for aboveground biomass† compartments.

Site‡	Crop	Soil type	Grain	Stk&Cha	Total	Yield§
			------------	kg N ha^{-1}	----------	Mg ha^{-1}
One	WW	Loamy Sand	139	60	199	6.9
	WW	Sandy Loam	169 **	58	227 **	8.7 *
Two	WCR	Loamy Sand			68	3.2
	WCR	Sandy Loam			134 **	5.2**

† Compartments are grain, stalk and chaff (Stk&Cha), and total aboveground biomass for winter wheat (WW); and winter cover rye (WCR).

‡ Within a site, *, ** show significant differences at $P < 0.05$, 0.01, respectively between the loamy sand and sandy loam.

§ Yield for WW is expressed as grain yield with 12% moisture. Yield for WCR is expressed as aboveground dry weight.

in the loamy sand (199 kg N ha^{-1}) ($P < 0.01$). Grain yield in the sandy loam (8.7 Mg ha^{-1}) was higher than (6.9 Mg ha^{-1}) in the loamy sand ($P < 0.05$). Nitrogen uptake by aboveground winter cover rye biomass of 134 kg N ha^{-1} was higher in the sandy loam than the 68 kg N ha^{-1} in the loamy sand ($P < 0.01$). Aboveground winter cover rye dry matter biomass production in the sandy loam (5.2 Mg ha^{-1}) was higher than (3.2 Mg ha^{-1}) in the loamy sand ($P < 0.01$).

For site one, in fall of 1993, initial soil NO_3^-–N of 86 kg NO_3^-–N ha^{-1} in the sandy loam was higher than the 74 kg NO_3^-–N ha^{-1} of the loamy sand ($P < 0.07$)(Table 3). After winter wheat harvest, the 76 kg NO_3^-–N ha^{-1} retained in the top 0.9 m of the finer sandy loam was higher than the 63 kg NO_3^-–N ha^{-1} in the loamy sand ($P < 0.05$). For site two, initial soil NO_3^-–N of 113 kg NO_3^-–N ha^{-1} in the sandy loam was higher than the 84 kg NO_3^-–N ha^{-1} of the loamy sand ($P < 0.01$). After lettuce harvest, the 164 kg NO_3^-–N ha^{-1} retained in the top 0.9 m of the finer textured sandy loam was higher than the 99 kg NO_3^-–N ha^{-1} retained in the loamy sand ($P < 0.05$). There were no significant differences in residual soil NO_3^-–N after winter cover rye kill. Similar responses were observed for site three, where the sandy loam had higher initial ($P < 0.01$) and higher final ($P < 0.05$) residual soil NO_3^-–N than the loamy sand.

The data suggest that under similar management, and similar amounts of N fertilizer applications and irrigation, the sandy loam will be a higher N use efficiency system than the loamy sand (Tables 2 and 3). These are important facts that can be applied with precision agriculture management practices to protect water quality.

The winter cover rye reduced the residual soil NO_3^-–N in the 0- to 0.9-m soil depth, by 148 kg N ha^{-1} in the sandy loam, higher than the reduction of the 88 kg N ha^{-1} in the loamy sand ($P < 0.05$). The scavenger winter cover rye showed the benefit and potential to protect water quality under this spatial variability due to soil type and residual soil NO_3^-–N within this center-pivot irrigation sprinkler.

Table 3. Soil NO_3^-–N content for the 0 to 0.9 m soil layer.

Site†	Crop‡	Soil type	Initial kg ha^{-1}	Final kg ha^{-1}
One	WW	Loamy Sand	74	63
	WW	Sandy Loam	86 ($P < 0.07$)	76 *
Two	Lettuce	Loamy Sand	84	99
	Lettuce	Sandy Loam	113 **	164 *
	WCR	Loamy Sand	99	11
	WCR	Sandy Loam	164 *	16
Three	Potato	Loamy Sand	110	106
	Potato	Sandy Loam	233 **	227 *

† Within a site, *, ** show significant differences at $P < 0.05$, 0.01, respectively between the loamy sand and sandy loam.
‡ Winter wheat (WW) and winter cover rye (WCR).

NLEAP Outputs

The NLEAP model predicted effects of management practices on residual soil NO_3^-–N ($P < 0.01$). A regression analysis of predicted *vs.* observed values was significant for the winter wheat and winter cover rye data set ($y = 1.9 + 1.2\,x$; $r^2 = 0.85$, $P < 0.07$). A regression analysis of predicted *vs.* observed values was significant for the lettuce and potato data set ($y = 2.0 + 1.1\,x$; $r^2 = 0.68$, $P < 0.01$). A regression analysis of predicted *vs.* observed values for the whole data set was significant ($y = 2.9 + 1.1\,x$; $r^2 = 0.77$, $P < 0.001$; Fig. 4). The b_0 and b_1 were not significantly different from zero and one, respectively for these three regression lines.

Although it is expected that the soil type properties and N use efficiency will not be uniform through the whole center-pivot area, these NLEAP simulations show the importance of soil type to simulate residual soil NO_3^-–N in an irrigated system (Fig. 4; Tables 1, 2, and 3). NLEAP 1.2 was able to simulate the trends in residual soil NO_3^-–N for the root system and below the root systems of these crops grown in irrigated sandy soils of the San Luis Valley (Fig. 4). We assumed that this NLEAP 1.2 version can describe the transport of NO_3^-–N between layers and the movement of NO_3^-–N out of the root zone.

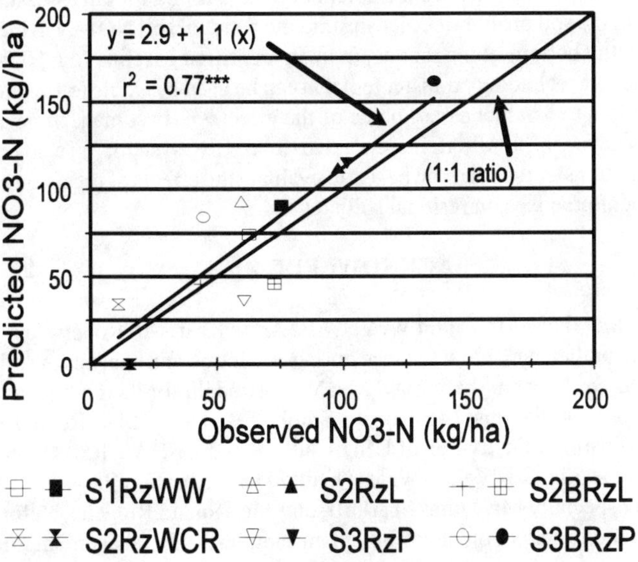

Fig. 4. Predicted *vs* observed residual soil NO_3^-–N for Sites 1 (S1), 2 (S2), and 3 (S3) for the root zone (Rz) and below the root zone (BRz) of winter wheat (WW), potato (P), lettuce (L) and winter cover rye (WCR). Solid symbols are for sandy loam and empty symbols are for loamy sand.

These simulations show the importance of using the bottom of the root depth and a similar base line for the whole system. This is necessary since winter cover rye can contribute significantly to the conservation of water quality by recovering NO_3^-–N remaining below the root zone of previous crops that otherwise could be susceptible to leaching (Fig. 4; Table 3).

Schaffer and Xu (1998) presented different flow charts showing the potential of linking different simulation models such as NLEAP to farm and/or yield data collected with global position systems (GPS). Their flow charts described how NLEAP simulations can be linked to GIS data layers to evaluate the yield responses and environmental impacts. Results from Fig. 4 show that NLEAP has the capability to evaluate the effects of soil type within a similar irrigation system. NLEAP simulated different fertilizer and irrigation management practices, crops and soils effects on residual soil NO_3^-–N for the root zone and below the root zone. These results suggest that if linked to a GIS system, NLEAP is a technology transfer tool capable of evaluating the effect of precision farming practices on residual soil NO_3^-–N transport in the soil profile. NLEAP can potentially be used with precision agriculture to increase N use efficiency and protect water quality.

CONCLUSIONS

The NLEAP 1.2 model has the potential to be a useful tool in assessing NO_3^-–N transport in the soil profile since it can simulate residual soil NO_3^-–N in the root zone and below the bottom of the root depth to an identified baseline (e.g., 0.9 m) depth. This model is a technology transfer tool that can be used by extension agents, farmers and educators to conduct assessments of the effects of best management practices on N budgets. NLEAP, if incorporated to a GIS system, is also a potential technology transfer tool that can be use to evaluate the effects of precision agriculture management practices on residual soil NO_3^-–N .

ACKNOWLEDGMENTS

The author thank Mr. David Wright, Ms. Anita Kear, Mr. Robert Lober, and Mr. Kevin Lee for their capable assistance during collection and analysis of soil and plant samples; to Dr. Marvin Shaffer and Ms. Mary Brodahl for their advice and support with NLEAP simulations; Mr. Donald Smokey Barker and Dr. Ronald Follett for their coordination of activities of USDA/NRCS and USDA/ARS personnel in this project; to the San Luis Valley Water Quality Demonstration Project (SLVWQDP) personnel especially Mr. James Sharkoff, and Mr. Ronald Riggenbach for technical assistance in the collection of soil and plant samples, and management information.

REFERENCES

Beckie, H.J., A.P. Moulin, C.A. Campbell, and S.A. Brandt. 1994. Testing effectiveness of four simulation models for estimating nitrates and water in two soils. Can. J. Soil Sci. 74:135–143.

Brinsfield, R.B., and K.W. Staver. 1991. Use of cereal cover crops for reducing groundwater nitrate contamination in the Chesapeake Bay region. p. 79–82. *In* W.L. Hargrove (ed.) Cover crops for clean water. Soil Water Conserv. Soc. Ankeny, IA.

Decker, A.M., A.J. Clark, J.J. Meisinger, F.R. Mulford, and M.S. McIntosh. 1994. Legume cover crop contributions to no-tillage corn production. Agron. J. 86:126–135.

Delgado, J.A., M.K. Brodahl, M.J. Shaffer, R.F. Follett, and J.L. Sharkoff. 1996. A list of definitions to consider when using the NLEAP model to evaluate N management practices in soils containing coarse fragments. Nitrogen Management NLEAP Facts Sheet 5/96. USDA-ARS and USDA-NRCS. USDA-ARS-SPNRU, Ft. Collins, CO.

Delgado, J.A. 1998. Sequential NLEAP simulations to examine effect of early and late planted winter cover crops on nitrogen dynamics. J. Soil Water Conserv. 53:(3) *(In print)*

Delgado, J.A., R.T. Sparks, R.F. Follett, J.L. Sharkoff, and R.R. Riggenbach. 1998. Use of winter cover crops to conserve soil and water quality in the San Luis Valley of South Central Colorado. *In* R. Lal (ed.) Soil quality and erosion. CRC Press, Boca Raton, FL. *(In press)*.

Follett, R.F., G.A. Reichman, E.J. Doering, and L.C. Benz. 1973. A nomograph for estimating evapotranspiration. J. Soil Water Conserv. 28(2):90–92.

Follett, R.F., M.J. Schaffer, M.K. Brodahl, and G.A. Reichman. 1994. NLEAP simulation of residual soil nitrate for irrigated and non irrigated corn. J. Soil Water Conserv. 49:375–382.

Holderbaum, J.F., A.M. Decker, J.J. Meisinger, F.R. Mulford, and L.R. Vough. 1990. Fall-seeded legume cover crops for no-tillage corn in the humid east. Agron J. 82:117–124.

Jensen, M.E., R.D. Burman, and R.G. Allen (ed.) 1990. Evapotranspiration and irrigation water requirements. ASCE Manuals and Rep. on Eng. Practice no. 70. Am. Soc. of Civil Eng., New York.

Khakural, B.R., and P.C. Robert. 1993. Soil nitrate leaching potential indices: Using a simulation model as a screening system. J. Environ. Qual. 22:839–845.

McCracken, D.V., M.S. Smith, J.H. Grove, C.T. MacKown, and R.L. Blevins. 1994. Nitrate leaching as influenced by cover cropping and nitrogen source. Soil Sci. Soc. Am. J. 58:1476–1483.

San Luis Valley Water Quality Demonstration Project (SLVWQDP). 1994. Best management practices for nutrient and irrigation management in the San Luis Valley. Colorado State Univ. Coop. Ext., Fort Collins, CO.

SAS Institute. 1988. SAS/STAT users guide. Ver 6.03. 3rd ed. SAS Inst., Cary, NC.

Shaffer, M.J., A.D. Halvorson, and F.J. Pierce, 1991. Nitrate leaching and economic analysis package (NLEAP): Model description and application. p. 285–322. *In* R.F. Follett et al., (ed.) Managing nitrogen for groundwater quality and farm profitability. SSSA, Madison, WI.

Shaffer, M.J., B.K. Wylie, and M.D. Hall. 1995. Identification and mitigation of nitrate leaching hot spots using NLEAP-GIS technology. J. Contaminant Hydrol. 20:253–263.

Schaffer, M.J., M.K. Brodhal, and J.A. Delgado. 1998. NLEAP Software version 1.2. USDA-ARS-GPRSU, Fort Collins, CO.

Schaffer, M.J., and C. Xu. 1998. Simulation modeling in precision agriculture: Soil fertility p. 55–60. *In* A.J. Schlegel (ed.) Proc. Great Plains Soil Fertility Conf., Denver, CO. 3–4 Mar., 1998. Kansas State Univ., Manhattan, KS.

Stogner, R.W. 1996. Nitrate monitoring of the shallow unconfined aquifer in the San Luis Valley. p. 123. *In* 13th Annual Potato/Grain Conf., Colorado State Univ. Coop. Ext., Fort Collins, CO.

USDA-SCS. 1973. Soil Survey of Alamosa Area, Colorado. USDA-SCS, Washington DC.

USDA-SCS. 1984. Soil Survey of Saguache County Area, Colorado. USDA-SCS, Washington DC.

USDA-SCS. 1988. National Agronomy Manual. 2nd ed., USDA-SCS, Washington DC.

Predicting the Effect of Nitrogen Deficiency on Crop Growth Duration and Yield

Upendra Singh and Paul Wilkens
Research and Development Division
International Fertilizer Development Center
Muscle Shoals, Alabama

Victor Chude, Sylvester Oikeh
Department of Soil Science
Ahmadu Bello University
Zaria, Nigeria

ABSTRACT

Manipulating agricultural systems for ecological, economic and agricultural gains is becoming increasingly important to agricultural policymakers and planners in both developed and developing countries and to the national agricultural commodity and trade industry. Systems tools such as crop growth simulation models are an important component in satisfying the above requirement. Crop growth duration is among the major determinants of yield. Existing crop growth simulation models have reliably predicted effects of temperature and photoperiod on crop duration. However, these models generally do not consider the effects of extremely high or low temperatures, drought stress, and nutrient deficiencies on crop duration. Drought stress and deficiencies of N and P during the vegetative phase have resulted in delayed tassel initiation and silking. Similarly, stresses during the ripening phase have resulted in early senescence and maturity. Thus, harvesting, yield forecasting, and planting of the following crop in a sequence, livestock rearing or fisheries would be influenced. A modified model that simulates the effect of N deficiency on phyllochron and phenological stages is presented. Simulation results are compared with field trials from Nigeria, Hawaii and Florida for tropical maize. Simulation studies showing importance of timely N management on yield and risk avoidance are presented.

INTRODUCTION

Simulation models of crop and soil processes are powerful tools because they encapsulate our understanding of the biophysical processes that underpin crop production; they serve to integrate existing scientific knowledge in a way that can be used to draw out additional understandings; and they can be used for priority setting by decisionmakers. These decision support tools are indeed indispensable for natural resource management and ecoregional studies. Increased degradation of the resource base in developing countries and excessive use of pesticides, irrigation and fertilizers in developed countries is making agricultural production unsustainable. Timely and accurate crop yield forecasts are becoming increasingly important to agricultural

Copyright © 1999 ASA-CSSA-SSSA, 677 South Segoe Road, Madison, WI 53711, USA. *Proceedings of the Fourth International Conference on Precision Agriculture.*

policymakers and planners in both developed and developing countries. Systems based effort is required to understand, predict, and manipulate outcomes from agricultural systems for ecological, agricultural, and economic gains.

The potential biomass yield of a crop can be thought of as the product of the rate of biomass accumulation times the duration of growth. The rate of biomass accumulation is dependent on the amount of radiation intercepted by the crop canopy and its efficiency of use in grain production. The development and maintenance of green leaf-area are important determinants of the proportion of the incident radiation intercepted (Muchow & Carberry, 1989). Phenological characteristics establish the pattern of leaf canopy development and maintenance, and the temporal framework within which biomass is distributed to various plant parts. The duration of growth for a particular cultivar is highly dependent on its thermal environment and photoperiod. Hence it is not surprising that the phenology of a crop is one of the major determinants of yield (Rabbinge et al., 1993). Accurate modeling of crop duration and growth are necessary for a reasonable yield prediction.

The essence of genotype by environment interaction is primarily associated with timing of phenological events and favorable environmental conditions (rainfall, temperature, radiation and photoperiod). Optimum timing of irrigation, fertilizer and pesticides is also dictated by a crop's phenology. In a farming system any delay or enhancement in phenology affects the above management inputs and harvest operation. As a consequence, the following activity—cropping, livestock or fisheries—is also influenced and in some instances such activities may not be possible at all. Thus, in a whole farm context the ability to accurately predict crop growth duration is the principal requirement of a reliable crop growth simulation model.

This paper deals with the use of decision support systems for developing country agriculture with the emphasis on the ability of the systems to predict crop growth duration. The effect of temperature, photoperiod, and nitrogen deficiency on crop development is discussed. A modification to the CERES-Maize model to simulate the effect of N deficiency on crop duration is presented. Examples from Nigeria, Hawaii, and Florida are used for validation and model application.

CROP DEVELOPMENT

As an approximation it is useful to consider crop development as independent of growth. The phasic development or phenology is an ordered sequence of processes, which take place over a period of time, punctuated by discrete events, such as sowing, emergence, floral initiation, anthesis and maturity. It is implicitly assumed that the plant, or part of the plant, possesses a development clock which proceeds at given rate constant (d^{-1}) for each of the above phases (Thornly & Johnson, 1990). Variation between modern annual cultivars within a species is usually most evident in the duration of growth, duration of the vegetative and reproductive phases, and the least in the rate of growth. Growth and development processes have different degrees of variation and sensitivity to environmental and management factors (Table 1). Crop development has two distinctly unique features -- phasic and morphological development. Phasic development has greater degree of genotypic diversity than morphological development (Table 1).

Table 1. Factors influencing plant growth and development processes in CERES models (adapted from Ritchie, 1991).

	Growth		Development	
	Mass	Expansion	Phasic	Morphological
Environmental Factor	solar radiation	temperature	temperature photoperiod	temperature
Degree of variation among cultivars	low	low	high	low
Sensitivity to plant water deficit	low- stomata moderate- leaf rolling and wilting	high- vegetative phase low- grain filling stage	low - delay in vegetative stage	low
Sensitivity to N and P deficiency	moderate	high	low	low - main stem high - tillers and branches

Effect of Temperature and Photoperiod

The effect of temperature on crop development rate has been studied extensively by several researchers (Gilmore & Rogers, 1958; Arnold, 1959; Tollenaar et al., 1979; Swan et al., 1987; Ritchie & NeSmith, 1991). Because a plant's time scale is closely coupled with thermal environment, it is appropriate to think of thermal time as plant's view of time. Likewise, if the photoperiod is used to modify thermal time, the resulting term is photothermal time. Thermal time accumulated per day (DTT) is simply mean air temperature minus the base temperature at which development stops. The base temperature (T_b) of maize in the CERES-Maize model is 8°C. The above DTT calculation is appropriate if (i) temperature response of the development rate is linear over the range of temperature experienced; (ii) the daily air temperature (T_a) does not fall below T_b; (iii) the daily temperature does not exceed an upper threshold temperature for a significant part of the day, for example 34°C in the maize model; and (iv) the growing region of the plant has the same mean temperature as T_a (Ritchie et al., 1998). The CERES-Maize model employs alternative methods when one of above assumptions are not correct. For example soil temperature is used up to the tenth leaf stage or tassel initiation instead of air temperature when the growing point is near the ground (Singh, 1985).

The duration of the juvenile phase, leaf-tip appearance rate, and the grain-filling duration in the CERES-Maize model is thermal time dependent. The cultivar differences in these growth stages are accommodated by cultivar-specific thermal time requirement coefficients. For maize the measured phyllochron (thermal time needed for leaf-tip appearance) is in the range of 38 to 45°C per leaf-tip appearance using base temperature of 8°C (Hesketh & Warrington, 1989). The phyllochron is also an input to the CERES-Maize model.

The length of the day (or night) can influence the rate at which maize plant switches from vegetative growth to reproductive growth. The threshold photoperiod below which there is no further photoperiod sensitivity in maize—a short day plant—is about 12.5 h. The rate of change of development, days delay per hour of photoperiod longer than 12.5 h, is expressed as cultivar specific coefficient (P2), in the CERES-Maize model (Ritchie et al., 1998). In addition to the delayed tassel initiation, photoperiod of > 12.5 h also resulted in increased number of leaves. The increased thermal time accumulation with the photoperiod delay (photothermal time) results in more leaves.

Model Performance

The comparison of simulated and observed phenological events using the CERES-Maize model has been done by several researchers (Singh, 1985; Ritchie et al., 1989; Singh et al., 1989, 1993; Elings et al., 1996). As evident from Fig. 1 the model in general was capable of reliably simulating days to silking and maturity for a wide range of environment and genotype. However, some of the simulated and observed days to silking and maturity differed by as much as 15 d. Under water- and nutrient - nonlimiting conditions such difference could translate to up to 3 t ha^{-1} dry matter production.

In the above example, field results were for irrigated maize with different N rates (0–200 kg N ha^{-1}). N stress slightly delayed the appearance of leaves, however, due to delayed silking, the final number of leaves were quite similar at low and high N rates (Bennett et al., 1989). Associated with the delayed silking the anthesis to silking interval (ASI) increased with N stress. It is well documented that drastic reduction in grains per ear occurs as the ASI goes above 5 d (Elings et al., 1996). Modifications were made to the CERES-Maize model to simulate the effect of N stress on crop development.

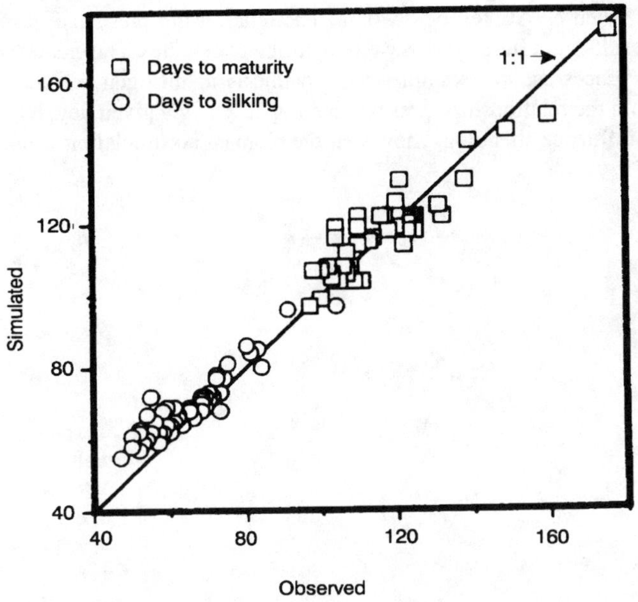

Fig. 1. Performance of CERES-Maize—without phenology modification—on days to silking and maturity (adapted from Singh, 1985).

SIMULATING NITROGEN RESPONSE

The current version of the CERES-Maize model, which is distributed in the DSSAT package (Tsuji et al., 1995) simulates N response on growth, leaf area, leaf weight, stem weight, grain yield and yield components (grain weight and grains per ear). The model uses two sets of N stress indices as a function of N deficiency factor to simulate the effect of N stress on leaf expansion and photosynthesis. The N deficiency factor (NFAC) is estimated as the ratio of N supply (from soil and fertilizers) to plant N demand. A complete description of the N model has been presented by Godwin and Singh (1998).

Effect of Nitrogen Stress on Phenology

The delay in phasic development prior to silking was modeled with severe N stress resulting in decreased rate of thermal time accumulation. For the same degree of N stress, the delay was more pronounced during the reproductive phase (tassel initiation to silking) with N deficiency factor (NFAC) below 0.5 resulting in reduced rate of thermal time accumulation (Fig. 2). The effect was manifested during the vegetative phase (emergence to tassel initiation) only under extremely severe conditions when NFAC fell below 0.38. The relationship presented in Fig. 2 was based on results from field trials as well as greenhouse studies. The greater sensitivity of growth processes, for example, photosynthesis to nitrogen deficiency is also evident from the relationships presented in Fig. 2. On any given day, NFAC of 0.6 would not influence phenology, however, the biomass accumulation would occur at 85% rate.

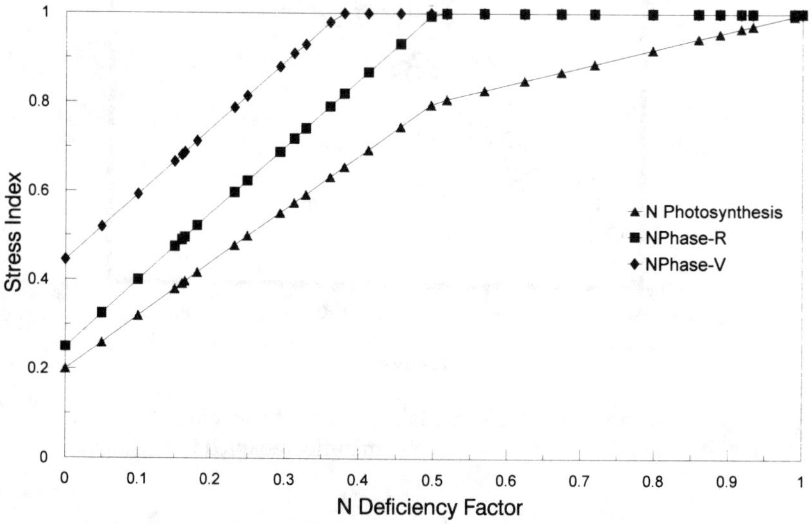

Fig. 2. Relationship between plant N deficiency factor and stress indices for growth (N photosynthesis), vegetative stage (NPhase-V) and reproductive stage (NPhase-R).

The average N stress effect over the reproductive period (tassel initiation to silking) was used by the model to modify the anthesis to silking interval (ASI). The ASI factor in turn determines the grain numbers per ear. As illustrated by the sensitivity analysis on maize cultivar Pioneer X304C sown during the dry season in Waipio, Hawaii (Singh, 1985), the days to silking increased from 78 to 108 d with increasing N deficiency in the plant (Fig. 3). In association with the delayed silking, the ASI increased.

The phyllochron, or the leaf appearance rate, is least affected by N stress. Over the range of NFAC given in Fig. 2, the phyllochron varied from 45°C with no N stress (NFAC = 1) to 47°C with extreme N stress (NFAC = 0). In general, the final leaf number would change a little. For the sensitivity analysis presented in Fig. 3, the final leaf number changed from 26 without any N stress during the reproductive stage (mean N stress = 1) to 23 leaves under extreme N deficiency. P deficiency, in contrast resulted in up to 32% increase in phyllochron (Rodriguez et al., 1998) and under increasing P stress conditions the final leaf numbers were reduced as well.

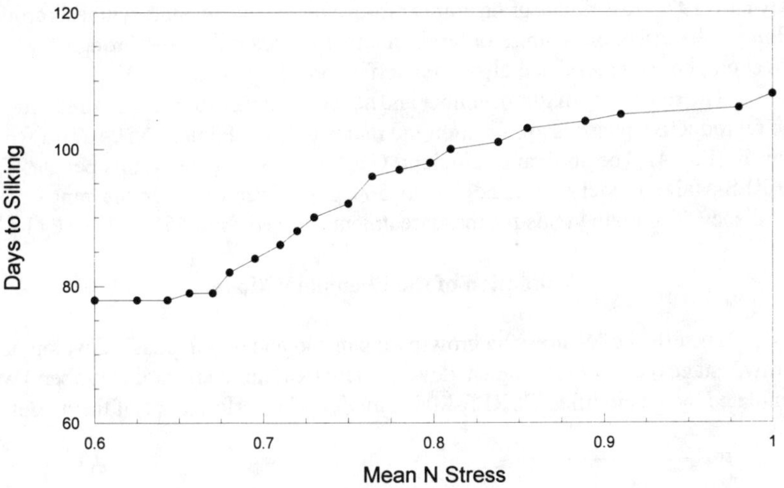

Fig. 3. Sensitivity of CERES-Maize to the effect of mean N stress during reproductive phase on silking date (days after sowing).

N stress during the grain-filling phase has the opposite effect compared with the pre-silking stress, that of reducing the grain-filling duration (Table 2). The combined effect of N stress on sowing to maturity duration may, however, be small. Under N limiting conditions, both the effect on growth and the shortened duration of the grain-filling stage contributes to lower grain yield.

Table 2. Effect of N application on phenological stages of two maize cultivars grown at three N rates (Singh, 1985).

Cultivar	N rate (kg N ha^{-1})	Days to silking	Grain filling duration (d)	Days to maturity
Pioneer X304C	0	83	53	136
	50	78	59	137
	200	78	59	137
H610 (Hawaii)	0	80	51	131
	50	76	57	133
	200	75	58	133

Effect of N Stress on Grain Number

Grain numbers per unit area are usually the most critical determinant of crop yield (Ritchie et al., 1998). It is well known that there is cultivar variability in grain numbers. To accommodate this effect the CERES-Maize model uses a cultivar coefficient, G2 (Ritchie et al., 1989). The model uses the concept of Edmeades &

Daynard (1979) to estimate grain number from average rate of photosynthesis around silking. The effect of N stress on grain number occurs indirectly through N stress effect on photosynthesis and also via its effect on silking date and ASI.

The sensitivity of grain number and hence grain yield to mean N stress during the reproductive phase is apparent on the maize cultivar Pioneer X304C, grown in Hawaii (Fig. 4). The cultivar coefficient, G2, for X304C is 690 grains per ear. The CERES-Maize model simulated 50- to 500 grains per ear over the range of N deficiency. The grain yields for these treatments ranged from 555 to 7180 kg ha^{-1}.

Validation of the Phenology Model

The effect of N stress on growth (expansion and mass), phasic development (growth stages) and morphological development (leaf numbers, grain numbers) was simulated by the modified CERES-Maize model. The performance of the model on

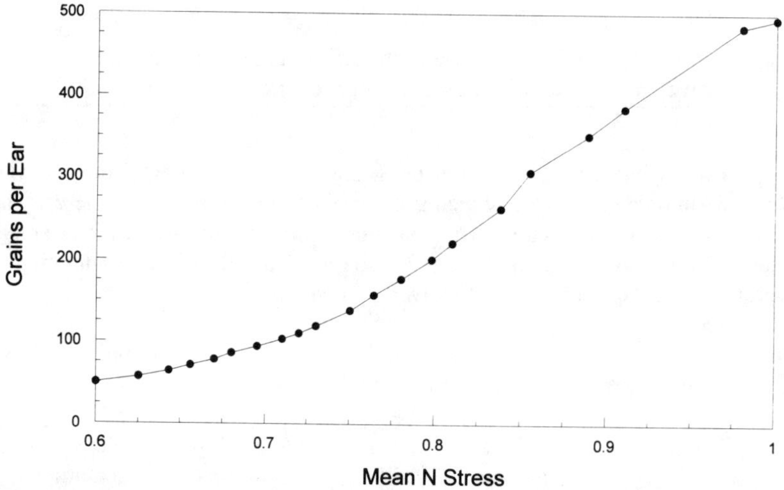

Fig. 4. Sensitivity of grain numbers per ear to the mean N stress during the reproductive phase as simulated by CERES-Maize.

actual field trials from Zaria, Nigeria (Oikeh, 1995), Gainesville, Florida (Bennett et al., 1989), and Waipio, Hawaii (Singh, 1985) is presented in Fig. 5. The model accurately captured the observed delays of up to 10 d to silking and reduced grain-filling duration of up to 14 d.

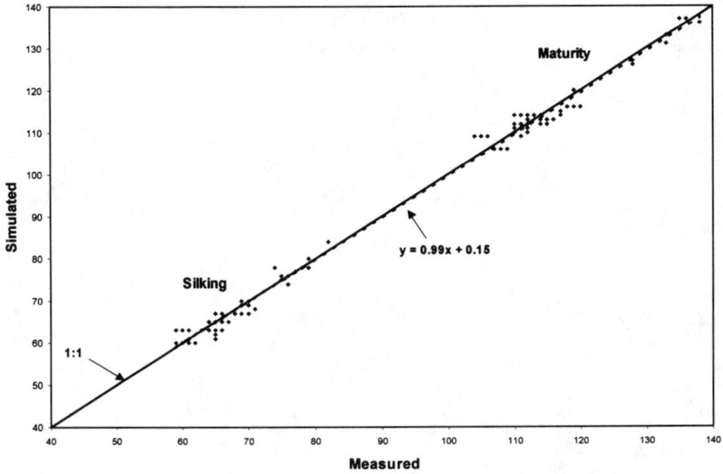

Fig. 5. Comparison of observed and simulated days to silking and maturity.

The phenology modification also improved the overall performance of the model. As evident from Fig. 6, the grain yield simulations were reliable for maize cultivar, Oba Super 2 grown over 2 yrs at five N rates (0, 30, 60, 90, and 120 kg N ha^{-1}) applied in two equal splits at Zaria in Nigeria. During 1993 trial, an additional treatment received all its 120 kg N ha^{-1} in a single application (Fig. 6). The effect of weather (years) was also accurately simulated with 1993 having higher grain yield than 1994. Rigorous validation of the model is planned with tropical maize data from' IFDC-Africa, CIMMYT and IITA.

DURATION BY ENVIRONMENT INTERACTION

In many rainfed maize growing regions, the window of opportunities for planting and growing season is limited. Under such conditions, any delay, whether environmental- or management- related could be costly. Grain yield would generally decrease due to the effect of N stress on phenology. However, a delay in time to tasseling or silking may help the plant avoid stress during a critical stage and thus result in higher grain yield. For example, on Millhopper sandy soil at zero N in Gainesville, Florida, the phenology model predicted 20 to 50% higher yield for 75% of the time than if the delay in phenology was not simulated. Much of the mid-season drought stress was avoided during the silking and the grain-filling stage due to the delayed development.

The year-to-year temperature variability in the tropics is lower compared to the temperate conditions. On the other hand variability associated with soil nutrient status is much greater in tropical soils. The effect of weather and N status on

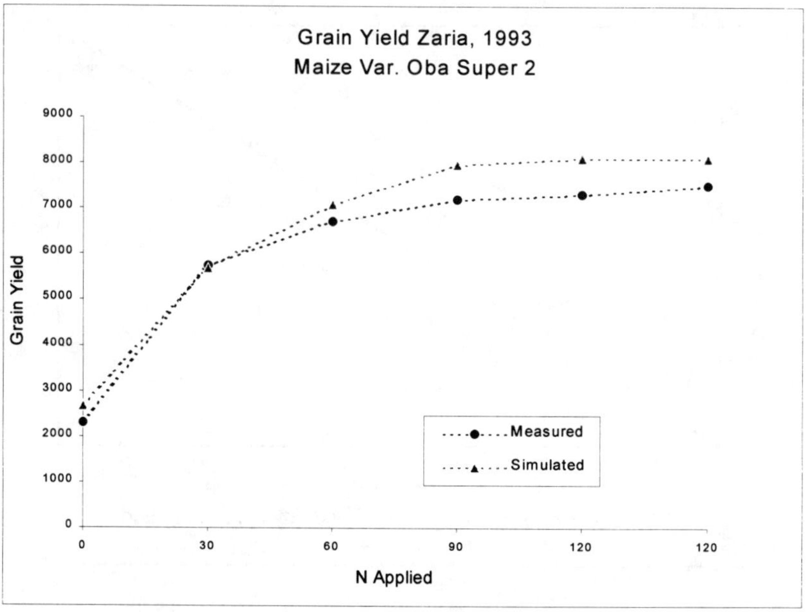

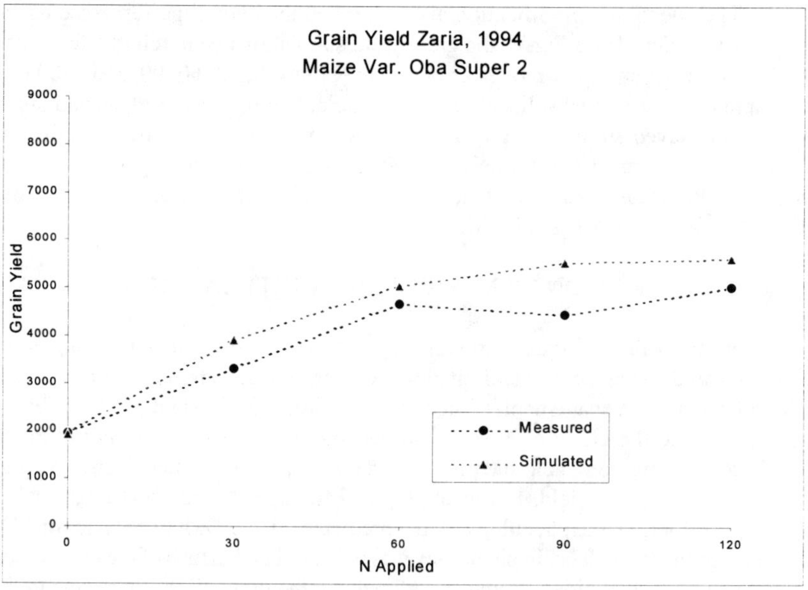

Fig. 6. Observed and simulated grain yield response to N application at Zaria, Nigeria during 1993 and 1994 season.

phenological development is illustrated using weather data for 7 yrs (1989–1995) and soil from Zaria, Nigeria (Fig. 7). The mean difference of about 10 days in silking date over 0 to 60 kg N ha^{-1} rate is much greater than the year-to-year variability associated with the weather. Much of maize in West Africa is grown without fertilizer application (Breman, 1990), hence nutrient status would be one of the key factors affecting phenological development.

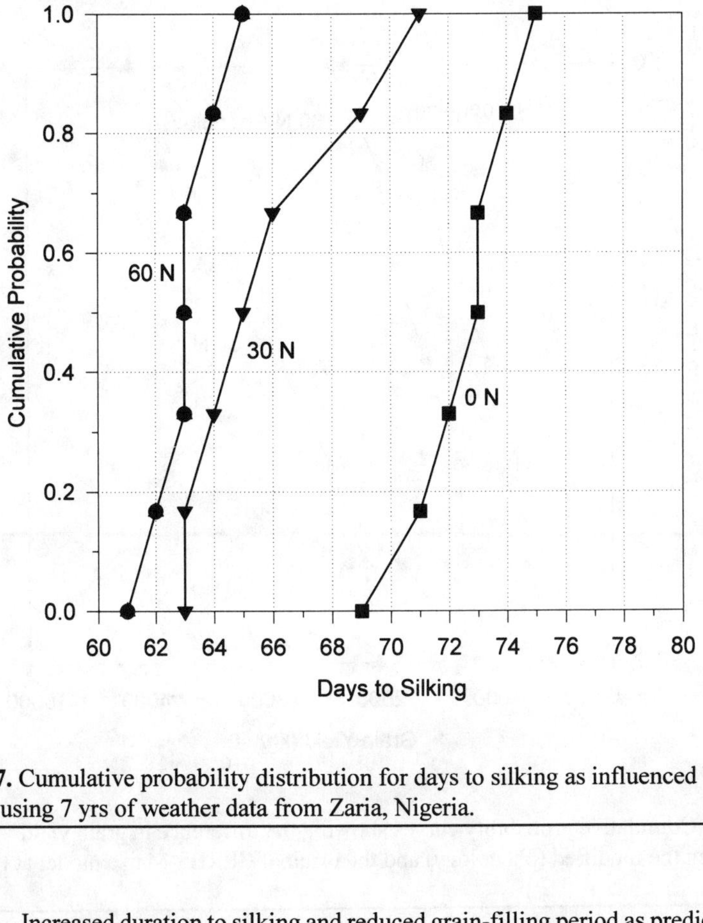

Fig. 7. Cumulative probability distribution for days to silking as influenced by N rates using 7 yrs of weather data from Zaria, Nigeria.

Increased duration to silking and reduced grain-filling period as predicted by the modified model, resulted in lower mean yields and increased variance when compared with the original version of the CERES-Maize model (Fig. 8). Soil fertility improvement thus would not only result in higher yields but reduced variance. The differences in timing of phenological events due to variable soil fertility would also make timing of field operations difficult. The continued mining of nutrients (negative annual balance of NPK) in agricultural soils of sub-Saharan Africa (with the exception of South Africa) implies that lower maize yields and increased variance would remain a critical factor.

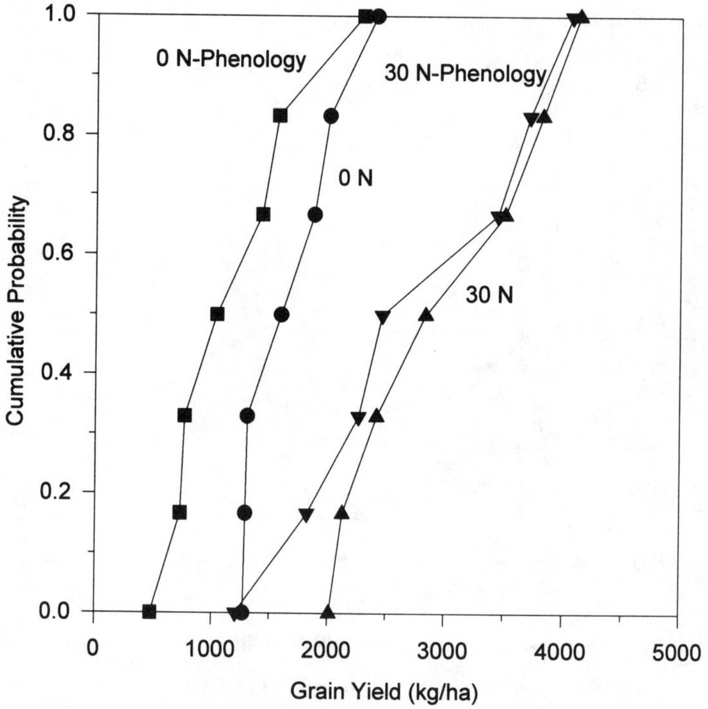

Fig. 8. Cumulative probability curves showing the difference in grain yield between the modified (phenology) and the original CERES-Maize model at two N rates.

CONCLUSIONS

As population pressures increase, the intensity of crop production in the developed and developing world will need to increase to keep pace. Nutrient and water availability is already a primary constraint to agricultural production in much of sub-Saharan Africa and these limitations will become even more pronounced. This critical challenge to the agricultural community may be addressed though dynamic crop growth simulation models such as CERES-Maize coupled with basic research for rapid assessment of nutrient management systems. Implicit with managing cropping systems under nutrient limited conditions is quantifying the duration and timing of phenological development to best utilize limited resources. As presented in this paper, the CERES-Maize model is sensitive to variation in the environment and nutrient status that modifies the duration of development stages and cropping season length. A validated crop model (such as CERES-Maize as demonstrated here) coupled with real-time weather will facilitate and enhance both the ability to forecast yields under limiting conditions and to fine-tune crop management to minimize deleterious effects.

ACKNOWLEDGMENTS

Part of this research was supported by the Soil, Water, and Nutrient Management Program of the CGIAR. Thanks to P. Stowe and L. Young for preparation of the figures.

REFERENCES

Arnold, C.Y. 1959. The determination and significance of the base temperature in a linear heat unit system. Proc. Am. Soc. Hortic. Sci. 74:430–455.

Bennett, J. M., L.S.M. Mutti, P.S.C. Rao, and J.W. Jones. 1989. Interactive effects of nitrogen and water stresses on biomass accumulation, nitrogen uptake, and seed yield of maize. Field Crops Res. 19:297–311.

Breman, H. 1990. No sustainability without external inputs. p. 124–133. *In* Beyond adjustment. African seminar, Maastricht, The Netherlands. Ministry of Foreign Affairs, the Hague, the Netherlands.

Elings, A., J. White, and G.O. Edmeades. 1996. Modelling tropical maize under drought and low N. p. 109. *In* Agronomy abstracts. 1996. ASA, Madison, WI.

Gilmore, E.C., Jr., and J.S. Rogers. 1958. Heat units as a method of measuring maturity in corn. Agron. J. 50:611–615.

Godwin, D.C., and U. Singh. 1998. Nitrogen balance and crop response to nitrogen in upland and lowland cropping systems. p. 55–77. *In* G.Y. Tsuji et al. (ed.) Understanding options for agricultural production. Kluwer Academic Publ., Dordrecht, the Netherlands.

Hesketh, J.D., and I.J. Warrington. 1989. Corn growth response to temperature: Rate and duration of leaf emergence. Agron. J. 81:696–701.

Muchow, R.C.,, and P.S. Carberry. 1989. Environmental control of phenology and leaf growth of maize in a semiarid tropical environment. Field Crops Res. 20:221–236.

Rabbinge, R., H.C. Van Latesteijn, and J. Goudriaan. 1993. Assessing the greenhouse effect in agriculture. p. 62–79. *In* Environmental change and human health. Ciba Foundation Symp. 1975. John Wiley & Sons, Chichester, England.

Ritchie, J.T. 1991. Specification of the ideal model for predicting crop yields. p. 97–122. *In* R.C. Muchow, and J.A. Bellamy (ed.) Climatic risk in crop production: Models and management for the semiarid tropics and subtropics. Proc. Int. Symp., St. Lucia, Brisbane, Queensland, Australia. CAB Int., Wallingford, England.

Ritchie, J.T., and D.S. NeSmith. 1991. Temperature and crop development. p. 5–29. *In* R.J. Hanks, and J.T. Ritchie (ed.) Modeling plant and soil systems. Agron. Monogr. 31. ASA, CSSA, and SSSA, Madison, WI.

Ritchie, J.T., U. Singh, and D.C. Godwin. 1989. A user's guide to CERES-Maize-V2.10. Int. Fertilizer Dev. Ctr., Muscle Shoals AL.

Ritchie, J.T., U. Singh, D.C. Godwin, and W.T. Bowen. 1998. Cereal growth, development, and yield. p. 79–98. *In* G.Y. Tsuji et al. (ed.) Understanding options for agricultural production. Kluwer Academic Publ., Dordrecht, the Netherlands.

Rodriguez, D., and J. Goudriaan. 1995. Effects of phosphorus and drought stresses on dry matter production and phosphorus allocation in wheat. J. Plant Nutr. 18:2501–2517.

Singh, U. 1985. A crop growth model for predicting corn (*Zea mays* L.) performance in the tropics. Ph.D. diss., Univ. of Hawaii, Honolulu.

Singh, U., D.C. Godwin, C.G. Humphries, and J.T. Ritchie. 1989. A computer model to predict the growth and development of cereals. p. 668–675. *In* Proc. 1989 Summer Computer Simulation Conf., Austin, Texas. 24–27 July, 1989. Soc. for Computer Simulation, San Diego, CA.

Singh, U., P.K. Thornton, A.R. Saka, and J.B. Dent. 1993. Maize modeling in Malawi: A tool for soil fertility research and development. p. 253–273. *In* F.W.T. Penning de Vries et al. (ed.) Systems approaches for agricultural development. Kluwer Academic Publ., Dordrecht, the Netherlands.

Swan, J.B., E.C. Schneider, J.F. Moncrief, W.H. Paulson, and A.E. Peterson. 1987. Estimating corn growth, yield, and grain moisture from air growing degree days and residue cover. Agron. J. 79:53–60.

Thornley, J.H.M., and I. R. Johnson. 1990. Plant and crop modelling: A mathematical approach to plant and crop physiology. Oxford Science Publ., Oxford, England.

Tollenaar, M., T.B. Daynard, and R.B. Hunter. 1979. Effect of temperature on rate of leaf appearance and flowering date in maize. Crop Sci. 19:363–366.

Tsuji, G.Y., G. Uehara, and S. Balas. 1994. DSSAT V.3: A decision support system for agrotechnology transfer. Univ. of Hawaii, Honolulu.

Potential Effects of Nitrogen-Fertilization Scenarios in Small Watersheds in Southern Ontario

G. Roloff
Departamento de Solos, Universidade Federal do Paraná (on leave)

K. B. MacDonald
Ontario Land Resource Unit, Greenhouse and Processing Crops Research Centre, AAFC

A. R. Couturier
Consultant
Guelph, Ontario

ABSTRACT

Modeling is an important tool for the decision-making process for site-specific management. An integrated model such as the Environment-Policy Integrated Climate (EPIC) can be used to assess the potential economic and environmental consequences of site-specific nitrogen fertilization. In this paper we test this application of EPIC to four small watersheds (280 to 380 ha) in southern Ontario, under intensive cropping dominated by soybeans, corn and winter wheat, and various tillage systems. In order to minimize data requirements, a simple area-weighting average will be used to produce watershed values from soil, crop and tillage-specific model runs. We will analyze the ability of the model, under this pooling strategy, to represent current conditions and two hypothetical 20-yr scenarios - N-fertilization using recommended rates or using site-specific N requirements. The analysis will focus on corn and winter wheat grain yields, and on total N discharge at the watershed outlet.

INTRODUCTION

Stream nutrient loads are controlled by environmental and management factors, which interact in complex ways. Therefore, studies dealing with the large set of interacting soil landscapes, crops and tillage systems must use computer models that replicate such interactions (e.g., Sugiharto et al., 1994; Fox et al., 1995). Modeling results can be used to advise land managers and policy makers to the potential benefits and consequences of strategies such as site-specific crop management. It has been suggested that the use of variable rate N-fertilization, based on early or within season soil test, can decrease N losses to the environment while maintaining crop yields (e.g., Meisinger et al., 1992).

Watershed-scale models deal with the source of the contaminants and their routing through the watershed, thus requiring detailed physiographic data to characterize individual watersheds, especially in relation to flow paths. Lack of such data can hamper the application of these models for large areas. The Environment Policy Integrated Climate (EPIC) model, on the other hand, was developed to produce estimates for a homogeneous small field (1–10 ha) using data that is usually readily

Copyright © 1999 ASA-CSSA-SSSA, 677 South Segoe Road, Madison, WI 53711, USA. *Proceedings of the Fourth International Conference on Precision Agriculture.*

available (Williams, 1995). However, EPIC has been extensively used to produce estimates for larger geographical areas (e.g., Bouzaher et al., 1993; Phillips et al., 1993) by linearly extrapolating estimates for specific conditions to large geographical areas by weight averaging.

EPIC's estimated average grain yields and annual stream loads of sediment and nutrients were found to be not significantly different than observed values for four monitored watersheds in southern Ontario (Table 1). A simple weight-averaging technique was used to generate the watershed-level values from selected EPIC output. This suggests that EPIC can potentially be used to indicate the direction and perhaps magnitude of long-term changes at the watershed scale brought about by alterations in management strategies.

In this paper, we report on a preliminary test of EPIC's ability in estimating the long-term effects of N-fertilization strategies over the stream N loads and net cash returns. We focused on the corn and winter wheat crops because they are important crops in southern Ontario and are suitable for variable rate N-fertilization. This test strives to use a minimum data set without specific calibration, in order to replicate conditions prevalent for large, regional studies.

MATERIALS AND METHODS

Watershed Data

Watershed data was obtained from two pairs of monitored watersheds in southern Ontario, Canada. One pair is located in Essex County (42°08' N, 82°42' W), while the other pair is near the City of London, part of the Kettle Creek watershed (42°53' N, 81°07' W). The watersheds are herein referred to as Essex East, Essex West, Kettle East and Kettle West. Main characteristics of the watersheds are described in Tables 2 and 3, and their crops and tillage distributions are shown in Table 4. Additional soil properties were also determined to a depth of 1.0 m, separated into three to five layers: bulk density, water retention at -33 kPa and -1.5 MPa water potential, cation-exchange capacity, pH, and the contents of sand, silt, organic C and calcium carbonate. The Essex watersheds were almost totally tile-drained at about 0.7-m depth, whereas the Kettle watersheds contained only localized tile drainage and, for modeling purposes, were assumed to contain no tile drainage.

EPIC modeling

EPIC input files for each combination of crop type (assuming monocultures), tillage system and soil type were prepared, with the soil and crop management data described above. Operation dates were those usual for the region (OMAFRA, 1994). Crop parameters for corn and winter wheat were those suggested by Kiniry et al. (1995).

The N-fertilization strategies tested were: (i) fixed recommended rates for expected grain yields of 7 Mg ha^{-1} of corn (175 kgN ha^{-1}) and 3.5 Mg ha^{-1} of winter wheat (30 + 50 kg N ha^{-1}) (OMAFRA 1994); (ii) one sidedressed application at least 20 d after seeding to bring the soil NO_3 concentration (0.2 m top layer) to 29 mg kg^{-1} for corn and 13 mg kg^{-1} for wheat (these concentrations correspond to the fixed recommended rates); and (iii) starter N fertilizer at 100 kg ha^{-1} for corn and 10 + 30 kg ha^{-1} for wheat, with one sidedressed application to bring soil NO_3 concentrations to the same levels as in the second strategy. Maximum annual N application for strategies two and three was limited to 200 kg ha^{-1}. These strategies will be referred to as *fixed*, *variable* and *starter+variable*, respectively.

Maximum and minimum daily air temperatures and daily precipitation for the Essex watersheds were obtained from a weather station distant about 21 km. For the Kettle watersheds, these data were from a station about 16 km away. These data were used to produce the weather statistics necessary for EPIC to generate the weather data needed for the arbitrary 22-yr EPIC runs. The first two years of the runs were considered initialization years and their results discarded. The PET method utilized was the Baier-Robertson, recently added to EPIC (Roloff et al., 1998).

The net return used as a measure of profitability, is the balance between direct costs (seed, fuel, fertilizers, chemicals) and harvested grain value. Costs and grain prices used were approximate and reflect the 1996–1998 crop years.

Analysis

For the tile-drained Essex watersheds, total N in the stream was taken as the sum of N in the surface runoff (soluble and within the sediment), and N in the subsurface and tile drain flows (soluble N). For the Kettle watersheds, total N was the sum of surface runoff and subsurface flow sources.

Each EPIC input file, and consequently each run, assumed that the entire watershed was of a specific combination of crop, tillage and soil. The relevant output was then area-weighted according to the relative proportion of the watershed for that combination of crop, tillage and soil (Tables 2 and 3), to produce an average value for the watershed. Estimated and observed values were compared and analyzed using comparison of means employing either a paired *t*-test or one-way analysis of variance.

RESULTS AND DISCUSSION

Nitrogen Loads

The relative contribution of corn and wheat to the stream N loads varied with fertilization strategy and watershed. In the Essex watersheds, variable and starter+variable caused small reductions in the relative contribution from corn (Fig. 1, left). However, they caused an almost doubling of the relative contribution of wheat, probably because the timing of simulated applications (mostly early to mid April) coincided with estimated high runoff and tile drain flows. For the Kettle watersheds, fertilization strategies made little difference in the relative contribution of corn (Fig. 2, left). For wheat, variable and starter + variable caused a small increase in Kettle East and a slight decrease in Kettle West.

Even though the changes in fertilization strategy caused some noticeable shifts in the relative contribution of corn and wheat, their effect on estimated total N stream loads was negligible (Fig. 1 and 2, right). The effects of these shifts became diluted in the watersheds because (i) the larger shifts were for wheat that occupied relatively small land surface (Table 4), and (ii) corn, which occupied substantial relative land surface, displayed only small shifts.

Certainly these modeling results are tentative and are influenced by the model and data limitations and simplifications, and by the assumptions embedded within the strategies tested. Nevertheless, the results suggest that the stream water quality effects of variable rate N fertilization may not be always or necessarily positive, and depend on the complex interactions between crop, soil and weather. For this environment, these effects may only be noticeable at the small watershed level in cases where the watershed is occupied almost entirely by one or more crops subjected to variable rate fertilization.

Net returns

Costs associated with variable N fertilization are usually higher than those for fixed rates. While we did not address costs such as additional soil sampling, we analyzed the estimated net cash returns, based on direct costs and grain prices, as a measure of profitability, one of the great motivators for change. For corn, there were small (2 to 3%) reductions in net return using variable or starter + variable, for both watersheds (Fig. 3 and 4). These reductions resulted from slightly smaller corn yields, which were at most 3% lower (Essex watersheds), and the additional operational costs, even though applied N rates were lower (96 to 154 kg N ha^{-1}) than the fixed rate (175 kg N ha^{-1}). These findings are consistent with the small monetary benefits for variable N fertilization rate for corn reported in other studies (e.g., Kitchen et al., 1995).

For wheat, the results indicate the reverse as for corn. The variable and starter+variable strategies resulted in significant gains in net returns (7 to 10%). Such gains were associated mainly with higher grain yields (7 to 14%), even though applied fertilizer rates (63 to 141 kg N ha^{-1}) were often higher than the fixed rate (80 kg N ha^{-1}). Wheat probably benefited more from the variable rate strategies because it has a longer season than corn, and has a stage of rapid vegetative growth (spring) closely following snowmelt, a period with high N leaching potential.

Even though again the modeling results should be interpreted with caution, they suggest that variable N fertilization had minimal effect on the economics of corn on the long term. Profits associated with wheat, on the other hand, may be somewhat increased. Further tests should be conducted to verify the monetary effects of rotations, long-term no tillage and timing of N application.

CONCLUSIONS

In this test, EPIC was able to generate varying N fertilization rates for corn and wheat, which resulted in relative contributions to stream N loads and net cash returns that were distinct than those obtained under a fixed rate. However, the effects over the stream N loads were negligible at the small watershed scale. Net cash returns were benefited by variable rates only for wheat.

These results certainly depend on the accuracy of EPIC and of the extrapolation procedure used to generate watershed-level data. While the absolute numbers may be meaningless, they were useful to demonstrate that the environmental effects of variable rate N fertilization at the watershed level are subtle and tend to be diluted within the watershed. These effects can probably only be noticeable in case of utilization of the variable rate over most of the watershed surface. In addition, it verified that economic effects of variable rates were essentially non-existent for corn, but may result in small gains in net cash returns for wheat.

Further tests of the various EPIC components involved with N dynamics are certainly necessary, but the model has demonstrated that it is sensitive enough to accompany and react to varying soil N concentrations in this environment. Therefore, it shows potential as a tool to help in the decision making process related to the application of variable N fertilization.

REFERENCES

Bouzaher, A., J.F. Shogren, D. Holtkamp, P. Gassman D., Archer, P. Lakshminarayan, A. Carriquiry, R. Reese, W.H. Furtan, R.C. Izaurralde, and J. Kiniry, 1993. Agricultural Policies and Soil Degradation in Western Canada: An Agro-Ecological Economic Assessment Rep. 2. The Environmental Modeling System. Techn. Rep. 5/93. Agric. Canada, Policy Branch.

Fox, G., G. Umali, and W.T. Dickinson, 1995. An economic analysis of targeting soil conservation measures with respect to off-site water quality. Can. J. Agric. Econ. 43:105–188.

Kiniry, J. R., D.J. Major, R.C. Izarraulde, J.R. Williams, P.W. Gassman, M. Morrison, R. Bergentine, and R.P. Zentner, 1995. EPIC model parameters for cereal, oilseed, and forage crops in the northern Great Plains region. Can. J. Plant Sci. 75:679–688.

Kitchen, N. R., D.F. Hughes, K.A. Sudduth, and S.J. Birrell, 1995. Comparison of variable rate to single rate nitrogen fertilizer application: Corn production and residual soil NO_3–N. *In* P.C. Robert et al., (ed.) Site-specific management for agricultural systems. ASA, CSSA, and SSSA, Madison, WI.

Meisinger, J.J., V.A. Bandel, J.S. Angle, O'Keefe, and C.M. Reynolds, 1992. Presidedress soil nitrate test evaluation in Maryland. Soil Sci. Soc. Am. J. 56:1527–1532.

OMAFRA (Ontario Ministry of Agriculture, Food and Rural Affairs). 1994. 1995–1996 Field Crop Recommendations. OMAFRA Publ. 296. The Queen's Printer for Ontario, Toronto.

Phillips, D.L., P.D. Hardin, V.W. Benson, and J.V. Baglio, 1993. Non-point source pollution impacts of alternative agricultural management practices in Illinois: A simulation study. J. Soil Water Conserv. 48:449–457.

Roloff, G., R. De Jong, C.A. Campbell, R.P. Zentner, and V.W. Benson, 1998. EPIC estimates of soil water and soil nitrogen under semiarid temperate conditions. Can. J. Soil Sci. (in press)

Sugiharto, T., T.H. McIntosh, R.C. Uhrig, and J.J. Lardinois, 1994. Modeling alternatives to reduce dairy farm and watershed non-point source pollution. J. Environ. Qual. 23:18–24.

Williams, J.R. 1995. The EPIC model. p. 909–1000. *In* V.P. Singh (ed.). Computer models of watershed hydrology. Water Resources Publ., Littleton, CO.

Table 1. Comparison of observed and estimated average annual water quality parameters and crop grain yields (standard deviations in parentheses) between observed and estimated values (period 1989–1994), for four watersheds in southern Ontario. *Source*: **G. Roloff, 1998**. Evaluation of EPIC at the small watershed scale in southern Ontario. Unpublished internal report. Agriculture and Agrifood Canada.

Parameter		Essex		Kettle	
		East	West	East	West
Stream discharge mm	observed	348 (219)	453 (194)	270 (197)	305 (252)
	estimated	215 (154)	319 (138)	235 (129)	224 (120)
Sediment Mg ha^{-1}	observed	0.53 (0.31)	0.68 (0.35)	0.31 (0.29)	0.75 (1.11)
	estimated	0.43 (0.35)	0.42 (0.38)	2.84 (2.30)	1.77 (1.46)
N kg ha^{-1}	observed	17.5 (9.0)	21.6 (7.1)	14.9 (11.4)	24.6 (22.5)
	estimated	14.9 (6.6)	30.0 (8.6)	21.6 (14.1)	19.6 (12.3)
P kg ha^{-1}	observed	1.4 (1.3)	2.1 (0.9)	1.2 (1.1)	1.9 (2.6)
	estimated	1.4 (0.8)	1.8 (1.0)	3.5 (2.1) *	2.7 (1.7)
Corn Mg ha^{-1}	observed	6.2 (1.8) ‡		6.1 (0.4)	
	estimated	6.7 (0.7)		6.6 (1.1)	
Winter wheat Mg ha^{-1}	observed	4.0 (1.3)		3.3 (1.4)	
	estimated	3.7 (0.6)		3.7 (0.4)	

* Significantly different at $P \leq 0.05$; absence of symbol indicates values are not significantly different. Comparison of observed with estimated values.

‡ Results pooled per watershed pair.

Table 2. Relevant watershed characteristics.

Watershed	Soil classification†	Texture‡	Slope steepness ($m\ m^{-1}$)	Fraction of total area
Essex East	Orthic HG (Brookston)	C L	0.01	1.0
Essex West	Orthic HG (Brookston)	C L	0.01	1.0
Kettle East	Gleyed Brunisolic GBL (Gobles)	L	0.03	0.14
	Gleyed Brunisolic GBL (Gobles)	Si L	0.03	0.23
	Orthic HG (Maplewood)	Si L	0.04	0.04
	Gleyed Brunisolic GBL (Tavistock)	Si L	0.03	0.35
	Gleyed Brunisolic GBL (Tuscola)	vf S L	0.03	0.04
	Orthic HG (Colwood)	Si L	0.01	0.09
	Orthic HG (Kelvin)	Si L	0.02	0.03
	Brunisolic GBL (Muriel)	Si L	0.04	0.07
Kettle East	Brunisolic GBL (Muriel)	Si L	0.04	0.05
	Brunisolic GBL (Muriel)	C L	0.04	0.04
	Orthic HG (Maplewood)	Si vf S	0.01	0.10
	Orthic HG (Kelvin)	L	0.02	0.13
	Gleyed Brunisolic GBL (Gobles)	L	0.03	0.42
	Brunisolic GBL (Brant)	L vf S	0.05	0.01
	Brunisolic GBL (Bennington)	L	0.03	0.06
	Gleyed Brunisolic GBL (Tavistock)	L	0.03	0.19

† HG – Humic Gleysol, GBL – Gray Brown Luvisolic; soil unit name in parentheses.
‡ C – clay, L – loam, S – sand, Si – silt, vf – very fine.

Table 3. Main EPIC input parameters that are watershed specific.

EPIC input parameter	Watershed			
	EE[†]	EW	KE	KW
Surface area (ha)	281	381	355	380
Slope length (m) [x]	40	40	35	35
Mainstem channel slope (m m^{-1})	0.008	0.008	0.01	0.01
Distance from outlet to most distant point (km)	3.0	2.9	3.3	2.3
Runoff curve number				
- corn, soybeans, sunflower (moldboard and disc)	81[§]		88[§]	
- corn, soybeans, sunflower (no tillage and chisel)	78		85	
- winter wheat (moldboard, cultivator and disc)	76		84	
- winter wheat (no tillage)	75		83	
- alfalfa (no tillage)	72		81	

[†] EE – Essex East, EW – Essex West, KE – Kettle East, KW – Kettle West.
[‡] Typical values, which are best estimates from contour maps and local knowledge.
[§] From EPIC User's Manual, assuming soil hydrological type B ([v]) or C ([w]).

Table 4. Average land fraction occupied by the various combinations of crop and tillage.

Crop	Tillage†	Average 1989-1994‡			
		EE§	EW	KE	KW
corn	MB	0.13	0.04	0.41	0.20
	DS	0.01	0.01	0.02	0.03
	CL	0.07	0.03	0.01	-
	CH	-	-		0.03
	NT	-	0.10	0.06	0.15
soybeans	MB	0.31	0.18	0.14	0.10
	DS	0.04	0.18	0.01	0.03
	CL	0.18	0.10	0.03	0.01
	CH	0.01	0.02	0.00	0.01
	NT	0.07	0.16	0.07	0.20
winter wheat	MB	0.02	-	0.03	-
	DS	0.04	0.06	-	0.02
	CL	0.07	0.04	0.04	0.01
	NT	0.01	0.04	0.01	0.08
field peas	MB	-	0.01	0.01	0.03
snap beans	MB	0.01	0.01	-	-
	CH	-	-	0.01	-
sunflower	MB	0.01	-	-	-
	NT	0.01	0.01	-	-
alfalfa	NT	0.03	0.01	0.14	0.10

† MB – moldboard plow, DS – disk harrow, CL – cultivator, CH – chisel plow, NT – no-tillage.

‡ Crop year starting in the 4th quarter of the previous year, and ending at the 3rd quarter of the year; no data available for 1992.

§ EE – Essex East, EW – Essex West, KE – Kettke East, KW – Kettle West.

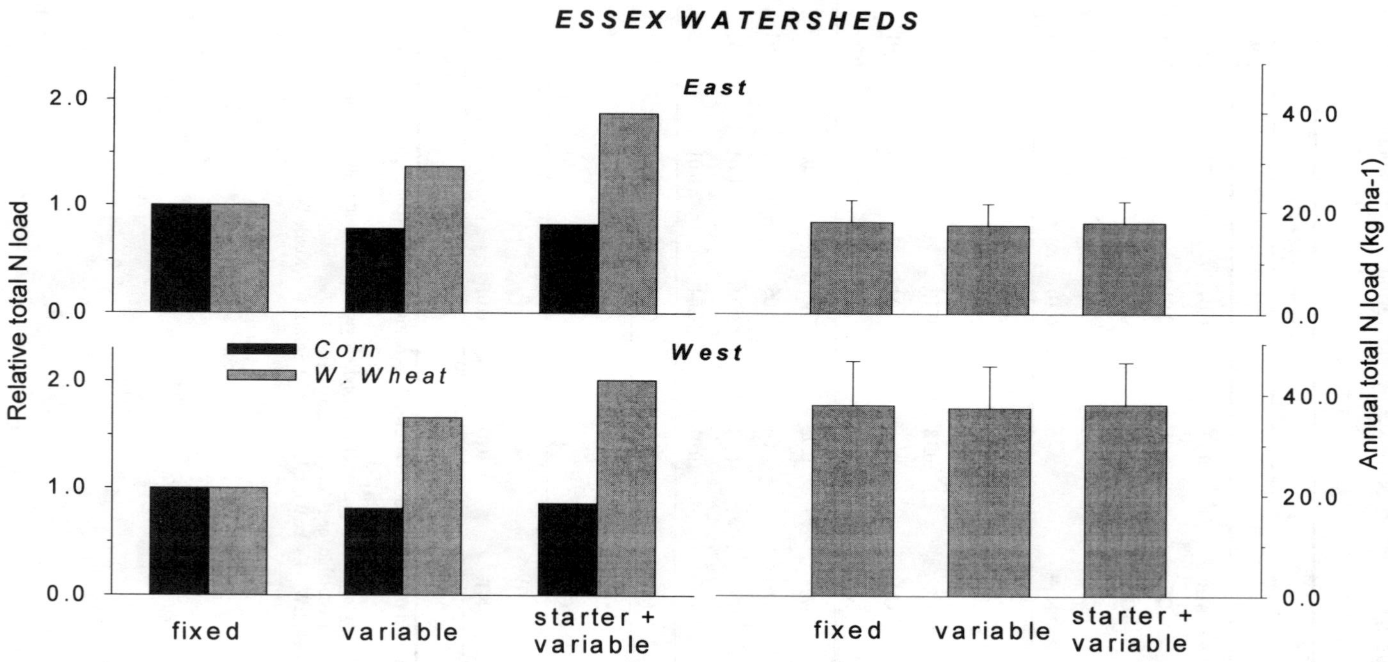

Fig. 1. Relative contribution to stream N loads (left) and estimated annual stream N loads (right) in relation to N fertilization strategy for the Essex watersheds.

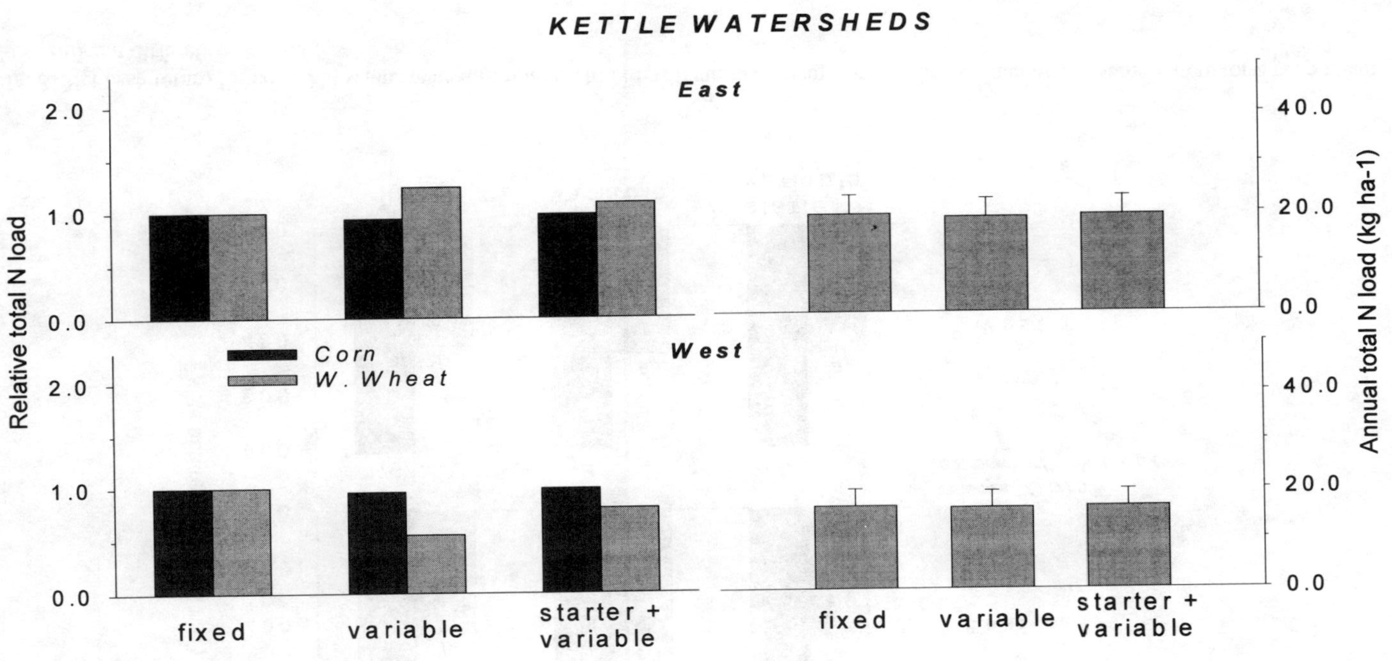

Fig. 2. Relative contribution to stream N loads (left) and estimated annual stream N loads (right) in relation to N fertilization strategy for the Kettle watersheds.

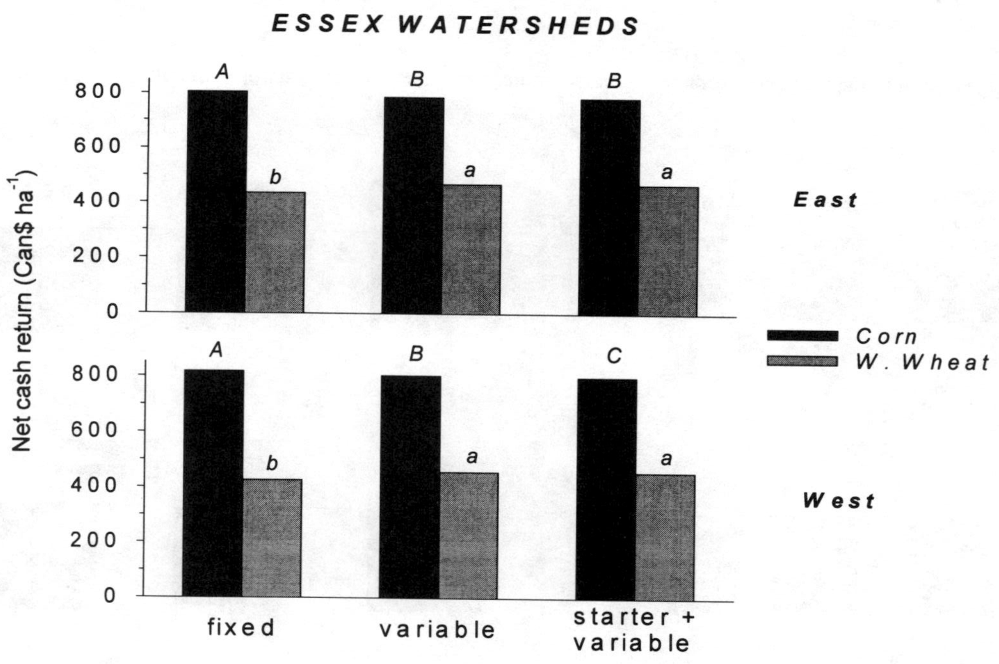

Fig. 3. Net cash return for corn and winter wheat in relation to N fertilization strategy for the Essex watersheds. Same letters for a crop mean no significant difference ($P > 0.05$).

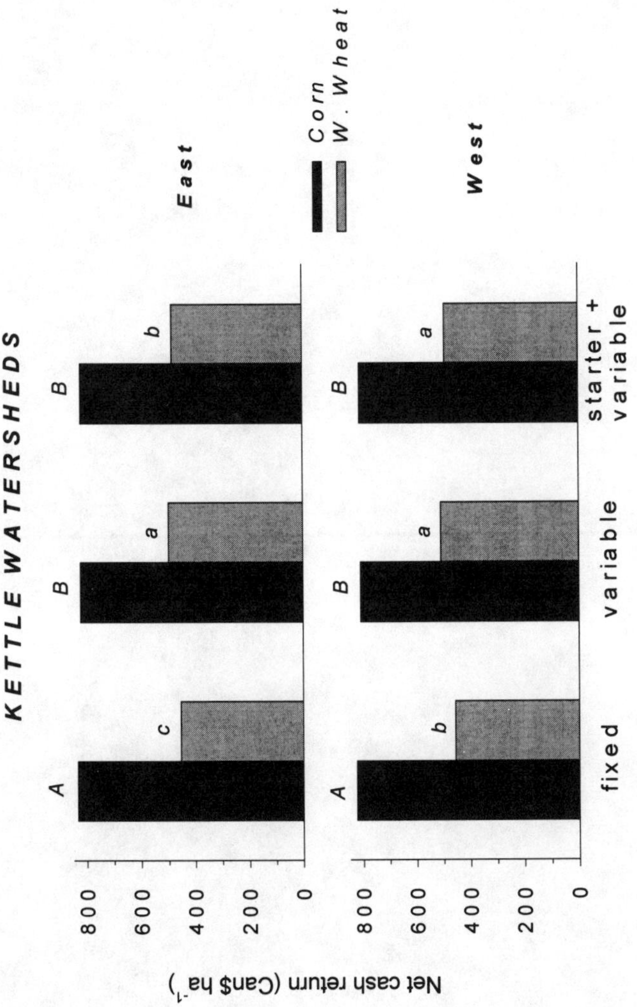

Fig. 4. Net cash return for corn and winter wheat in relation to N fertilization strategy for the Essex watersheds. Same letters for a crop mean no significant difference ($P > 0.05$).

Weed Detection in Field Corn Using High Resolution Multispectral Digital Imagery and Field Scouting

Brian L. Broulik
K. J. Dalsted
S. A. Clay
D. E. Clay
C. G. Carlson
M. M. Ellsbury
D. D. Malo

South Dakota State University
Plant Science Department
Brookings, South Dakota

ABSTRACT

Weed species occur in non-uniform Patches across agricultural fields with the amount of patchiness differing among weed species and field. This patchiness complicates herbicide recommendations; however, herbicide applications can be targeted to specific areas by identifying the locations and weed species in the field. The size, shape, and location of weed infestations can be determined by intensive field scouting, but this approach is expensive. Remote sensing information, combined with ground-truth data, may provide a useful method to solve this problem. A study was conducted to determine the feasibility of using remote sensing techniques to detect weed populations at several stages of corn growth. A charge-coupled device (CCD) with four spectral filters mounted in an airplane was used to obtain several near digital images with 1 m * 1 m resolution of a 65 ha no-till corn field from May through September 1997. The CCD sensor contained four spectral filters sensitive in the blue, green, red, and near infrared (NIR) wavelengths. Latitude and longitude coordinates of the field perimeter were integrated into a geographical information system so that coordinate of anomalous areas of the field could be identified. Using the coordinates of the anomalous areas, ground scouting from the aerial images was conducted to define the nature of the anomaly. Comparing the aerial images to the ground-truthed data indicated that the NIR wavelength showed the greatest differences between corn and weeds. A normalized difference vegetative index was generated using the image corresponding with maximum green canopy cover and was highly correlated with corn yield, indicating that future yield predictions may be possible using remote sensing, however, field scouting was necessary to distinguish weed species and densities. Remote sensing combined with ground scouting provided an excellent method to determine the location of weed infestations over an entire field, to create a database for site specific herbicide management, and to monitor changes in weed species density over time.

Copyright © 1999 ASA-CSSA-SSSA, 677 South Segoe Road, Madison, WI 53711, USA. *Proceedings of the Fourth International Conference on Precision Agriculture.*

REMOTE SENSING

Applying Remote Sensing Technology to Precision Farming

C. J. Johannsen
P. G. Carter

Laboratory for Applications of Remote Sensing
Purdue University
West Lafayette, Indiana

P. R. Willis

Resource21
Englewood, Colorado

E. Owubah
B. Erickson
K. Ross
N. Targulian

Agronomy Department
Purdue University
West Lafayette, Indiana

ABSTRACT

Major changes have occurred and continue to occur in remote sensing technology in recent years. Significant changes related to advances in spatial, spectral and temporal resolution. For precision farming, this means that remote sensing can provide (i) much greater detail of an individual field, (ii) much more precisely defined colors or delineations of variations of the vegetation, residues or surface soils, and (iii) repeat viewing of the same scene every 2 to 7 days. In effect, the farmer can detect missing plants or stress damage, assess the causality of a stress by looking more closely at the spectral responses, and receive data–information in a timely fashion so that corrective action may be taken. Within the next 10 yrs there will be more than 50 new land viewing satellites. The impact on precision farming means more data available, varying types of data such as wavelength bands that are specific to stress conditions, more timely information and costs which should be reasonable to the farmer because of the competition.

INTRODUCTION

Remote sensing technology is experiencing a resurgence of popularity, prominence and possibilities in the agriculture arena from the advancement of spatial, spectral and temporal resolution developments. For the precision farming

Copyright © 1999 ASA-CSSA-SSSA, 677 South Segoe Road, Madison, WI 53711, USA. *Proceedings of the Fourth International Conference on Precision Agriculture.*

revolution, this means availability of services and products that help to manage the farm operation more efficiently and profitably reducing some of the stresses of today's small, medium and large farming enterprises. This revolution began in the 1970s and went through a period of failures at delivering promises. Due to this sting of failure an aura of cautiousness yet exists. Research and evaluation by non-partisan entities still remains weak as innovations rapidly abound but the situation is much improved from previous years. Some of the visions agriculture users were left with following the early stages of the revolution have now been evaluated in detail showing considerable confidence. The development of newer, faster, stronger and cheaper electronic hardware and software as well as the release of farming practices to a newer generation (many with a lifetime of computer experience) will help to adopt this renewal of management opportunities. With the approach of using this technology as a management assistance tool and not with a problem solver attitude, the future looks brighter for precision farming and the ability of increased production for a hungry world while balancing with a sensitive environment.

CHANGES IN TECHNOLOGY

We have literally taken agriculture into the space age from this humble beginning. Farmers now have services available that involve satellites collecting data, transmitting locational information, and providing data from a variety of other sources. Some of these sources involve data from sensors on tractors, combines, and other equipment, sensors on airplanes to aid in crop scouting, and receiving or analyzing satellite information. For a fee, they can rely on consulting and ag supply companies to do much of this for them.

We will concentrate on remote sensing in this paper but want people to realize that other technologies are also involved in making the data collection, analysis and interpretation successful. Some of these technologies are GIS for assimilation of additional data layers and information (soil map or digital terrain aspect) to assist in the analysis of remotely sensed data, GPS for locating ground observations so computers can link them to other data, and improvements in communications such as the Internet for transferring information and other datasets. With all of these new innovations we are concerned that remote sensing might again be oversold by stating abilities that are beyond their present potential. If one has management style well defined, can absorb large volumes of data and information without becoming disgusted or confused, and can separate fact from speculation then there is a chance of successfully applying remote sensing technologies.

Remote sensing technology has seen many changes in the past five years. Because of improvements in sensors, computer chips, software and services, agriculture is seen as a potential application to reap benefits at ground and space altitudes. The term, precision farming has captured the essence of what is happening related to remote sensing and also that of other important technologies, namely geographic information systems (GIS) and global position systems (GPS). It is interesting that this term precision farming denotes a level of preciseness that is yet to be achieved. We would suggest that the term site specific farming is a better term since it considers what one can do at a specific location. There are other terms such as prescription farming, and variable rate technology that are also used while others have incorrectly called it, GPS when referring to this technology (Johannsen, 1997). Whatever it is called, we are seeing an information revolution occurring and once farmers have been provided this additional information about their crops, soil, and land, they will keep asking for more!

We should not think of remote sensing as obtaining data only from satellites (Lozano-Garcia et al., 1995). The tractor and other field equipment have and will continue to have a big role in the use of agriculturally oriented sensors. Purdue University researchers have devised a machine system to rapidly measure soil nitrate and pH using ion-specific, field-effect-transistors (ISFETS). The system consists of a rolling core sampler and a computer-controlled, automated analysis station mounted on a toolbar (Loreto & Morgan, 1996). Nitrate and pH measurements are made in <10 s and recorded with field location obtained from a GPS receiver. Mapping of the spatial distribution can be obtained using a GIS. ARS researchers at various locations (Moran et al., 1997) have also been developing and testing sensors with similar capabilities. In other Purdue efforts, the sound of a tillage tool pulled through the soil is being correlated to soil texture, specifically the percentage of sand and clay (Liu et al., 1993). These efforts are additions to the research which led to the development of a patented sensor for soil organic matter which enables site-specific herbicide applications (Shonk et al., 1991).

Other sensor technology has also seen many innovations. From the tractor, we are using remote sensors that measure soil and plant parameters; from an airplane, we are obtaining aerial photography and digital images showing anomalies within a field; from satellites we will be obtaining images with spatial resolutions that previously were top secret. The major changes are that from satellite altitudes we are or will be able to (i) image or see with more detail, a smaller piece of land, (ii) define more precisely the specific colors or light responses reflecting or radiating off of the field surface and, (iii) obtain data on a regular interval of every other day or every 2 to 7 d (Johannsen, 1994, 1996). These changes make for real advances to agriculture as we interpret remotely sensed data received on regular intervals. We have the need to be able to view those small portions in the field that are not performing well and determine what caused the anomaly. We will review these changes in more detail as they are important to the future of agriculture and the methods involved to gather data and information using this technology.

SPATIAL RESOLUTION

When you view yield image maps, you are looking at about 300 to 500 data points per hectare depending on how fast the harvester was traveling and how often (1, 2 or 3 s) the yield measurement was recorded. The specific area represented by one data point that can be seen on an image, whether yield monitor or remotely imaged, is called spatial resolution. Spatial resolution in satellite data collection is improving as current satellite resolution areas are 30 m $*$ 30 m (11 measurements hectare^{-1}), 20 $*$ 20 m (25 measurements hectare^{-1}) and 10 $*$ 10 m (100 measurements hectare^{-1}). With future satellites, we will be receiving data that have a variety of spatial resolutions or pixels (short for picture elements which are the smallest ground areas observable by the sensor) that in some cases will be as detailed as 1 $*$ 1 m or across 10 000 data points per hectare as shown in Table 1.

Table 1. Conversion of Spatial Resolution from Images to Data Points on the Ground.

Resolution (meters)	Data points (Pixels) Acre^{-1}	Points Ha^{-1}
1000	0.004	0.01
80	0.6	1.56
30	4.5	11.1
23.5	7.3	18.1
20	10	25.0
15	18	44.4
10	40	100
5	162	400
4	253	625
3	450	1 111
2	1012	2 500
1	4046	10 000

In terms of sensor technology, we are seeing improvements in spatial resolutions that allow one to see greater detail. At recent conferences, we have learned that there is a potential for over 50 land observing satellites to be launched before the year 2007 (Johannsen & Carter, 1997) that provide an interesting choice of data (Table 2). These satellites which are both government and commercial will have a large range of spatial resolutions from 1 m to 1 km. The highest resolution is proposed by Space Imaging Inc. and EarthWatch (QuickBird satellite) who plan to launch satellites in the next few years that will have 1 m panchromatic and 4 m digital data. Companies like Resource21 and GER are promising data in the 5 to 15 m spatial range. EarthWatch launched the EarlyBird satellite recently and it failed to reach the proper orbit. The satellite had a 3 m panchromatic and 15 m multispectral sensor, which would have been the first high resolution commercial satellite in space. The loss of this satellite has caused other commercial companies to be very cautious in their approach and advertising.

Table 2. Remotely Sensed Data Choices of the Future and Some Characteristics.

Satellite Platforms
Spatial Resolutions:
1, 2, 3, 4, 5, 10, 15, 20, 23.5, 30 m
Spectral Resolutions:
5, 10, 50, 100 nm
Temporal Resolutions:
1-5 d, 8-10 d, 14-16 d

Aircraft Platforms
Sensors similar to satellite platforms
Greater flexibility in obtaining coverage
Fly under the clouds!

Tractor Platforms
Potential sensors for organic matter content, pH, soil texture, soil moisture
Sensor measurements relate to specific tasks
Sensors will work under a variety of conditions
Spatial resolutions range from centimeters to meters
Usually better resolution in one field direction (direction of travel)

Note that we are not limited to thinking only about satellite coverage for remotely sensed data. Several companies are providing sensor coverage by airplane from a variety of altitudes at the time that the consultant and the farmer would like to see what is happening to the crop. The other choice is that companies are developing sensors that can provide specific measurements for specific elements or conditions. As mentioned previously, a number of these sensors involve making measurements from tractors or similar field vehicles and hand held units.

SPECTRAL RESOLUTION

Spectral resolution is the width of the individual bands. For example, band 1 of the Thematic Mapper measures reflected light energy from a wavelength of 450 to 520 nm. Thus the spectral resolution of that band is 70 nm. Spectral resolutions of 50 to 150 nm are typical for current satellite-based remote sensing. Multispectral sensors are so named because they measure incoming electromagnetic radiation in several discrete spectral bands. Landsat Thematic Mapper, for instance, measures seven spectral bands for each data point. The number of bands is known as the dimensionality of the data.

However, improved detector technologies have lead to sensors with spectral resolutions on the order of 5 to 15 nm. Two examples are the Airborne Visible/InfraRed Imaging Spectrometer (AVIRIS) (Vane et al., 1993), and the Hyperspectral Digital Imaging Collector Experiment (HYDICE) sensor (Mitchell 1995). Given that a sensor has a certain maximum range over which it may make measurements, we see that narrower spectral resolutions allow an instrument to measure more bands. In other words, narrower spectral resolution enables higher spectral dimensionality. As a result, AVIRIS has 224 bands and HYDICE has 210. Because of the higher dimensionality, this type of sensor is called hyperspectral.

At present, hyperspectral sensors are strictly airborne. The Lewis satellite, launched 23 August, 1997, had 384 bands and would have been the first hyperspectral instrument in space, but it failed to achieve orbit. Now NASA plans to launch the first hyperspectral sensor in 1999.

The reason hyperspectral data is important to farming is improved discrimination among various plants, soils, and their conditions. Lower dimensional data can be classified into rather gross distinctions like whether or not a crop is stressed, but analysis of hyperspectral data may be able to determine whether the stress is nitrogen deficiency, disease, drought, etc. (Moran et al., 1997). Why hyperspectral sensing should be so much better can be intuitively grasped. It is as if you went from coloring with 8 crayons to a box of 64; the sort of pictures you could color would be much more refined. Likewise, as illustrated here, the remote sensing picture of a particular image pixel goes from a rather blocky histogram to a figure approaching the shape of a continuous reflectance spectrum.

Due to the complexity of hyperspectral data, it may be some time before there are widespread, reliable techniques to interpret it. The high dimensional space of the data cannot be visualized. Classification techniques applied to lower dimensional data are inappropriate. Some researchers, drawing on the fact that the data now approximates the kind of spectra seen in chemistry, are applying the techniques of spectroscopy (Vane & Goetz, 1993; Martin & Aber, 1997). Others are taking the classification techniques of pattern recognition to a new level (Hoffbeck & Landgrebe, 1996). Some hope to find indices (band ratios) that outperform indices like the normalize difference vegetation index (NDVI) that has been widely used with lower dimensional multispectral data (Malthus et al., 1993). In any case, the analysis will probably so complex that individual farm managers will not wish to invest the time and resources needed to master it themselves. They will probably look to agribusiness to provide remote sensing specialists (Moran et al., 1997).

TEMPORAL RESOLUTION

The strength of remote sensing in precision farming management lies in the ability to learn more about crop growth variability while the crop is still growing. Temporal resolution implies looking at the same field or location on a repetitive sequence or change detection. The significance of temporal resolution for precision agriculture is evident, since it gives the grower a view of timely changes in crop condition, possible hazardous situations and yield variations. The temporal resolution or return frequency in an ideal remote sensing system for precision agriculture needs has to be approximately 3 d (Dudka, 1997).

A big concern is how reliably in time the data over the same agricultural location can be collected. The Landsat Thematic Mapper, for instance, provides data of the same scene every 16 days. Repeat viewing of the same scene every 2-7 days would appear to be possible with future sensors (Stoney, 1996, 1997). The importance of obtaining frequent data from a field or vegetable crop is the ability to detect changes that otherwise would not be readily seen on the ground. Variations in sensing responses for instance, could indicate some crop abnormality. Plant stress conditions, which may be due to drought, drainage, insect damage, nutrient deficiency, or weed competition are evident by a loss of infrared reflectance. Detection of nutrient deficiency early in the growing season may result in an easy diagnosis and management recommendation such as fertilizer application (National Academy of Science, 1997). With the commercial remote-sensing era coming up, the data reliability will be increasing rapidly. The number of commercial launches, planned to achieve in the next 10 years is around 30 (Johannsen & Carter, 1997).

The upcoming commercial data is of significance to all users especially if the images can reach the grower in time for them to take appropriate actions. However, the quick enough concept would be different for diverse agricultural crops. How often the imagery is needed also depends on what farmers are trying to monitor and how they are able to incorporate the information into their management scheme. For example, Nichols (1997) states that sensitive, high-value crops such as potatoes, cotton, or vegetables may need weekly or bi-weekly imaging depending on the growth stage. Other crops may require monthly imaging. A single image taken at peak of vegetative growth might be enough if a farmer is using it as a guide to check on certain areas throughout the growing season. For those farmers who are on weekly or monthly basis, the best choice would be to get the first image after 50% of the canopy covers the row (Nichols, 1997) to reduce background reflectance.

Plant stress conditions are usually evident by a change of infrared reflectance. Again, the availability of frequent data is useful in monitoring such practices, all of which adds to the crop history of the land (Barber, 1982). Such information will greatly enhance precision farming because it indicates field and crop problems before harvest, when they can potentially be corrected before yield losses occur. The temporal resolution concept is often underestimated, mostly due to potential problems with consistency of collection of remote sensing data, but temporal dimension is important for precision agriculture to be actually precise.

SUMMARY AND CONCLUSIONS

Remote sensing technology will improve by increasing spatial, spectral and temporal resolutions starting during this year. There also will be an effort to provide the data within 24 to 48 h after it has been acquired. Another technology besides remote sensing that will assist in improving answers and interpretation will be Geographic Information Systems (GIS). The ability to merge soil maps with remotely sensed data to understand crop variability is a great asset to interpretation. Many farmers have commented that they would like more detailed soil information than currently provided by the standard soil maps. The ability to take other data such as terrain data, slope, aspect or even other remotely sensed data and look at crop variability causes many people to see many opportunities for better understanding of what is causing the variability.

Where does (GIS) and remote sensing fit with precision farming? One answer is that one would have a better understanding of what is causing yield variation within a field and that one has better management over their farm records. Several companies are starting to market GIS record-keeping systems so farmers can record all of the field operations such as planting, spraying, cultivation and harvest (along with specific information such as type of equipment used, rates, weather information, time of day performed, etc.). Additionally, farmers are able to record observations through the growing season such as weed growth, unusual plant stress or coloring and growth conditions. Data collected by the GPS operations can be automatically recorded with the GIS program. Remotely sensed data can be analyzed and added to the GIS using soil maps, digital terrain and field operations information as ground truth.

More attention is being paid to the type of information that farmers will need. It would appear that most remote sensing vendors will not be delivering raw images directly to all farmers. Rather they will provide data/information to the information multipliers or the value-added vendors such as agricultural business dealers, extension personnel, crop consultants, and special agricultural information services who in turn will analyze and interpret the data and deliver it to the farmer. Farmers are collecting a lot of supporting data and those analyzing the remote sensing data will need to gain access to the farmer's data. Farmers will be in a position to perform their own image analysis but we must remember the needed training aspects for this to be successful.

Advances in remote sensing technology are changing the way we will look at agriculture. The success of remote sensing will be measured by the type of information that is provided to the farmer, how quickly the information is delivered and the fee that is charged for the information. Competition for the farmer's business should help in making the success a reality.

REFERENCES

Barber, J.J. 1982. Detecting crop conditions with low-altitude aerial photography. *In* C.J. Johannsen, and J.L. Sanders (ed.) Remote sensing for resource management. Soil Conserv. Soc. at Am., Ankeny, IA.

Dudka, M., 1997. Precision agriculture initiative. WWW documents of Environmental Remote Sensing Center (ERSC) at the University of Wisconsin, Madison.

Hoffbeck, J.P., and D.A. Landgrebe. 1996. Classification of remote sensing images having high spectral resolution. Remote Sens. Environ. 57:119–126.

Johannsen, C.J., 1994. Precision Farming: Farming by the Inch. The Earth Observer, NASA, 6:24–25.

Johannsen, C.J. 1996. Farming by the inch. The World and I. February.

Johannsen, C.J., 1997. Glossary of terms for precision farming. Modern Agric., Issue 1, January–February. p. 44–46. Also published on Ag Electronics Association's URL at: http://www.agelectronicsassn.org/.

Johannsen, C.J. and P.G. Carter, 1997. Who's who in commercial satellite remote sensing: A status check. p. 4. *In* Proc. of InfoAg Conf., 7 August and published and updated at URL: http://dynamo.ecn.purdue.edu/~biehl/LARS/.

Liu, W., L.D. Gaultney, and M.T. Morgan. 1993. Soil texture detection using acoustical methods. Paper no. 93–1015. ASAE, St. Joseph, MI.

Loreto, A.B., and M.T. Morgan. 1996. Development of an automated system for field measurement of soil nitrate. Paper no. 96–1087. ASAE, St. Joseph, MI.

Lozano-Garcia, D.F., R.N. Fernandez, K. Gallo, and C.J. Johannsen. 1995. Monitoring the droughts in Indiana, USA. Int. J. Remote Sens. 16:(7)1327–1340.

Malthus, T.J., B. Andrieu, F.M. Danson, K.W. Jaggard, and M.D. Steven. 1993. Candidate high spectral resolution infrared indices for crop cover. Remote Sens. Environ. 46:204–212.

Martin, M.E., and J.D. Aber. 1997. High spectral resolution remote sensing of forest canopy lignin, nitrogen, and ecosystem processes. Ecol. Appl. 7:431–443.

Mitchell, P.A. 1995. Hyperspectral digital imagery collection experiment (HYDICE). p. 70–95. *In* Proc. of SPIE The Int. Soc. for Optical Eng. Paris, France.

Moran, M.S., Y. Inoue, and E.M. Barnes. 1997. Opportunities and limitations for image-based remote sensing in precision crop management. Remote Sens. Environ. 61:319–346.

National Academy of Sciences. 1997. Precision Agriculture in the 21st Century. Geospatial and information technology in crop management. Natl. Acad. Press. Washington, DC.

Nichols, C. 1997. Satellite imagery detects in-field variability. Modern Agric. 1(3):19–20.

Shonk, J.L., L.D. Gaultney, D.G. Schulze, and G.E. Van Scoyoc, 1991. Spectroscopic sensing of soil organic matter content. Trans. ASAE. 34:1978–1984.

Stoney, W.E. 1996. The Pecora legacy—land observation satellites in the next century. Pecora 13 Symp., Aug. 22, Sioux Falls, SD.

Stoney, W.E. 1997. Data deluge, the satellites are coming. Geotimes. 42 (10):18–22.

Vane, G.T., R.O. Green, T.G. Chrien, H.T. Enmark, E.G. Hansen, and W.M. Porter. 1993. The airborne visible/infrared spectrometer (AVIRIS). Remote Sens. Environ. 44:127–143.

Vane, G.T., and A.F.H. Goetz. 1993. Terrestrial imaging spectroscopy: current status and future trends. Remote Sens. Environ. 44:117–126.

Radar Imagery for Precision Crop and Soil Management

M. Susan Moran, Daniel C. Hymer, and Jiaguo Qi
USDA-ARS U.S. Water Conservation Laboratory
Phoenix, Arizona

Yann Kerr
Centre d'Etudes Spatiales de la Biosphere
Toulouse, France

ABSTRACT

Studies during the past 25 yrs have shown that measurements of surface reflectance and temperature (termed optical remote sensing) are useful for monitoring crop and soil conditions. Far less attention has been given to the use of radar imagery, even though Synthetic Aperture Radar (SAR) systems have the advantages of cloud penetration, all-weather coverage, high spatial resolution, day–night acquisitions, and signal independence of the solar illumination angle. In this study, we obtained coincident optical and SAR images of an agricultural area to investigate the use of SAR imagery for precision farm management. Results showed that SAR imagery was sensitive to variations in field tillage, surface soil moisture and vegetation density. The coincident optical images proved useful in interpretation of the response of SAR backscatter to soil and plant conditions.

INTRODUCTION

By the year 2000, there will be about 10 earth-observation satellites supporting optical sensors with the spatial, spectral and temporal resolutions suitable for many farm management applications (Moran et al., 1997a). These optical sensors provide information in the reflective and thermal emissive portions of the electromagnetic spectrum. In a multitude of studies, this information has been used for such important farm applications as scheduling irrigations, predicting crop yields, and detecting certain plant diseases and insect infestations (see review by Hatfield & Pinter, 1993). Although optical remote sensing is a powerful farm management tool, there are some serious limitations that have restricted farm management applications. For example, acquisitions are limited to cloudfree sky conditions; the signal is attenuated by the atmosphere; and image interpretation is a complex function of the sun/sensor/target geometry. An alternative to the use of optical remote sensing for farm management is the use of radar backscattering data obtained from Synthetic Aperture Radar (SAR) sensors. There are currently four SAR sensors aboard polar-orbiting satellites, and there are plans for two more by the year 2000.

Copyright © 1999 ASA-CSSA-SSSA, 677 South Segoe Road, Madison, WI 53711, USA. *Proceedings of the Fourth International Conference on Precision Agriculture.*

SAR sensors measure the spatial distribution of surface reflectivity in the microwave spectrum. The radar transmits a pulse and then measures the time delay and strength of the reflected echo (i.e., amplitude and phase measurements), where the amplitude is called the radar backscatter (σ^o). The scattering behavior of the SAR signal is governed by the dielectric properties of both soil and vegetation, and the geometric configuration of the scattering elements (soil roughness, leaves, stalks and fruit) with respect to the wavelength, direction and polarization of the incident wave. SAR systems have the advantages of cloud penetration, all-weather coverage, high spatial resolution, day–night acquisitions, and signal independence of the solar illumination angle (Table 1). These advantages allow SAR images to meet the rigid data requirements involved with precision farm management (PFM) decisions. Furthermore, for PFM applications, several inherent disadvantages of SAR imagery (Table 1) are countered by the *a priori* information generally available from farm managers, such as cultivation practices, crop type, planting date, row direction, soil type, and topography (particularly with laser-leveled or terraced fields).

The greatest weakness of SAR data for precision farming is the poor understanding of the response of SAR σ^o to agricultural soil and plant conditions. Research in the optical region has benefitted from three fortuitous circumstances: (i) the LACIE and AgRISTARS Programs, (ii) availability of inexpensive, handheld optical sensors, and (iii) access to reliable optical images from orbiting sensors, particularly Landsat TM and SPOT HRV. The Large Area Crop Inventory Experiment (LACIE) and AgRISTARS Programs defined the physics of relations between optical measurements and biophysical properties of crop canopies and soils. These pioneering programs established the potential of optical remote sensing for crop management, and inspired many subsequent studies of agricultural remote sensing. Subsequent studies advanced the science based on easy and often-inexpensive access to optical data obtained with handheld, airborne or satellite-based sensors. SAR research has not had such advantages. First, there has not been a research effort of the magnitude of the LACIE and AgRISTARS Programs. Second, there are no commercially-available, inexpensive, ground- or aircraft-based SAR sensors for intensive field experiments. Third, up until this decade, there have been no SAR sensors aboard polar-orbiting satellites (Table 2). These limitations make field studies of SAR applications for agricultural management very difficult at best.

In the study presented here, we attempted to capitalize on the good understanding of the response of the optical data to plant–soil conditions in order to interpret SAR images of an agricultural region. For five dates in 1995 through 1997, we acquired pairs of images from the Landsat TM sensor and the ERS-2 SAR sensor covering the University of Arizona Maricopa Agricultural Center in central Arizona. The information obtained from multispectral reflectance (ρ) and temperature (T_s) measurements made with the TM sensor was used to interpret the signal received by the ERS-2 C-band SAR sensor. In particular, we focused on the determination of within-field variations in

- soil roughness (related to tillage, subsidence or erosion);
- vegetation density (related to seeding, crop vigor and pest infestations); and
- surface soil moisture condition (related to monitoring irrigation efficacy, soil texture).

Table 1. Complementarity of optical and SAR remote sensing. The pros and cons of these data as a source of crop and soil information for precision agriculture applications.

Issue	Optical	SAR
Atmospheric attenuation	**CON:** Limited to periods of cloudfree sky conditions	**PRO:** Characterized by cloud penetration and all-weather coverage
	CON: Sensitive to atmospheric scattering and absorption	**PRO:** Independent of atmospheric scattering and absorption
		CON: Characterized by speckle †
Sun–Sensor –Target Geometry	**CON:** Surface temperature (T_s) and reflectance (ρ) are a complex function of solar and viewing angles (θ_z and θ_v, respectively)	**PRO:** SAR provides its own energy source; it is independent of θ_z and maintains a relatively constant θ_v
	CON: Reflectance measurements are limited to daylight hours	**PRO:** SAR allows acquisition 24h day^{-1}
	CON: Measurements of surface T_s and ρ are a complex function of topography	**CON:** SAR backscatter is a function of topography, though the correction with available DEM data is relatively straight-forward
Sensitivity to soil and vegetation conditions	**PRO:** Nominally independent of small-scale soil roughness conditions	**CON:** Very sensitive to soil roughness conditions, particularly at roughness scales similar to the wavelength (2–6 cm)
	CON: The signal from soil is attenuated by the signal from overlying vegetation	**CON:** The signal from soil is attenuated by the signal from overlying vegetation
	PRO: A good theoretical and empirical basis for application in farm management	**CON:** Poor understanding of the response of SAR backscatter to agricultural soil/plant conditions
Data availability	**PRO:** Orbiting sensors characterized by high spatial resolution (10–120 m) and wide coverage (60–180 km swaths)	**PRO:** Orbiting sensors characterized by high spatial resolution (12–20 m) and wide coverage (25–300 km swaths)
	PRO: Most optical systems are multi-spectral, allowing use of multiple bands to discriminate crop and soil conditions	**CON:** Currently orbiting SAR sensors (i) are not multi-frequency and (ii) are only low-frequency (C- or L-band)

† Speckle is the combination of scattering from lots of small scatterers within a pixel that causes the "grainy" appearance of the radar images. This effect can be alleviated by averaging several radar measurements together (through multi-looking or post-processing) to reduce variation with a consequent reduction in spatial resolution.

Table 2. Characteristics of four orbiting SAR sensors.

	ERS-1, ERS-2	JERS-1	Radarsat
Wavelength	5.7 cm (C)	23.5 cm (L)	5.7 cm (C)
Polarization	VV	HH	HH
Resolution	25 m	25 m	10–30 m
Swath Width	100 km	80 km	50–170 km
Incidence Angle†	23°	38°	17°–43°

† SAR scattering is strongly dependent on incidence angle (θ_i), e.g., specular reflection occurs at θ_i=0° and small changes in surface elevation are more easily visible at near-grazing angles (θ_i~80°-90°). Smooth vs. rough surfaces are easier to detect at θ_i>20°.

BACKGROUND AND THEORY

In the reflective region of the optical spectrum, discrimination of crop growth and plant status is generally accomplished by assessing the reflectance of red and near-infrared (NIR) reflectance (ρ_{Red} and ρ_{NIR}, respectively) of the plant canopy. Simply put, plants absorb red radiation and scatter NIR radiation resulting in a large difference between ρ_{NIR} and ρ_{Red}; in contrast, for bare soil, $\rho_{NIR} \approx \rho_{Red}$. This difference between plant and soil reflectances is often enhanced by computing a ratio of visible and near-infrared reflectances, termed a Vegetation Index (VI). A commonly-used VI is the Soil Adjusted Vegetation Index

$$\text{SAVI} = (\rho_{NIR}-\rho_{Red})/(\rho_{NIR}+\rho_{Red}+L)(1+L), \quad [1]$$

where L is a unitless constant assumed to be 0.5 for a wide variety of leaf area index values (Huete, 1988). SAVI has been found to be sensitive to such vegetation parameters as green leaf area index (GLAI), fraction absorbed photosynthetically active radiation, and percentage of the ground surface covered by vegetation.

In the thermal region, remotely sensed measurements of soil and foliage temperature have been linked to soil moisture content, plant water stress, and plant transpiration rate (e.g., Jackson, 1982). The sensitivity of surface temperature to plant and soil moisture conditions is related primarily to the heat loss associated with evaporation and transpiration. As such, the thermal signal is related to the percentage of the site covered by vegetation and the water status of the vegetation and soil (i.e., EvapoTranspiration or ET).

In the microwave region, specifically the C-band SAR wavelength (Table 3), it is generally assumed that σ^o is directly related to surface roughness, soil moisture and vegetation density. This can be expressed by the water-cloud model, in which the power backscattered by the whole canopy σ^o is the sum of the contribution of the vegetation σ^o_v and that of the underlying soil σ^o_s. The latter is attenuated by the vegetation layer as a function of τ^2, the two-way attenuation through the canopy. Thus,

$$\sigma^o = \sigma^o_v + \tau^2\sigma^o_s, \quad [2]$$

where τ^2 is a function of green leaf area index (GLAI), σ_v^o is a function of τ^2 and GLAI, and σ_s^o is a function of volumetric soil moisture content (h_v) and surface roughness (Ulaby et al., 1984; Prevot et al., 1993).

It is apparent from this short discussion that there is a relation between the optical and SAR sensitivities to variations in soil surface roughness, vegetation cover, and soil moisture (Table 4). Theoretically, as the surface roughness increases, σ^o increases due to increased SAR scattering, ρ_{Red} and ρ_{NIR} decrease due to increased surface shadows, and T_s and SAVI remain relatively unchanged. As crop cover decreases, σ^o increases due to an increase in τ^2, T_s increases due to decreased transpiration rate, ρ_{Red} increases due to decreased leaf chlorophyll, ρ_{NIR} decreases due to decreased leaf scattering, and the SAVI decreases dramatically. As surface soil moisture increases, σ^o increases due to a change in the surface dielectric constant, T_s decreases due to increased evaporation rate, ρ_{Red} and ρ_{NIR} decrease due to water absorption, and the SAVI remains relatively unchanged.

Table 3. Specifications and characteristics of commonly-used SAR spectral bands.

Spectral Band	**Wave-length**	**Examples of SAR responses to agricultural targets**
X	~3 cm	**Shorter wavelengths** are sensitive to plant parameters such as GLAI, plant biomass, and % vegetation cover; **Longer wavelengths** are sensitive to surface (1–5 cm) soil moisture content and attenuated by increasing vegetation cover; **All wavelengths** are sensitive to variations in surface roughness and topography (Prevot et al., 1993; Ulaby et al., 1994; Moran et al., 1997b, 1998)
C	~6 cm	
L	~24 cm	
P	~70 cm	

Table 4. Theoretical response of Optical and SAR measurements to changes in plant–soil condition.

Change in Plant–Soil Condition	**σ^o**	**T_s**	**ρ_{Red}**	**ρ_{NIR}**	**SAVI**
Increase in surface roughness	↑	–	↓	↓	–
Decrease in vegetation biomass	↑	↑	↑	↓	↓
Increase in surface soil moisture content	↑	↓	↓	↓	–

↑ indicates an increase, ↓ indicates a decrease, and – indicates no substantial change. σ^o is backscatter, T_s is surface temperature, ρ_{Red} and ρ_{NIR} are surface reflectance in the Red and NIR spectrum, and SAVI is the soil adjusted vegetation index.

For analysis of the SAR information, we defined a set of normalized difference (Δ_N) indices, where

$$\Delta_N\sigma^o = (\sigma^o{}_1 - \sigma^o{}_2)/(\sigma^o{}_X - \sigma^o{}_M), \quad [3]$$

$$\Delta_N T_s = (T_{s1} - T_{s2})/(T_{sX} - T_{sM}), \quad [4]$$

$$\Delta_N\rho_{Red} = (\rho_{Red1} - \rho_{Red2})/(\rho_{RedX} - \rho_{RedM}), \quad [5]$$

$$\Delta_N\rho_{NIR} = (\rho_{NIR1} - \rho_{NIR2})/(\rho_{NIRX} - \rho_{NIRM}), \quad [6]$$

$$\Delta_N SAVI = (SAVI_1 - SAVI_2)/(SAVI_X - SAVI_M), \quad [7]$$

and the subscripts 1 and 2 refer to two locations within the field, and subscripts X and M refer to the maximum and minimum values within the entire farm. These indices range from -1 to 1, and are indicative of the optical and SAR responses to changes in plant/soil condition summarized in Table 4.

EXPERIMENT

The site of the Agricultural SAR/Optical Synergy (ASOS) study was the University of Arizona Maricopa Agricultural Center (MAC). MAC is a 770 ha research and demonstration farm located about 48 km south of Phoenix. The demonstration farm is composed of large fields (up to 0.27 * 1.6 km) in which alfalfa is grown year-round, cotton is grown during the summer, and wheat is grown during the winter. A data management system is in place to archive planting, harvesting and tillage information, and the times and amounts of water, herbicide and pesticide applications. Since the predominant irrigation method for the MAC demonstration farm is flooding, each field is dissected into level-basin borders.

The ASOS study was conducted in two parts. A retrospective study was conducted based on existing images in the European Space Agency (ESA) and EROS Data Center (EDC) archives. These images from 1995 and 1996 were ordered with the intent of determining field soil moisture, vegetation cover, and tillage conditions based on the response of the optical and SAR signals, and validating these determinations with the field notes archived by the MAC Farm Manager. A second study was conducted in which we ordered TM/SAR image pairs for three dates (May, June, and July) in 1997. During all three overpasses, we arranged for one field to be flood irrigated such that a large portion of the field was saturated, and, for contrast, a large portion was completely dry. A kenaf crop was planted in May, and by the June overpass dates, the GLAI was 0.3; by the July overpass, the GLAI was 1.5. We also monitored vegetation and soil moisture conditions in two fields of alfalfa at various growth stages with a variety of soil moisture conditions.

During each TM/SAR overpass in 1997, we made ~50 gravimetric measurements of soil moisture content to 5-cm depth in the dry and wet portions of the fallow field and in the two alfalfa fields. These were converted to volumetric soil moisture using estimates of field bulk density. We also measured GLAI in situ at multiple locations using a LICOR LAI2000 plant canopy analyzer.

The SAR raw data were averaged to one value for each field border (a minimum of 100 pixels) to minimize the speckle effect, and the mean was converted to values of σ^o according to Moran et al. (1997b). The TM raw data were converted to values of apparent reflectance and radiometric temperature according to Moran et al. (1995) and Markham and Barker (1986). The term apparent reflectance refers to reflectance factors derived from satellite images that have not been corrected for atmospheric effects. Considering that the TM data were acquired on days with clear, dry atmospheric conditions, the difference between apparent and surface reflectance in the Red and NIR wavelengths should be minimal. The radiometric temperature (T_r) was converted to surface kinetic temperature (T_s) based on measurements of surface emissivity (ε) using the relation $T_s = (T_r^4/\varepsilon)^{1/4}$, where $\varepsilon = 0.98$ for dense alfalfa, $\varepsilon = 0.95$ for rough bare soil and recently-harvested alfalfa, and $\varepsilon = 0.89$ for laser-leveled bare soil (Reginato & Jackson, 1988).

For a number of reasons, the ASOS study did not go as smoothly as planned. First, there were few TM/SAR pairs available in the ESA and EDC archives. We were only able to obtain images for November and December 1995 and 1996 (Table 5). During this time of year, there was very little farm activity, and the only crops were alfalfa and emergent wheat. Second, though we ordered the ERS-2 SAR and Landsat TM images for May, June, and July 1997, we only received one SAR/TM image pair (May 1997; Fig. 1). The reason for the failure to obtain the images as ordered is still unknown; however, such acquisition failure is not uncommon for satellite-based sensors, as reported by Moran (1994).

Table 5. ERS-2 SAR and Landsat TM scenes ordered for the 1995–1996 Agricultural SAR/Optical Synergy (ASOS) Study.

ERS-2 SAR	Landsat-5 TM	Notes
6 Nov. 1995	8 Nov. 1995	Wheat planted; cotton harvested; several disked fields; no irrigations
11 Dec. 1995	10 Dec. 1995	
25 Nov. 1996	26 Nov. 1996	
30 Dec. 1996	28 Dec. 1996	
19 May 1997	21 May 1997	Soil moisture study with bare soil conditions in Field 3
23 June 1997	22 June 1997	Soil moisture study with kenaf GLAI = 0.3 in Field 3 *SAR scene not acquired by ESA*
10 July 1997	9 July 1997	Soil moisture study with kenaf GLAI = 1.5 in Field 3 *Neither the SAR nor TM scene was acquired*

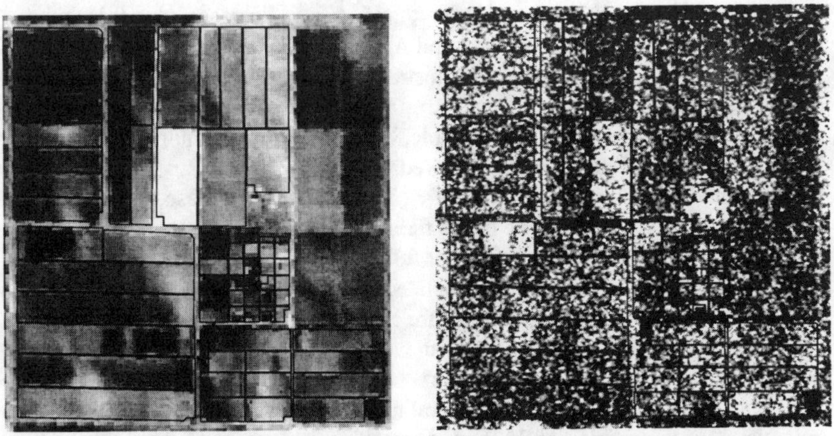

Fig. 1. Images of Landsat TM reflectance (left) and ERS-2 SAR backscatter (right) covering Maricopa Agricultural Center acquired on 21 May and 19 May 1997, respectively. The vector overlay designates the MAC field borders, and the total area covers 770 hectares.

RESULTS AND DISCUSSION

Retrospective ASOS Study 1995–1996

For this preliminary analysis, we selected all MAC fields in the four 1995–1996 images that had a record of distinctive within-field differences in tillage, soil moisture, and vegetation density. Since results were similar for fields of similar surface conditions, three fields were selected as examples for illustration in this section. According to field notes and on-site observations, Field 1 was fallow, but part of the field had been laser leveled and part was still rough due to cultivation; Field 2 was planted with alfalfa, but one-half of the field had been recently harvested, and Field 3 was also fallow, but part of the field had been flood irrigated.

All three fields (numbered 1 to 3 for reference herein) had a notable increase in the SAR σ^o ($\Delta_N\sigma^o \sim 0.2$) from one end of the field to the other (Fig. 2 and 3). The increase in $\Delta_N\sigma^o$ in Field 1 was due to the increased scattering of the SAR signal due to soil roughness. In Field 2, the increase in $\Delta_N\sigma^o$ resulted from a decrease in the alfalfa crop density due to a recent harvest, resulting in a larger τ^2 value in Eq. [2]. In Field 3, $\Delta_N\sigma^o$ increased due to the change in soil moisture and the sensitivity of the SAR signal to the dielectric constant of the surface. The dielectric constant of water is about 80 (in the C-band wavelength) and that of dry vegetation or soil is about 2 to 3.

The visual and quantitative assessment presented in Fig. 2 and 3 showed that the response of the optical data to the three different field conditions corresponded well with the theoretical hypotheses presented in Table 4. In Field 1, as the soil roughness increased, $\Delta_N\rho_{NIR}$ and $\Delta_N\rho_{Red}$ decreased by 0.2 due to increased surface shadows, and $\Delta_N T_s$ and Δ_NSAVI remained near zero for the two roughnesses. In Field 2, as the vegetation decreased due to harvest, $\Delta_N T_s$ increased by about 0.2 due to the decrease in transpiration, $\Delta_N\rho_{NIR}$ decreased by 0.5 and $\Delta_N\rho_{Red}$ increased by 0.4 due to the decrease in leaf area and photosynthetic activity, causing a decrease in Δ_NSAVI of 0.62. In Field 3, as the soil moisture increased, $\Delta_N T_s$ decreased by about 0.5 due to evaporative cooling, $\Delta_N\rho_{NIR}$ and $\Delta_N\rho_{Red}$ decreased by 0.1 and 0.2 respectively due to water absorption, and Δ_NSAVI remained near zero.

Based on data for fields not illustrated in Fig. 2 and 3, we found that the optical data were also useful for discriminating mixes of effects of roughness, vegetation and soil moisture. For example, in the SAR image acquired in November 1995, two adjacent fields of alfalfa showed no difference in SAR σ^o ($\Delta_N\sigma^o \sim 0$). Yet, we computed large negative values of $\Delta_N T_s$ and Δ_NSAVI. Based on the optical response, we postulated that one of the fields was recently harvested and had a low soil moisture content; the other was near full vegetation cover and had been recently irrigated. As a result, the high σ^o associated with low crop cover was offset by the low σ^o associated with high soil moisture content, and $\Delta_N\sigma^o \sim 0$.

Overall, the Δ_N indices worked well to discriminate the causal relation between surface conditions and SAR σ^o. Though results for only three fields are illustrated here, similar results for several more fields showed that this method has potential for interpretation of SAR imagery with coincident optical imagery. These results also illustrated the sensitivity of Landsat TM and ERS-2 SAR imagery to differences in tillage, surface soil moisture, and vegetation density.

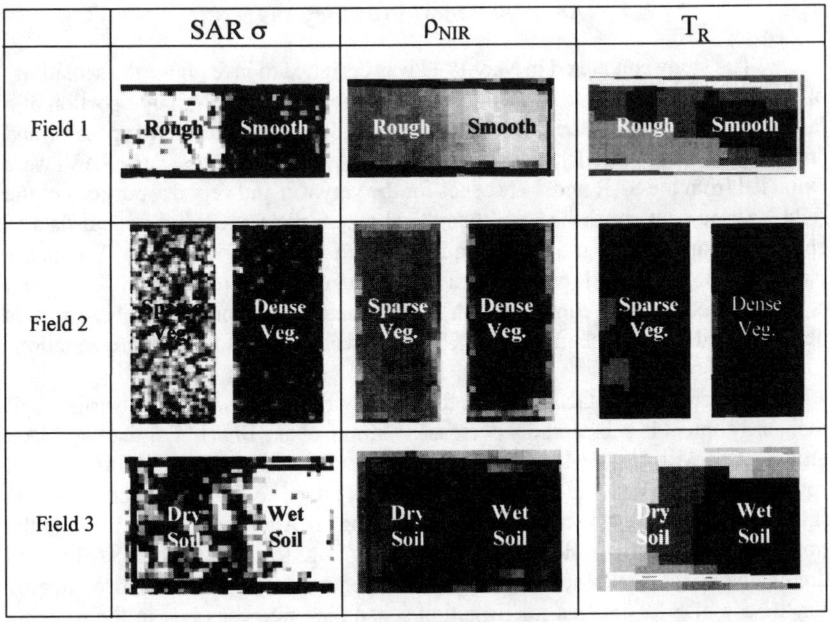

Fig. 2. Extracts of SAR and optical data for the three study fields, illustrating the differences in spectral response in SAR backscatter (σ), NIR reflectance (ρ_{NIR}), and radiometric surface temperature (T_r) to variations in field tillage, vegetation density, and surface soil moisture.

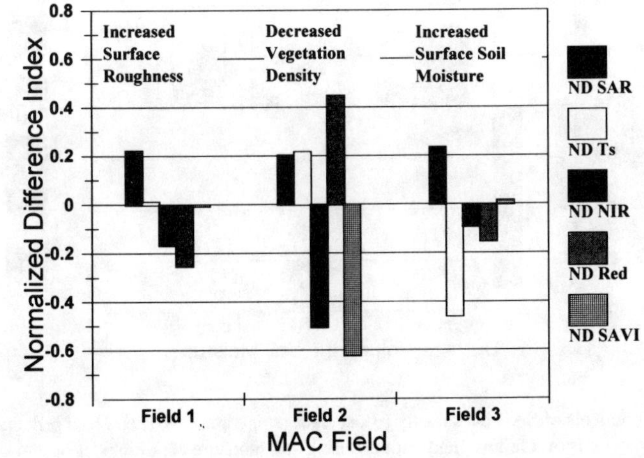

Fig. 3. The response of Δ_N indices (Eq. [3]-[7]) to variations in field roughness, vegetation density and surface soil moisture. The five legend captions refer to $\Delta_N\sigma^o$, $\Delta_N T_s$, $\Delta_N\rho_{Red}$, $\Delta_N\rho_{NIR}$, and Δ_NSAVI, respectively.

ASOS Soil Moisture Study 1997

The study conducted in May 1997 was designed to investigate the sensitivity of SAR and optical data to differing soil moisture conditions. A large portion of a fallow field was flood irrigated during the ERS-2 and Landsat overpasses, and another portion was left dry. Measurements of SAR σ^o, T_s, ρ_{Red}, ρ_{NIR}, and SAVI were extracted from the SAR and TM scenes for the very wet and very dry portions of the field. These data confirmed the theoretical response of SAR and optical data to changes in surface soil moisture conditions (Fig. 4). That is, for a soil moisture increase of 35%, the SAR σ^o increased by nearly 8 dB, T_s decreased by 8 °C, ρ_{Red} and ρ_{NIR} decreased by 0.07 each, and SAVI remained nearly constant. These results demonstrated the large changes in σ^o, T_s, ρ_{Red}, and ρ_{NIR} due to soil moisture variations for bare soil conditions.

For crops with GLAI > 1.0, the sensitivity of the SAR σ^o to surface soil moisture content is substantially decreased (Moran et al., 1998). For the two MAC alfalfa fields with GLAI~4.0, the σ^o was completely insensitive to the difference in soil moisture in the two fields, and instead, responded to the differences in GLAI (Fig. 5). That is, the σ^o increased with decreasing GLAI. According to Eq. [2], the transmittance through the dense alfalfa canopy (τ^2) was low, and thus the SAR σ^o was dominated by the backscatter signal from the vegetation (σ^o_v). This is discouraging for the use of SAR images for irrigation scheduling purposes late in the growing season. However, information about surface soil moisture conditions obtained early in the growing season will still be useful for monitoring irrigation efficacy, mapping precipitation events, and determining soil texture.

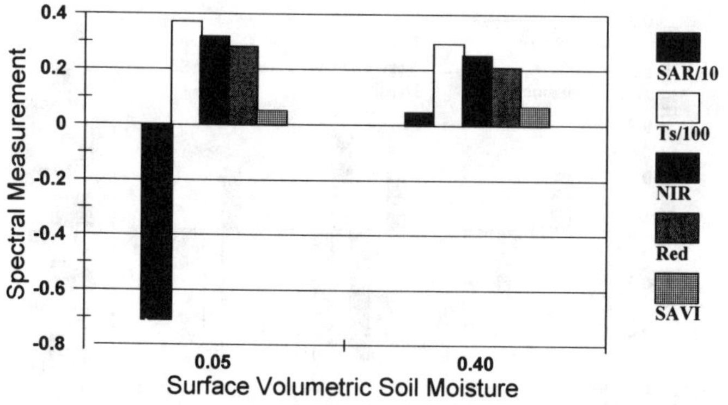

Fig. 4. Measurements of SAR backscatter (σ), surface temperature, NIR and Red reflectance and SAVI in two sections of a fallow field with differing soil moisture conditions. For purposes of graphic clarity, the SAR σ^o values were divided by 10 and the T_s values were divided by 100.

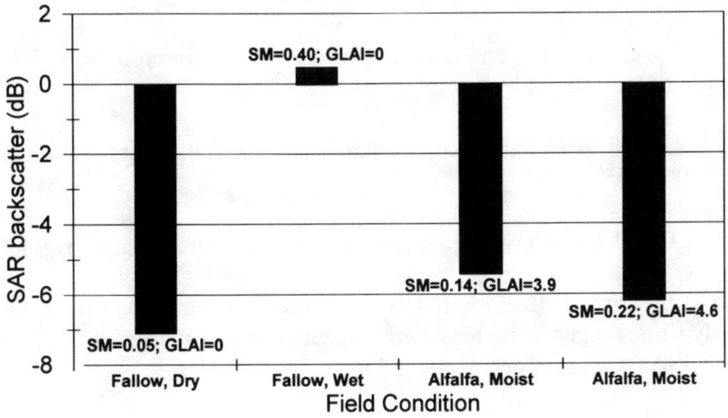

Fig. 5. A comparison of the sensitivity of SAR backscatter to soil moisture and vegetation density conditions for a fallow field and an alfalfa field. In the figure, the bars are labeled with measurements of volumetric soil moisture (SM) and green leaf area index (GLAI).

CONCLUDING REMARKS

The objective of this study was to investigate the utility of SAR images for precision farm management applications. These preliminary results showed that the SAR σ^o was sensitive to differences in field roughness (related to tillage), vegetation density, and surface soil moisture. Furthermore, we found that optical imagery obtained coincident with SAR imagery allowed a better understanding of the interactions of the SAR signal with soil and plant surfaces. Thus, it may be possible to model SAR σ^o based on optical measurements rather than the time-consuming *in situ* measurements of surface roughness and GLAI. Future work on this data set will be focused on compiling the SAR, optical and field information necessary to develop a relation to facilitate interpretation of the SAR image, which may take the form

$$\Delta_N\sigma = a + b\Delta_N T_s + c\Delta_N\rho_{NIR} + d\Delta_N\rho_{Red} + e\Delta_N SAVI, \quad [8]$$

where the parameters *a-e* are empirical coefficients determined by multiple regression analysis. Recognizing the limitations of optical remote sensing data due to cloud interference and atmospheric attenuation, the findings of this study should encourage further studies of SAR imagery for crop and soil assessment.

ACKNOWLEDGMENTS

We would like to acknowledge valuable help from Wanmei Ni, Chandra Holifield, Ross Bryant, Robin Marsett, and Michael Helfert, and the cooperation of personnel at the Univ. of Arizona Maricopa Agricultural Center (MAC). Financial support was provided by the NASA Landsat Science Team (NASA S-41396-F), NASA EOS IDS (NAGW-2425), and European Space Agency (ESA-AO2.F119).

REFERENCES

Hatfield, J.L., and P.J. Pinter, Jr. 1993. Remote sensing for crop protection. Crop Protec. 12:403–414.

Huete, A.R. 1988. A soil-adjusted vegetation index (SAVI). Rem. Sens. Environ. 25:89–105.

Jackson, R.D. 1982. Canopy temperature and crop water stress. Adv. Irrig. 1:43–85.

Markham, B.L., and J.L. Barker. 1986. Landsat MSS and TM post-calibration dynamic ranges, exoatmospheric reflectances and at-satellite temperatures. EOSAT Landsat Tech. Notes 1:3–8.

Moran, M.S. 1994. Irrigation management in Arizona using satellites and airplanes. Irrig. Sci. 15:35–44.

Moran, M.S., R.D. Jackson, T.R. Clarke, J. Qi, F. Cabot, K.J. Thome, and B.L. Markham. 1995. Reflectance factor retrieval from Landsat TM and SPOT HRV data for bright and dark targets. Rem. Sens. Environ. 52:218–230.

Moran, M.S., Y. Inoue, and E.M. Barnes. 1997a. Opportunities and limitations for image-based remote sensing in precision crop management. Rem. Sens. Environ. 61:319–346.

Moran, M.S., A. Vidal, D. Troufleau, J. Qi, T.R. Clarke, P.J. Pinter, Jr., T.A. Mitchell, Y. Inoue, and C.M.U. Neale. 1997b. Combining multi-frequency microwave and optical data for farm management. Rem. Sens. Environ. 61:96–109.

Moran, M.S., A. Vidal, D. Troufleau, Y. Inoue, and T.A. Mitchell. 1998. Ku- and C-band SAR for discriminating agricultural crop and soil conditions. IEEE Geosci. Rem. Sens. 36:265–272.

Prevot, L., I. Champion and G. Guyot. 1993. Estimating surface soil moisture and leaf area index of a wheat canopy using dual-frequency (C and X bands) scatterometer. Rem. Sens. Environ. 46:331–339.

Reginato, R.J., and R.D. Jackson. 1988. Emissivity measurements of soil and plant surfaces. p. 9. *In* The MAC III Experiment Report. USDA-ARS U.S. Water Conserv. Lab., Phoenix, AZ.

Ulaby, F.T., C.T. Allen, G. Eger III, and E. Kanemasu. 1984. Relating the microwave backscattering coefficient to leaf area index. Rem. Sens. Environ. 14:113–133.

Remote Sensing Research in Wisconsin Soybean and Corn Production Fields

R. T. Schuler
Biological Systems Engineering Department
University of Wisconsin
Madison, Wisconsin

S. J. Langton
Cooperative Extension Service
University of Wisconsin
Madison, Wisconsin

R. P. Wolkowski
Soil Science Department
University of Wisconsin
Madison, Wisconsin

M. Dudka and J. D. Gage
Environmental Remote Sensing Center (ERSC)
University of Wisconsin
Madison, Wisconsin

T. M. Lillesand
ERSC/Civil Engineering Institute of Environmental Studies
University of Wisconsin
Madison, Wisconsin

ABSTRACT

Remote sensing provides an opportunity to collect information about crop growth during the growing season. The ability to relate remotely sensed data to physical anomalies of the growing crop or soil will permit producers an opportunity to correct problems in a timely manner. Relationships between remotely sensed data and field anomalies are very limited. In this study, three production fields (16 to 24 ha), two in corn and one in soybeans, were used to compare ground and remotely sensed data. These data were collected four times during the 1997-growing season by low-altitude fixed-wing aircraft with instrumentation having a resolution of one meter. The wavelengths of the remotely sensed data were red, blue, green, infrared and near-infrared. Geospatial crop and soil data were collected on the ground throughout the growing season. At the times of remote sensing, intensive ground soil and crop data were collected. Some of the anomalies observed in these production fields were caused by white mold infestation in soybeans, weed infestations, wheel traffic, planter and sprayer malfunctions, tillage, topography and soil morphology. The results of these anomalies were observed in the remotely sensed data.

Copyright © 1999 ASA-CSSA-SSSA, 677 South Segoe Road, Madison, WI 53711, USA. *Proceedings of the Fourth International Conference on Precision Agriculture.*

INTRODUCTION

Recent technological developments have made remote sensing more available for agricultural applications. Remote sensing provides data to evaluate the spatial variability in agricultural fields. Crop scouts and producers can then use the data to more efficiently conduct their ground assessments and focus their efforts in the problem areas of a field. Also remote sensing can be used to identify not only the location but also the cause of the variability. Since a large number of factors that can cause spatial variability, identifying causes of variability with remote sensing has had a limited degree of success to date.

Advancements that attributed to the greater feasibility of remote sensing for agriculture include global positioning systems, improved low cost computers capable of managing large databases and improved equipment for collecting remote sensing data. In addition, the costs of these new technologies are at a level, which makes remote sensing feasible in many crop production systems, especially high value crops such as fruits and vegetables.

A field study was initiated to evaluate which factors most effectively describe the spatial variability in three crop production fields. During the 1997 growing season extensive spatial field data such as crop yield, soil fertility, crop growth and stand, were collected. Four times during the growing season, remotely sensed data were collected. At the time of remotely sensed data collected extensive spatial field data were collected.

In addition soils maps and digital elevation models data were obtained for additional analysis. The focus of this study was to identify the causes of any anomalies, which appeared in the remotely sensed and yield data and evaluate the ground problems that are observable in the remotely sensed data.

LITERATURE REVIEW

Remote sensing is the collection of data where the energy reflected or emitted from a surface or object. It is a non-contact method of collecting data. The research challenge is to relate this energy to a useful characteristic of the surface or object.

Frazier et al. (1997) divided remotely sensed data into direct and indirect as it is applied to site-specific crop management. Direct application is the use of remotely sensed data to control the application rate of a machine such as planter, sprayer and fertilizer applicator. The data may be from an image in a on-board computer or machine mounted sensor based the visible and/or infrared energy being reflected or emitted.

Another method of dividing the remotely sensed data as applied to site specific management into groups is based on ground or air transport of the sensor. The ground sensor lends more direct use of the data as in variable rate application.

Much of the site-specific research using remote sensing has been directed at fertility, primarily nitrogen. Using photometric sensors mounted on the application machine, Blackmer et al.(1995) studied a system to apply N to corn.

The system senses the corn canopy color, an indication of the nitrogen stress, and applies N based on a reference color for well-fertilized corn. Stone et al. (1996) studied the uses of remote sensing for correcting N fertilizer deficiencies in winter wheat during the growing season. Near infrared (774 to 786 nm) and visible red (665–677 nm) were used to developed a plant N spectral index which was an indication plant nitrogen stress. Their research demonstrated a significant fertilizer savings for variable rate application over a uniform application.

Wanjura and Hatfield (1987) evaluated the sensitivity of some spectral indices to crop biomass on the soil surface. Studying cotton, soybeans, sunflowers and grain sorghum with a tractor mounted sensor, they found some vegetation indices (ratio of visible red and near infrared) were more sensitive at higher levels of canopy biomass and leaf area. At lower levels of leaf area and biomass other vegetation indices (sum of visible red and near infrared) were more accurate.

Additional applications of vegetation indices listed by Moran et al. (1997) were crop phenology, crop growth, crop disease, crop evapotranspiration rate, and weed and insect infestations. Remote sense has the potential to provide information about the soil surface. Walters (1995) used remotely sensed data to divide a complex soil landscape into logical units. By dividing a agricultural field into areas based on soil properties, management zones can be defined for site specific management of crops. The remotely sensed data may also be used to improve the accuracy of the county soil maps currently published. In addition, this remotely sensed data may provide an opportunity to decrease the intensity of grid soil sampling and still create a accurate fertility map (Frazier et al., 1997). These management zones can also be established by using aerial remotely sensed data from the growing crop in a field (Yang et al., 1998).

The acquisition of remotely sensed data by crop producers is becoming more available (Davis, 1998). Although the producers can use the data to gain additional information about their fields that they will find to be useful, much research is needed to obtain the full potential this data can provide.

A National Research Council report describes the importance of site specific management and the area of needed research to insure successful and profitable adoption. Remote sensing was identified as an important element of site specific management where additional research was needed.

Although much research is needed, remotely sensed data with other data has the potential of identifying problems in crop fields during the growing season. This provides an opportunity for the producer to take corrective action during the growing season. Although yield data is very useful, but it is information after the problems have affected the yield. This is particularly important because many producers with crop yield monitors find large season to season variability in their fields.

STATEMENT OF GOALS

The primary goal of this study was to develop a systematic approach for determining the most important measurable factor limiting production in a site-specific managed field. Specific objectives were to

1. determine the level of variability and the relationships among multi-year site-specific data on production fields,
2. collect and evaluate topographic and soil morphology data more accurately than available in published data,
3. collect and evaluate multi-spectral remotely sensed data as a tool for assessing soil and crop variability and crop pest problems, and
4. study the influence of within field variability on com and soybean production.

METHODS

Three crop production fields located north of Madison, WI, in Dane County, 16 to 24 ha each, were identified for intensive geo-referenced data collection. During the 1997 growing season, two fields were in corn (Zea mays L.) production and one field was in soybeans (Glycine max L.). These fields were identified the names of current or previous owners. The Stone Corp was in soybeans for 1997 and was in corn production during 1996. The Caldwell and Watzke fields were in corn during 1997. In 1996, the Caldwell field was in corn and the Watzke field was split, one quarter of it was soybeans and the remainder was corn. All the data reported here were observed during the 1997 growing season. The soils in all three fields were primarily silt loam.

The fields were mapped with a Global Positioning System (GPS) receiver and grid soil sampled. The fields were soil sampled on 0.4–ha grids. Periodically during the growing season the crop plant and weed population and plant growth data were collected and geo-referenced. Crop yield data also were collected and geo-referenced with grain combine yield monitoring system recording at one second intervals.

As this field data were collected, several observations were made with respect to impact of equipment and disease infestations. Examples of equipment effect included planter skips, sprayer skips and wheel traffic during a late herbicide application. Using a global positioning system, these areas were mapped for later analysis. In the soybean field, white mold infestations were rated and mapped. Because of anomalies observed in the multi-spectral remote sensing data, additional observations were made in adjacent fields regarding tillage practices and pesticide applicator problems.

Remotely sensed data were collected by Airborne Data Systems. (Wabasso, MN) with their instrument known as SPECTRA-VIEW. The multi-spectral data were collected with CCD cameras with the appropriate filtering and with a spatial resolution of 1 m. The particular spectral coverage they provided is listed in Table 1. They collected data on 12 May, 5 July and 11 September. On 12 May, the soil was free of green vegetation and was partially covered with crop residue from the previous growing season. On 5 July, the crop canopy had reach complete closure, the surface of the soil was not visible from an aerial view. Seed development was occurring on 11 September. The data were transferred to CD-ROM and processed by Environmental Remote Sensing Center staff and visible and infrared maps were created.

Table 1. Airborne Data Systems sl2ectral coverage for the remote sensing data.

Visible Range				
Wavelength (μm)	0.4–0.5	0.5–0.6	0.6–0.7	
Color	Blue	Green	Red	
Infrared				
Wavelength(μm)	0.7–0.8	0.6–2.2	2.4–4.5	8.0–14
Type	Near	Near	Thermal	Thermal

In addition, US-NASA provided remotely sensed data from their ATLAS sensor for 24 September with a resolution of 3 m. On this date seed development was nearly complete. In some areas the crop was undergoing early senescence. With their airborne terrestrial applications sensor the spectral coverage included four wavelengths in the visible range, four in the near-infrared and six in the thermal infrared, Table 2. To minimize camera scale distortions created by cross scanning, using WiscImage, a software designed for this purpose (Scarpace, 1998).

Table 2. NASA Airborne Terrestrial Applications Sensor (ALTAS) spectral coverage for the remote sensing.

Visible range				
Wavelengths (μm)	0.45–0.52	0.52–0.60	0.60–0.63	0.63–0.69
Color	Blue	Green	Orange	Red
Infrared (Near)				
Wavelengths (μm)	0.69–0.76	0.76–0.90	1.55–1.75	2.08–2.35
Infrared (Thermal)				
Wavelengths (μm)	8.20-8.60	8.60-9.00	9.00-9.40	9.60-10.2
	10.2–11.2	11.2–12.2		

Topographic errors were corrected using ERDAS Imagine, which provided image to image registration of remote sensing data and Dane County, WI orthophotographs. A digital elevation models (DEM) of the Caldwell and Stone Corp fields were developed from the digitized aerial photographs using softcopy photogammetry and precise elevation data in the fields with survey quality GPS receivers. All data were registered to a common coordinate system using ARCinfo and ERDAS Imagine. The Dane County coordinate system was used due it its accuracy in the small geographic area of this study.

With the various wavelengths and dates, a large number of analyses can be made because the number of wavelengths can be studied and various combinations of wavelengths can be evaluated. For the discussion in this paper, the primary focus will be the data collected on 5 July and 24 September. The near infrared wavelengths were used in the analysis up to this point. All the wavelengths were listed to provide an indication of the potential for additional analysis. The following discussion is directed at the observations related to the machinery impact and other related data. White mold disease in the Stone Corp soybean field is discussed in detail in Use of Digital Imagery to Evaluate Disease Severity and Yield Loss Caused by White Mold of Soybeans, another report in these proceedings.

RESULTS AND DISCUSSION

No extensive statistical analysis was done to determine the degree of correlation between the various sets of data at this stage of this study. Qualitative observations were made when comparing the maps for infrared data and other ground geo-reference data. Discussion will be organized field by field. The remotely sense data collected on 12 May is not discussed even though it provided limited information. Weather and equipment limited the quality of the 11 September data.

Caldwell Corn Field

In the Caldwell com field, an anomaly appeared as a two east-west lines (255 and 131 m long and 1 to 2 m wide), oriented in the same direction as the planting pattern on the 5 July remote sensing imagery. Ground observations identified the lines were created by planter malfunction where one row was not planted for the distance of the anomaly. For the 24 September data, these lines were still observable. In addition, three areas of the field exhibited another anomaly for the 24 September data. The corn in these areas reached early senescence, which was closely associated with topographic and soil variability in the field. The corn grain yield map exhibited patterns similar to the 24 September remotely sensed data. Lower yields occurred at the planter skips and the areas of early senescence.

In a soybean field adjacent to the Caldwell field, the remote sensing data exhibited a pattern on the east headlands and northern edge. Upon further evaluation this pattern was caused inadequate cleaning of the spraying equipment when switching from a corn herbicide to the soybean herbicide used for this field.

In another soybean field adjacent to the Caldwell field, a research study was being conducted to evaluate soybean varieties and tillage. The 24 September data exhibited difference among the tillage treatments. This was believed to be related to degree of white mold infestations, most in the moldboard plow and least in the no-till.

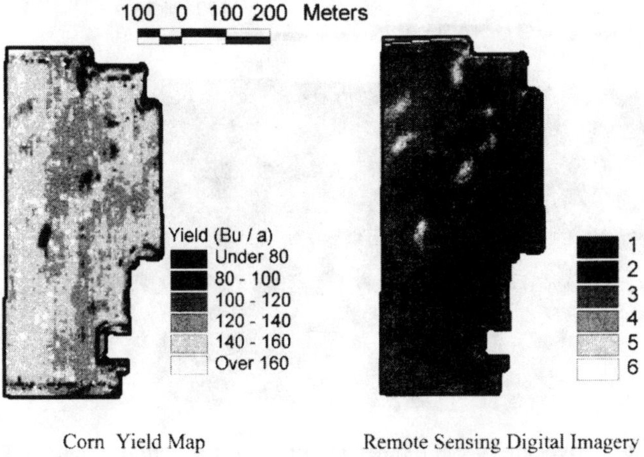

Fig. 1. Watzke corn field yield map and 24 September remote sensing imagery (Band 6, 0.76 to 0.90 μm).

Watzke Corn Field

The Watzke field had a different cropping pattern in 1996 when a portion of the field was in corn and the remaining portion was in soybeans. The boundary between these two areas of the field was apparent in all the remotely sensed data collected on this field in 1997 when corn was grown. In Fig. 1, the 1996 soybean area is seen in the western one quarter of the corn field. The corn growing on the 1996 soybean area was undergoing earlier senescence. The impact the soybeans is observed in the yield map, (Fig. 1).

The white spots in the remotely sensed imagery in Fig. 1 are areas of early senescence and the crop was moisture stressed in these areas. These spots are associated with gravelly eroded knolls. The crop yields were lower in these areas (Fig. 1). These areas were noticeable on the remotely sensed data from 12 May.

An additional observation from Watzke field included the headlands where higher levels of soil compaction took place. This is more apparent on the northern headlands (Fig. 1).

Stone Corp Soybean Field

Remotely Sensed Digital Imagery

The Stone Corp soybean field remotely sensed data exhibited a larger number anomalies than the other two fields. In Fig. 2, the white areas within the field in the remotely sensed digital imagery is associated with waterways. Although

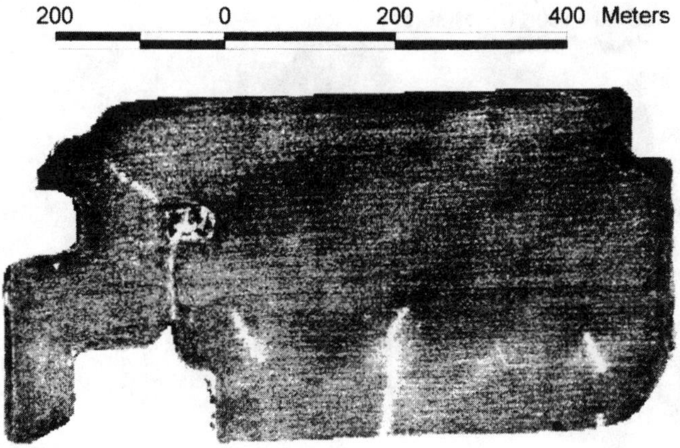

Fig. 2. Stone Corp soybean 24 September remote sensing digital imagery (Band 6, 0.76–0.90 μm).

soybeans were planted in the waterways in a east–west pattern, the soybean population was lower in this area and weed infestations were present in the waterways, primarily perennial weeds.

In Fig. 2, there appears to be east–west pattern or grain in the digital imagery. This was more apparent in the 5 July digital imagery because of its one meter resolution compared to three meters in Fig. 2. This is associated with late herbicide application when the tractor and applicator wheel traffic damaged the growing soybean plants. Scouting the field revealed that a herbicide was recently applied to the drilled soybeans. In addition, a sprayer skip was observed in the July imagery but it is not apparent in Fig. 2.

Other characteristics to note on the digital imagery are the compaction effects on the East and West headlands, similar to the Watzke corn field. The area west of the field center that has a number of anomalies is not in production. Several large trees are growing in that area.

This field had white mold infestations in the southern area of this field, Fig. 2. They occur east and west of the north–south waterway in the south–center area of the field. These areas are lighter in color. This issue will be discussed in much more detail in another paper in these proceedings, titled Use of Digital Imagery to Evaluate Disease Severity and Yield Loss Caused by White Mold of soybeans.

Stone Corp Soybean Yield

The patterns exhibited by the digital imagery in Fig. 2 are present in the soybean yield map (Fig. 3). The waterways in the southern area have higher

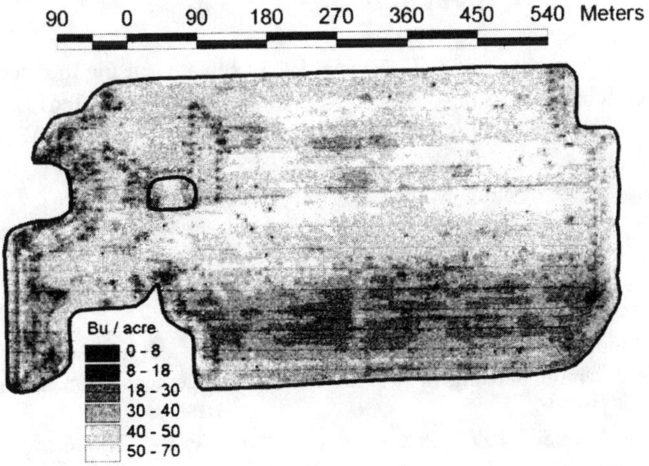

Fig. 3. Stone Corp soybean yield map.

yields than the adjacent areas that are associated with high levels of white mold infestation.

A east–west pattern believed to be created by the late pesticide application is present but different than the digital imagery. Some of this difference can be attributed to the spatial resolution of the data. The digital imagery had a 3 m resolution while the harvesting width of the combine was 4.6 m. And the resolution in the direction of travel was estimated to be 1.8 m with 1-s interval for yield data collection.

SUMMARY

Although no extensive statistical analysis has been done with these data, the patterns of the remotely sensed data are very similar to the patterns caused by diseases, equipment operations and weed infestations. The yield maps exhibit many of the same patterns as the digital imagery. This an indication of the potential for the use of remote sensing and it becomes more useful when combined with other data that is available from other databases found in county, state and federal agencies. Examples include topographic files and soil texture information.

The primary advantage of remote is obtaining information during the growing season. In some cases crop producers will have an opportunity to address the problems limiting growth immediately. As remote sensing becomes more available, the cost will go down and it will become more feasible to use by crop producers using site specific management.

REFERENCES

Blackmer, T.M., J.S. Schepers, and G.E. Meyer. 1995. Remote sensing to detect nitrogen deficiency in com. p. 505–512. *In* P.C. Robert et al. (ed) Site specific management for agricultural systems. ASA Misc. Publ. ASA, CSSA, and SSSA, Madison, WI

Davis, G. 1998. Space age farming. Modern Agric. 1(5):18–20.

Frazier, B.E., C.S. Walters, and E.M. Perry. 1997. Role of remote sensing in site-specific management. *In* The State of Site-Specific Management for Agriculture. F. J. Pierce and E. J. Sadler (ed.) ASA, CSSA, and SSSA, Madison, WI.

Moran, M.S., Y. Inoue, and E.M. Barnes. 1997. Opportunities and limitations for image-based remote sensing in precision crop management. Rem. Sens. Environ., 6:319–346.

National Resource Council. 1997. Precision agriculture in the 21st century: geospatial and information technologies in crop management. National Academy Press, Washington D.C.

Scarpace, F. 1998. Wisclmage. ERSC, Univ. of Wisconsin, Madison, WI.

Stone, M.L., J.B. Solie, W.R. Raun, R.W. Whitney, S.L. Taylor and J.D. Ringer. 1996. Use of spectral radiance for correcting in-season fertilizer nitrogen deficiencies in winter wheat. Trans. ASAE. 39(5):1623–1631.

Wanjura, D.F. and J.L. Hatfield. 1987. Sensitivity of spectral vegetative indices to crop biomass. Trans. ASAE. 30(3):810–816.

Yang, C., G.L. Anderson, and J.H. Everitt. 1998. A view from above: Characterizing plant growth with aerial videography. GPS World. 9(4):34–37.

Mapping In-Season Soil Nitrogen Variability Assessed through Remote Sensing

Kenan Diker
Department of Chemical and Bioresource Engineering
Colorado State University
Fort Collins, Colorado

Walter C. Bausch
USDA-ARS Water Management Research Unit
Fort Collins, Colorado

ABSTRACT

Conventional recommendations of N fertilizer are based on composited soil samples taken over the entire field. This may result in either underfertilization or overfertilization due to the neglected spatial variability in the field. This paper shows a methodology to estimate in-season soil N at various growth stages of irrigated corn. Soil N was estimated from plant N assessed through remote sensing. The nitrogen reflectance index (NRI) was employed to estimate plant N. Regression analysis between soil N and plant N showed good relationships at various growth stages. Geographic information system (GIS) mapping of measured and estimated soil N showed an agreement except in locations where hot spots were measured.

INTRODUCTION

Nitrogen is one of the essential nutrients for crop production. Eakin (1972) reported nutrient use of corn; N was removed in the largest quantity for both grain and straw yield. Excess amounts of N, however, lead to prolonged vegetative growth, delayed maturity, and it increases the danger of lodging (Mills & Jones, 1979). On the other hand, N deficiency results in small leaves, thin stems, fewer lateral branches, and a change in the color of leaves. Crop yield is affected in both cases.

Making N recommendations without knowing the N supply capability of a soil can lead to inefficient use of N and a potential pollution of the groundwater (Hong et al., 1990). There are many soil testing techniques such as the stalk NO_3 test (Iverson et al., 1985) and the pre-sidedress NO_3 soil test (Magdoff et al., 1984) that determine soil N availability; however, all these techniques are time consuming, laborious, and do not produce immediate results. In addition, these techniques could be economically prohibitive for precision farming.

In addition to soil N tests, plant N status is also used to determine the need of sidedress N fertilization. Plant N can be estimated from tissue sampling, chlorophyll meter (Wood et al., 1992), or remote sensing (McMurtrey et al., 1994;

Copyright © 1999 ASA-CSSA-SSSA, 677 South Segoe Road, Madison, WI 53711, USA. *Proceedings of the Fourth International Conference on Precision Agriculture.*

Bausch et al., 1994; Blackmer et al.,1994). Tissue sampling also is a very laborious and time-consuming process. Chlorophyll meter measurements provide quick estimates of the plant N status; however, this technique requires many measurements to determine the spatial distribution of N in the field. Bausch and Duke (1996) introduced the nitrogen reflectance index (NRI) to estimate the plant N status of corn. The index is calculated by a ratio of NIR/GREEN reflectance of an area of interest to NIR/GREEN reflectance of a reference, i.e. N stress free corn. This technique provided very fast estimates of plant N status and its spatial variability. The NRI may help determine "where and how much fertilizer should be applied?" to meet the needs of crops without applying excess N that may adversely impact the groundwater.

In this study, intensive soil and plant N sampling was conducted to determine the relationships between soil and plant N status during the growing season. These relationships were then used to estimate soil N status by remote sensing.

MATERIALS AND METHODS

The research was conducted at the Agricultural Research, Development and Education Center, Colorado State University, Fort Collins, CO. The research consisted of two different experiments. The first experiment was a fertilizer level study which established the interrelationships between soil and plant N and the canopy reflectance data. This experiment was conducted during the 1996 and 1997 growing seasons. Six N fertilizer levels of 0, 56, 84, 112, 168, and 224 kg ha^{-1} (east side of Fig. 1) were made available for the plant before planting. The zero level was residual N in the soil profile that was 26 kg ha^{-1}.

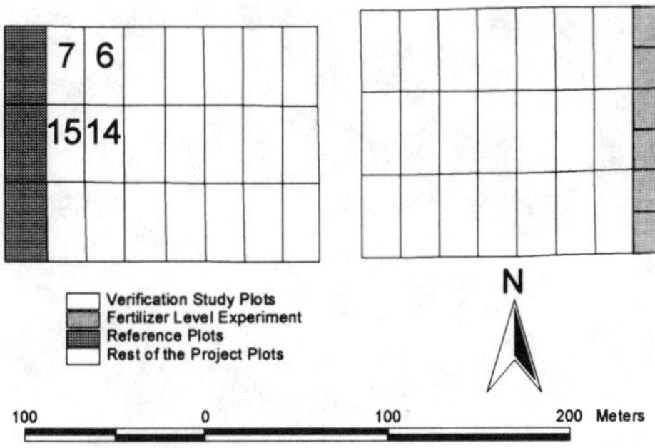

Fig. 1. Experimental layout.

Plot size was 11.5 m by 22 m in 1996 and 22 m by 22 m in 1997. Plot widths in 1996 were one-half of those in 1997 because two corn varieties were investigated in 1996 and the plots were split in half in the north–south direction.

The other experiment was conducted to verify the relationships developed from the fertilizer level study. In the second experiment, there were four N fertilizer treatments with plot dimensions of 22 m by 50 m. Two of the four plots (Plots 7 and 14) were fertilized when they needed fertilizer (spoon-fed) that was determined by a SPAD chlorophyll meter, one was pre-plant fertilized (Plot 6), and one was sidedress fertilized (Plot 15). The difference between the two spoon-fed N treatments was that Plot 7 had 100% of ET applied as irrigation whereas Plot 14 had 150%. All other treatments in both experiments had 100% of ET as irrigation water. Eight sampling locations (Fig. 2) were established in these four plots for soil and plant analysis. These sampling locations were established next to the rows from which reflectance data were collected so that no plant disturbance was occurred in the reflectance rows. Three reference plots for the verification study were also established. Corn hybrid Pioneer 3790 was planted in these experiments with row spacing of 0.76 m and row direction of north–south. Soil and plant N samples as well as reflectance data were collected at V6, V9, V12, V15, R1, and R2 growth stages.

Soil sampling was performed at soil depths of 0- to 30- and 30- to 60-cm. Soil samples were analyzed for NO_3–N and NH_4–N concentrations by the KCL extraction method. Tissue samples were collected by removing eight plants from the fertilizer level study and three plants from each location in the verification study. All the samples were composited for each sample plot in the fertilizer level

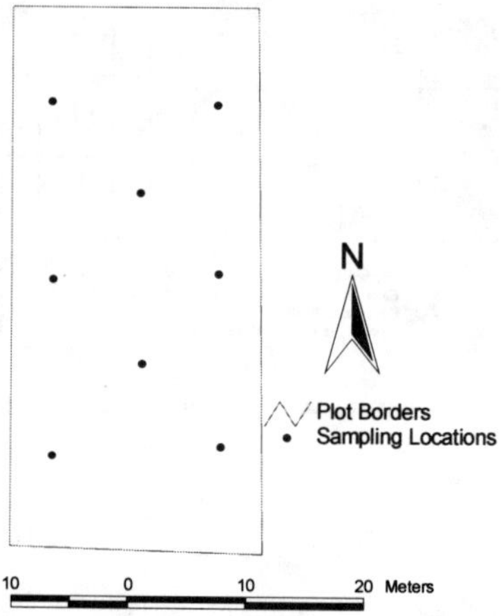

Fig. 2. Sampling locations in the verification study plots.

study and by sampling location in the verification study. The samples were dried at 55 °C for 24 h and ground in a dental mill. Samples were analyzed for total N content by a LICO N analyzer.

Corn canopy radiance and incoming irradiance were measured simultaneously with a mobile data acquisition system (Bausch et al., 1990). Three Exotech 100BX four-channel radiometers were used to collect the canopy radiance and incoming irradiance. These measurements were conducted around solar noon on the same day soil and plant samplings were undertaken. The canopy irradiance was measured from both nadir (10 m above the soil surface) and 75° (1 m above the plant canopy) view angles. The viewed spot on the ground by the nadir view radiometer was 2.6 m in diameter. The 75° view radiometer covered an ellipsoidal area with a length of 5.2 m and a width of 1.2 m. Bidirectional reflectance of the target was calculated using a procedure described by Neale (1987). Reflectance calculated for the various treatments and the reference plots in green and NIR wavebands was used to calculate the NRI.

The spatial variability of soil N was mapped using a GIS for both measured and estimated data. The inverse distance weighted (IDW) interpolation with a 16 m radius and a power of two was employed for mapping.

RESULTS AND DISCUSSION

Regression analysis was conducted between plant N and reflectance data for both nadir and 75° view angles for before and after tasseling stage. Before and after tasseling stage classification was made based on the magnitude of the plant

N at respective stages (Diker, 1998). Data collected at the V9, V12 and V15 growth stages were used to develop the before VT equation, whereas data from VT, R1, and R2 stages were used to develop the after VT equation. The V6 growth stage was excluded from the before tasseling equation because the plant N values were considerable larger than the rest. The before tasseling stage equations are given below for nadir and 75° views, respectively

$$PN = 0.44 + 2.10*NRI \ (r^2 = 0.77) \quad [1]$$

$$PN = 0.23 + 2.86*NRI \ (r^2 = 0.86) \quad [2];$$

corresponding equations for after tasseling were

$$PN = -0.18 + 2.64*NRI \ (r^2 = 0.80) \quad [3]$$

$$PN = -1.33 + 3.87*NRI \ (r^2 = 0.84) \quad [4]$$

where PN is plant N in %.

The 75° view reflectance data correlated better than nadir view data for before and after tasseling equations; however, after tassels develop, a 75° view radiometer sees mostly tassels rather than the green plant parts (Diker, 1998); therefore, it is not suggested that the 75° view be used after tasseling for plant N estimates. Figure 3 shows the spatial variability of plant N measured and estimated from both nadir and 75° view reflectance data at the V9 growth stage.

As seen in Fig. 3 the nadir view estimates of plant N are lower than those measured; however, 75° view estimates were very similar to measured plant N. This result was mainly because of the soil background effect because the soil background tends to increase green reflectance and reduce NIR reflectance, which greatly reduces sensitivity. The nadir view at the V9 growth stage saw significant soil background that was minimized by the 75° view (Diker, 1998). The largest plant N was observed in Plot 15 where sidedressing was applied at the V6 growth stage. This was evident in all three maps; however, in the nadir view map, the magnitude of plant N estimates was lower. The 75° view estimates eliminated the bullseyes that represent either a high or low plant N spot that possibly resulted from sampling errors in the measured plant N map. Figure 4 shows the plant N distribution at the R1 growth stage for measured and estimated data.

At the R1 growth stage, the estimated plant N from the nadir view also was lower than those measured. The estimates in Plots 6 and 14, however, were somewhat similar to measured plant N values in these plots. These lower estimates of the plant N may be because of the windy conditions on the day of sampling. A nadir view, according to Diker (1998), is more robust to use under windy conditions than an oblique view. Therefore, combining the windy conditions and the difficulties that the 75° view was having after tasseling no effort was made to map plant N estimates from the 75° view at the R1 growth stage. These results indicate that the 75° view can successfully be used to estimate the plant N status before the tassels developed; however, a nadir view should be employed for estimates after tasseling stage or when there are no 75° view reflectance data.

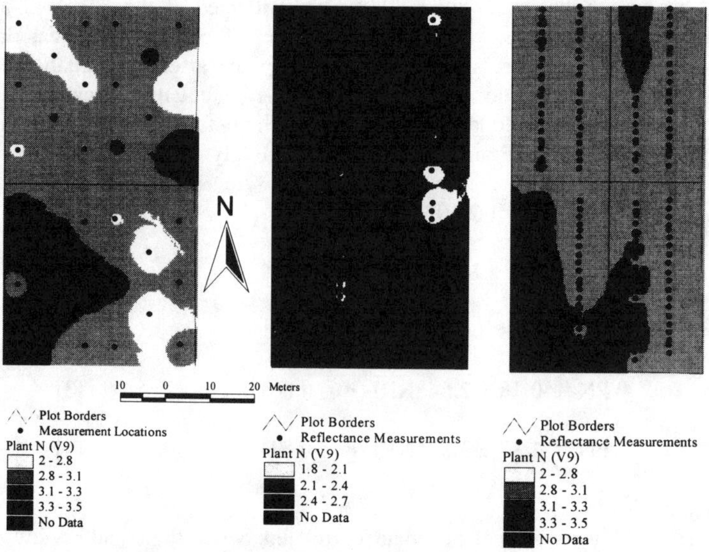

Figure 3. Spatial variability of measured (left), nadir view (middle) and 75° view estimated (right) plant N (%) at the V9 growth stage.

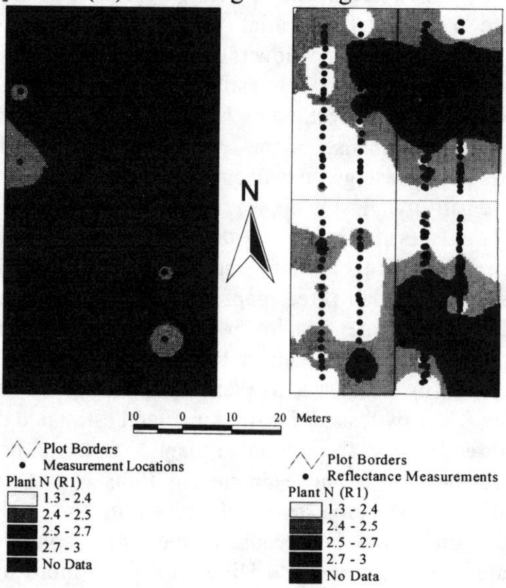

Figure 4. Spatial variability of measured (left) and estimated from nadir view (right) plant N (%) at the R1 growth stage.

Plant and soil N data collected from the fertilizer level study were used to develop the relationships between the parameters. Plant N was well regressed with the inorganic soil N (NO_3 + NH_4) rather than NO_3 itself. Therefore, the equations for various growth stages were developed between plant N and inorganic soil N. An equation for each growth stage under investigation was developed. For the relationships, data either from 1996 or from 1997 were used. The relationships developed are given below:

$$SN = -45.80 + 14.7*PN; r^2 = 0.78 \text{ at V6} \quad [5]$$

$$SN = 1.69 + 4.17*PN; r^2 = 0.56 \text{ at V9} \quad [6]$$

$$SN = -5.54 + 6.86*PN; r^2 = 0.62 \text{ at V12} \quad [7]$$

$$SN = -1.84 + 5.18*PN; r^2 = 0.79 \text{ at V15} \quad [8]$$

$$SN = -2.87 + 4.82*PN; r^2 = 0.74 \text{ at R1} \quad [9]$$

$$SN = 0.83 + 2.49*PN; r^2 = 0.76 \text{ at R2} \quad [10]$$

where SN is total soil N in ppm.

As indicated by the r^2 values, good relationships were observed between inorganic soil N and plant N. Using these equations, the inorganic soil N was estimated from the estimated plant N. The maps generated for measured and estimated inorganic soil N from both nadir and 75° view angles in kg ha^{-1} are given in Fig. 5.

As seen in Fig. 5, large variability in the maps of inorganic soil N occurred. In the measured inorganic soil N map, there were three very hot spots in Plot 6. These hot spots resulted in large areas of high soil N in the entire field. In estimated maps of inorganic soil N, the greatest inorganic soil N was in Plot 15. This also was the case for the measured map. The nadir view estimates of soil N were lower than those using the 75° view. The zonal statistics developed from these maps indicated that the inorganic soil N level was 202 kg ha^{-1} in Plot 6. This high soil N was a result of those three hot spots. In this plot, the estimated soil N was 100 and 126 kg ha^{-1} for nadir and 75° views, respectively. Zonal statistics for total soil N in Plot 15 were 129, 103 and 130 kg ha^{-1} for measured, nadir view and 75° view estimated maps, respectively. Corresponding values in Plot 7 were 132, 100 and 125 kg ha^{-1}. These results indicate that the inorganic soil N can be successfully estimated through remote sensing estimates of plant N status and that the 75° view is a better estimator of the inorganic soil N than the nadir view for the V9 growth stage. Similar results were also observed for other growth stages of corn. Figure 6 shows the measured and estimated inorganic soil N at the R1 growth stage. The estimates of inorganic soil N were 80, 76, 78, and 76 kg/ha for Plots 6, 7, 14, and 15, respectively while the corresponding values for measurements were 82, 55, 75, and 115 kg ha^{-1}. The high value in Plot 15 also was a result of a hot spot that can be seen in Fig. 6.

Our results indicated that there were some hot spots in measurements of the inorganic soil N at various growth stages. This may be because of the errors either in sample collection or in laboratory analysis of the samples; however, remote sensing estimates of plant and soil N data were much smoother than the measurements and the hot spots were minimized or eliminated. This means that human introduced errors in the measurements could be minimized by using remote sensing for monitoring plant and soil parameters.

Recommended N amounts also were estimated for the verification study at each of the growth stages of corn. For this process, the inorganic soil N amount in the 168 kg ha^{-1} N treatment of the fertilizer level study was assumed to be the optimal soil N level to support plant growth because it produced a grain yield that was statistically the same as the 224 kg ha^{-1} N treatment. These values for the 168 kg ha^{-1} N treatment were 127, 115, 101, 90, and 70 kg ha^{-1} for V9, V12, V15, R1, and R2 growth stages, respectively. The available N amounts were subtracted from the optimum soil N amount to obtain the suggested N fertilizer additions needed. The measured and estimated N fertilizer needs at the V9 growth stage were mapped and presented in Fig. 7.

In Fig. 7, it is clear that there was a small area in Plot 14 of the measured soil N map where there was a N shortage >30 kg ha^{-1} while some areas had a N deficiency of 0 to 30 kg ha^{-1}. There also was a larger area where there was no N deficiency. However, in the deficiency map estimated from the nadir view, large areas showed deficiencies of 0 to 30 and some 30 to 61 kg ha^{-1}. This indicates that from a nadir view angle, no N deficiency area was underestimated. The map generated from the 75° view agreed better with the map generated from the measurements by showing very small areas of N deficient areas. However, the 75° view

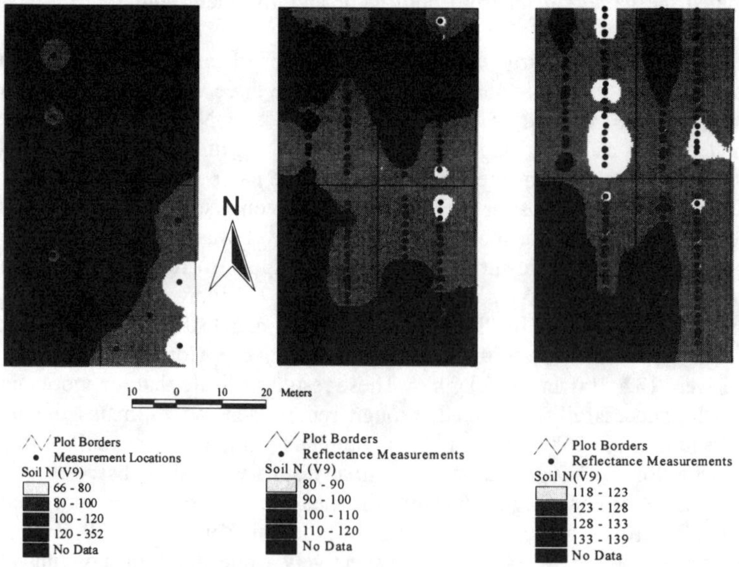

Fig. 5. Spatial variability of measured (left), nadir view (middle) and 75° view (right) estimated inorganic soil N (kg ha^{-1}) at the V9 growth stage.

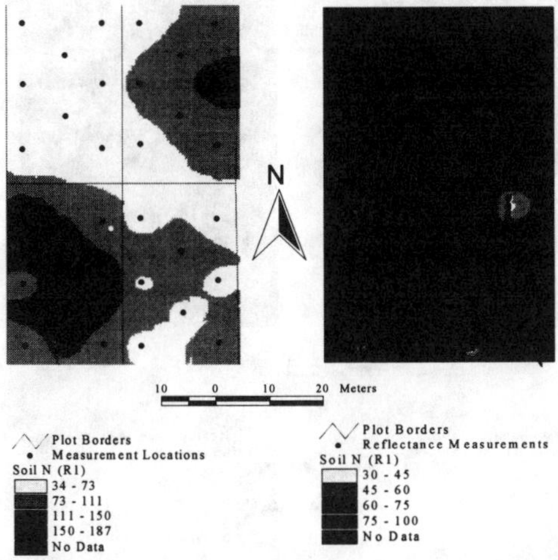

Fig. 6. Spatial variability of measured (left) and estimated (right) inorganic soil N (kg ha^{-1}) at the R1 growth stage.

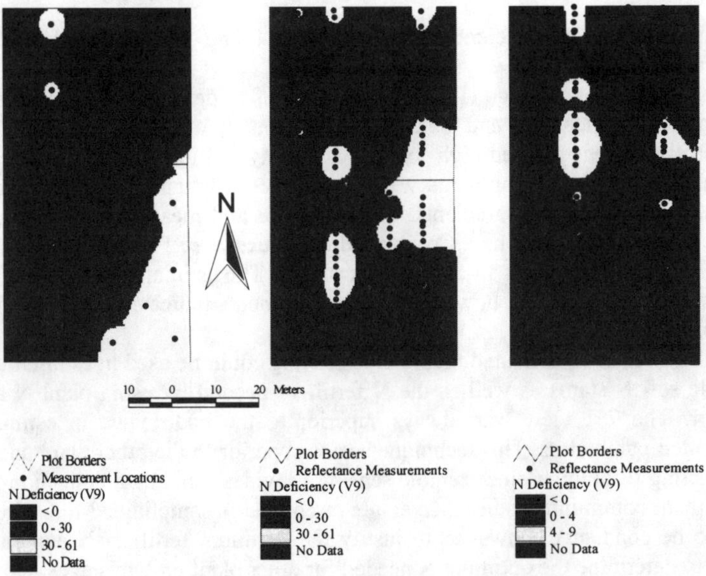

Fig. 7. Spatial variability of measured (left), nadir (middle) and 75° view (right) estimated N fertilizer needs (kg N ha^{-1}) at the V9 growth stage.

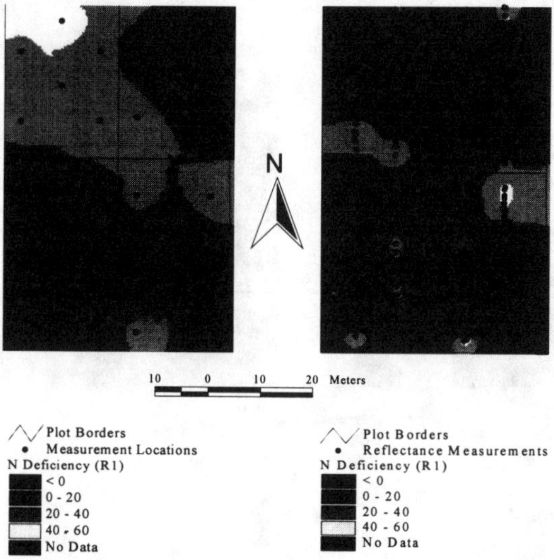

Fig. 8. Spatial variability of measured (left) and estimated (right) N fertilizer needs at the R1 growth stage.

overestimated the N deficient areas in Plots 14 and 15 compared with the measured map.

At the R1 growth stage (Fig. 8), the status of N deficiencies showed some difference in the measured and the estimated deficiency maps. The measured map of N fertilizer needs showed high soil N availability in Plots 15 and 6 whereas in these plots the excess N amounts were negligible. Estimates in Plots 6 and 14 were very similar to measurements. The estimated and measured prescription N for Plot 6 was 8 and 10 kg ha^{-1}. Corresponding values were 13 and 15 kg ha^{-1}, and 14 and 35 kg ha^{-1} for Plots 14 and 7, respectively. The estimated N fertilizer need for Plot 15 was 14 kg ha^{-1}, however, the measurements indicated that Plot 15 had 24 kg ha^{-1} available N.

This research indicated that remote sensing could be used in estimating the available soil N status as well as the N fertilizer needed by using plant N as an indicator. The 75° view was always superior to the nadir view in estimating investigated parameters. This technique seems superior to the other plant and soil N estimating techniques since remote sensing provides rapid data collection from a large plant community rather than single or countable samplings. More research needs to be conducted, however, to justify the optimum fertilizer N amounts as well as to determine the optimum N needed for other plant growth stages.

REFERENCES

Bausch, W.C., D.M. Lund, and M.C. Blue. 1990. Robotic data acquisition of directional reflectance factors. Rem. Sens. Environ. 30:159–168.

Bausch, W.C., H.R. Duke, and L.K. Porter. 1994. Remote sensing of plant nitrogen status in corn. *In* 1994 ASAE Summer Meeting, Kansas City, MO. 19–22 June, 1994

Bausch, W.C., and H.R. Duke. 1996. Remote sensing of plant nitrogen status in corn. Trans. ASAE. 39(5):1869–1875.

Blackmer, T.M., J.S. Shepers, and G.E. Varvel. 1994. Light reflectance compared with other nitrogen stress measurements in corn leaves. Agron. J. 86(6):934–938.

Diker, K. 1998. Use of geographic information management systems (GIMS) for nitrogen management. Ph.D. diss., Colorado State Univ., Fort Collins.

Eakin, J.H. 1972. Food and fertilizers. p. 1–21. *In* The Fertilizer Handbook. The Fertilizer Inst., Washington, DC.

Hong, S.D., R.H. Fox, and W.P. Piekielek. 1990. Field evaluation of several chemical indexes of soil nitrogen availability. Plant Soil. 123:83–88.

Iverson, K.V., R.H. Fox, and W.P. Piekielek. 1985. The relationships of nitrate concentrations in young corn stalks to soil nitrogen availability and grain yields. Agron. J. 77:927–932.

Magdoff, R.F., D. Ross, and J. Amadon. 1984. A soil test for nitrogen availability to corn. Soil Sci. Soc. Am. J. 48:1301–1304.

McMurtrey J.E., III, E.W. Chappelle, M.S. Kim, J.J. Meisinger, and L.A Corp. 1994. Distinguishing nitrogen fertilization levels in filed corn (*Zea mays* L.) with actively induced fluorescence and passive reflectance measurements. Rem. Sens. Environ. 47(1):36–44.

Mills, H.A., and J.B. Jones, Jr. 1979. Nutrient deficiencies and toxicities in plants: Nitrogen. J. Plant Nutr. 1:101–122.

Neale, C.M.U. 1987. Development of reflectance-based crop coefficients for corn. Ph.D. diss., Colorado State Univ., Fort Collins.

Wood, C.W, D.W. Reeves, R.R. Duffield, and K.L. Edmisten. 1992. Field chlorophyll measurements for evaluation of corn nitrogen status. J. Plant Nutr.. 15(4):487–500.

Developing Calibration Techniques to Map Crop Variation and Yield Potential Using Remote Sensing

G.A. Wood
G. Thomas
J.C. Taylor

School of Agriculture, Food and Environment, Cranfield University, Natural Resources Management Department
Silsoe, Bedfordshire, England

ABSTRACT

A significant problem with the site-specific management of crops is the amount of data collection required to confidently map crop conditions and related parameters. In Taylor et al (1997) high resolution remote sensing data using Airborne Digital Photography (ADP) was used successfully to map the within-field spatial distribution of crop parameters and yield potential for winter barley in the England. The calibration was based on the regression equation derived from field sampling vs. NDVI measurements. Our first approach used a large number of field observations to calibrate the ADP data, which inhibits the approach from being adopted for practical use. This paper explores the possibility of reducing field sampling and demonstrates a practical methodology for field calibration of ADP data as a source of management information for precision farming in wheat and barley. Data taken from 90 field quadrat observations is compared statistically to a reduced sample size (24 quadrats) that was extracted from the same data. Using parallel lines analysis the reduced sample is shown not to be significantly different from the larger sample. The reduced sample technique was put into practice in four fields growing cereals. In each field seven sample sites were selected representing the range of tiller variation. The crop was sampled at each measurement site, at two times during tillering, using three 50□50 cm quadrat samples, arranged in a triangular subset. The coefficients of determination were very high, e.g., 0.98, 0.95, 0.94, 0.78, with probabilities <0.002; in one case an apparent background soil effect reduced the r^2 to 0.63. The ability to map crop parameters and yield potential during a growing season provides a valuable source of management information to the farmer and agronomist. The development of a rapid methodology for field-based calibration will allow the use of ADP for the monitoring of crop parameters, bringing this technology closer to being used for monitoring both crop condition and the effects of managing variable inputs throughout, a growing season.

INTRODUCTION

To manage crops on a site-specific basis, by precision farming, it is first necessary to be able to confidently measure and map within-field spatial

Copyright © 1999 ASA-CSSA-SSSA, 677 South Segoe Road, Madison, WI 53711, USA. *Proceedings of the Fourth International Conference on Precision Agriculture.*

variability. Interpolation techniques and geostatistics offer a means of interpolating between sample points but rely on the quality and density of data for producing reliable maps. In some cases, geostatistics is inappropriate or misapplied where data are too few or not spatially correlated. The costs associated with some field measurements (e.g., soil related analysis) precludes the collection of a sufficiently large number of measurements necessary for mapping purposes.

In order to progress, precision farming must be based on quality maps which are timely and cost effective to produce. One way forward is to develop the use of surrogate data, which is cheaper and easier to collect.

High resolution remote sensing, using Airborne Digital Photography (ADP), offers precision farming a surrogate with the capability for real-time measurement and mapping of crop variation and ultimately the provision of crop management information during the season.

Remote Sensing

Remote sensing measures the energy reflected by the earth's surface. The origin of this is sunlight and measurements are made in a digital format by sensing devices that can be mounted on tractors, aircraft and satellites. Red reflected energy, which is visible to the eye, is absorbed by chlorophyll in healthy green plants for use in photosynthesis. Reflected near-infrared energy, which is not visible to the eye, is particularly sensitive to leaf cell structure and water content. Healthy dense crops are characterised, therefore, by strong absorption of red energy and strong reflectance of near-infrared energy. It is advantageous to combine these measurements into a single index that enhances the sensitivity to variation in the crop. Such mathematical combinations of the digital data are known as vegetation indices. One such index is the Normalised Difference Vegetation Index, Tucker (1979), defined as:

$$NDVI = (Infrared - Red) / (Infrared + Red) \quad [1]$$

In this case, the NDVI is the surrogate variable used to extrapolate a sample of ground measurements to all positions in the field. The actual ground observations will relate to specific plant parameters depending on the growth stage of the crop.

This paper extends our work, reported in Taylor et al (1997), by exploring the possibility of reducing field sampling to a practical level. In doing so it demonstrates a practical methodology for field calibration of remote sensing data as a source of management information for precision farming in wheat and barley.

Airborne Digital Photographic System

The ADP system consists of two Kodak Professional DCS420 digital cameras housed in a special mount designed to be carried on a light aircraft. Two cameras were used separately to take two high resolution digital images on 5 May 1996, of the red and near-infrared reflected energy. The flying height was selected to give a pixel size equivalent to 600 mm on the ground.

CALIBRATION METHODOLOGY

Digital Image Preparation

The red and near-infrared image bands are combined by digitally overlaying one image onto the other. The red and near-infrared images are processed into an NDVI image by applying Eq. [1] to the digital values for each pixel.

Yield Estimates from Crop Components

Estimates of crop yield can be made during the season by measuring plant components. Measurements were made of the number of grains per unit area after heading, four weeks prior to harvest and 6 wks after the ADP survey. The measurements consisted of the number of ears per unit area (N_e), and the average number of viable grains per ear (N_g). The TGW was unknown at the time, therefore, yield was estimated using an average value for the grain weight (w_g), equivalent to a TGW of 53.2 g - the average TGW for Cv. Intro (NIAB, 1996). Thus the forecast yield, Y, is given by:

$$Y = N_e . N_g . w_g \qquad [2]$$

Sampling Strategy

In order to estimate yield variation, measurements of crop components were made at different locations within the field. The aim was to use the ADP derived NDVI as an auxiliary variable for a regression estimator (Cochran, 1977) to extrapolate the sample observations to all locations within the field. Random sampling is not a requirement for this approach so the ADP images were used to select four locations within the field.

The four sites selected represented the range of the NDVI response, and by inference, the range of yield variation across the field (Fig. 4). The sampling pattern at each site consisted of five transects across the direction of the tramlines, with five 500 mm square sample quadrats spaced along the transect, as shown in Fig. 1. Subsequently, an additional position *f* was introduced in the ctr of the tramlines adjacent to each plot. This was because the original sample design resulted in two samples taken at positions close to the tramlines (a and e), two samples at an intermediate distance (b and d), but only one sample (c) representing the mid-position.

The locations of the sample sites were fixed by ground measurement relative to the tramline system which was also clearly visible on the ADP images. Thus, the ground survey sample sites were located in the ADP images by appropriate scaling of the ground measurements using the tramline spacing, verified by field measurement at several locations, to determine the image scale.

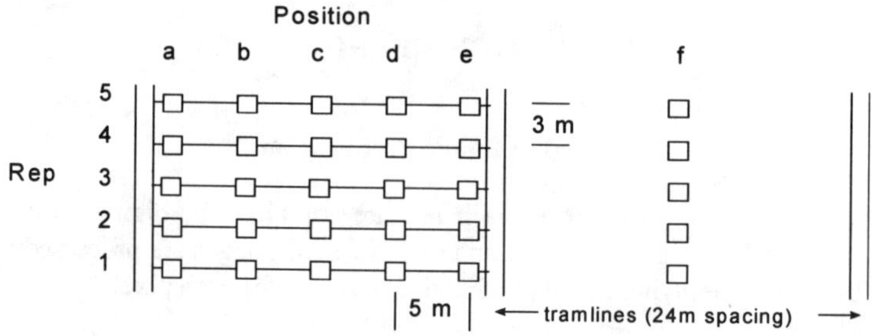

Fig. 1. Pattern of sample sites at each of four locations within the study field.

Measurements

On 19 June, 1996, a sub-sample of five ears, selected blindly, were taken, and the number of ears in each quadrat were recorded. The former were used to estimate the average number of grains per ear.

On 16 July, 1996, immediately prior to harvest, the crop was destructively sampled from quadrats positioned adjacent to those used for the ear counts. These samples were used to determine the actual TGW's and final yield - independently of the yield mapping combine systems.

Calibration

The crop yield at each of the 100 quadrats was forecast using Eq. 2. To reduce both the effects of highly localised crop variation on the estimates of crop components, and the effect of error in locating the quadrat sites on the imagery (estimated to be of the order of 600 mm), the yield forecasts and the NDVI values were aggregated for the five quadrats in each repetition (e.g., a1 to a5 in Fig. 1). The corresponding NDVI measurements, measured by the ADP images, were similarly aggregated.

Figure 2 shows that the yield potential from measured crop components is highly correlated to the NDVI (due to slug damage in site D, two data points were removed from the regression, leaving 18 observations). The regression equation was applied to the image data to make forecasts of the yield across the field on a 6 by 6 m grid. The grid was then used to produce a map indicating the within-field spatial distribution of yield potential (Fig. 3).

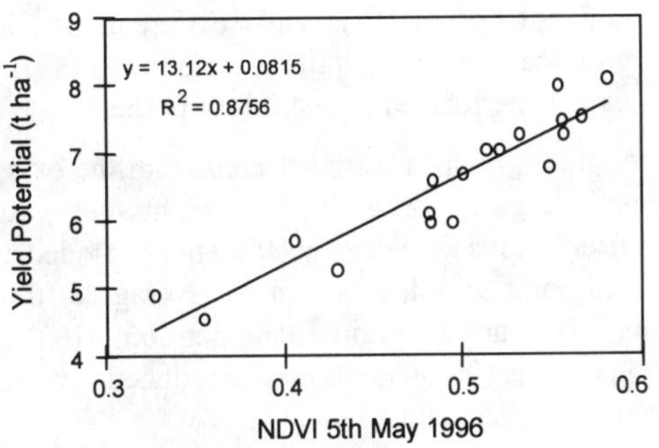

Fig. 2. Relationship between NDVI and yield forecast from ground data collected on 19/06/96.

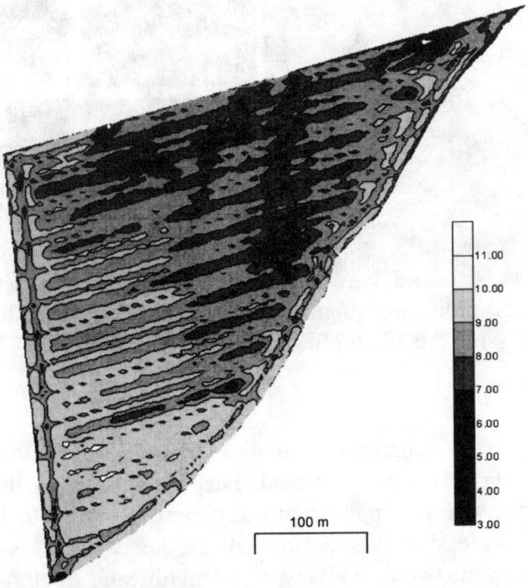

Fig. 3. Yield Forecast for Cv. Intro using May 5 ADP imagery calibrated with ground measurements made June 19.

Rapid Calibration Methodology

The above technique used a large number of field observations to calibrate the ADP data, inhibiting the approach from being adopted for practical use. The following section tests the possibility of further reducing field sampling demonstrating a practical technique for field calibration of ADP images.

The effect on the resulting regression relationship of choosing fewer sample sites was assessed statistically using the same data outlined above. It was hypothesised that the regression relationships using either a reduced data set or the above example would not be not statistically different, i.e., fewer observations could be used to produce the same calibration equation. A technique called, parallel lines analysis was performed to test this hypothesis.

By visual inspection, eight sites were selected from the existing data set using only the ADP image as a guide. At each of the eight sites a sub-sample of three localised quadrat observations were extracted. This provided a reduced data set comprising eight field point samples — each an average of three observations (a total of 24 quadrats). The larger, original data set comprised 18 points, each an average of five observations (90 quadrats). The reduced sample-size data set was compared with the full original data.

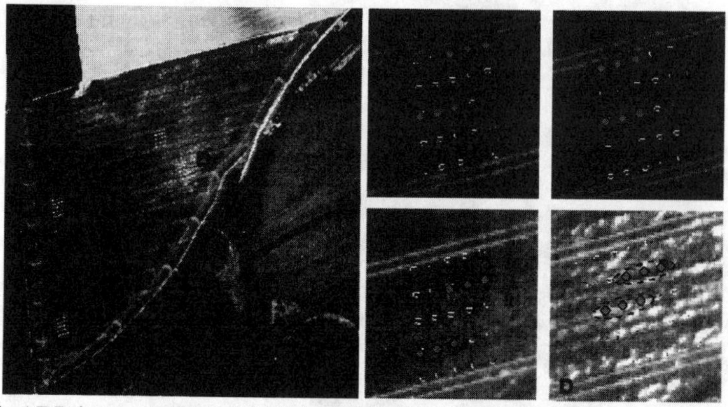

Fig. 4. ADP image indicating original sample sites along with the selected sites identified for assessing the effect of using a reduced sample.

The pooled *Y* on *X* regression was highly significant, $P < 0.001$ (Table 1). When the two data sets were regressed independently (with the slopes forced equal) the offsets were not significantly different, $P = 0.649$. Finally, when a normal *Y* on *X* regression was performed on the two data sets (Fig. 5) the differences between the two data sets were not significant, $P = 0.848$.

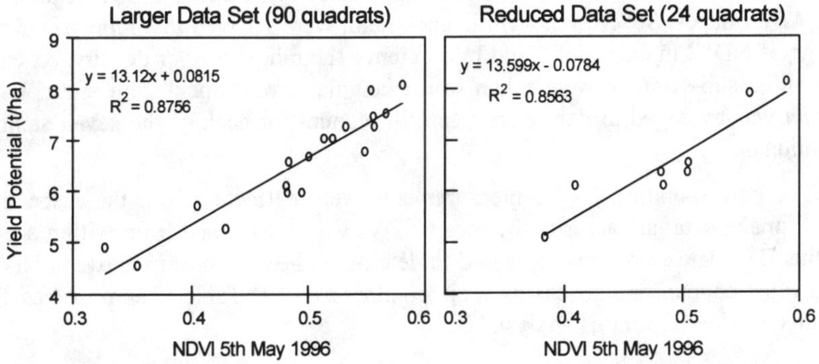

Fig. 5. Relationship between NDVI and yield forecast from ground data collected on 19 June 1996. The two graphs indicate the regressions based on two different sample sizes.†

Table 1. Accumulated analysis of variance comparing results from two regressions using Parallel Lines Analysis.

Change †	*d.f.*	*s.s.*	*m.s.*	*v.r.*	*Fprob*
NDVI	1	21.282	21.282	147.14	<.001
Treatment	1	0.031	0.031	0.21	0.649 N.S.
NDVI treatment	1	0.005	0.005	0.04	0.848 N.S.
Residual	22	3.182	0.145		
Total	25	24.5	0.98		

† The term *treatment r* efers to the two different sets of regression data: one regression (treatment) is based on 90 quadrat

The significance of this is that the reduced data set performed equally as well the larger data set allowing the development of a manageable, operational technique

Rapid Calibration Results

The rapid calibration technique was put into practice in four fields, three growing winter wheat, and one growing winter barley. At the time of writing assessments had been made at two times early in the development of the crop - both during tillering. The first was 22nd December and the second was 12th March. The NDVI measurements taken from ADP surveys prior to field sampling were expected to correlate with the number of tillers per unit area.

At each field, suitable sample locations were selected through interpretation of the ADP images. Seven sites, within each field, were selected that represented the range of NDVI in each field, and by inference the range in tiller density. At each site three sub-samples were taken in a triangular arrangement 2 m apart. Each *triplet* was averaged to derive an mean tiller count for each of the seven sample locations.

The corresponding NDVI measurements were extracted from the respective ADP image data: at each site the mean NDVI was determined from within a 2 m radius. The data were then regressed to determine the relationship between NDVI and tiller counts, and to assess their goodness of fit. Table 2 summarises the results obtained from this analysis.

Table 2. Summary of the coefficients of determination (r^2) resulting from the regression of tiller counts against NDVI for two dates in four fields

Field name	*Dec-97*	*Mar-97*
Trent (WB) †	0.94 (P < .001)	0.98 (P = 0.003)
Onion (WW)	0.94 (P < .001)	0.95 (P < .001)
12 Acres (WW)	0.57 (P = 0.028)	0.63 (P = 0.019)
Far Sweetbriar (WW)	(not surveyed)	0.78 (P = 0.008)

† WW, Winter Wheat; WB, Winter Barley.

CONCLUSIONS

ADP calibrated vegetation measurements for obtaining information, in near real-time, of the within-field variation of yield potential has great potential as a tool for providing timely management information. Our original investigative work used a large number of field observations, but served to demonstrate the principles and provide a basis for further development. It would not be practical to perform such a large amount of ground data collection, as this inhibits the approach from being adopted for practical use; however, the results of this investigation support the use of a reduced sample size

REFERENCES

Cochran, W. 1977 Sampling techniques. John Wiley & Sons, New York.

Taylor, J.C., G.A. Wood, and G. Thomas, 1997, Mapping yield potential with remote sensing, p. 713–720. *In* Stafford, J.V. (ed.) *Proc. of 1st European Conf. on Precision Agriculture,* SCI, London.

Tucker, C.J. 1979 Red and photographic infrared linear combinations for monitoring vegetation. *Rem. Sens. Environ.*, 8:127–150.

Assessing Yield Parameters by Remote Sensing Techniques

P. R. Willis

Resource21
Englewood, Colorado

P. G. Carter
C. J. Johannsen

Laboratory for Applications of Remote Sensing
Purdue University
W. Lafayette, Indiana

ABSTRACT

A case study is presented using 2 yrs of yield data in conjunction with high spectral resolution data, HYDICE and AVIRIS, and high spatial resolution imagery, simulated IKONOS. Procedures for integrating yield monitor data and remotely sensed data are discussed. There are many errors inherently associated with yield monitor data that need to be considered when using the data for scientific research. Correction methods which best suited this case study are presented. Statistical correlation between individual bands of hyperspectral imagery and yield data is analyzed on the basis of the entire agricultural scene, crop type, varieties, bare soil, vegetation cover, and sensor type. Averaging filters are used to study the effect of differing spatial resolutions on hyperspectral data correlation to yield. These results were then used to assess the performance of classifications to identify yield patterns prior to harvest. Traditional methods of detecting vegetation patterns are discussed, such as filters, linear stretches, tasseled cap, principal components, and classifications. An evaluation of hyperspectral bands with relation to yield for specific anomalous regions is given. Statistical correlation between specific anomalies identified in the field to the corresponding yield information was studied. It lays the groundwork for further research related to spectral responses of specific anomalies and their effects on yield. These results will be applied in other continuing studies to develop yield evaluation and anomaly detection models using varying spectral and spatial resolutions.

INTRODUCTION

The main focus of this study is to further the research in the area of anomaly detection and provide remote sensing input toward crop yield prediction. The most correlated hyperspectral bands to use with yield evaluation are explained. A simple linear regression using yield monitor data and corresponding hyperspectral data are used for this evaluation. This study sets the groundwork for improving our knowledge of spectral responses of specific anomalies and their

Copyright © 1999 ASA-CSSA-SSSA, 677 South Segoe Road, Madison, WI 53711, USA. *Proceedings of the Fourth International Conference on Precision Agriculture.*

effects on yield. Developing the methodology and procedures of how to get the raw yield data and the raster based remotely sensed data to interchangeable formats to work is a goal pursued in this research to provide assistance for other researchers. Anomaly detection and yield modeling are two areas where remote sensing has the greatest potential impact on agriculture. If a reliable methodology for estimating yield before harvest can be developed, the agriculture and remote sensing industries would be permanently linked together. Its impact would range from predicting food shortage crises to better management of information about food production.

The study site is the Davis-Purdue Agricultural Center, located in East Central Indiana, northeast of Muncie in Randolph County. Yield monitor data recorded with Rockwell Vision equipment is available for the three previous growing seasons. The crops in these fields are corn and soybean rotations.

SPECTRAL AND SPATIAL RESOLUTIONS

The spatial resolution for yield maps has a wide range of data points per acre depending on how fast the harvester is traveling and how often the yield measurements are recorded. For example, a harvester traveling at about 15 ft s^{-1} and recording yield measurements every second would have a spatial resolution of approximately 5 m.

Table 1. Conversion of spatial resolution from images to ground data points (Johannsen, et al., 1998).

Resolution (m)	Data Points (Pixels) ac^{-1}	Points ha^{-1}
1000	0.004	0.01
80	0.6	1.56
30	4.5	11.1
23.5	7.3	18.1
20	10	25.0
15	18	44.4
10	40	100
5	162	400
4	253	625
3	450	1111
2	1012	2500
1	4046	10000

Spatial and spectral advancements are improving many existing remote sensing applications and creating new possibilities. As many as 50 land observing satellites may be launched before the year 2007 (Johannsen, & Carter, 1997). Two examples of vastly increasing spectral resolutions are the AVIRIS and HYDICE sensors described in the next section. An example of increasing spatial resolution is the simulated IKONOS data also described in the next section.

REMOTELY SENSED DATA

Hyperspectral Digital Imagery Collection Experiment (HYDICE) is an airborne platform sensor, which collects data in 210 bands ranging from 400 to 2500 nm at approximately 10-nm increments. The actual bandwidths are affected by the air pressure (altitude) and temperature within the sensor and can vary from 3 to 16 nm. The spatial resolution is approximately 2 m. It is a military owned sensor, which is available for limited commercial use. For more detail on the specifications of the HYDICE sensor see Mitchell (1995). For this study an 7 Aug., 1997 HYDICE dataset was available.

Airborne Visible–infrared Imaging Spectrometer (AVIRIS) is an airborne platform sensor, which collects data in 224 wavelength bands recorded at 10 nm intervals from 400 to 2500 nm. The spatial resolution is reported at 20 meters although it depends on the customer's requested spatial resolution. We have a 5 July, 1996 dataset, which shows very little vegetation emergence. AVIRIS is a NASA owned sensor. For a more detailed description see Vane, et al., (1993).

One meter multispectral imagery provided by Space Imaging EOSAT simulates the yet to be launched IKONOS satellite. The actual satellite will be 1 m panchromatic and 4 m multispectral, consisting of four bands; blue, green, red, and near infrared. The data simulates a pansharpened multispectral image. Three dates of this imagery was used in this study, 7 Aug., 1997, 5 Sept., 1997, and 26 Sept., 1997.

YIELD DATA

When farmers look at yield maps, they are looking for general patterns and trends, such as unusually high or low yielding areas. Once these areas are identified, it is then the task of the farmer, who is familiar with the soils and field history, to suggest why these variations are occurring. Actions can then be taken to manage the lower yielding areas differently in order to obtain higher yields.

Yield monitors coupled with GPS units are a relatively new piece of equipment available to growers. Yield maps serve the needs of the farmers as described above and are designed to be readily understandable and visually pleasing, but for scientific research the data collection methods are lacking. The raw yield monitor data has a host of inherent errors, which one needs to understand in order to have enough knowledge to correct the errors as best as possible.

INHERENT ERRORS

There are many errors inherently associated with yield monitor data, which need to be considered when using the data for scientific research. Numerous studies have addressed some of the errors and various correction techniques such as Birrell et al. (1996) and Murphy et al. (1996). Errors include time lag of crop from intake to yield sensor, yield sensor calibration, GPS accuracy, uncertain crop width entering the header, surging grain, grain losses, and many more not listed (Blackmore & Marshall, 1996).

The time lag associated with the recorded latitude and longitude of crop yield measurement from intake to the actual yield sensor is a major problem in correlating yield to corresponding pixels in hyperspectral imagery. The Rockwell Vision software uses a default time shift correction of 12 s, as does most of the yield monitor industry. This number was determined somewhat arbitrarily as the average lag time. It will vary based on a vast set of operating and site variables and can vary anywhere from 8 to 30 s depending on how heavy the yield is and the speed of the combine. The next section describes how these errors were handled.

DATA PREPARATION

The National Agronomic Statistics Service of the USDA (NASS) conducted extensive hand-harvesting to calculate very precise yield measurements in two of the fields within the study site. The NASS data was used to determine the most appropriate time shift for the raw yield monitor data.

Fig. 1. Time shift correlation.

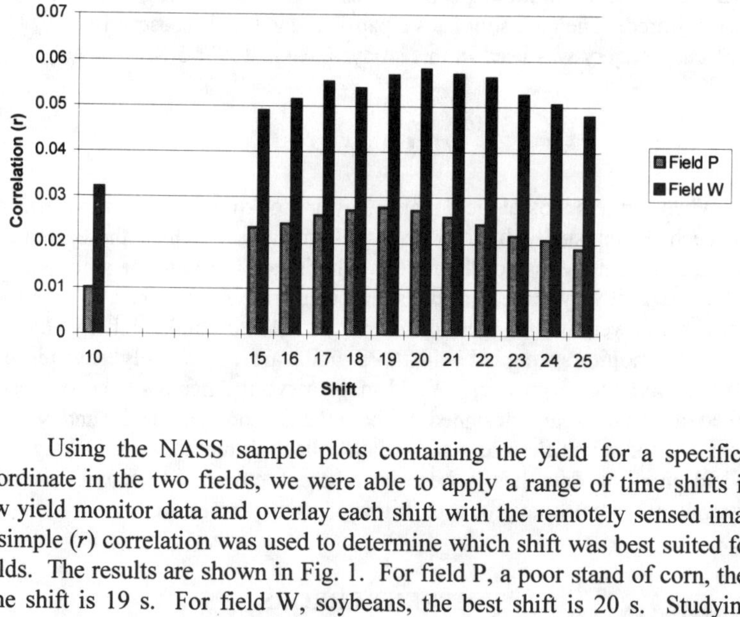

Using the NASS sample plots containing the yield for a specific X,Y coordinate in the two fields, we were able to apply a range of time shifts in the raw yield monitor data and overlay each shift with the remotely sensed imagery. A simple (*r*) correlation was used to determine which shift was best suited for the fields. The results are shown in Fig. 1. For field P, a poor stand of corn, the best time shift is 19 s. For field W, soybeans, the best shift is 20 s. Studying the trends in the (*r*) values of the shifts and assuming many of the calibration and operating variables remain constant throughout the entire dataset, an average 19 s shift was then applied to the rest of the sites without NASS sample plots.

Differing interpolation methods, such as various types of kriging and inverse distance weighting (IDW), were studied to determine which to use in the conversion of the yield point data to a raster based format. Kriging is the most acceptable way to grid data, but is very intensive in terms of computing times (Murphy, et al. 1996). IDW methods are best suited for very dense datasets, which are dense enough to represent the local variation of a surface. For this study we used an inverse distance method that gave very similar results to kriging methods, but requiring a fraction of the computing time.

The remotely sensed imagery was georeferenced and then resampled to a compatible spatial dimension as the yield data. The 1997 yield data was resampled to 2 m to match the HYDICE data, and the 1996 data was resampled to 18 m to match the AVIRIS data. For the spatial resolution comparisons the imagery and the yield data were down sampled together. Once in a raster based format, the geocoded hyperspectral imagery was registered to the yield layer. The corresponding pixels were run in a SAS correlation program on the basis of individual bands and spectrum ranges and on the basis of differing spatial resolutions.

Fig. 2. NASS, yield data, and image overlay.

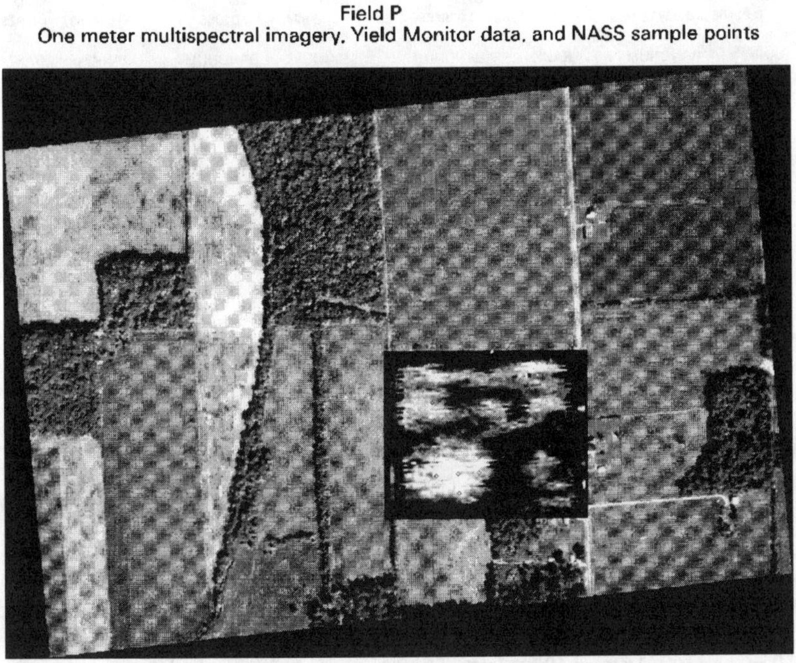

CORRELATION RESULTS AND DISSCUSSION

The results are currently being studied and interpreted. Only the results of the HYDICE data to yield is presented in this paper. At 2 m, the overall *r* correlation to yield is 0.50526. This is a relatively low number due to the large number of negatively correlated bands, generally in the visible wavebands and the water absorption regions. Depending on the strength of the negative correlation, these bands, used in conjunction with the most positively correlated bands could provide more accurate vegetation classifications and yield evaluations than previously available. HYDICE bands 630 to 700 nm, 1040 to 1100 nm, are consistently the most highly correlated to yield.

Table 2. Differing spatial resolutions correlation to yield for field N. The 20 most positively correlated bands are shown.

30 m		15 m		10 m		2 m	
CORR	MIN	CORR	MIN	CORR	MIN	CORR	MIN
0.84615	0.78147	0.82256	0.80533	0.79551	0.7656	0.50526	0.46539
MAX	STD	MAX	STD	MAX	STD	MAX	STD
0.90466	0.044122	0.83284	0.0091666	0.82434	0.020657	0.5371	0.023506
RANK	BANDNUM	RANK	BANDNUM	RANK	BANDNUM	RANK	BANDNUM
1	76	1	67	1	68	1	67
2	68	2	68	2	70	2	68
3	62	3	70	3	62	3	62
4	67	4	69	4	67	4	63
5	70	5	63	5	76	5	70
6	63	6	76	6	63	6	69
7	69	7	65	7	69	7	65
8	89	8	64	8	77	8	64
9	78	9	66	9	78	9	66
10	79	10	62	10	65	10	71
11	65	11	71	11	79	11	74
12	64	12	74	12	64	12	73
13	71	13	72	13	71	13	72
14	82	14	73	14	89	14	76
15	83	15	77	15	74	15	77
16	74	16	85	16	66	16	83
17	81	17	83	17	82	17	85
18	80	18	82	18	83	18	82
19	66	19	84	19	84	19	84
20	84	20	86	20	81	20	78

Increases in spatial resolution increase the positive correlation to yield as seen in Fig. 2. Interesting trends can be seen as the pixel size is increased. The negative correlation in the visible becomes positive. The overall correlation increases from 0.50526 for 2 m, 0.79551 for 10 m, 0.82256 for 15 m, 0.84615 for 30 m, and 0.94161 for 40 m data. The reason for this may be because many of the inherent yield monitor and image resampling errors are being generalized. Also, down sampled pixels containing averaged numbers result in a higher overall correlation. These results will be used to assess the performance of classifications to identify yield patterns prior to harvest.

Fig. 3. Correlation of yield to individual HYDICE spectral bands.

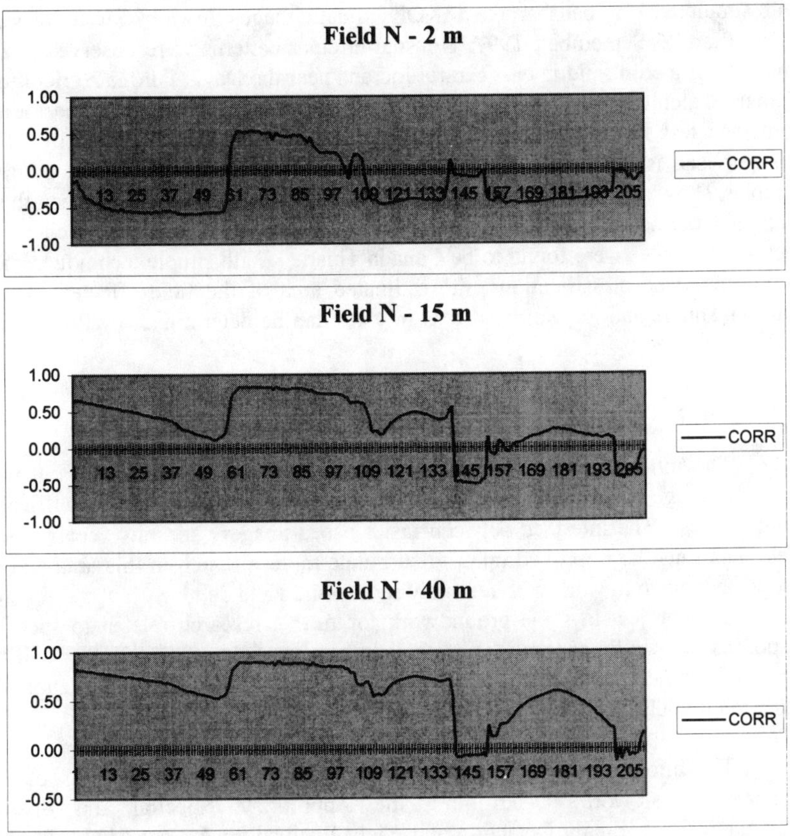

ANOMALY DETECTION

Similar techniques as described above are being used to identify anomalous regions, such as soil drainage problem areas and thistle plots. The initial attempts to identify a unique spectral signature for a specific weed type, such as thistle, using the hyperspectral data we have has proved to be very difficult. The dates of the imagery are not ideal for characterizing the thistle coupled with so much natural variation within the actual field as compared to signatures taken in labs or controlled studies.

Traditional remote sensing techniques can be used effectively to identify areas of anomalous conditions, visually and by using image processing with pattern recognition techniques. Much research needs to be completed before the process can be automated and before the computer can be used for anomaly distinction. We are to the point where, with the increasing spatial, spectral, and temporal resolutions, we can do a quick analysis and locate the anomalous areas with corresponding GPS coordinates so the growers can visually check for problems within their field.

The following example of an actual anomaly detection study was done with simulated 1 m, pansharpened IKONOS data images flown by Space Imaging EOSAT on 26 September, 1997. Unusual circular patterns were observed in the imagery for a corn field in east central Indiana near the Davis-Purdue Agricultural Center. Calculating the NDVI for the field showed that the circular areas contained less healthy vegetation (brighter areas are healthier vegetation). The unsupervised isodata classification, using the lower vegetation areas as training samples, showed that there was a significant difference statistically between these circular patterns and the surrounding corn. When the areas were field checked, these "low" spots were found to be Canada Thistle. With simple techniques such as unsupervised classifications, the estimated area of the field effected by the Canada Thistle and an estimated loss in yield can be determined (Willis, et al., 1997).

FURTHER RESEARCH

The inherent inaccuracies of yield monitor data as well as the inherent errors in image resampling must be understood in order to conduct scientifically valid studies. The interface between raster based imagery and raw vector based yield data must be improved upon to stimulate more research in this area and to allow the layman to work with remotely sensed and yield data.

This project lays the groundwork for further research related to spectral responses of specific anomalies and their effects on yield. These results will be applied in other continuing studies to develop yield evaluation and anomaly detection models using varying spectral and spatial resolutions.

The information presented in this paper is taken from a portion of the author's thesis work: Determining the Appropriate Spectral and Spatial Resolution for Anomaly Detection and Yield Evaluation: A Case Study. Purdue University, August 1998.

REFERENCES

Birrell, S.J., S.C. Borgelt, and K.A. Sudduth, 1996. Crop yield mapping: Comparison of yield monitors and mapping techniques. p. 15–32. *Proc. of 2nd Int. Conf., Site-Specific Management for Agricultural Systems, Minneapolis, MN. 27–30 Mar., 1996.*

Blackmore, B.S., and C.J. Marshall, 1996. Yield mapping: Errors and algorithms. p. 403–415. *In Proc. of the 3rd Int. Conf. on Precision Agriculture, Minneapolis, MN. 23–26 June, 1996.*

Johannsen, C.J., M.F. Baumbardner, P.R. Willis, and P.G. Carter, 1998. Advances in remote sensing technologies and their potential impact on agriculture. *In 1st Int. Conf. on Geospatial Information in Agriculture and Forestry, Lake Buena Vista, FL. 1–3 June, 1998.*

Johannsen, C.J., and P.G. Carter, 1997. Who's who in commercial satellite remote sensing: A status check. p. 4. *In InfoAg Conf., August 7, 1997.* Published and updated at URL: http://dynamo.ecn.purdue.edu/~biehl/LARS/

Mitchell, P.A. 1995. Hyperspectral digital imagery collection experiment (HYDICE). p. 70–95. *In Proc. of SPIE The Int. Soc. for Optical Engineers, Paris, France.*

Murphy, D.P., E. Schnug, and S. Haneklaus, 1996. Yield mapping - a guide to improved techniques and strategies. p. 33–48. *In Proc. of 2nd Int. Conf., Site-Specific Management for Agricultural Systems, Minneapolis, MN. 27–30 March 1996.*

Vane, G.T., R.O. Green, T.G. Chrien, H.T. Enmark, E.G. Hansen, and W.M. Porter. 1993. The airborne visible–infrared spectrometer (AVIRIS). Rem. Sens. Environ., 44:127–143.

Willis, P.R., C.J. Johannsen, and J.H. Arvik, 1997. High spatial resolution data for Precision Agriculture, Poster. *In ASA Annual Meetings, Anaheim, CA, October 27–30, 1997.*

Corn and Soybean Yield Indicators Using Remotely Sensed Vegetation Index

Minghua Zhang, Paul Hendley, and Dirk Drost
Zeneca Ag Products
Richmond, California

Michael O'Neill, and Susan Ustin
Department of Land, Air and Water Resources
University of California
Davis, California

ABSTRACT

Precision farming involves crop management in parcels smaller than field size. Yield prediction models based on early growth stage parameters are one desired goal to enable precision farming approaches to improve production. To accomplish this goal, spatial data at a suitable scale describing the variability of yield, crop condition at certain growth stages, soil nutrient status, agronomic factors, moisture status, and weed–pest pressures are required.

This paper discusses the potential application of aerial imaging to monitor and predict the potential yield for corn and soybean at various growth stages in the season. Included in the analyses were aerial images, yield monitor data and soil grid sampling. The relationship between remotely sensed Normalized Difference Vegetation Index (NDVI) and yield was best at 9 m spatial resolution. Preliminary results indicate that it is possible to use NDVI to estimate the potential yield for soybean and corn when canopy reaches full cover.

INTRODUCTION

Precision farming is a new agricultural system concept with the goals of optimizing returns in agricultural production and environment. This concept involves the development and adoption of remote sensing (Barnes et al., 1996), and Geographic Information System (GIS) technology applications, and knowledge-based technical management systems (NRC, 1997). With a refined GIS and spatial knowledge-based management system, farmers should have the ability to appropriately manage field operations at each location in the field, as well as the ability to predict likely yield from early season indicators.

Many studies have focused on variable rate applications (Gotway et al, 1996; Stafford and Miller, 1996) while some are focused on yield mapping (Sudduth et al., 1996). Yield mapping provides not only information about the yield itself, but it allows comparison to field conditions that may explain spatial yield variation. Yield mapping, however, is normally accomplished only once, at the end of a season. It is difficult to quantitatively evaluate efficacy of

Copyright © 1999 ASA-CSSA-SSSA, 677 South Segoe Road, Madison, WI 53711, USA. *Proceedings of the Fourth International Conference on Precision Agriculture.*

management because many factors that comprise the measured value change over successive seasons (Blackmore & Marshall, 1996; Davis, 1998). Over the longer term, the combined technologies of variable rate applications and yield mapping allows analyses of individual variables and their correlation to crop production between seasons. Remote sensing from airborne or spaceborne sensors can provide spatially distributed synoptic data acquired multiple times during the growing season. The normalized ratio of near-infrared reflectance to red reflectance, called the normalized difference vegetation index (NDVI) has been shown to be a sensitive indicator of biomass and leaf area in several crops, which can be used to track crop development over the season. Because crop yield is generally correlated with canopy development, this index can be used to develop a relationship to yield. Once a relationship between yield and NDVI is developed, then farmers can predict their yield earlier in the growing season and therefore, better harvest management, planning the following season's inputs, and other more effective management can be achieved.

To explore the potential for corn and soybean yield prediction, the study objectives were: (i) to monitor Midwest row crop growth during the growing season through aerial images, (ii) to understand the relationship between crop yield and NDVI, (iii) to investigate the optimum image spatial scale for the best performance in yield prediction, and (iv) to examine the relationships between NDVI and soil nutrients.

MATERIALS AND METHODS

Corn and soybean fields near Hills, IA (Fig. 1) were selected for study. Hills is situated in a small valley located south of Iowa City. Most agriculture in the region is rainfed and the growing season normally starts early in May with harvesting in late September or October. Soils are generally rich in organic matter. The selected cornfield, planted 24 April 1997, was approximately 46 acres in size. The selected soybean field, planted 10 May 1997, was approximately 110 acres in size. There were no specific pest pressures influencing crop growth in either field.

Aerial images of both fields were obtained three times (17 June, 16 July, and 18 Sept. in 1997) during the growing season using the ADAR 5500 4-band digital camera. The 4-spectral band intervals were filtered to match Landsat Thematic Mapper blue, green, red, and near- infrared bands. The airplane was flown at an altitude of approximately 7300 ft to provide ground spatial resolution of 1 m. Yield monitor data, recorded as a flow rate, and harvest speed were measured at two second intervals. Soil was sampled at 100 m by 100 m grid, and the major nutrients of N, P, K, and minor nutrients of Ca^{++}, Mg^{++} were analyzed.

The images, yield and soil data were georeferenced into a common coordinate, UTM zone 15 with no shift parameters. The NDVI was derived by ratio (NIR-R/NIR+R) from the aerial images for the selected fields. Various scales were resampled from the original 1 m resolution image and two second flow yield monitor point data in a raster environment. These scales included 1 m, 3 m, 6 m, 9 m, and 12 m.

ENVI image processing software (RSI), Arc/Info GRID and ArcView GIS software (ESRI, 1997) and SAS statistical software (SAS Institute 1996) were used for the data analysis. MS Excel was used for the graphs.

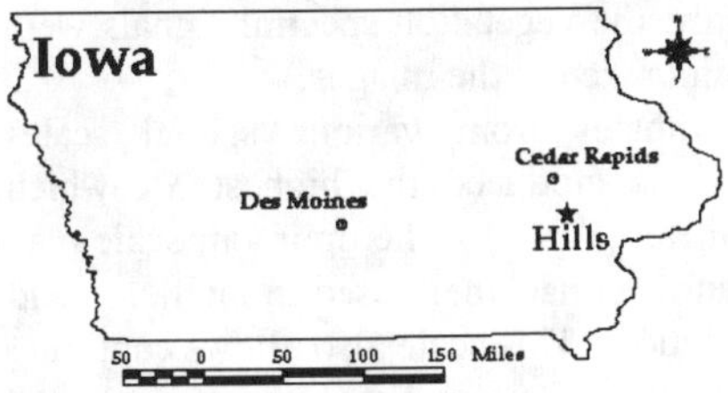

Fig. 1. Study location in Hills, IA.

RESULTS AND DISCUSSION

The NDVI maps and the false color composite images, which are comprised of green, red and near-infrared bands, clearly displayed anomalies in the corn and soybean fields at the 1 m spatial resolution. From the cropping history, these field patterns are associated with the agronomic variables, which are used to assist farmers in field scouting and farm management decisions. The analysis results indicated a strong correlation between corn and soybean yields and NDVI in both June and July images. Figure 2 displayed the similar spatial patterns in yield and NDVI for both corn and soybean crops.

a. b.

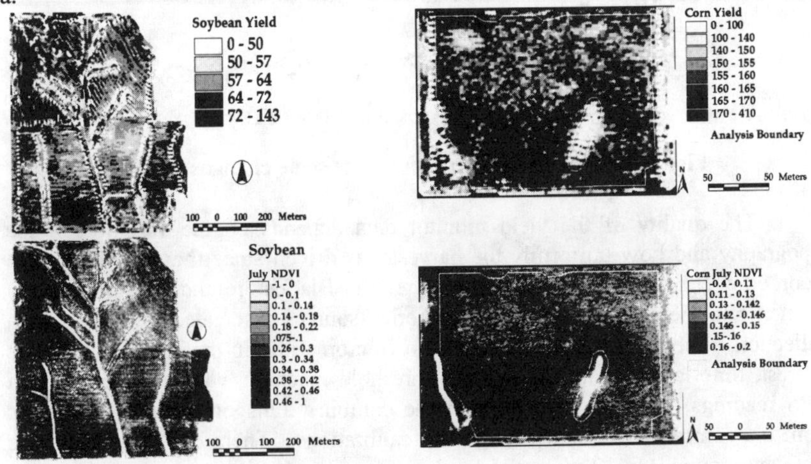

Fig. 2. (a.) Soybean yield map and (b.)
NVDI map for the field

Our analysis showed that more than 70% of yield variation can be explained by aerial imaged NDVI for both corn and soybean crops when the NDVI values were grouped into fine intervals. The relationships for NDVI yield estimates from both June and July images were significant; however, the estimates from the July images were better yield predictors. These results may be attributed to the facts that the crops had reached full cover in July, the vegetation spectral signals were maximized, and the soil exposures were minimized in the images.

Comparisons of NDVI yield estimates from various spatial scales indicated that 9 to 12 m spatial resolutions produced the highest R^2, which provided the best fit in both corn and soybean (Fig. 3). The optimum scale may be associated with the planting variables and the machinery used in the field yield mapping as well as the quality of data collection. This scale also allows capturing the soil anomaly variation present in these fields. Corn and soybean were planted in approximately 30-in row spaces. Most corn and soybean machinery cover >1 m in width, therefore it is difficult to precisely match field measurements to 1 m resolution imagery. In fact, the yield harvester was at least 6 m in width.

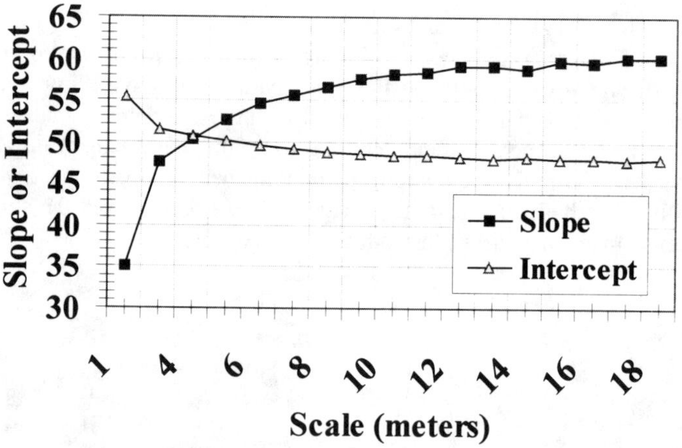

Fig. 3. Slope and interception vs. the scale changes in input data.

The quality of the yield monitor data depends on the variation of the topography and how uniformly the harvester is driven since the yield data were recorded every two seconds as a flow rate. As Blackmore and Marshall (1996) pointed out, there were seven potential error sources associated with yield data collection. These sources included unknown crop width in the header during the harvest, time lag of grain through the threshold machine, wandering error from GPS readings, surging grain through the combine transport system, grain loss from the combine, sensor accuracy, and calibrations. Therefore, it makes sense that data around 9 to 12 m spatial resolution would reveal the more robust results since the variations of yield recording can be averaged, and the random and systematic variations of both NDVI and yield can be reduced.

The results also showed that there was an increase in R^2 for yield estimates with the increased resampling from 1 m up to 9 m on NDVI while yield data stayed at 1 m. The magnitude of increased R^2 was larger for yield estimates when the yield data were re-sampled from 1 m up to 9 m while NDVI stayed at 1 m. The results indicated the sensitivity of scaling yield data to a proper scale in order to find the meaningful yield estimates using NDVI.

The residue map of the corn field using 9 m data inputs (Fig. 4) was based on the differences between the predicted yield map and actual yield map. The various ranges identified in the residue map show the spatial locations of the unmatched predictions. The histogram (Fig. 5) indicated that 70% of correct yield estimates were within ±5 bu ac^{-1}; 90% of correct yield estimates were within ±9 bu ac^{-1}.

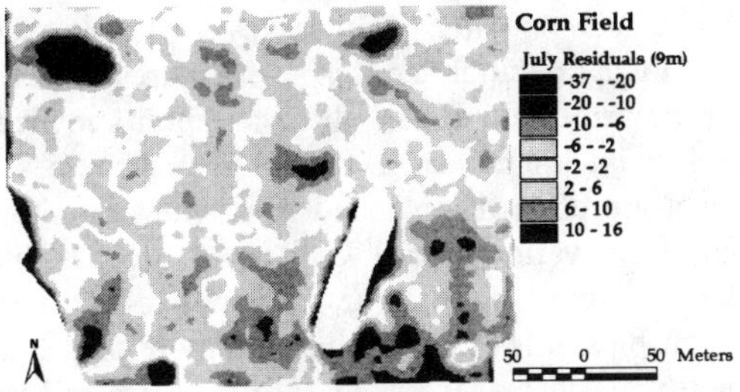

Fig. 4. Spatial distribution of the residue map for the corn field. The lighter colors represent well-matched yield predictions and the darker colors are the unmatched yield predictions.

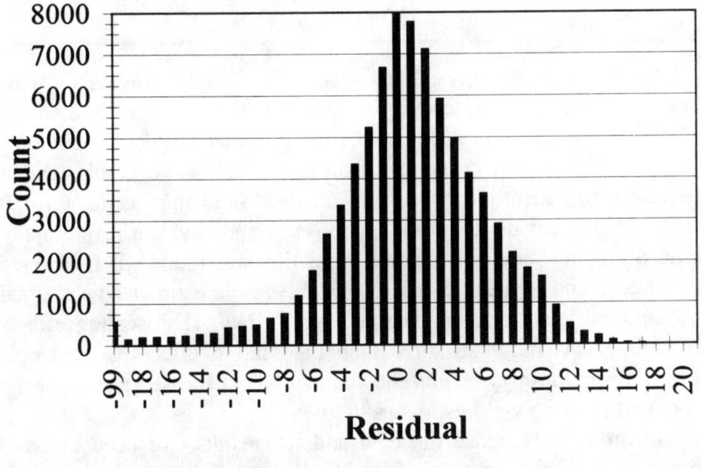

Fig. 5. Residue histogram. For most pixels, the difference between the predicted yield and the actual recorded yield was within 5 bu ac^{-1}.

There is a general hypothesis that yield is a function of soil nutrient variables in addition to environmental factors. Therefore, a relationship between yield and soil nutrients was expected; however, with the given data, we found no direct spatial relationship between yield and soil nutrients, and no direct relationship between NDVI and soil nutrients. This result may be due to the soil data quality and the inconsistent spatial scales of the yield, aerial images and soil nutrients. Hence, future analysis should include higher resolution of soil data, better accounting for the environmental factors, and middle season pest pressures.

Since field conditions and nutrients vary from field to field, one linear regression developed from one condition is not adequate to predict the yield for other fields. However, developing this type of relationship in multiple fields could provide farmers with information for site-specific management.

Understanding the field agronomic features is important in performing precision farming, and data quality is often another key to help us correctly understand spatial variation in agricultural landscape structures and temporal variations at a given field (Franzen et al., 1996). It is common knowledge that spatial data for precision farming is collected from various sources and in different formats. Yield data are often downloaded from yield monitors, soils from grid sampling, images from various aerial vendors and/or satellites, and agronomic features from farmer's growing history. To continue to improve our understanding of these relationships, it is critical to develop more precise location information for samples and to consider new strategies to optimize soil-sampling procedures. For future work, multiple fields should be studied and better image calibration using a common base should be developed to achieve the yield predictability.

CONCLUSIONS

From the analyses, we found that aerial images provide farmers with the knowledge needed to monitor crop growth, identify some of field anomalies, and efficiently direct the field scouting person to the anomalies. This knowledge is very important for short growing season crops such as vegetables and other high value cash crops in order to perform on time field management for the crops; however, it may not be as important for corn and soybean row crops. We also concluded that the better image yield prediction was found when crop reached full cover for corn and soybean crops, and the best yield estimates were obtained with 9 m resampled yield monitor data and NDVI data. The scaling effects of yield monitor data was more sensitive than the scaling effects of resampled NDVI images. Our preliminary results showed that NDVI can serve as an early indicator for corn and soybean yield estimates in Iowa.

In order to best use the data and the results, we need to have a better understanding of the impact of combining multiple errors into assessments. Though the data quality and potential error sources for each of the variables may potentially impede understanding of agricultural process and limit the rate at which technology applications are adopted in precision agriculture, the results from this study demonstrated yield prediction up to 3 mo before harvest, and addressed the choice of sampling scales for data collection. Nonetheless, more work is needed to determine the optimum strategy for correlating yield maps with NVDI to determine and verify apparent relationships. The image calibration and analysis methods need further research to improve the reliability of interpretations for technology applications in precision farming.

ACKNOWLEDGMENT

We would like to acknowledge Zeneca Ag products for providing the funding for the project and would like to thank useful comments from the colleagues in Zeneca and the University of California Davis.

REFERENCES

Barnes, E.M., M.S. Moran, P.J. Pinter, Jr., and T.R. Clarke. 1996. Multispectral remote sensing and site-specific agriculture: examples of current technology and future possibilities. p. 845–854. *In* Proc. of the 3rd Int. Conf.. 23–26 June.

Blackmore, B.S., and C.J. Marshall. 1996. p. 403–415. *In* Yield mapping: Errors and algorithms. Proc. of the 3rd Int. Conf.. 23–26 June.

Davis, G. 1998. Space age farming: Aerial sensor applications for precision farming and management. Modern Agric.. 1(5):18–20.

ESRI. 1996. ARC/INFO GIS 7.04 software manuals.

Franzen, D.W., L.J. Cihacek, and V.L. Hofman. 1996. Variability of soil nitrate and phosphate under different landscapes. p. 521–529. *In* Proc. of the 3rd Int. Conf.. 23–26 June.

Gotway, CA, R.B. Ferguson, and G.W. Hergert. 1996. The effects of mapping and scale on variable rate fertilizer recommendations for corn. p. 321–330. *In* Proc. of the 3rd Int. Conf.. 23–26 June.

NRC. 1997. Site-specific farming, information systems, and research opportunities. Natl. Academy Press.

SAS Institute, 1996. SAS/STAT user's guide. Statistics. Release 6.08 ed., SAS Inst., Cary, NC.

Sudduth, K.A., S.T. Drummond, S.J. Birrell, and N.R. Kitchen. 1996. Analysis of spatial factors influencing crop yield. p. 129–149. *In* Proc. of the 3rd Int. Conf.. 23–26 June.

Stafford, J.V., and P.C.H. Miller. 1996. Spatially variable treatment of weed patches. p. 465–474. *In* Proc. of the 3rd Int. Conf.. 23–26 June.

Tomato Yield – Color Infrared Photograph Relationships

G. Stuart Pettygrove
Department of Land, Air and Water Resources
University of California
Davis, California

Shrini K. Upadhyaya
Matthew G. Pelletier
Department of Biological and Agricultural Engineering
University of California
Davis, California

Timothy K. Hartz
Department of Vegetable Crops
University of California
Davis, California

Richard E. Plant
R. Ford Denison
Department of Agronomy and Range Science
University of California
Davis, California

ABSTRACT

Yield of processing tomato was measured in the Sacramento Valley, California, in three furrow-irrigated farm fields of 32 to 44 ha in size with a prototype weighing yield monitor mounted on a conventional single-row harvester. Normalized difference vegetation index derived from false-color infrared aerial photographs taken at full bloom was not related to fruit yield. However, fruit yield and vine dry weight just before harvest were closely related, suggesting that late-season factors caused yield variability. In one field, fruit yield was lowest both in the poorest-drained and the best-drained areas. In the best-drained area, soil texture was coarser, but yields were low, possibly because of inadequate lateral flow of water from the furrow to bed center. Yields were also low in the poorest-drained areas of the field due to prolonged saturation of soil following irrigation.

INTRODUCTION

Site-specific farming and its supporting technologies have been largely ignored by field crop farmers and researchers in the southwest USA. Grain crops, for which yield monitors are available, generally play a less central economic role than cotton, vegetables, and forages. Crop rotations are usually more diverse than those common in the U.S. Grain Belt. Variations in crop yield and quality are more likely to be influenced by irrigation system performance than by rainfall distribution. Characterizing the relationship of soil physical properties to plant

Copyright © 1999 ASA-CSSA-SSSA, 677 South Segoe Road, Madison, WI 53711, USA. *Proceedings of the Fourth International Conference on Precision Agriculture.*

growth and development is complicated by the fact that soil provides both the medium for root growth and the surface over which water flows across the field.

An important cropping system in California's Central Valley is the processing tomato-based rotation. We are not aware of any studies of field-scale variability in this system or other fruiting annuals such as pepper or cucurbits. Processing tomato varieties are determinate in fruiting and are mechanically harvested. The recent development of an improved tomato yield monitor at the University of California, Davis (Pelletier & Upadhyaya, 1999 this volume), allowed us to obtain yield maps in a tomato–wheat–sunflower rotation.

MATERIALS AND METHODS

Site Description and Crop Management

The three farm fields (42, 32, and 33 ha) chosen for this project are located near Davis, CA, in the southern Sacramento Valley. Rainfall averages 460 mm yr^{-1} and falls mostly from late fall to early spring. Soils are alluvial silty clay loams, clay loams, and silty clays. Fields are graded to uniform slopes for furrow irrigation. Field lengths were 400 m in the smaller two fields and 800 m in the larger field. To map soil texture, soil samples (0- to 15-cm depth) were collected on a 61-m square grid with 15 cores collected from the tops of raised beds in a 25-m^2 area at each grid location. Sand, silt, and clay content were determined by a modified pipette method.

Spring wheat (*Triticum aestivum* L.) was grown in the first year of the project, processing tomatoes (*Lycopersicon lycopersicum* L.) in the second year. Wheat was drill-seeded in the fall on raised beds with furrows spaced 1.52 m apart. The furrows provided surface drainage of heavy rain that occurred in early January and were used for a single irrigation in April at flowering. After June harvest, fields were disced and new raised beds were formed, also on a 1.52-m spacing. In two of the fields, tomatoes were direct-seeded in February. In the third field, transplants were used in April. In all three fields, the tomatoes were planted in a single row on each bed. Tomatoes were irrigated with hand-move impact sprinklers during the establishment stage, then by furrow irrigation. Water run lengths in two of the tomato fields were reduced by the use of flexible (lay-flat) gated plastic pipe in mid-field. The tomatoes were managed by the grower using standard practices for the area (Strand et al., 1990) and harvested in late July or early August.

Yield Measurements

Before harvest of wheat, the grower's combine (7.5-m swath) was retrofitted with an Ag Leader/GPS yield mapping system. The following year, tomato yield was measured with a prototype monitor mounted on one of the grower's harvest machines, which straddles a single bed of tomatoes. The yield monitor is described elsewhere (Pelletier & Upadhyaya, 1999 this volume). Briefly, it consists of a load cell to measure the weight of the fruit as it passes over a conveyor on the boom elevator just before discharge into the receiving trailer. Red fruit is separated from clods, vines, and most off-grade fruit before the weighing point. Weights were recorded once per second, but due to limitations of the differential GPS receiver and data logger, latitude and longitude were only sporadically recorded. Neither time of day nor harvester speed was recorded. The load cell readings were calibrated by use of a mobile weigh cart. To compile the yield data set, a four-point running average of yield was written to each data location possessing an actual GPS reading. Spacing between adjacent yield points in the final data set generally ranged from 3 to 12 m.

There were several inherent sources of error and smoothing in the yield monitoring procedure. Fruit followed a long, split flow path through the harvester before reaching the weighing point. Also at two locations in the harvester, a minimum mass of fruit was required to build up before any fruit was pushed beyond those points. Pelletier and Upadhyaya (1999 this volume) concluded that the yield monitor was not capable of detecting variability over a distance of less than nine meters. This is not greatly different from the resolution achieved by commercial grain yield monitors.

Commercial tomato fields are usually harvested with two or more machines operating at the same time. Because a machine harvests only one bed at a time, and there was only one yield monitor for our project, we were not able to obtain yield data on every bed. In two of the fields, the majority of the yield data are from every fifth bed. Due to the cooperation of the grower, on one field, yield was measured on every bed in several large areas.

To facilitate comparison of yields to aerial images, the tomato yield point data were converted in ArcView (ESRI, Redlands, CA) to a 9.15-m grid using inverse distance squared weighting of the 12 nearest neighbors. For geostatistical analysis, the original non-gridded data were used. Duplicate points and a small number (~5 per 1000) of high yield points (>130 Mg ha^{-1}) were removed. To examine anisotropy, we produced variogram surfaces (Isaaks & Srivastava, 1989) using Variowin 2.2 software (Pannatier, 1996). Geostatistical analyses were limited by available computer memory to 1,400 yield points at a time, which represented areas ranging from 0.7 to 1.5 ha.

Color Infrared Aerial Photography

Aerial photography and image processing methods are described by Denison et al. (1996). Color IR photos were taken at mid- to full-bloom stage from an altitude of 1200 m above the site using 70 mm Kodak Aerochrome 2443 film. Transparencies were scanned using a slide scanner with a resulting spatial resolution of about 0.74 m on the ground. Resulting 24-bit images in TIF format consisted of three colors corresponding approximately to the near-infrared, red, and green. These were analyzed using Image Pro for Windows (Media Cybernetics, Silver Springs, MD). The scanned images were ortho-corrected using GPS-determined locations of the field corners. NDVI values of <40 were removed as these were mostly associated with field edges and irrigation pipelines. Normalized Difference Vegetation Index (NDVI) was calculated as (IR-red)/(IR+red). Using the ARC/INFO resample function, NDVI grids of 3.04 and 9.15 m were created for geostatistical analysis and for comparison to tomato yield.

RESULTS AND DISCUSSION

Yield distribution of fresh fruit of the three fields (designated field 5, 14, and 58) is shown in Fig. 1. Mean yield ranged from 61 to 74 Mg ha^{-1}, but yield distributions in the three fields were similar. Based on yield values interpolated for 9.15 ∗ 9.15-m blocks, the least productive non-contiguous quarter of each field yielded 71 to 75 % of the field mean and 55 to 57 % of the most productive quarter. Based on a typical price paid to growers of \$55 Mg^{-1}, the lowest and highest yielding quartiles of a 32-ha (80-acre) field would provide approximate gross incomes of \$21,700 and \$38,600, respectively.

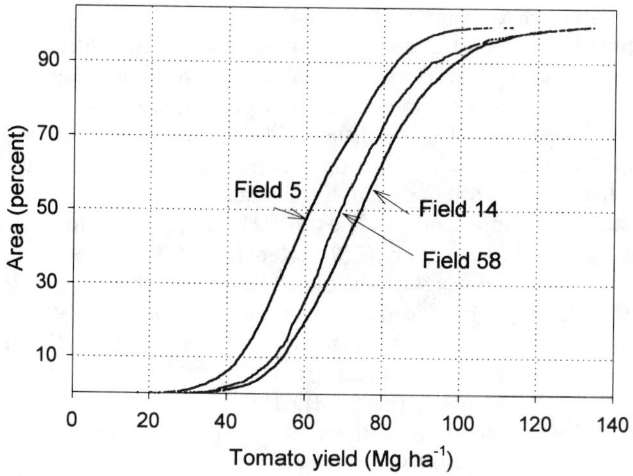

Fig. 1. Processing tomato fruit yield distribution in three commercial fields.

NDVI at full bloom and yield were unrelated in the three fields (Fig. 2). A more detailed analysis of NDVI-yield-soil texture relationships was conducted on one field (Field 5) because of its more detailed yield monitor data record. Tomato yield was lowest in the northern half of the field and in a small area in the southwest corner (Fig. 3, left). NDVI at full bloom was lowest in a large area just north of the center of the field and in the southwest part of the field (Fig. 3 right). Soil was sandiest in the southwest corner of the field and generally decreased in sand content in a north–northeast direction (Fig. 4). The use of gated pipe in the middle of two fields to reduce irrigation water run length did not produce any obvious effect on yield or NDVI.

Both yield and NDVI in Field 5 displayed somewhat greater spatial continuity parallel to the rows (east–west direction) than perpendicular to the rows. This is shown for four randomly selected areas of the field in Fig. 5. The variance surface graphs show that pairs of yield or NDVI values tended to be more alike (i.e., have lower semi-variance) when they were in the same row than when they were several rows apart. In one of the four areas (Fig. 5, third from top), yield showed a more complicated pattern of spatial dependence that did not match the row direction. The repeating pattern of the NDVI variance contours seen in the top three plots in Fig. 5 is probably a moiré pattern resulting from the fact that the tomato rows and NDVI pixels were not perfectly parallel.

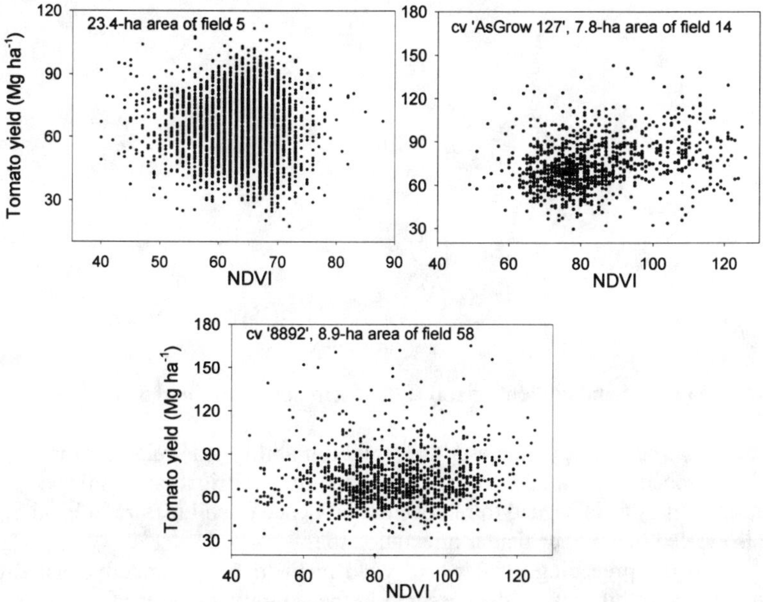

Fig. 2. NDVI at full bloom (19 June) vs. fruit yield of processing tomato. Each data point represents an area of 83.7 m^2.

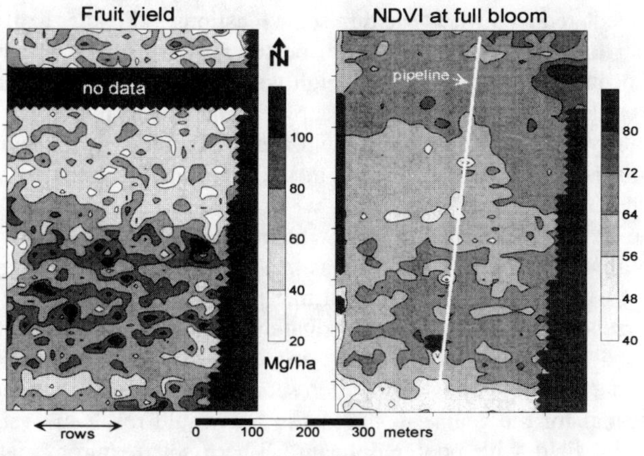

Fig. 3. Tomato yield and NDVI in Field 5.

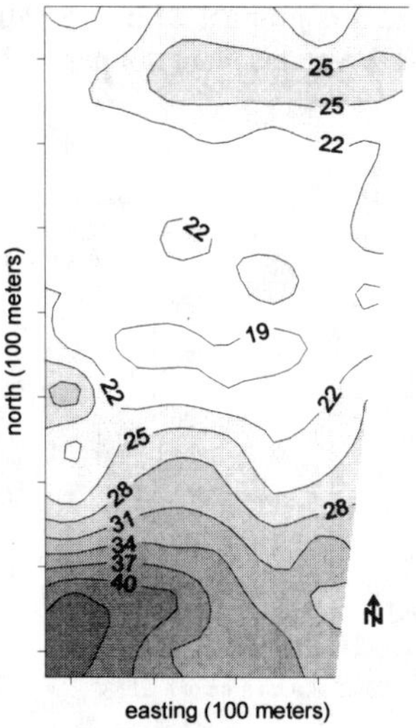

Fig. 4. Sand content of soil (0-to15-cm depth) in Field 5.

This anisotropy suggests that yield variability was related to management factors occurring parallel to rows, such as furrow irrigation or tillage; however, further analysis is needed to remove the influence of soil texture in Field 5, which also varied in a perpendicular orientation to rows.

In the preceding year, wheat yield in Field 5 was directly related to soil texture, with highest yields measured in the southwest corner of the field. In the northern half of the field where soil was finer-textured and drainage was slow, wheat tillered poorly following heavy rains in early January. Tomato yield showed a similar pattern in most of the field, but unlike wheat, yields were reduced in the southwest corner. Wheat and tomato yields are compared in Fig. 6. Grassy weeds reduced wheat yield in some locations in Field 5, and we compared wheat and tomato yield only at grid locations where weeds had been visually rated. Locations with high or very high weed density ratings in the wheat crop were discarded. The resulting graph shows a curvilinear relation between wheat and tomato yield, with high wheat yield and low tomato yield occurring in areas with coarser-textured soil, indicated in the graph as locations with silt content below 29%.

The grower's explanation for lower tomato yields on the better-drained soil in the southwest corner of the field was that the 1.5-m furrow spacing works well for the less well-drained Class II soil that predominates in Field 5; but in areas with coarser-textured Class I soil, irrigation water may not reach the center of each raised bed. This would result in under-irrigation and lower irrigation application efficiency. Longer irrigation set times or closer row spacing would be more optimal for the coarser-textured soil but would result in over-irrigation in areas of the field with poorer drainage. Where soil texture variation parallels furrows (as it does in Field 5), it should be possible to increase irrigation application efficiency and tomato yield by irrigating more often in the coarser-textured soil areas; however, rescheduling the irrigator's time, refilling head ditches, etc. in order to irrigate a small part of a field would require more intensive management.

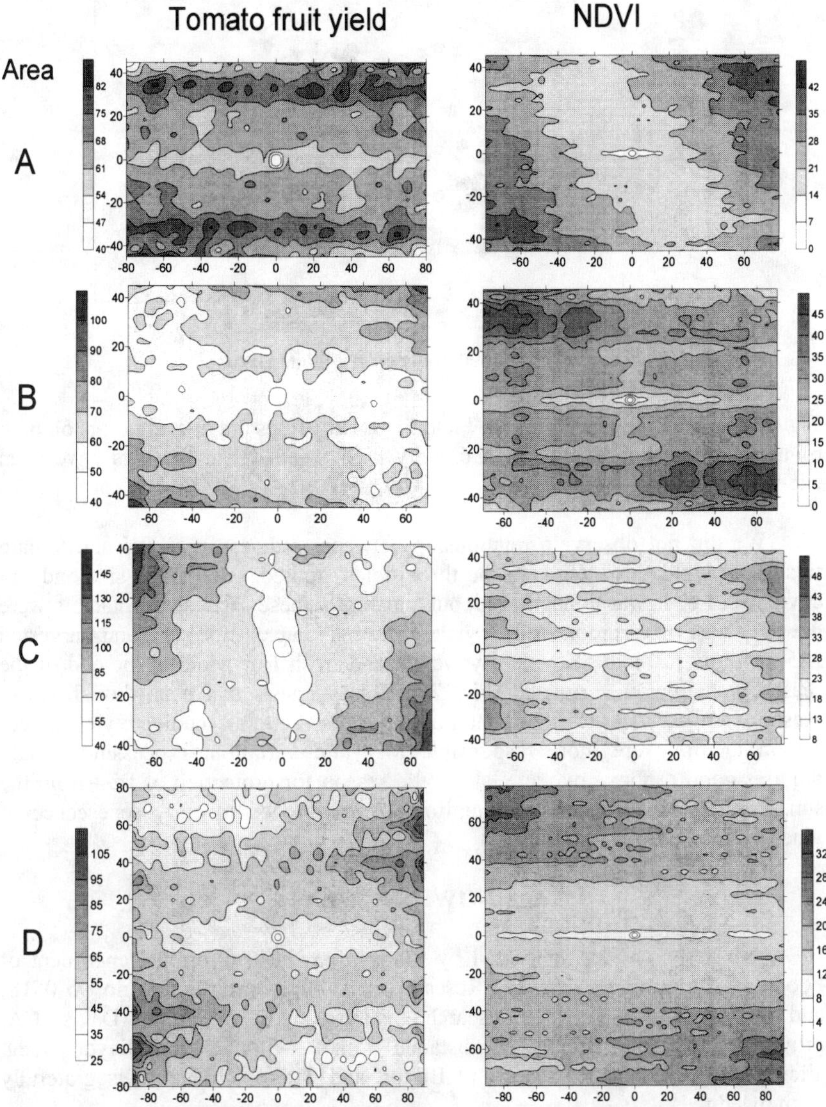

Fig. 5. Variogram surface of tomato fruit yield and NDVI in four areas of Field 5 showing greater spatial continuity parallel to rows. Yield in Area C is an exception. Graph axes are in units of meter.

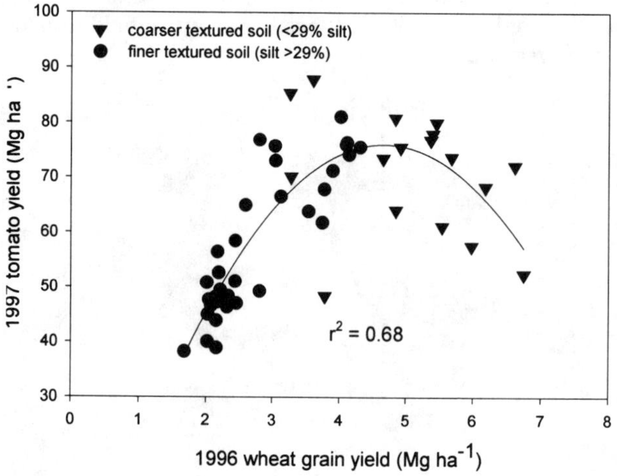

Fig. 6. Tomato vs. wheat yield in Field 5. Yield values are at grid points on 61 * 61-m spacing. Points with high or very high weed density ratings have been excluded.

We did not observe a relationship between mid-season NDVI and tomato fresh fruit yield. In general, in the three fields studied, good plant size and full coverage of beds was not achieved, but apparently these early-season factors were not the cause of variation in fruit yield six weeks later. In seven hand-harvested 225-m^2 plots in Field 58, vine dry weight and fresh fruit yield at the end of the season were highly correlated (Fig. 7). This suggests that even though early-season vigor was not very good and possibly caused overall mediocre yields, late-season factors were more responsible for yield variation. Processing tomato requires good canopy coverage late in the season for protection of fruit from the sun. Possibly, inadequate or nonuniform irrigation led to early senescence of vines in some parts of the field.

ACKNOWLEDGMENTS

This research was supported by funds from the California Department of Food and Agriculture's Fertilizer Research and Education Program (no. 95-0518) and the California Tomato Research Institute. W.E. Wildman, Davis, CA, provided aerial photographs. Assistance from Julie A. Young, Jiayou Deng, Victor Huey, Robert Rosseau, and Button and Turkovich Farms is gratefully acknowledged.

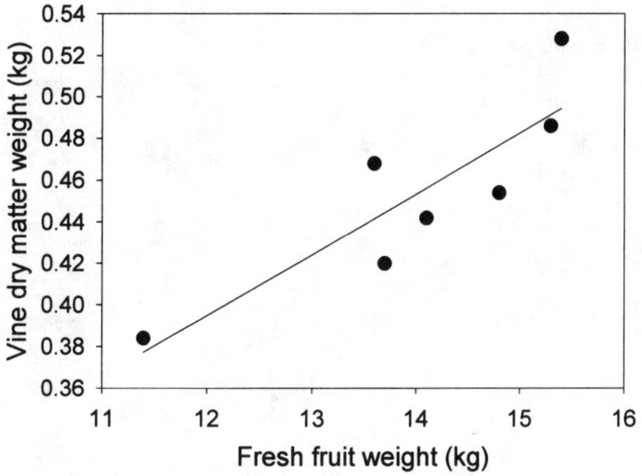

Fig. 7. Fresh fruit and vine dry matter yield just before harvest in hand-harvested plots in field 58. Each point is the average of eight 1-m ∗ 1 row subplots.

REFERENCES

Denison, R.F., R.O. Miller, D. Bryant, A. Abshahi, and W.E. Wildman. 1996. Image processing extracts more information from color infrared aerial photos. California Agric. 50(3):9–13.

Isaaks, E.H., and R.M. Srivastava. 1989. An introduction to applied geostatistics. Oxford Univ. Press, New York.

Pannatier, Y. 1996. Variowin: Software for spatial data in 2D. Springer-Verlag, New York.

Strand, L.L. et al. 1990. Integrated pest management for tomatoes. 3rd ed.. Publ. 3274. Division of Agric. and Nat. Resourc., Univ. of California, Oakland.

Evaluating Commercial Cranberry Beds for Variability and Yield Using Remote Sensing Techniques

Marilyn G. Hughes
Rutgers University Cooperative Extension–Center for Remote Sensing
Rutgers University
New Brunswick, New Jersey

Peter V. Oudemans
Plant Pathology–Blueberry and Cranberry Research Center
Rutgers University
Chatsworth, New Jersey

Joan R. Davenport
Crop and Soil Science
Washington State University
Prosser, Washington

Keri Ayres
Blueberry and Cranberry Research Center
Rutgers University
Chatsworth, New Jersey

Teuvo M. Airola
Center for Remote Sensing
Rutgers University
New Brunswick, New Jersey

Abbott Lee
Lee Brothers
Chatsworth, New Jersey

ABSTRACT

This study uses GPS–GIS–RS techniques to analyze cranberry (*Vaccinium macrocarpon* Ait.) crop health and yield. Extensive field sampling has been used in the past as a means of estimating potential bed yields. The major problem for predicting yield appears to be to high *intra-bed* spatial variability. For this study, color-IR photography from commercial cranberry beds (May 1996) was rectified to earth coordinates using GPS technology. An unsupervised multi-spectral classification and an NDVI were done to statistically group pixels in the image. Results indicate that a number of features within cranberry beds can be identified,

Copyright © 1999 ASA-CSSA-SSSA, 677 South Segoe Road, Madison, WI 53711, USA. *Proceedings of the Fourth International Conference on Precision Agriculture.*

including variations of vegetative cover, irrigation and drainage systems, and areas of beds damaged by insects and fungal disease (*Phytophthora cinnamomi).* In the future, remotely sensed imagery will be linked to ground based data to gain further insight into the spatial variation of factors affecting crop yield and health.

INTRODUCTION

Cranberries are a low growing perennial crop indigenous to the sandy wetland soils found in the Pine Barrens region of New Jersey. The berries develop on uprights along a network of vegetative runners that grow as a ground cover over the bed (Eck, 1990) and are harvested from late September to early November. Currently, New Jersey is home to >3 300 acres of cranberry beds valued between $6,000 and $30,000 per acre annually (Roper & Vorsa, 1997). Due to stringent wetland laws, expanding the cultivated acreage to meet an increasing demand for the crop is difficult. Therefore, cranberry growers are turning to new technologies such as precision agriculture to spatially map, predict, and ultimately increase current crop yield in existing cranberry beds.

Cranberry yield is influenced by the physical characteristics of the soil environment (i.e., soil profile, percentage of sand, percentage of organic matter) and management practices (i.e., irrigation, fertilizer, pesticide, and fungicide applications). These factors affect the fruit set, berry size, and number of flowers per upright (Eck, 1990; Baumann & Eaton, 1986). Currently, estimates of cranberry yield are made using a combination of two techniques: (i) determination of bed areas in production using aerial photography to define the boundaries, (ii) through intensive field sampling during the growing season. In the future prediction of yield and crop health using remote sensing (RS) techniques will provide a non-intrusive means of acquiring this information from individual sites as well as on a regional scale.

The detection and monitoring of crop health and soil drainage properties have been successfully undertaken using RS techniques in field crops (see Frazier et al., 1997 for review). Remote sensors are made up of detectors that record specific wavelengths of the electromagnetic spectrum (ERDAS, 1997). All types of land cover absorb a portion of the electromagnetic spectrum giving it a signature for identification. For example, areas of healthy green vegetation reflect strongly in the near-IR band and absorb in the red band, and areas covered by water absorb strongly in the near-IR part of the spectrum. Indicators of plant stress can be developed using a variety of techniques based upon the spectral reflectance properties of the crop of interest and the radiometric information available from the remote sensing instrument in use. To date, multi-spectral information available from satellites by virtue of their spatial resolution (5–80 m ground resolution), have little utility in monitoring cranberries that are grown in beds on the order of <10 ac. Recent advances in technology are leading to the development of new instruments that will allow access to a wide range of digital imagery from both aircraft and space borne platforms in the conversion of conventional imagery into digital format. The pending launch of a number of new commercial satellite remote sensing devices will provide significantly better spatial resolution data and a more frequent data acquisition in the future. In particular, the launch of a pointable instrument having spatial resolutions of 1 m panchromatic and 4 m multi-spectral is planned for this year by the SpaceImaging Corp. (Thornton, CO). At these higher resolutions satellite imagery will become a viable way to monitor cranberry production at regional and global scales.

AIM OF STUDY

For this study, we used the available 1:12000 color-IR aerial photography obtained and archived each May from Ocean Spray (Lakeville-Middleboro, MA). The CIR photography is currently used to catalogue production acreage. Data were collected in three bands, the green (0.5-0.6), red (0.6-0.7), and near-IR (0.7-2.0). The overall goal of this study was to determine the feasibility of using remotely sensed data in conjunction with ground-truthed field data to map spatial variations in cranberry yield. To that end the following approach was taken. (i) image processing techniques were used on the color-IR photography to map bed features and variability, (ii) spectral information was used to derive statistical relationships to predict bed yield, and (iii) the information was used to develop a protocol for future studies incorporating both traditional field sampling and remote sensing techniques for mapping, understanding and predicting cranberry yield.

METHODOLOGY

Color infrared (CIR) photography from 20 May, 1996 providing coverage of the Lee Brother's cranberry farm was scanned in using a high-resolution color scanner at 600 dots per inch (dpi). This resulted in a ground resolution of approximately 2 ft and provides information about reflectance in the green, red, and near infrared portions of the electromagnetic spectrum. Using a series of ground control points obtained in the field with a GPS unit, the image was rectified and re-projected into the New Jersey State Plane Coordinate System. In addition, this imagery serves as a backdrop for information surveyed in the field, and as the source for multi-spectral data for remote sensing classification and yield estimation. The location of irrigation sprinkler heads within two of the beds provided the initial data sample locations in 1996 (Fig. 1). These locations were digitized on screen by scanning in a picture of their precise positions, and later refined using GPS data from within the beds. Based on a preliminary study using this data set it was found that the distance between the irrigation sprinkler heads as sampling points (50–80 ft) did not reflect the actual spatial variability within the beds. Large areas of low yield occurred between sampling points, leading to the consensus that a new sampling method was needed. In 1997, ground sampling was modified to better reflect the heterogeneous conditions of the cranberry beds. This procedure, called smart sampling was laid out using flags and the coordinates for each point were determined using a GPS unit (Fig. 1). Field data collected at each point include parameters such as number, weight and condition of fruit, upright density, canopy height, soil characteristics, and pH.

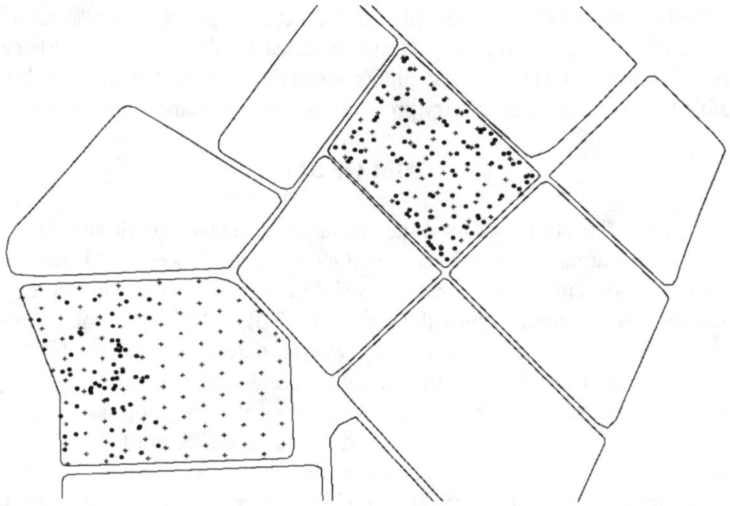

Fig. 1. Locations of sampling points on two cranberry beds in Speedwell, NJ. Crosses identify sprinkler heads, and filled circles identify smart sampling points. The bed on the lower left is planted with the cu. Stevens and the bed in the upper center is planted with the cu. Early Black.

An unsupervised multi-spectral classification was performed on the color-IR imagery to cluster the digital reflectance numbers into 10 statistically based classes. The percentage of each class (%CL) within each bed for the whole farm was then computed. A correlation analysis between %CL and total bed yield at harvest was done to determine significant relationships between the spectral data and yield. A stepwise multiple regression was done using the spectral classes with the highest correlation coefficients to predict bed yield. In addition, a normal difference vegetative index (NDVI) was calculated for each pixel based on the formula NDVI=(NIR-red)/(NIR+red) to determine variations in vegetative health over the beds. *In-situ* ground data for two test beds were overlain onto the NDVI surfaces to visually assess variations in observed data within the beds.

RESULTS

Initial studies using the color-IR photography show that remote sensing techniques have great potential for mapping bed boundaries, estimating yield and providing valuable information to growers regarding crop health. The results of the unsupervised classification on all beds on this farm for 1996 are shown in Fig. 2. Major yield components based on this technique are identifiable on the imagery as verified by field observations. For example, a close-up of the Stevens bed from 1996 (Fig. 3) reveals many significant features. The dark gray pixels shown in the lower left-hand part of the bed are lower lying areas with poor drainage that are wet. Inspection of this bed in the field revealed the presence of acute symptoms of *Phytophthora* infestations, specifically, *P. cinnamomi* that is introduced into the bed via irrigation (Oudemans, 1998). The areas surrounding the dark, wet areas are also poorly drained and appear to be affected by, but not killed by *P. cinnamomi.* Based on these results, the grower placed several under drains in the same bed prior to the 1997-growing season. The white pixels located above the water in Fig. 2 represent areas of low vine coverage and lower yield possibly due to insect damage.

The percent of area in the 18 beds occupied by pixels for each of the 10 statistical clusters was computed (%CL). Results of the correlation analysis between %CL and yield at harvest are shown in Table 1. Four of the %CL classes gave correlation coefficients over 0.4 and were selected for use in a multiple regression analysis (Eq. [1]). The yield values predicted from (Eq. [1]) compare favorably to the actual yields determined at harvest (Table 2). These four classes explain approximately 85% of the variation in yield. Residuals range from a low of -7100 $lb.acre^{-1}$ to a high of +4600 $lb.acre^{-1}$. Average deviation is 1900 $lb.acre^{-1}$. These results support development of a method for performing supervised classifications whereby the operator delineates the ground features on the imagery.

Fig. 2. Results of an unsupervised classification on a CIR image taken on 20 May, 1996.

Table 1. Correlation coefficients between yield in individual beds and 10 spectral classes resulting from unsupervised classification

	Spectral class defined in the unsupervised classification									
	1	2	3	4	5	6	7	8	9	10
Correlation coefficient (r)	0.05	0.02	-0.18	-0.25	-0.40	-0.14	0.09	0.41	0.48	-0.58

Yield = 263.022-0.21*%CL5+1.87*%CL8+1.54*%CL9-45.0*%CL10
(r^2 = 0.85) Eq. [1]

Fig. 3. Results of an unsupervised classification on a CIR image taken on 20 May, 1996.

The above analysis shows that RS information may be used to predict yield. Ground observations based on this map were made in an attempt to identify these spectral classes in the field. Although significant work remains to be done, it appears that the classes contributing positively to yield reflect areas in the bed that have high vine density. Areas that contribute negatively to yield appear to be those high in weeds, and damaged by root disease and/or insects.

The results of the NDVI (data not shown) indicate that areas within beds of low or no vegetation are aligned with areas of high water/poor drainage, root damage and disease. Of the two test beds, the Early Black bed had the higher overall NDVI, less disease and insect damage and produced approximately 50% more fruit than the Stevens bed. The associations between the yield in the bed and the other sampled variables indicate that yield at harvest is strongly associated with upright density similar to results found by Baumann and Eaton (1986).

Table 2. Raw data used to calculate multiple regression from unsupervised multi-spectral classification of image. %CL represents area for each statistical class defined in the unsupervised classification. Actual is the measured yield at harvest. Prediction is the yield computed using the regression equation (Eq. [1]). Resid. is the residual difference between the actual and computed yields. Raw data are measured in bbl/acre, where 1 bbl $acre^{-1}$=100 $lbs.acre^{-1}$.

Bed#	%CL5	%CL8	%CL9	%CL10	Actual	Pred.	Resid.
1	55.5	1.4	1.6	2.1	181	162	19
2	13.8	2.5	0.32	0.1	294	261	33
3	33.4	1.3	1.52	1.3	206	202	4
4	7.2	3.8	0.08	0.02	261	267	-7
5	15.6	0.4	0.1	0.1	233	256	-23
6	1.8	0.4	0.5	0.81	272	228	46
7	0.9	0.06	0.06	0.03	253	262	-9
8	0.9	0.28	0.4	0.29	264	251	13
9	7	5.2	1.4	1.18	252	220	32
10	1.6	57.8	8	0.01	388	382	5
11	0.02	13.6	85.8	0.52	400	397	3
12	41	0.73	0.24	0.05	251	253	-2
13	22	0.2	0.06	0.01	258	258	0
14	13.8	0.8	0.48	0.03	189	260	-71
15	14.2	18.8	11.3	4.11	98	127	-29
16	0.19	0	0	0	264	262	2

1998 GROWING SEASON

The above study shows that spectral analysis of color-IR imagery provides valuable insights into the composition of, and variability within cranberry beds. Results suggest that there are at least four significant factors affecting cranberry yield that can be mapped and identified using CIR imagery. For the upcoming season the following objectives are planned.

1. Digital color-IR imagery of the cranberry acreage will be obtained in early May 1998.
2. For selected beds, ground data will be collected at smart sampling sites and irrigation sites.
3. An unsupervised classification will be performed on the images to identify the spectral clusters in the beds.
4. Ground based data will be collected within the clusters to identify the contributing factors for each spectral class.
5. The *in-situ* data will be linked to the remotely sensed imagery via geostatistical techniques within a GIS.

SUMMARY

Current agricultural methodology is aimed at maximizing productivity while minimizing the area of cultivated land. This is extremely important in cranberry production because strict federal guidelines prevent new cranberry acreage from being developed in the wetland environment. Thus, with little opportunity to expand the area of production, cranberry growers must use new technologies to decrease the impact of farming on the environment and maximize yields. These preliminary results indicate that color infrared photography provides useful information regarding crop health and yield. Using an unsupervised classification, a number of features within the beds are identifiable, including variations in vegetative cover within and between beds, irrigation system features, and areas of the beds impacted by both root disease and insect damage. Results from the correlation analysis suggest that aerial photography flown early in spring may be used to predict yield at harvest. In addition, it may be possible to identify areas within beds vulnerable to fungal disease, insect damage, and drainage problems early enough in the growing season that mitigating actions could be taken. Further research linking the remotely sensed data with the site-specific data will aid in understanding the complex and interrelated factors that contribute to crop yield and should help to improve current agricultural practices for cranberry.

REFERENCES

Baumann, T.E., and G.W. Eaton. 1986. Competition among berries on the cranberry upright. J. Am. Soc. Hort. Sci. 111(6):869–872.

Eck, P. 1990. The American Cranberry. Rutgers Univ. Press, New Brunswick, NJ.

ERDAS, 1997. Remote sensing. p. 5–10. *In* ERDAS Imagine Field Guide. 4th ed. ERDAS, Atlanta, GA.

Frazier, B.E., C.S. Walters, and E.M. Perry. 1997. Role of remote sensing in site-specific management. p. 149–160. *In* The Site-Specific Management for Agricultural Systems. ASA, CSSA, and SSSA, Madison, WI.

Oudemans, P.V. 1998. Detection and monitoring of Phytophthora species in cranberry irrigation water by lupine baiting. Plant Dis. 82:(accepted).

Roper, T.R., and N. Vorsa. 1997. Cranberry: Botany and horticulture. Hort. Rev. 21:215–249.

Remote Sensing as an Aid for the Spatial Management of Nutrients

K. Panten
S. Haneklaus
D. Schroeder
E. Schnug

Institute of Plant Nutrition and Soil Science
Federal Agricultural Research Center
Braunschweig, Germany

M. Vanoverstraeten

Kemira S.A. /N.V.
Wavre, Belgium

ABSTRACT

The number of suppliers of remote sensing images taken from airborne platforms and space vehicles will increase in future and thus extend the limited availability of this information tool. Remote sensing enables a fast and economic data acquisition for the efficient investigation of soil and crop features. This paper presents existing and future tools for the collection of remotely sensed data, alternatives of data processing and examples of how this information can be efficiently used for the coordination and optimization of soil and plant sampling strategies. Special emphasis is put on the potential misinterpretation of vegetation indices, which leads to an inappropriate evaluation of the plant nitrogen status of the crop.

INTRODUCTION

When the first aerial photographs were taken at the beginning of the 20th century (Loeffler, 1994), the potential benefits of remote sensing for agricultural purposes were already obvious. Relationships between image features and soil fertility parameters or nutritional disorders of crops were then determined (Schnug, et al. 1998).

Comprehensive knowledge of the interpretation of aerial photographs already exists, but the lack of experienced interpreters has been probably the greatest barrier for a widespread use of imagery for crop management (Pelzmann, 1997). Until today the application of remote sensing in agriculture was hampered by high costs, insufficient standardization (e.g., lack of geocoding and import into GIS) and failure to develop practical strategies to implementing the information of remote sensing into agricultural production.

Copyright © 1999 ASA-CSSA-SSSA, 677 South Segoe Road, Madison, WI 53711, USA. *Proceedings of the Fourth International Conference on Precision Agriculture.*

Soon the number of suppliers of remote sensing images taken from airborne platforms and space vehicles will increase. Among other features they will probably offer geocoded image data which are ready to go into existing GIS and especially for precision agriculture designed software such as LORIS (Schroeder et al., 1997). But besides a reasonable price for these services, interpretation procedures including ground truth campaigns and the final transformation of results into digital agro resource maps is essential for the acceptance of remotely sensed data as a source for gathering geo-referenced information about the spatial variability of soil and plant features.

In principle remotely sensed data offers a fast and economic way to optimize soil and plant sampling strategies and to design variable field management. Therefore the purpose of this paper is to give an overview about existing and future tools to collect remotely sensed data, to provide guidelines how to interpret them and finally to highlight possibilities of how to integrate them into spatially variable outputs of agro-resources.

APPLICATION OF REMOTE SENSING IN PRECISION AGRICULTURE WITHOUT GROUND TRUTH

Information on spatially variable soil and plant features that can be gathered without ground truth is limited. Water or heat-stressed areas as well as weed-controlled zones in the field can be identified (Davis 1997–1998). A discrimination of weed types from remote observation is, however, not possible and therefore requires direct verification in the field. For navigation, the coordinates of the image are used and for fast and economic positioning in the field a GPS is recommended.

In fields with regular flooding of certain areas, these zones can easily be identified and excluded from further management practices. In irrigated land use systems, the amount of irrigation water rates could possibly be adjusted according to drought stress.

APPLICATION OF REMOTE SENSING IN PRECISION AGRICULTURE WITH GROUND TRUTH

Bare soil images are required for the investigation of key soil parameters such as organic matter content, soil texture and geomorphology. A ground campaign is necessary because of the superposition of the reflectance characteristics of soil texture, organic matter content, water content and ferric oxides (Barnes et al. 1996). One of the main advantages of remote sensing images is the efficient coordination of the soil sampling strategy by directing the sampling locations according to 'visible' differences in the field which can reduce sampling efforts drastically (cp. Schnug et al., 1998).

Images taken during the vegetative period have a wide operational area. They reveal different patterns in the field but causal factors are unknown, so that a ground campaign is inevitable. The calibration of field and image data allows an evaluation of crop growth differences and nutrient supply or a prognosis of expected yield.

AVAILABLE AND FUTURE TOOLS TO COLLECT REMOTE SENSED DATA - AIRBORNE IMAGES

Advantages of airborne images are their high ground resolution, and the free choice of wavelengths. Disadvantages are that the timing for collecting images is dependent on weather conditions and the time and labor consuming organization and availability of an airplane and pilot.

At the moment panchromatic color and near infrared (NIR) images offer the fastest, most flexible and economic method to collect remote information. The desired wavelength and ground resolution depends on flight height, film and filter type (Ladouceur et al., 1986; Smith et al., 1992) and thus can be adopted according to individual demands. Distorted images require an individual post-processing procedure (Anon, 1992).

The latest generation of digital cameras operating without a film medium offers the following advantages: the actual ground resolution is less than 1 meter which is satisfying for agricultural demands, but less processing time for developing and scanning is the main advantage of these cameras. A major disadvantage is the time needed for saving a single image. It is expected that this problem will be overcome in the near future. Taylor et al. (1997a, b) describe the use of digital cameras for the determination of within-field variations of crop growth and the prognosis of expected yield differences.

Aircraft-based sensors offer an interim solution between photographs and satellite-based sensors. The ground resolution is satisfactory and depends on the height of flight. The fixed bandwidths limit the collection of reflection information. Cloutis et al. (1996) found that multi-spectral optical images could be used to monitor variations in crop condition parameters during the growing season for different crop types. Disadvantage of aircraft-based sensors is the time consuming post-processing and last but not least its dependency on equipment availability and weather conditions.

AVAILABLE AND FUTURE TOOLS TO COLLECT REMOTELY SENSED DATA - SATELLITE IMAGES

The identification of land use systems from satellite images are widely known (Vanoverstraeten & Trefois, 1993), but the use of this data source in agriculture for gathering spatial information about soil and crop parameters is just beginning. A major advantage of satellite images is that post-processing rules are established. Disadvantages are the restricted ground resolution, the delivery time span from acquisition to use and the fixed date of recording and price.

Some of these problems will be overcome with a new generation of satellites, which have been launched since last year. With higher resolution, reduced repeating time, fast delivering time (<48 h) and lower costs, the new satellite-based sensors will match most agricultural demands and compete with common photographs or aircraft-based sensors. Table 1 includes a list of the main important satellite-based sensors with their spectral regions, pixel resolution, and delivery characteristics.

The processing of SAR (Synthetic Aperture Radar) data for thematic mapping applications of soil moisture content or soil texture differs fundamentally from that of optical imagery and is much less understood (Williams, 1995). Many factors like surface roughness, incidence angle, resolution, wavelength, frequency and polarization influence the results. The future potential of radar sensors depends on the possibilities to optimize the regulation of factor combinations. Investigations on the correlation of the backscattered SAR signal (long wavelength, 5.7 or 21 cm) and the soil moisture content showed that an estimation was only possible for the surface soil (<10 cm, Moran et al., 1997; Jackson & O'Neill 1987).

AVAILABLE AND FUTURE TOOLS TO COLLECT REMOTELY SENSED DATA - SURF-EYES

Decision-making for sampling based on remote sensing images requires the availability of these images at the right moment. Technically against this requirement are the dependence on both time and availability of equipment. Images from satellites and airborne platforms both bear risks of either availability or unsuitable survey conditions mainly due to surface coverage by clouds. In future, images obtained by unmanned air vehicles (Anonymous, 1998a, b; Suplee, 1997) and partly also surf-eyes may overcome these limitations. Surf-eyes are fixed scanner systems on elevated positions scattered in the landscape within sight of each other that collect images of crops and soil surfaces at very low maintenance costs. The automatically rectified and geocoded images can be stored in a GIS (Geographical Information System) and will deliver real time images of land or crops enabling the processing of spectral information for physiological crop features for identifying areas of homogeneity or inhomogeneity respectively for directing sampling.

The area covered by a surf-eye mainly depends on the height of the pylon, or tower the camera is mounted to. For example a stepwise rotating digital camera equipped with a 28 mm wide-angle lens in an angle of 45°, fixed in 250 m height to the pylon of an overhead line, will cover an area of about 167 ha. A flatter angle will enlarge the area proportionally. By employing several of these, it is possible to gather continuously remote sensed images of large areas.

USE OF AIRBORNE IMAGES FOR THE EVALUATION OF THE SPATIAL VARIABILITY OF THE CLAY CONTENT IN SOIL AND YIELD OF WINTER BARLEY

The problem related to non-thematic soil information like physical properties or organic matter content (Pitts et al., 1986), however is that the same variability observed in an image from remote sensing may have several individual reasons. For instance: differences in soil darkness can be due to variations in soil organic matter but also due to differences in soil moisture content or soil mineral composition or simply due to variable shadowing of the ground by clouds.

Remote sensing applied to crops has the general problem that differences in the spectral reflectance of the canopy can have more than one reason. Changes in the "green" color of crops, preferably identified by means of the reflectance of near infrared are certainly most likely due to changes in the chlorophyll content and the nitrogen content of the plant. This is most often attributed to the nitrogen supply of the crop and thus claimed as a suitable information for the design of nitrogen dressings (Baret & Fourty, 1997). But there are many more reasons why the canopy's green or infrared reflectance can change - for instance sulfur or other nutrient availability as well as from shortage–excess of water or from pests and diseases. Most recently, for instance, the sulfur supply of crops has become the major explanation for reasons of "green" color intensity of crops in the Northern Hemisphere (Schnug & Haneklaus, 1998).

On a 8.8 ha field in Mariensee (52.704° N, 9.476° E) soil samples were taken in a 30∗30 m grid. The spatial variability of the clay content in soil and yield of winter barley in 1996 was set in relation to three 6∗6 cm color slides, each of them classified with the supervised and unsupervised classification method. The basis of the supervised classification method is the definition of training areas whereby each area represents one class. After that, all pixels of the image are run through the so-called Maximum Likelihood classification, which assumes that each class is normally distributed. In comparison the unsupervised classification method is a cluster analysis method which works without prior defined training areas. The user decides into how many classes the image will be classified. During the clustering process the pixels are attributed and after statistical analysis the different classes are established. For both classification methods the same numbers of classes (4) were chosen in order to enable a comparison of results. In total six classified images were set in relation to the clay content in the soil and yield of winter barley. Two of the images were shot during the vegetation period of winter barley in June and July 1996 and one of the bare soils in fall 1997. The images were scanned in a ground resolution ≤1-m and then geo-referenced using the ARC/INFO Software (Anonymous, 1992). The significance of results was evaluated by the *F*-test (Table 2).

The *F*-test results in Table 2 illustrate the discrepancy between surface reflection, soil analysis parameters and yield using different methods. The closest relationships between image and clay content could be proved for the unsupervised method at both dates in June and July. None of the images was related to the spatial variability of the soil organic matter content. Yield of winter barley and the unsupervisedly established classes of the image in early June correlated significantly, while no relationship existed in late July, independent of the method used.

Table 2. Significance (*F*-test) for selected soil features and yield in locations of a field attributed to spectral classes retrieved by different classification methods from remote images of a barley crop taken at different growth stages and bare soil (year: 1996–1997; location: ′Mariensee′ 52.704° N 9.476° E).†

Method and date	Clay	OMC	pH	P HNO_3	P CAL	K CAL	Mg $CaCl_2$	Zn HNO_3	Cu HNO_3	Yield
UN0706	*	-	-	-	-	-	-	-	***	**
SU0706	-	-	-	***	**	**	-	-	*	-
UN2107	**	-	-	-	-	**	-	*	-	-
SU2107	-	-	-	-	*	*	-	-	-	*
UN2509	-	-	-	-	-	-	-	-	*	*
SU2509	-	-	-	-	-	-	-	-	-	-

† UN, Unsupervised classification; SU, Supervised classification; alphanumeric are abbreviations for campaign dates. OMC; organic matter content; nutrients extracted by HNO_3 (0.43 m nitric acid), CAL (lactic acid), $CaCl_2$ (0.01 m calcium chloride).

Despite principal reservations against remote sensing images (see above), orthogonal crop images are a valuable tool to direct soil and plant sampling. Remote sensing images reveal obvious differences in soil and plant characteristics. Figure 2 and 3 demonstrate how the unsupervised classified image of 7 June 1996 can be used for the development of soil or plant sampling strategies. The relative clay content and yield of winter barley are shown in the form of isolines superimposed on the classified images (Fig. 2 and 3).

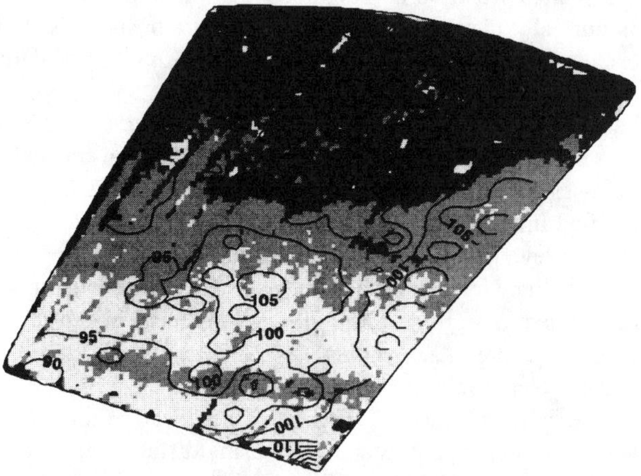

Fig. 1. Unsupervised classified image of 7 June 1996 overlaid with the spatial variation of clay contents in the soil.

The relative clay content in this field varies between 75 and 130%. In the images, four classes were build corresponding approximately with clay contents of ≤90, 90–100, 100–105, and >105%. The results reveal that directed soil sampling in the different classes of the remote sensed image (cp. Haneklaus et al., 1998; Schnug et al., 1998) will provide information about the spatial variation of clay content in the soil. Zheng and Schreier (1988) describe how soil color can be differentiated from an image, and set different classes in relation to soil features such as the organic C and moisture content.

Fig. 2. Unsupervised classified image of 7 June 1996 overlaid with the spatial variation of yield of winter barley.

The winter barley yield in t ha^{-1} varies in this field between 3.5 and 8 t ha^{-1}. The 4 classes build from the image correspond approximately with yields of ≤5.5 t ha^{-1}, 5.5-6.5 t ha^{-1}, 6.5-7.0 t ha^{-1} and >7.0 t ha^{-1}. Remotely sensed data can also be used for the determination of the spatial growth differences, whereby the prognosis of quantified yield data requires a ground truth (Haneklaus & Schnug 1993).

CONCLUSIONS

GPS technology and fast and powerful PCs provide the requirements for a successful implementation of remote sensing in precision agriculture. The use of remotely sensed data will become an important future tool for the efficient collection of spatially variable crop and soil data. The acceptance of this source depends on the accuracy of the information itself and its price in relation to other approaches such as self-surveying, directed sampling based on equifertiles or grid sampling (Haneklaus et al., 1997, and 1998; Schnug et al., 1998).

ACKNOWLEDGMENT

The authors wish to express their most heartily thanks to Dr. Kerr C. Walker (SAC, Aberdeen) for all his efforts in improving the language of this paper.

REFERENCES

Anonymous 1992. Image integration-incorporating images into your GIS. ESRI. Redlands, CA.

Anonymous 1998a. http://www.sikorsky.com/programs/cypher/index.html.

Anonymous 1998b. http://www.dasa.com/dasa/g/ri/orbit/inspect/inspector.htm.

Baret, F., and Th. Fourty. 1997. Radiometric Estimates of Nitrogen Status of Leaves and Canopies. p. 201–227. *In* G. Lemaire (ed.) Diagnosis of the nitrogen status in crops. Springer-Verlag, Heidelberg.

Barnes, E.M., M.S. Moran, P.J. Pinter, Jr., and T.R. Clarke. 1996. Multispectral remote sensing and site-specific agriculture: Examples of current technology and future possibilities. p. 845–854. *In* Proc. of the 3rd Int. Conf. on Precision Agriculture, Minneapolis, MN. 23–26 June 1996.

Cloutis, E.A., D.R. Connery, D.J. Major, and F. J. Dover 1996. Airborne multispectral monitoring of agricultural crop status: effect of time of year, crop type and crop condition parameter. Int. J. Remote Sensing. 17(13):2579–2601.

Davis, G. 1997–1998. Space age farming: Aerial sensor applications for precision farming and management. Modern Agric.. 1(5):18–20.

Haneklaus, S., and E. Schnug, 1993. Höhe, spatiale Variabilität und geostatistische Analyse der Erträge von Miscanthus auf den Versuchsflächen Rheinberg-Bernshof, Rheinberg-Gänsewiese und Scholven-Buer. ILLIT Report 10.

Haneklaus, S.I. Ruehling, D. Schroeder, and E. Schnug. 1997. Studies on the variability of soil and crop fertility parameters and yields in different landscapes of Northern Germany. p. 821–826. *In* J.V. Stafford (ed.) Proc. of the 1st European Conf. on Precision Agriculture, Warwick, England. 7–10 Sept. 1997. Vol II: BIOS Scientific Publishers Ltd., England.

Haneklaus, S., D. Schroeder, and E. Schnug. 1998. Decision making strategies for fertiliser use in precision agriculture. This Volume.

Jackson, T.J., and P. O'Neill. 1987. Temporal observations of surface soil moisture using a passive microwave sensor. Remote Sens. Environ. 21:281–296.

Ladouceur, G., R. Allard and S. Ghosh. 1986. Semi-automatic survey of crop damage using color infrared photography. Photogram. Eng. Remote Sensing. 52(1):111–115.

Loeffler, E. 1994. Geographie und fernerkundung. Teubner Studienbücher der Geographie, Stuttgart.

Moran, M.S., Y. Inoue and E.M. Barnes. 1997. Opportunities and limitations for image-based remote sensing in precision crop management. Remote Sens. Environ. 61:319–346.

Pelzmann, R.F., Jr. 1997. Using imagery in field management. Modern Agric.. 1(2):17–19.

Pitts, M.J., J.W. Hummel, and B.J. Butler. 1986. Sensor utilizing light reflection to measure soil organic matter. Trans. ASAE 29:422–428.

Schnug, E., and S. Haneklaus. 1998. Diagnosis of sulphur deficiency. E. Schnug and H. Beringer (ed.) Verlag Kluwer, Dordrecht (in press).

Schnug, E., K. Panten, and S. Haneklaus. 1998. Soil sampling and nutrient recommendations- the future. Comm. Soil Sci. Plant Anal. (in press).

Schroeder, D., S. Haneklaus, and E. Schnug. 1997. Information management in precision agriculture with LORIS. p. 821–826. *In* Proc. of the 1st European Conf. on Precision Agriculture, Warwick (England). 7–10 Sept. 1997,

Smith, J.L., J.A. Logan, and T.G. Gregoire. 1992. Using aerial photography and geographic information systems to develop databases for pesticide evaluations. Photogram. Eng. Remote Sensing. 58(10):1447–1452.

Suplee, C. 1997. Robot evolution. National Geographics. 192:76–95.

Taylor, J.C., G. Thomas, and G.A. Wood. 1997a. Diagnosing sources of within-field variation with Remote Sensing. p. 705–712. *In* Proc. of the 1st European Conf. on Precision Agriculture, Warwick (England). 7–10 Sept. 1997.

Taylor, J.C., G.A. Wood, and G. Thomas. 1997b. Mapping yield potential with remote sensing. p. 713–720. *In* Proc. of the 1st European Conf. on Precision Agriculture, Warwick (England). 7–10 Sept. 1997.

Vanoverstraeten, M., and P. Trefois. 1993. Detectability of land systems by classification from Landsat Thematic Mapper data. Virunga National Park (Zaïre). Int. J. Remote Sensing. 14(15):2857–2873.

Williams, J. 1995. Thematic information from space. p. 95–143. *In* Geographic Information from Space. Processing and Application of Geocoded Satellite Images. John Wiley & Sons, Chichester.

Zheng, F., and H. Schreier. 1988. Quantification of soil patterns and field soil fertility using spectral reflection and digital processing of aerial photographs. Fertil. Res.. 16:15–30.

Table 1. Satellite-based sensors with their spectral regions, pixel resolution and delivery characteristics (Moran et al., 1997).

Satellite	Sensor	Spectral region µm	Pixel resolution	Repeat cycle
Landsat-5	Thematic Mapper (TM)	0.45–0.52 0.52–0.60 0.63–0.69 0.76–0.90 1.55–1.75 2.08–2.35 10.4–12.5	30 m (visible, IR) 120 m (thermal IR)	16 d
SPOT-1 to SPOT-3	High Resolution Visible (HRV)	0.50–0.75 0.52–0.59 0.62–0.66 0.77–0.87	10 m (panchromatic) 20 m (multispectral)	26 d
IRS-1C	Panchromatic Linear Imaging Self-Scanning III (LIS-III)	0.50–0.75 0.52–0.59 0.62–0.68 0.77–0.86 1.55–1.70	5.8 m (panchromatic) 23 m (visible) 70 m (IR)	24 d
RADARSA T	Synthetic Aperture Radar (SAR)	5.3 Ghz (C-band)	28 m	24 d
Earth Watch	Early Bird (launch 1997)	0.45–0.80 0.50–0.59 0.61–0.68 0.79–0.89	3 m (panchromatic) 15 m (visible, NIR)	3 d
	Quick Bird (launch 1998)	0.45–0.90 0.45–0.52 0.53–0.59 0.63–0.69 0.77–0.90	1 m, 2 m (panchromatic) 4 m (visible, NIR)	3 d
Space Imaging	Space Imaging System (SIS) (launch 1997)	0.45–0.90 0.45–0.52 0.52–0.60 0.63–0.69 0.72–0.90	1 m (panchromatic) 4 m (visible, NIR)	1-3 d

Application of Remote Sensing to Irrigation Management in California Cotton

Richard E. Plant
Department of Agronomy and Range Science and Biological and Agricultural Engineering
University of California
Davis, California

Daniel S. Munk
University of California Cooperative Extension
Fresno, California

ABSTRACT

One of the most important irrigation management decisions in California Cotton Production is the timing of the last irrigation. We analyzed data obtained by scanning a sequence of false color infrared aerial photographs of an experiment at two field sites. The experiment involved a comparison of different dates on which the last irrigation was applied. Our preliminary results, based on one year of data, indicate a strong relationship between yield and vegetation index. They also indicate that remote sensing can detect differences in crop development due to soil texture differences within the field. Thus, remote sensing may be of value in assisting in timing the final irrigation and in determining whether within-field differences in soil texture are sufficient to economically justify a switch from furrow irrigation to other more precisely controlled systems.

INTRODUCTION

Remote sensing provides an important component of the information used in developing site-specific crop management strategies and tactics. In particular, it can be effectively combined with yield monitor data and point sample data to determine the factors underlying yield spatial variability. Knowledge of the relationship between crop stress status and remote sensing data is necessary to interpret these data for management purposes. The most commonly used format of remotely sensed data is in the form of a raster-based arrangement of cells, or "pixels" (Lillesand & Kiefer, 1994). A quantity called a digital number, taking on integer values between 0 and 255, represents radiation intensity in a particular wavelength band in each pixel. The most important radiation wavelength bands for interpreting plant production are the red (R) and infrared (IR) bands. A very large body of literature exists on the relationship between remotely sensed data and plant production, but much of this is based on satellite data that has a spatial resolution of no less than about 20- to 30-m. Although commercially available satellite data will soon be available at much higher resolution, for the present data at a resolution suitable for precision agriculture must come from aircraft-based

Copyright © 1999 ASA-CSSA-SSSA, 677 South Segoe Road, Madison, WI 53711, USA. *Proceedings of the Fourth International Conference on Precision Agriculture.*

systems, either through direct digital imaging or through film-based infrared aerial photography.

Both direct digital imagery and aerial photography have advantages and disadvantages in this application. Direct digital imagery has the very important advantage that it avoids one level of manipulation between the acquisition and the analysis of the data, namely, the conversion of an image recorded on film to a digital image through film processing and scanning. A primary disadvantage of direct digital imagery is that at present the cost of the data acquisition systems is very high. Moreover, the relationship between light intensity and digital number value, while probably more straightforward that that of a scanned aerial photograph, is still complex (Price, 1987). Because of its relatively low cost and widespread availability, it is likely that film-based aerial photography will play an important role in the initial application of aerial remote sensing to precision agriculture.

There have been relatively few studies relating aircraft-based remote sensing data to crop production. Moran et al. (1997) give a general review of applications of remote sensing in production agriculture. Wiegand et al. (1994) showed a strong correlation between the normalized difference vegetation index (NDVI, Tucker, 1979) and lint yield in cotton. Denison et al. (1996) showed similar relationships between NDVI and corn grain yield. Denison also used the concept of NDVI-days to interpret corn grain yield data. NDVI-days are computed by integrating daily NDVI over time, or estimating this integral based on available data. The theoretical argument in favor of NDVI-days is that NDVI has been related to photosynthetic rate (Wiegand et al., 1991; Sellers, 1989) and to biomass production rate (Tucker, 1979). Therefore, the time integral of NDVI should represent total biomass. To the extent that the harvest index is constant, this quantity should then be related to crop yield. Wiegand et al. (1991) show a correlation between yield and PVI (described below) summed over the season. Prince (1990) shows a high correlation between biomass and summed NDVI.

In the application of aerial remote sensing to site-specific field crop management in California it is natural to begin by investigating cotton production systems. Cotton is California's highest value field crop, both in terms of number of acres planted and dollar value of harvest. More than 95% of California cotton acreage is in the San Joaquin Valley. Since this region receives negligible rainfall during the cotton growing season, cotton production in California is fully irrigated, mainly by furrow irrigation. Virtually all California cotton fields have been laser leveled, so that infiltration is not influenced by topographical differences. Nevertheless, substantial differences in applied water may occur due to soil texture differences within the field (Or & Hanks, 1992).

Among the management practices that may be influenced by film-based remote sensing are irrigation and defoliation. The two most important decisions in cotton irrigation management are the scheduling of the first and the last irrigation of the season. In particular, the timing of the last irrigation is crucial to effective harvesting. Cotton is chemically defoliated prior to harvest to facilitate mechanical picking. If the last irrigation is too early the crop will be excessively vigorous and hard to defoliate. If, on the other hand, the final irrigation is too late then yield will be reduced due to water stress. Timing of defoliation is important since early defoliation will kill the plant before all harvestable bolls are set while late defoliation may delay harvest and increase the risk of being affected by winter rains.

In this paper we give a preliminary report based on one year's data of a study of the relationship between remotely sensed information and crop growth and yield. The overall objective of the study is to determine how remote sensing may be used effectively in crop management decision making. During 1997 we took aerial infrared photographs of field experiments at two locations in the San Joaquin Valley. Both experiments were replicated designs investigating the effect of date of final irrigation on yield. We present here results demonstrating the effect of water stress and variety on the time course of a vegetation index, on the relationship between vegetation index and crop phenology, and on the relationship between vegetation index and yield.

MATERIALS AND METHODS

A sequence of false color infrared aerial photographs was taken of irrigation timing experiments conducted at two sites in 1997. The first site was located on an 86.8 m by 325.2 m plot at the University of California West Side Research and Extension Center (WSREC) near Five Points, CA (lat. 36.3°N, long. 120.1°W). The soil at this site is relatively uniform and is classed as a Panoche clay loam that is a member of the nonacid thermic family of Typic Torriorthents. The experiment involved four different dates of the final irrigation (18 June, 21 July, 8 August, and 29 August) and four varieties (Acala Maxxa, Acala Phytogen 33, Acala GC-510, and Pima S-7). The experiment was laid out as a split plot with final irrigation date as the main plot factor and variety as the subplot factor. There were 4 replications. Each plot consisted of 6 1.02-m rows. The field was maintained in as stress-free a manner as possible according to University of California practices.

The second site was located in eastern Fresno County, CA (lat. 36.4°N, long. 119.9°W) in a 60 ha commercial cotton field. The variety was Acala Maxxa. The experiment involved three final irrigation dates (11 August, 25 August, and 5 September). The experiment was laid out as a randomized complete block with four blocks and four treatments per block. There were two 5 September treatments in each block. Figure 2 shows a black and white image of a false color infrared aerial photograph of the site taken on 26 August. Each plot spanned the length of the field (approximately 780 m) and consisted of 16 0.76 m rows. The field soil type was predominantly Traver sandy loam with two large sandy streaks, appearing as the lighter regions in Fig. 2. The sandy soils are classed as Hesperia sandy loam. A portion of each experimental plot was located in a sandy streak as shown in the figure. The field was maintained according to normal commercial production practices.

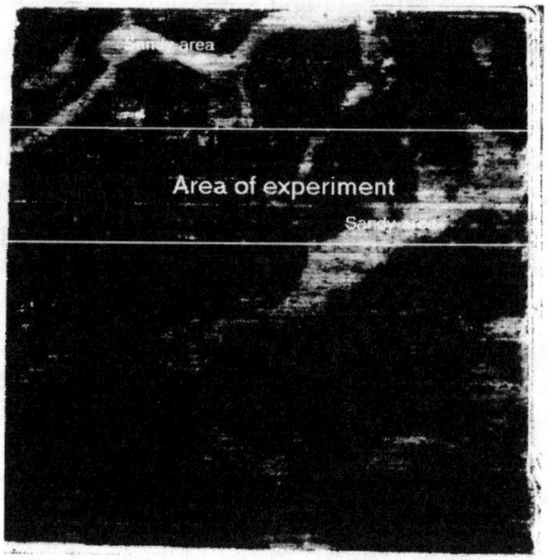

Fig. 1. Black and white reproduction of a false color aerial photo of the Fresno County site.

False color infrared aerial photographs were taken of the sites using Kodak 2443 film. Photographs were taken from an altitude of approximately 850 m at the WSREC site and 1525 m at the Fresno County site. The WSREC site was photographed on 28 July, 13 August, 26 August, 15 September, and 30 September. The Fresno County site was photographed on each of these dates and also on 19 June. Photos were taken at two aperture settings in rapid sequence. Photographs were taken at mid-day, and skies were generally cloud-free (at this time of year cloud cover is rare in the San Joaquin Valley). Positive images were scanned at 600 dpi using an Agfa Arcus II scanner. The resulting image files were imported into the Idrisi geographic information system (Clark University, Worcester, MA) where they were band separated and georegistered. Georegistration was carried out using reference points obtained with a Trimble Pro-XL global positioning system (Trimble Navigation, Sunnyvale, CA). During georegistration the images were resampled so that the number of image cells per crop row width was adjusted to the whole number nearest to the original image. Data were analyzed using Idrisi, Microsoft Excel (Microsoft Corp, Redmond, WA), and Minitab (Minitab, State College, PA). In statistical analysis of plot data only the cell values from the middle rows of each plot were used.

RESULTS

There are many vegetation indices that could be used to interpret aerial photographs (e.g., Wiegand et al., 1991). Almost all of these are based on formulas involving the magnitude of the number representing the red band intensity, denoted R, and that representing the IR band intensity, denoted IR. In this notation the formula for the NDVI is NDVI = (IR - R)/(IR + R). Many vegetation indices are algebraically related to each other. Perry and Lautenschlager (1984) have shown that most vegetation indices can be divided into two broad classes: ratio-based indices represented by NDVI, and distance-based indices represented by the perpendicular vegetation index (PVI, Richardson and Wiegand, 1977). Within these classes the indices are generally equivalent for management purposes. The PVI is calculated based on the soil line, i.e., the line in (R,IR) plane formed by the locus of points representing bare soil at different moisture contents. For the soils common in the Central Valley of California the soil line is approximately the 45° line in the (R,IR) plane. In this case the formula for the PVI can be shown through a little algebra to be $PVI = |IR - R|/\sqrt{2}$.

While the magnitude of the digital numbers can be calibrated using a reference surface, maintaining such a surface near an agricultural crop is probably not practical. Therefore one must assume that the aerial images will not include a calibrated surface. Since the NDVI involves a ratio of the digital numbers it is not sensitive to changes in absolute magnitude provided that the red and infrared change in the same proportion. Therefore if this were the case then a ratio image such as the NDVI would not suffer as much from the lack of a calibrated surface in the image. A distance-based image, on the other hand, would change in proportion to the change in absolute radiation intensity and thus would seem less appropriate for this application. Based on these considerations, our first objective was to determine the extent to which the NDVI is independent of site conditions.

Effect of Image Variability

Scanned photographic images can vary for a wide variety of reasons including scanner settings, exposure, incident solar radiation intensity, solar angle, cloud cover, and camera altitude. Of these, the effect of scanner setting and aperture setting could be readily controlled since aperture setting could be rapidly changed while other sources of photographic variability were constant. We therefore compared NDVI computed from images taken of the WSREC site at two different image scans and also two different aperture settings. We used the photographs taken on 26 August, and 15 September at aperture settings F6.8 and F8. Images at different aperture settings were scanned using the same scanner settings. Images at the same aperture were also scanned on two different days to test the effect of differences in scanner settings. We georegistered the images and computed the NDVI in Idrisi. We then used standard map algebra (Tomlin, 1990) to compute the ratios of NDVI values computed from the different image treatments.

Table 1 shows the mean and standard deviations of image cell values computed by dividing the image computed using the F8 aperture setting by the that computed using the F6.8 setting. The first column of the table shows these computations for the NDVI images. Since the means of the F8-based NDVIare:

Table 1. Mean and standard deviations of pixel value ratios of images from photos at F8 aperture divided by images from F6.8 aperture.

Date	NDVI ratio ± S.D.	IR ratio ± S.D.	R ratio ± S.D.
26 Aug.	1.15 ± 0.09	0.75 ± 0.06	0.57 ± 0.05
15 Sept.	1.32 ± 0.20	0.78 ± 0.05	0.59 ± 0.10

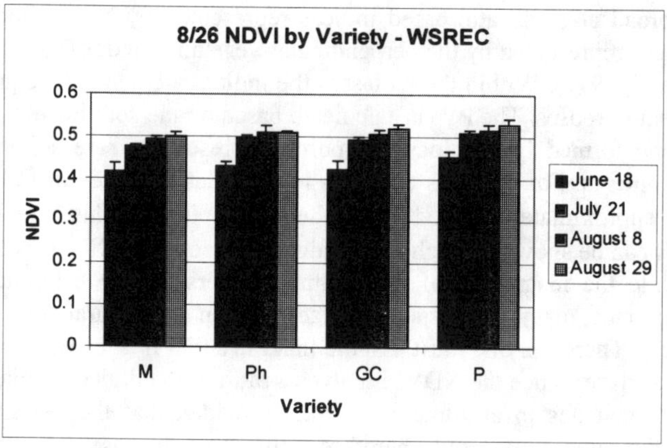

Fig. 2. Histograms of 26 August NDVI for each variety and each date of final irrigation. Error bars are 95% confidence regions.

substantially higher than the means of the F6.8-based NDVI, it follows that the R and IR bands are not proportionately affected by the aperture change. The second and third columns of Table 1 show the mean and standard deviation obtained by dividing the F8-based image band values by the F6.8-based values. The R image values are reduced proportionately more than the IR image values when the aperture is reduced. Scanning the positives on different days had a smaller but non-negligible effect on the NDVI. The ratio of NDVI values for the 26 August image was 1.06 (SD = 0.10) and the ratio for the 15 September image was 1.07 (SD = 0.28).

NDVI vs. Variety and Time

NDVI were computed for each experimental plot on each date in each of the two sites. ANOVA tables were generated for each date with NDVI as the response

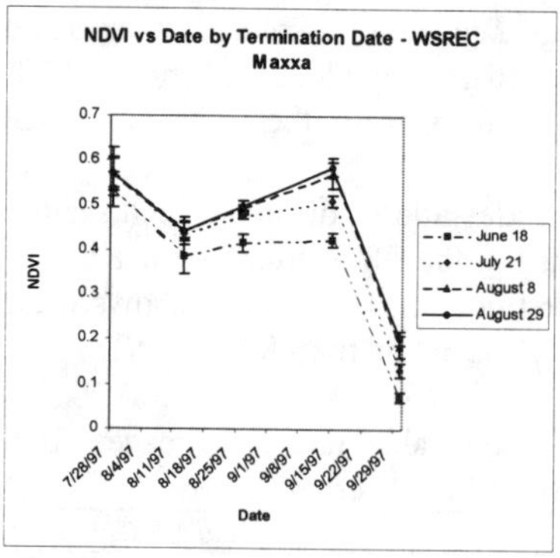

Fig. 3. NDVI vs. date, WSREC Maxxa

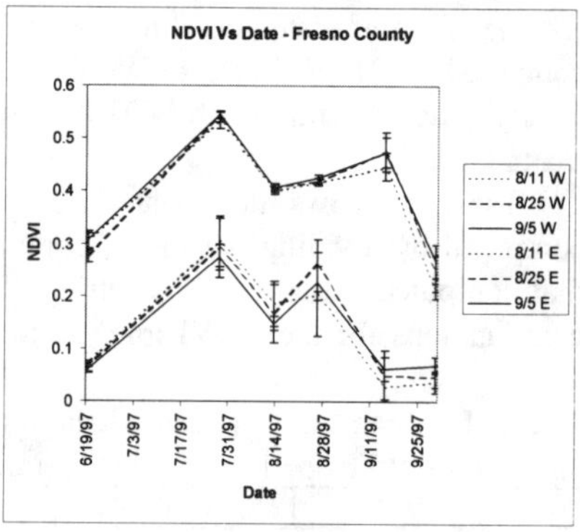

Fig. 4. NDVI vs. date, Fresno County

variable. Results were the same for all dates at the WSREC site: effects of both last irrigation date and variety were significant ($P < 0.01$), but their interaction is not ($P > 0.1$). At the Fresno County site last irrigation date did not have a significant effect on NDVI ($P > 0.1$) except for a marginally significant effect ($P = 0.084$) on September 30. Fig. 2 shows histograms of NDVI vs. variety for the August 26 WSREC image. The error bars represent 95% confidence intervals. Fig. 3 shows the time course of plot mean NDVI vs. time for the variety Acala Maxxa for each irrigation treatment at the WSREC. The error bars represent 95% confidence intervals based on between-plot variability. Because of the substantial aperture effect shown in the previous subsection we attempted to select images with consistent overall photographic density for the plot.

Figure 4 shows the time course of plot mean NDVI vs. time for each irrigation treatment at the Fresno County site. NDVI was computed separately for the east (sandy) and west (loamy) ends of the plots. The computations on the east side of the field were based on the average value in the sandy section of each individual plot. The computations on the west side were based on strips of approximately the same length.

Yield vs. NDVI-days

Lint yields were measured for each plot in both the WSREC and Fresno County experiments. NDVI-days were estimated for each plot using by integrating mean NDVI against time using the triangulation method. This method estimates the integral by connecting data points with straight lines and computing the integral of the resulting curve. Lint yields at the WSREC are plotted against NDVI-days in Fig. 5. All varieties except Acala GC-510 lie approximately on the

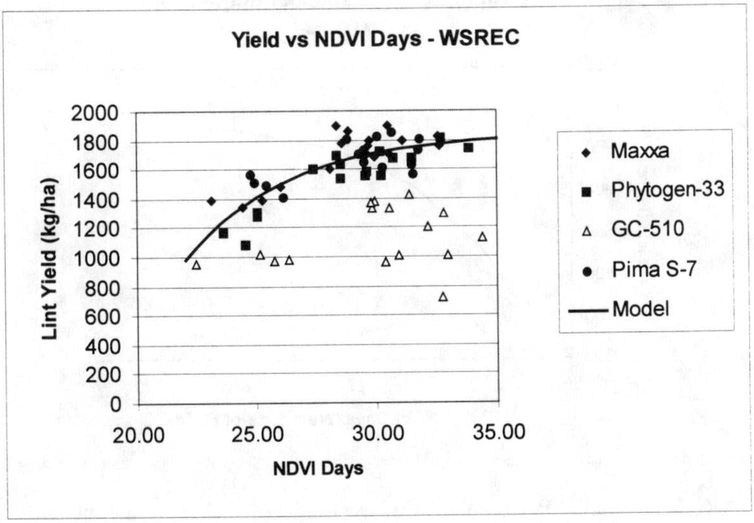

Fig. 5. Yield vs. estimated NDVI-days, WSREC.

same curve. The reason for the deviation in GC-510 data may be that this variety had to be replanted due to poor establishment of the original stand. A least-squares regression curve was computed for all varieties except GC-510. The figure also shows the regression curve

$$\text{Yield} = 1852.1(1 - e^{-0.227(\text{NDVI} - 18.67)}),$$

which provided the best least-squares fit to the data ($R^2 = 0.65$).

Based on the expectation that the loamy soils would contribute more to yield than the sandy soils, NDVI-days were computed only for the west side of the Fresno County site. Figure 6 shows the plot of lint yield vs. west side mean NDVI-days for each of the plots in the site. Also shown is the least-squares regression curve ($R^2 = 0.61$)

$$\text{Yield} = 991.27(1\text{-}e^{-1.003(\text{NDVI} - 24.02)}).$$

DISCUSSION

The most important and obvious conclusion from the analysis of NDVI computed at different aperture settings and scanner settings is that caution must be used in comparing data from aerial photographs taken on different days. Although the scanner setting did not have a very large effect on computed NDVI, the effect of aperture setting was substantial, ranging from 15 to 32%. The data in Table 1 indicate that this shift is due to the greater reduction in red than infrared digital number value with reduced aperture. It is possible that this effect is due to the density vs. exposure properties of the film (Heller, 1970). The response of the

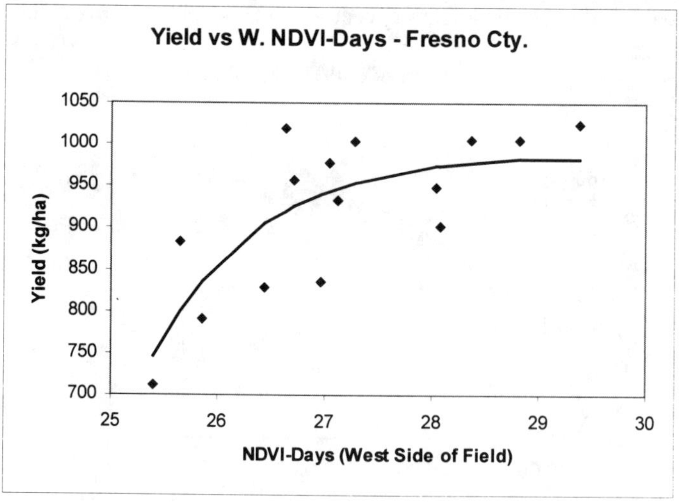

Fig. 6. Yield vs. NDVI days on the west side of the experimental plot, Fresno County site.

individual film layers is not necessarily in proportion, nor is it linear throughout the range of exposures.

Despite the problems with aerial photos taken at different dates, it is evident from inspection of the yield vs. time plots in Fig. 3 and 4 that these plots contain potentially useful information. Indeed, the pattern in these plots is generally consistent with expected behavior if NDVI reflects crop photosynthetic rate. It generally rises early in the season, peaks, and then declines as the crop senesces. A second, lower peak follows this, and then a final decline. It is tempting to try to correlate this time course with known end-of-season behavior of cotton. As the cotton crop matures, it shifts carbohydrate allocation from vegetative to reproductive structures (Kerby & Hake, 1986). This results in leaf senescence, which begins at mid-season and accelerates as the season progresses. The combined effects of decline in carbohydrate production caused by leaf senescence and increased demand by fruiting structures cause the crop to reach a point at which no carbohydrates can be allocated to new bolls. At this point, called cutout, no further production of harvestable bolls occurs. A second manifestation of the decline in carbohydrate availability at the end of the season is reduced plant growth, reflected in a decline in the number of main-stem nodes above the highest (newest) flower (which is colored white - lower, more mature flowers are red). Empirical observations indicate that in Acala cotton cutout occurs when there are five nodes above the white flower.

Figure 7 shows plots of the number of nodes above the white flower for the Fresno County field. Based on this plot, the loamy area of the field reached cutout between 8 August (based on least-squares regression) and 11 August (based on

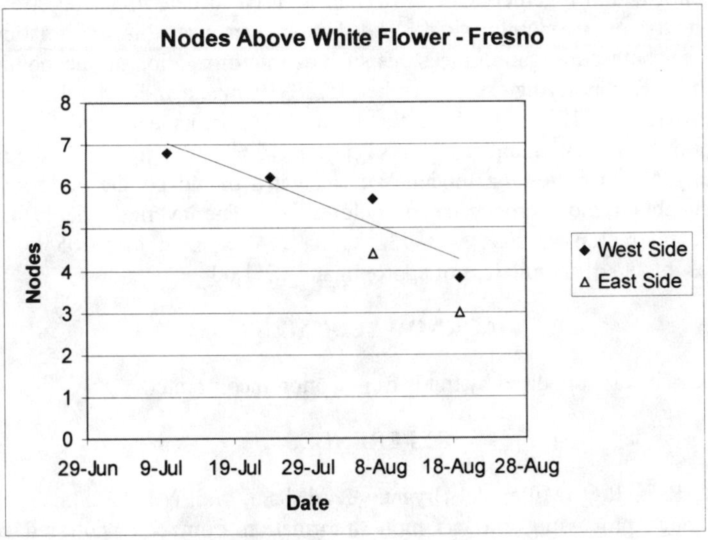

Fig. 7. Nodes above white flower on the east and west sides of the experimental plot at the Fresno County site. Regression line is the least squares fit to the west side data.

interpolation between the last two data points). Based on extrapolation between the two sandy soil data points, the sandy area reached cutout on about 1 August. This earlier date is consistent with increased plant stress and lower soil water holding capacity in the sandy region. From Fig. 4, the cutout dates fall between the higher NDVI 28 July image and the lower NDVI 13 August image. It is therefore possible that there is a relation between the time course of the NDVI plots and the occurrence of cutout, but the temporal resolution is insufficient to tell for sure. It is likely that the steep decline in NDVI observed at the end of the season corresponds with boll opening. The fact that NDVI in the east end of the Fresno County field declines before NDVI in the west end is again consistent with the more rapid maturity in the sandy soil. The reason for the late season increase is unknown.

Remote sensing time course data may potentially assist in two important end-of-season management decisions, applying the final irrigation and scheduling chemical defoliation. Determining the date and amount of the last irrigation requires a knowledge of the cutout date. Therefore, if a connection can be established between landmarks on the NDVI vs. time curve and cutout, this information can be used to assist in scheduling the final irrigation. Scheduling the date of defoliant application may be done by determining the number of nodes above the highest cracked boll (NACB) (Kerby & Hake, 1986). Future research will be done to determine the correlation between NACB and the end-of-season decline in NDVI.

In highly heterogeneous fields such as the Fresno County site, remotely sensed information together with a GIS can be used to determine the effects of heterogeneity on economic yield. The farmer can use this information to determine whether mitigation measures such as microirrigation are economically worthwhile. Reclassifying the September 15 NDVI image indicates that on that date approximately 12% of the field, that is, about 7.2 ha, have an NDVI <0.25. In this region the average number of NDVI-days is 16.6, which is outside the range of validity of the regression model. We can get a bound on the yield loss by assuming this region produced no yield at all. The average yield for the experimental region in 1997 was 922 kg ha^{-1}. At a price of \$1.65 kg^{-1} this translates into an economic loss of approximately \$11,000.

ACKNOWLEDGMENT

This research was supported by funds from Cotton Incorporated.

REFERENCES

Denison, R.F., R.O. Miller, D. Bryant, A. Abshahi, and W.E. Wildman. 1996. Image processing extracts more information from color infrared aerial photographs. California Agric. 50(3):9–13.

Heller, R.C. 1970. Imaging with photographic sensors. p. 35–72. *In* Remote sensing with special reference to agriculture and forestry. Natl. Academy of Sci., Washington, DC.

Kerby, T.A., and K. Hake. 1996. Monitoring cotton's growth. p. 335–355. *In* Cotton Production. Publ. 3352. Div. of Agric. and Nat. Resour., Univ. of California, Oakland.

Lillesand, T.M., and R.W. Kiefer. 1994. Remote sensing and image interpretation. John Wiley, New York.

Moran, M.S., Y. Inoue, and E.M. Barnes. 1997. Opportunities and limitations for image-based remote sensing in precision crop management. Rem. Sens. Environ. 61:319–346.

Or, D., and R.J. Hanks. 1992. Soil water and crop yield spatial variability induced by irrigation nonuniformity. Soil Sci. Soc. of Am. J. 56:226–233.

Perry, J.R., and L.F. Lautenschlager. 1984. Functional equivalence of spectral vegetation indices. Rem. Sens. Environ. 14:169–182.

Price, J.C. 1987. Calibration of satellite radiometers and the composition of vegetation indices. Rem. Sens. Environ. 21:15–27.

Prince, S.D. 1990. High temporal frequency remote sensing of primary production using NOAA AVHRR. p. 169–184. *In* Applications of remote sensing in agriculture, Butterworths, London.

Richardson, A.L., and C.L. Wiegand. 1977. Distinguishing vegetation from soil background information. Photogrammetric Eng. Rem. Sens. 43:1541–1552.

Sellers, P.J. 1989. Vegetation-canopy spectral reflectance and biophysical processes. p. 297–335. *In* Theory and applications of optical remote sensing. Wiley Interscience, New York.

Tomlin, C.D. 1990. Geographic information systems and cartographic modeling. Prentice-Hall, Englewood Cliffs, NJ.

Tucker, C.J. 1979. Red and photographic infrared linear combinations for monitoring vegetation. Rem. Sens. Environ. 8:127–150.

Wiegand, C.L., J.D. Rhoades, D.E. Escobar, and J.H. Everitt. 1994. Photographic and videographic observations for determininng and mapping the response of cotton to soil salinity. Rem. Sens. Environ. 49:212–223.

Wiegand, C.L., A.J. Richardson, D.E. Escobar, and A.H. Gerbermann. 1991. Vegetation indices in crop assessments. Rem. Sens. of Environ. 35:105–119.

Spectral Reflectance Pattern Recognition for Segmenting Corn Plants and Weeds

John W. Hummel
Crop Protection Research Unit
USDA-ARS
Urbana, Illinois

Jing Yu
Agricultural Engineering Department
University of Illinois
Urbana, Illinois

ABSTRACT

Spectral reflectance sensors have been developed that can detect the presence of plant material against a background of bare soil. These systems are able to identify areas of weed infestation in fallow fields or in the area between the rows of a row crop and control the application of herbicides. In typical U.S. corn production, a number of options are available for controlling weeds in between the rows, e.g., broadcast herbicide application or mechanical cultivation. In-row weed control might be possible with nonselective herbicides if detected plant material areas can be segmented into crop and weed subsets. Accomplishing this task in real time would allow application of the herbicide to only the weeds, which could significantly reduce the amount of herbicide used. This study reports on the use of an industry-developed spectral reflectance sensor to identify and locate plant material and the experimental C-language software that was developed to segment the corn plants and weeds.

INTRODUCTION

Uncontrolled weed populations can substantially reduce corn yields. Weeds can decrease the moisture and nutrients available to crops and reduce productivity. Also, weed abundance within corn fields can affect the severity of infestation by corn pests (Weber et al., 1990), again contributing to reduced productivity.

Weed control strategies have evolved for many years. Before the development of soil-applied herbicides in the 1960s, the basic tools for a producer to control weeds were mechanical and cultural practices. A cultural practice, such as crop rotation, can limit a buildup of weed population and prevent shifts in weed species composition. During the 1960s and 1970s, weed control consisted of combinations of cultivation and herbicide applications. Most herbicides were soil-applied, i.e., either tilled into the soil before planting or applied to the soil surface prior to crop emergence. During the mid-1970s and early 1980s, post emergence herbicides were commercialized for weed control (Steckel et al., 1990).

Copyright © 1999 ASA-CSSA-SSSA, 677 South Segoe Road, Madison, WI 53711, USA. *Proceedings of the Fourth International Conference on Precision Agriculture.*

Many farmers now rely primarily on herbicides for weed control. In 1988, a survey of U.S. farms in the Midwest showed that 96% of all U.S. corn fields received herbicide treatment (Lybecker et al., 1991). An average of 1.3 herbicide applications per hectare were applied. Herbicide usage in corn fields was about 60% of the total pesticide applied by crop farmers in the USA. Although herbicide use for weed control is important for successful corn production, this quantity of herbicide applied to corn crops may increase the negative impact of herbicides on nontarget organisms, on development of weed resistance, and on groundwater contamination. Scientists and farmers are searching for balanced use of agri-chemicals to provide economic weed control while also protecting groundwater resources (Lybecker et al., 1991). Production equipment and practices are needed that are beneficial to the environment and, at the same time, economical for agricultural producers.

The advent of site-specific crop management and the increased availability of foliar-applied herbicides has heightened interest in applying herbicides only to specific plants (weeds). In fields where weeds are not uniformly distributed, herbicide usage could be greatly reduced (Stafford & Miller, 1996).

REVIEW OF LITERATURE

Researchers have shown that spectral reflectance can be used to distinguish plant material from soil and crop residue (Hooper et al., 1976; Haggar et al., 1983; Bargen et al., 1993). Others have investigated the use of machine vision and image analysis (Everitt et al., 1987) and color-infrared photography (Brown et al., 1994) to identify plant species and to differentiate weeds from crop plants.

Jia and Krutz (1992) investigated the feasibility of using machine vision to locate corn plants. The centers of corn plants were located by photographing the top and side views of plants. The difference in reflectance properties of main veins and leaves was detected. Two algorithms were developed, i.e., the main leaf vein detection algorithm that extracted main vein lines, and the central location algorithm that located the center of a corn plant. The results showed that the corn plant can be identified and located by applying image processing techniques by detecting a specific leaf feature. However, many factors, such as shape, reflectance properties, and the distribution of light sources needed to be considered since changes in any of these factors could directly affect the algorithm. In addition, the computation intensity and cost of image analysis systems discourage their use in commercial herbicide applicators. Spectral reflectance-based sensors have been incorporated into sprayers to apply pesticides only on green vegetation. Felton and McCloy (1992) used a reflectance-based system, which automatically controlled the nozzles on a spray boom through a microprocessor. On most of the farms where tests were conducted, 95% of weeds were killed. Shropshire et al. (1991), as Bargen et al. (1993) reported, developed a ratio reflectance photo detector system for a spot agricultural sprayer. The desired accuracy and reliability for the detection of a plant were not consistently obtained. Stone (1994) developed a concept sprayer. The sprayer was triggered with a computer module which had its own optical sensors to detect weeds. The module detected weeds while monitoring applicator speed and determined when to activate the nozzles.

Beck (1996) developed a sensor based on differences in spectral reflectance at the visible and near infrared wavelengths exhibited by chlorophyll-bearing plant tissue and soil. The system used solid-state, frequency-modulated light emitters for weed detection, and could discriminate plants from soil. The sensor was incorporated into an eight channel device called the Patchen[1] WeedScanner (Patchen California, Los Gatos, CA) with each channel equipped with a solenoid-valve-controlled spray nozzle.

OBJECTIVES

This study was concerned with the post-emergent application of herbicides only to the site-specific region of a corn field that needed weed control. Lowered herbicide usage in corn production using site-specific application should reduce production costs and contamination of ground and surface water.

The overall objective of the research was to investigate the ability of the Patchen Weedscanner to identify and locate corn plants and, through software, to distinguish them from weeds. The specific objectives of the research were: (i) to identify distinguishable patterns and threshold values representing corn plants in the data stream from the Patchen Weedscanner, (ii) to develop appropriate algorithms to identify corn plants and weeds, (iii) to test the algorithms on data streams collected from field plots, and (iv) to evaluate the efficacy of the system.

ALGORITHM DEVELOPMENT

Visual observation suggested that the pattern of a corn plant, in the horizontal plane, might be sufficiently different in size and shape from that of co-occurring weeds that an algorithm might be able to differentiate between them. Initially, the objective was to develop appropriate algorithms to segment corn plants from weeds, although an additional requirement was to be able to make the determination rapidly enough to permit real-time control of pesticide application when weeds were identified. Algorithm development was carried out in the C and C++ languages, and a listing of the code is available in Yu (1997).

Four standard patterns of corn plants were developed for use in evaluating corn plant recognition algorithms. Preliminary data of a corn row were collected with the Weedscanner at a sensor scan rate of 200 scans s^{-1}, producing 4 096 lines of data representing a total row length of 25 m. When corn plants were at the 3 to 4 leaf growth stage, the leaves usually had a distinguishing orientation direction when viewed from above. Corn plants, with the leaves oriented parallel to the row, had a maximum pattern dimension of 15 cm and occupied about 24 lines of data (Pattern A). Corn plants oriented perpendicular to the row had a pattern that occupied only six lines of data (Pattern B). Two additional patterns (Patterns C and D) were

---

[1]Mention of a trade name, proprietary product, or specific equipment does not constitute a guarantee or warranty by the USDA-ARS or the University of Illinois and does not imply the approval of the named product to the exclusion of other products that may be suitable.

developed, representing corn plants whose leaves were oriented at 45° clockwise and counterclockwise, respectively, from being parallel to the corn row. These patterns consisted of 20 lines of data.

Algorithm One

Algorithm One used the relative horizontal area that a corn plant covered to segment corn plants from weeds. Based on an average travel speed and average size of corn plants, a minimum relative area (percentage of the area scanned by the eight sensor channels) that a corn plant covered was estimated. This relative area was set as the initial threshold value to segment corn plants from weeds. For example, if the relative area threshold value was set at 15%, then any pattern of contiguous data cells that resulted in a relative area of >15% of the whole view (eight channels) exceeding the threshold value was segmented as a corn plant. The initial area threshold value could be set to different values for different growth stages and different size corn plants. Corn plant recognition was based on the premise that, once a green plant was encountered, each successive scan would contain at least one value in excess of the threshold value until the plant was completely scanned. The logic accommodated situations in which green plant material was initially encountered by the sensor as well as situations where bare background soil was initially scanned.

Algorithm Two

Algorithm Two was developed for use in situations where weeds were similar in size to corn plants or where weeds grew in patches. The relative area that a patch of weeds covered could be larger than, or equal to, the relative area a corn plant covered. In these situations, relative area alone could not separate corn plants from weeds.

This second algorithm included a measure of the distance between two corn plants, since modern corn planters accurately place seeds to produce a relatively constant distance between corn plants in the row. Assuming constant sensor travel speed, the center-to-center distances between two corn plant patterns in a data set would be constant. As a result, the distance between two corn plants could be used to find the accurate positions of corn plants and help in distinguishing corn plants from weeds.

Algorithm Two used both the relative horizontal area that a corn plant covered and the distance between two corn plants to distinguish corn plants from weeds. These relative area and distance values were set as the initial threshold values to distinguish corn plants from weeds. For example, if the relative area threshold value for a corn plant was set at 15% and the distance threshold value was set at 32 lines (20 cm), then any pattern that resulted in a relative area of >15% of the whole view and a distance of 32 ± 4 lines (20 ± 2.5 cm) from the previous pattern was treated as a corn plant. Only plants whose relative area and distance met both the relative area condition and the distance condition could be treated as corn plants.

EQUIPMENT AND PROCEDURE

A prototype WeedScanner system, which was introduced in 1992, was used in this study. The system had eight 5.1-cm wide channels with each channel consisting of a modulated gallium aluminum arsenide light source, a spray nozzle, and a detector. The silicon photo-detector sensed the spectral reflectance at two wavelengths, approximately 660 nm and 960 nm, and the system combined the two signals into a single response ranging from 0 V to 4.0 V. Higher output values correlated with higher percentages of the sensed area (approximately 10 mm $*$ 5.1 cm) covered by chlorophyll-containing material.

A Gateway 486 laptop computer was interfaced with the Weedscanner via the computer serial port for system control and for archiving data. In the SCAN mode, detector output voltage values were saved to an array in the static RAM. After a data set had been collected, the operator could write the data to a comma-delimited file of 16 columns and 4 096 lines. The voltage levels representing the area of green chlorophyll-bearing material present in the scanned area for each of the eight sensors were stored in the first eight columns, while the gain data, which was the sum of the calibration values and the voltage levels, were stored in the last eight columns.

The Weedscanner sensor was cantilever-mounted on the side of a John Deere Gator transport vehicle when collecting field plot data. Power to operate the sensor, computer, etc. was obtained from the transport vehicle's 12 VDC system. The height of the Weedscanner detector was 50 cm above the ground.

Algorithm validation data were collected in a no-tillage field on the Agricultural Engineering Research Farm at Urbana, Illinois. Two 15-m rows of corn were planted. The centerline of the sensor was positioned over the row centerline during data collection. Using a frame time of 5 ms and a forward speed of the transport vehicle of approximately 4.8 km h^{-1}, each frame was approximately 7 mm in length. The distance between the corn plants in each row varied from 20 to 40 cm. When the corn plants were 16 d old, they had 3 to 6 leaves. Neither corn plants nor weeds overlapped at this growth stage. The maximum dimension of an individual corn plant ranged from 7 to 15 cm. Weeds were smaller, with a maximum dimension ranging from 2 to 5 cm.

RESULTS AND DISCUSSION

The two algorithms were tested by using the same data sets, which included the four standard pattern data sets of corn plants, modified pattern data sets of corn plants, and field data validation sets of corn plants and weeds collected from two rows of corn plants. The computation and comparison portions of the two algorithms took only a few nanoseconds. The threshold value of the relative area a corn plant covered was set to different values according to different data sets.

Four Standard Pattern Data Sets

These four standard pattern data sets only contained corn plants, with a constant center-to-center distance between corn plant patterns of 48 lines (30 cm). The threshold value of the relative area was initialized at 15% for all four data sets. Both algorithms were successful in identifying all plant patterns in a 10-plant data set when the distance between two corn plants was 30 cm (Table 1). After the 10 corn plants were scanned, the average values of the relative area of the ten corn plants in the four data sets were 17.7, 20.8, 19.9, and 19.9%, respectively. As the spacing between adjacent corn plants was reduced, the algorithms became less effective in identifying corn plants (Table 1). When the distance between adjacent plants was 15 cm, the algorithms could correctly calculate the pattern area for all four patterns, but could not correctly identify plants in the Pattern A data set. At a spacing of 12.5 cm, the pattern area was incorrectly calculated for the Pattern A data set, and corn plants in Pattern A,

Table 1. The effect of plant spacing on the ability of the algorithm to identify corn plant patterns.[1]

	Pattern area				Plants correctly identified			
	A[2]	B[3]	C[4]	D[5]	A	B	C	D
	percentage							
Algorithm One								
Spacing 1[6]	17.7	20.8	19.9	19.9	100.0	100.0	100.0	100.0
Spacing 2[7]	17.7	20.8	19.9	19.9	0.0	100.0	100.0	100.0
Spacing 3[8]	18.8	20.8	19.9	19.9	0.0	100.0	0.0	0.0
Spacing 4[9]	25.0	20.8	44.8	44.8	0.0	0.0	0.0	0.0
Algorithm Two								
Spacing 1	17.7	20.8	19.9	19.9	100.0	100.0	100.0	100.0
Spacing 2	17.7	20.8	19.9	19.9	0.0	100.0	100.0	100.0
Spacing 3	18.8	20.8	19.9	19.9	0.0	100.0	0.0	0.0
Spacing 4	25.0	20.8	44.8	44.8	0.0	0.0	0.0	0.0

[1]Area threshold value of 15%.
[2]Leaves of corn plants oriented parallel to the row.
[3]Leaves of corn plants oriented perpendicular to the row.
[4]Leaves of corn plants rotated 45° clockwise from parallel to the row.
[5]Leaves of corn plants rotated 45° counterclockwise from parallel to the row.
[6]Spacing between centers of two adjacent corn plants of 30 cm.
[7]Spacing between centers of two adjacent corn plants of 15 cm.
[8]Spacing between centers of two adjacent corn plants of 12.5 cm.
[9]Spacing between centers of two adjacent corn plants of 3.75 cm.

C, and D data sets could not be correctly identified. When the spacing was reduced further to 2.5 cm, the algorithms could not correctly identify plants in any of the four data sets.

Four Standard Pattern Data Sets with Weeds between Corn Plants

Modified pattern data sets were formed by adding some weed patterns to the four standard pattern data sets. The distance between adjacent corn plants and the pattern shapes of the corn plants were unchanged. Since weeds grow in no predicted location, weeds were added randomly at a rate of one weed pattern per corn plant. The pattern sizes of the weeds were smaller than the pattern size of the corn plants. The maximum horizontal distance of the weeds was 10 cm. The weed patterns did not have uniform shapes and did not overlap each other nor did the weeds overlap the corn plants. Weeds with four different sizes (2.5, 6.25, 10 cm, and a patch of weeds) were considered. The threshold values of the relative area of corn plants for all four data sets were set to 15%. The average values of the relative area for Patterns A, B, C, and D after 10 corn plants were scanned and when weeds were 2.5 cm were 17.7, 20.8, 19.9, and 19.9%, respectively, for both algorithms (Table 2). All corn plants were

Table 2. The effect of weed size on the ability of the algorithm to identify corn plant patterns in the presence of weed patterns.[1]

	Pattern area				Plants correctly identified			
	A[2]	B[3]	C[4]	D[5]	A	B	C	D
	percentage							
Algorithm One								
Weed 1[6]	17.7	20.8	19.9	19.9	100.0	100.0	100.0	100.0
Weed 2[7]	17.9	20.8	19.9	19.9	50.0	100.0	100.0	100.0
Weed 3[8]	19.0	20.8	20.1	20.1	50.0	100.0	50.0	50.0
Weed 4[9]	26.2	27.8	27.3	27.3	50.0	50.0	50.0	50.0
Algorithm Two								
Weed 1	17.7	20.8	19.9	19.9	100.0	100.0	100.0	100.0
Weed 2	17.7	20.9	19.9	19.9	100.0	100.0	100.0	100.0
Weed 3	17.8	20.8	19.9	19.9	100.0	100.0	100.0	100.0
Weed 4	17.7	20.8	19.9	19.9	100.0	100.0	100.0	100.0

[1]Area threshold value of 15%.
[2]Leaves of corn plants oriented parallel to the row.
[3]Leaves of corn plants oriented perpendicular to the row.
[4]Leaves of corn plants rotated 45° clockwise from parallel to the row.
[5]Leaves of corn plants rotated 45° counterclockwise from parallel to the row.
[6]Weeds with size of 2.5 cm grew between two corn plants.
[7]Weeds with size of 6.25 cm grew between two corn plants.
[8]Weeds with size of 10 cm grew between two corn plants.
[9]A patch of weeds grew between two corn plants.

correctly identified, and pattern areas were correctly calculated by both algorithms. The average values of the relative area for Patterns A, B, C, and D data sets after 10 corn plants were scanned and when weeds were 6.25 cm were 17.9, 20.8, 19.9, and 19.9%, respectively. Weeds in the Pattern A data set were incorrectly treated as corn plants by Algorithm One, even though pattern areas were correctly calculated for all data sets; Algorithm Two was able to correctly calculate relative area and to correctly identify all corn plants. When weeds with a maximum horizontal distance of 10 cm were included in the data set, all weed patterns in Patterns A, C, and D data sets were treated as corn plants by Algorithm One, while Algorithm Two accurately identified the corn plants. When patches of weeds were interspersed among the corn plants, weeds in all of the data sets were incorrectly treated as corn plants by Algorithm One, while Algorithm Two continued to correctly identify corn plants.

Four Standard Pattern Data Sets with Varying Distances between Plants

A second group of modified pattern data sets were formed to investigate the algorithms' ability to accommodate different intervals between corn plants. The pattern shapes of corn plants in these four data sets were the same as the data sets in Table 1, but the spacing between corn plants was allowed to vary by ±2.5 cm. The threshold values for the relative area of the corn plants were set to 15% for all of the data sets. The average values of the relative area of Patterns A, B, C, and D data sets were 17.7, 20.8, 19.9, and 19.9%, respectively, for both algorithms when the distances were 30 ± 2.5 cm (Table 3). All corn plants in the four data sets were distinguishable. When the spacing between two corn plants was reduced to 15 ± 2.5 cm, neither algorithm could correctly identify corn plants for the Pattern A data set, as well as for the parts of Patterns C and D data sets in which corn plants were closer than 15 cm. When the plant spacing was 12.5 ± 2.5 cm, corn plants in Patterns A, C, and D data sets were not identifiable by either algorithm. At a spacing of 3.75 ± 2.5 cm, all corn plants overlapped each other in the four data sets and could not be correctly identified by either algorithm.

Four Pattern Data Sets of Larger Corn Plants

A set of modified pattern data sets were formed using plant patterns representing large corn plants, and the effect of plant spacing on plant identification was investigated. The threshold values of the relative areas of the four data sets were 15% for both algorithms. The average values of the four data sets were 38, 46.9, 29.4, and

Table 3. The effect of plant spacing on the ability of the algorithm to identify corn plant patterns when the patterns are nonuniformly spaced.[1]

	Pattern area				Plants correctly identified			
	A[2]	B[3]	C[4]	D[5]	A	B	C	D
	percentage							
Algorithm One								
Spacing 1[6]	17.7	20.8	19.9	19.9	100.0	100.0	100.0	100.0
Spacing 2[7]	18.8	20.8	19.9	19.9	0.0	100.0	75.0	75.0
Spacing 3[8]	25.0	20.8	33.6	33.6	0.0	100.0	0.0	0.0
Spacing 4[9]	25.0	20.8	33.6	33.6	0.0	0.0	0.0	0.0

Algorithm Two								
Spacing 1	17.7	20.8	19.9	19.9	100.0	100.0	100.0	100.0
Spacing 2	18.8	20.8	19.9	19.9	0.0	100.0	75.0	75.0
Spacing 3	25.0	20.8	33.6	33.6	0.0	100.0	0.0	0.0
Spacing 4	25.0	20.8	33.6	33.6	0.0	0.0	0.0	0.0

[1]Area threshold value of 15%.
[2]Leaves of corn plants oriented parallel to the row.
[3]Leaves of corn plants oriented perpendicular to the row.
[4]Leaves of corn plants rotated 45° clockwise from parallel to the row.
[5]Leaves of corn plants rotated 45° counterclockwise from parallel to the row.
[6]Spacing between centers of two adjacent corn plants of 30 cm.
[7]Spacing between centers of two adjacent corn plants of 15 cm.
[8]Spacing between centers of two adjacent corn plants of 12.5 cm.
[9]Spacing between centers of two adjacent corn plants of 3.75 cm.

29.4%, respectively, and neither algorithm had difficulty identifying the corn plant patterns in Patterns B, C, and D data sets when the distance between two corn plants was 30 cm (Table 4). The patterns in the Pattern A data set overlapped each other, however, and the corn plants in the first data set were not distinguishable by either algorithm. When plant spacing was reduced to 15 cm and less, neither algorithm could correctly identify corn plants in any of the four data sets, since all of the plants overlapped.

Four Pattern Data Sets - Sensor not Scanning along a Straight Line

A set of modified patterns was formed to investigate the effect of driving accuracy on the algorithms' ability to identify corn plants. When the sensor was scanning a curved line data set, the patterns of corn plants were not always in the same column in the data set. However, assuming constant sensor speed, the pattern shapes would not change. The threshold values of the relative area of corn plants were all set to 15% for both algorithms. The average values of the relative areas of Patterns A, B, C, and D data sets were 17.7, 20.8, 19.9, and 19.9%, respectively, when the spacing between corn plants was 30 cm (Table 5) for both algorithms. The corn plants in all four data sets were correctly identified; however, when spacing was reduced to 15 cm,

Table 4. The effect of plant spacing on the ability of the algorithm to identify corn plant patterns when large corn plants are present.[1]

	Pattern area				Plants correctly identified			
	A[2]	B[3]	C[4]	D[5]	A	B	C	D
	percentage							
Algorithm One								
Spacing 1[6]	38.0	46.9	29.4	29.4	0.0	100.0	100.0	100.0
Spacing 2[7]	50.0	54.2	50.0	50.0	0.0	0.0	0.0	0.0
Spacing 3[8]	50.0	60.0	60.0	60.0	0.0	0.0	0.0	0.0
Spacing 4[9]	50.0	75.0	75.0	75.0	0.0	0.0	0.0	0.0
Algorithm Two								
Spacing 1	38.0	46.9	29.4	29.4	0.0	100.0	100.0	100.0
Spacing 2	50.0	54.2	50.0	50.0	0.0	0.0	0.0	0.0
Spacing 3	50.0	60.0	60.0	60.0	0.0	0.0	0.0	0.0
Spacing 4	50.0	60.0	60.0	60.0	0.0	0.0	0.0	0.0

[1]Area threshold value of 15%.
[2]Leaves of corn plants oriented parallel to the row.
[3]Leaves of corn plants oriented perpendicular to the row.
[4]Leaves of corn plants rotated 45° clockwise from parallel to the row.
[5]Leaves of corn plants rotated 45° counterclockwise from parallel to the row.
[6]Spacing between centers of two adjacent corn plants of 30 cm.
[7]Spacing between centers of two adjacent corn plants of 15 cm.
[8]Spacing between centers of two adjacent corn plants of 12.5 cm.
[9]Spacing between centers of two adjacent corn plants of 3.75 cm.

Table 5. The effect of plant spacing on the ability of the algorithm to identify corn plant patterns when the sensor path does not parallel the corn row.[1]

	Pattern area				Plants correctly identified			
	A[2]	B[3]	C[4]	D[5]	A	B	C	D
	percentage							
Algorithm One								
Spacing 1[6]	17.7	20.8	19.9	19.9	100.0	100.0	100.0	100.0
Spacing 2[7]	17.7	20.8	19.9	19.9	0.0	100.0	100.0	100.0
Spacing 3[8]	18.8	20.8	19.9	19.9	0.0	100.0	0.0	0.0
Spacing 4[9]	25.0	20.8	33.6	33.6	0.0	0.0	0.0	0.0
Algorithm Two								
Spacing 1	17.7	20.8	19.9	19.9	100.0	100.0	100.0	100.0
Spacing 2	17.7	20.8	19.9	19.9	0.0	100.0	100.0	100.0
Spacing 3	18.8	20.8	19.9	19.9	0.0	100.0	0.0	0.0
Spacing 4	25.0	20.8	33.6	33.6	0.0	0.0	0.0	0.0

[1]Area threshold value of 15%.
[2]Leaves of corn plants oriented parallel to the row.
[3]Leaves of corn plants oriented perpendicular to the row.
[4]Leaves of corn plants rotated 45° clockwise from parallel to the row.
[5]Leaves of corn plants rotated 45° counterclockwise from parallel to the row.
[6]Spacing between centers of two adjacent corn plants of 30 cm.
[7]Spacing between centers of two adjacent corn plants of 15 cm.
[8]Spacing between centers of two adjacent corn plants of 12.5 cm.
[9]Spacing between centers of two adjacent corn plants of 3.75 cm.

neither algorithm could correctly identify the corn piants for the Pattern A data set. At a spacing of 12.5 cm, corn plants in Patterns A, C, and D data sets were not identifiable by either algorithm, and at a spacing of 3.75 cm, neither algorithm could correctly identify corn plants for any of the four data sets.

Two Data Sets of Corn Plants and Weeds Collected in the Field

The algorithms were tested using data collected from two corn plant rows in the field. Due to late-summer planting that resulted in poor germination and emergence, the horizontal area of the corn plants varied considerably. The threshold value of the relative area was 13% for both algorithms. The average values of the relative area for Row 1 and Row 2 after 10 corn plants were scanned was 16.6 and 20.5%, respectively, for Algorithm One (Table 6). There were 65 corn plants in the first row. Algorithm One was able to correctly identify 55 of the 65 corn plants in Row 1 and 48 of the 66 corn plants in Row 2, for accuracies of 85 and 77%, respectively. For Algorithm Two, the average values of the relative area for Row 1 and Row 2 after 10 corn plants were scanned was 13.0 and 13.0%, respectively. Algorithm Two was able to correctly identify only a few of the corn plants (Table 6). Nonuniform travel speed

Table 6. The effect of plant spacing on the ability of the algorithm to identify corn plant patterns in field data sets where both corn plants and weeds are present.[1]

	Pattern area	Plants correctly identified
	percentage	
Algorithm One		
Row 1[2]	16.6	85.0
Row 2[3]	20.5	77.0
Algorithm Two		
Row 1	13.0	3.0
Row 2	13.0	3.0

[1]Area threshold value of 13%.
[2]The first corn plant row in the field.
[3]The second corn plant row in the field.

of the sensor and nonuniform plant spacing produced a data set in which large variation existed in the number of lines between adjacent plants. Even when the relative area of a corn plant was larger than the calibration value, the distance to the previous corn plant was outside the range of the calibration value, and the corn plant was not identified.

Comparison of Algorithm One and Algorithm Two

Both algorithms could distinguish corn plants from weeds when corn plants did not overlap each other, and when weeds were small and did not overlap with corn plants. Algorithm One, however, could not correctly identify corn plants when weeds were as large as corn plants or when weeds grew in patches since the algorithm treated big weeds and patches of weeds as corn plants. Algorithm Two overcame this disadvantage and could correctly identify corn plants when large weeds or a patch of weeds grew between two corn plants if the intervals between two corn plants were uniform. The accuracy of Algorithm Two was higher than the accuracy of Algorithm One when large weeds or patches of weeds were present. If the distance between two corn plants was not constant, Algorithm Two had difficulty in correctly identifying corn plants.

CONCLUSIONS

The Patchen Weedscanner system, in conjunction with appropriate software, can correctly distinguish corn plants from some weed sizes and patterns. The following conclusions can be made: (i) distinguishable corn plant patterns and threshold values can be identified in data streams from the Patchen Weedscanner; (ii) two algorithms were developed for identifying corn plant and weed patterns; (iii) tests were completed showing that the algorithms can, for specific corn plant sizes and weed sizes and densities, identify corn plants and distinguish them from weeds; and (iv) Algorithm One, in conjunction with the Patchen Weedscanner system, was able to successfully identify corn plants with an accuracy of at least 75%. Algorithm Two was unable to successfully identify corn plants in the field data sets due to nonuniform plant spacing.

ACKNOWLEDGMENT

The authors gratefully acknowledge the support of Patchen California, Los Gatos, CA, provider of the WeedScanner equipment used in this study.

REFERENCES

Bargen, K.V., G.E. Meyer, D.A. Mortenson, S.J. Merritt, and D.M. Woebbecke. 1993. Red-near infrared reflectance sensor system for detecting plants. p. 231–238. *In* J.A. DeShazer and G.E. Meyer (ed.) Proc. SPIE Conf. on Optics in Agriculture and Forestry. SPIE Volume 1836. SPIE, Bellingham, WA.

Beck, J. 1996. Reduced herbicide usage in perennial crops, row crops, fallow land and non-agricultural applications using optoelectronic detection. SAE Tech. Pap. Ser. 961758. SAE, Warrendale, PA.

Brown, R.B., J.P. G.A. Steckler, and G.W. Anderson. 1994. Remote sensing for identification of weeds in no-till corn. Trans. ASAE 37(2):297–302.

Everitt, J.H., R.D. Pettit, and M.A. Alaniz. 1987. Remote sensing of broom snakeweed (*Gutierrezia sarothrae*) and Spiny Aster (*Aster spinosus*). Weed Sci. 35(2):295–302.

Felton, W.L., and K.R. McCloy. 1992. Spot spraying. Agric. Eng. 73 (1):9–12.

Haggar, D.E., C.J. Stent, and S. Isaac. 1983. A prototype handhold patch sprayer for killing weeds, activated by spectral differences in crop/weed canopies. J. Agric. Eng. Res. 28:349–358.

Hooper, A.W., G.O. Harries, and B. Ambler. 1976. A photoelectric sensor for distinguishing between plant material and soil. J. Agric. Eng. Res. 21:145–155.

Jia, J., and G.W. Krutz. 1992. Location of the maize plant with machine vision. J. Agric. Eng. Res. 52:169–181.

Lybecker, D.W., E.E. Schweizer, and R.P. King. 1991. Weed management decisions in corn based on bioeconomic modeling. Weed Sci. 39(1):124–129.

Shropshire, G.J., K.L. Von Bargen, and D.A. Mortensen. 1991. Optical reflectance sensor for detecting plants. p. 222–235. *In* J.A. DeShazer and G.E. Meyer (ed.) Proc. SPIE Conf. on Optics in Agriculture. SPIE Volume 1379. SPIE, Bellingham, WA.

Stafford, J.V., and P.C.H. Miller. 1996. Spatially variable treatment of weed patches. p. 465–474. *In* P.C. Robert, et al. (ed.) Proc. 3rd Int. Conf. on Precision Agriculture. ASA, CSSA, and SSSA, Madison, WI.

Steckel, L.E., M.S. Defelice, and B.D. Sims. 1990. Integrating reduced rates of postemergence herbicides and cultivation for broadleaf weed control in soybeans (*Glycine max*). Weed Sci. 38(6):541–545.

Stone, M.L. 1994. Embedded neural networks in real time controls. ASAE Pap. 94–1067. ASAE, St. Joseph, MI.

Weber, D.C., F.X. Mangan, and D.N. Ferro. 1990. Effect of weed abundance on European corn borer (*Lepidoptera: Pyralidae*) infestation of sweet corn. Environ. Entomol. 19:1858–1865.

Yu, J. 1997. Corn plant identification using spectral reflectance pattern recognition. M.S. thesis. Univ. of Illinois, Urbana, IL.

Sensor for Weed Detection Based on Spectral Measurements

F. Feyaerts
P. Pollet
L. Van Gool
P. Wambacq

Dept. of Electrical Engineering, ESAT-PSI,
Katholieke Universiteit Leuven,
Belgium

ABSTRACT

A sensor is proposed here which distinguishes between crop and weed based on their different spectral reflectances. The sensor is built upon an imaging spectrograph. We chose this spectral reflectance sensor because of the fast spectral imaging and possible high spatial and spectral resolution. Parameters like the angle-of-view and the quality of the optics were optimized for maximal performance within reasonable cost.

Classification success rates depend not only on spatial and spectral filtering, both characteristics of the device, but also on the number of wavelengths and the crop itself. Under controlled conditions, corn and sugar beet can be separated from weed with a success rate of at most 90, respectively 80%. Herbicide savings which depend on weed density, the nozzle activation frequency and the spray resolution (width), are maximal with the MLNN classifier.

Keywords: Weed sensor - Multi-spectral imaging - Herbicide reduction

INTRODUCTION

Growing environmental consciousness and increased competition are the driving forces behind *precision farming*. One of the most promising techniques to decrease the use of herbicides is place-specific spraying, i.e. to only spray where needed. Initially, spot spraying was focused on distinguishing vegetation from soil. Spraying would then be restricted to patches covered with vegetation, i.e. weeds and/or crop. The distinction is usually based on the so-called `red-edge', the big difference in canopy/leaf reflectance for red and near infrared. The ratio of the reflectance at these wavelengths is an effective cue to distinguish between soil and vegetation [Haggar et al. 1983, Felton et al. 1991].

The goal of the reported work is to go a step further and also distinguish between the crop and the different weeds. It goes without saying that this is a much more subtle distinction to make.

A few approaches in that direction have already been suggested in the literature. Some are based on analyzing the shapes and sizes of leaves [Gerhards et al. 1993, Franz et al. 1995]. For the time being, the computational requirements seem beyond what is economically feasible for the online application envisaged in this

Copyright © 1999 ASA-CSSA-SSSA, 677 South Segoe Road, Madison, WI 53711, USA. *Proceedings of the Fourth International Conference on Precision Agriculture.*

study. Other research has shown that differences in spectral reflectance characteristics of different plants may suffice to tell them apart [Knipling 1970, Nitsch et al. 1991, Hahn and Muir 1993, Price 1994]. Classical multi-spectral measurement

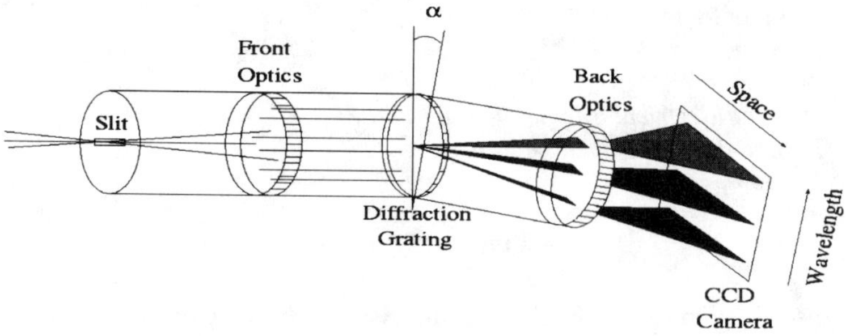

Figure 1: Working principle of imaging spectrograph

devices suffer from low spatial and spectral resolution [Felton et al. 1991] and are too expensive and/or too vulnerable to be mounted on a spray boom.

This paper presents a setup for the recognition of crop and weed that reduces these restrictions. Our aim is to design a vegetation sensor that is sufficiently cheap and rugged to be put on spray booms for analysis of lines with a length of approximately 2 meters and a width of 1-2 cm.

SENSOR SETUP

Proposed sensor

The proposed weed sensor yields the reflectance spectrum of each point on a narrow linear stripe [Herrala et al. 1994]. It consists of 3 parts: an objective lens, an imaging spectrograph, and a camera. The principle is shown in Figure 1.

The objective lens (not shown in Figure 1) projects the image of a field patch on the slit aperture of the spectrograph. This slit extracts a small stripe from the

Table 1: low cost, low resolution and high cost, high-resolution spectrographs

Parameter	Experimental spectrograph	Imspector
F-number	3	4 – 8
Spectral resolution	35	1.5 – 5 nm
Available spectral range	400 – 1000 nm	435 – 855 nm
Free spectral range	1 octave in available range	Idem
Slit width	200 μm	80 μm
Slit length	8 mm	8.8 mm
CCD	½" (4.8 x 6.4 mm)	½" (4.8 x 6.4 mm)
Spectral bands	150	200

patch on the ground. The front doublet in the spectrograph collimates the light coming from the slit. The light is then split into its spectral components by diffraction. The second doublet forms an image of the diffracted light on a (monochrome) camera. In this way, one of the axes of the camera act as a spatial axis while the other axis is a spectral axis. We have constructed such a device, which we will call the ***experimental spectrograph***, with parameters as given in Table 1. It should be emphasized that the design parameters of this spectrograph were chosen to obtain a low cost device. Its spectral range coincides with the typical spectral range of a low cost CCD or CMOS camera. The spectral resolution of this sensor (determined mainly by the slit width and the number of grooves in the diffraction grating) is limited to about 150 bands. Using an objective with f = 7 mm, the device has a spatial resolution of approximately 3 mm, when looking from a height of 1 meter. These are approximately the working conditions we envisage.

By using additional or more expensive components (lenses, grating, ...), a system with a finer spectral resolution (5 nm in the spectral range of 435 to 855 nm) and a higher spatial resolution (3 mm when placed at the same height as before), was designed and made commercially available by Specim [Herrala et al. 1994]. The number of spectral bands of this device (further referred to as the ***Imspector***) is approximately 200. tracts a small stripe from the
Table 1 summarizes its most important design parameters. The spectrograph with the highest resolution - the Imspector - probably shows the practical limitations of these devices. The other one gives the performance drop one has to reckon with when using cheaper optics and a wider angle of view, as to minimize the number of sensors on the spray boom.

Feature selection

There is no point in using all wavelengths' reflectance. Computation times would be prohibitive and there is substantial redundancy in the data. The optimal set of wavelengths depends on the plants to be distinguished. One can take the wavelengths that maximize the following class-to-class separation function:

$$F(\lambda) = \left|X(\lambda) - Y(\lambda)\right| \Big/ \sqrt{\sigma_X^2(\lambda) + \sigma_Y^2(\lambda)}$$

Equation 1: Class-to-class separation evaluation function

in which X(λ) and Y(λ) are the mean values of the reflected light for class X, respectively class Y at wavelength λ and $\sigma^2{}_X(\lambda)$ and $\sigma^2{}_Y(\lambda)$ are the standard deviations of the measurements of the reflected light at the same wavelength λ, also for classes X and Y. Local extremes of this function are found at certain wavelengths for which the corresponding spectral irradiants are used as features for the classification. The differences in reflectance of the plants reflect underlying, physical differences between the plants [Hahn and Muir 1993]. The spectral differences in the visible region are mainly determined by the production of chlorophyll. Chlorophyll production in plants depends on their response to several external factors such as competition, disease, weather, soil status,... At least as important for the discrimination of plant species are the differences in the near infrared. These dif-

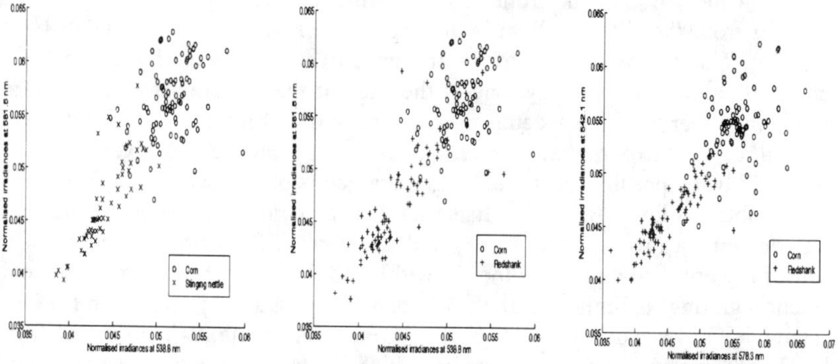

Figure 2: left: Normalized spectral irradiance values: wavelength selection for corn and stinging nettle, first wavelength couple (i.e. combination of the two best discriminating wavelengths); middle: Corn and redshank, for wavelength couple of Figure 2 [left], which is suboptimal for this crop-weed combination; Right: Corn and redshank, first wavelength couple for this combination of crop and weed

ferences depend rather on the internal structure of the plant. The number of cell layers, the size of cells and the orientation of the cell walls strongly determine the reflectance in the near infrared. Also leaf hairs and waxes that are characteristic for some plant species influence the infrared spectral reflectance.

As we are dealing with more than two classes, wavelengths are selected for the combination of the crop class with each one of the weed classes separately. The aforementioned comparison of their reflectance spectra yields a selection of wavelengths for each pair. The best discriminating wavelengths for each pair of classes are selected automatically. Lower ranking wavelengths are only added if they remain sufficiently far from those already selected (higher ranked extremes for each combination of the crop and the weeds). This procedure carries on towards lower ranking wavelengths up to rank 5 for every crop-weed pair that is being added. Figure 2 [left] shows the normalized reflectance values at the 2 best separating wavelengths for corn and stinging nettle. Although the clusters overlap, their centroids are reasonably apart. At these same wavelengths, redshank would be much more difficult to tell apart from corn, however (Figure 2 [middle]). The two best wavelengths for this pair and the corresponding clusters are shown in the right of the same figure. As these examples show, there is little chance of telling a crop like corn apart from a weed mix if only a couple or so wavelengths would be used.

The result section gives the wavelength selections that have been used. They typically contain around 10 different wavelengths. The feature space is analyzed to reduce the original feature space to the smallest set of orthogonal directions that contain at least 95 percent of the original variance.

APPLICATION: WEED-CROP SENSOR

Measurement conditions

Tests were performed under semi-field conditions: indoor measurements for actual plants. Two economically important crops, corn and sugar beet, were selected for the experiments. They were intermingled with weeds that are found among these crops in real fields. Special care was taken to gather healthy looking samples of the same age. In the following list, the number of samples that were gathered with the Imspector, respectively with the experimental spectrograph are given for the selected vegetation classes. With the Imspector, systematically more samples could be extracted from the scene. The study included corn [Zea Mais] (66-33), sugar beet [Beta vulgaris](91-56), buttercup [Ranunculus Pepens](65-37), Canada thistle [Cirsium Arvense](64-45), charlock [Sinapis Arvensis](81-32), chickweed [Stellaria media](66-62), dandelion [Tarraxacum Officinale](58-36), grass [Poa annua](63-43), redshank [Poligonum Persicaria](77-41), stinging nettle [Urtica dioica](48-44), wood sorrel [Onalis europaea](60-35) and yellow trefoil [Medicago Lupulina](62-39).

The scene illumination consisted of a controlled light source only: a broadband 100-Watt halogen lamp. Intensity variations, although not intended, are observed due to the spatial non-uniformity of the emission pattern of the light source, due to shadows and due to different degrees of specular reflection by leafs with different orientations. For this reason, the normalization (features normalized to norm 1) was introduced. Measurements are performed using a monochrome 1/2" (4.8 x 6.4 mm) CCD (MX5 of Adimec) in combination with either the Imspector or the experimental spectrograph. The processing unit consists of a 166 MHz Pentium PC with 32 MByte RAM.

It should be emphasized that under real field conditions the situation gets more complicated, e.g. due to the presence of sunlight and the variations thereof. As one should be able to use the weed sensor also at night, dusk, or dawn, a light source has to be provided. To fully exploit the imaging spectrograph's spectral range, the light source should have a broad and preferably flat spectrum in the visible and near-infrared bands. Usable sources are e.g. halogen lamps and light bulbs. A halogen lamp is preferred for its higher intensity at smaller wavelengths (visible part of the spectrum). Specular reflections are another important source of differences and variations in the spectral composition of the reflected light. Shadows also cause spectral variations, as Rayleigh scattering becomes more important in shaded areas and yields a higher share of shorter wavelengths. These considerations called for additional refinements like software-based illumination normalization.

Classification strategy

As we were dealing with a high dimensional feature space, dividing the original dataset into representative subsets was not possible. Therefore, we used a bootstrap sample generator method as described in [Efron and Gong 1983]. Following the bootstrap procedure, the estimated (reliable) classification success rates are obtained as the averages over a large number of experiments. Classification success rates in these experiments are calculated on synthetic dataset derived from the original dataset by drawing samples with replacement of the sample. As a result, this dataset can contain the same sample more than once.

Classification results in the rest of this paper were obtained using a leave-one-out strategy on dataset that were generated with a bootstrap method: one sample is eliminated from the dataset; the classifier is trained based on the remaining samples; the eliminated sample is assigned to a class by the classifier. This procedure is repeated for all the samples of the dataset. As the classifier varies from dataset to dataset (even from sample to sample), a leave-one-out strategy will give a strong indication on the performance the classifier would achieve. However, *it does not give the classifier parameters that can be used afterwards on the field.*

Implemented classifiers

The paper discusses the following classifiers:

1. Mahalanobis Distance (e.g. see [Fukunaga 1972]), further referred to as MD.

This simple parametric statistical method assumes the samples have Gaussian distributions. A sample x is then assigned to class A if the feature is closest in a Mahalanobis sense to the expected feature of class A. The number of dataset generated with the bootstrap method and analyzed with this classifier was limited to 100.

2. k-Nearest Neighbor (e.g. see [Longman 1977]), further referred to as k-NN.

With this method, no prior knowledge is needed about the class probability. Additionally, features don't need to have Gaussian distributions. Classes are characterized by a set of selected reference samples. An incoming sample is then assigned to the class with the closest reference samples, where simple Euclidean distances are used. The number of (smallest) distance k on which the decision was based, was tested to get best crop-weed discriminating results. The k's for the highest averages of the crop and minimal weed classification success rates in the range of $k=1 \rightarrow 30$ (for each combination of crop-weed and imaging spectrograph) will be given when the results are discussed. For computational reasons, the number of dataset generated with the bootstrap method and analyzed with this classifier was limited to 50.

3. Multi-layer neural network with nonlinear differentiable transfer functions (e.g. see [Rumelhart et al. 1986]), further referred to as MLNN.

The neural network consists of three layers in which each layer is fully connected to the next one. First, comes an input layer with a fixed number of 6 neurons, regardless the data source and feature dimension. Next, the hidden layer is found in which the number of neurons depends on the crop-weed combination and the type of imaging spectrograph. The exact numbers will be given when the results are discussed. Finally, the output layer consists of 2 neurons: one for each class (crop or weed). The output of these neurons is trained to be high when an input vector of the corresponding class is fed into the network. The higher the output, the more probable the input vector belongs to the class, corresponding to the neuron. The transfer functions of the neurons of layer one and two were a log-sigmoid with bias. This function maps the input of the neuron from the infinite interval (-∞,∞) to the finite interval (0,1). The third layer neurons had a tan-sigmoid transfer function, also with bias. The neuron input is mapped from (-∞,∞) to (-1,1). The training procedure was implemented with a back-propagation learning rule using momentum and an adaptive learning rate. The former minimizes the learning time. The latter minimizes the risk to get stuck in a sub-optimal local minimum of the error function. For computational reasons (training phase) the number of dataset generated with the bootstrap method and analyzed with this classifier was limited to 15.

Classification results

Ideally, every sample would be assigned to its own class enabling plant-specific herbicide spraying. Some weeds, like Canada thistle are treated that way. Current practice in field spraying is different however as most of the herbicides will kill several weeds class, e.g. the grasses. Mixing herbicides makes it possible to kill all the weeds in one run. Therefore, when assigning the results, the samples are in the end classified into only two classes: crop and weed. While weed samples need not to be assigned to the same weed class for the classification to be nevertheless `correct', the crop should only be assigned to its own class. As a result, crop samples can be expected to be classified 'erroneously' in a larger percentage of cases.

Table 2 lists the wavelengths that were selected according to Equation 1. It also gives the number of principal components that contained at least 95 percent of the original variance for four cases: each of the two imaging spectrographs was used for classification of either corn or sugar beet in combination with all the selected weed classes.

The average classification success rate for each of the classifiers is calculated as the mean of the classification success rates for the different classes. In this way, classification results are only based on how the classifier manages to discriminate

Table 2: Feature for each combination of crop and imaging spectrograph: wavelength selection based on Equation 1 (F), feature dimension after principal component analysis (PCA)

Combination	F	PCA
Corn – Imspector	539, 540, 542, 545, 549, 557, 565, 578, 585, 596, 605, 639, 675, 687, 703, 814, 840	5
Corn – Exp. Spectr.	478, 542, 550, 558, 562, 602, 666, 754, 822	3
Beet – Imspector	535, 542, 545, 554, 565, 578, 585, 595, 610, 628, 657, 666, 680, 690, 699, 720, 778, 804	8
Beet – Exp. Spectr.	478, 534, 590, 654, 666, 714, 754, 770, 774, 778, 798, 806	3

Table 3: Detailed classification success rates (bootstrap variances) for each of the classifiers, using the high resolution Imspector and the low resolution experimental spectrograph

	Imspector			Experimental spectrograph		
	MD	k-NN	MLNN	MD	k-NN	MLNN
Corn	87 (3)	88 (3)	97 (2)	61 (11)	84 (6)	89 (3)
Buttercup	98 (2)	99 (1)	95 (4)	95 (4)	95 (3)	85 (8)
Thistle	98 (2)	99 (1)	94 (4)	100 (0)	100 (0)	100 (1)
Charlock	91 (4)	91 (3)	83 (6)	99 (2)	97 (3)	89 (7)
Chickweed	100 (0)	100 (1)	100 (1)	100 (1)	94 (5)	86 (10)
Dandelion	96 (2)	95 (3)	93 (4)	93 (5)	92 (4)	72 (11)
Grass	100 (0)	99 (1)	99 (3)	99 (2)	98 (3)	93 (6)
Redshank	100 (0)	99 (1)	98 (3)	100 (0)	99 (2)	96 (4)
Stinging nettle	98 (2)	98 (2)	94 (3)	99 (2)	99 (1)	94 (4)
Wood sorrel	99 (2)	94 (3)	90 (6)	88 (8)	82 (8)	68 (9)
Yellow trefoil	100 (1)	100 (0)	100 (1)	91 (6)	98 (3)	87 (6)

between crop and weeds and not on the number of samples of the different classes.

Case 1

In a first case study, it is investigated how well *weeds and corn* can be told apart (under semi-field conditions as explained above). We refer to Table 2 for an enumeration of the wavelengths that were selected (different sets of wavelengths for each of the imaging spectrographs). Table 3 shows the detailed classification results for the discrimination of corn and the different weeds for each of the classifiers. At this point, the reader is reminded of the fact that for the weeds, success rates indicate whether that weed's samples were assigned to its own or any other weed class.

Table 4 summarizes the success rates for classification as corn and weed. Each row gives the estimated percentages for classification to crop and weed respectively. Consider first the classification of the samples that were gathered with the Imspector (cf. Table 4). With the MD classifier, an average of 98 percent of the weed samples were recognized as weed while also 13 percent of the crop was classified as weed. With the k-NN classifier with k = 5, 98 percent of the weed was correctly classified as weed but some 12 percent of the crop was classified as weed as well. With the MLNN classifier, 94 percent of the weed samples and only some 3 percent of the crop were classified as weed.

Table 4: Average classification success rates (bootstrap variances) for crop and weed

	Imspector			Experimental spectrograph		
	MD	k-NN	MLNN	MD	k-NN	MLNN
Corn	87 (3)	88 (3)	97 (2)	61 (11)	84 (6)	89 (3)
Weed	98 (2)	98 (2)	94 (4)	96 (3)	95 (3)	87 (7)

Using the experimental spectrograph (cf. Table 4), classification was systematically a bit worse. With the MD classifier, still 96 percent of the weed was correctly classified as weed, but almost 40 percent of the crop samples were classified as weed as well. With the k-NN classifier with k = 4, 95 percent of the weed was recognized correctly, while some 16 percent of the crop was misclassify as weed. Using the MLNN classifier would result in only about 87 percent of the weed being classified as weed. More than 10 percent of the crop would be misclassify as weed.

Case 2

In this second case study, it is investigated how well weeds and sugar beet can be told apart under semi-field conditions (as explained above). We refer again to Table 2 for an enumeration of the wavelengths that were used with each of the two imaging spectrographs.

Table 5 gives the detailed classification success rates using each one of the classifiers. Table gives the overall classification success rates for sugar beet and weed. Consider first the classification of the samples that were gathered with the Imspector (cf. Table 6). With the MD classifier, 89 percent of the weeds were correctly recognized as weed while 22 percent of the crop was erroneously classified as weed as well. With the k-NN classifier with k = 5, 93 percent of the weed samples were correctly recognized, but as much as 35 percent of the crop samples were classified as weed as well. With the MLNN classifier, 81 percent of the weed was classified correctly while 17 percent of the crop is erroneously classified as weed. Again, the results were a bit worse using the experimental spectrograph (cf. Table 6). Although 92 percent of the weed samples were correctly recognized as weed with the MD classifier, 34 percent of the crop was erroneously classified as weed. Some slightly better results were obtained using the k-NN classifier with k = 4. As much as 94 percent of the weed samples were correctly recognized while only slightly more than 25 percent of the crop samples were erroneously classified as weed.

Table 5: Detailed classification success rates (bootstrap variances) using a Mahalanobis distance based classifier

	Imspector			Experimental spectrograph		
	MD	k-NN	MLNN	MD	k-NN	MLNN
Sugar beet	78 (6)	65 (5)	83 (4)	66 (9)	74 (6)	88 (4)
Buttercup	81 (8)	90 (4)	73 (6)	77 (7)	89 (6)	72 (9)
Canada thistle	90 (5)	96 (3)	84 (7)	96 (4)	97 (2)	96 (3)
Charlock	73 (7)	84 (5)	56 (8)	88 (8)	95 (4)	69 (12)
Chickweed	96 (3)	96 (3)	88 (5)	100 (0)	100 (1)	99 (2)
Dandelion	82 (8)	92 (4)	71 (11)	98 (5)	98 (2)	95 (7)
Grass	91 (4)	94 (4)	82 (4)	100 (0)	100 (1)	100 (1)
Redshank	93 (5)	96 (2)	89 (4)	90 (8)	90 (6)	73 (11)
Stinging nettle	86 (5)	92 (4)	76 (7)	84 (9)	85 (7)	64 (9)
Wood sorrel	97 (2)	97 (2)	91 (10)	91 (5)	91 (5)	75 (8)
Yellow trefoil	97 (3)	97 (2)	95 (3)	99 (2)	100 (1)	97 (4)

Table 6: Average classification success rates (bootstrap variances) for crop and weed

	Imspector			Experimental spectrograph		
	MD	k-NN	MLNN	MD	k-NN	MLNN
Sugar beet	78 (6)	65 (5)	83 (4)	66 (9)	74 (6)	88 (5)
Weed	89 (5)	93 (4)	81 (7)	92 (5)	94 (4)	83 (7)

With the MLNN classifier, 83 percent of the weed samples were classified as weed together with only some 12 percent of the crop samples.

CONCLUSIONS

For both spectrographs and irrespective of the classifier, appreciably more crop will be classified as weed than vice-versa. This is desirable, because weed must be sprayed quasi completely, whereas some herbicides reaching crop is unavoidable anyway with current and foreseeable resolutions of the spraying nozzles. Moreover, not all of the weeds equally damage the selected crops. Dandelion, for instance, is far less noxious than Canada thistle. Classification results are therefore rather `conservative': some of the crop was assigned to the weed class because this rather unimportant vegetation class was present in the set of weeds.

The difference between the classifiers lies not only in their success rates, but also in the variances on the estimated success rates. The classifier with the highest success rates, coupled to the lowest variances will perform better than other classifiers. Another influencing parameter of practical import is the computational cost.

It is difficult if not impossible to draw final conclusions on `the best' classifier. This depends also e.g. on the effectiveness, harmfulness and price of the herbicides for the different weeds. Nevertheless, looking only at the different classification success rates, the k-nearest neighbor with k = 5 (Imspector) or k = 4 (the Experimental Spectrograph) comes out to be a good option.

Figure 3 gives the estimated herbicide savings [Pollet et al. 1998] (corn and weed) that could be obtained with the aforementioned classifiers for feasible working conditions (spray width of 50 cm, spray length (direction of motion) of

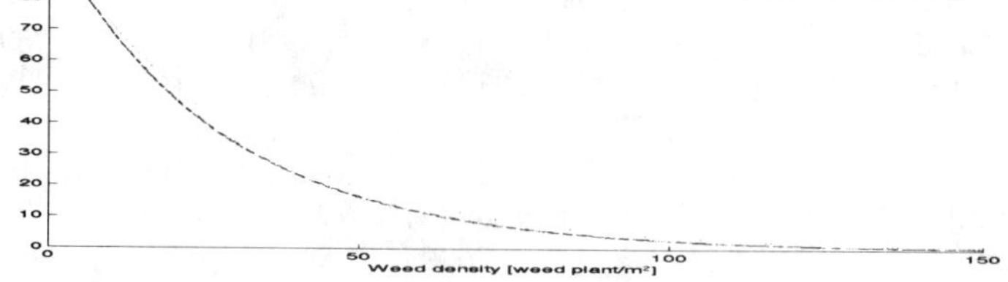

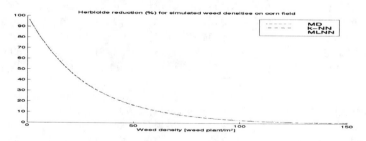

Figure 3: Herbicide savings for varying weed densities in the spray rectangle

10 cm), i.e. driving at a speed of 7.2 km/h and activating the spray nozzles at a frequency of 20 Hz. The number of weeds in the spray rectangle, on which the herbicide reduction is calculated, is estimated from a Poisson distribution. The number of crop plants is estimated, based on the known row and intra-row distances. The highest herbicide savings are obtained with the MLNN classifier: less classification errors are made for the crop samples (cf. Table 4).

ONGOING WORK

At this moment the speed of analysis is mainly determined by the (CCD) camera frame rate. By reading out only the selected camera lines (wavelengths) of a random addressable pixel CMOS-imager like the Fuga15 [Dierickx et al. 1996], higher frame rates (more spectral analyses) can be obtained. Another advantage of the CMOS sensor is its present on-chip A/D conversion and the possibility to implement parts of the algorithm on the camera chip. At this moment, we are experimenting with this type of camera to investigate feasibility.

When leaving semi-field conditions, extra research must be done to obtain a stable algorithm under changing sunlight conditions, using motion-blurred spectral images of plants. Hardware, optics and software may require some changes to meet the necessary performance. Additional structural information about the crop will be used to improve the system's performance on the field [Pollet et al. 1998]. To be economical, production cost must be reduced further, even for the experimental spectrograph: there will be a trade off between the price of the optical components and the overall performance of the sensor.

ACKNOWLEDGEMENTS

Our special thanks go to the Flemish Community (VLIM-project), the Belgian Ministry of Agriculture (IWONL-project) and Ecospray, our industrials partner, for their financial support.

REFERENCES

Dierickx B., Scheffer D. Meynants G., Ogiers W. Vlummens J., Random addressable active pixel image sensors, Invited paper AFPAEC Europe to Berlin, October 9 1996, SPIE proceedings vol. 2950 p.1, 1996

Efron B., Gong G., A Leisurely look at the Bootstrap, the Jackknife and Cross-Validation, The American Statistician, vol. 37(1), pp. 36-48, 1983.

Felton W.L., Doss A.F., Nash P.G., McCloy K.R., A Microprocessor controlled Technology to selectively spot spray weeds, 6 pp., 1991.

Franz E., Gebhardt M.R., Unklesbay K.B., Algorithms for extracting leaf boundary information from digital images of plant foliage, Transactions of the ASAE, vol. 32(2),pp. 625-633, 1995.

Fukunaga K., Introduction to statistical pattern recognition, Electrical Science, Academic press New York (N.Y.), ISBN/ISSN: 0-12-269850-9, p. 369, 1972.

Gerhards R., Nabout A, Sökefeld.M. , Kühbauch W., Nour Eldin H.A., Automatische Erkennung von zehn Unkrautern mit Hilfe digitaler Bildverarbeitung und Fouriertransformation, Journal of Agronomy and Crop Science, vol. 171, pp. 321-328, 1993.

Haggar R.J., Stent C.J., Isaac S., A Prototype Hand-held Patch Sprayer for Killing Weeds, Activated by Spectral Differences in Crop/Weed Canopies, Agricultural Engineering Research, vol. 28, pp. 349-358, 1983.

Hahn F., Muir A.Y., Weed-crop discrimination by optical reflectance, IV Int. Symposium on Fruit, Nut, and Vegetable Production Engineering, 8 pp., 1993.

Herrala E., Okkonen J., Hyvarinen T., Aikio M., Lammasniemi J., Imaging spectrometer for process industry applications, SPIE vol. 2248, pp. 33-40, 1994.

Knipling E.B., Physical and Physiological Basis for the Reflectance of Visible and Near-Infrared Radiation from Vegetation, Remote Sensing of Environment, American Elsevier Publishing Company.Inc., 5 pp., 1970.

Longman, Nearest neighbor analysis, Mathematics in geography 1, Longman London, *p. 27*, 1977.

Nitsch B.B., Von Bargen K., Meyer G.E., Visible and Near-Infrared Plant, Soil and Crop residue. Reflectivity for weed sensor design, ASAE International Summer Meeting Presentation, paper No.913006, 31 pp., June 1991.

Pollet P., Feyaerts F., Wambacq P., Van Gool L., Weed detection based on strucctural information using an imaging spectrograph, Proceedings of the 4th Int. Conf. On Precision Agriculture (in press), 1998.

Price J.C., How Unique are Spectral Signatures? Remote Sensing of Environment, American Elsevier Publishing Company.Inc., 6 pp., April 1994.

Rumelhart D.E., Hinton G.E., Williams R.J., Learning internal representations by error propagation, Parallel Data Processing, vol. 1(8), pp. 318-362, MIT Press, Cambridge, 1986.

Use of Digital Imagery to Evaluate Disease Incidence and Yield Loss Caused by Sclerotinia Stem Rot of Soybeans

M. Dudka
Environmental Remote Sensing Center
University of Wisconsin
Madison, Wisconsin

S. Langton
R. Shuler
Department of Biosystems Engineering
University of Wisconsin
Madison, Wisconsin

J. Kurle
C. R. Grau
Department of Plant Pathology
University of Wisconsin
Madison, Wisconsin

ABSTRACT

Remotely sensed spectral data were used to assess the incidence of Sclerotinia stem rot of soybean caused by the fungus *Sclerotinia sclerotiorum* and to determine its effect on variability of soybean yields. Multispectral data were obtained with an ATLAS sensor (Airborne Terrestrial Applications Sensor), yields were mapped with a combine-mounted yield monitor, and field disease assessments made both visually and by means of spectral reflectance observations obtained with a handheld radiometer. Limitations in data obtained during the ground truth survey prevented use of multispectral data for disease assessment. However, our results indicate that disease incidence and crop yield can be estimated from spectral reflectance data, that plant disease can explain a high percentage of yield variability in a production soybean field, and that diseased areas can be mapped using precision agricultural techniques. This information will enable growers to use variable rate technologies to control Sclerotinia stem rot.

INTRODUCTION

Precision agricultural techniques allow management of spatial variability in crop performance. GPS and Geographic Information Systems (GIS) provide accurate maps of seasonally stable (Moran et al., 1997) abiotic factors; soil fertility, soil type, depth of soil layers, and topographic characteristics that are static or change only slowly in a particular field. In conjunction with yield maps these techniques provide a means of prescribing soil amendments that have the potential of increasing crop yields. Although edaphic factors have been the focus of site specific management research, seasonally variable (Moran et al., 1997) biotic factors such as weeds and plant diseases that can limit crop yields also may occur spatially in a predictable manner. Although their occurrence is strongly influenced by weather conditions, plant diseases can occur predictably at locations that are determined by topography, edaphic factors, previous crops, or earlier disease outbreaks. These characteristics may allow their occurrence to be predicted and mapped using the techniques of site specific crop management. The usefulness of variable rate management technologies has been demonstrated for controlling aggregated weed populations (Mortenson et al., 1995). Similar approaches using site specific management practices also offer the possibility of limiting yield loss to plant diseases such as Sclerotinia stem rot that have an aggregated occurrence determined by edaphic factors, topography, or inoculum distribution.

Sclerotinia stem rot of soybean, caused by the fungus *Sclerotinia sclerotiorum* has become increasingly important as a limitation on increased soybean yields in the north central USA. Few of the soybean varieties that are currently grown are resistant to the disease. Numerous management changes, including narrower row widths, higher seeding rates, earlier planting dates, and shorter crop rotations, have contributed to the increased occurrence of this disease. Other practices applied to increase soybean yields also increase the likelihood of Sclerotinia stem rot. Increased soil fertility, liming, higher seeding rates, rhizobium inoculants, or selection of varieties resistant to other soybean diseases, such as soybean cyst nematode, may increase the occurrence and severity of Sclerotinia stem rot outbreaks by increasing stand density and canopy cover. Although fungicides can control Sclerotinia stem rot, their expense limits the use of broadcast applications on production soybean fields. Paradoxically, the most effective current Sclerotinia stem rot control recommendations include management changes such as lower planting rates or wider row widths that lower potential soybean yields (Grau, personal communication).

A number of characteristics of the pathogen interact with management practices to increase the frequency of disease outbreaks. Many dicotyledonous plant species are hosts to the stem rot fungus so that numerous crops and weeds are sources of soilborne inoculum. Soilborne sclerotia are highly persistent and may remain viable for several years, resulting in inoculum accumulation under crop rotations that include nonhost crops. Airborne inoculum formation and host infection are induced by high soil moisture and high humidity, conditions that are favored by the dense canopy of soybeans managed for maximum yields (Boland & Hall, 1988a).

In a field, the occurrence of Sclerotinia stem rot is aggregated and occurs first in those areas where inoculum availability interacts with both weather conditions and host plant development to produce infection (Boland & Hall, 1988b; Hartman et al., 1998). Topographic, edaphic and historical factors contribute to the aggregated occurrence of the disease in predictable hotspots (Boland & Hall, 1988a). Features such as bordering fencerows that reduce air circulation or depressions where fog collects produce environments favoring disease development. Areas of high soil fertility with favorable soil texture and moisture can promote soybean germination, dense stands, and the early formation of dense plant canopies. Areas of past disease occurrence have a much higher potential for renewed disease occurrence because of soilborne inoculum accumulation. This can be especially important in fields where susceptible host crops such as dry beans (*Phaseolus vulgaris* L.), canola (*Brassica napus* L var. *oleifera*), or sunflowers (*Helianthus annus* L.) have been grown in the past or dicotyledonous weeds such as lambsquarter (*Chenopodium album* L.) or redroot pigweed (*Amaranthus retroflexus* L.) are a problem.

Because Sclerotinia stem rot occurrence can be so highly aggregated, precision agricultural techniques may have several important applications for dealing with the disease. Remote sensing combined with precision mapping could enable earlier detection of disease outbreaks and can be used to provide more rapid and more accurate estimates of disease incidence and yield loss. More importantly, precision mapping of disease aggregation may enable the use of variable rate planting and pesticide application techniques for disease control.

OBJECTIVES

The objectives of this study were to determine if:

1. Spectral reflectance observations obtained with a handheld radiometer were correlated with visual estimates of the incidence of Sclerotinia stem rot in a production soybean field.
2. To determine if multispectral imagery obtained by a NASA Airborne Terrestrial Sensor (ATLAS) was correlated with yield data obtained from a combine mounted yield monitor.
3. To determine if the results of ground surveys and handheld radiometer observations could be combined with imagery obtained by the ATLAS system to provide an estimate average Sclerotinia stem rot incidence and disease related yield losses for the soybean field.
4. To determine what proportion of variation in soybean yield was explainable by the incidence of soybean Sclerotinia stem rot.

MATERIALS AND METHODS

The study area was an 18.3 ha production soybean field near Waunakee WI. The field had been cropped in a corn–soybean–alfalfa rotation following a single year of dry beans planted in 1991. In 1997 the field was planted to Kaltenberg soybean variety KB241. The field was mapped using a Global Positioning System (GPS) receiver equipped with differential correction, grid soil sampled, and surveyed for weed and crop plant populations. Visual disease assessments of Sclerotinia stem rot incidence were made at eleven locations in the field during the R6 growth stage and the locations mapped using a backpack GPS receiver. Disease assessment was made by counting four groups of 100 soybean plants in an area of approximately 0.8 m^2. Simultaneously with the visual disease assessments, reflectance observations were made at the same points with a Cropscan handheld radiometer (MSRWYS92 Multispectral radiometer, Cropscan, Rochester, MN). These reflectance observations represent the percentage of incoming sunlight reflected from plants and the soil surface in eight wavelengths (460, 510, 560,610, 660, 710, 760, and 810 nm with a 30 nm bandwidth). Aerial spectral observations made at the late R6 to early R7 growth stage were obtained by an ATLAS sensor in 14 bands in the visible, near IR, mid IR, and thermal wavelengths. Pixel size of these observations was 8 m^2. Yield data was obtained with a combine mounted yield monitor.

The multispectral data was obtained as a subset of an original multispectral image provided by NASA. Tangential distortion was removed using WiscImage processing software (algorithm developed by Frank Scarpace, Environmental Remote Sensing Center, Univ. of Wisconsin, Madison) to minimize scale distortions inherent to across-track scanner imagery. The ATLAS image for the study area was rectified to Dane County orthophotograph using ERDAS Imagine software (ERDAS Imagine, 1997). All datasets including ground truth data: yield, disease incidence, and radiometer observations were registered to the common coordinate system using ArcInfo and ERDAS Imagine. In order to increase positional accuracy in relatively small geographic area surveyed in this study, the local reference system (Dane Coordinate System) was applied.

Reflectance data obtained using the handheld radiometer data was analyzed statistically using correlation and regression. Because reflectance at wavelengths of 706 and 760 nm was most highly correlated with disease incidence ($r = 0.88$, $P < 0.001$) in preliminary analysis and because these wavelengths coincided with near-IR bands of the multispectal data, these wavelengths were emphasized in subsequent analyses of reflectance and multispectral data.

PC analysis was used to reduce the complete 14 band multispectral dataset to two bands that contained most of the spectral information (>95%). Subsequently, the first spectral band (blue) was excluded from the PC transformation since it contained a great deal of systematic noise. Moreover, as expected all thermal bands were highly collinear and were also removed from analysis. Unsupervised classification using ERDAS Imagine ISODATA clustering method was performed on both original and PC transformed imagery (ERDAS Imagine, 1997). Resulting spectral classes were converted into information classes using both yield monitor data and disease severity estimations. Attempts to use supervised classification failed due to difficulty in developing sufficient training sets based on available ground truth data.

Yield surfaces were interpolated through kriging from the original point data and the position of disease incidence assessments was converted from the original GPS referenced reflectance data to the GIS coverage. The yield GIS layer was re-coded into six yield ranges for stratified sampling. Selected points falling into each yield range were sampled for yield and spectral reflectance.

Statistical analysis relating ground truth data and reflectance from multispectral images was conducted in three steps: (i) Point files consisting of coordinates of the sample points were generated. (ii) Data for yield, reflectance data, disease incidence, and multispectral reflectance was obtained at the sample points using the corrected and transformed GPS coordinates. (iii) Data relating the four sample variables was analyzed using regression analysis to describe the observed relationship among the variables.

RESULTS AND DISCUSSION

In the study area soybean yields averaged 3695 to 672 kg ha^{-1} while Sclerotinia stem rot incidence ranged from 0 to 90%. Sclerotinia stem rot was concentrated on the southern half of the field on a portion of the field that had been the planted to dry beans in 1991. The bean crop had been seriously infected with Sclerotinia stem rot during that year. The field was also infested with Venice mallow (*Hibiscus trionum* L.) a possible

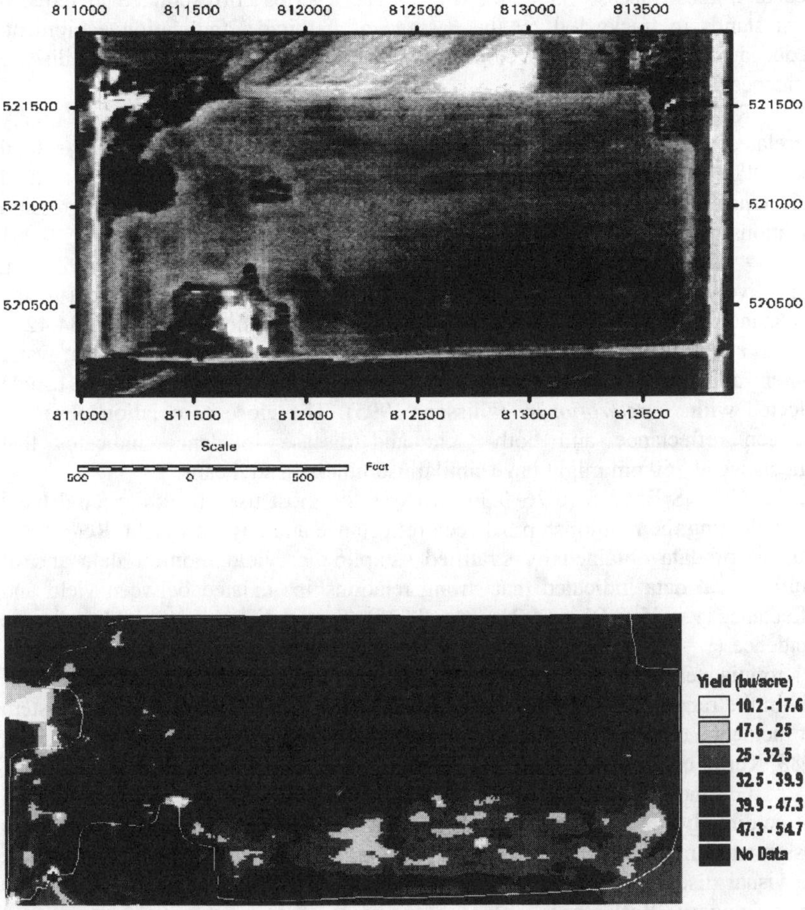

Fig. 1. Images of the study field representing color composite of ATLAS Bands 6, 4, 2 (RGB) (upper image) and yield response surface produced from point data obtained from combine mounted yield monitor (lower image). Images are oriented with upper edge to the North. Areas of white mold coincided with areas of low yield on south half of field (a) and paralleling east and west edges of the field where overlap occurred at headlands (b) in the area of highest white mold incidence reduced white mold incidence occurred in wheeltracks from herbicide application (c) and in eroded waterways (d).

host of *S. sclerotiorum*. This is infestation was most severe in the southern one-half of the field early in the growing season, however, it was controlled with herbicides.

Although Sclerotinia stem rot incidence was concentrated on the southern one-half of the field its occurrence was modified by management practices or by edaphic factors within this area and at several other locations in the field. Smaller diseased areas occurred in headlands where turning and overlapped planting had resulted in dense stands along the east and west edge of the field (Fig. 1). Strips of lower disease incidence that paralleled the south edge of the field ran through the area of highest disease incidence (Fig.1) These strips corresponded to reduced plant stands in tracks left by the passage of herbicide application equipment. Reduced plant stands that developed in waterways also resulted in lower disease incidence (Fig. 1).

Reflectance data obtained with the handheld radiometer was highly correlated with disease incidence at two wavelengths, 706 and 760 nm (r = 0.90 and 0.88 respectively) and the relationship between disease incidence (i) and reflectance (x) could be represented by highly significant linear regression equations for both wavelengths (i = -1.80339 + 0.478456x, r^2 = 0.81, P 0.001 and i = 82.86 – 186.12E-.02x, r^2 = 0.779, P 0.001 for 706 and 760 nm, respectively. There also was a strong relationship between final yield (y) and reflectance at 760nm measured with the handheld radiometer (y = 634.42 – 48.37x, r^2 = 0.81, P 0.001). Reflectance at 760 nm is associated with leaf water content and has also been reported to be a measure of plant health in canola infected with *S. sclerotiorum* (Nilsson, 1995). The close association observed between reflectance and both yield and disease incidence indicates that reflectance at 760 nm might have similar usefulness in soybeans.

ATLAS Band 6 (0.76–0.90 nm) was the most useful multispectral band for evaluating the relationship between reflectance and soybean yield. Regression analysis of data obtained by stratified sampling of yield monitor data and of multispectral data indicated that strong relationships existed between yield and reflectance (y =-2592.91 + 65.31x, r^2 = 0.838). The relationship of yield to disease incidence (y = 2633.18 – 13.48x, r^2 = 0.64), - 13.48 kg ha^{-1} per 1% increase in Sclerotinia stem rot incidence, at the 11 points of the ground truth assessment, lies within the range of -12.09 to -16.80 kg ha^{-1} per 1% increase in Sclerotinia stem rot incidence reported for susceptible soybean varieties (Chun et al., 1987; Grau, 1988; Kurle, unpublished data).

Ground truth data obtained by visual disease surveys and radiometry could not be directly linked to the ATLAS imagery because of shortcomings in survey design and limitations of the ATLAS sensor. The size of each point surveyed in the visual disease estimates was approximately $1/10^{th}$ that of each pixel in the aerial imagery. Reflectance values obtained by the aerial sensor represented averages of disease incidence that may have differed considerably from the much more detailed observations obtained with by visual observation and radiometry. In addition the distance between some ground survey points was less than the dimension of an individual pixel so that a single pixel may have overlapped two or more of the ground truth observations. This spatial overlap was further complicated by the possibility that slight errors existed in registration of different datasets.

The ATLAS sensor is characterized by relatively low spectral resolution and has a considerably wider bandwidth than the high spectral resolution bandwidths sensed with the handheld radiometer. Consequently, the much lower spectral resolution provided higher signal-to-noise ratio, while sacrificing the sensor's ability to discriminate fine spectral differences (Lillesand & Kiefer, 1994). The radiometer wavelengths that were most highly correlated with yield and disease incidence, 706 and 760 nm, yielded regression slopes of opposite sign (Fig. 2). The ATLAS band that most closely approximated these wavelengths; Band 5: 690 to 760 nm, overlapped both bands and as a result would have included reflectance values that were both increased and decreased by the same levels of disease incidence or yield.

The value of the results obtained in this study is reduced by the mismatch between the spatial resolution of the ground truth survey and aerial reflectance data and the inadequate spectral resolution of the aerial sensor when compared with that of the handheld radiometer. However, with appropriate modifications to equipment and survey methods multispectral images can be used to evaluate Sclerotinia stem rot incidence and its effect on crop yield. Equipment modifications would include increasing the spatial and spectral resolution of the aerial sensor. Narrower bandwidths of higher spectral resolution would be less likely to contain the conflicting spectral responses that we observed within the bandwidth of Band 5. Spatial resolution greater than the 3 m resolution of the ATLAS sensor would be necessary if there was considerable variability in disease severity over short distances. In this study, however, the 3-m resolution would have been adequate if the survey design had included disease incidence assessments made over larger areas, more observations in each disease incidence class, and had involved assessments from the entire field. An additional factor that is not addressed in this work is timeliness of the disease observations. The aerial observations were made when the soybean crop had progressed to the R7 stage of development. Leaf senescence occurring during this stage has limited the usefulness of multispectral data as a tool for disease assessment.

CONCLUSION

Spectral data obtained with a handheld radiometer was highly correlated with disease incidence at 706 and 760 nm and yield at 760 nm. Multispectral data obtained with an airborne sensor was correlated with yield. Limitations in data obtained during the ground survey prevented use of the ground truth data to calibrate the multispectral data for disease assessment. The high correlation of yield with reflectance data in ATLAS band 6 and yield with disease incidence obtained in the ground survey indicate that classification reflected disease severity and indicate that yield was strongly affected by Sclerotinia stem rot. With proper modification of ground survey methods and possibly the aerial sensor system, multispectral data in combination would yield monitor data that could be use for rapid assessment and mapping of Sclerotinia stem rot in areas as large as this production soybean field.

Based on this preliminary research, Sclerotinia stem rot occurrence explained a high percentage of yield variation occurring on this field. If Sclerotinia stem rot infection occurs repeatedly and predictably in areas determined by topography, edaphic factors, management practices and inoculum concentration, then those areas could be mapped using precision agricultural techniques. Growers could then use variable rate planting or pesticide applications to control Sclerotinia stem rot by reducing planting rates or patch spraying of fungicides or Sclerotinia stem rot suppressing herbicides in areas with a history of white mold.

Regression of disease incidence on reflectance at 706 nm

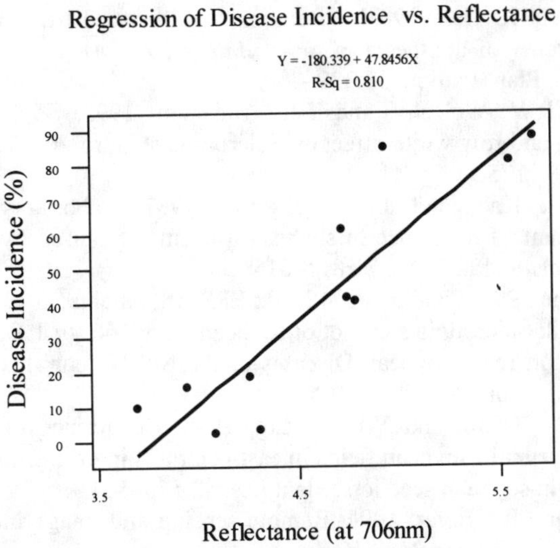

Regression of disease incidence on reflectance at 760 nm:

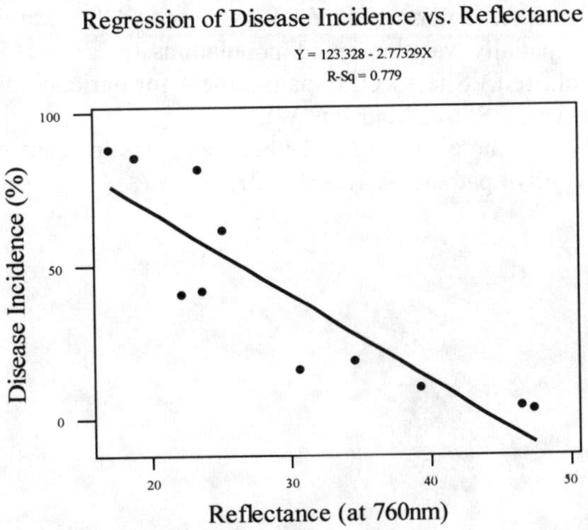

Fig. 2. Regression of disease incidence on reflectance obtained with handheld radiometer at 706 and 760 nm. Graphs show counteracting relationship that exists between two bands of the handheld radiometer that are contained within the wider bandwidth of Band 5 of the ATLAS sensor.

REFERENCES

Boland, G.J., and R. Hall. 1988a. Epidemiology of Sclerotinia stem rot of soybean in Ontario. Phytopathology. 78:1241–1245.

Boland, G.J., and R. Hall. 1988b. Relationships between the spatial pattern and number of apothecia of *Sclerotinia sclerotiorum* and stem rot of soybean. Plant Pathol.. 37:329–336.

Buzzell, R.I., T.W. Welcaky, and T.R. Anderson. 1993. Soybean cultivar reaction and row width effect on Sclerotinia stem rot. Can. J. Plant Sci. 73:1169–1175.

Chun, D., L.B. Kao, and J.L. Lockwood. 1987. Laboratory and field assessment of resistance in soybean to stem rot caused by *Sclerotinia sclerotiorum*. Plant Dis. 71:811–815.

ERDAS Imagine 1997. Field guide. 4th ed., ERDAS, Atlanta, GA.

Grau, C.R. 1988. Sclerotinia stem rot of soybean. p. 56–66. *In* T.D. Wyllie and D.H. Scott (ed.) Soybean Diseases of the North Central Region. APS Press, St. Paul, MN.

Hartman, G.L. L. Kull, and Y.H. Huang, 1998. Occurrence of *Sclerotinia sclerotiorum* in soybean fields in east-central Illinois and eumeration of inocula in soybean seed lots. Plant Dis. 82:560–564.

Lillesand, T., and R. Kiefer. 1994. Remote sensing and image interpretation. John Wiley & Sons. New York.

Moran, M.S., Y. Inoue, and E.M. Barnes. 1997. Opportunities and limitations for image-based remote sensing in precision crop management. Remote Sens. Environ. 61:319–346.

Mortenson, D.A., G.A. Johnson, D.Y. Wyse, and A.R. Martin. 1995. Managing spatially variable weed populations. p. 397–415. *In* P.C. Robert et al., (ed.) Site specific management for agricultural systems. ASA, CSSA, and SSSA, Madison, WI.

Nilsson, H.E. 1995. Remote sensing and image analysis in plant pathology. Annu. Rev. Phytopathology. 15:489–527.

Directed Sampling Using Topography as a Logical Basis

David W. Franzen
North Dakota State University
Fargo, North Dakota

Ardell D. Halvorson
USDA-ARS
Fort Collins, Colorado

Joseph Krupinsky
USDA-ARS
Mandan, North Dakota

Vern L. Hofman
North Dakota State University
Fargo, North Dakota

Larry J. Cihacek
North Dakota State University
Fargo, North Dakota

ABSTRACT

Some nutrient levels are often related to topography in the Northern Great Plains, particularly N. The use of topography to direct sampling is attractive because it may require fewer samples to reveal fertility patterns compared with dense grid-sampling methods. However, the use of topography-based sampling requires an additional measure of agronomic knowledge from the sampler. Determining where to establish soil-sampling zones may be difficult using only a topography map. Yield mapping, satellite imagery, and soil conductivity may be useful in determining sampling zone boundaries. When similar patterns to topography are produced using yield mapping, satellite imagery or soil conductivity sensors, soil sampling zone boundaries may be more confidently established or refined.

INTRODUCTION

Grid soil sampling has been used to reveal soil fertility patterns because it is a sampling system that reduces sampler bias (Linsley & Bauer, 1929; Peck & Melsted, 1973; Wollenhaupt et al., 1994; Franzen & Peck, 1995), however, there may be logical reasons that explain nutrient variability in some fields. One of the more compelling reasons for similar nutrient patterns between years is the effect of topography, or landscape, on the accumulation or depletion of plant nutrients. Landscape features have been described and categorized by Ruhe, (1960) and Jones et al. (1989). Topography influences the movement and accumulation of

Copyright © 1999 ASA-CSSA-SSSA, 677 South Segoe Road, Madison, WI 53711, USA. *Proceedings of the Fourth International Conference on Precision Agriculture.*

soil water (Fiez et al., 1994; Miller et al., 1988; Sinai et al., 1981; Stone et al., 1985). Several studies have shown a relationship between soil physical and fertility factors and topography (Brubaker et al., 1993; Fiez et al., 1994; Jones et al., 1989). Additional studies have described a relationship of soil nitrogen content with topography (Bruulsema et al., 1996; Cassel et al., 1996). Residual soil N may be influenced by topography due to the possible transformations it is subject to and its movement within the soil. Topography and its associated soil water content influences mineralization (Stevenson et al., 1995), nitrification, and denitrification (Stevenson, 1982). Differences in crop removal of nutrients due to topography and crop yield variability may also influence residual nutrient levels.

Topography has been shown to be related to fertility patterns, especially N, in North Dakota (Franzen et al., 1998). Topography-based sampling, although requiring fewer samples than a densely grid-sampled field, also requires careful consideration of where to establish management zone boundary lines. The dense grid approach, which relies on a high sample density backed by statistical analysis, requires less interpretation and a lower level of agronomic expertise than the management zone approach. Using a dense grid, the sampler is relatively confident that the boundaries initially laid out in the nutrient availability map are correct. Directed sampling, on the other hand, should be considered an iterative process, where additional layers of information are considered to refine and readjust boundary lines, increasing accuracy over time. The directed sampling approach requires a higher level of agronomic education than the grid approach because of the ability needed to find relationships between topography and soil nutrients and because of the interpretation of field observations taken over time. Determining where to draw management zone boundary lines is difficult using only an elevation map. Several methods for defining fertility management zones, including soil conductivity, satellite imagery, and yield mapping may be useful in helping to produce a more accurate representation of fertility levels.

MATERIALS AND METHODS

Two 16.2 ha sites were sampled in a 33 m grid. The Valley City, North Dakota site was sampled each fall from 1994 to 1997. The field was seeded to spring wheat in 1994, sunflower in 1995, spring wheat in 1996 and barley in 1997. A uniform rate of N and P was applied each year based on a composite soil test, except in 1997 when variable-rate treatments of N as anhydrous ammonia were applied. The site near Colfax, ND, was sampled each fall from 1995 to 1997. The Colfax site was in corn in 1995, spring wheat in 1996, and corn in 1997. A variable-rate N and S application was made in 1996 and a variable-rate N application was made in 1997.

A third site is located near Mandan, ND, at the USDA-ARS Northern Great Plains Research Laboratory, and was sampled from 1995–1997. The 31.6 ha site is divided into three fields; west, center, and east. The east field was sampled in a 33 foot grid. The west and center fields were sampled in a 45 m grid. The Mandan site is in a winter wheat, spring wheat, and sunflower rotation.

At each site, five to eight soil cores were taken at each sample location at two depths, 0–15 cm and 15–60 cm. Nitrate-N, P, S, and Cl were analyzed on the

0- to 15-cm depth. Nitrate-N, S, and Cl also were analyzed on the 15- to 60-cm depth. Nitrate-N, S, and Cl were reported as the total of two depths (0- to 60-cm). Elevations were measured using a laser-surveying device in a 33 m grid.

The SPOT satellite image of the Colfax site was taken late in the 1995 corn-growing season prior to the first soil sampling. Pixel size is approximately 10 m square. Electrical conductivity mapping at Valley City was conducted using a Veris Corporation (Salina, KS) soil conductivity instrument and an EM-38 (Geonics, Ltd., Missisauga, ON). The Veris recorded readings 1 second apart, each representing an area of about 42 m^2. EM-38 measurements were made in a 15 m grid.

The yield monitor used at Mandan was an Ag-Leader. The yield monitors used at Colfax and Valley City were John Deere Greenstar systems. All combines recorded yield at 1-s intervals. Some post-processing of yield information was required before the final map was produced.

Mapping was conducted using Surfer for Windows (Golden Software, Golden CO). Inverse distance squared was the interpolation method employed, using eight nearest neighbors and a simple search. Correlation was conducted using SYSTAT for Windows, Evanston, IL.

RESULTS AND DISCUSSION

Relationship of Nutrient Levels with Topography at Valley City

Nitrate-N levels from 1995, superimposed on the topography map in Fig. 1, illustrate how the higher N levels were found in depressions and convergent topography at this location. Correlation of N, P, and Cl between the 33 m grid and topography samplings were similar in 1994–1997 except for N and Cl in 1997 following the variable-rate N treatments (Table 1).

Table 1. Comparison of topography-based and 66 m grid-based N, P, S, and Cl with a 33 m grid at Valley City, ND, 1994–1997.

	Correlation (r)							
	Topography				66 m grid			
Year	N	P	S	Cl	N	P	S	Cl
1994	0.29*	0.27*	0.25*	0.37*	0.18	0.29*	0.37*	0.24*
1995	0.38*	0.34*	0.02	0.33*	0.50*	0.75*	0.72*	0.04
1996	0.49*	0.53*	0.14	0.53*	0.34*	0.51*	0.33*	0.49*
1997	0.16	0.39*	0.11	0.01	0.22*	0.56*	0.60*	0.11

* denotes correlation is significant at $P < 0.05$.

Fig. 1. Overlay of NO_3-N, 1995 with topography, Valley

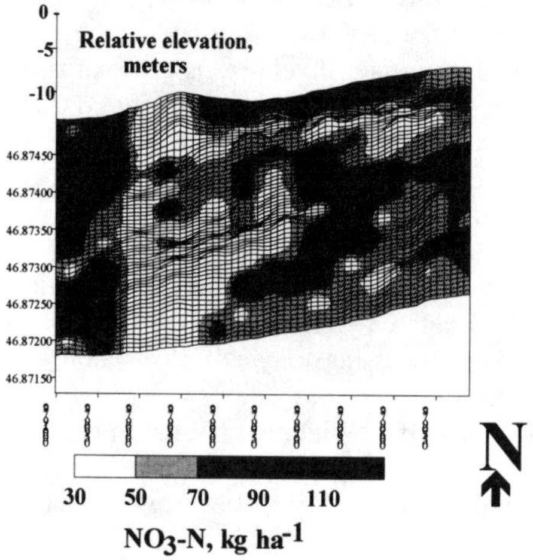

Fig. 2. Valley City hard red spring wheat yields,

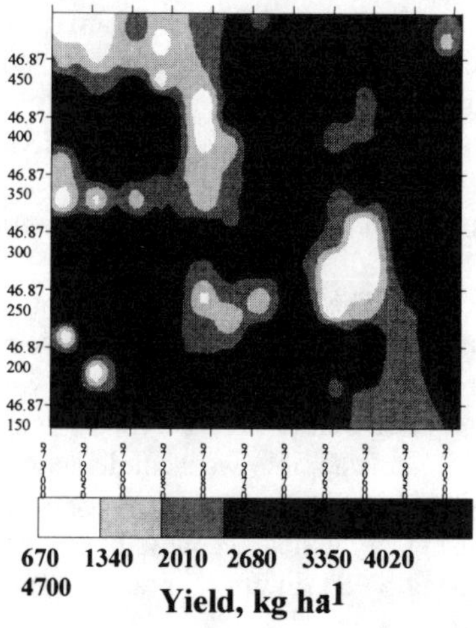

The correlation of S was higher for the 33 m grid sampling than the topography-based sampling.

Alternative Methods of Management Zone Definition at Valley City

In 1996, the yield map (Fig. 2) showed a low yielding area in the northwest corner of the field. By separating out this area as an additional sampling zone, 1996 correlation of topography N with the 33 m grid sampling was improved compared to 1995 (Table 1). Portions of the yield map mistakenly show zero yields, because in areas around the low yielding northwest corner the grain lodged and the grower left the swath at the site to pick up later, but did not turn the yield monitor off when moving through the area. It is very important when something unusual occurs in the field to note the variation so that the yield map is properly interpreted. Because of the many factors producing yield differences, using a yield map alone as the primary basis for directed sampling is not suggested. Yield was not related to N, P, or Cl, however the correlation with S was significant (Table 2).

Table 2. Correlation of Valley City yield with 33 m grid N, P, S, and Cl.

Comparison	---------- r--------	P value
Yield/ N	0.046	0.586
Yield/ P	0.001	0.986
Yield/ Cl	0.057	0.504
Yield/ S	0.175	0.038 *

*Significant at $P < 0.05$

In 1997, soil conductivity sensors were used to map salinity (Fig. 3).

Figure 3. Electrical conductivity patterns using a Veris instrument and an EM-38 instrument, Valley City, 1997.

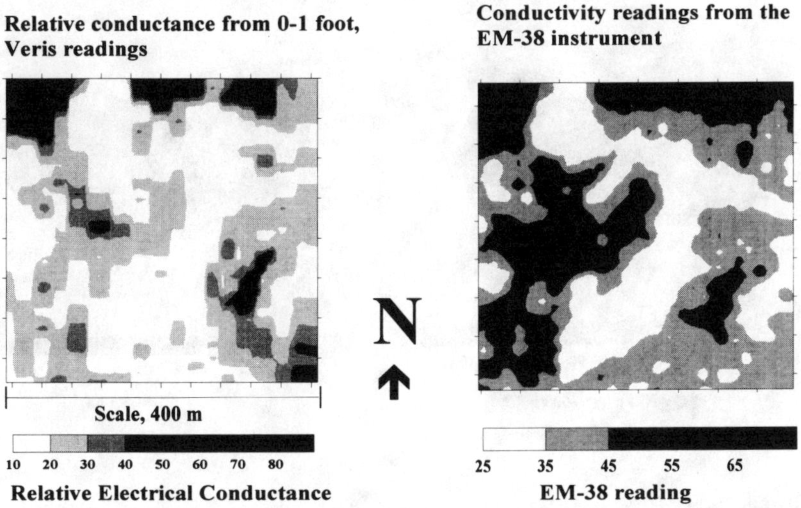

Patterns of conductivity are very similar to NO_3–N patterns, however, conductivity was not directly related to NO_3–N levels (Table 3). Correlation of conductivity with NO_3–N levels would result in overestimation of some NO_3–N levels. By using the patterns of conductivity to define sampling zones, however, conductivity-based zone sampling would be expected to give similar boundaries and correlation as topography sampling.

Table 3 Correlation of Veris and EM-38 conductivity with N, P, Cl, and S.

Comparison	---------- *r* ----------	*P* value
Veris / N	0.086	0.31
Veris / P	0.344	0.00 *
Veris / Cl	0.233	0.00 *
Veris / S	0.789	0.00 *
EM-38 / N	0.011	0.90
EM-38 / P	0.109	0.20
EM-38 / Cl	0.245	0.00 *
EM-38 / S	0.467	0.00 *

* Significant at $P < 0.05$

Relationship of Topography and Nutrient Levels at Colfax

Nitrate–N levels at Colfax from 1996 are shown in Fig. 4. Similar patterns of NO_3–N between years suggest that NO_3–N levels are related to landscape. Correlation of NO_3–N levels between 33 m and 66 m grid sampling and

Figure 4. Colfax NO3-N levels, 1996 over topography.

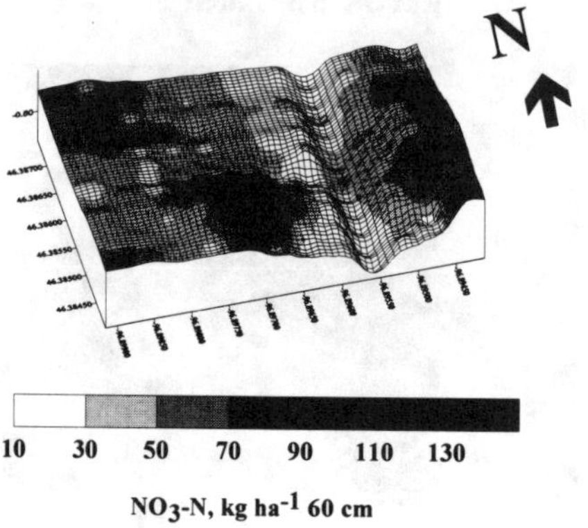

Figure 5. Colfax 1997 corn yield.

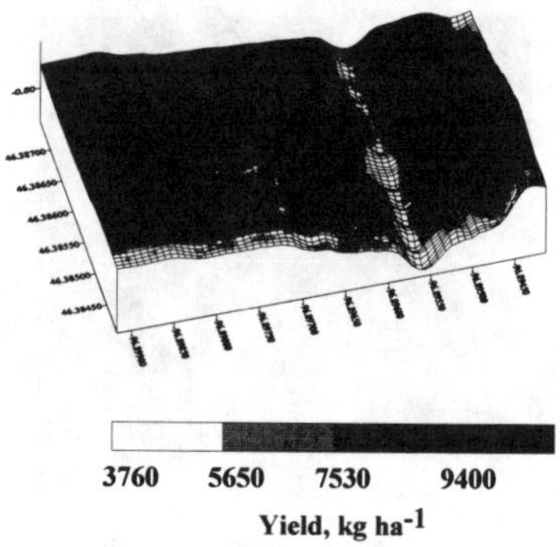

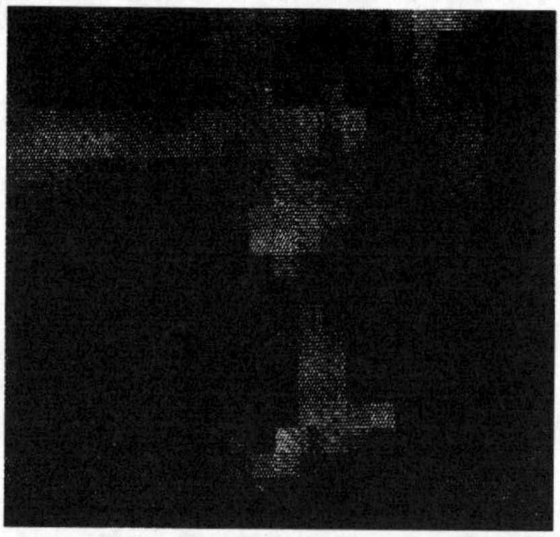

<---------------scale, 400 m --------------->

Fig. 6. SPOT satellite image, Colfax, corn, 1995.

topography sampling is shown in Table 4.

The 1997 yield map reveals a low yielding area in the center of the field (Fig. 5). This area also is consistently higher in P, S, and Cl and lower in N than the surrounding area. By using the yield map to define an important management zone, the topography relationship is supported.

The SPOT satellite image of the Colfax site in 1995 is shown in Fig. 6. Using the map to define management zones based on patterns in the image result in patterns similar to those defined by topography. In 1994, the poor corn growth areas in lighter toned areas were the result of excessive rainfall. The areas of better corn growth (darker tones) were at higher elevations. By identifying areas of varying plant health, management zones related to topography were defined in this relatively wet year.

Table 4. Comparison of topography-based and 66 m grid-based N, P, S, and Cl with a 33 m grid at Colfax, ND.

	Correlation (r)							
	Topography				66 m grid			
Year	N	P	S	Cl	N	P	S	Cl
1995	0.27*	0.21*	0.51*	0.09	0.62*	0.62*	0.68*	0.56*
1996	0.39*	0.53*	0.03	0.04	0.41*	0.25*	0.37*	0.24*
1997	0.26*	0.55*	0.19*	0.17	0.33*	0.54*	0.28*	0.16

* Significant at $P < 0.05$.

Figure 7. Mandan center field sunflower yield, 1995.

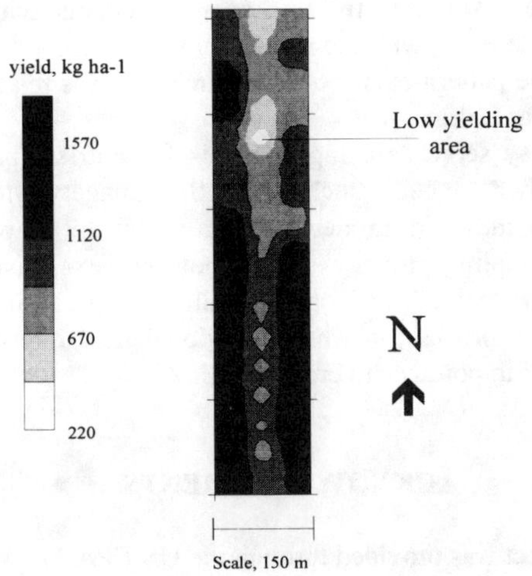

Relationship of Topography and Nutrient Levels at Mandan

Nitrate-N, P, S, and Cl levels were related to topography from 1995–1997 (Table 5). The yield mapping shows an area of relatively low yield in the center field (Fig. 7). This area was did not appear important when the topography map was produced, however, following the yield map, this area was separated out and improved correlation of topography and N levels from r = 0.70 to r = 0.83 in 1995.

Table 5. Comparison of topography-based and 66- to 90-m grid-based N and P with a 33-45 m grid at Mandan, ND .

	Correlation (r)							
	Topography				66 m grid			
Year	N	P	S	Cl	N	P	S	Cl
1995	0.83*	0.70*	0.13	0.22*	0.29*	0.58*	0.48*	0.56*
1996	0.29*	0.64*	0.36*	0.39*	0.20	0.72*	0.04	0.17
1997	0.83*	0.35*	0.63*	0.82*	0.81*	0.39*	0.59*	0.69*

* Significant at $P < 0.05$

SUMMARY

Soil nutrient levels were related to topography at three sites in North Dakota, except for Cl at Colfax. Yield mapping was related to topography and nutrient levels at Valley City, Colfax, and Mandan in some low yielding areas, however, unless there is further knowledge of the field to interpret maps, yield mapping should be used with caution to determine soil management zones. Soil conductivity measurements defined zones similar to topography, however, they were not related to N levels directly. Use of a soil conductivity map should be considered a pattern detector, with the reason for the pattern left up to the sampler to investigate why the pattern exists. Satellite imagery was related to topography and nutrient patterns at Colfax.

Landscape may serve as a logical basis for nutrient patterns especially NO_3-N, however, other methods, such as satellite imagery, yield data and soil conductivity sensors may aid in determining where to draw sampling zone boundaries. Zone sampling strategy should not rely exclusively on elevation mapping as its only method of zone definition. Zone sampling should be considered an iterative process, in which additional field information is used to further define areas of importance to crop growth.

ACKNOWLEDGMENTS

Funding for this project was provided through the US-EPA 319 Water Quality grant program, the Sugarbeet Research and Education Board of Minnesota and North Dakota, Agrium, the North Dakota Soil Conservation Districts of Stutsman Co., Cass Co., and Wild Rice, and the Potash & Phosphate Institute.

REFERENCES

Brubaker, S.C., A.J. Johnes, D.T. Lewis, and K. Frank. 1993. Soil properties associated with landscape positions. Soil Sci. Soc. Am. J. 57:235–239.

Bruulsema, T.W., G.L. Malzer, P.C. Robert, J.G. Davis, and P.J. Copeland. 1996. Spatial relationships of soil nitrogen with corn yield response to applied nitrogen. p. 505–512. *In* Proc. of the 3rd Int. Conf. Precision Agriculture, Minneapolis, MN. 23–26 June,1996. ASA, CSSA, and SSSA, Madison, WI.

Cassel, D.K., E.J. Kamprath, and F.W. Simmons. 1996. Nitrogen–sulfur relationships in corn as affected by landscape attributes and tillage. Agron. J. 88:133–140.

Fiez, T.E., B.C. Miller, and W.L. Pan. 1994. Winter wheat yield and grain protein across varied landscape positions. Agron. J. 86:1026–1032.

Franzen, D.W., and T.R. Peck. 1995. Field soil sampling density for variable rate fertilization. J. Prod. Agric. 8:568–574.

Franzen, D.W., L.J. Cihacek, V.L. Hofman, and L.J. Swenson. 1998. Topography-based sampling compared to grid sampling in the Northern Great Plains. J. Prod. Agric. In Press.

Jones, A.J., L.N. Mielke, C.A. Bartles, and C.A. Miller. 1989. Relationship of landscape position and properties to crop production. J. Soil Water Conserv. 44:328–332.

Linsley, C.M., and F.C. Bauer. 1929. Test your soil for acidity. Univ. of Illinois. College of Agriculture and Agriculture Exp. St. Circ. 346.

Miller, M.P., M.J. Singer, and D.R. Nielsen. 1988. Spatial variability of wheat yield and soil properties on complex hills. Soil Sci. Soc. Am. J. 52:1133–1141.

Peck, T.R., and S.W. Melsted. 1973. Field sampling for soil testing. p. 67–75. *In* Soil testing and plant analysis. 2nd ed.. SSSA, Madison, WI.

Penney, D.C., R.C. McKenzie, S.C. Nolan, and T.W. Goddard. 1996. Use of crop yield and soil landscape attribute maps for variable rate fertilization. pp. 126–140. *In* J. Havlin, (ed.) Proc. 1996 Great Plains Soil Fertility Conf. Denver, CO. 5–6 Mar., 1996. Kansas State Univ., Manhattan.

Ruhe, R.V. 1960. Elements of the soil landscape. p. 165–170. *In* Trans. of the 7th Int. Congress of Soil Sci.. Vol. 4. Int. Soc. of Soil Science. Madison, WI.

Sinai, G., D. Zaslavsky, and P. Golany. 1981. The effect of soil surface curvature on moisture and yield-Beer Sheba observations. Soil Sci. 132:367–375.

Stevenson, F.J. 1982. Origin and distribution of nitrogen in soil. p. 1–42. *In* F.J. Stevenson, (ed.) Nitrogen in Agricultural Soils. Agron. Monogr. 22. ASA, CSSA, and SSSA, Madison, WI.

Stevenson, F.C., J.D. McKnight, and C. van Kessel. 1995. Dinitrogen fixation in pea:controls at the landscape- and micro-scale. Soil Sci. Soc. Am. J. 59:1603–1611.

Stone, J.R., J.W. Gilliam, D.K. Cassel, R.B. Daniels, L.A. Nelson, and H.J. Kleiss. 1985. Effect of erosion and landscape position on the productivity of Piedmont soils. Soil Sci. Soc. Am. J. 49:987–991.

Wollenhaupt, N.C., R.P. Wolkowski, and M.K. Clayton. 1994. Mapping soil test phosphorus and potassium for variable-rate fertilizer application. J. Prod. Agric. 7:441–448.

A Digital Camera System for Weed Detection

T. Heisel
S. Christensen

Danish Institute of Agricultural Sciences
Research Centre Flakkebjerg
Slagelse, Denmark

ABSTRACT

Development of new methods for weed control is one of the biggest scientific and technological challenges in the future. Sensor steered sprayers and prototypes of robots steered by digital video cameras have been developed for non-chemical weed control. In 1997, a collaboration project between the Danish National Railways Agency, the Danish Institute of Agricultural Sciences and Hardi International was initiated. The project aimed at developing a digital camera system (WeedEye) and a spot sprayer to reduce herbicide usage to locations with weeds. The results of the first experiments show that WeedEye detects small weed seedlings at a speed of 45 km h^{-1}. Further, the results show that WeedEye provides a reasonable estimate of the percentage leaf area m^{-2} at this speed. The perspectives of using sensors and digital video cameras in agriculture are discussed.

BACKGROUND

One of the biggest challenges in agriculture is development of new tools and methods to control weed nonchemically or position specifically. Mechanical weed control is already an alternative to the chemical weed control in cereal crops, unless you have very aggressive weed species (Rasmussen & Rasmussen, 1995). In row crops band spraying and inter-row weeding could considerably reduce the consumption of chemicals. Advanced tools with automatic weed detection systems and precise chemical or nonchemical control measures on spots or patches of weeds are important for the goal of reducing or eliminating herbicide use in most crops.

During the last 10 yrs sensors and robots have taken over very tedious and loading industrial work. For several reasons this technique has not been implemented so rapidly in agriculture. First of all, the cost of chemical weed control has been low. Thus the engineering industry has had no demand for new technology. Another reason is that it is more difficult to recognize, e.g., plants than standardized objects in the industrial process. Plants vary in shape and color, and at the same time it is necessary to take into account the varying backgrounds and lighting conditions. During the last 3 to 4 yrs the situation has changed. Sensor steered sprayers and prototypes of non-chemical weed control tools

Copyright © 1999 ASA-CSSA-SSSA, 677 South Segoe Road, Madison, WI 53711, USA. *Proceedings of the Fourth International Conference on Precision Agriculture.*

steered by video cameras have appeared and are now being developed for commercial use.

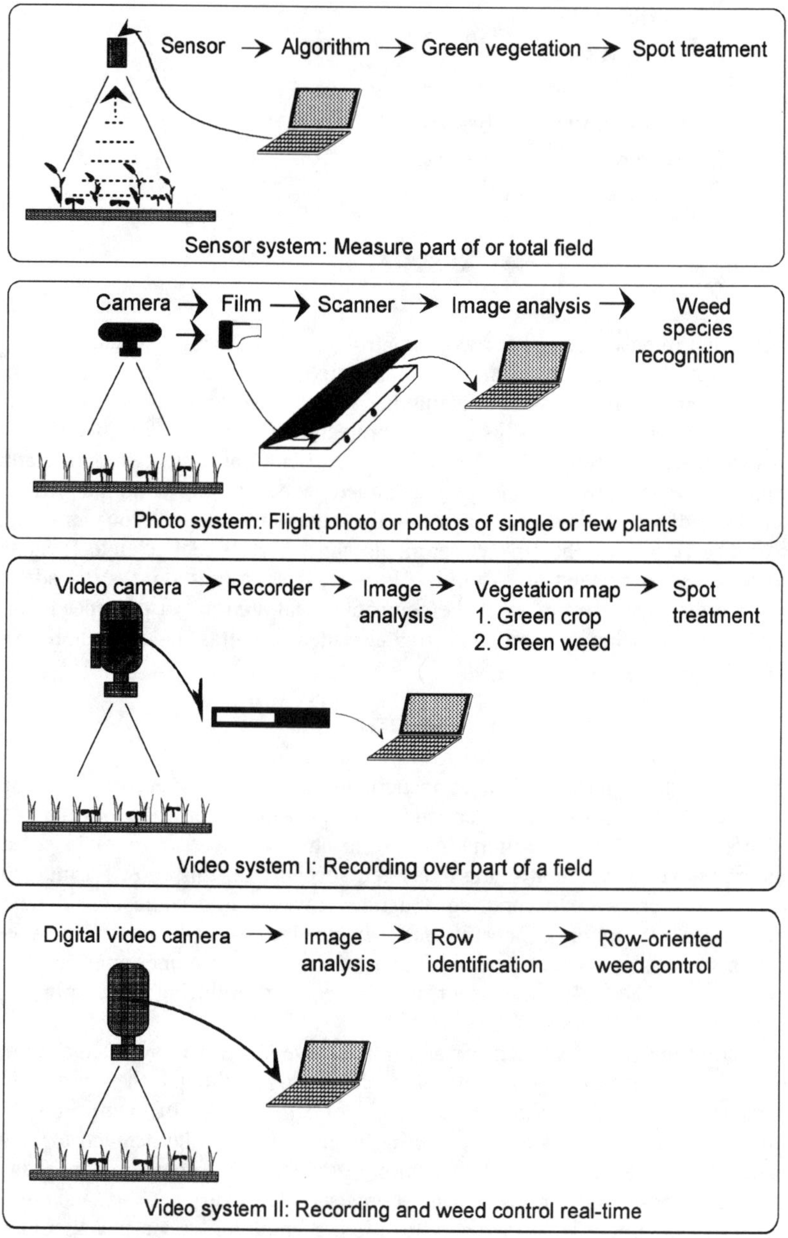

Fig. 1. Sensor and camera systems for weed detection and site specific weed control.

Several systems to measure crop and weed are available. The simplest system is a sensor system (Fig. 1) that measures the quantity of reflecting light from plants and soil by means of simple sensors. Green plants reflect visible and infrared light, whereas soil background and stone have different reflectance characteristics in these two wave bands. There is a difference between the visible and the infrared light of green plants and of soil and stone. The system has been used to investigate whether it is possible to recognise weed types on the light reflection of leaf (Borregaard, 1996). The results show that when plants are green it is impossible to measure the difference between species. However, there is a big difference between the colour of weed species at flowering. The sensor system is fitted to measure the quantity of green vegetation and can e.g. be used to screen weeds in stubble fields (Felton et al., 1991). The sensor system is simple and relatively cheep, thus it is the only system that for the time being is used on sprayers.

A photo system is another possibility where plants are photographed with a simple camera (Fig. 1). Each picture is digitized by means of a scanner, to make it possible to analyze the details of the photo in a computer. Several scientists have tried the photo system to identify weed types at early growth stages (Petry & Kühbauch, 1989; Woebbecke *et al.*, 1995; Andreasen *et al.*, 1997). The trials have been successful with separate weed plants; however, under natural conditions, where weed plants often overlap and vary in shape and colour, the results were not satisfactory. The photo system is useable to map out weed problems that can be seen from flight and satellite photos (Brown et al., 1994). This approach is only possible when the weed is dominating or when the colour of the weeds distinctly differs from the colour of the crop. Christensen et al. (1998) showed that weed leaf area measured with an image analysis technique could be used to map and stratify weed sampling.

At least two video systems are available (Fig. 1). Video system I uses videotape, preferably S-VHS. The pictures are digitized and analyzed on the same principles as the photo system. So far, video photos do not have the same quality as a common photo and can therefore only be used to separate crop and, e.g., dicotyledon weed (Zhang & Chaisattapagon, 1995; Benlloch *et al.,* 1996). In principle Video system II is the same as video system I. The difference is that the pictures are transferred directly to the computer and can be used to steer, e.g., a tool or a robot (Guyer et al., 1986).

This article describes a camera system that can function both as video system I and II. The camera system has been developed to detect weeds on the Danish National Railways Agency's track areas at a speed of 45 km h^{-1}. The Danish National Railways Agency aims to decrease the consumption of pesticides considerably, partly by reducing the consumption of pesticides and partly by limiting the control of weeds to the positions where the need for control is real. Therefore, a collaboration project between the Danish National Railways Agency, the Danish Institute of Agricultural Sciences and Hardi International was initiated in 1997. The purpose was to develop a digital camera system (WeedEye) and a spot sprayer that limits chemical control to the track area (Fig. 2), where green vegetation appears. The ultimate goal of the Danish National Railways Agency is to develop nonchemical methods and only use chemicals where nonchemical methods are impossible to apply or are insufficient.

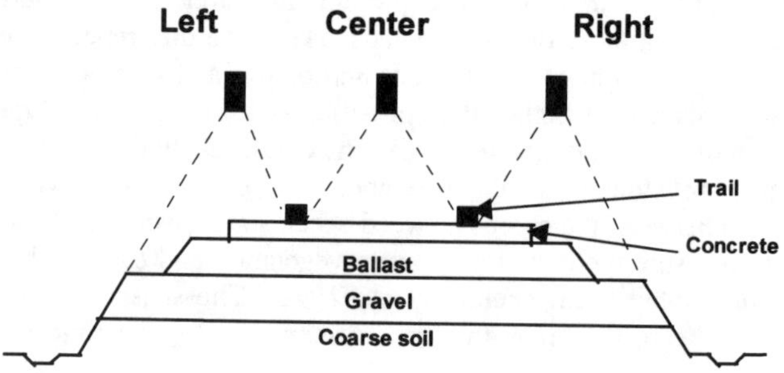

Fig. 2. Three detection areas–cameras on The Danish National Railways Agency trail.

MATERIALS AND METHODS

The project was established in 1997 by development of camera and computer techniques to automatically detect weeds. The demand was a system to function also at night at a speed of 45 km h^{-1}. The idea was to use that green plants have a characteristic spectral reflectance in the blue, red and infrared wavebunds that is different from the broken stones, rails, gravel etc. To detect weeds in the three control areas shown in Fig. 2, a steel box with three identical camera systems was constructed. The steel box has three windows, one at each side and one in the driving direction. Each camera system contains of three black-white digital cameras locked to each other. They function as a colour camera (Fig. 3).

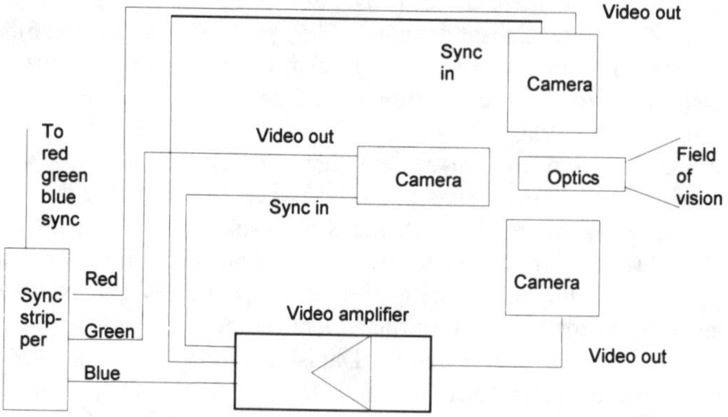

Fig. 3. Diagram of WeedEye that consists of three black and white digital video cameras.

The camera system (WeedEye) has been constructed to make it possible for each camera to see through a specific colour filter to be penetrated only by red, blue or infrared light, respectively. By means of a video amplifier and a syncstripper the cameras are synchronized to give the same signal as a RGB (red, green, blue) colour camera. Furthermore, the signal can be treated as a common RGB signal (Fig. 3) although the green signal is actually infrared. Each grabbed photo is treated by means of a common image analysis routine through linear combination of the three channels (RGB) and thresholding.

Each camera system is connected to a framegrabber with a dedicated 150 MHz Pentium PC with flashdisc (Fig. 4). The computer calculates the leaf area of a defined visual field. The three camera computers transfer the information of the leaf area to the front-end computer, which combines data with specifications of position from a GPS. For each desired distance the average leaf area is registered.

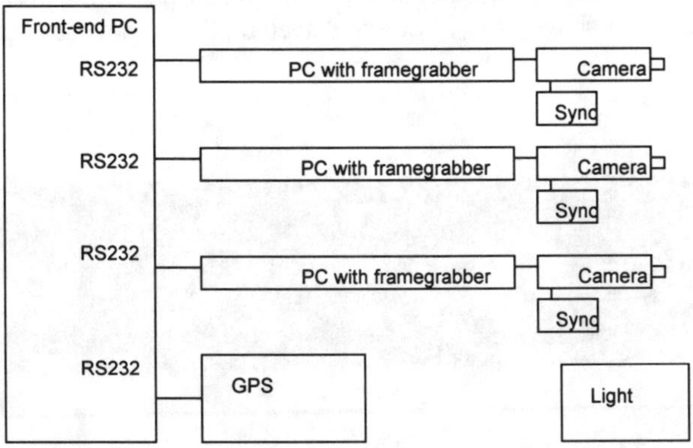

Fig. 4. Diagram of hardware linking the three camera systems.

To function independently of light conditions a halogen bulb was installed for each camera system. The halogen bulb sends out a continuous spectrum of light in the visible and infrared spectrum. Furthermore, auto iris lenses are used for each camera to reduce problems with varying lighting conditions.

WeedEye's detection of weeds was tested with transplanted wheat and charlock at 2 to 4 leaf stage. 189 plants were transplanted in nine spot areas to investigate how efficiently WeedEye detects different quantities of weed (Fig. 5). Varying numbers of wheat and charlock according to Table 1 were transplanted in spots at one side of the bed. The distance between the spots was approximately 20 m. The leaf area coverage in percent is approximately proportional with the number of plants at the spot. The leaf area of charlock and wheat was measured in the laboratory.

RESULTS

Leaf area of wheat and charlock was measured from 15 to 20 cm^2 per plant. Leaf area and the calculated degree of coverage of each spot are listed in Table 1.

Table 1. Weed plants and measured leaf area of nine weed patches.

	Spot								
No. plants	*1*	*2*	*3*	*4*	*5*	*6*	*7*	*8*	*9*
Charlock	**4**	**10**		**4**	**10**	**15**	**8**	**10**	**30**
Wheat			**15**	**4**	**10**	**15**	**4**	**20**	**30**
Leaf area (cm^2)	111	277	293	189	472	708	300	667	1416

In the lower part of Fig. 5 results from WeedEye are shown as a function of the distance travelled. Notice that there are nine peaks with almost the same distance between the peaks, which correspond to the nine spots with a distance of 20 m above. This result shows that WeedEye detected all the spots at a speed of 45 km h^{-1}.

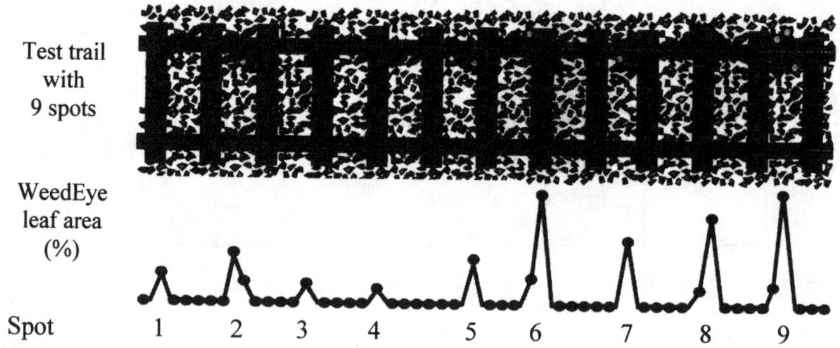

Fig. 5. Plan of the detection trials with WeedEye on a 200m railway track. The 9 weed patches (from left to right) are described in Table 1. Below the result of test run with WeedEye shown in a two-dimensional diagram.

Figure 6 shows calculated versus WeedEye measured percent leaf area. A correlation coefficient of 0.65 is found. WeedEye underestimates the leaf area compared to the calculated figures based on manual leaf area measurements. This is due to the fact that the calculations in Table 1 do not take into account the upright positioning of wheat leaves. However, the test shows that WeedEye provides an acceptable estimate of percentage leaf area per $meter^2$ at a speed of 45 km h^{-1}.

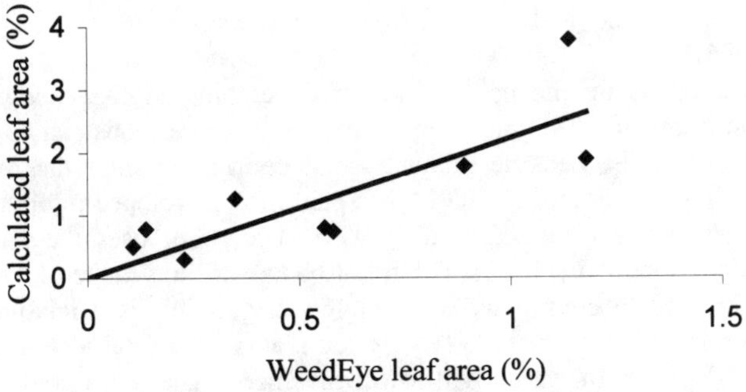

Fig. 6. The relationship between WeedEye measured and calculated leaf area in percentage.

DISCUSSION

Today three types of sensor steered sprayers are available: Detect Spray, WeedSeeker, and Bauman Scan-Ray. All three types use the simple sensor system, which can only screen for green vegetation. Detect Spray and WeedSeeker are primarily used for total control of vegetation in stubble fields or weed between rows in soybeans and maize. At each nozzle there is a sensor that opens the nozzle when passing green vegetation. Bauman Scan-Ray sees vegetation of a certain height. Bauman Scan-Ray is therefore primarily used to control weed species that are much higher than the crop. The simple sensors need exact adjustment of height, calibration for different types of soil or condition of humidity and cannot be used for detecting weed in crops. The German company BAYER developed a sensor-technique-system connecting the detection of plants and the nozzle function of the spraying equipment. According to Keilholz et al. (1998) it is possible to apply herbicide precisely at night with a train speed of 50 km h^{-1}.

The use of digital cameras and image analysis makes it possible to develop more intelligent weed detection algorithms. At the Danish National Railways Agency's track areas the simple sensors would immediately be able to divide and work up each pixel (element) of the photo. The presented digital camera system makes it possible to develop algorithms that continuously adjust image analysis for dissimilarities of the colors of broken stones etc. The system makes it possible to measure area and other geometrical items. Furthermore, filters and adjusted linear combinations of RGB makes it possible to discriminate weed species that have a characteristic reflecting spectre.

In agriculture WeedEye can be developed to monitor crop and weeds. WeedEye can be used as the simple sensors measuring vegetation indices by the infrared and the red reflection in a defined area. Both in Denmark and abroad several trials have shown that the relation between infrared and red reflection provides precise information about the physiologic status of a crop (Christensen & Goudriaan, 1993; Christensen et al., 1997). At early crop stages WeedEye could be used to measure the status of the crop and at the same time detect and evaluate the degree of weed leaf area between the rows by image analysis (Benlloch et al., 1996).

WeedEye also opens up the possibilities of developing high technology implements for nonchemical weed control measures. High value crops can cover the cost for advanced tools, because manual weed control is often the only possibility. CEMAGRAF in France and IVIA in Spain have developed a robot to control weeds in artichokes (Stafford et al., 1997). The robot uses the latest technology and video system II (Fig. 1). The robot has a working range of 1.6m and has been constructed to control weed by high voltage. This is a relatively slow method, which limits the control to three weed plants per m^2, at a speed of 0.8 km h^{-1}. Higher densities of weeds will lower the speed. The robot could be equipped with other controlling tools, i.e., cutting or defoliating, which may improve speed.

In row crops the consumption of herbicides can be reduced considerably by band spraying and inter-row weeding. Silsoe Research Institute in England has developed an autonomic machine, which is able to work in a cabbage field (Stafford et al., 1997). The cabbage plants and the distance between the rows must be unmistakable. The machine turns automatically by the end of a row and it is equipped with a video system II (Fig. 1).

Perspectives for use of high technology weed control equipment depend heavily on prices (herbicides, taxes, computer, and control technology) and on environmental politics. The political trend in the world in general and in Denmark in particular is pointing towards environmentally friendly solutions. Basis for research is obvious since the advancing information technology field is constantly getting cheaper.

ACKNOWLEDGMENTS

The Danish National Railways Agency funded the project.

REFERENCES

Andreasen C., M. Rudemo, and S. Sevestre 1997. Assesment of weed density at an early stage by use of image processing. Weed Res. 37(1):5–14.

Benlloch J.V., A. Sánchez, S. Christensen, and A.M. Walter 1996. Weed mapping in cereal crops using image analysis tecniques. AgEng Madrid 1996. Paper 96G-0.47.

Borregaard T. 1996. Application of imaging spectroscopy and multivariate methods in crop-weed discrimination. Ph.D. diss., Dep. of Agricultural Sci., Tåstrup.

Brown R.B., J.P.G.A. Steckler, and G.W. Anderson 1994. Remote sensing for identification of weeds in no-till corn. Trans. ASAE. 37(1):297–302.

Christensen S., and J. Goudriaan 1993. Deriving light interception and biomass from spectral reflectance ratio. Rem. Sens. Environ., 43:87–95.

Christensen S., T. Heisel, B.M.J. Secher, A. Jensen, and V. Haahr 1997. Spatial variation of pesticide doses adjusted to varying canopy density in cereals. 1st European Conf. on Precision Agriculture, Warwick. 8–10 Sept. 1997.

Christensen S., T. Heisel, and J.V. Benlloch 1998. Patch spraying and rational weed mapping in cereals. Proc. of the 4th Int. Conf. on Precision Agriculture, Minneapolis 19–22 July 1998.

FeltonW.L, A.F. Doss, P.G. Nash, and K.R. McCloy 1991 A microprocessor controlled technology to selectively spot spray weeds. p. 427–432. *In* Proc. ASEA Symp. 1991: Automated agriculture for the 21st century.

Guyer D.E., G.E. Miles, M.M. Schreiber, O.R. Mitchell, and V.C. Vanderbilt 1986. Machine vision and image processing for plant identification. Trans. ASAE 29(6):1500–1507.

Keilholz G., W. Benz, M. Kleinert, and H.M. Schloms 1998. Einsatz der Sensortechnik zur Optimierung der chemischen Vegetationskontrolle auf Gleisen der Deutschen Bahn AG. Z. PflKrankh. PflSchutz, Sonderh. XVI, 243–247.

Petry W., and W. Kühbauch 1989. Automatisierte Unterscheidung von Unkrautarten nach Formparametern mit Hilfe der quantitativen Bildanalyse. J. Agron. and Crop Sci., 163:345–351.

Rasmussen J., and K. Rasmussen 1995. A strategy for mechanical weed control in spring barley. p. 557–564. *In* Proc. from the 9th EWRS Symp., Budapest. 1995.

Stafford, J.V., J.V. Benlloch, E. Monto, S. Christensen, and J.M. Roger 1997. Reducing or eliminating agro-chemical inputs in efficient production of high quality produce with conventional, sustainable and organic farming systems. Final Project Report for The European Commission (Contact No. AIR3-CT93-1299).

Woebbecke D.M., G.E. Meyer, K. Von Bergen, and D. Mortensen (1995) Shape features for identifying young weeds using image analysis. Trans. ASAE 38(1):271–281.

Zhang N., and C. Chaisattapagon. 1995. Effective criteria for weed identification in wheat fields using machine vision. Trans. ASAE 38(3):965–974.

Weed Detection Based on Structural Information Using an Imaging Spectrograph

P. Pollet
F. Feyaerts
P. Wambacq
L. Van Gool

Department ESAT-PSI
Katholieke Universiteit Leuven
Belgium

ABSTRACT

This paper presents a system for the detection of weed amongst crop based on the extraction of structural field information. Data is gathered online with a sensor built upon an imaging spectrograph optimized for this purpose. The optical sensor splits the light from a line on the ground parallel with the spray boom in its spectral components that are projected on a camera. One of the main advantages of this sensor type is the high spatial resolution (up to a few mm). The weed detection algorithm uses the reflectance differences between the red and the near-infrared bands to make the distinction between plants and soil. Detection of weed amongst crop is based on the detection of weed patches and the extraction of the crop rows. A mathematical model for the herbicide reduction and the hit rate is presented that shows the relations between the different system parameters such as the spray resolution and the performance of the classifier.

INTRODUCTION

Today, environmental pressure is increasingly changing the farming practice towards a more ecologically acceptable situation. One of the hot topics is the demand to reduce the herbicide use. Therefore a growing demand has arisen for new application techniques that reduce the herbicide use and still keep the yield losses low. One of the most promising techniques is the place specific application of herbicides. Place specific herbicide application however is only possible with reliable and affordable plant and weed detection methods. Only a few plant sensors such as Patchen and Detectspray are now on the market (Felton, 1995; Beck et al., 1995). These systems are both based on the spectral differences between plant and soils.

Weed detection is much more difficult than plant detection and is still under investigation. Lots of research has been done in the area of digital image recognition (Gerhards et al., 1993; Guyer et al., 1986). These systems are often too slow or not good enough to be implemented in an online application. Other groups have investigated the spectral properties of the plants and concluded that it should be possible to make a distinction between crops and weeds based on small differences in the spectral signature (Hahn & Muir, 1993).

Copyright © 1999 ASA-CSSA-SSSA, 677 South Segoe Road, Madison, WI 53711, USA. *Proceedings of the Fourth International Conference on Precision Agriculture.*

This paper presents a sensor that can collect the spectral signatures of the plants online on the field. The output of this sensor can be used to detect weeds based on their spectral signature (Feyaerts et al., 1999). Here, an algorithm for detecting weeds in crop rows using this sensor is described. Finally, a mathematical model is presented to calculate the herbicide reduction and the hit rate based on the system parameters.

SENSOR PRINCIPLE

The weed sensor is based upon an imaging spectrograph (Herrala et al., 1994; Battey & Slater, 1993). An imaging spectrograph splits the light from a line in its spectral components and projects it on a camera. The principle of this sensor is shown in Fig. 1. The optical sensor consists of three parts: an objective lens, an imaging spectrograph, and a camera. The objective lens projects the reflected light from the ground on a slit in the spectrograph. With an objective lens of 3.5 mm it is possible to analyze a line of 3 m from a height of 1 m above the ground. The slit lies in the focal plane of the first lens, in our case a doublet. The doublet with a focal length of 75 mm collimates the light on a transmission grating with 75 grooves mm^{-1}. This transmission grating splits the light in its spectral components according to:

$$md = \lambda\,(\sin\theta - \sin\varphi)$$

With d: the distance between the grooves, m: the diffraction order, λ: the wavelength, θ: the angle between the normal of the transmission grating and the diffracted light, and φ: the angle between the normal and the incident light. The second doublet projects the light on the camera. The camera surface lies in the focal plane of the second doublet. The end optics and the camera are tilted to minimize the angle between the diffracted rays and the optical axis of the doublet in order to minimize the aberrations. The spectral resolution is mainly determined by the slit width and the number of grooves. The spatial resolution is determined by the quality of the front and end optics.

The spectral range of the sensor (400–1000 nm) coincides with the spectral sensitivity of a standard CCD/CMOS-camera. A larger spectral range is only possible with a more expensive type of camera. Some typical parameters for our sensor are given in Table 1.

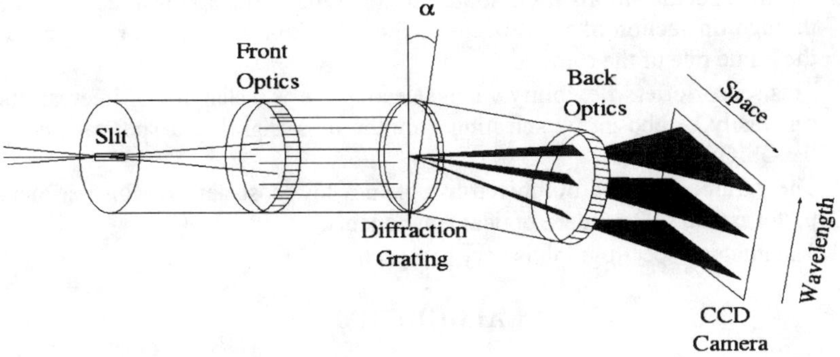

Fig. 1. Principle of the imaging spectrograph.

Table 1. Spectrograph parameters.

f-number	3
Available spectral range	400–1000 nm
Free spectral range	1 octave from the available range
Spectral resolution	35 nm
Slit width	200 μm
Slit length	8 mm
Size spectrograph	135∗7060∗ cm
Spatial resolution	depends on height and objective lens

One of the axes of the camera acts as the spectral axis while the other acts as the spatial axis. This can be seen in Fig. 2 that shows an example of a spectral image obtained from some plants. Two camera types were used in the setup:

- The MX5, a standard CCD-camera from ADIMEC with a 8-bit linear output.
- The FUGA15C, a CMOS camera from CCAM technologies has several advantages over the standard CCD-camera. First of all CMOS-technology is cheaper than CCD-technology. The CMOS camera also has a random pixel access capability enabling to read out only the spectral lines of interest. This property gives the CMOS camera a superior read-out speed compared with a standard CCD-camera (Dierickx et al., 1996). The CMOS-camera with 512∗512 pixels is capable of reading out 2.4M pixels per second. Plant detection only needs two lines (one red typically 740 nm and one infrared typically 660 nm). For a stripe on the ground with a width of 2 cm and a length of 2 m, the sensor should be able to detect plants with a size of 80 mm^2 at a speed of 170 km h^{-1}. This makes the sensor ideal for road and railway spraying.

This sensor has several advantages over other multi-spectral techniques:

- The imaging spectrograph has no moving parts, resulting in higher robustness with respect to vibrations of the spray boom.

- All the spectral information about the analyzed surface is available at once, through diffraction of the reflected light. The limiting factor for the speed is the frame rate of the camera.
- This sensor offers flexibility when selecting the wavelengths of interest: they can easily be chosen by selecting only the most significant spectral lines on the CMOS-camera.
- The parameters were optimized to obtain a low cost device with reasonable performance and standard optical components.
- The imaging spectrograph is very compact.

ALGORITHM

Weed detection using the sensor described above can be based upon the spectral differences between the crops and weeds. (Feyaerts et al., 1998). It turns out that this technique, however, is not capable of detecting every single weed in a field. Therefore we have chosen to combine this spectral technique in with additional field information to improve the performance of the detector.

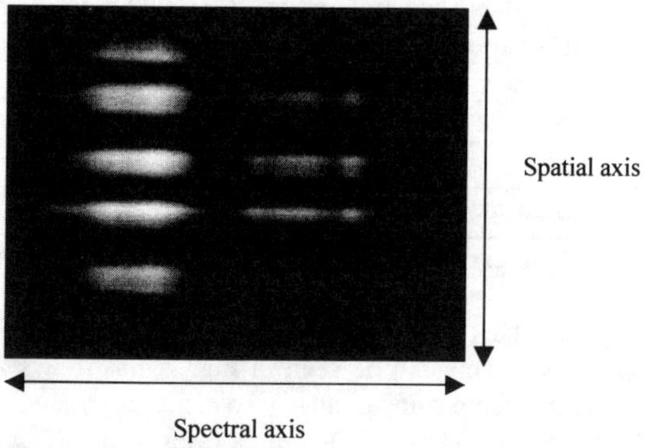

Fig. 2: Imaging spectrograph image of five plants.

We can for example take into account that crops like beets and corn are planted in with additional field information to improve the performance of the detector. We can for example take into account that crops like beets and corn are planted in rows. Other additional information can be the expected maximum crop sizes in a certain crop stadium or the fact that weeds often grow in patches. Therefore an algorithm was developed that takes these additional factors into account. It is important to realize that the combination of two nonperfect algorithms can lead to a highly performant system. When for example each algorithm can recognize 85% of the weeds, both algorithms combined will be able to recognize almost 98% of the weeds!

The algorithm is basically a row detection algorithm enriched with information on the plant sizes. While other row detection algorithms (Hague et al., 1997) can often use two-dimensional information (an image), this sensor provides us only with one-dimensional information (a line). To minimize the computation time, the algorithm uses as much information as possible to detect the weeds that are growing between the crop rows.

The parameters used by the algorithm are the height of the sensor, the angle of view of the objective lens, the crop row distance and the maximum crop size. The height and the angle of view determine the analyzed distance on the ground. This distance (typical 1.5 to 2 m) is crucial information for any algorithm because this is necessary to determine which spray nozzle must be activated. The accuracy of the algorithm and the herbicide application depend on the accuracy of this distance. It is therefore important to have a stabilized spray boom and/or to measure the distance from the sensor to the ground e.g. with an ultrasonic sensor.

The algorithm starts by reading out a red line and an infrared line. Every point of the infrared line is divided by its corresponding point of the red line. The resulting vector with the vegetation index of each point is thresholded to segment the green parts from the soil. To extract reliable row information, 10 subsequent stripes are analyzed and combined by the OR-operation. So instead of looking at only an area of $2 * 150\ cm^2$, the sensor looks at an area of $10–20 * 150\ cm^2$. The width of the area should be at least the crop distance within a row.

The algorithm then calculates the positions, the sizes and the number of the green areas. Succeeding plants with a total size (= size plant + size next plant + space between them) smaller than the row width are first glued together and considered as one plant. This is done to avoid problems in the case a crop is seen as two small separate green pieces. All the plants bigger than the maximal tolerated size (row width) are immediately removed from the green vector and considered as weeds. The evaluation for the remaining plants is based on the distances between the succeeding plants. Two distances are calculated for each plant: the distance from the examined plant to a plant on the left side of the examined plant and the distance from the examined plant to a plant on the right side of the examined plant. If both these distances are smaller than the (crop row distance – row width distance) or larger than the crop row distance, the plant will be sprayed. The second condition is necessary in case a crop in a row is missing. For the leftmost and the rightmost plant, the evaluation criterion will be the same but only based on one plant.

It is clear that the above criterion for spraying will always be fulfilled in highly populated weed patches. This means that all the weed patches will be effectively sprayed.

HERBICIDE REDUCTION AND HIT RATE

Herbicide Reduction

Herbicide reduction depends on several factors such as spray resolution, weed density, and performance of the detector. To evaluate the influence of these parameters, we have developed a mathematical model that describes the relation between these parameters and the herbicide reduction.

The model uses the following parameters: W weed density (number of weeds m^{-2}), l length of the spray rectangle, and b width of the spray rectangle, p is the probability that a crop is correctly classified as a crop and q is the probability that a weed is correctly classified as weed.

The size of the spray rectangle is determined by the space between the nozzles (typical 50 cm) and the dynamic behavior of the nozzles. At a tractor speed of 7 km h^{-1} and an opening/closing frequency of 20 Hz of the nozzle valve, it is possible to spray in a rectangle of $10*50$ cm^2. We assume that the plants are small enough so that it is only necessary to spray in only one spray rectangle to destroy a weed.

The herbicide reduction in an area in a field with a certain weed density equals the probability:

$$\text{Herbicide reduction} = \text{P (no spraying in a rectangle)}$$

The following two conditions must be satisfied if no herbicide may be applied:

1. All present crops in the spray rectangle must be recognized as crops.
2. All present weeds in the rectangle must be classified as crops.

Or,

$$P(\text{no spraying}) = P(\text{no weed activated spraying}).\text{P(no crop activated spraying)}$$

For the first part, we get:

$$\text{P(no weed activated spraying)} =$$

$$\sum_{k=0}^{\infty} \text{P(k weeds in rectangle).P(all k weeds classified as crops)}$$

The number of weeds in a rectangle can be described by the Poisson distribution. This leads to:

$$\text{P(no weed activated spraying)} = \sum_{k=0}^{\infty} \frac{(\text{lbW})^k}{\text{k!}} e^{-\text{lbW}} (1-\text{q})^k$$

This can be simplified to:

$$\text{P(no weed activated spraying)} = e^{-\text{lbW}} e^{\text{lbW(1-q)}} = e^{-\text{lbWq}}$$

The probability that k crops are in the spray rectangle cannot be described by a Poisson function because crops are planted in a regular way and not in a locally random pattern. This makes the mathematical description much more complicated. Therefore we have chosen to limit ourselves to realistic cases instead of using a generalized mathematical model. The limitation of the model lies in the fact that we only allow spray rectangles with in one direction a spray distance smaller than the typical crop distance and in the other direction a spray distance smaller than two times the crop distance. Due to this limitation the maximum number of crops in a spray rectangle will be two.

For example, in a corn field with a row distance of typical 75 cm and an intrarow distance of typical 13 cm, the maximum size of the spray rectangle will be $150 * 13$ cm^2 or $75 * 26$ cm^2. This spray rectangle is larger than the spray rectangles in which we want to apply the herbicides, so the model is acceptable. As stated before the typical dimension of our spray rectangles would be 50 ∗ 10 cm^2 or smaller.

In the calculations that follow *R* will be the row distance and *S* will be the intra-row distance. Three cases are now possible.

1) $R \geq l$ and $S \geq b$

$$P(0\ \text{crops}) = 1 - \frac{lb}{RS} \qquad P(1\ \text{crop}) = \frac{lb}{RS} \qquad P(2\ \text{crops}) = 0$$

2) $R < l < 2R$ and $S \geq b$

The first condition $R < l < 2R$ implies that one or two crop rows will lie in the spray rectangle. This leads to the following probabilities:

$$P(\text{1 row in the spray rectangle}) = \frac{2R - l}{R} \qquad P(\text{2 rows in the spray rectangle}) = \frac{l - R}{R}$$

This gives us for the 'crop probabilities':

$$P(0\ \text{crops}) = \frac{2R - l}{R}\left(\frac{S - b}{S}\right) + \frac{l - R}{R}\left(\frac{S - b}{S}\right)^2 \qquad P(1\ \text{crop}) = \frac{2R - l}{R}\frac{b}{S} + \frac{l - R}{R}\frac{2b(S - b)}{S^2}$$

$$P(2\ \text{crops}) = \frac{l - R}{R}\left(\frac{b}{S}\right)^2$$

3) $S < b < 2S$ and $R \geq l$

similar as under 2)

So we get:

$$P(\text{no crop activated spraying}) = \sum_{k=0}^{2} P(k\ \text{crops}).p^k$$

Finally we obtain:

$$\text{Herbicide reduction} = e^{-lbWq}\left(\sum_{k=0}^{2} P(k\ \text{crops}).p^k\right)$$

In Fig. 3 we have plotted the herbicide reduction in a corn field vs. the weed density in the case of a spray rectangle of 50 ∗ 10 cm^2 and a smaller rectangle of 50 ∗ 5 cm^2. The classification parameters were $p = 0.9$ and $q = 65/75$. The latter means that we assumed a row width of 10 cm and a row distance of 75 cm. A herbicide reduction of 70% can be achieved in areas with a low weed density (<10 weeds m^{-2}). Figure 3 also illustrates the importance of a small spray rectangle. In an area with a weed density of 50 weeds m^{-2} a reduction of 36% can be achieved with the smallest rectangle, while a reduction of 13% is possible with the larger rectangle. The size of the spray rectangle, however, is less important in areas with very high or very low weed density. Similar conclusions are also valid for beets.

Figure 4 shows the effect of a realistic classifier (p =0.8, q=0.8) on the herbicide reduction in a corn and with a spray rectangle of 50 x 10 cm^2. The

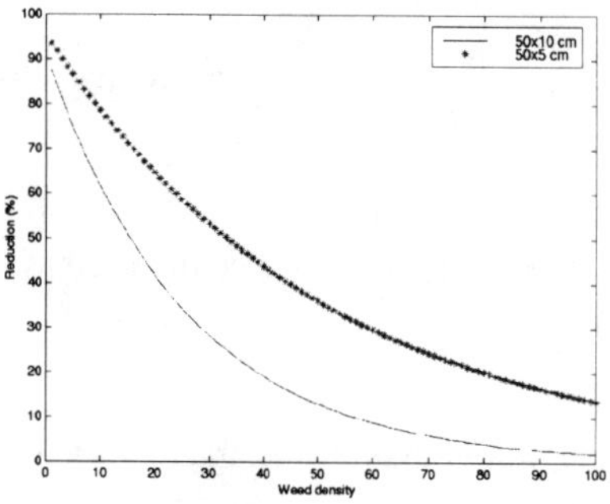

Fig. 3: Herbicide reduction with a small and a large rectangle.

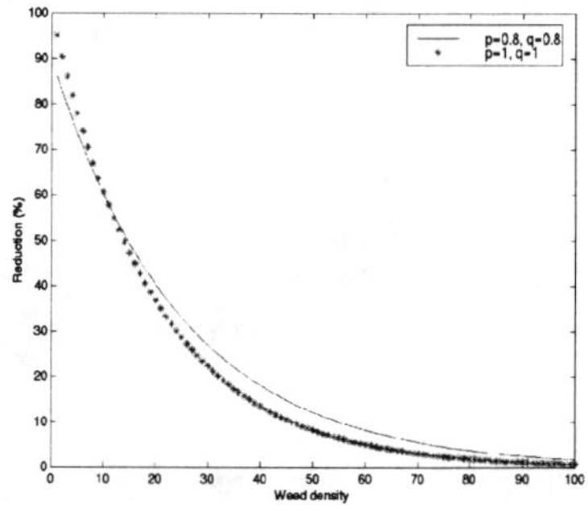

Fig. 4 : Reduction with a perfect and non-perfect classifier.

herbicide reduction of a perfect classifier will be higher in areas with a low weed density and lower in areas with a high weed density. In areas with a low weed density the non-perfect system will often be activated by the crops, which explains the lower reduction. In highly infested areas, the opposite effect becomes more important.

Some weeds will be misclassified and not sprayed, which results in a higher herbicide reduction. The net effect of this phenomenon on the herbicide reduction in a field depends on the spatial variability of the weed density.

Hit rate

Another important parameter in the design of a weed sensor is the hit rate. The hit rate is the probability that a weed in a field will be sprayed. It is important to realize that the hit rate is not the same as the classifier performance. The hit rate depends also on the weed density and the size of the spray rectangle. A weed will be sprayed if the weed itself, or any other weed in the rectangle, or a crop in the spray rectangle is classified as a weed. To calculate this, we will first calculate the weed activated spraying part and then the crop activated spraying part of the hit rate formula.

In a given rectangle with at least one or more weeds, the probability that the rectangle contains exactly k weeds is:

$$P(\text{k weeds in a rectangle with weeds}) = \frac{P(k)}{1 - P(0)}$$

P(k) is the Poisson distributed probability. The probability that the sprayer isn't activated in a rectangle with one or more weeds is:

$$\text{P(no weed activated spraying)} = \sum_{k=1}^{\infty} \frac{P(k)}{1-P(0)}(1-q)^k$$

After substituting the Poisson formula, we get:

$$\text{P(no weed activated spraying)} = \frac{e^{-lbW}}{1-e^{-lbW}}(e^{lbW(1-q)}-1)$$

For infinitesimally small spraying rectangles, we get:

$$\text{P(no\_weed\_activated\_spraying)} = \lim_{lb\to 0} \frac{e^{-lbW}}{1-e^{-lbW}}(e^{lbW(1-q)}-1) = 1-q$$

Which is in accordance with what is expected. Very small rectangles will contain at most one weed and the hit rate will be uniquely determined by the classifier performance.

The crop part is the same as before, therefore we get:

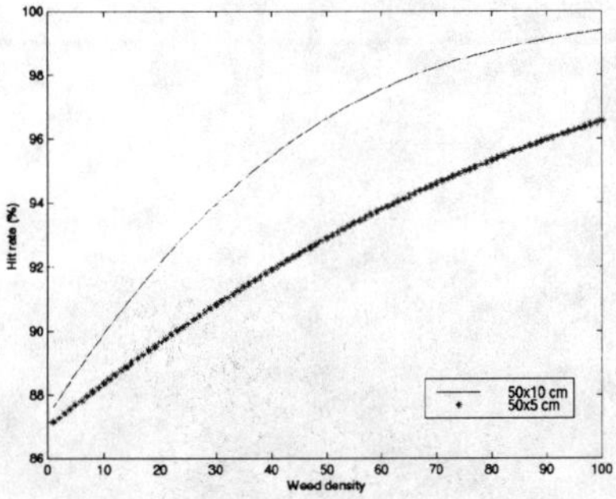

Fig. 5. Hit rate vs. weed density for two spray rectangles.

$$\text{Hit rate} = 1 - \frac{e^{-lbW}}{1-e^{-lbW}}(e^{lbW(1-q)}-1).\left(\sum_{k=0}^{2} P(\text{k crops}).p^k\right)$$

The hit rate is plotted in Fig. 5 for a corn field with a nonperfect classifier with $p = 0.9$ and $q = 65/75$ and a spray rectangle of $50*10\ \text{cm}^2$. As expected the hit rate is higher in larger spray rectangles. Large spray rectangles will thus result in lower herbicide reductions, but a better hit rate. The hit rate also increases with the weed density. For weed patches with a weed density of 50 weeds m^{-2} the hit rate will be >90%.

It is important to notice that this mathematical model is only correct in Poisson distributed weed areas. Weeds however often occur in patches, resulting in a non-uniform weed distribution. Herbicide reduction and hit rate in a real field can easily be obtained by calculating the mean of the herbicide reduction and the hit rate in every grid point using the formulas above.

RESULTS

Test Setup

A mobile test unit was built for the evaluation of the sensor and the algorithm. This test unit is shown in Fig. 6. This test unit consists of the weed sensor (spectrograph + camera), a standard color camera, a Pentium 166 MHz with 32 MB RAM, a video mixer, a video recorder and a power supply.

The data of the spectrograph are collected and processed by the PC. A color camera positioned next to the imaging spectrograph records the plants passing under the weed sensor. The processing output of the PC results in a display with green stripes for the weeds and red stripes for the crop. This output display is then mixed with the signal of the color camera and recorded on videotape for further evaluation. After appropriate positioning and calibrating of the color camera and

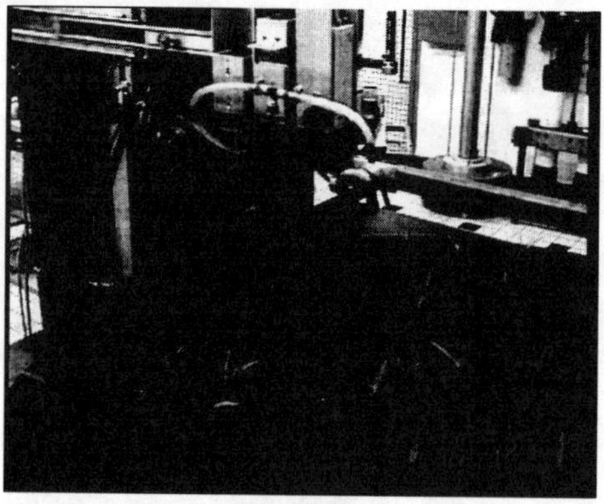

Fig. 6. Weeds and corn on the turning table with sensor.

the sensor, the recorded image on the tape will show us the plants together with their quotation (soil (no stripe), weed (green stripe) or crop (red stripe)).

The initial tests were done on a large turning table illuminated by a halogen lamp. The turning table was also equipped with two spray nozzles. A Visual C++ program was written to evaluate the different algorithms.

Evaluation of the Sensor

The sensor was first calibrated with interference filters (bandwidth of 10 nm) to determine the spectral range of the sensor. A spectral range from 400 to 1000 nm was achieved with the CCD-camera, but only when a very strong light source (1000 W) was used. The low sensitivity of the camera in the IR-range restricts the sensor to a range from 400 to 850 nm in more practical cases. The sensor showed also a very good linearity between wavelength and pixel position. The sensor equipped with an objective lens of 3.5 mm was placed 30 cm above the turning table to determine the spatial resolution of the green detection. The sensor was able to analyze a line of 80 cm on the ground. The smallest plant that the sensor could detect had a size of 0.5 cm^2. From experiments we found out that a threshold between two and three on the vegetation index was most suited for detecting plants.

Evaluation of the Row Detection Algorithm

Corn was planted in three concentric circles on the turning table together with some weeds to test the row detection algorithm. The distance between the corn was 25 cm. The performance of the detector strongly depended on the parameter for the crop width. Crops were often wrongly recognized as weeds when this parameter was too small. When the parameter was too large the opposite happened and weeds close to the crops were not detected. Weeds in the crop rows were of course never detected. A good estimation of this parameter is therefore crucial. A typical value for this parameter in our experiments was 7 cm. This parameter also depends on the size of the crop and the angle of view of the sensor. The wider the angle of view, the larger that the (highest) plants will be seen by the the sensor. It is therefore necessary to attach the sensor mask as high as possible on the spray boom.

CONCLUSION

We have described in this paper a system for weed detection based upon an imaging spectrograph. The sensor splits the reflected light from the ground in its spectral components and projects it to the surface of a camera. The detector was especially designed to be cheap, robust and compact. The sensor is capable of detecting plants as small as 0.5 cm^2. In combination with a CMOS-camera such as the FUGA15C it is possible to detect plants at a speed of 170 km h^{-1}. This makes the sensor ideal for railroad spraying.

An algorithm for the detection of weeds between crop rows was presented. The algorithm only uses the one-dimensional spatial information of the sensor. The detection is based on the distances between subsequent plants. Tests on a turning table showed that this algorithm is capable of detecting weeds between crop rows.

Finally, a mathematical analysis was carried out to evaluate the parameters that are crucial in a weed detection system. The herbicide reduction is determined by the size of the spray rectangle, the weed density and the performance of the detector. The herbicide reduction will decrease if the weed density or the size of the spray rectangle increases. The effect of the performance of the detector is subtler. In areas with a high weed density, herbicide reduction is even better with a less performing sensor while in areas with a low weed density herbicide reduction can be lower. Another important parameter is the hit rate: this is the probability that a weed will be sprayed. The hit rate also depends on the size of the spray rectangle, the weed density and the performance of the detector. The hit rate will increase if the weed density or the size of the spray rectangle increases. Large herbicide savings and a good hit rate are therefore two opposite demands for a nonperfect weed detection system.

ACKNOWLEDGMENTS

We wish to thank the Flemish Community (VLIM-project), the Belgian Ministry of Agriculture (IWONL-project) and Ecospray, our industrial partner, for their financial support.

REFERENCES

Battey, D.E., and J.S. Slater, 1993. Compact holographic imaging spectrograph for process control applications. SPIE Vol. 2609 (Optical methods for chemical process control).

Beck, J., and T. Vyse, 1995. Structure and method usable for differentiating a plant from soil in a field, U.S. Patent No. 5,389,781.

Dierickx B., D. Scheffer, G. Meynants, W. Ogiers, and J. Vlummens, 1996. Random addressable active pixel image sensors. *In* AFPAEC Europto proc., Berlin. 9 Oct. 1996. SPIE, Vol. 2950.

Felton, W.L., and K.R. McCloy, 1992. Spot spraying, microprocessor controlled, weed-detecting technology helps save money and the environment, Agric. Eng./November, 4 pp.

Feyaerts F., P. Pollet, L. Van Gool, and P. Wambacq, 1998. Sensor for weed detection based on spectral measurements, Proc. of the 4th Int. Conf. On Precision Agriculture.

Gerhards, R., A. Nabout, M. Sökefeld, W. Kühbauch, and H.A. Nour Eldin, 1993. Automatische Erkennung von zehn Unkrautern mit Hilfe digitaler Bildverarbeitung und Fouriertransformation, J. Agron. Crop Sci. 171:321–328.

Guyer, D.E., G.E. Miles, M.M. Screiber, O.R. Mitchell, and V.C. Vanderbilt, 1986. ASAE Machine vision and image processing for plant identification, Trans. 29(6):1500–1507.

Hague, T., J.A. Marchant, and N.D. Tillet, 1997. A system for plant scale husbandry. p. 635–642. *In* Proc. of the 1st European Conf. Precision Agriculture.

Hahn, F., and A.Y. Muir, 1993. Weed-crop discrimination by optical reflectance, IV Int. Symposium on Fruit, Nut, and Vegetable Production Engineering, 8 pp.

Herrala, E., J. Okkonen, T. Hyvarinen, M. Aikio, and J. Lammasniemi, 1994. Imaging spectrometer for process industry applications. SPIE. 2248:33–40.

Spectral Properties of Sugarbeets Related to Sugar Content and Quality

D. S. Humburg
Department of Agricultural Engineering
South Dakota State University
Brookings, South Dakota

K. W. Stange
Department of Agricultural Engineering
South Dakota State University
Brookings, South Dakota

ABSTRACT

Sugarbeets stand to benefit from site-specific management to a greater extent than many other crops. Yield maps for sugar production have been unavailable as no effective method of spatially quantifying sugar content has been available. An experiment was conducted in an attempt to relate spectral properties of beet canopy with sugar content and quality at harvest. Detailed spectral measurements, taken in August, September, and October of 1997, were related to beet quality. Models were developed using canopy reflectance at three and four spectral bands combined in indices and related to beet quality. The best model involved spectral bands at 500, 550, and 830 nm and was able to account for just >50% of the variation in sugar content of the sampled beets. The model was also able to predict one half or more of the variation in sodium, amino N, and recoverable sugar per ton.

INTRODUCTION

The development of sensors for yield and quality variables in rowcrops, combined with the availability of inexpensive positioning systems, and controls for the application of inputs has provided the required components for site-specific management systems in some crops. Such management systems are often characterized by a cyclical series of steps where spatially variable management decisions are implemented by control and positioning systems followed by, or interspersed with, information gathering steps such as remote sensing and yield monitoring. Information gathered is either implemented immediately or archived for use in formulating management decisions for the next crop cycle.

It is not surprising that the most complete of these systems center around the mainstream rowcrops with the largest total acreages such as corn and soybeans. The potential market for hardware, software, and services encourages the direction of resources to these crops. However, from the economic standpoint of producers these crops are not necessarily optimum targets for site-specific management. The costs of hardware, software, services, and management effort must be recouped through changes in the crop management that result in higher total production, lower input

Copyright © 1999 ASA-CSSA-SSSA, 677 South Segoe Road, Madison, WI 53711, USA. *Proceedings of the Fourth International Conference on Precision Agriculture.*

costs or combinations of these two (Lowenberg-DeBoer & Swinton, 1997). In the case of inputs, such as fertilizer, sophisticated producers are often reluctant to take a chance on operating below the point where the marginal cost of the input is equal to the marginal value of crop response. Simply put, the only cost to the producer of excessive fertilizer application is the value of the fertilizer itself. Indirect costs such as water quality impacts are largely not borne by the farmer. This makes the recovery of the costs of implementing site-specific technology more difficult.

In contrast to crops such as corn, growers of sugarbeets face a direct economic penalty for excessive fertility, as well as for inadequate fertility. Quality and quantity of sugar produced are closely linked to fertility variables and management decisions. It is not surprising that beet growers have been early adopters of site-specific technology; however, the systems for precision management in sugarbeets lack key components. Yield monitors are typically the entry point into site-specific operation for grain producers, and knowledge of yield variability is a motivating factor in the adoption of additional technology. Technology for mapping of yield in beets has not been widely available. Tonnage monitors have been developed (Hofman et al., 1996) but these still face problems of accuracy and difficulty in accounting for tare (Hall et al., 1998). Even assuming that the technical problems with tonnage monitors are solved they do not reflect spatial sugar production to the extent that sugar content and quality are not uniform across the field. Without a means of quantifying spatial variation in sugar content it is not possible to map true yield in the crop. Also, since sugar content and quality are directly related to compensation for the crop, a map of profitability is not possible without measuring or estimating sugar content. Maps of yield are only retrospective tools in the management system. They do however provide insight and motivation for management variation in subsequent crop years.

Nitrogen is well established as the fertility component most directly affecting sugar content and quality in beet roots. The color of the canopy also has been linked to the N remaining in the beet canopy. Moraghan (1998) proposed a methodology for using beet canopy spectral properties (color) as a spatial indicator of the residual N returned to the soil and available for the subsequent crop. Beet tops with a Yellow or Yellow-Green color were found to contribute appreciably less N to the soil than Green tops (Moraghan & Smith, 1996). The characteristic yellowing of the beet canopy late in the season is well recognized by growers as a generally desirable sign associated with higher sugar content. This visible change occurs as an increase in canopy reflectance at wavelengths of 500 to 600 nm. It is also possible that reflectance changes occur outside of the visible wavelengths that might be associated with variation in sugar content at harvest. Xie (1997) conducted an experiment in which both visible and near infrared (NIR) wavelengths were examined for relationships to sugar content and the quality variables of potassium, sodium, harmful amino N (HAN), and recoverable sugar (RST). Statistically significant links were identified for a number of wavelengths and sugar content and HAN. However, the correlation for a multiple linear regression using several of these wavelengths, and sugar content was weak ($R^2 = 0.27$).

Because of the potential to provide useful maps of sugar content and quality variables from multispectral aerial images, a more extensive experiment was conducted during the 1997 growing season with the following objectives:

1. Sample beet canopy reflectance spectra for beets of varying sugar content across a wide range of wavelengths.
2. Collect sample spectra at several points in time during the latter half of the growing season to test for a temporal effect of spectral changes.
3. Identify spectral wavelengths most likely to be useful in an imaging system for quantification of Sugarbeet quality variables.
4. Establish predictive models for sugar quantity and quality variables for the sample data and field.

MATERIALS AND METHODS

A sugarbeet field was selected for the study in August of 1997 southwest of Raymond, MN. The crop appeared to be very healthy with a consistent stand and no obviously diseased areas. What were apparent, however, were areas of deep green beet canopy and areas of yellower canopy. The regions were oriented spatially so that it appeared that the cause of the variation was not a result of disease or the result of random fertility variation. The visible differences occurred in regular stripes suggesting variation in fertility caused by characteristics of the machine used to apply the fertilizer. A series of 50 sample sites was established across the center of the field. Selected sites alternated between darker and lighter areas of canopy. In addition to these alternating sites an additional 10 sites were established in other areas of the field that showed noticeable variation in color. Each site was flagged and its coordinates recorded with the use of a differentially corrected GPS receiver. The coordinates of identifiable points on the perimeter of the field, such as field boundary and road junctions were also recorded. Four blue nylon tarps were staked down in areas adjoining the field boundaries also, to be used as additional ground control points for aerial images of the field. Coordinates of these markers were also recorded.

Spectral data were collected on 15 August, 17 September, and 30 September. Ground truth spectra were collected with a FieldSpec FR portable spectrograph. The spectrograph samples and measures light from wavelengths of 350 to 2500 nm at intervals of 3 nm. The procedure at each field sample site involved the collection and recording of five spectra files. Three spectra, typically the first, third, and fifth recorded, were taken of the beet canopy from a height of approximately 2 m. No fore optic was used on the collection fibre, which resulted in an angle of view for the sensor of 25°. The second and fourth spectra collected at each site were taken from a reflectance panel. The panel was a calibrated reflectance surface with a nominal reflectance of 50%. It was mounted on a tripod and leveled at each site to obtain a consistent measurement of the solar radiation spectra present at the time each site was sampled.

Airborne images were also collected on 24 August, 17 September, and 30 September, 1997. These images were acquired by Airborne Data Systems of Wabasso, MN using a four-camera system from an altitude of approximately 1524 m (5000 ft). Four images were acquired, using filters centered at the following wavelengths: 500, 570, 670, and 870 nm. The three visible wavelength filters used passbands of 50nm width. The NIR band (870 nm) had a passband of 400 nm width.

The four band images were spatially registered to one another and combined into a single image file prior to delivery.

After the collection of spectral data on 30 September beet root samples were taken from each sample site. At this point the preharvest lift had removed three of the sample sites so that 57 remained for analysis. Two tare bag samples were taken from each site with roots of similar size and number divided between the two bags at each site. The roots were analyzed for sugar content and quality at the Southern Minnesota Beet Sugar Cooperative Tare lab.

Data collected with the spectrograph were processed to provide a single reflectance curve for each sample site. The three spectra representing the beet canopy were averaged for a site. The two reflectance panel spectra also were averaged. The panel spectrum at a site was adjusted for the calibrated panel reflectance at each wavelength to provide the solar power spectrum for that time and sight. The average of the beet canopy spectra was divided by the solar power spectrum to obtain the leaf reflectance curve for that site. Prior to this process each of the beet spectra was graphed and reviewed. It is possible to collect erroneous spectra in which a combination of beet canopy and reflectance panel are sampled. This is identifiable in a review of the raw canopy spectra as curves that are intermediate between pure canopy and pure sunlight. A number of these data collection errors were identified in the files and removed from the process. In those cases where an error was made the data representing the beet canopy may no longer be the average of three spectra, but instead may be represented by one or two sample spectra from that site.

The reflectance spectra were studied for identifiable links to the sugar content and quality of the associated locations in the field. Combinations of wavelengths, or indices, and changes occurring between sample times were used to develop a predictive model for sugar in this field.

RESULTS

A set of typical spectra from a test site is given in Fig. 1. The two spectra taken of the 50% reflectance panel nearly coincide and represent the sunlight reflected from that panel. The beet spectra also nearly coincide and are lower than, or equal to, the panel spectra. The water absorption bands near 1400 and 1800 nm were removed to prevent the noise in these wavelengths from artificially adjusting the scale of the graphs. The reflectance of the canopy is obtained by dividing the beet spectra by the panel spectra and adjusting for the calibrated reflectance of the panel at that wavelength.

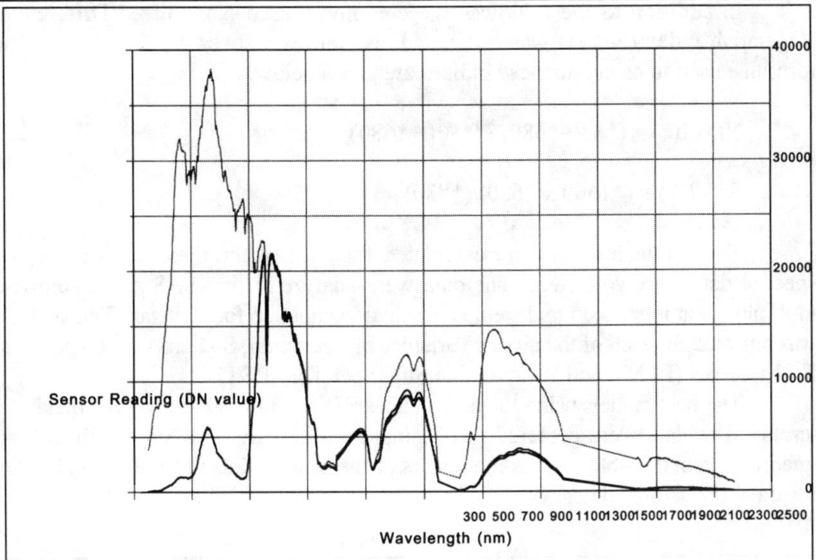

Data representing the five sites in the field with the highest sugar content were averaged. Similarly, the reflectance spectra from five sites with the lowest sugar content were averaged. These reflectance curves, shown in Fig. 2, indicate areas in which differences occur that might be associated with changes in beet sugar content and quality. The average spectra for the high sugar content beets have a higher reflectance between 550 and 650 nm than the spectra representing the low sugar beets. This corresponds to the visible yellowing that is observable in the field. These spectra also have a much lower reflectance in the near infrared from 750 to 1400 nm than those representing the beets with lower sugar. The high sugar average spectra also appear to have a lower reflectance at 350 and 1680 nm.

Based on the trends visible in Fig. 2 two new vegetative indices were composed in an attempt to maximize the predictability of sugar content from the spectral data. An index denoted as BDVI (Beet Differential Vegetative Index) was calculated as follows:

BDVI(*) = (*550-*500) / (*830-*500)

The numbers represent wavelengths at which reflectance measurements are taken. The (*) in this equation represents the time (August[A], September[S], or October[O]) for the particular calculation of the index. The second index was composed of reflectance measurements from the same wavelengths with the addition of a value for 360 nm. The index B2DVI (2nd Beet Differential Index) has the following composition:

B2DVI(*) = ((*550-*500)+(*500-*360)) / (*830-*500)

In addition to these indices the commonly used Normalized Differential Vegetative Index (NDVI) was calculated, as well as a stress index (STVI). The formulae used to calculate these indices are given below.

NDVI(*) = (*830-*680) / (*830+*680)

STVI(*) = (*1660 x *680)/(*830)

These four indices were calculated for each of the three dates on which spectral data were collected. The data were analyzed using SAS and a stepwise multiple linear regression to develop a prediction equation for % sugar. The analysis was repeated for each of the quality variables of Potassium, Sodium, Amino N, Loss-To-Molasses (LTM), and Recoverable Sugar per Ton (RST).

The best models using the new indices BDVI and B2DVI involved quadratic terms. The data were centered to eliminate co-linearity between the linear and quadratic terms. Table 1 gives the results of the best models fit to the BDVI index for the three sample dates.

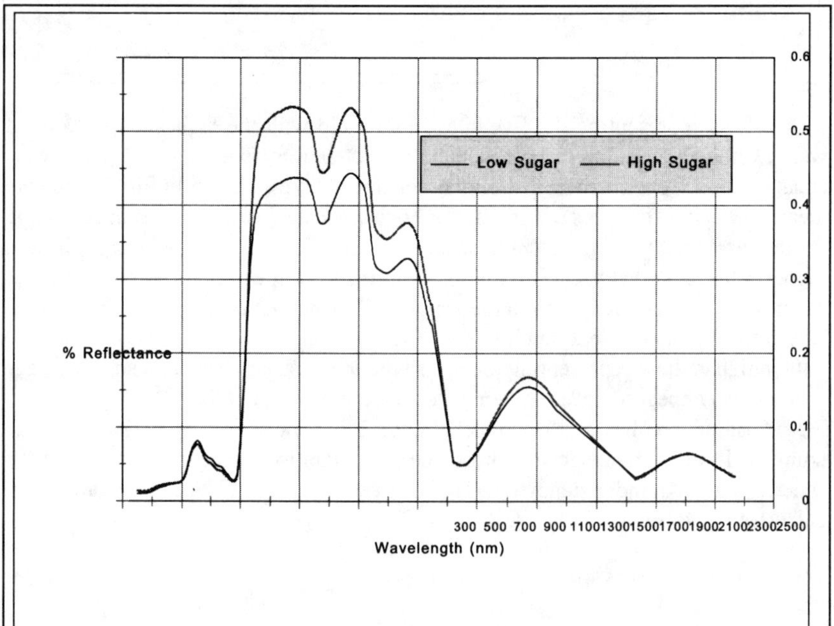

Table 1. MLR coefficient values and correlation coefficients for the BDVI index and the sugar quality variables. Note that CBDVI represents the *centered* BDVI index data set.

Quality Variable	Int.	CBDVI	CBDVI²	Time (Int)	Time (CBDVI)	R^2
%Sugar	19.07	22.58	-968.7	-0.21(S) -0.58(O)	32.08(S) 19.81(O)	**0.522**
Potassium	1709.2	NA	257,260	60.0(O)	-9922(O) -7007(S)	**0.254**
Sodium	153.8	-1535.8	54,588	14.14(S) 35.27(O)	-1607(S) -835(O)	**0.503**
Amino N	142.1	-2884	65,953	26.14(S) 52.77(O)	-1042(O)	**0.613**
LTM	0.86	-3.26	164.2	0.070(O)	-3.02(S) -5.09(O)	**0.332**
RST	364.5	530.7	-22,578	-4.93(S) -13.3(O)	442.5(S) 728.0(O)	**0.528**

The models derived from this analysis are of quadratic form. The first three terms of each equation would represent the model if the August data were used to predict the quality variable. The **Time**, and **Time(CBDVI)** columns modify the intercept and linear terms of the respective model for the sample data from September and October. Absence of either of the time terms from the table indicates that the modifying term was not statistically significant and was excluded.

Results for the multiple linear regression using the B2DVI index are given in Table 2. The formats for the table and the models predicting sugar and quality are the same as given for Table 1.

The analysis for the Normalized Differential Vegetative Index (NDVI) and the stress index, (STVI) did not produce significant quadratic terms in their models. The linear models produced had very low correlations to sugar, with a maximum R^2 of 0.209 for percentage of sugar from the NDVI linear model. Other correlations were lower still and the results are not tabulated here.

DISCUSSION

The field in which the experiment was conducted was a well-managed crop, and the sugar analysis results verified this. The sugar content for the 113 samples processed ranged from 20.11 to 14.54%. The average for the 113 samples was 18.60% sugar, with a standard deviation of 0.913%. Only seven of the 113 tare samples had a sugar content below 17%. Two tare samples were collected at each of 57 sites for which spectral data were also acquired. A single sample from one set was lost.

In a few of the sample pairs there is a marked difference in the sugar analysis. Five of the sample pairs had differences in sugar content of 1% or more.

Table 2. MLR coefficient values and correlation coefficients for the B2DVI index and the sugar quality variables. Note that CB2DVI represents the *centered* B2DVI index data set.

Quality Variable	Int.	C2BDVI	C2BDVI^2	Time (Int.)	Time (C2BDVI)	R^2
%Sugar	19.09	15.94	-470.7	-0.29(S) --.65(O)	16.07(S) 23.57(O)	**0.514**
Potassium	1717.0	NA	123,459	68.54(O)	-5,405(S) -7,485(O)	**0.250**
Sodium	151.7	-1056	27,990	18.55(S) 39.33(O)	-734.4(S) -1247(O)	**0.501**
Amino N	136.7	-2106	33,390	34.40(S) 57.71(O)	-614.9(O)	**0.585**
LTM	0.827	-3.95	70.83	0.065(S) 0.116(O)	-1.73(O)	**0.318**
RST	365.1	373.7	-11,004	-6.74(S) -14.91(O)	360.7(S) 535.1(O)	**0.519**

These are potentially problematic as the difference represents nearly 20% of the range of the sugar values and a single set of spectral data is used to represent that site. It is not known whether the variation in sugar between sample bags for the same site is genuine, or if it is an artificial result of the sampling process, such as inadequate or inconsistent topping of the beets. Sample weights were deliberately kept high, between 25 and 30 lbs, to avoid having the sample dominated by a single plant. No attempt was made to adjust or remove the sample pairs with the wide variation in analysis.

The results indicate a measurable relationship between spectral properties of the beet canopy and the quantity and quality of the sugar in the beet roots. The model developed from the BDVI index explains just over one half of the variation in sugar content and a similar percentage of the variation in recoverable sugar. This model also explains just over 50% of the variation in sodium for these samples, and 61% of the variation in amino N. The model does not do very well at predicting the potassium as used in the calculation of loss-to-molasses. The correlation for LTM is correspondingly low at 0.318, primarily because of the poor predictability of potassium.

The variation in reflectance in the NIR bands is much greater in magnitude than the variation in the visible wavelengths (Fig. 2); however, the differences in the visible bands *may* be both consistent and valuable in contributing to the models. It is possible that the small variation in the visible bands may be overwhelmed by the variation in the NIR making the index dominated by the NIR reflectance. It may be possible to normalize the numerator of the BDVI (*830–*500) for its range within the field, and similarly normalize the denominator (*550–*500) for its range within the field. This approach will be tested in the near future to see if it improves the performance of the selected indices.

The models developed used data collected by a portable instrument and wavelengths were selected for the indices from a near continuum. In an application of remote sensing limited wavelengths are available. It would be possible to select filters with center pass bands at wavelengths of 360, 500, 550, and 830 nm if the index were useful as developed. Landsat TM bands 1, 2, and 5 could be used to calculate the BDVI from a satellite platform.

The time effect was found to be a significant portion of the models in this experiment. The first data were collected on 15 August, followed by the second and third events on 17 September, and 30 September. In subsequent experiments data will be collected beginning in July and the effect of time will be explored in more detail. It is possible that the models could be enhanced by additional temporal data particularly if that data were collected before spectral changes become visible within the field.

The model developed for the BDVI index could be used to predict spatial variation in sugar content and quality of the beets in the sample field. Airborne, or satellite data, at appropriate wavelengths would be necessary to do this. The model is less than perfect, but is a beginning point in producing a sugar map as feedback to producers adopting site-specific technology. It is unlikely, however, that the model developed would be widely applicable in predicting actual sugar content values for fields outside of this study. What is more likely is that the model could be modified to predict the relative sugar content within the field, indicating the variation of sugar within the field while not specifying absolute values. Those values could ostensibly be predicted if ground truth beet quality sampling were to be done at appropriate sites within the field. A map of BDVI modeled sugar quality, that was geo-referenced to points on the perimeter of the field, could be used to identify the coordinates of points for optimum ground truth sampling.

CONCLUSIONS

Sample spectral data were collected that represent a range of healthy sugarbeet canopy. The spectra covered wavelengths of 350 to 2500 nm and represent the range of wavelengths that might be expected to be linked with sugar content. Beet samples were also collected and showed the field to have beets with sugar contents ranging from 20 to 14.5%. Most of these values were concentrated in the upper portion of that range.

Examination of the reflectance spectra for high and low sugar content beets indicated apparent differences at 360, 550, and 700 to 1200 nm. The reflectance at 500 nm seemed to be a crossover point in which no differences between high and low sugar content beets occurred and this wavelength was used also as a basis. Two indices were proposed from the reflectances at these wavelengths. These, and two other commonly used indices were evaluated for links to sugar content and quality.

The strongest predictive model used the BDVI index comprised of reflectance values at 500, 550, and 830 nm. The model was quadratic in BDVI and also included time effects for the September and October collection events. This model produced an R^2 value of 0.522 for sugar content. It also was useful in predicting sodium, amino N, and recoverable sugar per ton. R^2 values for these variables were 0.503, 0.613, and 0.528 respectively. The model was less useful in predicting potassium with an R^2 of 0.254 for this data set. The model using the B2DVI as the independent variable gave similar, but slightly lower performance. The other indices did not produce correlations >0.2 for any of the variables.

The model developed based on the BDVI index is one that has potential for the mapping of relative sugar content in beet fields from remotely sensed image data. It may be that small improvements in the model may be made by normalizing some of the spectral differences so that they are not dominated by the NIR variation that occurs. It will be necessary to develop a means of normalizing the model within a field. It is unlikely that any absolute numerical model such as that developed for BDVI in this work would be applicable across field boundaries and varietal differences. The potential exists, however for reasonably inexpensive mapping of the relative sugar content within field boundaries. This would be a useful retrospective tool for growers adopting site-specific fertility management as a means of maximizing sugar production.

ACKNOWLEDGMENTS

This work was supported by the Sugarbeet Research and Education Board of Minnesota and North Dakota, and conducted with the assistance of the Southern Minnesota Beet Sugar Cooperative.

REFERENCES

Hall, T.L., L.F. Backer, V.L. Hofman, and L.J. Smith. 1997. Evaluation of sugarbeet yield sensing systems operating concurrently on a harvester. *In* 1997 Sugarbeet Res. and Ext. Rep. Vol 28. North Dakota State University Ext. Serv. Fargo.

Lowenberg-DeBoer, J, and S.M. Swinton. 1997. Economics of site-specific management in agronomic crops. p. 369–396. *In* F.J. Pierce and E.J. Sadler (ed.) The state of site specific management. ASA, CSSA, and SSSA, Madison, WI.

Moraghan, J.T. 1997. Testing a new precision farming technique involving Sugarbeet tops in grower's fields. *In* 1997 Sugarbeet Res. and Ext. Rep. Vol 28. North Dakota State University Ext. Serv. Fargo.

Moraghan, J.T., and L.J. Smith. 1996. Nitrogen in sugarbeet tops and the growth of a subsequent wheat crop. Agron. J. 88:521–526.

Walter. J.D., V.L. Hofman, and L.F. Backer. 1996. Site-specific sugarbeet yield monitoring. *In* P.C. Robert et al. (ed.) Precision Agriculture Proc. 3rd Intl. Conf. 1996. Minneapolis, MN. 23–26 June 1996. ASA, CSSA, and SSSA, Madison, WI.

Xie, B. 1997. Spectral properties of Sugarbeet canopy related to sugar content and quality. M.S. thesis. South Dakota State Univ., Brookings. South Dakota.

Spatial High Resolution Crop Measurements with Airborne Hyperspectral Remote Sensing

J.C. Deguise
H. McNairn
K. Staenz
Canada Center for Remote Sensing
Natural Resources Canada
Ottawa, Ontario, Canada

M. McGovern
Central Experimental Farm
Agriculture and Agri-Food Canada
Ottawa, Ontario, Canada

ABSTRACT

With the future launch of high spectral and spatial resolution satellites, we foresee that that it will be possible to use the information from these satellites in a timely fashion for the detection of stress, the prediction of yield and as a diagnostic tool for recurring low productivity areas for numerous crops. Preliminary results indicate a correlation between the vegetation fraction of airborne hyperspectral pixels and ground truth validation measurements. The advantages of the pixel-unmixing algorithm for within-field crop properties mapping will be demonstrated.

INTRODUCTION

The important role of remote sensing in precision agriculture was well summarized in a recent paper by Moran *et al.* (1997). With the long list of environmental satellites announced for launch in the coming years, and the increasing use of digital databases for farm management, the role of remote sensing cannot be ignored. In particular many authors have demonstrated relationships between plant reflectance and agricultural parameters of crops (Hinzman *et al*., 1986; McMurtrey *et al*., 1994; Blackmer *et al*., 1995). Previous studies were based mostly on hand held spectro-radiometer reflectance measurements above individual plants or small portions of a field. In particular, Ma *et al.* (1996) suggested that remotely sensed reflectance is a more efficient way of determining plant nitrogen status over large areas than taking costly and time consuming field measurements. This paper will report on preliminary results obtained from visible and near infrared airborne hyperspectral imagery.

The Canada Centre for Remote Sensing, in collaboration with Agriculture Canada and the private sector, is leading a study on the applications of remote sensing to precision agriculture in Canada. An important component of this program is the use of high spectral and spatial resolution remote sensing data for agricultural information extraction. In the context of this multi-year precision

Copyright © 1999 ASA-CSSA-SSSA, 677 South Segoe Road, Madison, WI 53711, USA. *Proceedings of the Fourth International Conference on Precision Agriculture.*

agriculture program, agricultural biophysical and hyperspectral airborne data were collected in July 1997 in the region of Carman, Manitoba (Canada).

DATA ACQUISITION AND PROCESSING

The field of interest for this report was planted with potatoes in regular rows oriented perpendicular to the direction of the image acquisition flight. The plants in this field were at the early flowering stage.

Location of a regular grid of 24 sites in this field was predetermined. A real time differential GPS was used to locate these sites in the field. Each site consisted of a 0.5 m by 0.5 m area. On 13 July 1997 the mean height of the plants was measured, photographs were taken and personal observations–comments were recorded at each site. The total mass of vegetation contained within each site was then cut and bagged for immediate total weight determination (wet biomass). The sealed bags were sent to a laboratory, the content was oven dried and reweighed (dry biomass) to determine the water content and the total leaf area of the samples.

On 15 July 1997, this potato field was imaged with a Compact Airborne Spectrographic Imager (*casi*). The hyperspectral data set was acquired at an altitude of 2600 m above ground. At this altitude and with the selected data acquisition configuration of the sensor, each pixel of the image covered a ground area of 4 m by 4 m and measured radiance in 96 narrow spectral bands (6.6 nm full width at half-maximum) ranging from 413 nm to 956 nm.

Processing of the *casi* hyperspectral radiance data has been carried out with the "Imaging Spectrometer Data Analysis System" (ISDAS) developed at the Canada Centre for Remote Sensing in collaboration with industry (Staenz *et al.*, 1996). To avoid pixel resampling, only a line shift correction was applied to the imagery to remove the aircraft roll effects. The pixel shift of each line was calculated using the roll data recorded during the flight with a precise inertial aircraft attitude measurement system. Atmospheric correction using the MODTRAN3 radiative transfer model was then applied to the imagery converting the at-sensor radiance into surface reflectance (Staenz and Williams, 1997). A post-processing module of ISDAS was finally used to remove remaining atmospheric and calibration effects from the reflectance (Staenz & Williams, 1997).

A constrained linear unmixing procedure was applied to the reflectance data. This spectral analysis method expresses the reflectance spectrum of an image pixel as a linear sum of individual spectra from "pure pixels" (Shimabukuru & Smith, 1991; Boardman, 1995). These pure pixels were selected by performing a principal component (PC) analysis on the reflectance data set. Scatter plots of a pair of PCs were generated. Three endmembers were selected from averages of those pure pixels located in the extremities of the scatter plot: high and low density vegetation and bare soil. For each pixel of the image, the spectral unmixing procedure calculates fraction of the total reflectance spectra contributed by each endmember.

In order to demonstrate the use of remotely sensed crop reflectance for within-field variability measurements, the location of the ground sampling sites had to be established in the roll-corrected image. As mentioned previously, in order to preserve the radiometric integrity of the data, a full image-to-map registration was not considered. Our experience also showed that polynomial fit tools currently available in commercial image analysis packages do not lead to a satisfactory registration accuracy for non-orthogonal data from airborne line scanner and push-broom type scanners.

A specialized software system for correcting airborne imagery developed at the Canada Centre for Remote Sensing called GEOCOR (Gibson, 1994) has been used to correct and georeference another similar *casi* image of this field with sufficient accuracy. This system, however does not permit the usage of the transformation equations backward to transpose the geolocation of our sampling sites into the nongeometrically corrected image.

Given this major constraint, the 24 sites were located using their geographic coordinates in the *casi* image corrected with GEOCOR. Based on the location of these sites in the geometrically corrected image, the sampling points were visually located in the noncorrected image using relative distance to natural features visible in both images. To account for the location uncertainty, analysis results were computed and averaged across a 3 by 3 pixel matrix centered on these estimated locations.

RESULTS

Table 1 shows the Pearson correlation coefficients of the wet biomass of the plants with the other biophysical parameters (crop height, dry biomass, and leaf area) measured at all the sites.

Table 1. Correlation of biophysical parameters.

	Wet biomass
Crop height	0.65
Dry biomass	0.88
Leaf area	0.91

Because of the strength of the correlation between the biophysical parameters in Table 1, only the wet biomass was used for further investigation purposes.

Figure 1 shows the high-density vegetation endmember fraction compared to the wet biomass data collected at each site of the potato field. The high-density vegetation endmember fraction for each of the nine reflectance pixels surrounding the estimated location, their mean and the wet biomass are plotted for each site.

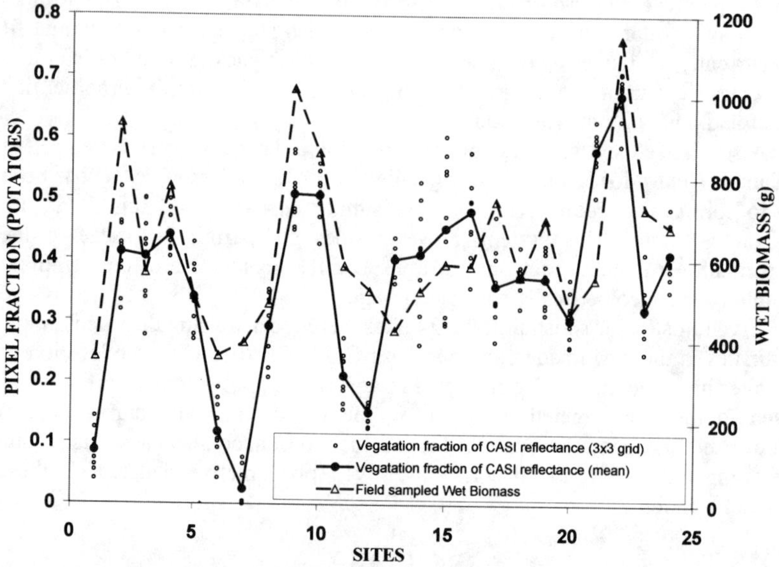

Fig. 1. Endmember fraction (potato) from pixel unmixing and wet biomass of ground samples at each site.

Figure 2 is a scatter plot of these two quantities showing a linear regression with a coefficient of determination (R^2) of 0.476.

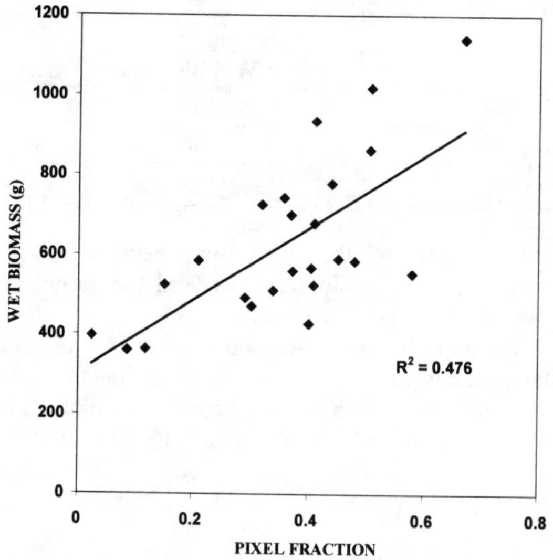

Fig. 2. Relationship between crop wet biomass and endmember fraction (potato) from pixel unmixing.

DISCUSSION

Figure 1 shows an evident relationship between the image pixel vegetation fraction retrieved from hyperspectral data and the measured biomass. For most of the sites, the upward and downward trends are the same for both data sets. At many sites, however the amplitude of the change in these two curves is quite different which partially explains the lower R^2 of the linear relation in Fig. 2. At this stage of our analysis we showed that hyperspectral pixel unmixing procedures can map relative within-field variability of potato biomass. We cannot, however make a quantitative estimate of agricultural parameters with only this method. Further empirical or physical modeling, or a combination of both is required for this purpose (Staenz et al., 1997). The main reasons for this are the location uncertainty of the field samples in the image, the uncertainty of the representation of these field measurements taken from a small site relative to the pixel size of the image, and the heterogeneity of the crop reflectance in the area selected for sampling.

This method would gain enormously if the location of the ground sampling sites in the geometrically noncorrected image could be improved. Future studies should be done in more homogeneous areas with well-marked sampling sites that can be easily identified in the imagery. Field sampling of biophysical parameters should be done on a larger area to better match the image pixel size.

The potential for pixel fractions to map within-field biophysical variability cannot be denied. The advantage of this method, compared to other techniques such as vegetation indices, is its capability to estimate the contribution of each component to the total reflectance of a pixel. This feature is most advantageous in the case of row crop where an important proportion of the pixel reflectance is due to soil or where weed infestation is important. This method also takes full advantage of the spectral information in the image compared with empirical methods establishing a relationship between biophysical data and a limited number (combination) of spectral bands. Contrary to empirical methods, pixel unmixing does not require ground reference data to create a mapping function involving the sensor signal.

CONCLUSION

Preliminary validation tests indicate that the use of linear spectral unmixing to extract endmember fractions has the potential to permit within-field mapping of potato crop properties and eventually some other similar row crops. This method can provide data over an entire field that can easily be compared with other biophysical data in a GIS database. The use of hyperspectral data may not be cost-effective now, but should become a valuable tool for precision agriculture when satellite-based hyperspectral data are available in the near future.

ACKNOWLEDGMENT

The authors wish to thank Tom Szeredi for his patient explanations of pixel unmixing and Christine Burke for preparing the final layout of this document.

REFERENCES

Blackmer, T.M., J.S. Schepers, and G.E. Meyer. 1995. Remote sensing to detect nitrogen deficiency in corn. *In* Proc. Site-Specific Management for Agric. System, Minneapolis, MN. pp. 505–512.

Boardman, J.W. 1995. Analysis, understanding and visualization of hyperspectral data convex sets in N-space. pp.14–20. *In* Proc. SPIE's International Conf. on Imaging Spectrometry, Orlando, FL. Vol. 2480.

Gibson, J.R. 1994. Photogrammetric calibration of a digital electro-optical stereo imaging system. Geomatica. 48(2):95–109.

Hinzman, L.D., Bauer, M.E., Daughtry, C.S. 1986. Effects of nitrogen fertilization on growth and reflectance characteristics of winter wheat. Remote Sens. Environ. 19:47–61.

Ma, B.L., Morrison, M.J., Dwyer, L.M. 1996. Canopy light reflectance and field greenness to assess nitrogen fertilization and yield of maize. Agron. J. 88:915–920.

McMurtrey III, J.E., Chappelle, E.W., Kim, M.S., Meisinger, J.J., Corp, L.A. 1994. Distinguishing nitrogen fertilization levels in field corn (Zea mays L.) with actively induced fluorescence and passive reflectance measurements. Remote Sens. Environ. 47:36–44.

Moran, M.S., Inoue, Y., Barnes, E.M. 1997. Opportunities and limitations for image-based remote sensing in precision crop management. Remote Sens. Environ. 61:319–346.

Shimabukuru Y.E., Smith J.A. 1991. The least squares mixing models to generate fraction images from remote sensing on multispectral data. IEEE Trans. Geosci. Remote Sens. 29:16–20.

Staenz, K., Deguise, J.-C., Chen, J.M., McNairn, H., Brown, R.J., Szeredi, T., and McGovern, M. 1998. The use of hyperspectral data for precision farming. Proceedings of the ISPRS ECO BP'98. Budapest, Hungary (in press).

Staenz, K., and Williams D.J. 1997. Retrieval of surface reflectance from hyperspectral data using a look-up table approach. Can. J. Remote Sens. 23(4):354–368.

Staenz, K., Szeredi, T., Brown, R.J., McNairn, H., and Van Acker, R. 1997. Hyperspectral information techniques applied to agricultural *casi* data for detection of within-field variations. Intl. Symp. on Geomatics in the Era of Radarsat. Ottawa, Ontario. (CD-ROM).

Staenz, K., Schwarz, J., and Cheriyan, J. 1996. Processing/analysis capabilities for data acquired with hyperspectral spaceborne sensors. Acta Astronautica. 39(9–12):923–931.

Application of Remote Sensing to Irrigation Management in California Cotton: Factors Influencing Vegetative Growth

Daniel Munk
University of California Cooperative Extension
Fresno, California

Richard E. Plant
Department of Agronomy and Range Science and Biological and Agricultural Engineering
University of California
Davis, California

ABSTRACT

An improved understanding of linkages between plant crop stress level, actual canopy vegetation, and vegetation index is needed to properly evaluate the relationship between false color infrared aerial photographs and yield. In Part II of this paper, we identify some of the key plant and soil based parameters useful in the evaluation of vegetative indices and yield prediction for irrigated cotton. The accumulation of crop water stress over time is partially related to climatic conditions and soil characteristics such as water holding capacity and rooting depth. These soil water storage parameters were found to be highly effective in predicting an ideal irrigation termination date for cotton and associated canopy development. Our preliminary analysis of one-year results suggests a strong correlation between plant water stress and crop canopy size in the late-season for fields having highly contrasting soil moisture regimes.

Interest of Superspectral Remote Sensing Data and Agronomic Reflectance Models for Crop Management

Bernard Coquil

Matra Marconi Space
Toulouse, France

ABSTRACT

A Critical issue for Crop Management is to get the variability at the subfield level. Ground measurements and yield monitoring techniques are one method to achieve these goals. But remote sensing also can be a very efficient way to reach these objectives with the use of hyperspectral (superspectral) data. MATRA MARCONI SPACE (MMS) one of the largest space companies worldwide has been involved since 3 yrs in the definition of a new commercial satellite system able to generate such superspectral remote sensing data with a high revisit frequency and fast delivery. Working with the National French Agronomic Research Institute and using agronomic reflectance models it was possible to generate from the superspectral images biophysical indexes very important to detect stress (water, disease) or to create application maps (fertilizer). Since the last 2 yrs MMS has performed a lot of commercial campaigns in the USA and in Europe to validate this concept and test on different types of crops. The robustness of the agronomic algorithms and demonstrate the accuracy of the method.

Copyright © 1999 ASA-CSSA-SSSA, 677 South Segoe Road, Madison, WI 53711, USA. *Proceedings of the Fourth International Conference on Precision Agriculture.*

Remote Sensing as a Guide to Directed Sampling for Improved Site-Specific Management

Devon D. Liston
J. S. Schepers

Department of Agronomy
University of Nebraska
Lincoln, Nebraska

ABSTRACT

Now that remotely sensed images of agricultural fields are commercially available, questions arise about proper use of this technology and interpretation of the data. A ground-truthing guide for remotely sensed fields is in progress. One of the two methods studied for this guide includes interpretation of weekly images to guide sampling to characterize soil variability and crop stresses. This is in contrast with pre-season determination of sampling sites based on GIS guided classification that utilizes available image data. The focus on the latter is a nontraditional approach to ground truthing stresses based on GPS guidance to predetermined points. The methodology and timeliness of sampling for soil variability and crop stresses are of interest to the users of remotely sensed products.

Copyright © 1999 ASA-CSSA-SSSA, 677 South Segoe Road, Madison, WI 53711, USA. *Proceedings of the Fourth International Conference on Precision Agriculture.*

Evaluation of General Weed Management (GWM) Model: A Computerized Bioeconomic Weed Decision Aid for Row Crops

Brian L. Broulik
C. L. Reese
S. A. Clay
D. E. Clay
M. M. Ellsbury

South Dakota State University
Plant Science Department
Brookings, South Dakota

ABSTRACT

With the large number of herbicide options available, deciding which herbicide is best for a given situation may be overwhelming resulting in poor or incorrect decisions. GWM, General Weed Management, a bioeconomic model, has been proposed as an approach to simplify herbicide selection. Estimated net return is based on predicted yield and commodity price minus the loss from weed competition and control cost. Weed species and density in the evaluation areas are used as input parameters and a list of herbicide options based on predicted net return and efficacy is generated. The experiment evaluated different weed management strategies at different landscape positions. Five different strategies were tested in a RCBD at three locations in two fields in 1996 and 1997. Locations within fields were selected based on landscape position and previously mapped weed populations. Treatments included a weedy control, the weed management strategy selected by the producer, and three GWM generated treatments (preemerge, postemerge and preemerge plus post emerge). Preliminary results suggest that GWM may perform as good or better than the producer blanket herbicide treatment. Model generated decisions with regards to economics may play a role in the future of weed management as cropping practices move to more intensive management practices with precision farming techniques, and therefore need to be further evaluated. Specific results will be presented later.

Copyright © 1999 ASA-CSSA-SSSA, 677 South Segoe Road, Madison, WI 53711, USA. *Proceedings of the Fourth International Conference on Precision Agriculture.*

Image Analysis of 35-mm Aerial Photography for Estimating Various Plant Parameters

Walter Bausch
Kenan Diker

USDA-ADS Water Management Research
AERC-CSU Foothills Campus
Fort Collins, CO

ABSTRACT

Plant parameters such as leaf area, plant N status, and crop coefficients for crop Et have been successfully estimated from canopy reflectance measured in discrete, narrow spectral wavebands. Site specific water and N management for irrigated corn will require at least weekly inputs from spectral data to assist with management decisions. Unfortunately, weekly acquisition over large fields with narrow-band spectral data is still not feasible. Therefore, the objective of this paper is to determine the usefulness of 35-mm broad-band spectral data for estimating the plant N status of irrigated corn. Thirty-five mm color plus black and white infrared as well as false color infrared aerial photography was acquired at several growth stages over corn plots with established N treatments. The slide film was developed and scanned at five resolutions and written to CD ROM media by a commercial source. Ground-based canopy reflectance was measured with Exotech four band radiometers filtered in the blue, green, red, and near infrared portions of the electromagnetic spectrum. SPAD chlorophyll meter and leaf area measurements as well as plant samples for N analysis were also taken. All of these measurements were made on the same day of the over flights.

Copyright © 1999 ASA-CSSA-SSSA, 677 South Segoe Road, Madison, WI 53711, USA. *Proceedings of the Fourth International Conference on Precision Agriculture.*

PROFITABILITY

Evaluating the Economics of Precision Agriculture

Carol Snyder
Farmer's Software Association
Fort Collins, Colorado

John Havlin
Department of Soil Science
North Carolina State University
Raleigh, North Carolina

Gerard Kluitenberg
Department of Agronomy
Kansas State University
Manhattan, Kansas

Ted Schroeder
Department of Agricultural Economics
Kansas State University
Manhattan, Kansas

ABSTRACT

This study was conducted to evaluate the economics of site-specific N management for irrigated corn in central Kansas. Spatially detailed yield mapping and soil sampling were used to impose spatially variable N rates from which economic comparisons of variable and uniform N management was conducted. Less N was applied using variable-rate rather than uniform N management in every year studied. Variable-rate management was determined to be economically viable in some years while not in others. Additional research is necessary in order to generalize these results for application at other sites.

INTRODUCTION

Site-specific and related information technologies enable one to spatially measure, monitor, and manage factors and inputs that influence crop yield, input efficiency, profitability, and environmental quality. As applications of site-specific technology are evaluated, adoption will be limited by uncertainties in economic returns. The economic benefit from site-specific farm management depends on identifying places in a field where additional input use will increase revenue by a magnitude greater than the added costs, and in identifying places where reduced input use will reduce costs by more than possible revenue reduction attributed to the same or lower yield potential.

Copyright © 1999 ASA-CSSA-SSSA, 677 South Segoe Road, Madison, WI 53711, USA. *Proceedings of the Fourth International Conference on Precision Agriculture.*

Economic returns to site-specific technology are essential for adoption, yet are difficult to quantify (Snyder, 1996). The economic value of spatial information is unknown due to the newness of the technology and limited availability of resources. Since economic and agronomic results are highly dependent on soil, crop, and environmental characteristics in each field, it is difficult to draw generalized conclusions regarding profitability. Positive economic returns to site-specific technology may be achievable, however, management of implicit interactions that are difficult to isolate may be necessary. Economic returns will be difficult to realize if efforts are placed on managing variability that is unmanageable or not a significant of yield variability. Therefore, identifying relevant soil and crop characteristics that limit yield is imperative. It is important for producers to recognize that several years may be required to accurately identify all of the relevant yield limiting factors due to complexity of in-field interactions and temporal influences.

With the elimination of farm programs, commodity markets will become increasingly more volatile due to our global economy while input costs continue to rise. Producers will have to become more efficient and perhaps more specialized to remain competitive and to meet consumer demands. Before the profit potential of site-specific management can be addressed, the spatial interrelationships of all factors that influence the productive capacity of the field must be understood. Economic analysis of crop production, regardless of the technology, requires determining the responsiveness of crop yield to inputs (Lowenberg-DeBoer et al., 1994). The interaction between soil, applied inputs, and the environment must be managed to optimize yield. The success of site-specific management is contingent on the reliability of identifying, quantifying, and effectively managing the spatial distribution in crop yield as well as those factors influencing yield.

This paper examines the economics of site-specific technology relative to variable N management and is based on the field research of Redulla (1998). Benefits from variable application of N are expected to be derived from savings in N costs and/or a positive yield response.

MATERIALS AND METHODS

A study was conducted to compare variable and uniform N management on continuous, center pivot irrigated corn in central Kansas. Five years of information are available at Site 1(1993–1997), and three years at Site 2 (1993–1995). The soil at Site 1 (123 acres) is mapped as Pratt loamy fine sand, and the soil at Site 2 (161 acres) is mapped at Pratt-Tivoli loamy fine sand. The Pratt series (sandy, mixed, slightly acid, thermic Psammentic Haplustalfs) and the Tivoli series (sandy, mixed, slightly acid, thermic Typic Ustipsamments) consists of deep, well drained, rapidly permeable soils formed in sandy eolian deposits on uplands (Redulla et al., 1997; Redulla, 1998). The coarse-textured soil at these sites makes N management important for efficient N utilization.

Spatially distributed yield goals and soil profile NO_3 content were used to prescribe spatially variable N fertilizer rates. Less N has been consistently recommended under variable than uniform application. Yield goals were based on the actual yield maps obtained with minor adjustments for previous yield history. Grid-based soil samples (0-4ft) taken as a single core within each cell (0.75 ac) , were collected to determine profile NO_3, (Table 1 and 2). Point data was block kriged to come up with whole cell estimates. Spring NO_3 levels were low averaging 6.8 ppm at Site 1 and 6.0 ppm at Site 2.

The Kansas State University Soil Testing Laboratory N recommendation model (Lamond, 1994) was used to determine the N application rate for the VRT cells and is described by:

$$N_{rec} = (1.35 * YG * 1.1) - (7.5 * NO_3) \quad [1]$$

where:
N_{rec} = N fertilizer recommendation (lbs N ac^{-1})
YG = yield goal (bu/ac)
1.1 = textural adjustment factor for sandy soil
NO_3 = profile soil nitrate content (ppm N ac^{-1}, 0-2ft soil depth)

A new N_{rec} was calculated each spring based on yield goal and spring profile NO_3. As information for additional years became available, revisions were made to the grid-based yield goal which had an impact on N_{recs}. Environmental conditions had a significant influence on the differences in yield between years at both sites. Excessive rain in 1993 and 1995 reduced yield at each site an average of 40 to 50 bu ac^{-1} relative to years with more ideal growing conditions.

Although Nrecs for the VRT treatments represent a continuous range of N rates, VRT Nrecs were grouped into six N rates to facilitate N application. The VRT N rates were 130, 160, 190, 220, 250, and 280 lbs N ac^{-1} (1994-1995), 160, 190, 220, and 250 lbs N ac^{-1} (1996) and 180, 195, 210, 225, and 255 lbs N ac^{-1} (1997). Application of N was allocated to VRT cells based on the N_{rec} for that cell. The uniform N rate was a field average recommendation of 230 lbs N ac^{-1} at Site 1 (1994–1996) and at Site 2 (1994–1995) and was adjusted up to 240 lbs N ac^{-1} at Site 1 in 1997. A completely randomized block design with 12 replications was used to divide the fields into uniform and variable N treatments. Each experimental plot was composed of six 180 by 180 ft (0.75 acre) contiguous cells. Treatment effects on grain yield were measured in the fall with a yield monitoring combine equipped with a global positioning system (GPS).

To evaluate the influence of N and other soil and field characteristics on grain yield, a production function was developed following procedures outlined in Snyder et al. (1997). A quadratic function was estimated using data from both uniform and variable treatments (Eq. [2]). A separate production function was estimated for each site and each year. The quadratic function has commonly been used in quantifying corn yield response to N (Bundy & Andraski, 1995; Cerrato & Blackmer, 1990; Featherstone et al., 1991; Schlegel & Havlin, 1995, Snyder et al., 1997).

The quadratic model used was:

$$Y = \beta_0 + \beta_1 NAVAIL + \beta_2 NAVAIL2 + \beta_3 NOVER + \beta_4 NUNDER + \beta_5 ELEV + \beta_6 CHGELEV + \beta_7 OMA + \beta_8 PHA + \beta_9 PHA2 + \beta_{10} SANDA + \beta_{11} CLAYA + \beta_{12} CLAYA2 + \beta_{13} DPTHCLAY + \beta_{14} CLAYA \quad (2)$$

Definitions and summary statistics of variables in Eq. [2] are presented in Tables 1 and 2.

The relationship of most interest in the evaluation of VRT N management is the interaction of N and yield. Total N available (NAVAIL) was calculated as the sum of N applied and NO_3 present in the soil profile (0–2 ft). The difference between N applied and the N_{rec} represents the amount of over or under application of N relative to the recommended application (Eq. [1]) for each cell. The average absolute value of misapplication on the VRT cells is small (15 lbs N ac maximum^{-1}) since the exact N_{rec} was not applied when grouping the N_{rec} into six N rates. Variation in the magnitude of misapplication on the uniform cells is greater due to differences in the field average N recommendation and the actual N_{rec} for each uniform treatment cell. A differentiation was made between over and under application of N to evaluate the yield response to misapplication of N. From these estimates, a distinction was made between results of farming the whole field using VRT or uniform N management and the benefits and costs examined.

RESULTS AND DISCUSSION

The production functions were estimated using ordinary least squares regression. The parameter estimates by site and year are presented in Table 3. The models explain from 17 to 48% of per acre yield variability within each field for a particular year. Site 1 had coefficients of determination (R^2) that increased each year (0.17–0.43) indicating that yield variability was explained better each year as more information became available to set yield goals and therefore N recommendations. This suggests the importance of conducting such analysis on a year-by-year and field-by-field basis since additional variability was explained as more information became available to improve estimated N_{recs} based on productive information.

The amount of N available (NAVAIL) was positively correlated with yield at Site 1 in 1994 and 1997 (Table 3) and at Site 2 in 1994 and 1995. The nonlinear NAVAIL2 term was negatively correlated with yield and statistically significant in 1994 and 1997 at Site 1 and in 1996 at Site 2. Nitrogen over-applied (NOVER) was negatively correlated to yield and statistically significant at each site and for each year (Table 3). Nitrogen under-applied (NUNDER) was negatively related to yield in 1994 and 1995 at each site. The impact of misapplication of N on corn yield at these sites is significant and reflects the importance of proper determination of N fertilizer recommendations. Based on the N recommendation model (Eq. [1]), primary emphasis should be placed on setting accurate yield goals given the low spring NO_3 levels (Tables 1 and 2).

The key to this study was to examine the net returns to VRT over uniform N management. Net return is defined as the additional return from VRT less additional VRT costs. Additional return is the change in VRT income derived from a change in yield and changes in fertilizer cost as compared with uniform N application. Additional returns are calculated for each cell and totaled for the field. To estimate the effects of VRT and uniform N application on yield for the entire field, the prescribed N application rates along with the parameter estimates from the production function (Table 3) were used to determine expected yields for each cell in the years relevant to each site. Additional returns to VRT N management can be determined from the expected total N application and predicted yield response.

Comparing variable and uniform N management treatments, the predicted amount of total N applied at each site was always less under VRT N management than uniform N management (Table 5). Average savings in N applied using VRT over uniform application was 10% although variations occur by year based on the determination of yield goal and spring NO_3. With less N applied, overall N expense was reduced at both sites under VRT N management for the time periods evaluated. These savings will continue to be significant as N fertilizer prices increase.

Net returns per acre were calculated as the difference between VRT and uniform N management (Table 6). Changes in returns were based on predicted differences in corn yield and N use under each N management treatment. At Site 1, positive net returns to VRT N management over uniform N management were observed. Assuming a \$2.70/bu corn price and \$0.29/lb N, average net returns at Site 1 ranged from \$8.88/ac to \$33.78/ac. At Site 2, differing results were obtained between 1994 and 1995. At the same price levels, Site 2 net returns to variable N management over uniform N management averaged \$42.38/ac in 1994, but -\$6.37/ac in 1995. These results may reflect the difference in growing conditions between years.

The additional costs of VRT consist of additional N application, soil sampling, laboratory analysis, labor, yield monitor/GPS receiver and data management costs (Table 7). These costs are based on surveys and estimates from site-specific farming users. Given an estimated additional cost of \$13.05/ac, VRT N management was an economically viable option at Site 1 in 1995 and 1996 and at Site 2 in 1994. Those years where costs weren't recovered were years with unique environmental conditions that may have affected the outcome of the results. When evaluated across years, however, the average return to VRT management was \$4.11/ac at Site 1 (1994–1997) and \$11.23/ac at Site 2 (1994–1995).

CONCLUSION

The ability of precision farming technology to monitor and manage factors that cause yield variability provides farmers with a higher level of management information and control. Increased efficiency obtained through increased knowledge will boost profitability and drive technological adoption. Motivation for adoption of site-specific farming depends on variability present in a field. The key to accurate site-specific analysis is the identification of variability that is within a relevant range and is manageable. Variable N management is expected to reduce the number of areas in the field to which N is either over or under applied. Results from this study consistently indicate that less total N fertilizer was used with VRT than uniform N management. The potential for net returns to VRT N management to provide at least profitable results were illustrated, although not always positive, in predictions from each field during the years studied. The importance of setting accurate yield goals is essential on these fields, since spring NO_3 levels are low. The results confirm the potential for profitable use of precision N management. Recording environmental impacts and timing of weather events is crucial in effective analysis on a year-by-year basis. For the adoption of site-specific farming technologies to be economically viable for producers, benefits of using the system must outweigh the costs. Benefits from site-specific agriculture are expected to be derived from savings in input costs and/or increased yields. Coupled with future reductions in technology costs associated with site-specific technology, VRT N management will continue to have promise of being a profitable investment over traditional N management. Additional studies are warranted to document a methodology to be used to define manageable and economically variable factors in order to draw generalized conclusions regarding profitability of VRT N management across all crop and field conditions.

Table 1. Definition and summary statistics of variables used in the production function at Site 1.

Variable	Definition	Site 1 Mean	St. Dev.	Min.	Max.
Y93	1993 corn yield (bu ac^{-1})	166.5	21.6	91.6	201.4
Y94	1994 corn yield (bu ac^{-1})	187.4	20.6	132.0	247.0
Y95	1995 corn yield (bu ac^{-1})	136.3	29.9	4.8	181.2
Y96	1996 corn yield (bu ac^{-1})	206.4	15.7	146.6	231.5
Y97	1997 corn yield (bu ac^{-1})	214.7	11.6	134.7	233.8
NAVAIL94†	1994 total N available (N applied+spring NO_3) (lbs N ac^{-1})	299.6	24.5	219.7	368.7
NAVAIL95†	1995 total N available (N applied+spring NO_3) (lbs N ac^{-1})	276.5	19.7	202.8	322.2
NAVAIL96†	1996 total N available (N applied+spring NO_3) (lbs N ac^{-1})	291.2	27.7	199.6	345.2
NAVAIL97†	1997 total N available (N applied+spring NO_3) (lbs N ac^{-1})	273.9	16.5	208.8	313.6
NOVER94‡	1994 N over-applied (lbs N ac^{-1})	10.5	21.8	0.0	141.9
NOVER95‡	1995 N over-applied (lbs N ac^{-1})	13.5	20.5	0.0	100.1
NOVER96‡	1996 N over-applied (lbs N ac^{-1})	25.3	27.5	0.0	130.6
NOVER97‡	1997 N over-applied (lbs N ac^{-1})	10.4	19.0	0.0	90.5
NUNDER94§	1994 N under-applied (lbs N ac^{-1})	-8.7	12.6	-52.2	0.0
NUNDER95§	1995 N under-applied (lbs N ac^{-1})	-3.0	5.5	-22.9	0.0
NUNDER96§	1996 N under-applied (lbs N ac^{-1})	-1.6	3.8	-15.9	0.0
NUNDER97§	1997 N under-applied (lbs N ac^{-1})	-2.5	4.1	-21.5	0.0
NO3S94	Spring 1994 soil nitrate (ppm)	6.1	1.5	4.1	9.4
NO3S95	Spring 1995 soil nitrate (ppm)	6.5	1.3	3.9	9.8
NO3S96	Spring 1996 soil nitrate (ppm)	9.8	1.8	6.8	14.4
NO3S97	Spring 1997 soil nitrate (ppm)	5.0	1.1	3.2	9.2

† N available is calculated as the sum of N applied plus NO_3 present in the soil

‡ N over-applied is positive difference between N applied and the N recommendation

§ N under-applied is negative difference between N applied and the N recommendation

Table 2. Definition and summary statistics of variables used in the production function at Site 2.

Variable	Definition	Site 2 Mean	St. Dev.	Min.	Max.
Y93	1993 corn yield (bu ac^{-1})	123.5	26.9	62.0	179.9
Y94	1994 corn yield (bu ac^{-1})	174.3	25.3	91.0	233.0
Y95	1995 corn yield (bu ac^{-1})	114.5	20.9	68.6	169.8
NAVAIL94†	1994 total N available (N applied+spring NO_3) (lbs N ac^{-1})	248.0	32.1	154.2	311.8
NAVAIL95†	1995 total N available (N applied+spring NO_3) (lbs N ac^{-1})	235.8	21.3	162.1	297.3
NOVER94‡	1994 N over-applied (lbs N ac^{-1})	23.1	31.8	0.0	134.1
NOVER95‡	1995 N over-applied (lbs N ac^{-1})	17.9	23.8	0.0	119.7
NUNDER94§	1994 N under-applied (lbs N ac^{-1})	-7.2	15.3	-74.4	0.0
NUNDER95§	1994 N under-applied (lbs N ac^{-1})	-6.4	11.0	-64.9	0.0
NO3S94	Spring 1994 soil nitrate (ppm)	6.5	1.3	3.9	9.8
NO3S95	Spring 1995 soil nitrate (ppm)	5.5	3.6	1.0	23.6

† N available is calculated as the sum of N applied plus NO_3 present in the soil.
‡ N over-applied is positive difference between N applied and the N recommendation.
§ N under-applied is negative difference between N applied and the N recommendation.

Table 3. Corn yield production function estimates, by site and year†.

	Site 1				Site 2	
Variable	1994	1995	1996	1997	1994	1995
Intercept	41.99	-141.89	357.37*	10.06	85.54	144.97
NAVAIL	1.81*	-0.93	-0.35E-01	2.78**	1.10**	0.44
NAVAIL2	-0.30E-02*	0.198E-02	0.46E-03	-0.47E-02**	-0.18E-02*	-0.30E-04
NOVER	-0.29**	-0.30**	-0.22**	-0.29**	-0.43**	-0.34**
NUNDER	-0.70E-01	-0.88*	0.17	0.11	-0.17	-0.66**
ELEV	-2.01**	-0.28	-0.66	-0.74E-01	3.94**	1.00
CHGELEV	1.26	1.85	-1.30	-0.31	-7.63**	0.43
OMA	-0.46	-31.37**	-4.57	5.05	-8.89*	0.97
PHA	-29.63	135.32	-33.38	-60.98	50.58	5.06
PHA2	2.11	-10.56	2.53	4.82	-4.39	-0.19
SANDA	-0.26	-0.12	-0.72	-0.15	-1.97**	-1.63
CLAYA	0.25	-1.59	0.77	0.81	-2.25	-0.31
CLAYA2	-0.65E-01	0.33E-01	-0.12	-0.49E-01	0.34E-01	0.65E-01
DPTHCLAY	-0.12	0.23	-0.21	0.36E-01	0.98E-01	0.52
CLAY	4.92	6.89	8.30*	0.68	0.73	-6.27
R^2	0.17	0.22	0.28	0.43	0.48	0.38
Obs.	161	161	161	161	162	162

*, ** Significant at 0.10 level, at 0.05 level

† Estimates obtained using the entire data set for each site where ½ the field was farmed with uniform N management and ½ with VRT N management.

Table 4. Definition and summary statistics of predicted N management variables (lbs N ac^{-1}).

Variable	Definition	Site 1				Site 2			
		Mean	St. Dev.	Min.	Max.	Mean	St. Dev.	Min.	Max.
Site 1									
NREC94	1994 N available under VRT management	290.5	33.7	210.7	359.7	232.1	50.6	117.2	336.5
NREC95	1995 N available under VRT management	269.8	24.8	202.8	322.2	224.3	32.4	139.2	305.3
NREC96	1996 N available under VRT management	267.5	25.2	183.8	324.4	n/a	n/a	n/a	n/a
NREC97	1997 N available under VRT management	265.9	19.9	202.8	298.8	n/a	n/a	n/a	n/a
NUNI94	1994 N available under uniform management	300.5	11.6	286.5	329.2	258.1	2.82	252.5	263.1
NUNI95	1995 N available under uniform management	282.7	9.4	261.3	302.0	240.9	7.24	232.1	277.2
NUNI96	1996 N available under uniform management	308.0	14.6	284.4	345.2	n/a	n/a	n/a	n/a
NUNI97	1997 N available under uniform management	279.7	9.12	265.6	313.6	n/a	n/a	n/a	n/a
NOVRUNI4	1994 N over-applied on uniform plots	16.1	27.9	0.0	141.9	35.8	37.2	0.0	138.1
NOVRUNI5	1995 N over-applied on uniform plots	21.2	24.8	0.0	100.1	23.3	24.9	0.0	99.7
NOVRUNI6	1996 N over-applied on uniform plots	40.7	26.5	0.0	131	n/a	n/a	n/a	n/a
NOVRUNI7	1997 N over-applied on uniform plots	16.8	20.2	0.0	90.5	n/a	n/a	n/a	n/a
NUNDUNI4	1994 N under-applied on uniform plots	-12.5	15.2	-52.2	0.0	-9.83	17.9	-74.4	0.0
NUNDUNI5	1995 N under-applied on uniform plots	-2.2	5.4	-22.9	0.0	-6.6	12.9	-64.9	0.0
NUNDUNI6	1996 N under-applied on uniform plots	-0.2	1.75	-20.0	0.0	n/a	n/a	n/a	n/a
NUNDUNI7	1997 N under-applied on uniform plots	-2.9	5.6	-25.6	0.0	n/a	n/a	n/a	n/a

Table 5. Total N applied (lbs) under different N management scenarios.

Location	Method of N management		Change in total VRT N applied over uniform application	
	VRT	Uniform		
Site 1				
1994	36 334	37 260	(926)	-3%
1995	34 625	37 260	(2 635)	-8%
1996	30 652	37 260	(6 608)	-22%
1997	36 608	38 640	(2 032)	-5%
Site 2				
1994	33 229	37 490	(4 261)	-13%
1995	34 750	37 490	(2 740)	-8%

Table 6. Net returns ($/ac) to VRT N over uniform N management. †

	1994	1995	1996	1997
Site 1:	$10.13	$33.78	$15.85	$8.88
Site 2:	$42.38	($6.37)	n/a	n/a

† Assuming $2.70/bu corn price and $0.29/lb N

Table 7. Additional per acre costs associated with VRT N Management.

Additional Spreader Charge	$2.00
Soil Sampling Costs	$7.00
Labor ($8 h^{-1} @ 6 samples h^{-1})	$1.30
Yield Monitor/GPS Receiver	$2.00
Data Management($0.50 ac^{-1} maps + $0.25 ac^{-1} record keeping)	$0.75
Total Additional Cost of VRT Practices (per acre):	$13.05

REFERENCES

Bundy, L.G., and T.W. Andraski. 1995. Soil yield potential effects on performance of soil nitrate tests. J. Prod. Agric. 8:561–567.

Cerrato, M.E., and A.M. Blackmer. 1990. Comparison of models for describing corn yield response to nitrogen fertilizer. Agron. J. 82:138–143.

Featherstone, A.M., J.J. Fletcher, R.F. Dale, and H.R. Sinclair. 1991. Comparison of net returns under alternative tillage systems considering spatial weather variability. J. Prod. Agric. 4:166–173.

Lamond, R.E. 1994. Nutrient management. p. 12–15. *In* Corn production handbook. Agric. Exp. Stn. and Coop. Ext. Serv. Publ. C-560 Rev., Kansas State Univ., Manhattan.

Lowenberg-DeBoer, J., S. Hawkins, and R. Nielsen. 1994. Economics of precision farming. Purdue Univ., West Lafayette, IN.

Redulla, C.A. 1998. Variable-rate nitrogen management for irrigated corn. Ph.D. diss. Kansas State Univ., Manhattan.

Redulla, C.A., J.L. Havlin, G.J. Kluitenberg, N. Zhang, and M.D. Schrock. 1997. Variable nitrogen management for improving groundwater quality. p. 1101–1110. *In* Proc. 3rd Int. Conf. on Precision Agriculture, Minneapolis, MN. 23-26 June, 1996. ASA, CSSA, and SSSA, Madison, WI.

Schlegel, A.J., and J.L. Havlin. 1995. Corn response to long-term nitrogen and phosphorus fertilization. J. Prod. Agric. 8:181–185.

Snyder, C.J. 1996. An economic analysis of variable-rate nitrogen management using precision farming methods. Ph.D. diss. Kansas State Univ., Manhattan.

Snyder, C., T. Schroeder, J. Havlin, and G. Kluitenberg. 1997. An economic analysis of variable-rate nitrogen management. p. 989–998. *In* Proc. 3rd Int. Conf. on Precision Agriculture, Minneapolis, MN. 23-26 June, 1996. ASA, CSSA, and SSSA, Madison, WI.

Spatial Break-Even Variability for Variable Rate Technology Adoption

B. C. English
R. K. Roberts
S. B. Mahajanashetti
Department of Agricultural Economics
University of Tennessee
Knoxville, Tennessee

ABSTRACT

Adopting variable-rate technology (VRT) by agricultural producers depends on several factors including spatial variation in yield potential. Many farmers currently use consulting firms to implement VRT. A farmer breaks even using the technology when there exists a minimum yield variability such that the VRT custom cost is offset by additional returns. A farmer with greater spatial variability than the break-even variability will employ VRT. In this analysis, we have first presented a methodology for determining this spatial break-even variability and then applied it to a hypothetical 30-ac corn field consisting of two land qualities. Break-even variability was determined for a range of nitrogen and corn prices by varying land quality proportions. For the response functions assumed and a custom VRT rate of $4.67/ac, VRT was economically feasible within range of 15 to 70% of the field in good quality land.

INTRODUCTION

Farm fields are characterized by areas with varying potential to produce crop output. Across a given field, one could find variability with respect to soil type, nutrient status, landscape position, organic matter content, water holding capacity and so on. Variation in these factors leads to variation in the potential of different areas to utilize applied inputs and produce crop output (Carr et al., 1991; Hibbard et al., 1993; Malzer et al., 1996; Sawyer, 1994; Snyder, 1996; Wibawa et al., 1993).

Precision farming or site-specific farming uses site-specific information to identify homogeneous subareas within a field, make site-specific prescriptions, and apply inputs at spatially variable rates. Since precision technology helps in matching input application to crop and soil needs, under- and over-application of inputs like seeds, fertilizers and pesticides is reduced. As a result, the adoption of this technology is expected to result in greater yields and/or reduced inputs (Morgan and Ess, 1997; Sawyer, 1994; Snyder, 1996). Though management of within-field variability holds the promise of economic and environmental benefits, voluntary adoption of variable rate technology (VRT) is likely to be most dependent on profitability (Daberkow, 1997; Sawyer, 1994). Profitability is

Copyright © 1999 ASA-CSSA-SSSA, 677 South Segoe Road, Madison, WI 53711, USA. *Proceedings of the Fourth International Conference on Precision Agriculture.*

one favorable outcome desired by virtually all producers in a market economy for voluntary technology adoption (Lowenberg-DeBoer-J & Swinton, 1995).

Economic benefits from switching to VRT from uniform application methods depend on the economic value of assessing and treating within-field variability (Forcella, 1993; Hayes et al., 1994). This value in turn depends upon the extent of within-field variability. In nearly homogeneous fields, the cost incurred to assess variability and use the VRT might not be offset by additional returns generated. As a result, economic returns to VRT could be negative (Sawyer, 1994). From a purely economic perspective, therefore, there are two factors driving adoption of precision farming technology -- 'spatial variability' (Morgan and Ess, 1997) and the magnitude of spatial yield differences.

Forcella (1993) presented a nice hypothetical illustration of how within-field variability could influence the economic outcomes of VRT. He created 11 hypothetical fields, each field composed of 10 hectares with two types of soil in different proportions. Assuming that corn yield response to nitrogen (N) was characterized by a linear response and a plateau (LRP) function for both soil types, he calculated misapplication costs that would be incurred with average rate application. Misapplication costs were equivalent to money saved if soils were managed by prescription and they increased with increasing within-field variability. The analysis also showed changes in how the input commodity-price ratio would impact misapplication costs. The cost of technology, however, was not explicitly considered in the analysis.

Many VRT farmers currently use consulting firms to develop nutrient application maps and to custom apply nutrients (Morgan & Ess, 1997; Lowenberg-DeBoer & Swinton, 1995; Swinton & Ahmad, 1996). A farmer seeking to purchase VRT services on a custom-hire basis is primarily faced with the question, "Do additional returns generated at least cover custom charges?" Since the extent of economic returns from VRT is dependent on the extent of spatial variability within a field, the farmer ultimately needs to know the minimum spatial variability at which he/she can break even using the technology. A farmer with within-field variability that is higher than this minimum will employ VRT; one with lower variability will prefer the uniform application method. Knowledge of this minimum variability can help the farmer make appropriate economic decisions about the adoption of VRT.

We are not aware of any analytical illustrations that explicitly address the issue of minimum spatial variability at which a farmer can break even with the cost of using VRT. The objective of this analysis was to develop a hypothetical model and present a methodology for determining this minimum variability, which we refer to as *spatial break-even variability,* for VRT adoption. The role of input and crop prices and custom charges in determining spatial break-even variability will also be examined.

METHODS

In this section, we first illustrate a methodology for determining spatial break-even variability and later apply the methodology to a hypothetical field situation. Assume a corn field consisting of two land types: one with a high yield response to N covering an area of A_h acres and the other with a low yield response occupying A_l acres. Economic analysis of crop production, regardless of the technology used, requires determining the responsiveness of crop yields to inputs (Snyder, 1996). Say, the yield responses for the two land types in the field are represented by the following quadratic response functions.

$$Y_h = \alpha_h + \beta_h*N_h + \gamma_h*N_h*N_h \quad [1]$$

and,

$$Y_l = \alpha_l + \beta_l*N_l + \gamma_l*N_l*N_l \quad [2]$$

Where Y_h is estimated corn yield (bu ac^{-1}) and N_h is N applied (lb ac^{-1}) on the high yield potential land; Y_l is estimated corn yield (bu ac^{-1}) and N_l is N applied (lb ac^{-1}) on the low yield potential land; and α_h, β_h, γ_h, α_l, β_l, and γ_l are estimated parameters for the two response functions. When using VRT, the farmer applies N to each land type based on the individual response functions. On the other hand, he/she relies on an average yield response function when average rate application methods are used. The average yield response function for the field can be represented as

$$Y_a = \alpha_a + \beta_a*N_a + \gamma_a*N_a*N_a \quad [3]$$

Where Y_a is estimated average corn yield (bu ac^{-1}) for the field; N_a is uniform rate of N applied (lb ac^{-1}) on the field; and α_a, β_a, and γ_a are the estimated parameters. The parameters of the average function are weighted averages of the parameters of the two individual response functions, with proportions of the two land types in the field as weights. Thus, for a given proportion of the two kinds of land in the field, the parameters of the average response function can be calculated as

$$\alpha_a = \alpha_h* Pr^h + \alpha_l* Pr^l \text{ ; } \beta_a = \beta_h* Pr^h + \beta_l* Pr^l \text{ ; and } \gamma_a = \gamma_h*Pr^h + \gamma_l* Pr^l \quad [4]$$

Where Pr^h and Pr^l are the proportions of high and low yield potential areas in the field, respectively; $Pr^h = A_h / (A_h + A_l)$, $Pr^l = A_l / (A_h + A_l)$; and $Pr^h + Pr^l = 1$.

Economically Optimum Nitrogen Application

Variable Rate Application Technology

A profit-maximizing farmer tries to optimize the use of an input in crop production by using it up to the point at which the additional returns equal the additional costs. In our example, precision application of nitrogen on the corn field assumes that the farmer follows this principle for each land type, using the information provided by the individual response functions (Eq. [1] and [2]). Denoting the price of N as P_n and the price of corn as P_c, the profit-maximizing conditions under VRT are

$$(\beta_h + 2\gamma_h*N_h) = P_n / P_c \quad [5]$$

for the high yielding area and,

$$(\beta_l + 2\gamma_l*N_l) = P_n / P_c \quad [6]$$

for the low yielding area. Expressions on the left hand side (LHS) of Eq. [5] and [6] are marginal physical productivities of nitrogen derived from the individual response functions. Solving Eq. [5] and [6] provides the optimum N quantities, N_h^* and N_l^*, such that

$$N_h^* = [(P_n / P_c) - \beta_h] / 2\gamma_h \quad [7]$$

$$N_l^* = [(P_n / P_c) - \beta_l] / 2\gamma_l \quad [8]$$

Optimum good and poor land yields (Y_h^* and Y_l^*) are obtained by substituting N_h^* and N_l^* into Eq. [1] and [2], respectively. The optimum returns above nitrogen costs for the field, R^*_{VRT}, can then be calculated

$$R^*_{VRT} = A_h*(P_c*Y_h^* - P_n*N_h^*) + A_l*(P_c*Y_l^* - P_n*N_l^*) \quad [9]$$

Average Rate Application

When the farmer instead prefers use uniform rate application technology, he/she is assumed to base input optimization on the average response function represented by Eq. [3]. The profit-maximizing level of N_a^* can be calculated from the optimality condition

$$\beta_a + 2\gamma_a*N_a = P_n / P_c \quad [10]$$

Solving the above equation gives

$$N_a^* = [(P_n / P_c) - \beta_a] / 2\gamma_a \quad [11]$$

Optimum return above nitrogen cost from the field, R^*_{ART}, can then be calculated as $R^*_{ART} = (A_h + A_l)*(P_c*Y_a^* - P_n*N_a^*)$ [12]

Where Y_a^* is the optimum yield obtained by substituting N_a^* into Eq. [3].

Spatial Break-Even Variability

Whether the farmer adopts VRT, is shown by the relationship: $(R^*_{VRT} - R^*_{ART})$ O C, where R^*_{VRT} and R^*_{ART} are the returns from VRT and average rate application as given by Eq. [9] and [12] and C is the additional custom charge the farmer has to pay for implementing VRT. The LHS being larger than the right hand side of the equation (RHS) implies preference for VRT; the RHS being larger implies the preference for uniform application. When within-field variability leads to equality between the LHS and the RHS in the above relationship, the farmer breaks even using the technology. It is this within-field variability that we refer to as *spatial break-even variability*. In a field setting like the one we are considering, the spatial break-even varibility proportions are defined as the minimum and the maximum values of *Pr* so the farmer can at least cover the additional cost incurred with VRT implementation. Values of Pr^l and Pr^h beyond this range result in negative economic returns from VRT adoption.

To determine the spatial break-even variability proportions, different proportions of low and high yielding lands (Pr^l and Pr^h) in the field can be simulated. Each time the land proportions are changed, R^*_{VRT} (Eq. [9]) changes. Further, there will be a change in average response function (Eq. [3]) leading to changes in N_a^* (Eq. [10]), Y_a^* and R^*_{ART} (Eq. [11]). For every change in land proportions, the difference between R^*_{VRT} and R^*_{ART} can be compared with C and spatial break-even variability proportions identified.

Application to a Hypothetical Field

For purposes of exposition, we assumed a 30-ac corn field. The corn yield response functions for nitrogen on the high and low yield potential lands in the field (Eq. [1] and [2]) were represented as

$$Y_h = 120 + 1.04*N_h - 0.0025*N_h* N_h \quad [12]$$

$$Y_l = 75 + 0.33*N_l - 0.00102*N_l* N_l \quad [13]$$

First, break-even analysis was conducted for the base scenario with P_n = \$0.33/lb, assuming that the source of nitrogen is urea-ammonium nitrate (Duke, 1997) P_c = \$2.56/bu, a three year average corn price (Anonymous, 1997); and C = \$4.67/ac, an estimated cost of VRT (Caldwell, 1997).

For determining spatial break-even variability, the proportion of the poor land in the field was varied from 0 to 100 % in 5% increments. Sensitivity of break-even variability proportions to the changes in nitrogen and corn prices (P_n and P_c) as well as the additional cost associated with VRT (C) was also analyzed.

RESULTS

Economic outcomes of precision and average rate applications of nitrogen to corn changed depending on the proportion of the two kinds of land in the field. Table 1 shows optimum variable rate N applications for the two types of land and the optimum average rate N application for the entire field, for all land proportions considered. Table 1 also displays optimum returns above the cost of nitrogen under the two application methods. The optimum N application under the average rate application method varied with changes in land proportions. The last column shows that differences in returns above the cost of nitrogen between the two application methods increased with increases in the proportion of the low yielding land in the field. Conversely, when the field was more than 60% low yielding land, differences in returns declined as the proportion of low yielding land increased.

For this 30-ac field, the additional cost of custom hiring precision farming services is \$140.10 or \$4.67 per acre times 30 acres. This cost would be covered if the field were >30% and <85% low yielding land. These two percentages of low yielding land are the spatial break-even variability proportions. A farmer seeking to implement VRT in a field with the yield characteristics portrayed in equations 12 and 13 should determine whether the yield variability in the field falls between the minimum and maximum spatial break-even variability proportions. If the field's yield variability is within these limits, the farmer can increase net returns by hiring precision farming services; otherwise, the farms would be better off using uniform rate technology.

Sensitivity of Break-Even Variability

Change in Nitrogen to Corn Price Ratio

Spatial break-even variability changes depending on the shapes of the response functions and input and output prices. For given response functions, changes in the input-output price relationship lead to changes in optimum quantities of N and optimum returns above the cost of N; therefore, spatial break-even variability proportions also change with changes in the input-output price ratio. For purposes of illustration, the N-corn price ratio, P_n / P_c, was varied upward and downward by 5% by (i) changing P_n, keeping P_c constant, and (ii) changing P_c, keeping P_n constant. Table 2 presents the impact of 5% price changes on the spatial break-even variability proportions. With a rise in the price of nitrogen, the minimum proportion requirements decrease for both the high and low yielding areas. The opposite happens when the price of nitrogen falls. The corn price, similarly, influences these spatial break-even variability proportions but in the opposite direction.

Table 1. Optimum returns above N cost under variable rate and average rate technology, hypothetical 30-ac corn field.

Proportion of Low Yield Area (Pr^l)	Optimum N†			Optimum Returns‡ Above N Cost		$R^*_{VRT}-R^*_{ART}$
	N_h^*	N_l^*	N_a^*	R^*_{VRT}	R^*_{ART}	
	…… (lb/ac) ……			………. (dollars /acre)…………		
0.00	182.36	-	182.36	15588.45	15588.45	0.00
0.05	182.36	98.91	180.60	15135.09	15108.41	26.68
0.10	182.36	98.91	178.74	14681.73	14629.59	52.14
0.15	182.36	98.91	176.75	14228.37	14152.11	76.27
0.20	182.36	98.91	174.63	13775.01	13676.10	98.92
0.25	182.36	98.91	172.37	13321.66	13201.71	119.95
0.30	182.36	98.91	169.94	12868.30	12729.12	139.18
0.35	182.36	98.91	167.32	12414.94	12258.53	156.40
0.40	182.36	98.91	164.51	11961.58	11790.18	171.40
0.45	182.36	98.91	161.47	11508.22	11324.33	183.88
0.50	182.36	98.91	158.17	11054.86	10861.31	193.55
0.55	182.36	98.91	154.59	10601.50	10401.47	200.03
0.60	182.36	98.91	150.67	10148.14	9945.27	202.87
0.65	182.36	98.91	146.38	9694.78	9493.23	201.56
0.70	182.36	98.91	141.66	9241.42	9045.97	195.46
0.75	182.36	98.91	136.43	8788.06	8604.26	183.80
0.80	182.36	98.91	130.61	8334.70	8169.04	165.67
0.85	182.36	98.91	124.10	7881.34	7741.46	139.88
0.90	182.36	98.91	116.77	7427.98	7322.99	105.00
0.95	182.36	98.91	108.44	6974.62	6915.46	59.16
1.00	-	98.91	98.91	6521.27	6521.27	0.00

† N_h^* is the optimum N/ac for the high yield area and N_l^* is the optimum N/ac for the low yield area for VRT. N_a^* is the optimum N/ac for average rate technology.
‡ R^*_{VRT} is the optimum return from VRT and R^*_{ART} is the optimum return for average rate technology.

Table 2. Impact of changes in nitrogen and corn prices on spatial break-even variability proportions

Change in price ratio	Spatial break-even variability proportion	
	Minimum. share of low yield response area	Maximum. share of low yield response area
	.... (percent)	....
Increasing Price Ratio by 5% through a:		
Rise in P_n	28.7	85.9
Fall in P_c	30.5	84.8
Decreasing Price Ratio by 5% through a:		
Fall in P_n	32.0	84.0
Rise in P_c	30.0	85.1

Return-Difference Contour Analysis

In our example, we assumed that the corn grower knew the price of nitrogen and VRT custom charges. Figure 1 shows differences in returns per acre between variable and uniform technologies for various combinations of N prices and land proportions assuming a constant corn price. The axis labeled LANDRAT measures the proportion of the low yield area in the field. The axis labeled PIN measures N price and the vertical axis labeled NRTNDIFF measures the per acre difference in returns.

Figure 2 presents the contours of return differences. Each contour is a locus of points representing different combinations of N price (PIN) and the proportion of low yield area in the field (LANDRAT), holding the return constant. Consider the contour representing a return difference of $4.67/ac. If custom charges are set at $4.67/ac as in our previous example, the points A and B indicates spatial break-even variability proportions, when N price is $0.33/lb. Any proportion of the two land qualities between these two points enables the farmer to have surplus revenue after covering VRT costs. The spatial break-even variability proportions are 30 and 85% poor land. When N price increases to $0.49/lb, the spatial break-even variability proportion changes to points C and D or 24 and 89% poor land. The range of variability over which the farmer enjoys surplus has now widened. Assume now that N price remained the same at $0.488/ac, but the custom charge increases to $8/ac. In this case, the spatial break-even variability proportions are given by the points E and F and the spatial break-even variability proportions are 43 and 80% for poor land. In this case, the VRT adoption range has narrowed.

CONCLUSIONS

This analysis simulates net returns to VRT for a hypothetical corn field. The impact on a farmer's decision to adopt VRT depends on the proportions of soils in a field, along with expected nitrogen and corn prices. The analysis provides a mechanism to access the effect VRT will have on net returns considering the yield potential variability of the soil, the response of the crop to the input being applied on the different soils, and changing input and output prices. Net returns for both VRT and uniform rate technology are compared and spatial break-even variability proportions for the field are determined. These spatial break-even variability proportions are calculated as the proportions of the different soils within a field for which the use of VRT just pays for itself.

The analysis suggests for our hypothetical field additional returns from VRT adoption more than covers the additional costs when the field is between 30 to 85% poor land. These boundaries are defined as the spatial break-even variability proportions.

Farmers can use the concept presented in this paper to evaluate whether to enlist the services of VRT. The analysis demonstrates that the net return differential that occurs when using VRT when compared with uniform rate technology depends on field variability, crop response to the input being applied, and input and output prices.

Further analysis on the variability of input availability in the soil prior to application should be incorporated into the analysis. Also, the analysis should be expanded to introduce more than two soils.

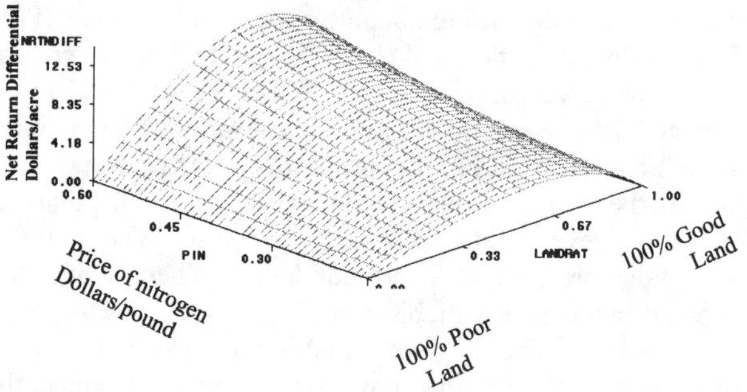

Fig. 1. Spatial break-even variability analysis for corn with two land qualities and changing relative input-output prices

Fig. 2. Per acre net return analysis for corn.

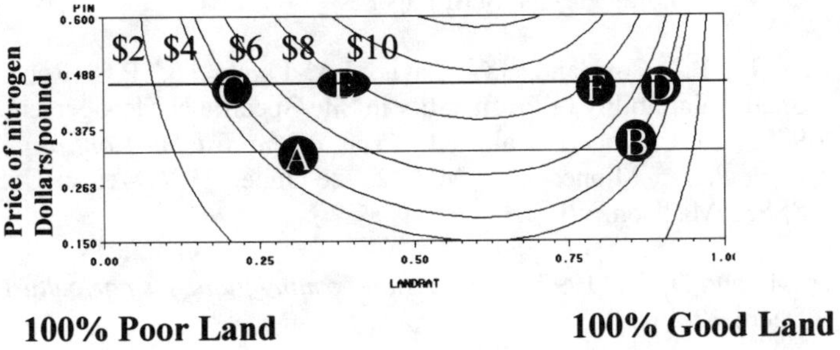

REFERENCES

Annonymous, 1997. Tennessee agriculture 1996. Dep. of Agric.. Nashville, TN.

Caldwell, Joey, 1997. Personal Communication. Tennessee Farmers Cooperative, Covington.

Carr, P.M., G.R. Carlson, J.S. Jacobson, G.A. Nielsen, and E.O. Skogley. 1991. Farming soils, not fields: A strategy for increasing fertilizer profitability. J. Prod. Agric. 4 (1):57–61.

Daberkow, S. 1997. Adoption rates for selected crop management practices-implications for precision farming. Choices, Third Quarter 1997

Duke, John, 1997. Personal Communication. Tennessee Farmers Cooperative

Forcella, F. 1993. Value of managing within-field variability. p. 125–132. *In* P.C. Robert et al., (ed.) *Proc. of Soil Specific Crop Management: A workshop on Research and Development Issues*. Minneapolis, MN. 14–16 Apr., 1992. ASA, CSSA, and SSSA, Madison, WI.

Hayes, J.C., A. Overton, and J.W. Price. 1994. Feasibility of site-specific nutrient and pesticide applications. p. 62–68. *In* K.L. Campbell et al., (ed.) *Environmentally Sound Agriculture*. Proc. of the 2nd Conf., Orlando, FL. 20–22 Apr. ASAE, 1994. St. Joseph, MI.

Hibbard, J.D., D.C. White, C.A. Hertz, R.H. Hornbaker, and B.J. Sherrick. 1993. Preliminary economic assessment of variable rate technology for applying P and K in corn production. Presented at the *Western Agricultural Economics Association Meeting*, Edmonton, Alberta.

Lowenberg-DeBoer, J., and S.M. Swinton. 1995. *Economics of site-specific management in agronomic crops.* Dep. of Agric. Economics, Purdue Univ.. W. Lafayette, IN. Staff Paper 95–14.

Malzer, G.L., P.J. Copeland, J.G. Davis, J.A. Lamb, and P.C. Robert. 1996. Spatial Variability of Profitability in Site-Specific N Management. p. 967–975. *In* P.C. Robert et al., (ed.) *Proc. of the 3rd Int. Conf. on Precision Agriculture*, Minneapolis, MN. 23–26 June, 1996. ASA, CSSA, and SSSA, Madison, WI.

Morgan M., and D. Ess. 1997.*The precision-farming guide for agriculturists*. John Deere. Moline IL.

Sawyer, J.E. 1994. Concepts of variable rate technology with considerations for fertilizer application. *J. Prod. Agric.* 7(2):195–201.

Snyder C.J. 1996. An economic analysis of variable-rate nitrogen management using precision farming methods. *Ph.D. diss.*, Kansas State Univ., Manhattan.

Swinton, S.M., and Ahmad, Mubariq. 1996. Returns to farmer investments in Precision Agriculture Equipment and Services. p. 1009–1018. *In* P.C. Robert et al., (ed.) *Proc. of the 3rd Int. Conf. on Precision Agriculture*, Minneapolis, MN. 23–26 June, 1996. ASA, CSSA, and SSSA, Madison, WI.

Wibawa, W.D., D.L. Dludlu, L.J. Swenson, D.G. Hopkins, and W.C. Dhanke. 1993. Variable fertilizer application based on yield goal, soil fertility and soil map unit. *J. Prod. Agric.* 6(2):255–261.

Economics of Variable Rate Planting for Corn

J. Lowenberg-DeBoer
Agricultural Economics Department
Purdue University
West Lafayette, Indiana

ABSTRACT

This analysis works out the economic implications of the corn (*zea mays L.*) population response curves by yield potential category estimated by Pioneer Hi-Bred agronomists. Variable rate seeding examples are developed for various mixes of low, medium and high yield potential soil, as well as for a range of seed costs and variable rate equipment costs. The strategies analyzed are: variable rate planting using agronomic recommendations for each yield potential zone, variable rate planting using an economic decision rule, and two information strategies that set a uniform planting rate based site specific yield potential information. The general conclusion is that variable rate seeding by yield potential zone has profit potential only for farmers with some low yield potential land (<100 bu a^{-1}). Farmers with mix of medium and high potential land are better off with uniform rate seeding. The surprise is that variable rate seeding is potentially profitable when the proportion of low yield land is small. In the example, the farm with 10% low yield potential soil shows positive returns to variable rate planting. The results are not particularly sensitive to seed cost, corn price or variable rate equipment investment cost.

INTRODUCTION

Varying plant populations within fields is an old idea that has been given new life by the availability of GPS technology. Intuitively, it makes sense to reduce plant populations on soils that have lower yield potential. From the investment cost perspective, variable rate planting is relatively inexpensive, especially for a producer who already has invested in GPS. A variable rate planter controller sells for about $3500 to $4000. But intuitive appeal and low investment cost do not necessarily generate profit. The objective of this analysis is to examine the profitability of variable rate planting for corn given available information on crop response to plant populations. This will be done using crop responses estimated by Pioneer Hi-Bred agronomists (Pioneer Hi-Bred International, 1997) and a spreadsheet model.

Operator managed variable rate planting systems have been around for about 20 yrs, but failed to catch on because they depended on the alertness of the operator. Farmers who used manual systems often say that they worked when the operator was fresh, but as fatigue crept in operators would often forget to switch populations. All the gains from lower populations for the low yield potential areas could be lost if a

Copyright © 1999 ASA-CSSA-SSSA, 677 South Segoe Road, Madison, WI 53711, USA. *Proceedings of the Fourth International Conference on Precision Agriculture.*

few rounds were made at the lower population in higher yield areas. Precision technology automates the process and reduces switching errors.

Numerous agronomic studies have considered variable rate planting. Doerge (1997) summarizes the results of several Pioneer Hi-Bred studies by saying that optimum seeding rates do not vary much across a wide range of soil and yield conditions in the U.S. Corn Belt. He also states that while seeding rates below the optimum can reduce yields; higher than optimum seeding rates carry little penalty. He concluded that variable rate seeding may be profitable on farms with some areas with yield potential below 100 bu a^{-1}. Barnhisel et al. (1996) provide data from a 3 yr study in Kentucky that indicates a modest return to variable rate planting on fields which include some soils with below 100 bu a^{-1} yield averages. Kinsella (1997) reported a soybean response to higher populations on lighter, lower yield potential soils. Some farmers have reported success in variable rate planting by soil type, instead of by yield potential.

This analysis will address the following specific questions for variable rate planting based on yield potential:

1. What proportion of corn area must be below 100 bu a^{-1} yield potential to justify variable rate planting?
2. Does the profitability of variable rate planting depend on the productivity of the better soils in the mix? Is variable rate planting profitable for only a mix of low and high productivity soils or also for a mix of medium and low productivity land?
3. How sensitive are returns to variable planting to seed and equipment costs?

This analysis will not deal with the risk considerations related to seeding rate decisions, interaction of row width and variable rate planting, or use of variable rate planting to increase populations in low germination areas. These are important issues, but beyond the scope of this analysis and of the data available. This analysis assumes that the low yield potential land will be farmed. It does not deal with alternate uses for that land in forage, forestry or other crops.

METHODS

The approach was to develop spreadsheet budget examples that estimate returns to various seeding rate strategies and then to vary certain parameters parametrically to determine the sensitivity of results to baseline assumptions. The population response functions by yield potential estimated by Doerge (1998, personal communication) were used to estimate U.S. corn yields because they are the best available for Corn Belt conditions. The focus is on corn because no soybean population response functions are available.

One of the key questions raised by Doerge (1997) and others is that if planting rates are varied, how will the changes be determined. Accurately mapping yield potential is in itself a major problem. It is assumed that the yield potential zones are relatively small, irregularly shaped and interspersed, so that management by field or other unit would be difficult. It is assumed that yield potential zones are accurately mapped. For simplicity this study assumes that we know the proportion of the corn area that has high, medium, and low yield potential and that variable rate equipment can accurately change populations given a map of the zones. The zones are defined as:

High: over 180 bu a^{-1} expected yields
Medium: between 120 and 140 bu a^{-1} expected yields
Low: under 100 bu a^{-1} expected yields

Five strategies were evaluated based on approaches suggested in the literature for managing variable rate inputs (Lowenberg-DeBoer & Boehlje, 1996):

- **Uniform seeding** to achieve a population of 28 000 plants/a at harvest - this is the control to which other strategies are compared.
- **VRT, agronomic rule** - Variable rate seeding based on Pioneer agronomic recommendations.
- **VRT, economic rule** - Variable rate seeding based on the economic criteria that the marginal value of the additional product be equal to the marginal cost of the additional input in every management zone.
- **Information strategy, agronomic rule** - An information strategy based on agronomic recommendations, which identifies the highest optimal plant population for any site in the field and plants all the corn acres uniformly at that rate.
- **Information strategy, economic rule** - An information strategy that uses site specific yield potential information to determine the population at which the marginal value equals marginal cost criteria for all the corn acres.

The plant populations for each management strategy are listed in Table 1. Because the crop response functions were estimated on a limited range of populations, no populations below 18 000 were used.

Price assumptions are listed in Table 2. The $67/bag seed price is the average price for 1997 with all relevant quantity and other discounts. The other costs include hired labor, chemicals, fertilizer, equipment, interest, crop insurance, machinery, drying equipment, and storage for average soils from Doster et al. (1996). Other assumptions include: 1000 acres of corn planted, to allow for germination and

Table 1. Plant populations for variable rate and information strategies by yield potential zone.

Yield potential zone	Variable rate, agronomic recommendation	Variable rate economic rule	Information, highest agronomic recommendation	Information, economic rule
		plants $acre^{-1}$ at harvest		
Low	18 000	20 000	30 000	28 000
Medium	28 000	26 000	30 000	28 000
High	30 000	30 000	30 000	28 000

Table 2. Price assumptions for estimation of returns to variable rate corn planting.

Item	Unit	Price
Corn at harvest	bu	$3.00
Corn seed	bag, 80 000 kernels	$67.00
Dryer fuel	gal	$0.50
Variable rate planter controller & monitor	controller and monitor	$3500
Discount rate	year	10%
Other costs	acre	$168.55

other problems the planted populations should 10% higher than desired population at harvest and the useful life of the planter controller is 5 yrs. The 10% discount rate is used to estimate an annualized cost for the VRT equipment.

In the examples, only the seeding rate was influenced by the site specific information. All other inputs were held constant. Larger gains are to be expected in an integrated system which manages several inputs site specifically.

To keep the example simple the baseline farm was assumed to have only two yield potential zones: high and low. Scenarios were also developed with varying proportions of high and medium yield potential land and of medium and low yield potential soil. Other scenarios include:

- varying seed price from $50 to $110 per 80 000 kernel bag and
- changing the variable rate equipment cost from $2000 to $8000 per farm.

The baseline scenario assumes that the producer's only investment is the variable rate planter controller and monitor. The sensitivity testing considers the case of producers who must also purchase GPS units, computers or other equipment to implement variable rate planting.

RESULTS AND DISCUSSION

Consistent with the agronomic studies variable rate planting shows economic advantage on farms with some land with under 100 bu a^{-1} yield potential (Table 3). The benefits vary with the proportion of low potential soil, but are of similar magnitude for mixes of high and low or medium and low potential land. Surprisingly, benefits occur for a very modest proportion of low yield land. Both variable rate strategies show small positive returns at 10% of land being low yielding.

The benefits are highest when a small part of the farm has high potential soil. In the baseline example, when 95% of the farm is low yield potential (5% high yield potential), the gain from variable rate is more than $4 a^{-1} for both variable rate strategies. This large benefit when a high proportion of the land is low yielding, depends on the assumption that under a uniform rate strategy producers would maintain a desired harvest population of 28 000.

With the variable rate strategies, the source of economic benefits depends on the proportion of low yield land. When proportion of low yield land is small, yield gains provide most of the benefits and seed savings are small. When the proportion of low yield land is large, the largest source of benefits is seed savings.

Compared with the variable rate approaches, the information strategy using the economic rule appears to offer some promise. For a mix of low and high yield potential land, the net gains compared to uniform planting are slightly greater than those of the variable rate strategies for small percentage of low yield land (Table 3). The information strategy might have consistently greater return if the costs of installing variable rate equipment, learning to operate it, mapping yield potential zones and additional field time are factored in. With a mix of high and medium yield potential land, the information strategy with the economic rule shows small gains relative to uniform planting for situations in which variable rate planting shows losses (Table 4). The populations for the information strategy with the economic rule drop to near 20 000 when a high proportion of the land is low yielding, and rise to near 30 000 when most of the farm is in high yield potential zones.

The information strategy with the agronomic rule seems to offer small benefits compared to either uniform or variable rate planting when small parts of the farm are in low yield potential zones (Tables 3 and 4). When the farm has some high yield

Table 3. Gains from using site-specific information in corn plant population decisions for farms with varying percentages of low yield potential land.†

Percentage of land with low yield potential	Variable rate seeding, agronomic recommendation	Variable rate seeding, economic decision rule	Information strategy, agronomic recommendation	Information strategy, economic decision rule
	----------------	---------------- $ a^{-1}	----------------	----------------
5%	0.26	0.28	0.77	0.82
10%	0.47	0.50	0.57	0.70
25%	1.10	1.19	-0.04	0.34
50'%	2.16	2.33	-1.06	0.00
75%	3.22	3.46	-2.08	0.66
90%	3.85	4.15	-2.69	2.44
95%	4.06	4.37	-2.89	3.68

† Compared with returns with a uniform population of 28 000 plants acre^{-1}. Remainder of the farm is high yield potential.

Table 4. Gains from using site-specific information in corn plant population decisions for farms with varying percentages of medium yield potential land.†

Percentage of land with medium yield potential	Variable rate seeding, agronomic recommendation	Variable rate seeding, economic decision rule	Information strategy, agronomic recommendation	Information strategy, economic decision rule
	-------------	-------------	$ a^{-1} -------------	-------------
5%	0.00	0.02	0.87	0.87
10%	-0.05	-0.00	0.77	0.80
25%	-0.19	-0.08	0.46	0.59
50%	-0.44	-0.22	-0.05	0.24
75%	-0.68	-0.35	-0.56	0.00
90%	-0.83	-0.43	-0.87	0.09
95%	-0.87	-0.46	-0.97	0.19

† Compared with returns with a uniform population of 28 000 plants $acre^{-1}$. Remainder of the farm is high yield potential.

potential land, this strategy sets plant populations at the high yield potential rate of 30 000. When >50% of the farm is in the low yield zones, uniform planting is preferred to the information strategy with the agronomic criteria.

Results for the information strategies appears to be quite sensitive to the exact population response functions used. Early response function estimates reported by Doerge (1997) show less favorable results for the information strategy.

If the farm has a mix of medium and high productivity land, the cost of variable rate planting is greater than the yield or seed savings benefits (Table 4). For the mix of medium and high yield soils, the information strategy that uses the highest agronomic rate shows consistently lower returns than the uniform rate strategy. For that soil mix, the information strategy with the economic decision rule shows modest positive returns; probably too modest to justify a change in management.

Seed cost increases augment the value of site-specific variable rate management (Table 5), but they do not change the general management advice. Farmers with a mix of high and low productivity land may benefit from variable rate seeding at any seed cost in the range from $50 to $110 per bag. Farmers with a mix of medium and high yield potential soil are better off with uniform rate planting.

Table 5. Gains from using site-specific information in corn plant population decisions across a range of seed corn prices for farms with 50% low yield potential land.†

Seed corn prices, \$ bag^{-1}	Variable rate seeding, agronomic recommendation	Variable rate seeding, economic decision rule	Information strategy, agronomic recommendation	Information strategy, economic decision rule
	-----------------	---------- \$ a^{-1} ----------	-----------------	
50	1.23	1.70	-0.59	0.00
60	1.78	2.05	-0.87	0.00
70	2.33	2.45	-1.14	0.00
80	2.88	2.98	-1.42	0.00
90	3.43	3.57	-1.69	0.04
100	3.98	4.19	-1.97	0.18
110	4.53	4.81	-2.24	0.31

† Compared with returns with a uniform population of 28 000 plants $acre^{-1}$. Remainder of the farm is high yield potential.

Similarly, a higher investment cost for variable rate equipment reduces the benefit (Table 6), but the farmer with a mix of low and high yield potential soil still shows positive returns even with an \$8000 investment on 1000 acres of corn. The farmer with a mix of medium and high potential soils does not show positive returns from variable rate even at a \$2000 investment cost.

CONCLUSIONS AND IMPLICATIONS

The general conclusion is that variable rate seeding by yield zone has profit potential only for farmers with some low yield potential land (<100 bu a^{-1}). The examples indicate that profitability does not depend on the productivity of the other land in the mix, as long as it is substantial higher than that of the low productivity soil. Variable rate seeding can be profitable for mixes of high and low or of medium and low yield potential soils. Farmers with a mix of medium and high potential land are better off with uniform rate seeding.

Sensitivity tests indicate that variable rate seeding is potentially profitable when the proportion of low yield land is small. In the example, the farm with 10% low yield potential soil shows positive returns to variable rate planting. The results are not particularly sensitive to seed cost or variable rate investment cost. This result suggests that variable rate seeding may be most valuable in areas on the fringe of the U.S. Corn Belt where some of the low yield potential soils are farmed.

Table 6. Gains from using site-specific information in corn plant population decisions across a range of VRT equipment investment levels for farms with 50% low yield potential land.†

Variable rate equipment investment $ farm^{-1}	Variable rate seeding, agronomic recommendation	Variable rate seeding, economic decision rule	Information strategy, agronomic recommendation	Information strategy, economic decision rule
	----------	----------	$ a^{-1} ----------	----------
2000	2.56	2.72	-1.06	0.00
3000	2.29	2.46	-1.06	0.00
4000	2.03	2.19	-1.06	0.00
5000	1.77	1.93	-1.06	0.00
6000	1.50	1.67	-1.06	0.00
7000	1.24	1.40	-1.06	0.00
8000	0.97	1.14	-1.06	0.00

† Compared with returns with a uniform population of 28 000 plants acre^{-1}. Remainder of the farm is high yield potential.

The information strategies appear to offer some promise, especially for farms with small areas of low yield potential. This is especially true if the time and effort required to implement variable rate planting is factored in. The information strategies have the advantage of being simpler to implement and do not require special variable rate equipment.

This analysis is only as good as the corn population response functions used. The functions used in this analysis were estimated with pooled data from around the Corn Belt. Site-specific population response functions would probably improve the performance of both the variable rate and information strategies. This type of function could be estimated with small plot data or yield monitor data from several years of strip plot population trials.

REFERENCES

Barnhisel, R.I., M.J. Bitzer, J.H.Grove, and S.A. Shearer. 1996. Agronomic benefits of varying corn seed populations: A central Kentucky study. *In* P.C. Robert et al., (ed.) Proc. of the 3rd Int. Conf. on Precision Agriculture. ASA, CSSA, and SSSA, Madison, WI.

Doerge, T. 1997. Variable rate seeding of corn. Crop Insights. Pioneer Hi-Bred Int., http://www.pioneer.com.

Doster, D.H., S.D. Parsons, E.P. Christmas, D.B. Mengel, and R.L. Nielsen. 1996. 1997 Purdue Crop Guide. ID-166. Purdue Univ., West Lafayette, IN.

Kinsella, J. 1997. What my farm plots say, presentation at the top farmer crop workshop. Purdue Univ., West Lafayette, IN.

Lowenberg-DeBoer, J. and M. Boehlje. 1996. Revolution, evolution or deadend: economic perspectives on precision agriculture. *In* P.C. Robert et al., (ed.) Proc. of the 3rd Int. Conf. on Precision Agriculture. ASA, CSSA, and SSSA, Madison, WI.

Pioneer Hi-Bred Int., 1997. Pioneer Agronomy Research Update. Pioneer.

Economics of Variable Rate Lime in Indiana

Rodolfo Bongiovanni
National Institute for Agricultural Technology
Manfredi, Córdoba
Argentina

James Lowenberg-DeBoer
Department of Agricultural Economics
Purdue University
West Lafayette, Indiana

ABSTRACT

In Indiana, variable rate application of lime is often considered a good place to start site-specific management (SSM). This is because pH is one of the most variable of manageable soil characteristics and because spreaders can be retrofitted relatively inexpensively to do variable rate application (VRA). The objective of this study is to evaluate the profitability of VRA for lime as a stand-alone activity. The methodology involves a spreadsheet model using corn and soybean pH response functions estimated with small plot data. The overall results indicate increased annual returns to corn and soybean production with site-specific pH management strategies. On average, SSM with agronomic recommendations provides an increased annual return of \$2.93 acre^{-1} (+1.78%). SSM with the economic decision rule provides an average increase in annual return of \$7.91 acre^{-1} (+4.82%).

INTRODUCTION

Soil Acidity and Liming

Soil acidity is commonly indicated by soil pH, a measure of hydronium ion activity in a soil suspension. Acidity may be created by a removal of bases by harvested crops, leaching, and an acid residual that is left in the soil from N fertilizers, and it has long been recognized as one reason soils become unproductive. Liming to correct soil acidity has been practiced for a long time, but during the last several years, limestone use has tended to decrease while crop yield and N fertilizer use have increased markedly.

Most crops require some lime if pH falls below 5.0, most require no lime if pH is 7.0 or higher. If pH exceeds 7.5, soils become basic and present problems of nutrient availability, plant nutrient uptake, microorganism activity, potential herbicide damage, and weed control deficiency.

From an economic point of view, lime is a crop-production input that provides certain benefits at a cost. If liming increases crop yield (or reduces the requirements for other inputs) and if the value of the increased yield (savings in the cost of other inputs) exceeds the cost of lime, then liming is profitable. A farmer who wants to maximize the net returns will increase his lime rate as long as the value of the benefits exceed the cost of the lime (Hall, 1983).

Copyright © 1999 ASA-CSSA-SSSA, 677 South Segoe Road, Madison, WI 53711, USA. *Proceedings of the Fourth International Conference on Precision Agriculture.*

How much lime will pay for itself at any specific location? Answering that question requires knowledge of how crop yields respond to lime applications. In this study, yield-response functions are fitted to field-plot data from controlled lime rate experiments reported in the literature.

Site-Specific Lime Management

Beyond the determination of optimal lime rates, this study addresses the decision to adopt site-specific management (SSM). Variable rate application (VRA) of lime is often considered a good place to start SSM, since pH is one of the most variable of manageable soil characteristics, and because spreaders can be retrofitted relatively inexpensively to do VRA. Theoretically, this practice should allow growers to manage field variability while improving profitability in a manner that is environmentally responsible.

If fields were uniform, there would be no need for SSM. Since fields contain a complex of soils and landscapes, extensive spatial variability in soil properties and crop productivity is the norm rather than the exception in most fields (Mulla & Schepers, 1997). With such extensive variations, potential exists for adoption of SSM.

Production Strategies

Different SSM strategies can be implemented, but only two SSM strategies and one intermediate approach will be considered here. First, SSM using agronomic recommendations (SSM-Agronomic), because is the most common current practice for those beginning precision farming. This approach grid samples the field, and applies the recommended rate of lime to the individual grid using the agronomic recommendation rules (e.g., Vitosh et al., 1995).

Second, SSM with the economic rule (SSM-Economic) is considered. This approach is similar to SSM-Agronomic, but uses the economic rule, marginal value product equal to marginal factor cost (MVP = MFC), to determine the recommended rate of lime to the individual grid. This strategy calculates a rate of lime that is wealth maximizing in a 4-yr period. Wealth is maximized when the marginal response to one unit of lime equals the number of units of output that must be sold to pay for that unit of lime. SSM with the economic rule is considered because it is the approach recommended in the production economics literature and as a test of the sensitivity of the SSM adoption choice to the lime rate decision rule.

Third, an Information Strategy is examined. It applies a uniform rate of lime to bring the grid cell with the lowest pH up to full production. The information strategy considered here uses agronomic recommendations. It is the case of the farmer who has grid soil information, but does not have equipment available to apply at a variable rate. Information strategies are considered because some farmers do not have access to variable rate equipment and because they have been found potential profitable in management of P and K (Schnitkey et al., 1996).

In order to evaluate profitability, these three SSM strategies are compared to whole field management (WFM) strategy, considered here as the base case because is the most common current practice in the Midwestern U.S.. WFM makes a composite soil test, resulting in an average pH for the field and a single rate for lime application. The option of doing nothing also is reported for comparison purposes as the control case.

Problem

Soil variability within farm fields has long been recognized by soil scientists as well as farmers, and pH is one of the soil characteristics with the highest spatial variability (Cline, 1944). Lime used to correct pH is an important cost for Indiana farmers. The per unit cost is low relative to other fertilizer, but the application rates are comparatively high. Lime application is measured in tons, instead of the pounds used for P, K and nitrogen (N). High uniform rates can leave underlimed and overlimed spots in the field. Unlike some other fertilizers, lime can produce a negative effect on crop response if it is applied in excess. If the soil becomes basic, it will affect nutrient availability, plant nutrient uptake, and microorganism activity. Moreover, lime will potentially affect herbicide choice, herbicide damage and weed control (Childs et al., 1997).

There are no published economic evaluations of SSM of lime. Most existing studies of the economics of SSM focus on variable rate application of N, P, and K because this was the first SSM technology that was technically feasible (Lowenberg-Deboer & Swinton, 1997). Published studies on the economics of precision farming report mixed results. Farmers and agribusinesses are currently beginning to invest on SSM for lime and there is an urgent need for information on the profitability of the practice to guide decisions.

Objectives

The main objective of this research is to evaluate the profitability of custom-operated VRA for lime as a stand-alone activity, using data for Indiana farms that produce grain in a corn-soybean rotation. Profitable, in the sense used in this paper, means that switching to SSM yielded higher net present value (NPV) of returns than WFM, according to the definition of Lowenberg-Deboer and Swinton (1997).

The specific objectives were to determine (i) if SSM is profitable using agronomic recommendations for lime rates, compared to WFM; (ii) to determine if SSM is profitable using the economic rule (MVP=MFC) for lime rates, compared to WFM; (iii) to determine if the information strategy presents any advantage over VRA and (iv) to compare profitability with 1-acre and 2.5-acre soil sampling systems.

Hypothesis

The general hypothesis is that site-specific soil pH correction is wealth maximizing over a 4-yr soil sampling cycle, and that a 2.5-acre grid is the most profitable for SSM with current technology.

MATERIALS AND METHODS

Simulation Model

The methodology involves a spreadsheet simulation model using corn and soybean pH response functions estimated with experimental data. The ideal data for this study would be results from long-term field trials in several locations comparing SSM and WFM for lime. Unfortunately, such data is not available and many farmers and agribusiness people will need to make decisions about SSM for lime before such data could be collected. Simulation is a way to use available data to provide preliminary results. A complete description and other aspects of the methodology is given by Bongiovanni (1998).

The planning period for the simulation is four years, a typical soil sampling cycle and consistent with a decision to maintain an adequate pH level. In multiperiod cases it is useful to cast the problem in terms of maximizing net present value (NPV), also referred to in the economics literature as wealth. NPV allows for the time preferences of decision makers who often value returns in the future differently from current returns. For a 4 yr corn–soybean rotation, the objective function is the NPV of returns:

$$W = \frac{P_{corn} Y_{1corn}}{(1+d)} - P_l g(pH) - OtherCosts + \frac{P_{soy} Y_{2soy}}{(1+d)^2} + \frac{P_{corn} Y_{3corn}}{(1+d)^3} + \frac{P_{soy} Y_{4soy}}{(1+d)^4}$$

where: W = wealth; P_{corn} = price of corn; P_{soy} = price of soybean; P_l = cost of lime; $Y_{i\,corn}$ = Yield of corn in year i, (i = 1,3); $Y_{i\,soy}$ = Yield of soybean in year i, (i = 2,4); d = discount rate; $g(pH)$ = lime requirement (LR) as a function of initial pH.

This function represents the NPV of returns above variable costs (e.g., hired labor, fertilizer, seed, chemicals, machinery repairs, fuel, dryer fuel, interests, storage, crop insurance, and miscellaneous), taken from Doster (1998). Because returns to lime vary from year to year, average returns are summarized for communication purposes in dollars per acre per year, using the annualized value estimated with the function PMT of the Quattro-Pro spreadsheet. This function calculates the even periodic payment over *Nper* periods with the NPV of *Pv* at a discount rate of *Rate*. PMT uses the following formula:

$$\frac{Pv * Rate}{1-(1+Rate)^{-Nper}}$$

Estimation of the Yield Response to pH

Crop response to pH is a complex phenomena dependent on soil characteristics and other factors. Because of its complexity and its long term nature, few researchers have attempted to estimate pH response functions and no such functions have been published for Indiana conditions. To allow an initial evaluation of SSM profitability for lime, a general corn and soybean response function was estimated with data pooled over several locations in the United States and Canada.

Agronomic evidence indicates that corn and soybean response to pH should exhibit (a) decreasing marginal returns to lime application (Woodruff et al., 1987) and (b) a yield plateau beginning somewhere below pH 7.0 (Adams, 1969). A quadratic response and plateau (QRP) function was chosen because it fits the biological response and because it is widely used in the agricultural literature (Mengel et al., 1987; Gotway Crawford et al., 1997).

Mathematically, the QRP can be expressed as:

$$Yield = \begin{Bmatrix} a + bX + cX^2, X < X^* \\ m, otherwise \end{Bmatrix}$$

where X is the input level and X^* is some specified value of X, and the parameters are restricted, so that the curve and its first derivative are continuous at X^*.

The form of the initial diminishing returns is unknown, so the quadratic can be thought of as a Taylor series approximation of that unknown function. Using the grafted polynomial technique (Fuller, 1969), the QRP functions were estimated with linear ordinary least squares (OLS).

Because available pH response data did not include yields at high pH levels, the estimated response function did not include the negative effects of high pH. These negative effects were modeled using a weed science rule of thumb of a 1% decline in yields for each one tenth (0.1) increase in pH over 7.5 (Jordan, 1997).

Experimental Data

Published data from controlled experiments were compiled from experiment stations in Alabama, Delaware, Florida, Georgia, Illinois, Indiana, Iowa, Kentucky, Minnesota, Mississippi, Missouri, Nebraska, Ohio, Ontario, Pennsylvania, South Carolina, Virginia, and Wisconsin. The data are for corn and soybeans. Data from experimental plots were sought for at least two variables: soil pH and crop yield.

Experimental data is converted and reported as relative yields (percentage of maximum), following the methodology used by Mc Lean and Brown (1984). This methodology allows the pooling of data from diverse areas and various sources without significant bias. The next step was to convert the relative yields reported in the tables to representative yields for Indiana. The values for representative yields were taken from Doster et al. (1998), corresponding to average yields of 132.6 bu acre^{-1} for corn and 42.4 bu acre^{-1} for soybeans. These are the yield values used to convert the estimated relative yields to absolute yields.

Corn Production Function

In terms of relative yields, the pH response function for corn is:

$$Relative\ Yield = \begin{cases} -0.8287 + 0.5318 * pH - 0.0391 * pH^2, pH \le 6.8 \\ 97.96\%, otherwise \end{cases}$$

According to the estimated regression, corn yield increases to a pH of 6.8 and then approaches a plateau with a relative yield of 97.96% (129.89 bu acre^{-1}). The reason why the plateau level is reached at 97.96% (and not at 100% of relative yield), is because the regression estimates average response with error in both directions. This means that there are observations above and below the OLS estimator.

Soybean Production Function

The estimated production function in terms of relative yields is:

$$Relative\ Yield = \begin{cases} -0.7510 + 0.5003 * pH - 0.0367 * pH^2, pH \le 6.8 \\ 95\%, otherwise \end{cases}$$

According to the estimated regression, soybean yield increases to a pH of 6.8 and then approaches a plateau with a relative yield of 95% (40.28 bu acre^{-1}).

Field Soil Test Data

Field soil test data for this thesis was taken from three sources: (i) the Ph.D. thesis of Karr (1988), which studied six fields in southwestern Indiana, sampled on a 0.11-acre grid; (ii) data from seven fields of Purdue's Davis Farm, as reported in Top-Soil (1994), sampled on a 2.5 acre grid; and (iii) data provided by Lynn Harvest Land Co-op, from nine fields grid sampled on a 2.5 acre grid. Lynn, Indiana is located in the east central part of the state. For the Karr fields the 0.11-acre cells were grouped into roughly rectangular 2.5- and 1-acre grids for the analysis. The soil sample for the 0.11-acre cell nearest to the center of the 2.5- or 1-acre grid was used to represent that grid.

Lime Requirement

Lime requirement (LR) for agronomic recommendations was calculated with the method from Watson and Brown (1997), who provide a table with the lime required to reach a desired pH level, according to the present lime indexes. This table is the same used by Vitosh et al. (1995). Thus, LR was estimated for each grid with a spreadsheet equation, based on the original pH (pH_0).

LR for economic recommendations was calculated using marginal analysis, which states that when the value of increasing the amount of lime to be applied by unit equals the cost of applying one additional unit, profit is maximized; or when marginal value product equals marginal factor cost (MVP = MFC). In this case the MVP is the NPV of marginal returns during the 4-yr planning period.

In order to calculate LR for a 4-yr period, lime carryover was estimated through the change in pH. According to Mengel (1997), there are two factors which cause pH to drop. One is the removal of calcium and magnesium through cropping and the other is N fertilizer application and N transformation processes in the soil that add hydronium ions (H^+), which must be neutralized. In other words, with normal cropping, the decrease in soil pH is a function of crop uptake, of nitrogen acidification, and of the buffering capacity of the soil. Using the values cited by Potash and Phosphate Institute (1997), total lime consumption was estimated as 0.35 tons acre^{-1} yr^{-1} for corn and 0.15 tons acre^{-1} yr^{-1} for soybeans, or as stated by Mengel: one ton of lime every 4 yrs in a corn–soybean rotation.

Having this value of lime consumption, change in pH was estimated for each individual cell by modifying the equation from Black (1993):

$$pH_i = B_s\left(\sqrt{LA - \sum_{j=1}^{i-1} LCj + 1} - 1\right) + pH_0$$

where: pH_i is the pH in year i from lime application in year 0, $i=1,2,3,4$; B_s = pH buffering capacity of soils; LA= lime applied (megagrams of limestone per hectare) and LCj= lime consumed in the previous period of time considered (megagrams of limestone per hectare).

Assumptions, Prices, and Costs Used

A few general assumptions are necessary to complete the study.

1. All other production factors are constant. The only factor considered in determining yields is the pH of the soil.
2. All applications and soil testing is done by custom operators. All pH corrections occur in 1 yr, in a single application.
3. LR assumes the use of agricultural limestone with a neutralizing value of 90%, which is incorporated in the first 4 in of soil.
4. Is assumed that the applicator is capable of delivering the required rates with VRA. Rates can be varied in every grid of 1 or 2.5 ac. No interpolated maps are used because the number of 2.5-acre grids per field is relatively small, especially in the Karr data.
5. Inflation is not considered in this study. Therefore, all the values are given in real terms.
6. Corn price: \$2.60 bu^{-1}. and soybean price: \$6.35 bu^{-1}. (Doster et al., 1998).
7. Cost of aglime (90% neutralizing value): \$22.00/ton, including transportation (Delphi Limestone, 1997).
8. Conventional lime spreading \$3.00 per acre; extra cost of lime spreading with VRA: \$3.00 per acre over conventional spreading cost; cost of soil sampling: \$1.00 per acre; cost of soil analysis: \$5.50 each (Swaim, 1997).
9. Other costs of production were taken from Doster et al. (1998) and interpolated in the spreadsheet for the estimated yields within each grid.

RESULTS

Average Annualized Returns for the Three Locations

To observe behavior of the five strategies for the three locations, average results are shown in Table 1 and in Fig. 1. The overall mean represents the case in which the farmer has no information on the pH of the fields tested. For instance, this may be the case when a producer rents or buys land. In many cases, farmers have some information about pH and need for lime because of cropping practices and time since last lime application. In those cases they can concentrate test on fields that they suspect need lime. The average of the fields needing lime represents the case in which the farmer can focus testing on fields that probably need lime.

In general, the most profitable option is SSM-Economic, which gives a 4.82% improvement in return above variable cost or \$7.91 acre^{-1} yr^{-1} more. SSM-Agronomic is the second most profitable option, with a 1.78% improvement in return above variable cost or \$2.93 acre^{-1} yr^{-1} more. WFM provides economic benefits over the option of doing nothing.

Information Strategy is less profitable than whole field strategy due to the high rate of lime required to bring the lowest grid cell up to the plateau yield level. Some 873 out of 1413 cells (62%) are pushed up over a pH of 7.5. Since this model includes a yield discount for values of pH higher than 7.5, it makes sense to have a lower return above variable cost when pH exceeds 7.5. SSM-Economic gives the highest return above variable cost over time: \$172.05 acre^{-1} yr^{-1}.

Table 1. Average annualized returns of the five strategies for the three locations.

	Strategy ($ acre^{-1} yr^{-1})				
	Do-nothing	WFM	SSM-Ag	SSM-Econ	Information
All fields:					
Average	$161.88	$164.14	$167.07	$172.05	$147.45
%over/below WFM	-1.38%	0.00%	1.78%	4.82%	-10.17%
$ over/below WFM	-$2.26	$0.00	$2.93	$7.91	-$16.69
Fields receiving lime:					
Average	$157.51	$163.65	$167.91	$173.15	$147.56
%over/below WFM	-3.75%	0.00%	2.60%	5.80%	-9.83%
$ over/below WFM	-$6.14	$0.00	$4.26	$9.49	-$16.09

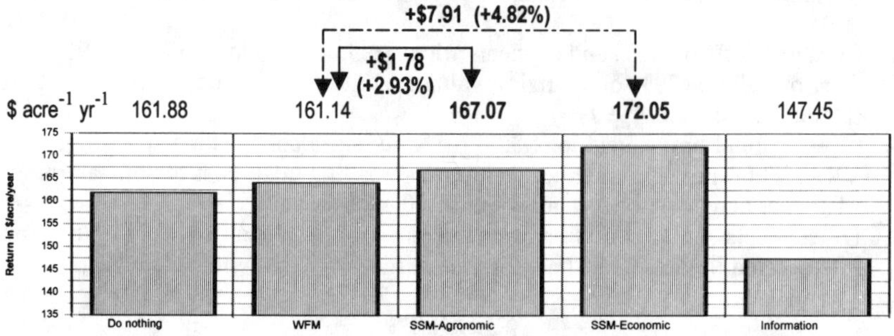

Fig. 1. Overall average annual returns by strategy for all fields.

If only fields where lime was applied are considered, the benefits of the SSM strategies are higher: the most profitable option is SSM-Economic, which gives a 5.80% improvement in return above variable cost or $9.49 acre^{-1} yr^{-1}. SSM-Agronomic is the second most profitable, with 2.60% improvement or $4.26 acre^{-1} yr^{-1} more. SSM-Economic gives the highest return above variable cost over time: $173.15 acre^{-1} yr^{-1}.

In order to observe the variability in pH and yields, the change in the spatial standard deviation of the samples was estimated. Average initial pH and average pH of year four were compared for each strategy and data set. For yield, corn was chosen as the reference crop, and yield without lime application was compared with yield of the third year under each strategy.

On average, SSM strategies tend to decrease estimated spatial variability in yields. The decrease in the standard deviation of the yield was 2.85 bu acre^{-1} for both SSM strategies, while WFM decreased by 0.89 bu acre^{-1}. This decrease in yield variability may be linked to a decrease in the variability of returns, as reported by Lowenberg-DeBoer and Aghib (1997). On the other hand, the information strategy increases the standard deviation of the yield by 0.97 bu acre^{-1}.

Karr Data. 1-Acre Grid with Grid Center Sampling

Karr data set was used for this sensitivity analysis because it is sampled at 0.11-acre. With 1-acre grids four fields out of six require some lime. The results of benefit or loss are presented with respect to WFM with 2.5-acre grid (Fig. 2).

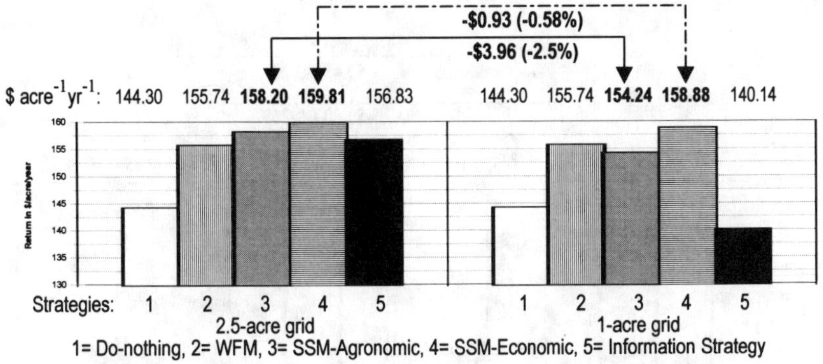

Fig. 2. Comparative annual returns of soil sampling methods.

The 1-acre grid does not show any economic benefits over the 2.5-acre grid, because of the cost involved on sampling at a higher intensity. All strategies show some differences in terms of dollars and of percentages, when compared with the original model. Although the overall mean of SSM is profitable with respect to WFM (+$3.14), the overall annual return for SSM-Economic is 0.58% (-$0.93) lower than the baseline model.

With the 1-acre grid, information strategy has lower values than WFM and SSM strategies. This can be explained by the high rate of lime required to bring the lowest grid cell up to the plateau yield level. Some 254 out of 365 0.11-acre cells (70%) are pushed up over a pH of 7.5.

Increase in the Charges for Variable Rate Application

Complete sensitivity testing results are given by Bongiovanni (1998). Given uncertainty about appropriate VRA fees, one sensitivity test of particular interest is the effect of fee increases. The baseline assumption is a $3/acre additional fee for VRA.

With a $6 acre^{-1} charge for VRA, the most profitable option continues to be SSM-Economic, which gives a 4.54% average improvement in return above variable cost, or $7.46 acre^{-1} yr^{-1} more. SSM-Agronomic is the second best choice, with a 1.52% improvement, or $2.50 acre^{-1} yr^{-1} more. SSM-Economic gives the highest return above variable cost over time: $171.60 acre^{-1} yr^{-1}, $0.45 below the annual return with the assumed $3 acre^{-1} extra cost of VRA (Fig. 3). The decrease in the annual return is less than with the $3 VRA cost because the fee is spread over the 4 yrs of the soil sampling cycle.

Because VRA is still being perfected as a new technology and its market is still growing, the costs are not strongly established. As precision farming technologies evolve, the costs associated with VRA will probably decrease and VRA may eventually be provided as the standard practice at no extra cost. Currently, extra costs of VRA and grid sampling vary from region to region. This makes it imperative to obtain current costs from local companies when it comes to estimating potential economic benefits of VRA of lime in a specific area. In addition, if the cost of soil sampling and analysis could be spread out over more inputs, the benefits of intensive sampling and VRA would increase.

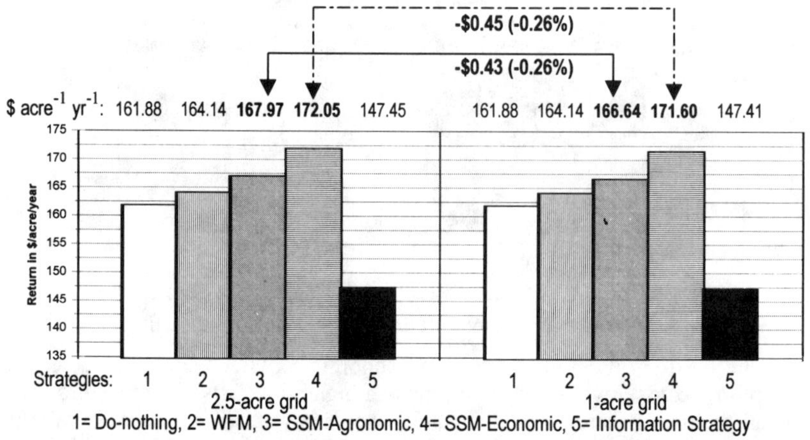

Fig. 3. Comparative annual returns of increasing the extra charge for VRA.

GENERAL CONCLUSIONS

The baseline results indicate that with either the SSM agronomic recommendations or the economic optimization, VRA of lime is profitable as a stand-alone technology. The overall results indicate that SSM-Agronomic provides an increased annual return of \$2.93 acre^{-1} (+ 1.78%). SSM-Economic provides an increased annual return of \$7.91 acre^{-1} (+ 4.82%). The information strategy is not profitable because it tends to over lime certain areas, leading to herbicide damage and other ill effects of alkalinity. Because of the extra cost of sampling, returns are higher on a 2.5-acre grid than on a 1-acre grid.

Sensitivity testing indicates that SSM for lime with a 2.5-acre grid is more profitable than WFM over a wide range of prices and conditions. A \$3 change up or down in the cost of VRA services did not have a significant effect on profitability, compared to the baseline model. Estimates indicate that spatial variability of yields decreases with SSM strategies.

Further Research

Given the experience in this study, the priority research needs in the area of SSM for lime include: better methods for testing spatial variability of soil pH, improved understanding of crop response to lime and site-specific field testing of the agronomics and economics of variable rate lime. Accurate sensors for soil pH will probably be developed eventually, but until that time we will rely on soil sampling. For grid sampling, grid size and alignment of the sample points are important questions. Guided sampling designs using soil type, topography, remote sensing and other information also need to be considered.

The results of this study depend to a large degree on the reliability of the crop response functions estimated with sparse data collected under a wide range of conditions. This is barely adequate for a first approximation study and inadequate to capture the specifics of the range of conditions under which lime might be used. Improved lime response models are needed that include the effect of soil characteristics as well as interactions with other nutrients and management practices. The long run benefit of such improved models would be decision support systems that would help producers improve productivity and profitability.

Simulation studies can only do so much. They are useful in providing timely answers to urgent questions. They do not replace gathering and analyzing field data. Yield monitors and other precision farming technology has made it easier and cheaper for farmers and researchers to collect that field data. On-farm trials are needed to verify the economics of SSM of lime and to fine tune lime requirement estimates for site-specific conditions.

ACKNOWLEDGMENTS

Authors would like to acknowledge John Trott, director of Purdue Agricultural Centers, and Doug Biehl, Precision Farming Specialist, Lynn Harvest Land Co-op, for providing soils data used in this work. The research was made possible by an assistantship funded by Rotary International and the National Institute for Agricultural Technology (INTA) of Argentina.

REFERENCES

Adams, F. 1969. Response of Corn to Lime in Field Experiments on Alabama Soils. Auburn University Agric. Exp. Station, Bulletin 391, June 1969, 18 p.

Black, C.A., 1993. Soil testing and lime requirement. p. 693–694. *In* Soil fertility evaluation and control. Iowa State Univ., Ames, IA.

Bongiovanni, R. 1998. Economic Evaluation of Site-Specific Lime Management. M.S. thesis. Dep. of Agric. Economics, Purdue Univ.. West Lafayette, IN.

Childs, D., T. Jordan, T. Bauman, and M. Ross. 1997. 1997 Weed Control Guidelines for Indiana. Purdue Univ., West Lafayette, IN.

Cline, M. 1944. Principles of soil sampling. Soil Sci. 58:275–288.

Delphi Limestone. 1998. Delphi, IN. Phone (765) 564-2580.

Doster, H.; S. Parsons, E. Christmas, D. Mengel, and R. Nielsen. 1998 Purdue Crop Guide. Document: ID-166-Rev. Purdue Univ., West Lafayette, IN.

Fuller, W. 1969. Grafted polynomials as approximating functions. Aust. Journal of Agricultural Economics, Volume 13, 1969, 35-46.

Gotway Crawford, C.; Bullock, D.; Pierce, F.; Stroup, W; Hergert, G.; and K. Eskridge. 1997. Experimental Design Issues and Statistical Evaluation Techniques for SSM. p. 301–335. .*In* F.J. Pierce and E.J. Sadler, (ed.) 1997. The state of site-specific management for agriculture. ASA, CSSA, and SSSA, Madison, WI.

Hall, H. 1983. Economic Evaluation of Crop Response to Lime. Paper no. 81-1-153, Kentucky Agric. Exp. Stn.. Am. Agric. Econ. Assoc.

Jordan, Thomas N. Personal Communication, 1997. Professor of Weed Science. Purdue University, 1155 Botany and Plant Pathology, West Lafayette, IN 47907-1155, Office: LILY G-312, Phone: (765) 494-4629, FAX: (765) 494-0363, Email: jordan@btny.purdue.edu.

Karr, M.C., 1988. Spatial and temporal variability of soil chemical parameters. Ph. D. diss. Purdue Univ., West Lafayette, IN.

Lowenberg-DeBoer, J., and S.M. Swinton, 1997. Economics of site-specific management in agronomic crops. p. 369–396. *In* F.J. Pierce and E.J. Sadler, (ed.) The state of site-specific management for agriculture. ASA, CSSA, and SSSA, Madison, WI.

Lowenberg-DeBoer, J., and T. Aghib. 1997. Average returns and risk characteristics of site-specific P and K management: On-farm-trial results from Indiana. Purdue Univ., W. Lafayette, IN.

Mc Lean, E., and J. Brown. 1984. Crop response to lime in the midwestern United States. p. 267–303. *In* soil acidity and liming. St. Joseph, WI., USA: ASAE. Series: Agron. no. 12.

Mengel, D; W. Segars, and G. Rehm. 1987. Soil Fertility and Liming. p. 461–496. *In* ASA-CSSA-SSSA: Madison, Wisconsin, USA. Soybeans: Improvement, Production and Uses, 2nd ed.. Agron. Monog. 16. ASA, CSSA, and SSSA, Madison, WI.

Mengel, D.B. 1997. Unpublished information. Purdue Univ., West Lafayette, IN. Ph: 765-494-4801, Email: dmengel@dept. agry.purdue.edu

Mulla, D., and J. Schepers, 1997. Key processes and properties for site-specific soil and crop management. p. 1–18. *In* F.J. Pierce and E.J. Sadler, (ed.) 1997. The State of Site-Specific Management for Agriculture. ASA, CSSA, and SSSA, Madison, WI.

Potash & Phosphate Institute. 1997. Manual internacional de fertilidad de suelos. PPI, Norcross, GA.

Schnitkey, G., J. Hopkins, and L. Tweeten. 1996. Information and application returns from precision fertilizer technologies: a case study of eighteen fields in northwest Ohio. Dep. of Agric. Ohio State Univ.

Swaim, Dave. 1997. Personal Communication. Crop Consultant. West Lafayette, IN. Phone: (765) 362-4946.

Top-Soil Testing Service, Co. 1994. "Soil Test Report. Spring, 1994. Purdue University. Davis Purdue Ag. Center Farm", Frankfort, IL 60423.

Vitosh, M, J. Johnson, and D. Mengel. 1995. Tri-state fertilizer recommendations for corn, soybeans, wheat and alfalfa. Michigan State Univ., Ohio State Univ., Purdue Univ. Extension Bulletin E-2567, July 1995, p. 3.

Watson, M., and J. Brown. 1997. pH and lime requirement. p. 15. *In* Recommended Chemical Soil Test Procedures. North Central Regional Res. Publ. no. 221.

Woodruff, J., F. Moore, and H. Munsen. 1987. Potassium, boron, nitrogen, and lime effects on corn yield and earleaf nutrient concentrations. Agron. J. 79:520–524.

Economic Returns and Environmental Impacts of Variable Rate Nitrogen Fertilizer and Water Applications

K. B. Watkins
Y. C. Lu

USDA-ARS
Beltsville, Maryland

W. Y. Huang

USDA-Economic Research Service
Washington, District of Columbia

ABSTRACT

Several studies have evaluated the profitability of variable rate application of one type of input, but few have investigated the profitability of variable rate application of two or more types of inputs. In addition, most studies ignore the impact of variable rate input application on the environment. This study evaluates the profitability and environmental outcomes associated with spatial variation of nitrogen fertilizer and irrigation water in seed potato production. Seed potato yields and nitrogen losses are simulated for four different areas of a 63 ha field using the EPIC (Environmental Policy Integrated Climate) crop growth model. A dynamic optimization model is used to determine optimal levels of N fertilizer for each area of the field. Average nitrogen losses and economic returns are evaluated for both uniform and variable rate application of nitrogen and water. The results indicate greater economic and environmental benefits may be achieved from varying water application than from varying nitrogen application across the field.

INTRODUCTION

Variable rate application (VRA) refers to the application of agricultural inputs in specific and changing rates throughout the field. The goal of VRA is to apply a precise amount of fertilizers, pesticides, water, seeds, or other inputs to specific areas in the field where and when they are needed for crop growth. VRA has the potential to increase both agricultural productivity and environmental stewardship. It must be shown to be profitable, however, before farm operators will adopt it.

The economic feasibility of VRA has not been fully explored. The economic studies that have been done have produced mixed results (Lowenberg-DeBoer & Boehlje, 1996; Lu et al., 1997). Most economic studies focus on only one type of input, usually fertilizer, and use a partial budgeting framework to evaluate the returns and costs associated with variable rate application (Carr et al.,

Copyright © 1999 ASA-CSSA-SSSA, 677 South Segoe Road, Madison, WI 53711, USA. *Proceedings of the Fourth International Conference on Precision Agriculture.*

1991; Feiz et al., 1994; Wibawa et al., 1993). In addition, most earlier studies ignore the effects of variable rate input application on the environment.

The objective of this paper is to determine the long-term economic and environmental feasibility of variable rate N fertilizer and water application for a specific field situation and seed potato rotation considering N carryover effects. A dynamic optimization model is used to determine optimal steady-state N fertilizer rates for different parts of a field near Ashton, ID, exhibiting spatial seed potato yield variability. The EPIC (Environmental Policy Integrated Climate) crop growth model is used to simulate crop yields and N losses for different parts of the field under conventional and variable rate N fertilizer and water application. Discounted net present values are calculated for each management strategy and are compared with simulated N losses for the field.

SITE DESCRIPTION OF FIELD

This study is based on farm level production data from a seed potato operation near Ashton, ID. The field under study is a 63 ha field composed of Kucera silt loam with bedrock substratum and Lostine silt loam soils. The typical rotation used on this field is a three-year seed potato-spring wheat-feed barley rotation, with seed potatoes planted in early May and harvested in early October and the two grain crops planted in early April and harvested in late August. Seed potato yields were monitored across the field in October, 1995 using a load cell type yield monitor system similar in design to that in Rawlins et al. (1995). The resulting data indicated that seed potato yields for 1995 ranged from <6 Mg ha^{-1} in some areas to >22 Mg ha^{-1} in other areas. This information was used to group seed potato yields into the following four ranges.

- Yield Range 1 - below 6 Mg ha^{-1} (5 ha);
- Yield Range 2 - between 6 and 11 Mg ha^{-1} (21 ha);
- Yield Range 3 - between 11 and 17 Mg ha^{-1} (31 ha); and
- Yield Range 4 - >17 Mg ha^{-1} (6 ha).

A dynamic optimization model was used to determine optimal steady-state nitrogen fertilizer levels for each yield range and the field during the potato year of the rotation under both conventional and variable rate water application.

THE DYNAMIC OPTIMIZATION MODEL

Potatoes are grown in rotation with other crops for pest management reasons. In Idaho, seed potatoes may be rotated with as many as two or three other crops in 3- and 4-yr rotations. Small grain crops such as wheat and barley are the typical crops used in these rotations. These crops often represent expenses to seed potato production and are included primarily for pest management benefits rather than profitability.

Historically, a 3-yr seed potato–spring wheat–feed barley rotation has been used on the 63 ha study field. Assume the farm operator uses variable rate N fertilizer application during the seed potato year but uses conventional N fertilizer application during the wheat and barley years. Taylor (1983) shows how dynamic optimization models can be used to determine optimal fertilizer application rates for continuous cropping systems with fertilizer carry-over. We have modified the Taylor model to determine optimal N fertilizer application rates for the 3-yr seed potato rotation. For the sake of simplicity, assume the amounts of N applied in the wheat and barley years are held constant at A^W and A^B, respectively. Given the decision maker desires to maximize the expected value of profit for the rotation, the recursive equation for optimal N application during the potato crop year can be expressed as

$$F_k = \text{MAX}\, [(P_t^P\, Y_{k,t}^P\{N_{k,t}^P\} - rSUBt\ A_{k,t}^P) + \alpha\,(P_{t+1}^W\, Y_{k,t+1}^W\{N_{k,t+1}^W\} - r_{t+1}\, A^W) + \alpha^2\,(P_{t+2}^B\, Y_{k,t+2}^B\{N_{k,t+2}^B\} - r_{t+2}\, A^B) + \alpha^3\, F_{k+1}]$$

subject to

$$N_{k,t}^P = A_{k,t}^P + R_{k,t}^P$$
$$N_{k,t+1}^W = A^W + R_{k,t+1}^W$$
$$N_{k,t+2}^B = A^B + R_{k,t+2}^B$$

$$R_{k,t+1}^W = V_{k,t}^P\{N_{k,t}^P\}$$
$$R_{k,t+2}^B = V_{k,t+1}^W\{N_{k,t+1}^W\}$$
$$R_{k+1,t+3}^P = V_{k,t+2}^B\{N_{k,t+2}^B\}$$

$$A_{k,t}^P \geq 0$$
$$R_{11}\ \textit{given}$$

where F_k = the expected present value of rotation k; t = the crop year; P_t^P, P_{t+1}^W and P_{t+2}^B = expected crop prices; $Y_{k,t}^P$, $Y_{k,t+1}^W$, and $Y_{k,t+2}^B$ = expected yield response functions for potato, wheat, and barley, respectively; $N_{k,t}^P$, $N_{k,t+1}^W$, and $N_{k,t+2}^B$ = available N fertilizer identities for each crop year; $R_{k,t+1}^W$, $R_{k,t+2}^B$, and $R_{k,t+3}^P$ = residual soil N carry-over levels for each crop year; $V_{k,t}^P$, $V_{k,t+1}^W$, and $V_{k,t+2}^B$ = expected N carry-over functions; $A_{k,t}^P$, A^W, and A^B = amounts of N fertilizer applied in each crop year; r_t = the expected price of N fertilizer in crop year t; and α = the time preference discount factor = $1/(1+d)^t$, where d equals the discount rate. The dynamic economic model above is designed for a risk-neutral farmer who wants to determine the optimal N fertilizer application rate that maximizes the present value of expected net farm income. The optimal steady-state level of N available for plant uptake in the potato year (N^{*P}) is determined by holding expected N and crop prices constant over time, and assuming all expected yield and expected N carry-over functions are invariant over time (e.g., $Y^P = Y_{k,t}^P$, $Y^W = Y_{k,t}^W$, and $Y^B = Y_{k,t}^B$; $V^P = V_{k,t}^P$, $V^W = V_{k,t}^W$, and $V^B = V_{k,t}^B$). A soil test for residual soil N would be required each potato crop year to determine the amount of applied N necessary to maintain N^{*P}. The EPIC crop growth model was used to simulate the data required to estimate yield response functions and N carry-over functions for each yield range and the field.

DESCRIPTION OF EPIC

EPIC is a biophysical process simulation model that is capable of simulating many different crops in either single cropping systems or crop rotations. The model can simulate a wide variety of environmental and biological processes including erosion-sedimentation, nutrient cycling, crop growth, and pesticide and nutrient movement with water and sediment (Mitchell et al., 1996; Williams, 1989). In this study, we use EPIC to simulate seed potato yields and N losses for each yield range under uniform and variable rate N and water application.

EPIC CALIBRATION

Two steps were taken to calibrate EPIC to simulate seed potato yields for the four yield ranges. The first step was to calibrate EPIC to simulate a conventional three-year seed potato-spring wheat-feed barley rotation. This step was accomplished using soil data, farm production and management data, and daily weather data from the Ashton seed potato operation for the period 1987 through 1994. The second step involved calibrating EPIC to simulate 1995 seed potato yields falling within each of the four yield ranges. In this step, EPIC simulations were made for the period 1987 through 1995 and were sequenced so that a seed potato yield observation would be simulated for 1995. The simulations were made using actual daily weather data for 1987 through 1995 and soil data for Kucera silt loam with bedrock substratum and Lostin silt loam soils. Calibrations for Yield Ranges 1, 2, and 3 were made using Kucera silt loam soil data. The seed potato operator indicated that soil depths are variable across the field due to the bedrock substratum being close to the surface in many areas. Thus, the model was calibrated for the first three ranges by reducing soil depths. Lostine silt loam soil data were used to simulate 1995 seed potato yields for Yield Range 4.

WEATHER AND IRRIGATION SIMULATION

EPIC was used to simulate crop yield and N loss data for each yield range during a 30-yr period. Daily weather observations were generated by EPIC using monthly weather parameters calculated from Ashton daily weather data for the period 1988 through 1995. All irrigation water was applied automatically using a center pivot system. Automatic irrigation of seed potatoes began on 16 June and continued until 21 August, while automatic irrigation of spring wheat and feed barley began on 6 June and continued until 24 July. The center pivot systems on the seed potato operation apply up to 1.9 cm of water per application and require 3 d to make one complete application (or circuit). Based on this information, EPIC was programmed to apply 1.9 of water in 3-d intervals whenever soil water tension measured in kilopascals (kPa) was above a preset critical level. Different critical soil water tension levels were determined by parameterization for each yield range and the field, with each critical level representing the level of soil water tension necessary to achieve maximum average seed potato yields for the rotation during thirty simulation years. Kucera silty loam with bedrock substratum was used to determine the critical soil water tension level for the field, since the majority of the field is composed of this soil type. The critical soil water tension levels were 75 kPa for Yield Range 1, 55 kPa for Yield Range 2, 40 kPa for Yield Range 3, 48 kPa for Yield Range 4, and 34 kPa for the field. The maximum amount of water applied to each crop in a growing season was 30.5 cm, with an additional 2.5 cm applied to seed potatoes on September 21 to soften dirt clods for potato digging. The maximum irrigation water depth of 30.5 cm represents that used by the farm operator.

FUNCTIONS OF YIELD RESPONSE TO NITROGEN

Yield response functions were estimated for all three crops by yield range. Sixteen nitrogen application levels were specified for potatoes in 11 kg increments ranging from 0 to 168 kg ha^{-1}. The rotation was simulated during a 30-yr period for each potato N application holding N applied to wheat and barley constant at their historical averages of 105 and 87 kg ha^{-1}, respectively. Simulated potato yields and soil NO_3–N carry-over at the end of April in the potato year were collected and averaged for each simulation. We assumed that a preplant N soil test would be used to determine the amount of NO_3–N in the soil at the end of April. The average soil NO_3–N values were added to the amount of nitrogen applied to provide the average amount of available nitrogen for potato production for each simulation. The average yield data and the average available nitrogen data were then used to estimate potato yield response functions for each yield range and the entire field. Yield response function data for wheat and barley were generated in a slightly different manner. Nitrogen applications to both wheat and barley were held constant rather than parameterized. Thus, only soil N carry-over was allowed to vary from year to year. Average available N for both crops was calculated as average soil NO_3–N carry-over at the end of March plus the fixed amount of N applied to each crop. A quadratic function was found to provide the best fit for the data.

NITROGEN CARRY-OVER FUNCTIONS

Nitrogen carry-over (NCO) to the current year was defined as some proportion of available nitrogen from the previous crop year. For example, NCO into the potato year was calculated as some proportion of available N in the barley year, and NCO into the barley year was calculated as some proportion of available nitrogen in the wheat year. The data used to estimate yield response functions were also used to estimate NCO rate functions.

NCO to the present period can vary depending on the crop grown in the previous period. Therefore, NCO rate functions were estimated for all three crops within each yield range. NCO rates were calculated as the ratio of average soil NO_3–N carry-over in the present period to average available N in the previous period. The rates were then specified as linear functions of available N in the previous crop year.

MANAGEMENT STRATEGIES AND ECONOMIC DATA

The EPIC model was used to simulate crop yields and annual N losses in sediment, runoff, percolation, and subsurface flow for four different management strategies during a 30-yr period. In all simulations, seed potatoes were rotated with spring wheat and feed barley. The four management strategies are as follows:

1. **CONV** Conventional uniform N fertilizer application and conventional water application during all crop years;

2. **VRAN** Variable rate N fertilizer application during the potato year, conventional irrigation during all crop years;

3. **VRAW** Variable rate water application during all crop years with conventional N fertilizer application during the potato year; and

4. **VRANW** Variable rate N fertilizer application during the potato year and variable rate water application during all crop years.

Applied N in the potato year of the rotation was calculated as the difference between the optimal steady-state level of available N obtained from the DP model, N^{*P}, and the amount of N in the soil at preplant simulated by EPIC for the end of April in the potato year, $R_{k,t}^{P}$. We assumed the farm operator would determine $R_{k,t}^{P}$ using a soil N test. We assumed water would be applied variably every crop year for the variable rate strategies, since the irrigation system is a fixed investment and remains stationary on the field.

The economic value of each N application strategy was calculated using the field-level net present value approach proposed by Hewitt and Lohr (1995). The discrete form of the basic model for calculating net present value (*NPV*) for a rotation is as follows:

$$NPV = \sum_{i=1}^{I} \frac{P_i Y_i - VC_i - AC_i}{(1+d)^i}$$

where i equals the year in the rotation crop sequence, I equals the number of years in the complete crop sequence, P_i is the market price of the crop in the ith year (\$/Mg), Y_i is the annual yield of the crop in the ith year (Mg/ha), VC_i is the variable cost of producing the crop in the ith year (\$/ha), AC_i is the annualized capital cost of yield monitor and center pivot irrigation equipment in the ith year (\$/ha), and d is the farmer's real discount rate. Equation [5] may be used to compare rotations of equal length, as in the case of this study. Hewitt and Lohr (1993) show how Eq. [5] may be modified to evaluate rotations of unequal length (e..g., 2-yr, 3-yr, 4-yr, etc.).

Average crop yields from each 30-yr simulation were used as steady state yield estimates for each N application strategy. Market prices for each crop were calculated as 5-yr averages for the period 1991 through 1995. The average price used for seed potatoes was \$105 Mg^{-1} (\$4.76 cwt^{-1}), the average price used for spring wheat was \$130 Mg^{-1} (\$3.54 bu^{-1}) and the average price used for feed barley was \$106 Mg^{-1} (\$2.31 bu^{-1}). Government program payments were excluded from the analysis as the 1996 Farm Bill legislation calls for a phase out of farm income support after the year 2002.

A real discount rate of 4.2% was used for the net present value calculations. It was calculated as the average prime rate of 7.3% for the period 1991 through 1995 less an average inflation rate of 3.1% for the same period. Average inflation was calculated based on the Consumer Price Index.

Variable costs for each crop were obtained from University of Idaho Cooperative Extension Service enterprise budgets and were supplemented with cost data from the Ashton seed potato operation. All variable cost data were in 1995 dollars. Variable costs included machinery operating costs, irrigation costs, costs of materials used in production (fertilizer, insecticide, seed), custom costs, consultation fees, crop insurance, operator labor costs, and interest on operating capital.

All fertilizer was custom applied. For **CONV** and **VRAW**, the custom charge for fertilizer application was \$16.91 ha^{-1} and represented a \$12.35 ha^{-1} charge for conventional fertilizer application plus a \$4.56 ha^{-1} fee for soil testing. The latter charge was estimated assuming four soil samples were taken for the 63 ha field (one every 16.2 ha) at a cost of \$72 per sample (\$9 multiplied by 8 core samples per sample). Fertilizer application costs for wheat and barley remained the same for **VRAN**, and **VRANW**. However, the fertilizer application charge for variable rate fertilizer application in the potato year was \$34.58 ha^{-1}. This charge was obtained from personal communication with fertilizer dealers in Southern Idaho and includes \$29.64 ha^{-1} for variable rate fertilizer application and \$4.94 ha^{-1} for map making.

Grid sampling is used in most instances to determine spatial soil nutrient content across the field. We separated the field into four yield ranges to reduce the burden of simulating several grid cells; however, we account for grid sampling on the cost side. A grid sampling fee of \$44.45 ha^{-1} was charged to the **VRAN** and **VRANW** strategies. This charge was also obtained from personal communication with Southern Idaho fertilizer dealers, and includes \$24.70 ha^{-1}

for grid soil sampling and $19.75 ha^{-1} for soil analysis. All custom grid sampling charges were based on a grid size of 0.567 ha (1.4 acres).

For the variable rate strategies, we assumed the farm operator owns a potato yield monitor with a DGPS (Differential Global Positioning System) receiver. The cost of the yield monitor was $2,700, while the cost of the DGPS receiver was $3,500. The former fee represented a cost estimate from the Ashton, Idaho, farm operator for a new seed potato yield monitor, while the latter fee was the midpoint of the cost range for DGPS receivers reported in Lu et al. (1997). We assumed the farm operator obtains differential correction free of charge from a Coast Guard or Corps of Engineers station. An additional fee of $500 was added to cover training costs. The total capital cost of equipment plus training was annualized to a per year expense of $1513 using the 4.2% real discount rate and a 5-yr replacement period. This expense was then converted to a per hectare expense of $7.80 assuming the yield monitor would be used on 194 ha of a typical 582 ha seed potato operation. We assumed the farm operator uses a 0.4 km (0.25 mile) center pivot sprinkler system to irrigate each crop on the field. The total cost of a conventional center pivot system was assumed to be $43,000, while the cost of a variable rate center pivot system was assumed to be 1.5 times that of the conventional system ($64,500). Both expenses were annualized to $16,063 per year and $24,095 per year, assuming the 4.2% discount rate and a 10-yr replacement period for each system. The annualized capital costs of the two systems were converted to per hectare expenses of $82.80 and $124.20, assuming the farm operator would purchase three systems to irrigate 194 ha of a typical 582 ha seed potato operation.

RESULTS

Optimal Steady-State Nitrogen Levels

Optimal steady-state levels of N available, N applied, and residual N in the potato crop year of the rotation were determined for this study using the DP optimization model presented by (1) above. The price of N fertilizer was held constant at $0.66 kg^{-1}, while crop prices were calculated as the average market prices reported above less per unit operating costs ($13.45 Mg^{-1} for seed potato storage; $18.74 and $5.51 Mg^{-1} for custom hauling of wheat and barley). The 4.2% real discount rate was used to calculate the time preference discount factor α. The model was solved for each yield range using GAMS (General Algebraic Modeling System) Version 2.25 (Brooke et al., 1992). The output of the model for the potato year of the rotation are presented in Table 1.

Conventional and variable rate water application had little impact on the optimal steady-state levels of N available and N applied for YR3 and YR4. However, the two irrigation strategies had much impact on steady-state N available and N applied for YR1 and YR2, the two least productive yield ranges. The steady-state N available for YR1 was the smallest of the four yield ranges when water was applied conventionally (192 kg ha^{-1}). However, the steady-state potato yield on YR1 was considerably lower than that on the other yield ranges under conventional water application (14.4 Mg ha^{-1}), indicating the production efficiency of N was very small on this yield range under uniform water application. The productive efficiency of N fertilizer was increased when water was applied variably to each yield range based on critical soil water tension levels. The steady-state N available and the steady-state yield for YR1 were both increased with variable rate water application (212 kg ha^{-1} and 23.7 Mg ha^{-1}, respectively). For YR2, variable rate water application resulted in reduced steady-state levels of N available (194 kg ha^{-1}), N applied (160 kg ha^{-1}) and increased steady-state seed potato yield (25.2 Mg ha^{-1}). Thus, variable rate water application tended to have a yield enhancing effect on the two less productive yield ranges, which was a result of increased N fertilizer utilization.

Nitrogen Losses by Management Strategy

Nitrogen losses associated with each management strategy are presented by yield range in Table 2. Average N losses for the field were nearly the same for both **CONV** and **VRAN** (14 and 16 kg ha^{-1}, respectively). Similarly, average N losses for the field were equal for **VRAW** and **VRANW** (7 kg ha^{-1} each). Thus, there was no environmental benefit achieved from variable rate application of N fertilizer for the field. Variable rate water application, however, resulted in reduced N loss for the field relative to conventional water application. Most of the reduction in N loss occurred on YR1 and YR2, the least productive yield ranges. Much of the N applied to these yield ranges percolated below the root zone under conventional uniform water application. Variable rate water application reduced N percolation and allowed for more N applied to be made available to the plant on both yield ranges.

Net Present Values by Management Strategy

Discounted net present values (NPVs) for the rotation are presented for each management strategy by yield range in Table 3. **VRAN** produced the smallest average NPV for the field ($375 ha^{-1}), followed by **CONV** ($418/ha). The NPV of **VRAN** was smaller than that of **CONV** because of the additional cost of variable rate N fertilizer application under the former strategy. This cost outweighed any benefits achieved from maintaining the steady-state level of available N for each yield range in the potato year of the rotation. **VRAW** produced the largest average NPV for the rotation ($528 ha^{-1}), followed by **VRANW** ($478 ha^{-1}). The larger NPVs for the rotation under the two variable rate water application strategies were primarily due to the increased NPVs to YR1 and YR2 resulting from improved N fertilizer use and increased yields. The additional cost of variable rate application of N fertilizer made **VRANW** less profitable than **VRAW**. The greater profitability of **VRAW** and **VRANW** relative to **CONV** and **VRAN** depended largely on the cost charged for the variable rate center pivot sprinkler system, which was assumed would be 1.5 times that of a conventional system. Sensitivity analysis indicated that **VRANW** would be nearly as profitable as **CONV** if the cost of the variable rate system were two times that of the conventional system.

CONCLUSIONS

The results of this study indicate variable rate water application may be more important than variable rate N application, both economically and environmentally. Economic returns were greatest for the field when water was applied variably, whereas economic returns were smallest when N fertilizer was applied variably. Variable rate N application did not reduce the level of N loss from the field when compared with conventional N application. However, variable rate water application cut N losses by half for the field when compared with uniform water application. Areas of the field with low yield productivity appear to have the greatest potential to benefit from variable rate water

Table 1. Optimal steady-state N levels and yields for the potato year of the rotation.

Yield range	N available	N applied	Residual N	Potato yield
Conventional water application				
		kg ha^{-1}		Mg ha^{-1}
YR1	192	169	23	14.4
YR2	219	186	33	22.8
YR3	198	158	40	28.1
YR4	202	130	77	29.2
Field	197	152	45	28.7
Variable rate water application				
Yield range	N available	N applied	Residual N	Potato yield
		kg ha^{-1}		Mg ha^{-1}
YR1	212	185	27	23.7
YR2	194	160	33	25.2
YR3	199	161	38	29.0
YR4	200	131	69	29.2
Field	196	154	42	28.7

Table 2. Average N loss for the rotation by management strategy and yield range.

Management strategy	YR1	YR2	YR3	YR4	Average †
			kg ha^{-1}		
CONV	36	18	7	12	14
VRAN	41	24	7	11	16
VRAW	15	7	5	10	7
VRANW	20	7	5	9	7

† Weighted by the number of hectares within each yield range.

Table 3. Discounted net present value for the rotation by management strategy and yield range.

Management strategy	YR1	YR2	YR3	YR4	Average †
			$ ha^{-1}		
CONV	-641	128	708	832	418
VRAN	-663	122	639	771	375
VRAW	87	319	700	753	528
VRANW	84	263	648	697	478

† Weighted by the number of hectares within each yield range.

application. The productive efficiency of applied N was smaller on the low yielding areas than on the higher yielding areas of the field. Thus, much applied N on the low yield areas was lost by percolation when water was applied uniformly across the field. The timely application of irrigation water under variable rate application reduced N percolation losses, increased the availability of N to the plant, and enhanced crop yields.

Some caution must be used in interpreting the results of this study. The results apply to a specific field situation and cannot be generalized to a whole region. Also, several limitations need to be mentioned. One is that the model was calibrated to simulate crop yields for a seed potato operation, but was not calibrated to simulate N losses. Thus simulated N losses must be interpreted in relative terms (e.g., one strategy produces more N loss than another) as actual levels of N loss for the field are unknown. A second limitation is that we used only one year of yield map data to measure spatial yield variability across the field. Much of the literature indicates that yields may vary from year to year as well as spatially across the field. Four to five years of yield map data would be more appropriate for determining spatial patterns in yields for the field. Only one year of data were available for this study. Finally, we vary only two inputs, N fertilizer and water. Profitability may be improved by variable application of other inputs such as pesticides and seeds.

REFERENCES

Brooke, A., D. Kendrick, and A. Meeraus. 1992. GAMS Release 2.25. A user's guide. The Scientific Press Series. Boyb & Fraser Publ. Company. Danvers, MA.

Carr, P.M., G.R. Carlson, J.S. Jacobsen, G.A. Nelson, and E.O. Skogley. 1991. Farming soils, not fields: A strategy for increasing fertilizer profitability. J. Prod. Agric. 4:57–61.

Fiez, T.E., B.C. Miller, and W.L. Pan. 1994. Assessment of spatially variable nitrogen fertilizer management in winter wheat. J. Prod. Agric. 7:86–93.

Hewitt, T.I., and L. Lohr. 1995. Economic and environmental simulation of alternative cropping sequences in Michigan. J. Sustain. Agr. 5:59–86.

Lowenberg-DeBoer, J., and M. Boehlje. 1996. Revolution, evolution or dead-end: Economic perspectives on precision agriculture. p. 923–944. *In* P.C. Robert et al., (ed.) Proc. of the 3rd Int. Conf. on Precision Agriculture, Minneapolis, MN, 23–26 June, 1996, ASA, CSSA, and SSSA, Madison, WI.

Lu, Y., C.C. Daughtry, G. Hart, and B. Watkins. 1997. The current state of precision farming. Food Rev. Int. 13:141–162.

Mitchell, G., R.H. Griggs, V. Benson, and J. Williams. 1996. EPIC user's-guide-draft. Version 5300. The EPIC model. Environmental Policy Integrated Climate, formerly Erosion Productivity Impact Calculator. USDA-ARS, Grassland Soil and Water Res. Lab., Temple, TX.

Rawlins, S.L., G.S. Campbell, R.H. Campbell, and J.R. Hess. 1995. Yield mapping of potato. p. 59–68. *In* P.C. Robert et al., (ed.) Proc. of Site-Specific Management for Agricultural Systems, 2nd Int. Conf., Minneapolis, MN, 27–30 March, 1994. ASA, CSSA, and SSSA, Madison, WI.

Taylor, C.R. 1983. Certainty equivalence for determination of optimal fertilizer application rates with carry-over. West. J. Agric. Econ. 8:64–67.

Wibawa, W.D., D.L. Dludlu, L.J. Swenson, D.G. Hopkins, and W.C. Dahnke. 1993. Variable fertilizer application based on yield goal, soil fertility, and soil map unit. J. Prod. Agric. 6:255–256.

Williams, J.R. 1989. EPIC: The Erosion-Productivity Impact Caculator. p. 676–681. *In* J.K. Clema (ed.) Proc. of the 1989 Summer Computer Simulation Conf., Austin, TX. 1989.

From Data to Information: Adding Value to Site-Specific Data

Scott M. Swinton and Kezelee Q. Jones
Department of Agricultural Economics
Michigan State University
East Lansing, Michigan, USA

ABSTRACT

A conceptual model is developed to measure the value of information from in-field soil sensing technologies as compared with grid and other soil sampling methods. Soil sensing offers greater spatial accuracy and the potential to apply inputs such as nitrogen fertilizer immediately, avoiding changes in nutrient status that occur with delays between soil sampling and fertilizer application. By contrast, soil sampling offers greater measurement accuracy, because it does not rely on proxy variables such as electrical conductivity to infer nutrient status. The average profitability and relative riskiness of soil sensing versus sampling depend upon (i) the trade-off between, on the one hand, the spatial and temporal accuracy of sensing and, on the other hand, the measurement accuracy of sampling; (ii) the cost of data collection, and (iii) input and product prices. Similar trade-offs govern the relative riskiness of sensing versus sampling.

PROBLEM STATEMENT

Site-specific agricultural information technologies are being tested by farmers across the USA. How and where site-specific farming (SSF) is adopted over the long term depends on the value of supplementary information it offers. Information has value when it leads to improved management decisions (Davis & Olson, 1985). In crop production, this may mean better crop yields, better crop quality, lower costs, or reduced agricultural pollution, all leading to more profitable, more sustainable farming.

For an information technology to be profitable to a business (and hence worth adopting without public intervention), the information value to the firm must outstrip its cost. Most current information to guide variable-rate application of agricultural inputs comes from manual soil sampling. Most commonly available as grid sampling, such soil sampling is typically done on 2- to 3- acre grids on Midwestern farm fields. At prices of $5 to 10 per acre, it is viewed as expensive by many farmers who are otherwise interested in site-specific farming (Wehrspann, 1996; Swinton & Ahmad, 1996). Apart from cash costs, there also is an opportunity cost to the time delay between sampling and treatment.

Sensing technologies — both remote and in-field — represent an emergent alternative to soil sampling (Sudduth et al., 1997; Hummel & Birrell, 1997; Frazier et al., 1997). They have the potential to offer a much lower marginal cost of data collection, better geographic accuracy, more intensive sampling and sometimes also the opportunity for immediate input application. Sensing technologies, however, tend to observe a proxy for the real crop input of interest. For example, crop N deficiency

Copyright © 1999 ASA-CSSA-SSSA, 677 South Segoe Road, Madison, WI 53711, USA. *Proceedings of the Fourth International Conference on Precision Agriculture.*

might be measured by infrared reflectance from leaves or from soil electrical conductivity, rather than a direct measure of soil nitrogen available for crop uptake. Apart from yield monitoring, agricultural sensing technologies have not yet seen widespread adoption by U.S. farmers.

This paper aims to develop a conceptual model of information economics related to site-specific crop management. That model is examined for indicators of:

1. How does the value of soil nutrient information vary between sampling and sensing technologies?
2. When could sensing be more profitable than sampling?
3. When could sensing yield less variable net returns than sampling?

REVIEW OF THE LITERATURE

Information Value and How to Measure It

Existing literature offers helpful context on how information is valued, how information is used in a spatially variable setting, and how much spatial information is required for optimal decision making.

Marschak (1968) defines information as a message that alters probabilistic perceptions of random events. This definition is used in statistical decision theory and has a great appeal among researchers of different disciplines (Chavas & Pope, 1984). Information in this sense acts as a means to decrease risk and has little or no value in the absence of risk. In general, risk characterizes those uncertain events whose outcomes alter the decision maker's welfare (Robison, 1987) as measured, for example, by the net income of farmers.

Given the link between information and risk, it is important to define how risk is measured. A practical working definition of risk can be based on mean and variance of the risky outcome variable under one of two broadly applicable conditions. The first is that the decision maker does not care about statistical moments of the probability distribution other than mean and variance. The second is that two or more risky outcome variables have probability distributions that are equal except for location (mean) and scale. This condition applies if "there exists some random variable X such that each [variable] Y_i is equal in distribution to $\mu_i + \sigma_i X$," where X is normalized to have mean 0 and variance 1 (Meyer, 1988). Under this definition, given two random variables Y_1 and Y_2 with the same mean, Y_2 is said to be riskier than Y_1 if Y_2 has a greater variance than Y_1 (Meyer, 1987; Pratt, 1964). This implies that probabilistic information is more valuable for managing Y_2 than Y_1.

Value of Spatial Information in Agriculture

Having explored briefly how information is valued, we now turn to information use in a spatially variable agricultural setting. Feinerman et al. (1989) studied the stochastic optimization of irrigation water in a spatially variable agricultural field. Their main objective was to predict mean yield based on available information and to determine which information level led to highest welfare for the decision maker. They found that spatial sampling (conditional analysis) significantly improved yield prediction and efficiency of irrigation water use, as compared with assuming a uniform probability density function (pdf) for the whole field (unconditional analysis).

Increased welfare under the conditional approach has also been found for temporal information embodied in pre-sidedress NO_3 samples in corn (Babcock et al. 1996). They used field plot data in estimating a corn production function, and showed that pre-sidedress NO_3 sampling could reduce average fertilizer application rates between 15 and 41%.

Chavas and Pope (1984) showed that the value of costless information is always nonnegative. The more samples taken on a given field, the less uncertainty there is about the random variables and therefore the smaller the deviation between the nutrient that should truly be applied and what is applied based on measurements; however, sampling is not costless. Denser sampling costs more, and if the added value is less than the information cost, the decision maker's welfare will be lower. As Feinerman et al.(1989) note, determining the optimal number of observations and the optimal spread over a field is a difficult statistical problem.

MODEL DEVELOPMENT AND DISCUSSION

Sampling Versus Sensing

As a prelude to the discussion that follows, we make a distinction between accuracy and precision. Accuracy in measuring a physical attribute or process describes how well the measurement reveals the true level of the attribute or process. Precision, on the other hand, refers to the resolution of the measurement instrument. Therefore a measure can have high precision but may be highly inaccurate if the measurement instrument is not well calibrated.

Many data collection tools are available for agricultural field management (Fig. 1). Broadly speaking, data can be collected by sampling or sensing. Sampling refers to collecting individual observations from the population of interest, and using them to make inferences about the population as a whole. Importantly, sampling entails direct measurement of the attribute(s) of interest (e.g., NO_3–N in a soil sample) for making management inferences. Whole-field soil sampling has been used for half a century to generate fertilizer recommendations at the field level. Site-specific approaches, such as sampling on a grid or by soil type within a field, use a denser sampling approach to make inferences about smaller areas within a field. Our focus in this paper will be on grid soil sampling to represent all classes of in-field soil sampling.

Sensing refers to automated data collection using intensive sampling. Sensing frequently relies on a proxy variable that is correlated with the attribute of management interest. Remote sensing, for example, may entail measures of near-infrared reflectance from a cropped field as imaged from an airplane or satellite. Inferences for crop management must be based on extrapolating from the proxy

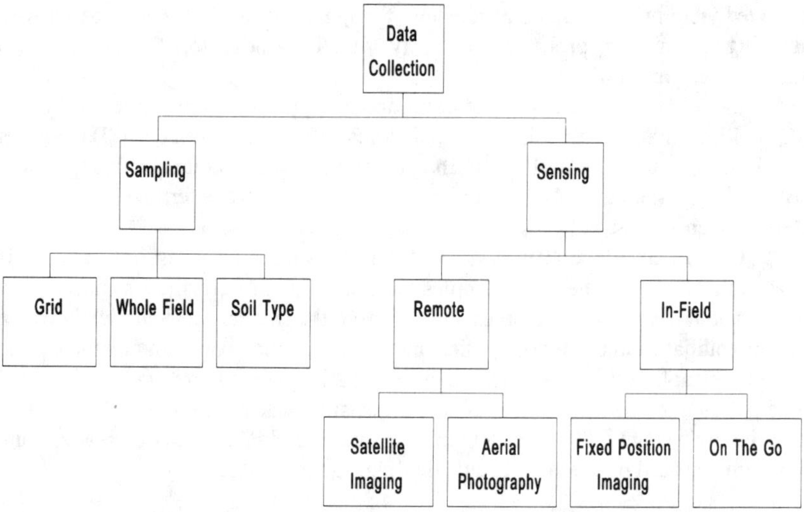

variable to predicted levels of the management variable of interest. This paper will focus on in-field soil sensing (IFSS), rather than remote sensing. A helpful illustration of IFFS is use of the soil electrical conductivity to predict the presence of soil NO_3^-. NO_3^- ions conduct electricity, but so can other soil media. Therefore electrical conductivity is an imperfect proxy for NO_3–N. There is always a margin of error, however, fine-tuned or calibrated the measurement instrument. The automated data collection inherent in sensing yields a frequency of measurement that typically leads to high spatial accuracy. Because IFSS allows immediate variable rate technology (VRT) nutrient application, there also is potentially less time for nutrient status to change, a real problem with leachable NO_3–N. Hence, IFSS may offer greater temporal accuracy as well as spatial accuracy to compensate for its potential proxy variable inaccuracy.

Conventional soil sampling has a higher level of measurement accuracy than IFSS (since the true attribute is being measured and not a proxy). Grid sampling, however, has relatively lower spatial and temporal accuracy since the measured values represent a larger area, and the elapsed time between measurement and application is relatively long. Ideally a farmer would like accurate location, accurate measurements and instantaneous application. What we have instead are relatively precise location and relatively imprecise measurements for IFSS. If IFSS is linked to variable rate N fertilizer application on-the-go, then IFSS offers relatively short elapsed time before fertilizer application. For conventional sampling, we have relatively imprecise location, relatively precise measurements combined with relatively long elapsed time to application. This tradeoff between spatial and temporal accuracy and measurement accuracy is illustrated in Fig. 2.

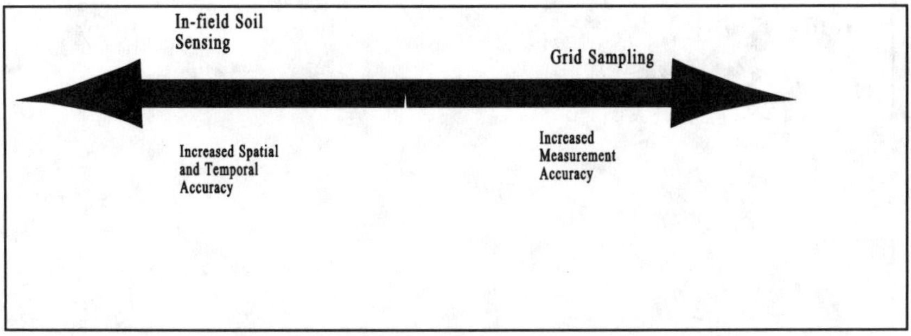

Which sampling approach best suits farmer needs depends on a number of factors including cost of information acquisition, ease of use, and clarity of interpretation. The two key factors will probably be (i) expected profitability, and (ii) variability of profits, especially if the farmer is risk-averse. In the remainder of this paper we seek to identify conditions under which sensing leads to higher expected net income or lower variance of net income than grid sampling.

A Model of Spatial Nitrogen Management

Consider the management of soil N in corn production. Nitrogen is a spatially and temporally random input to the corn production process. Following Babcock and Blackmer (1992) we assume that the mean yield y is a function only of the mean concentration of available soil N N_s at the time and location where the application decision is being made. Therefore the production relationship is represented by $f(N_s) = E(y \mid N_s)$. Binford et al. (1992) found that corn yield response to N is approximately linear up to a plateau level where N is no longer limiting. So the corn production function can be modeled by a linear response and plateau (LRP) function (Fig. 3) as:

$$y = \begin{cases} \alpha N_T + \varepsilon^y \text{ if } N_T \leq N_r \\ Y_r + \varepsilon^y \text{ if } N_T > N_r \end{cases} \qquad [3]$$

where Y_r is maximum yield (at and above recommended N rate N_r), N_T is the total N in the soil after application, and N_r is the recommended level of N. ε^y is the yield disturbance, and is independent of any spatial, temporal and measurement disturbances in N measurement. We assume that the yield disturbance is normally distributed such that $\varepsilon^y \sim N(0,\sigma_y^2)$. $N_a = N_r - \hat{N}$ is N applied where $\hat{N}$ is a measure of the available N before application. We define $\hat{N}$ by Eq. [4], decomposing the associated sampling error into three components:

$$\hat{N} = N_s + \varepsilon^t[\text{weather}(t^* - t^0)] + \varepsilon^a(\text{area}) + \varepsilon^p(\text{proxy}) \qquad [4]$$

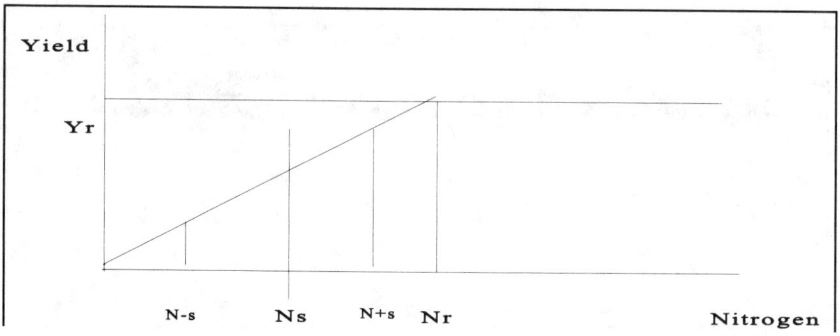

where N_s is the true available N level in soil at time of testing; t^*, t^0 are respectively time of nutrient intake by plant and initial sampling time; ε^t is error due to passage of time between sampling and application; ε^a is spatial error, and ε^p is error due to measurement by proxy instead of the true attribute. We assume that all the disturbances are normally distributed according to: $\varepsilon^t \sim N(0,\sigma_t^2)$, $\varepsilon^a \sim N(0,\sigma_a^2)$, and $\varepsilon^p \sim N(0,\sigma_p^2)$, implying that $E[\hat{N}] = N_s$.

With one variable input (N), profit is given as:

$$\pi = P(\alpha N_T + \varepsilon^y) - w(N_r - \hat{N}) - \text{Fixed Cost} \qquad [5]$$

where P is product price and *w* is N fertilizer cost.

Mean Profitability of Sensing Versus Sampling

For mean profitability, we consider two scenarios for errors in measuring soil N. Under the first scenario, suppose the measured value of the available soil N is overestimated at N^+_s (Fig. 3). Based on this measured value, the farmer under applies $N_r - N^+_s$ that falls short of recommended rate N_r by the amount $N^+_s - N_s$. Yield falls short of Y_r leading to a loss of potential profit.

Suppose instead the measured value of available soil N is underestimated at N^-_s. The farmer over applies $N_r - N^-_s$, with a surplus $N_s - N^-_s$. The over application does not increase yield above Y_r; however, there is waste of N fertilizer, leading to reduced profitability. Any deviation from the true N_s leads to a loss in profitability. The greater the deviation, the more drastic the loss in profits.

The error band $N^+_s - N^-_s$ shrinks as the number of sampling points increases over a given area (i.e., the denser the sampling procedure). By assumption, IFSS offers more spatial accuracy than grid sampling. Hence if we consider IFSS as denser sampling (i.e., we are still sampling but the area that each sample represents under IFSS is much smaller), we can conclude intuitively that IFSS has a smaller error band leading to smaller overestimates and underestimates of available N and therefore smaller expected loss in profitability relative to conventional grid sampling.

Mathematically, we observe the same result. We first assume that a rational farmer will apply soil N as long as the marginal value product ($MVP = P*df(N)/dN$) is greater than the cost of the input (w). In our LRP corn production model, this implies that $P\alpha > w$ up to point N_r. If the error band, N^+_s - N^-_s, is narrower for sensing, then $N^+_{sen} < N^+_{sam}$. Now consider the overestimate scenario described above. We introduce two new symbols: N^+_{sam} (overestimated measure of N_s under sampling) and N^+_{sen} (overestimated measure of N_s under IFSS). We also define $\pi_j = PY_j - w(N_r - N^+_j)$ (j = sen, sam). Then the difference in expected profitability between sensing and sampling depends on whether the value of increased yield from sensing is greater than the increased cost of fertilizer, $\pi_{sen} - \pi_{sam} = P(Y_{sen} - Y_{sam}) + w(N^+_{sen} - N^+_{sam})$. From our assumptions, $P(Y_{sen} - Y_{sam}) = P\alpha\,(N^+_{sam} - N^+_{sen}) > w\,(N^+_{sam} - N^+_{sen})$. So IFSS leads to a higher expected profitability given overestimation of N_s.

Scenario two is more straightforward. Underestimation of soil N leads to over application of N; however over application leads to yield of $Y_{sam} = Y_{sen} = Y_r$. We observe that the wasted fertilizer, $N_r - N^-_s$, is smaller for IFSS then for sampling. Therefore $\pi = PY\text{-}w(N_r - N^-_s)$ is higher under IFSS.

Variance of Profitability in Sensing Versus Sampling

The analysis of risk effects from soil sensing versus sampling depends upon the probability distributions of the input of interest–N in this case. In the mean-variance framework developed above, risk effects hinge on variances and covariances. Building on the profit expressing in Eq. [5], we assume that the time-lag variability of soil N, N_s is unrelated to either spatial accuracy or proxy measurement variability($cov(\varepsilon^t,\varepsilon^a) = cov(\varepsilon^p,\varepsilon^t) = 0$), but that the proxy measurement variability increases with spatial variability ($cov(\varepsilon^p,\varepsilon^a) > 0$). These generic assumptions hold for both conventional sampling and IFSS.

The three components of soil attribute assessment risk are assumed to vary systematically between grid sampling and soil sensing. For grid sampling, we assume there exists no proxy error ($\sigma_p^2 = 0$) because soil analysis laboratories directly measure NO_3–N. We also assume that $\sigma_a^2>0$ and $\sigma_t^2>0$, and that both variances are large. The assumption concerning spatial variability is reasonable considering that a particular sample represents a grid cell area typically 2.5 acres. Such an area may contain great agronomic variability (Pierce & Mueller, 1997). The assumed variability due to time lags is based on how water-soluble soil NO_3 levels fluctuate in response to precipitation during the weeks that typically elapse between soil sampling and pre-sidedress N application.

Given these assumptions, the variance of $\hat{N}$ with grid soil sampling is:

$$\sigma^2_{\hat{N}}(sam) = var(N_s + \varepsilon^t + \varepsilon^a + \varepsilon^p) = \sigma_t^2 + \sigma_a^2 . \quad [6]$$

In general with the LRP corn production function, the variance of profit is:

$$\sigma^2_\pi = \begin{cases} var(\alpha PN_T + P\varepsilon^y - wN_r + w\hat{N}) & \text{if } N_T < N_r \\ var(PY_r + P\varepsilon^y - wN_r + w\hat{N}) & \text{if } N_T \geq N_r \end{cases} \quad [7]$$

Inserting the variance of $\hat{N}$ with grid soil sampling yields the variance of profit under soil sampling:

$$\sigma_\pi^2(sam) = \begin{cases} P^2\sigma_y^2 + \sum_{j=a,t}[(w-\alpha P)^2\sigma_j^2 + 2P(w-\alpha P)cov(\varepsilon^y,\varepsilon^j)] & if\ N_T < N_r \\ P^2\sigma_y^2 + \sum_{j=a,t}[w^2\sigma_j^2 + 2Pw\ cov(\varepsilon^y,\varepsilon^j)] & if\ N_T \geq N_r \end{cases} \quad [8]$$

The risk effects of in-field soil sensing arise from the distinct nature of sampling disturbances associated with sensing. To begin, assume that input application on-the-go in response to the sensory input makes temporal disturbance negligible, so $\sigma_t^2 = 0$. Likewise, since sensing involves very dense sampling, we assume that although locational disturbance is nonzero ($\sigma_a^2 > 0$), it is small. Finally, we assume that there is measurement error due to use of a proxy for the true attribute of interest ($\sigma_p^2 > 0$), but that there is medium to high correlation between the proxy measurement and the true level of N.

The variance of $\hat{N}$ with sensing is thus given as:

$$\sigma_{\hat{N}}^2(sen) = var(N_s + \varepsilon^t + \varepsilon^a + \varepsilon^p) = \sigma_p^2 + \sigma_a^2 + 2cov(\varepsilon^p,\varepsilon^a) \quad [9]$$

Inserting this expression into Eq. [7] yields the variance of profit with in-field soil sensing:

$$\sigma_\pi^2(sen) = \begin{cases} P^2\sigma_y^2 + 2(w-\alpha P)^2 cov(\varepsilon^p,\varepsilon^a) + \sum_{j=a,p}[(w-\alpha P)^2\sigma_j^2 + 2(w-\alpha P)Pcov(\varepsilon^y,\varepsilon^j)] \\ \quad if\ N_T < N_r \\ P^2\sigma_y^2 + 2w^2 cov(\varepsilon^p,\varepsilon^a) + \sum_{j=a,p}[w^2\sigma_j^2 + 2wPcov(\varepsilon^y,\varepsilon^j)] \\ \quad if\ N_T \geq N_r \end{cases} \quad [10]$$

Variability of Profits Under the Two Scenarios

Locational error is likely to be higher under grid sampling than under sensing [σ_a^2(sam) > σ_a^2(sen)], as assumed above. The general form for the difference in profit variability under the two scenarios is therefore:

[11]

$$\sigma^2_\pi(sam) - \sigma^2_\pi(sen) = \begin{cases} (w-\alpha P)^2 [\sigma^2_{\hat{N}}(sam) - \sigma^2_{\hat{N}}(sen)] + 2P(w-\alpha P)[cov_{sam}(\varepsilon^y, \hat{N}) - cov_{sen}(\varepsilon^y, \hat{N})] & if\ N_T < N_r \\ w^2 [\sigma^2_{\hat{N}}(sam) - \sigma^2_{\hat{N}}(sen)] + 2Pw\,[cov_{sam}(\varepsilon^y, \hat{N}) - cov_{sen}(\varepsilon^y, \hat{N})] & if\ N_T \geq N_r \end{cases}$$

Given (11), we have equal variability of net income for sampling and sensing in two instances. The first is if $P\alpha = w$ when total nitrogen (N_T) is less than the recommended rate (N_r) or if $w = 0$ when total nitrogen exceeds N_r. These conditions imply that the value of the yield gain just equals the cost of N, so any N rate would be equally profitable. The second instance is if the variances of the sampled nitrogen levels are equal for sampling and sensing, [$\sigma_N^2(sam) = \sigma_N^2(sen)$]. However, grid sampling will lead to riskier profit if it has higher variance in sampled soil nitrogen

[12]

$$\sigma^2_\pi(sam) > \sigma^2_\pi(sen)\ if$$
$$\{\sigma^2_{\hat{N}}(sam) > \sigma^2_{\hat{N}}(sen)\}\ and\ \{cov_{sam}(\varepsilon^y, \hat{N}) \geq cov_{sen}(\varepsilon^y, \hat{N})\}$$

level and if that soil N level covaries with crop yield:

This general form; however, does not reveal much. We therefore use the profit variability expressions in Eq. [8] and [10] and add the simplifying assumption that yield variability (ε^y) is independent of spatial, temporal and measurement disturbances,

$$\sigma^2_\pi(sam) - \sigma^2_\pi(sen) = \begin{cases} (w-\alpha P)^2 [(\sigma^2_{a(sam)} - \sigma^2_{a(sen)}) + (\sigma^2_{t(sam)} - \sigma^2_{p(sen)})] - 2(w-\alpha P)^2 cov(\varepsilon^p, \varepsilon^a) & if\ N_T < N_r \\ w^2 [(\sigma^2_{a(sam)} - \sigma^2_{a(sen)}) + (\sigma^2_{t(sam)} - \sigma^2_{p(sen)})] - 2w^2 cov(\varepsilon^p, \varepsilon^a) & if\ N_T \geq N_r \end{cases}$$

[13]

to obtain:

By assumption, the covariance between the proxy error and spatial error is positive. However, we cannot unambiguously determine the sign of the difference between temporal variability in N_s and proxy measurement variability. We can say that if temporal variability is sufficiently high and spatial variability sufficiently low relative to proxy measurement variability, then profit variability under sampling is higher than profit variability under sensing. This amounts to a trade-off between the proxy measurement variability of sensing, on the one hand, and the spatial and temporal soil variability of grid sampling, on the other. If total variability is higher under grid sampling than under in-field soil sensing, then sensing information has greater value to a risk-averse decision maker.

CONCLUSIONS

This paper has used a conceptual model to show that both expected profit and variance of profit from site-specific input management depend on the accuracy of spatial information. The relative merits of sensing information are that it tends to be more accurate spatially and temporally (if used for VRT input control) than soil sampling information. On the other hand, soil sampling tends to give more accurate measurements. Which of the two is preferable depends upon the relative importance of location, timing, and measurement error.

This result has important implications for profitable site-specific input management today, as well as for the development of new site-specific technologies in future. Regarding profitable management, the payoffs to sensing are highest (i) when sensor equipment gives a fairly accurate measure of desired attributes, (ii) when timeliness matters (e.g., plant growth hormone or N fertilizer application), (iii) when in-field micro-variability is great. By contrast, the payoffs to in-field soil sampling methods (by grid, soil type, or other) are highest when (i) sensor equipment is not reliable, (ii) timeliness does not matter (e.g., P or K fertilizer application), and (iii) spatial variability occurs on a larger scale. Grid sampling tends to impose only a variable cost for data collection labor and laboratory analysis. On the other hand, in-field sensor equipment is a capital good with a high fixed cost for equipment and operator learning, so its cost per acre declines with the land area over which it is used.

From the standpoint of technological innovation, the spatial and temporal accuracy advantages of sensing combined with far lower cost-per-measurement suggest continuing intensive research and development of better sensor technologies, both in-field and remote. The focus will be on reliable equipment that minimizes proxy measurement error. In the meantime, there remains large scope for empirical measurement of the relative profitability and riskiness of sensing versus sampling technologies in the market.

REFERENCES

Auld, D.A. 1972. Imperfect knowledge and the new theory of demand. J. Polit. Econ. 80:1287–94.

Babcock, B., A. Carriquiry, and H. Stern. 1996. Evaluation of soil test information in agricultural decision-making. App. Stat. 45(4):447–461.

Babcock, B., and A.M. Blackmer. 1992. The value of reducing temporal input nonuniformities. J. Agric. Resour. Econ. 17:335–347.

Binford, G., A. Blackmer, and M. Cerrato. 1992. Relationships between corn yields and soil nitrate in late spring. Agron. J. 84(1):53–89.

Chavas, J., and R. Pope. 1994. Information: its measurement and valuation. Am. J. Agric. Econ. 66:705–710.

Davis, G.B., and M.H. Olson. 1985. Management information systems: Conceptual foundations, structure, and development. 2nd ed. McGraw-Hill, New York.

Feinerman, E., E. Bresler, and G. Dagan. 1989. Optimization of inputs in a spatially variable natural resource: Unconditional vs. conditional analysis. J. Environ. Econ. Mgmt. 17(1):140–154.

Frazier, B.E., C.S. Walters, and E.M. Perry. 1997. Role of remote sensing in site-specific management. p. 149–160. *In* F.J. Pierce and E.J. Sadler (ed.) The state of site-specific management for agriculture. ASA, CSSA, and SSSA, Madison, WI.

Hummel, J.W., and S.J. Birrell. 1997. Sensors and the future of site-specific nutrient management. p. 55–58. *In* D.D. Warncke (ed.) Managing diverse nutrient levels: Role of site-specific management. ASA, CSSA, and SSSA, Madison, WI.

Marschak, J. 1968. Economics of inquiring, communicating, deciding. Am. Econ. Rev. 58(1):1–18.

Meyer, J. 1987. Two moment decision models and expected utility maximization. Am. Econ. Rev. 77:421–430.

Meyer, J. 1988. Two moment decision models and expected utility maximization: Some implications for applied research. p. 170–183. *In* G. Carlson (ed.) Risk analysis for production firms: Concepts, informational requirements and policy issues. Proc. of Southern Regional Research Project S-180, July 1988. Dep. of Agric. Econ., North Carolina State Univ., Raleigh.

Pierce, F.J., and T.G. Mueller. 1997. The expectations and realities of site-specific nutrient management. p. 1–10. *In* D.D. Warncke (ed.) Managing diverse nutrient levels: Role of site-specific management. ASA, CSSA, and SSSA, Madison, WI.

Pratt, J.W. 1964. Risk aversion in the small and in the large. *Econometrica.* 32(1):12–36.

Robison, L.J., and P. Barry. 1987. *The competitive firm's response to risk.* MacMillan, New York.

Sudduth, K.A., J.W. Hummel, and S.J. Birrell. 1997. Sensors for site-specific management. p. 183–210. *In* F.J. Pierce and E.J. Sadler (ed.) The state of site-specific management for agriculture. ASA, CSSA, and SSSA, Madison, WI.

Swinton, S.M., and M. Ahmad. 1996. Returns to farmer investments in precision agriculture equipment and services. p. 1009–1018. *In* P.C. Robert et al., (ed.) Proc. of the 3rd Int Conf. ASA, CSSA, and SSSA, Madison, WI.

Wehrspann, J. 1996. When grids don't fit. Farm Ind. News. 29(Oct.):4–7.

Additional Analysis Tools Based on Yield Data

Hal Watkins

GROWMARK
Bloomington, Illinois

ABSTRACT

Objective: To offer farmers an effective analysis tool to evaluate profitability of a field or farm using yield data. While there are numerous options in global information system (GIS) software for the presentation of visualized data, the actual tools to analyze are more limited and certainly less used. One aspect of the analysis of data is most often overlooked by commercial providers is the analysis of profitability. This analysis will require the confidence and cooperation of the producer to be accurate and useful as a management tool. Algebraic formulas that convert yield data to gross and net profit are not that difficult to write, but most yield oriented GIS packages lack the ability to apply them. In our presentation we will demonstrate the practicality and necessity of such financial analysis. We will make a practical application that a Precision Farmer might use in analyzing items such as cash rent, input costs, drainage cost effectiveness and tillage practices. Results: By applying algebraic formulas to yield data we can represent yield as revenue per acre as well as a profitability by acre, hybrid or soil type. We believe that better managers will use this data to decide whether to continue to rent some farms and use the data to re-negotiate other cash rents. Conclusions: Farmers who use yield data for a variety of decision making processes will be the farmers of the next century. Yield analysis comparing varying crops is another area that could be like comparing apples and oranges. By taking yields to an index value based on yearly averages we believe farmers can make financial decisions more readily. Financial analysis will also be performed on this data, demonstrating the practical application of yield data in the determination of profitability for various farms and fields.

Impact of Spatial Resolution on the Potential Profitability of Site-Specific N and P Management of Corn

Gary Malzer
S. Braum
A. Hopkins
D. Mulla
D. Huggins
J. Bell
P. Robert

University of Minnesota
St. Paul, MN

ABSTRACT

The profitability associated with site-specific nutrient management will be impacted by the spatial scale that is used for the interpolation of management decisions. A 12 ha research site near Windom, MN was established to quantify the impact of sampling scale on potential increased return. The field was soil sampled on a 0.06 ha grid for routine soil test values (P, K, pH, etc) to a depth of 15 cm and to a depth of 60 cm for nitrate-N. Three replications of five N rate treatments (0, 67, 112, 156, and 202 kg ha^{-1}) and three P rate treatments (0, 56, and 112 kg P205 ha^{-1}) were applied as constant rate strips, randomized within a split block arrangement with P rate as the main block. Grain yields were obtained every 15 m along each treatment transect. Multiple regression techniques, using a split-block-in-space approach were used to determine yield response and economic optimum rates of N and P fertilization at different locations within the field (minimum resolution was 0.18 ha). The field was subdivided into several different management areas based upon grid cell size (0.36, 0.52, 1.04, 2.08, and 12 ha) and potential management areas based on soil test values, organic matter and topographic features. Each management area was described according to the appropriate soil testing procedure for making fertilizer recommendations. The impact of that fertilizer application was evaluated economically utilizing the response equations determined for that sub-region of the field.

ENVIRONMENT

Targeting Precision Agrichemical Applications to Increase Productivity

S. A. Clay
G. J. Lems
D. E. Clay

Plant Science Department
South Dakota State University
Brookings, South Dakota

M. M. Ellsbury

USDA-ARS
Northern Grain Insect Laboratory
Brookings, South Dakota

F. Forcella

USDA-ARS
Morris, Minnesota

ABSTRACT

Spatial variability of weeds in a field was used as input information for a bio-economic weed control model to generate preemergence, pre- plus postemergence, and postemergence herbicide strategies at three field locations. Weed control effectiveness, crop production, and profitability were estimated and compared with a producer's blanket herbicide application at each site. The recommendations from the bio-economic model were less expensive than the producer's treatment and resulted in similar or better weed control, yield, and net return. Using site-specific herbicide application and placement would optimize economic returns and environmental safety, benefiting the producer and society.

INTRODUCTION

Weeds and insects have been shown to occur in patches across field landscapes (Cardina et al., 1995; Mortensen et al., 1995; Ellsbury et al., 1996; Lems, 1998). This patchiness causes excessive or inappropriate application of pesticides in production agriculture. Prophylactic blanket pesticide treatments normally are applied to all areas even if low or no pest infestations are present. Management of spatial heterogeneity would lead to better use of pesticides (Mortensen et al., 1995). Knowledge of species present in specific areas of a field would be useful to better target pesticide applications.

Copyright © 1999 ASA-CSSA-SSSA, 677 South Segoe Road, Madison, WI 53711, USA. *Proceedings of the Fourth International Conference on Precision Agriculture.*

Several approaches can be used to obtain site-specific recommendations. The first requirement is to define the heterogeneity of the species and density in the field. This has been done using grid sampling approaches (Brown & Steckler, 1995; Gerhards & Wyse-Pester, 1997) coupled with kriging (Heisel et al., 1996). Another method that is being investigated is using remote sensing (to determine anomalies in the field) coupled with site specific field scouting (Broulik et al., 1997). The next requirement is to generate a site-appropriate recommendation that incorporates information on the crop, soil properties, pest species, pest density, management options, and control cost. To this end, several computer based decision support systems have been developed including GWM (Wiles et al., 1996) and WEEDSIM (Swinton & King, 1994).

GWM is classified as a bio-economic model because it generates weed treatment decisions based on economic return from weed control strategies (Wiles et al., 1994). This model is flexible to the user. Several user-defined characteristics including crop, weed, field specific soil properties, herbicide cost, and weed efficacy can be incorporated easily into the model. Preemergence applications are based on weed seed counts derived from counting seeds in known diameter soil cores whereas postemergence recommendations are based on weed counts and size from sampled areas. GWM predicts net return from the available user-defined weed management options based on expected yield and crop price minus treatment and applications costs and yield loss from weed competition.

Growers are reluctant to move from a uniform application of herbicides to a site-specific approach. The objective of this research was to compare model approach recommendation scenarios for pre- and postemergence herbicides based on a decision support system with a grower treatment based on the grower's knowledge of the field, weed problems, and other decisions based on price, rebates, and other factors. A field in soybean–corn was chosen that had a history of intensive field scouting and a wide diversity of weed species, densities, and weed patchiness.

MATERIALS AND METHODS

A study was conducted in 1996 in a soybean field in Moody Co., Standard Deviation (44.17^0 N lat., -96.63^0 W long.) that had an average elevation of 523 m above sea level with 16.5 m of relief. Three trials were located in this field. At the summit position the soil was a fine-loamy, mixed, Udic Haploboroll (Venagro series), on the shoulder position the soil was a fine-loamy, mixed, Udic Haploboroll (Doland series), and at the toeslope position the soil was a fine-silty and fine-loamy, frigid, Typic Calciaquoll (Vallers series).

Each trial was a randomized complete block design with five treatments replicated four times. Individual plot size was 3 by 9 m. The five treatments included a weed control, grower's treatment, and three GWM generated recommendations, preemergence only, preemergence plus postemergence treatments, and a postemergence treatment only. For the preemergence and pre + postemergence weed seed inputs, one soil core, 10-cm diam by 5-cm depth was collected from each of the four replicates and weed seed extracted using a root-washing system (Gross

& Renner, 1989). Seeds were manually enumerated by species and composite results of the four cores were used for model input. For postemergence treatments, weed seedling counts by species were taken in a 20 by 50-cm area in each of the four replications at each location. Growth stages of the weed seedling including leaf number and height were recorded. A composite average per unit area was calculated and used for model input. Weed competitive index, using expert knowledge to rank the species competitiveness in the local area and where the most competitive weed has a value of 1, (Table 1) was also used a model input. Treatment cost was based on herbicide prices at local dealerships and expected yield for each field location was based on historical field records.

The top-ranked or highest profit treatments generated by GWM were chosen for the three GWM treatments. Herbicides were applied with a bicycle sprayer using 4-8002 flat fan nozzles on 76 cm spacing calibrated for 134 L ha^{-1} output at 274 kPa and 4.8 kph. Preemergence herbicides were applied in late May and post emergence herbicides were applied June 20.

Herbicide efficacy was evaluated 8 and 27 d after postemergence treatment by comparing the number and vigor of grass and broadleaf weed species to the untreated check on a 0 to 100 scale. Weed and crop biomass samples were clipped from 4-0.1 m^{-2} areas of the plots at soybean full bloom and mid-pod fill development, separated into weed and crop components, dried at 60°C to constant weight, and weighed. The center 1.5 by 4.6 m area of the plots was harvested with a plot combine at physiological maturity.

Data were analyzed using analysis of variance techniques and when the *F*-test was significant, least significant differences were calculated at $P = 0.05$. Net profit was calculated using harvested yield data, soybean price in the fall of 1996 ($238 Mg^{-1}) minus cost of herbicide and other expenses including land rent($150 ha^{-1}), seed and planting costs ($65 ha^{-1}), and custom combining ($50 ha^{-1}).

Table 1. Weed competitive index used in the GWM model for soybean trial.

Common name	Scientific name	Competitive index
Pigweed sp.	*Amaranthus* sp.	1.00
Common ragweed	*Ambrosia artemisiifolia*	0.20
Common milkweed	*Asclepias syriaca*	0.25
Common lambsquarter	*Chenopodium album*	0.45
Canada thistle	*Cirsium arvense*	0.75
Pensylvania smartweed	*Polygonum pensylvanicum*	0.28
Foxtail sp.	*Setaria* sp.	0.09
Black nightshade	*Solanum nigrum*	0.35

Table 2. Weed seed bank sampling results for preemergence only and preemergence plus postemergence treatments.

Area	Weed species	Preemergence	Pre + post emergence
		---------------seeds m^{-2}---------------	
Summit	*Amaranthus* sp.	555 (383)[†]	400 (465)
Shoulder	*Amaranthus* sp.	832 (825)	585 (692)
Toeslope	*Amaranthus* sp.	801 (555)	1109 (1692)
	Ambrosia artemisiifolia	893 (1786)	524 (1047)
	Polygonum pensylvanicum	524 (623)	--
	Setaria spp.	277 (554)	--

[†] Numbers in parentheses are the standard deviation of the mean.

RESULTS AND DISCUSSION

Only seeds of *Amaranthus* sp. (pigweeds) were found in the summit and shoulder slope positions with numbers ranging from about 400 to 850 seeds m^{-2} (Table 2). At the toeslope position, *Setaria* sp. (foxtails) seeds numbered about 300 m^{-2}, *Polygonum pensylvanicum* numbered about 525 seeds m^{-2}, *Amaranthus* sp. number about 900 seeds m^{-2},and *Ambrosia artemisiifolia* ranged about 500 to 900 seeds m^{-2}. Based on the weed species and numbers of seeds present, GWM recommended different herbicide strategies for each area (Table 3), treatments ranged from herbicide applied both pre- and postemergence to no herbicide applied at either time. Cost of treatments based solely on weed seed counts ranged from \$0 to \$59.50 ha^{-1}. The grower's treatment was a postemergence treatment only of imazethapyr at a cost of \$63.20 ha^{-1}.

There were more weed species present in each area than the weed seed evaluation would have predicted, although total weed seedlings were less dense than the weed seeds. At the summit and shoulder positions, *Setaria* was the dominant species, with 13 and 19 plants m^{-2}, respectively, although other species were present (Table 4). At the toeslope position, *Ambrosia artemisiifolia* had 55 plants m^{-2} and *Cirsium arvense* had 17 plants m^{-2} while *Setaria* plants were scarce. Although there were differences in weed species and densities, GWM recommended chlorimuron at each position at a cost of \$40.25 ha^{-1} (Table 3).

At the summit, the producer treatment of imazethapyr had the lowest weed biomass at the full bloom stage of soybean growth (Table 5). Soybean biomass was less in the control and preemergence treatment (trifluralin) at full bloom. By mid pod fill, there were no differences in either soybean or weed biomass. Yields among treatments were similar, however, net return for the post (chlorimuron) treatment was significantly greater than the producer treatment of imazethapyr.

There were no differences either in soybean or weed biomass, yield, or net return at the shoulder position among treatments. Average soybean yield over all treatments was slightly greater than at the summit with slightly higher net return. Applying herbicides at either of these positions did not improve yield over untreated controls.

At the toeslope position, soybean biomass was greatest in the preemergence clomazone treatment at full bloom, while all treatments had less weed biomass than the untreated control. By mid pod fill, there were no differences in soybean biomass among all treatments, but weed control was best in the postemergence chlorimuron treatment. Yields differed among treatments. The preemergence clomazone treatment had the greatest yield and the untreated control and producer imazethapyr treatments had the lowest yields. Net returns in this area were very poor with both the control and producer treatments losing money. In the control plots, there was no weed control and yield losses resulted because of competition. In the producer treatment, the herbicide cost did not offset the gain in yield. The best return was when clomazone was applied preemergence.

GWM treatments generally had similar or higher net returns than the producer's treatment. In all instances, GWM herbicide tactics at the three locations were less expensive than the producer's herbicide of choice. The reduced herbicide cost contributed towards increased profits of the GWM vs producer treatment. At the shoulder location, the model recommended no weed control for the pre + postemergence treatment based on seed bank data and low weed infestation levels. Hence, the producer's treatment had better weed control but the no herbicide decision had higher net profit because the weed infestation level was not high enough to cause significant yield reductions.

These results suggest that using a model such as GWM to obtain site specific recommendations or making herbicide decisions based on the actual weeds present in the field can improve profitability. While it is perceived as too risky for a grower not to apply any herbicide, changing the herbicide recommendation for the site has positive outcomes. For example in this model field, 13 ha would have received an application of clomazone due to infestations of *Ambrosia artemisiifolia* while 42 ha would have received trifluralin to control *Setaria* sp. These treatments would have cost $700 for clomazone and $800 for trifluralin, a total cost of $1500, and a savings of $2600 over the producer treatment. In addition, net returns would have increased to about $2200 in areas covered by *Setaria* and about $2300 in areas infested with *Ambrosia artemisiifolia*. Thus, net returns, using combinations of site specific applications, may have increased $4500 over the entire field while achieving similar or better weed control.

Table 3. Herbicide treatment recommendations for three areas in a soybean field.

	Summit			Shoulder			Toeslope		
Treatment	Herbicide	Rate	Cost[a]	Herbicide	Rate	Cost	Herbicide	Rate	Cost
		a.i. ha^{-1}	\$ ha^{-1}		a.i. ha^{-1}	\$ ha^{-1}		a.i. ha^{-1}	\$ ha^{-1}
Preemerge	trifluralin	0.84 kg	19.25	trifluralin	0.84 kg	19.25	clomazone	1.12 kg	54.25
Pre + Postemerge	none none	-- --	-- --	trifluralin + chlorimuron[b]	0.84 kg 13.4 g	19.25 40.25	none + bentazon[c]	-- 1.12 kg	-- 51.85
Postemerge	chlorimuron[b]	13.4 g	40.25	chlorimuron[b]	13.4 g	40.25	chlorimuron[b]	13.4 g	40.25
Producers treatment	imazethapyr[d]	71 g	63.20	imazethapyr[d]	71 g	63.20	imazethapyr[d]	71 g	63.20

[a]Cost of treatment include application cost of \$9.80 ha^{-1}, herbicide, and surfactants. Prices obtained from local dealers.
[b]Chlorimuron mixed with crop oil concentrate at 1 L ha^{-1} and 28% N at 1 L ha^{-1}.
[c]Bentazon mixed with crop oil concentrate at 2.37 L ha^{-1} and 28% N at 4.47 L ha^{-1}.
[d]Imazethapyr mixed with nonionic surfactant at 2.37 L ha^{-1} and 28% N at 2.37 L ha^{-1}.

Table 4. Weed densities, height, and number of leaves used for postemergence recommendations using GWM.

		Seedling		
Area	Weed species	Density	Height	Growth stage
		no. m^{-2}	cm	no. leaves
Summit	*Setaria* sp.	13	9	3
	Chenopodium album	2	9	4
	Asclepias syriaca	1	8	2
	Cirsium arvense	1	10	4
Shoulder	*Setaria* sp.	19	9	3
	Chenopodium album	3	8	3
	Asclepias syriaca	3	11	4
Toeslope	*Setaria* sp.	2	8	3
	Chenopodium album	3	10	4
	Asclepias syriaca	3	10	4
	Cirsium arvense	17	13	4
	Solanum nigrum	1	5	4
	Ambrosia artemisiifolia	55	12	4

Table 5. Crop and weed biomass at full bloom and mid-pod fill stages of soybean development, yield, and net return all treatments.

Area	Treatment	Biomass full bloom		Biomass mid pod		Yield	Net return
		Soybean	Weed	Soybean	Weed		
		-----g m^{-2}-----		-----g m^{-2}-----		Mg ha^{-1}	\$ ha^{-1}
Summit	Control	350.5	159.5	766.5	271.5	2.30	245
	Pre	301.0	258.5	575.0	151.3	2.42	255
	Pre+Post	482.3	160.8	762.8	32.8	2.37	262
	Post	502.5	167.3	617.0	176.5	2.64	286
	Producer	579.8	77.5	742.3	2.8	2.27	175
	$LSD_{(0.05)}$	269.7	122.9	ns	ns	ns	107
Shoulder	Control	502.8	78.0	650.3	187.3	2.60	317
	Pre	462.3	105.0	862.3	188.0	2.70	322
	Pre+Post	508.8	74.0	814.3	66.5	2.73	288
	Post	531.8	106.8	644.8	134.5	2.53	260
	Producer	648.8	16.0	888.0	30.3	2.27	270
	$LSD_{(0.05)}$	ns	ns	ns	ns	ns	ns
Toeslope	Control	181.5	566.3	308.3	663.8	1.08	-46
	Pre	635.3	108.8	520.0	353.3	2.11	143
	Pre+Post	264.8	220.3	404.0	562.8	1.59	24
	Post	299.0	66.0	393.3	10.6	1.68	57
	Producer	303.0	129.8	525.3	382.8	1.39	-35
	$LSD_{(0.05)}$	183.8	259.9	ns	505.8	0.42	101

REFERENCES

Brown, R.B., and J.P.G.A. Steckler. 1995. Prescription maps for spatially variable herbicide application in no-till corn. Trans. ASAE. 38:1659–1666.

Broulik, B., S.A. Clay, D.E. Clay, and C.G. Carlson. 1997. Weed detection in field corn using high resolution multispectral digital imagery and field scouting. Proc. of North Central Weed Sci. Soc. 52:53.

Cardina, D. Sparrow, and E.L. McCoy. 1995. Analysis of spatial distribution of common lambsquarters (*Chenopodium album)* in a no-till soybean (*Glycine max*). Weed Sci. 43:258–268.

Ellsbury, M.M., W.D. Woodson, S.A. Clay, and C.G. Carlson. 1996. Spatial characterization of corn rootworm population in continuous and rotated corn. p. 487–494. *In* Robert P.C. et al., (ed) Proc. of the 3rd Int. Conf. on Precision Agriculture. ASA, CSSA, and SSSA, Madison, WI.

Gerhards, R., and D.Y. Wyse-Pester. 1997. Characterizing spatial stability of weed populations using interpolated maps. Weed Sci. 45:108–119.

Gross, K.L., and K.A. Renner. 1989. A new method for estimating seed numbers in the soil. Weed Sci. 37:836–839.

Heisel, T., C. Andreasen, and A.K. Ersboll. 1996. Annual weed distributions can be mapped with kriging. Weed Res. 36:325–337.

Lems, G.J. 1998. Weed spatial variability and management on a field-wide scale. M.s. thesis. South Dakota State Univ.

Mortensen, D.A., G.A. Johnson, D.Y. Wyse, and A.R. Martin. 1995. Managing spatially variable weed population. p. 397–415. Proc. soil specific crop management for agricultural systems. ASA, CSSA, and SSSA, Madison, WI.

Swinton, S.M., and R.P. King. 1994. A bioeconomic model for weed management in corn and soybean. Agric. Syst. 44:313–335.

Wiles, L.J., R.P. King, E.E. Schiweizer, D.W. Lybecker, and S. M. Swinton. 1996. GWM: General weed management model. Agric. Syst. Technol. 50:358–376.

REFERENCES

Influence of Sewage Sludge Application on Soil Quality: I. Organic Matter, pH, Phosphorus, Potassium, and Inorganic Nitrogen

C. D. Tsadilas and V. Samaras

National Agricultural Research Foundation
Institute of Soil Classification and Mapping
Greece

ABSTRACT

In a 2-yr field experiment the influence of sewage sludge application to some soil properties was investigated. The soil was a Typic Xerochrept and it was cultivated with cotton after amendment with sewage sludge rates ranging from 0 to 300 ton ha^{-1} yr^{-1} in a completely randomized blocks experimental design. The results showed that sewage sludge application significantly increased cotton yield. Its implication on soil properties was as follows: Organic matter content was significantly increased while soil pH was slightly decreased. Available phosphorus was highly increased in the surface layer. Nitrate N was highly increased in the whole soil depth. Phosphorus was strongly correlated with organic matter content.

INTRODUCTION

Treatment of sewage sludge effluent leads to the production of a considerable amount of sewage sludge that requires management both economically efficient and environmentally safe. The main options available for sewage sludge management includes sea disposal, landfill, incineration, land reclamation, forestry and application to agricultural land. In Europe sea disposal ends this year according to the directive concerning the urban wastewater treatment (CEC, 1991). Landfill includes a significant percentage of sewage produced covering in Europe about 40%. The practice of incineration covers about 11% while land reclamation, forestry and some other options cover together a small percentage, i.e., about 6% (Smith, 1996). The latter option, i.e., application of sewage sludge to agricultural land has been widely adopted in almost all over the world. In Europe it includes about 40% of the sewage sludge produced and this percentage is expected to increase. That is because this option offers the possibility of exploration of some beneficial properties of sewage sludge such as the considerable organic matter content and the high concentration of some elements that are essential for plant growth. In Greece during the last ten years many wastewater treatment plants were established and the number of them is anticipated to be dramatically increased in the next few years. Many cities, even the small ones, taking advantage of the European Union legislation and the motivations offered, are now interested in managing their urban wastes in an

Copyright © 1999 ASA-CSSA-SSSA, 677 South Segoe Road, Madison, WI 53711, USA. *Proceedings of the Fourth International Conference on Precision Agriculture.*

environmentally safe way establishing contemporary wastewater treatment plant units. So, it is expected that the total amount of the sewage sludge will rise.

Sewage sludge application to the land has direct or indirect effects on soil quality. Many characteristics of sewage sludge are among those proposed to be included as basic indicators of soil quality, e.g., total organic C and N content, pH, electrical conductivity and mineral nutrients such as potassium and phosphorus (Doran & Parkin, 1994). Therefore sewage sludge application to the land must be considered from the point of view of the consequences caused to the soil quality by this practice. In Greece during the few last years farmers has shown interest in using sewage sludge as a substitute for mineral fertilizers for the crops grown. The purpose of the present study was to investigate the consequences of sewage sludge application on some soil properties included in soil quality, i.e., organic matter content, pH, P, K and N.

MATERIALS AND METHODS

Soil and Sewage Sludge

A two year (1996, 1997) field experiment was conducted with cotton (*Gossypiunm hirsutum.* L.) in the area of Rizomilos of Magnesia prefecture, central Greece. The soil developed on alluvial deposits was a deep, well drained, clay loam, flat, with a slope 2% classified as Typic Xerochrept. Selected soil properties of the soil studied are given in Table 1. This soil is a typical soil of the wider Thessaly Plain area in which cotton crop is cultivated. In this area the mean annual precipitation is about 450 mm. Sewage sludge used came from the wastewater treatment plant of Volos, central Greece, which was recently established. Performance of this treatment plant includes a biological treatment by activated sludge processes or percolating filters after an initial screening of litter and grit removal. The basic physicochemical properties of the sewage sludge used are shown in Table 2.

Experimental Design

The experimental design was completely randomized blocks with the following treatments:

C, control, no sewage sludge, no mineral fertlizers,
CF, no sewage sludge but 160 kg N ha^{-1} yr^{-1} and 80 kg P_2O_5 ha^{-1},
SS1, no minearl fertilizers but 10 ton dry sewage sludge (105 ^oC) ha^{-1} yr^{-1},
SS2, no mineral fertlizers but 30 ton dry sewage sludge ha^{-1} yr^{-1}, and
SS3, no mineral fertilizers but 50 ton dry sewage sludge ha^{-1} yr^{-1}. Each treatment was replicated four times.

The experimental plots were 5 $*$ 5 m wide and between them a corridor was left 2 m wide. The seeds were planted on lines in a distance between each other 0.98 m and in a density of 18 to 20 seeds m^{-1}. Plantings were carried out about April 20 in both years. The first year, the variety was a Greek one named *Gossipium hirsutum, var. Corina* and the second year an American one named

Gossipium hirsutum, var. Pioneer-50. Sewage sludge was applied to the soil about 30 d before planting and it was incorporated to a depth of about 25 cm. The whole mineral fertilization was applied basically at the same time with the sewage sludge. The plants were irrigated using a trip irrigation system. Cotton harvesting was made by hand. The external lines of the plot experiments were not harvested. The experiments were carried out at exactly the same places using the same treatments in both years of experimentation.

Samplings

Before the experiment establishment, soil sampling was carried out to a depth of 0 to 90 cm for soil characterization. Soil samples were air dried, crushed to pass a 2-mm sieve and analyzed for texture using the hydrometer method (Gee & Bauder, 1986), pH in a soil solution ratio 1:2 (McLean, 1982), calcium carbonate volumetrically by the calcimeter method (Allison & Moodie, 1965), exchangeable cations by 1*M* NH_4OAc extraction (Thomas., 1982), and cation-exchange capacity by 1 *M* NaOAc, pH 8.2 method (Rhoades, 1982). Organic matter content was measured by the wet oxidation procedure of Walkley and Black (Nelson & Sommers, 1982) and available P by the 0.5 *M* $NaHCO_3$ (pH 8.5) extraction (Olsen & Sommers, 1982). Total N was determined by the Kjeldhal method (Bremenr & Mulvaney, 1982).

At the beginning of September in the second year of experimentation composite soil samples from each experimental plot were selected from a depth 0 to 100 cm. Soil samples after the proper preparation, were analyzed for pH, organic matter, available P, and exchangeable K using the above mentioned procedures. In addition, total, ammonium and NO_3–N also were determined. Total N was determined by the Keldhal method (Bremner & Mulvaney, 1982), while ammonium and NO_3–N were determined colorimetrically after extraction with 2 *M* KCl using the indophenol blue method for ammonium N and the cooperized Cad reduction method for NO_3–N, respectively (Keeney & Nelson, 1982).

Sewage sludge was analyzed for pH in a 1:5 sludgewater suspension, free Ca carbonate volumetrically by the calcimeter method, organic matter content by the wet oxidation procedure used for the soils, total N with the Kjeldhal method, total P colorimetrically after a wet digestion with perchloric acid (Olsen & Sommers, 1982), and total Mn, Cu, Zn, Cd, Pb, and Ni by atomic absorption spectrophotometry after a digestion procedure described by Baker & Amacher (1982).

RESULTS AND DISCUSSION

Soil and Sewage Sludge Characteristics

Some physical and chemical properties are shown in Table 1. Clay content ranged in the whole profile around 300 g kg^{-1}, pH was neutral to slightly alkaline, equivalent $CaCO_3$ was low showing a tendency to increase with soil depth, cation-exchange capacity was high, about 30 cmol kg^{-1} in the whole profile. Soil was relatively well supplied by exchangeable K especially in the upper horizons, Mg was around 5 cmol kg^{-1} and exchangeable Na was very low in relation to the cation exchange capacity. Available P was relatively low in the root zone. Organic matter content was about 25 g kg^{-1} in the plough layer and tended to decrease with soil depth. These values of organic matter content are considered high for Greek soils in which typical values range from about 10 to 15 g kg^{-1} soil.

Sewage sludge characteristics are presented in Table 2. It was rich in organic matter content, total N and P and relatively poor in K, not permitting to make anyone significant difference to the recommended quantities for fertilizer K necessary for most crops. The latter one finding is in accordance to those reported by many other workers (Smith, 1996). Heavy metal concentrations were very low and well below the limits stipulated in the 86/278/EEC Directive of the European Union or in the USEPA (1993).

Sewage sludge application to the soil resulted in a significant increase of cotton yield (Table 3). Seed cotton was raised in the first year (1996) from 3505 kg ha^{-1} in the control treatment (C) to 4115 kg ha^{-1} in treatment with the highest sewage sludge rate (SS3). In the second year the yield was lower because both of the unfavorable weather conditions and the different variety of cotton used. In the first year the variety was *Gossipium Hirsutum, var. Corina* and in the second year *Gossipium Hirsutum var. Pioneer-50.*

Table 1. Selected soil properties of the soil studied

Property	Depth, cm		
	0–30	30–60	60–90
Clay, g kg^{-1}	310	290	290
pH (soilwater 1:1)	7.62	7.78	7.76
$CaCO_3$, g kg^{-1}	trace	26.0	62.0
Organic matter, g kg^{-1}	15.4	13.0	10.8
Cation-exchange capacity, cmol kg^{-1}	33.04	31.46	33.57
Exchangeable cations, cmol kg^{-1}			
K^+	0.68	0.60	0.40
Mg^{2+}	4.79	5.31	4.79
Na^+	0.34	0.54	0.52
Ca^{2+}	27.9	43.5	48.8
Available P, mg kg^{-1}	9.7	11.6	-
Total N, g kg^{-1}	1.40	1.10	0.80

However, also in this second year sewage sludge significantly increased cotton yield (Table 3).

Table 2. Some physics-chemical properties of the sewage sludge used and limit values for heavy metals of Directive 86/278/EEC

Property	1996	1997	86/278/EEC limits
Organic matter, g kg^{-1}	366.3	447.7	-
Total N, g kg^{-1}	26.5	26.5	-
pH (H_2O 1:5)	6.89	6.91	-
$CaCO_3$, g kg^{-1}	53.4	86.4	-
Total P, g kg^{-1}	33.5	244.4	-
Mn, mg kg^{-1}	260	199	-
Cu	224	289	1000–1750
Zn	1812	1714	2500–4000
Cd	5.24	2.8	20–40
Pb	442	525	750–1200
Ni	37.56	76.65	300–400

Table 3. Influence of sewage sludge on cotton yield

Treatments	Cotton seed yield, kg ha^{-1}	
	1996	1997
C, control (no fertilizer, no sewage sludge - SS)	3495a†	2732a
CF, mineral fertilizer, no SS	3978ab	2967ab
SS1, no fertilizer, 10 ton ha^{-1} dry SS	4062ab	2862ab
SS2, no fertilizer, 30 ton ha^{-1} dry SS	3978ab	3125ab
SS3, no fertilizer, 50 ton ha^{-1} dry SS	41115b	3237b

† Values into the same column followed by different letters differ significantly ($P<0.05$) according to the LSD test.

Influence of Sewage Sludge on Soil Properties

Soil pH

Soil pH was slightly but statistically significantly decreased because of the sewage sludge application (Table 4), The decrease was about 0.3 units in the treatment with the highest sludge rate (SS3) in the surface soil and about 0.2 units in the subsurface layer. Similar observation of sewage sludge effect on soil pH were also reported by Tsadilas et al. (1985) and Wen et al. (1997). In addition, the previously authors referred to a buffering action of sewage sludge on soil pH. However, soil pH was maintained well above the lowest value of 6.0 stipulated in the 86/278/EEC. Directive for soils to which sewage sludge is to be applied. Decrease in soil pH was attributed to the mineralization of organic matter contained in sewage sludge.

Organic Matter

Soil organic matter content was substantially increased by the sewage sludge. From 24.7 g kg^{-1} in the control treatment (C) it was raised to 30.7 g kg^{-1} in the treatment with the higher sewage sludge rate (SS3) (Table 4). This contribution of sewage sludge to the soil organic matter content is of significant importance for Greek soils in which after thousand years of intensive cultivation, organic matter content level has been dramatically decreased.

Available Phosphorus

Available P was also significantly affected by the sewage sludge in both surface and subsurface layer. In the surface layer it was raised from 4.0 mg kg^{-1} soil in the control treatment to about 25 mg kg^{-1} in the treatment SS3. In the subsurface layer (25–50 cm) the respective increase was from about 2 mg kg^{-1} in the C treatment to about 12 mg kg^{-1} in the SS3 treatment. Phosphorus is an immobile element in the soil because of a variety of fixation and precipitation processes. Furrer and Staufer (1986) with lysimeters studies have not found any P leaching in soils amended with sewage sludge; however the results of the present study showed that P can move in soil even if in small distances. From the data of Table 4 it is obvious that this move was observed in the treatments with including sewage sludge but not in the treatment with mineral fertilizer. So, it seems that organic matter of sewage sludge may affect P transportation. This is supported by the strong relationship found between available P and organic matter content (r = 0.88***). More research on P forms in sewage sludge is required for explaining P leaching in soils amended with sewage sludge.

Exchangeable Potassium

Exchangeable K was not significantly affected by sewage sludge application. It ranged from about 360 to 400 mg kg^{-1} in the surface layer and between 260 to 300 mg kg^{-1} soil in the subsurface layer. Similar observations were reported by several workers (Smith, 1996). Sewage sludge contains small quantities of K that do not increase K concentration, especially in soils well supplied with K.

Nitrogen

Total N ranged from 0.12 to 0.27 while NH_4^+ was very low ranging from about 2.53 to 5.5 mg kg^{-1}, in the various treatments. Both total and NH_4^+ did not significantly differ between the various treatments (data not presented). NH_4^+ life span is very short in the soil ranging from 1 to 3 wks or more depending on the nitrification rate (Evangelou et al. 1994).

Table 4. Influence of sewage sludge application on soil pH, organic matter, available P, and exchangeable K

Treatments	pH		Organic matter		Available P		Exchangeable K	
	H_2O 1:1		g kg^{-1}		mg kg^{-1}		mg kg^{-1}	
	Depth, cm							
	0–30	30–60	0–30	30–60	0–30	30–60	0–30	30–60
C	7.97c†	7.97b	21.57a	21.70a	3.98a	2.10a	397a	295a
CF	7.84bc	7.93b	22.15a	20.42a	6.53a	1.41a	382a	258a
SS1	7.90bc	7.93b	23.15a	19.20a	6.85a	3.50a	362a	264a
SS2	7.78ab	7.78a	25.95b	21.07a	19.20b	7.69ab	386a	281a
SS3	7.66a	7.75a	26.97b	20.60a	25.83c	12.28b	368a	261a

† values in the same columns followed by different letters differ significantly ($P < 0.05$) according to the LSD test.

NO_3–N was strongly influenced by the treatments applied in the whole depth of the soil profile (Fig. 1). In the control treatment it was maintained almost at a level of 5 to 10 mg kg^{-1} soil. In the treatment CF (no sewage sludge, mineral fertilizer) was found in the surface soil to be about 65 mg kg^{-1} being reduced gradually with depth to about 20 mg kg^{-1}. It seems that NH_4–N applied with mineral fertilization was nitrified but was not entirely absorbed by the plants. An amount of the NO_3–N moved down the profile beyond the root depth. Rainfall distribution in the area studied is such that a considerable rainfall takes place during the winter. So, NO_3^- is subject to leaching through the following winter rainfall. Treatment SS1 (no mineral fertilization, sewage sludge application at the lowest rate) showed the most preferable pattern of NO_3–N distribution in the soil profile. It remained low and almost constant in the whole profile except in the depth of 25 to 50 cm where it was higher (about 25 mg kg^{-1} soil). This means that from the point of view of NO_3^- leaching this sewage sludge rate (10 ton dry sewage sludge per ha per year) was the best among those studied. In the treatment SS2, NO_3–N was higher than in the SS1 but it was reduced with depth and remained at a lower level than in the treatment CF. Finally in the treatment SS3 with the highest sewage sludge rate, NO_3–N concentration remained high although it was reduced with depth. This means that high sewage sludge rates increase the hazard of NO_3^- leaching and ground water contamination.

CONCLUSIONS

Sewage sludge from the treatment plant of Volos, central Greece is found suitable for cotton fertilization being able to entirely substitute mineral fertilizers. Sewage sludge application increases organic matter content as well as available P. However, sewage sludge rates higher than 10 ton ha^{-1} yr^{-1} may cause NO3 leaching beyond the root depth and effect groundwater pollution.

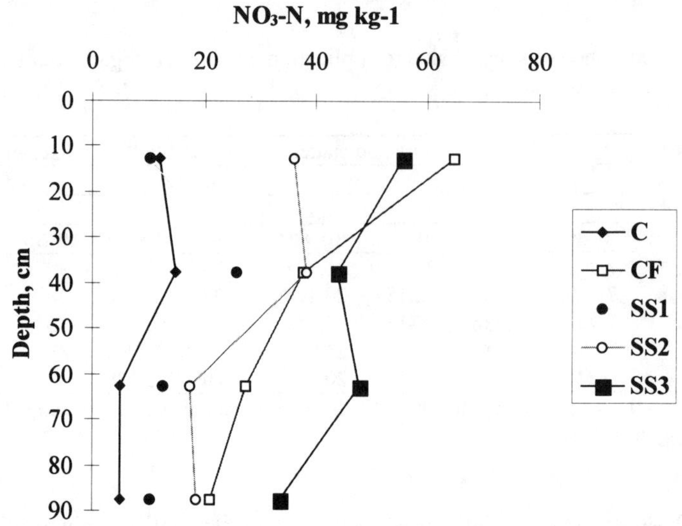

Fig. 1. Influence of sewage sludge on the NO_3–N distribution in the soil profile.

ACKNOWLEDGMENTS

The authors wish to express their thanks to the Municipality of Volos, Greece, for financing this work. Special thanks also are expressed to Ms Vaso Gianni and Mr. Savas Papadopoulos for the laboratory measurements.

REFERENCES

Allison, L.E., and C.D. Moovie. 1965. Carbonate. p. 1379–1396. *In* C.A. Black et al., (ed.) Methods of soil analysis. Part 2. Agron. Monog. 9, ASA, CSSA, and SSSA, Madison, WI.

Baker, D.E., and M.C. Amacher. 1982. Nickel, copper, zinc, and cadmium. p.323–336. *In* A.L. Page et al., (ed.) Methods of soil analysis. Part 2. 2nd ed. Chemical and Microbiological Properties. ASA, and SSSA, Madison, WI.

Bremenr, J.M., and C.S. Mulvaney. 1982. Nitrogen-total. p.595–624. *In* A.L. Page et al., (ed.) Methods of soil analysis. Part 2. 2nd ed. Chemical and Microbiological Properties. ASA, and SSSA, Madison, WI.

CEC, Council of the European Communities. 1991. Council Directive of 21 May 1991 concerning urban wastewater treatment (91/27/EEC). Official J. European Comm. no. L 135/40–52.

Doran, J. W. and T. B. Parkin, 1994. Defining and assessing soil quality. p. 3–21. *In* J.W. Doran et al., (ed.) Defining soil quality for a sustainable environment. SSSA, and ASA, Madison, WI.

Evangelou, V.P., J. Wang, and R.E. Philips. 1994. New developments and perspectives on soil potassium quantity/intensity relationships. Adv. Agron., 52:173–226.

Furrer, O.J., and W. Stauffer. 1986. Influence of sewage sludge and slurry application on nutrient leaching losses. p. 108–127. *In* A. Dam Kofoed et al., (ed.) Efficient land use of sludge and manure. Elevier Applied Sciences Publ., London.

McLean, E.O. 1982. Soil pH and lime requirement. p. 119–224. *In* A.L. Page et al., (ed.) Methods of soil analysis. Part 2. 2nd ed. Chemical and Microbiological Properties. ASA, and SSSA, Madison, WI.

Gee, G.W., and J.W. Bauder. 1986. Particle Analysis. p. 383–412. *In* A. Klute (ed.) Methods of soil analysis. Part I. 2nd Ed. Physical and Mineralogical Methods. Agron. Monog. 9, ASA, and SSSA, Madison, WI.

Keeney, D.R., and D.W. Nelson. 1982. Nitrogen-inorganic forms. p. 634–698. *In* A.L. Page et al., (ed.) Methods of soil analysis. Part 2. 2nd ed. Chemical and Microbiological Properties. ASA, and SSSA, Madison, WI.

Nelson, D.W., and L.E. Sommers. 1982. Total carbon, organic carbon, and organic matter. p.539–580. *In* A.L. Page et al., (ed.) Methods of soil analysis. Part 2. 2nd ed. Chemical and Microbiological Properties. ASA, and SSSA, Madison, WI.

Olsen, S.R., and L.E. Sommers. 1982. Phosphorus. p. 403–430. *In* A.L. Page et al., (ed.) Methods of soil analysis. Part 2. 2nd ed. Chemical and Microbiological Properties. ASA, and SSSA, Madison, WI.

Rhoades, J.D. 1982. Cation exchange capacity. p.149–158. *In* A.L. Page et al., (ed.) Methods of soil analysis. Part 2. 2nd ed. Chemical and Microbiological Properties. ASA, and SSSA, Madison, WI.

Smith, S.R. 1996. Agricultural recycling of sewage sludge and the environment. CAB Int., London.

Thomas, G.W. 1982. Exchangeable Cations. p. 159–166. *In* A.L. Page et al., (ed.) Methods of soil analysis. Part 2. 2nd ed. Chemical and Microbiological Properties. ASA, and SSSA, Madison, WI.

Tsadilas, C.D., Theodora Matsi, N. Barbayiannis, and D. Dimoyiannis. 1995. Influence of Sewage Sludge Applications on Soil Properties and on the Distribution and Availability of Heavy Metal Fractions. Commun. Soil Sci. Plant Anal. 26(15&126): 2603–2619.

US EPA. 1993. Part 503. Standard for the use or disposal of sewage sludge. Federal Register 58:9387–9404.

Wen, G., J.P. Winter, R.P. Voroney, and T.E. Bates.1997. Potassium availability with application of sewage sludge, and sludge and manure composts in field experiments. Nutr. cycling agroecosyst. 47:233–241.

Site-Specific Herbicide Management for Preserving Water Quality

B. R. Khakural
P. C. Robert
D. J. Mulla
R. S. Oliveira

Department of Soil, Water, and Climate
University of Minnesota
St. Paul, Minnesota

G. A. Johnson

University of Minnesota, Southern Experiment Station
Waseca, Minnesota

W. C. Koskinen

USDA-ARS
St. Paul, Minnesota

ABSTRACT

Soil properties that affect the fate and transport of herbicides and weed density were studied across a soil-landscape in Blue Earth County, MN. Soil properties such as organic matter, texture, pH, and adsorption coefficients of herbicides and weed density varied spatially. Adsorption coefficient (K_d) of imazethapyr was strongly correlated with soil pH while K_d of alachlor was strongly correlated with organic matter. Distribution of broad leaf weeds were related to soil-landscape characteristic. Preliminary results of this research suggest that site-specific application of herbicide (pre- or post-emergence) based on soil properties and weed density can reduce herbicide use.

INTRODUCTION

Contamination of ground and surface waters by agri-chemicals is a matter of growing public concern. Soil properties that affect the fate and transport of herbicides varies spatially. Weed populations are also spatially heterogeneous (Gerhards et al., 1995; Johnson et al., 1995) resulting in aggregated weed patches of varying density and areas with few or no weeds. A conventional herbicide management system uses a uniform herbicide rate for the entire field. Site-specific management of herbicide is a new alternative to conventional herbicide management. In site-specific management, rates of herbicide can be varied according to crop and soil needs to avoid over-application. It has great potential

Copyright © 1999 ASA-CSSA-SSSA, 677 South Segoe Road, Madison, WI 53711, USA. *Proceedings of the Fourth International Conference on Precision Agriculture.*

for reducing ground and surface water pollution while maintaining or increasing net returns (Larson et al., 1997; Marks & Ward, 1993).

Few studies have dealt with environmental aspects of site-specific herbicide management. Khakural et al. (1994) have reported a decrease in alachlor concentrations in surface runoff as a result of site-specific management in a fine loamy catena in southwestern Minnesota. By adopting site-specific rates of alachlor application instead of applying a uniform rate in the entire field, alachlor concentration in runoff water, sediment and water + sediment was reduced by 10, 24 and 22%, respectively. Weed mapping and site-specific application of herbicide can substantially reduce herbicide use (Mortensen et al, 1995). A 40 to 60% reduction in herbicide use was reported in cereal crop by targeting herbicide application to grass weed patches (Stafford & Miller, 1996).

Spatial patterns in soil properties that affect the fate and transport of herbicides and spatial distribution of weeds need to be studied for determining optimum site-specific herbicide rates. The objectives of this research were: (i) to study variations in soil properties that affect the fate and transport of herbicides and weed distribution across paired mini-watersheds in southern Minnesota and (ii) to develop a site-specific herbicide management map based upon pattern or variation in soil properties and grass weed distribution, and (iii) to study back ground water quality data.

MATERIALS AND METHODS

Paired mini-watersheds (32 h) from Blue Earth County, were used for this study. The mini-watersheds consist of wide range of soil and landscape characteristics. There is a 6.6 m difference in elevation between the lowest and highest points in the landscape (Fig. 1). Poorly drained soils with tile lines occur at the toeslope position. Well and moderately well drained soils occur at the summit, side slope and shoulder positions. Slope gradient ranges from 0 to 6%. Major soils at the study site were Lester L (fine-loamy, mixed, mesic, Mollic Hapludalf), Shorewood SiCL (fine, montmorillonitic, mesic Aquic Argiudolls), Cordova CL (fine-loamy, mixed, mesic Typic Argiaquoll), Waldorf SiCL (fine, montmorillonitic, mesic Typic Haplaquolls), Lura SiCL (fine, montmorillonitic, mesic, Cumulic Haplaquoll), and Blue Earth SiCL (fine-silty, mixed (calcareous), mesic Mollic Fluvaquents). Auto samplers and area-velocity sensors were installed in each section of the mini-watershed (Fig. 1). Tile-line drainage flow and surface runoff are being monitored, samples collected, and analyzed for background information.

Elevation survey was conducted using a geodimeter (Geodimeter Model 126, Geotronics AB, Danderyd, Sweden). A total of 234 soil surface (0–15 cm) samples were collected from nine south-north oriented parallel transects (806 m long and 45 m apart) at 30-m intervals. Each grid point was georeferenced using a global positioning system (GPS).

Soil properties that affect the fate and transport of herbicides such as particle size, organic matter, pH, and adsorption coefficient (K_d) were determined. Particle-size analysis was performed using a hydrometer method (Gee & Bauder,

1986). Soil organic matter content was determined using a modified Walkley Black method (Nelson & Sommers, 1986). Soil pH was measured in a 1:1 soilwater suspension. Forty-two samples from selected sites representing the entire field were used to determine K_d of imazethapyr (Pursuit) and alachlor (Lasso) herbicides. Batch equilibration technique was used to determine K_d.

Weed scouting was done to study spatial patterns in weed distribution before application of post-emergent herbicide in soybean crop (1997). Common weeds were recorded from a 0.304 m * 0.304 m area at a 15 m * 15 m grid. Each grid point was georeferenced using a global positioning system (GPS). Spatial variability in K_d values of alachlor and grass weed density (1997) were used for preparing site-specific herbicide application map of acetochlor (Harness, a pre-emergence herbicide) for corn in 1998.

RESULTS AND DISCUSSION

Adsorption co-efficient of imazethapyr was the most variable soil property (Table 1). Surface sand content and K_d of alachlor also had relatively high coefficient of variability. Soil texture varied from loam to silty clay. Surface sand, silt and clay contents varied from 1 to 48%, 22 to 59%, and 26 to 65%, respectively. Soils at the steeper sideslope had the greatest sand content and least clay content (Fig. 2).

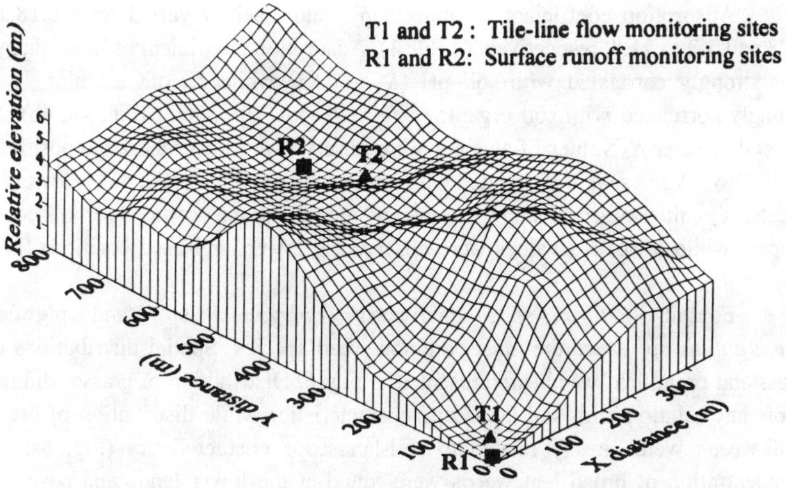

Fig. 1. Relative elevation map of the research site showing locations of subsurface tile line flow and surface runoff monitoring sites.

Table 1. Summary of simple statistics for selected soil (0–15 cm) properties at the experimental watershed.

Variable	No. of samples	Mean	Std Dev	Range	CV (%)
Sand (%)	234	14.26	8.12	1.1-48.2	57.0
Silt (%)	234	41.98	5.50	21.6-58.9	13.1
Clay (%)	234	43.76	7.20	25.6-65.1	16.5
Organic matter (%)	234	7.10	2.07	2.2-10.0	29.4
pH	234	6.20	0.80	4.9-7.7	12.2
Adsorption coefficient for imazethapyr (K_d)	42	1.56	1.08	0.18-3.78	69.0
Adsorption coefficient for alachlor (K_d)	42	10.34	4.19	3.31-21.82	40.0

Soils at the closed depression (low land) had the greatest silt content. Soil organic matter ranged from 2.2% in steeper sideslope to greater than 10% in lower landscape positions (closed depression) (Table 1; Fig. 3a). Surface soil pH ranged from 4.9 at the steep sideslope/backslope to 7.7 at the lower landscape positions (closed depression).

Adsorption coefficient of imazethapyr and alachlor varied from 0.18 to 3.78 and 3.3 to 21.8, respectively (Table 1). Adsorption coefficient of imazethapyr was strongly correlated with soil pH (R^2 = 0.83) while K_d of alachlor was strongly correlated with soil organic matter (R^2 = 0.70) content (Fig. 4a, 4b). In general, greater K_d values of alachlor were observed at lower landscape positions while lower values were associated with upper slope positions (Fig 5a). However, greater K_d values of imazethapyr were found at upper slope positions (or at steeper slopes) while lower Kd values were observed at lower landscape positions (Fig. 5b).

Foxtail (*Setaria* sp.), smartweed (*Polygonum* sp.) and pigweed (*Amaranthus* sp.) were the most common weed species. Spatial distributions of grass and broad leaf weeds are presented in Fig. 6. Distribution of grasses did not show any relation with soil-landscape characteristics while distribution of broad leaf weeds were closely related to soil-landscape characteristics (Fig. 6a, b). Concentration of broad leaf weeds were found at the lower landscape positions with greater organic matter contents (Fig. 6b). Weed populations were significantly aggregated with large areas being weed free or with few weeds. Broad leaf weeds were concentrated in <20% area of the field while 80% of the area was weed free. Grass weed density was relatively high [>4 weeds/ (0.304 m ∗ 0.304 m area)] in 35% of the field area while 65% of the field area had low grass weed density [≤ 4 weeds/0.304 m ∗ 0.304 m area)].

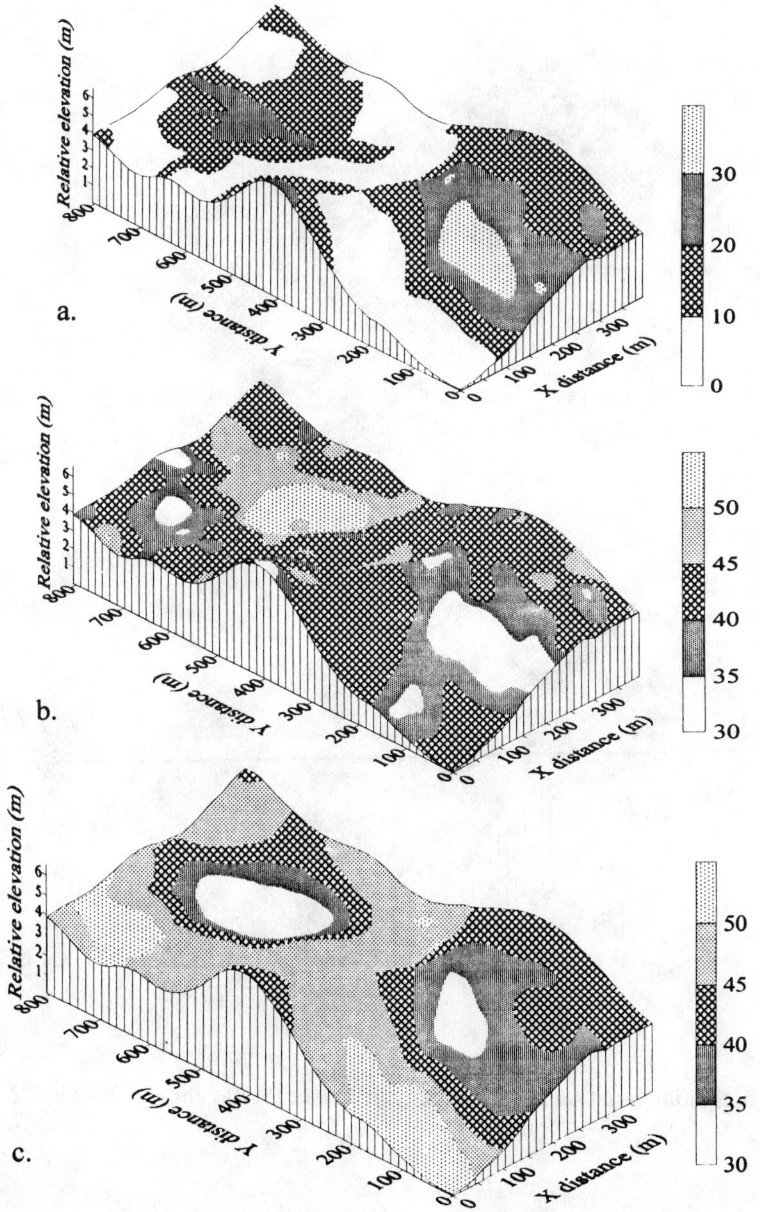

Fig.2. Spatial variation in surface sand (a), silt (b), and clay (c) contents (%).

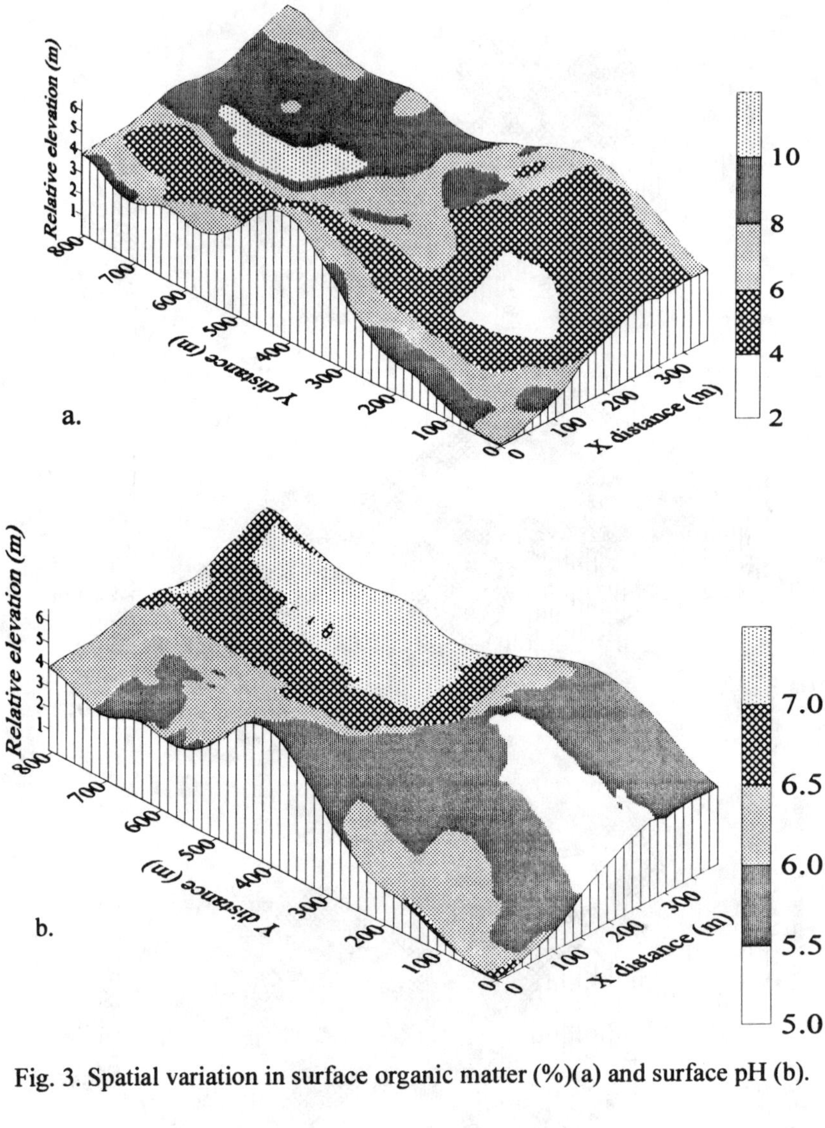

Fig. 3. Spatial variation in surface organic matter (%)(a) and surface pH (b).

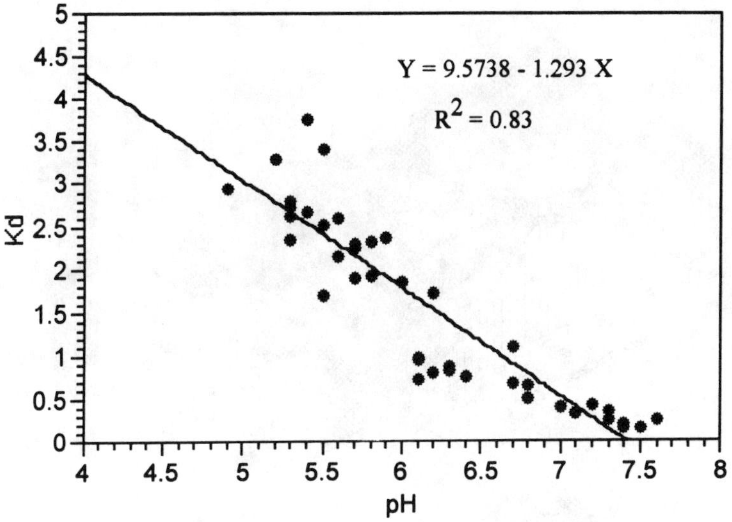

Fig. 4a. Relationship between adsrption coefficient (Kd) of imazethapyr and soil pH.

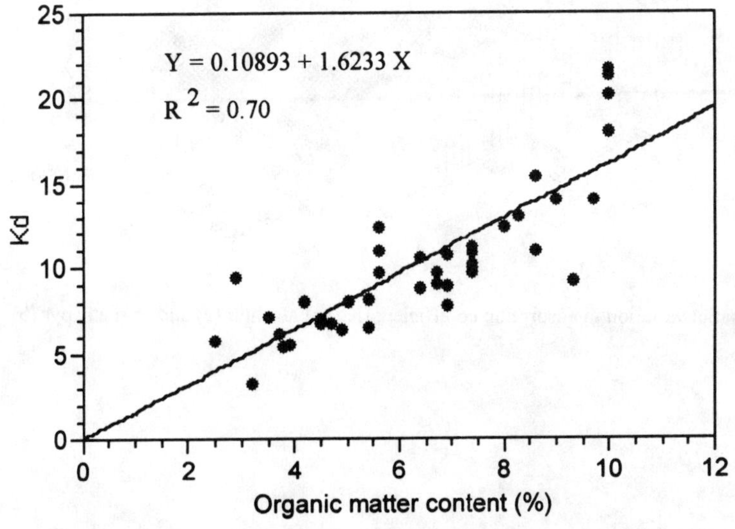

Fig. 4b. Relationship between adsorption coefficient (Kd) of alachlor and soil organic matter.

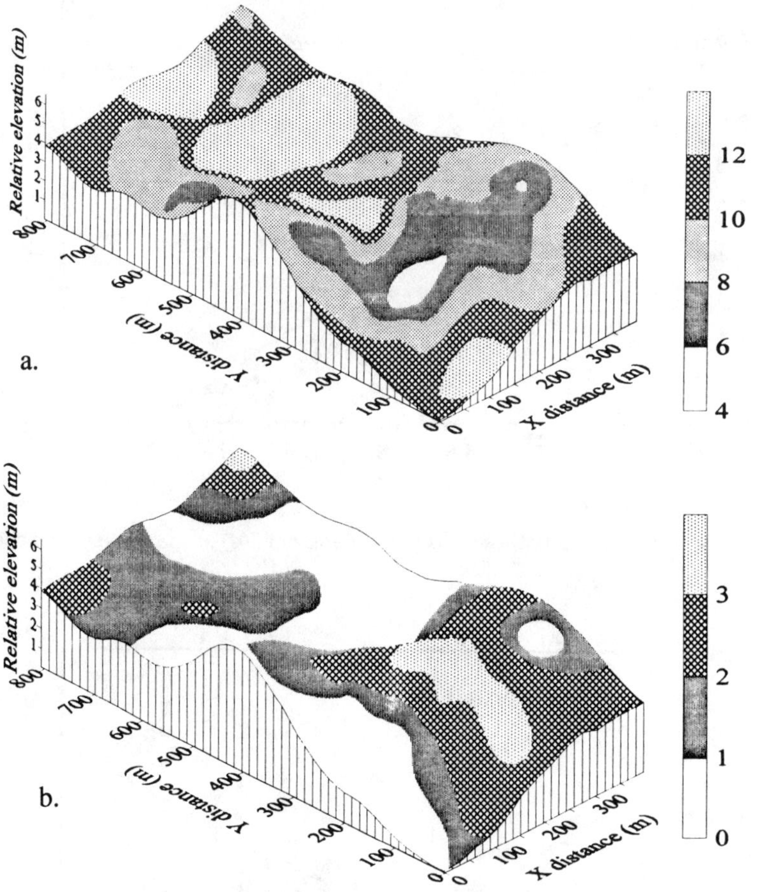

Fig.5. Spatial variation in adsorption co-efficient (Kd) of alachlor (a) and imazethapyr (b).

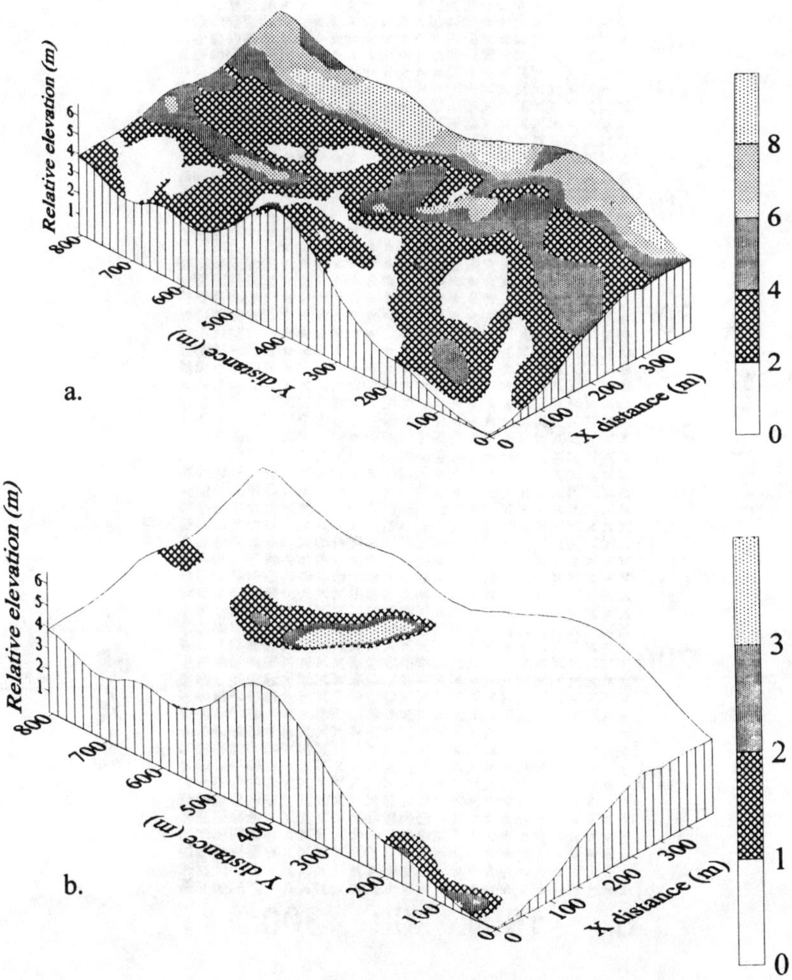

Fig. 6. Spatial variation in grass (a) and broad leaf (b) weeds (number in a 0.304 m X 0.304 m area).

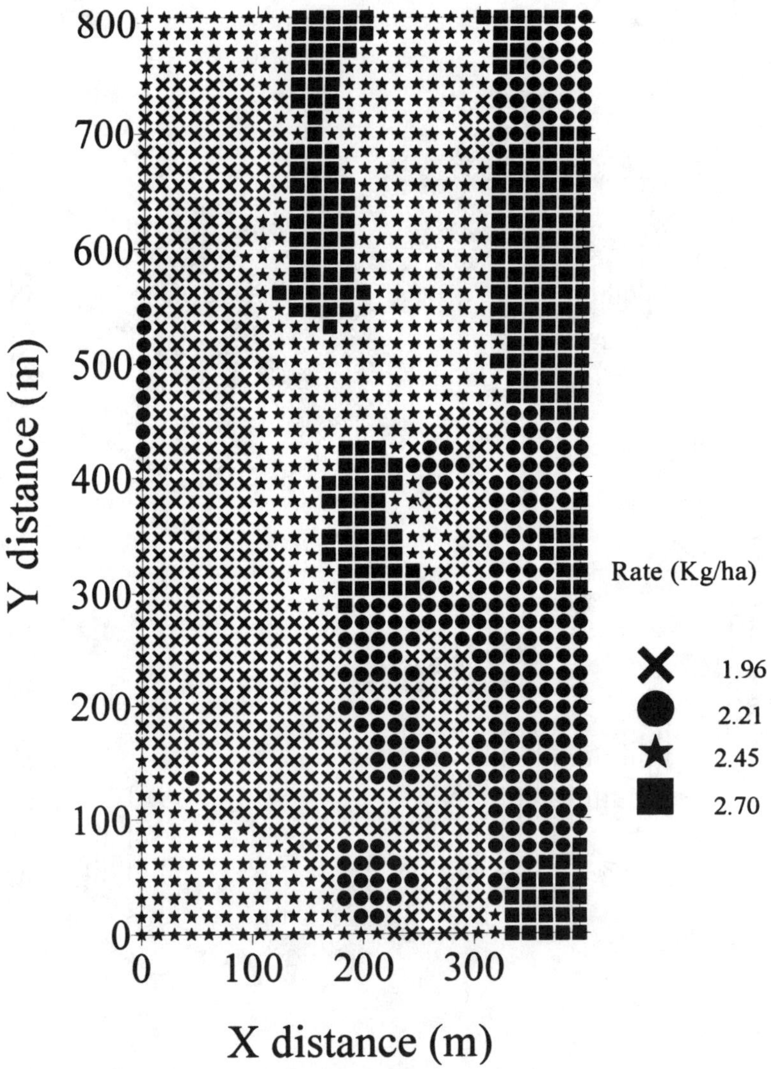

Fig.7. Site-specific herbicide management map used for applying acetochlor (1998).

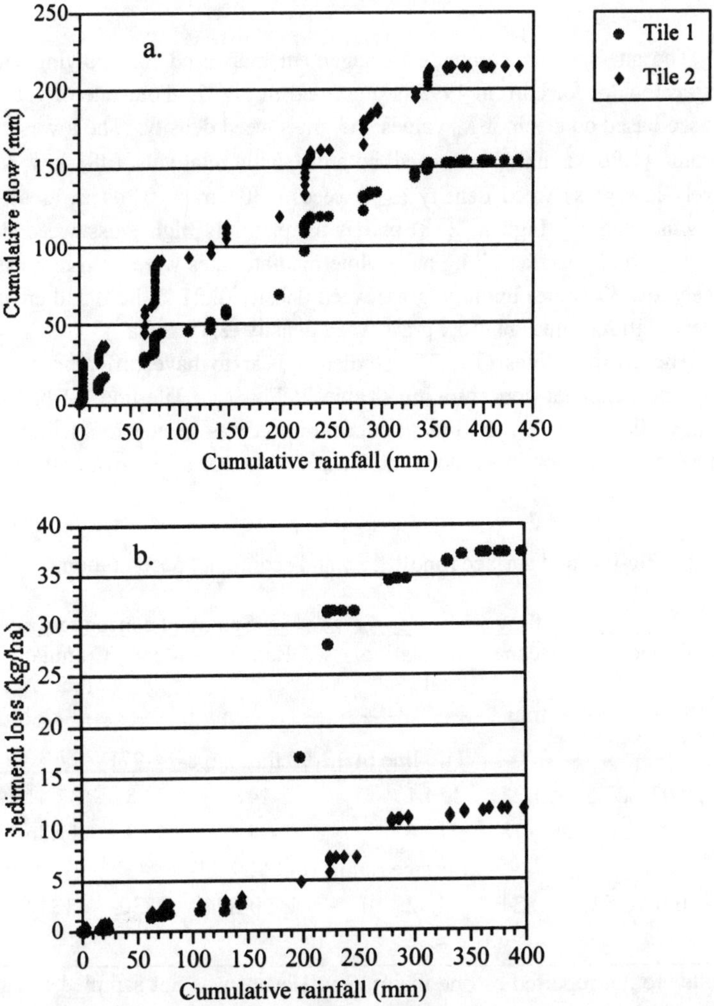

Fig. 8. Relationship between cumulative rainfall and cumulative tile-line flow (a) and cumulative sediment loss (b).

The site-specific herbicide management map used for applying variable rate of acetochlor for corn in 1998 is presented in Fig. 7. Four rate of acetochlor were used based on alachlor K_d values and grass weed density. The lowest rate of acetochlor (1.96 kg ha^{-1}) was used in areas with relatively (≤ 8) low K_d and relatively low grass weed density ($\leq$ 4 weeds/0.304 m $*$ 0.304 m area). Field areas with relatively high K_d (>8) and with relatively high grass weed density received the highest rate (2.7 kg ha^{-1}). Intermediate rates were used in areas with relatively low K_d values but high grass weed density (2.21 kg ha^{-1}) and areas with relatively high K_d values but low grass weed density (2.45 kg ha^{-1}).

The two tile-lines (T1, T2) studied appear to have differences in flow volume and sediment concentration (Table 2; Fig. 8). Tîle-line T2 had higher cumulative flow than T1. Sediment concentration was greater in T1 than in T2. Sediment concentration in surface runoff was 500 to 800 times greater than in tile-line flow.

Table 2. Tile-line and surface runoff flow and sediment concentration.

Site	flow			Sediment concentration		
	Range	Mean	Cumulative total	Range	Mean	Cumulative loss total
	mm			mg L^{-1}		kg ha^{-1}
			Tile-line (4-17-97 through 8-14-97)			
T1	0.03 - 27.5	1.28	154.1	2-147	5.08	37.41
T2	0.03 -13.6	1.78	214.4	2-94	3.38	11.86
			Surface runoff (6-28/29-97)			
R1	0.04-0.52	0.24	4.92	1500-5940	2830	131.9
R2†	-	-	-	-		-

†Results are not reported as one major runoff event was not sampled because of equipment problem

CONCLUSION

Soil properties that affect the fate and transport of herbicides and weed density varied spatially. Adsorption coefficient (K_d) of imazethapyr was strongly correlated with soil pH while adsorption coefficient (K_d) of alachlor was strongly correlated with organic matter. Soil-landscape characteristics influenced the distribution of broad leaf weeds but not grass weeds. Preliminary results of this research suggest that weed scouting and spot application of post-emergence herbicide can reduce herbicide use.

ACKNOWLEDGMENT

This project was funded in part by the USDA-CSREES grant no. 95-37102-2174 and the Minnesota Soybean Research and Promotion Council. We thank Monsanto Co. for providing the imazethapyr and acetochlor herbicides and Nyle Wollenhaupt from Ag-Chem Equipment Co. for his help in arranging site-specific herbicide applicator and in map preparation.

REFERENCES

Gee, G.W., and J.W. Bauder. 1986. Particle-size analysis. p. 383–441. *In* A. Klute (ed.) Methods of soil analysis. Part 1. 2nd ed. Agron. Monogr. 9. ASA and SSSA, Madison, WI.

Gerhards, R., D.Y. Wyse-Pester, and D.A. Mortensen. 1996. Spatial stability of weed patches in agricultural fields. p. 495–504. *In* P.C. Robert et al. (ed.). Proc. of the 3rd Int. Conf. on Precision Agriculture. Minneapolis, MN. 23–26 June, 1996. ASA, CSSA, and SSSA Madison, WI.

Johnson, G.A., D.A. Mortensen, and A.R. Martin. 1995. A simulation of herbicide use based on weed spatial distribution. Weed Res. 35:197–205.

Khakural, B.R., P.C. Robert, and W.C. Koskinen. 1994. Runoff and leaching of alachlor under conventional and soil specific management. Soil Use Manage. 10:158–163.

Larson, W.E., J.A. Lamb, B.R. Khakural, R.B. Ferguson, and G.W. Rehm. 1997. Potential of site-specific management for nonpoint environmental protection. p. 339–369. *In* F.J. Pierce and E.J. Sadler (ed.) The State of Site Specific Management for Agriculture. ASA Misc. Publ., ASA, CSSA, and SSSA, Madison, WI.

Marks, R.S., and J.R. Ward 1993. Nutrient and pesticide threats to water quality. p. 301–307. *In* P. C. Robert et al. (ed.) Proc. of 1st Workshop on Soil Specific Crop Management. A Workshop on Research and Development Issues. Minneapolis, MN. 14–16 April, 1992.

Mortensen, D.A., G.A. Johnson, D.Y. Wyse, and A.R. Martin 1995. Managing spatially variable weed population. *In* P.C. Robert et al. (ed.). Site Specific Management of Agricultural Systems. ASA, CSSA, and SSSA, Madison, WI.

Nelson, D.W., and L.E. Sommers. 1986. Total carbon, organic carbon, and organic matter. p. 539–577. *In* A.L. Page (ed.) Methods of soil analysis. Part 2. 2nd ed., Agron. Monogr. 9. ASA, and SSSA, Madison, WI.

Stafford, J.V., and P.C.H. Miller. 1996. Spatially variable treatments of weed patches. pp. 465-473. In P.C. Robert et al. (ed.). Proc. of the 3rd Int. Conf. on Precision Agriculture. Minneapolis, MN. 23-26 June, 1996. ASA, CSSA, and SSSA, Madison, WI.

Influence of Sewage Sludge Application on Soil Quality: II. Heavy Metals

V. Samaras and C. D. Tsadilas

National Agricultural Research Foundation
Institute of Soil Classification and Mapping
Larissa, Greece

ABSTRACT

In a 2 yr field experiment the influence of sewage sludge application on the total and available forms of heavy metals, i.e., Zn, Mn, Cu, Fe, Pb, Cd, and Ni was studied. The soil was a Typic Xerochrept and it was cultivated with cotton after amendment with sewage sludge rates ranging from 0 to 30-ton $ha^{-1}yr^{-1}$. The experimental design was completely randomized blocks with five treatments each replicated four times. After 2 yrs of sewage sludge application, soil samples were taken from all the plots from a depth 0 to 50 cm and the concentrations of the total and available forms (DTPA extractable) of the above mentioned heavy metals were determined. The results showed that sewage sludge application significantly increased the total concentration of Cu, Zn, and Pb in the surface layer. DTPA exctractable forms were increased significantly in all the metals studied except Mn. Total concentration of Cu, Zn, and Cd were strongly correlated with organic matter content. DTPA extractable Cu, Zn, Cd, and Fe were also strongly correlated with organic matter content positively and negatively with soil pH. For all the metals studied except Cu and Fe, there was no evidence of leaching in the deeper layers.

INTRODUCTION

Recycling of sewage sludge in agriculture is a very widely adopted practice in all over the world since it provides an outlet for solving a serious problem in both an economically beneficial and environmentally safe way. That is because sewage sludge contains several elements most of which are essential for plant growth. These elements basically come from domestic, road-run-off and industrial inputs combining in the sewerage systems. Some of these elements in high concentrations, however, may become toxic to the plants and harmful to human health through the food chain. To avoid the harmful effects coming from sewage sludge to the human health and the environment the respective authorities in many countries have adopted a series of measures that users of sewage sludge must follow (CEC, 1991; USEPA, 1993). The practice of sewage sludge recycling in agriculture is new in Greece and started in recent years after the establishment of a large number of wastewater treatment plants in many cities, even the small ones. Therefore research related to the impacts of sewage sludge application to the soil is of significant importance in Greece. In the first part of this work (Tsadilas &

Copyright © 1999 ASA-CSSA-SSSA, 677 South Segoe Road, Madison, WI 53711, USA. *Proceedings of the Fourth International Conference on Precision Agriculture.*

Samaras, 1999, in this publication) we presented the results of a research on the impacts of sewage sludge application to a soil cultivated with cotton on some basic physicochemical properties, i.e., organic matter content, pH, P, K, and inorganic N concentration, and on cotton yield. In this second part, we report the impacts of sewage sludge on some heavy metals in soil and cotton plants, i.e., Cu, Zn, Cd, Fe, Ni, Pb, and Mn.

MATERIALS AND METHODS

Soil and Sewage Sludge

Soil and sewage sludge characteristics as well as details of the experimentation are given in the first part of the work (Tsadilas & Samaras, 1999, in this publication). The soil was a Typic Xerochrept cultivated in the area mainly with cotton, while sewage sludge came from the treatment plant of Volos, Magnesia, central Greece. The experimental design was completely randomized blocks with five treatments including control (**C**), no sewage sludge, no mineral fertilization, **CF**, no sewage sludge but the conventional mineral fertilization with N and P, **SS1**, no mineral fertilization but 10 ton ha^{-1} yr^{-1} dry sewage sludge (105 ^{o}C), **SS2**, no mineral fertilizers but 30 ton ha^{-1} yr^{-1}dry sewage sludge, and **SS3**, no mineral fertilizers but 50 ton ha^{-1} yr^{-1}dry sewage sludge. Each treatment was replicated four times.

Samplings

At the beginning of September in the second year of experimentation (1997) composite soil samples from each experimental plot were selected from a depth 0 to 50 cm. Soil samples after the proper preparation, were analyzed for pH, organic matter, and heavy metals as follows: surface samples were analyzed for total heavy metal concentration after a wet digestion according to Baker and Amacher (1982). Both surface and subsurface soil samples were extracted with a 0.005M DTPA 0.01M $CaCl_2$, and 0.1M TEA (pH 7.3) solution (Baker & Amacher, 1982). In all the extractions Cd, Pb, Zn, Ni, Mn, and Cu were determined by atomic absorption spectrometry. In all the samples collected pH was measured in a soil solution ratio 1:2 (McLean, 1982), while in the surface samples organic matter content was also determined by the wet oxidation procedure of Walkley and Black (Nelson & Sommers, 1982). At the full bloom stage (end of July) composite leaf samples were also collected from each experimental plot. Leaf samples after the appropriate preparation, were dried at 75 ^{o}C for 24 h, ashed at 500 ^{o}C and after a dilution with 0.1 *M* HCl were analyzed for the above mentioned metals using atomic absorption spectrometry.

RESULTS AND DISCUSSION

Physical and chemical properties of the soil and sewage sludge used are given in the first paper (Tsadilas & Samaras, 1999, this publication). The soil used was a typical soil cultivated in the area with cotton. Sewage sludge was characterized by low heavy metal concentrations well below the limits stipulated in the 86/278/EEC Directive of the European Union or in the USEPA (1993).

Sewage sludge application to the soil significantly increased cotton yield in both years of experimentation (Tsadilas & Samaras, 1999, this publication).

Influence of Sewage Sludge on Soil Heavy Metals

Sewage sludge application resulted in a significant increase in the surface soil of total concentration of Cu, Zn, Pb, and Cd. In contrast, Ni and Mn tended to decrease in the treatment with the highest sewage sludge rate and Fe was not significantly affected (Table 1). In more detail the influence of sewage sludge on the metals studied is discussed below.

Copper

Copper concentration was slightly increased with the sewage sludge rates from about 59 mg kg^{-1} in the treatments C and CF to 62 mg kg^{-1} in the treatment SS3. The same was true also for Cu extracted by DTPA ($Cu\text{-}_{DTPA}$). Both total as well as $Cu\text{-}_{DTPA}$ were significantly correlated with organic matter content. In addition $Cu\text{-}_{DTPA}$ was significantly correlated with total Cu (Fig. 1). $Cu\text{-}_{DTPA}$ was also significantly increased in the subsurface. From 1.90 mg kg^{-1} in the C1 treatment it was raised to 2.74 in the treatment SS3 while organic matter content remained unaffected in subsurface. Increase in $Cu\text{-}_{DTPA}$ in subsurface layer may be attributed to the decrease of soil pH to which $Cu\text{-}_{DTPA}$ was negatively correlated (Fig. 1).

Zinc

Zinc, in general, showed similar behavior to Cu. Total Zn as well as DTPA extractable Zn ($Zn\text{-}_{DTPA}$) were significantly increased with increasing sewage sludge rate. Total Zn as well as $Zn\text{-}_{DTPA}$ was strongly correlated with organic matter content while $Zn\text{-}_{DTPA}$ was also significantly correlated with total Zn and negatively with soil pH (Fig. 2). In both cases, i.e., Cu and Zn, sewage sludge did not affect their leaf concentration instead increased their soil concentration. DTPA method was found suitable for predicting available heavy metal forms for several crops (Mitchell et al., 1978). However for cotton crop it seems that for both elements DTPA method is not a good indicator for predicting available forms in soil since the respective correlation between leaf and soil concentration were not statistically significant (data not presented). $Zn\text{-}_{DTPA}$ in subsurface soil was not affected by sewage sludge addition (Table 1).

Cadmium

Cadmium concentration both total and DTPA extractable were significantly increased because of the sewage sludge application, although increase was not high and in any case well below the limits imposed by the EU with the Directive 86/278/EEC. However, besides the significant increase in total and DTPA

Table 1. Influence of sewage sludge application on total concentration of soil heavy metals (mg kg^{-1}) in surface soil

Metal	Treatments†				
	C	CF	SS1	SS2	SS3
Cu	58.37a‡	59.00a	60.75ab	61.87b	62.38b
Zn	75.00a	79.00a	87.38b	110.12c	103.5c
Cd	0.10a	0.10a	0.10a	0.57b	0.77c
Fe	3,705ab	3,850b	3,802ab	3,837ab	3,637a
Ni	237.25ab	243.50b	240.75b	245.25b	228.00a
Pb	31.87a	32.62a	33.12ab	36.37b	36.00b
Mn	862.25b	859.00b	844.75b	847.50b	780.75a

† for symbol explanation see text, ‡ numbers in the same line followed by different letters differ significantly ($P < 0.05$) according to LSD test

Table 2. Influence of sewage sludge application on DTPA extractable heavy metal concentration (mg kg^{-1}) in surface and subsurface soil

Metal	Treatments†				
	C	CF	SS1	SS2	SS3
0–25 cm					
Cu	1.90ab‡	1.87a	1.93ab	2.33bc	2.74c
Zn	1.84a	2.86ab	2.30ab	6.09bc	9.32c
Cd	0.035ab	0.035ab	0.033a	0.041bc	0.043c
Fe	2.34a	2.29a	2.53a	3.78ab	5.54b
Ni	1.42ab	1.32a	1.33a	1.38ab	1.56b
Pb	1.35a	1.28a	1.29a	1.57ab	1.67b
Mn	7.79a	7.04a	6.36a	6.31a	9.24a
25-50 cm					
Cu	1.88a	1.79a	1.92ab	2.07ab	2.19b
Zn	1.35a	2.08a	2.03a	2.52a	3.77a
Cd	0.027a	0.024a	0.027a	0.029a	0.029a
Fe	2.44ab	2.32a	2.28a	3.03ab	3.32b
Ni	1.39c	1.20a	1.27ab	1.37bc	1.37bc
Pb	1.30a	1.23a	1.24a	1.35a	1.37a
Mn	7.95b	6.08a	9.09a	7.07ab	7.40ab

† for symbol explanation see text, ‡ numbers in the same line followed by different letters differ significantly ($P < 0.05$) according to LSD test

extractable Cd, Cd concentration in cotton leaves was not also significantly affected. As in the Cu and Zn, DTPA was not found a good indicator for assessing available Cd in soil. As in the previously mentioned elements, both total and DTPA extractable Cd were strongly correlated with organic matter content (Fig.

3). As in the case of Zn, Cd in the subsurface layer was not significantly affected.

Iron

Total Fe content was not significantly influenced by sewage sludge probably because of the high Fe content of the soils that was around 3.8% Fe. However Fe extracted by DTPA was highly increased because of the sewage sludge addition. From 1.61 it raised to about 5.50 mg kg^{-1} in the treatment SS3. DTPA Fe extractable was strongly correlated with organic matter content, positively, and with soil pH, negatively (Fig. 4). DTPA method failed also for Fe to extract available Fe forms for cotton plants. As in the case of Cu, Fe_{DTPA} was significantly increased also in subsurface layer (Table 1).

Nickel

Total Ni was not significantly affected by the sewage sludge while Ni extracted by DTPA was significantly increased and followed the trend of organic matter content with which it was significantly correlated. pH negatively influenced Ni_{DTPA} (Fig. 5). Ni_{DTPA} tended to increase with increasing sewage sludge rate also in subsoil like Cu and Fe (Table 1).

Lead

Lead showed similar behavior to Ni. Total Pb was not influenced by the sewage sludge addition while Pb_{DTPA} was significantly increased being positively correlated with organic matter content and negatively with soil pH (Fig. 6). In subsurface soil Ni_{DTPA} was not significantly influenced by the sewage sludge addition.

Manganese

Finally, Mn was not at all affected by the sewage sludge application. Total and DTPA extractable forms were found to be independed on organic matter content or soil pH (data not presented). The same was true for both surface and subsurface soil.

From all the above mentioned it is clear that all the metals studied, except Mn, are strongly associated with the organic matter content of sewage sludge. Concerning the organic matter role in the protection of the heavy metal hazard when we recycle sewage sludges in land there are two hypotheses (McBride, 1994): The first one usually called sludge time bomb hypothesis suggests that mineralization of organic matter could release metals in more soluble forms increasing the hazard of heavy metal toxicity. The second one called sludge protection hypothesis claims that the residuum of sludge decomposition can for a very long maintain heavy metal solubility's at very low levels based on field observations that plant uptake reaches a maximum as sewage sludge rates increase (Chaney & Ryan, 1993). Sludge protection hypothesis was criticized by McBride

(1994) who suggested that, because of the saturation effect, uptake of metals into plant tops cannot reflect solubility of metals. On the other hand, because of the capacity of soils to immobilize metals is finite, a sharp increase in heavy metal uptake by plants would be expected. Therefore, the problem of sewage sludge management is not so simple. It requires long term experiments and a very careful interpretation of the results in order to avoid unsafe conclusions for the long term effects of waste recycling in soils.

ACKNOWLEDGMENTS

The authors wish to express their thanks to the Municipality of Volos, Greece, for financing this work. Special thanks are expressed to Vaso Gianni and Savas Papadopoulos for their laboratory work.

REFERENCES

Allison, L.E., and C.D. Moovie. 1965. Carbonate. p. 1379–1396. *In* C.A. Black et al. (ed.) Methods of soil analysis. Part 2. Agron. Monog. 9. ASA, and SSSA, Madison, WI.

Baker, D.E., and M.C. Amacher. 1982. Nickel, Copper, Zinc, and Cadmium. p.323–336. *In* A.L. Page et al., (ed.) Methods of soil analysis. Part 2. 2nd ed. Chemical and Microbiological Properties. ASA, and SSSA, Madison, WI.

Bremenr, J.M. and C.S. Mulvaney. 1982. Nitrogen-total. p.595–624. *In* A.L. Page et al., (ed.) Methods of soil analysis. Part 2. 2nd ed. Chemical and Microbiological Properties. ASA, and SSSA, Madison, WI.

CEC, Council of the European Communities. 1991. Council Directive of 21 May 1991 concerning urban waste water treatment (91/27/EEC). Official J. Euro. Commun. L 135/40–52.

Chaney, R.L., and J.A. Ryan, 1993. p.451–506. *In* H.A.J. Hoitink and H.M. Keener (ed.) Science and engineering of composting: Design, environmental, microbiological and utilization spects. Renaissance Publications, Worthington, OH.

Doran, J.W., and T.B. Parkin, 1994. Defining and assessing soil quality. p. 3–21. *In* J.W. Doran et al., (ed.) Defining soil quality for a sustainable environment. SSSA, and ASA, Madison, WI.

Evangelou, V.P., J. Wang, and R.E. Philips. 1994. New developments and perspectives on soil potassium quantity/intensity relationships. Adv. Agron. 52:173–226.

Furrer, O.J., and W. Stauffer. 1986. Influence of sewage sludge and slurry application on nutrient leaching losses. p. 108–127. *In* A. Dam Kofoed et al., (ed.) Efficient land use of sludge and manure. Elsevier Applied Sciences Publ., London.

Gee, G.W., and J.W. Bauder. 1986. Particle analysis. p. 383–412. *In* A. Klute (ed.) Methods of soil analysis. Part I. 2nd ed. Physical and Mineralogical Methods, Agron. Monog. 9, ASA, and SSSA, Madison, WI.

Keeney, D.R., and D.W. Nelson. 1982. Nitrogen-Inorganic Forms. p. 634–698. *In* A.L. Page et al., (ed.) Methods of soil analysis. Part 2. 2nd ed. Chemical and Microbiological Properties. ASA, and SSSA, Madison, WI.

McBride, M.B. 1994. Toxic metal accumulation from agricultural use of sewage sludge: Do USEPA regulations ensure long-term protection of soil? Composing Frontiers II(4):4–16.

McLean, E.O. 1982. Soil pH and Lime Requirement. p. 119–224. *In* A.L. Page et al., (ed.) Methods of soil analysis. Part 2. 2nd ed. Chemical and Microbiological Properties. ASA, and SSSA, Madison, WI.

Nelson, D.W., and L.E. Sommers. 1982. Total carbon, organic carbon, and organic matter. p.539–580. *In* A.L. Page et al., (ed.) Methods of soil analysis. Part 2. 2nd ed. Chemical and Microbiological Properties. ASA, and SSSA, Madison, WI.

Olsen, S.R., and L.E. Sommers. 1982. Phosphorus. p. 403–430. *In* A.L. Page et al., (ed.) Methods of soil analysis. Part 2. 2nd ed. Chemical and Microbiological Properties. ASA, and SSSA, Madison, WI.

Rhoades, J.D. 1982. Cation exchange capacity. p.149–158. *In* A.L. Page et al., (ed.) Methods of soil analysis. Part 2. 2nd ed. Chemical and Microbiological Properties. ASA, and SSSA, Madison, WI.

Smith, S.R. 1996. Agricultural recycling of sewage sludge and the environment. CAB Int. London.

Thomas, G.W. 1982. Exchangeable Cations. p. 159–166. *In* A.L. Page et al., (ed.) Methods of soil analysis. Part 2. 2nd ed. Chemical and Microbiological Properties. ASA, and SSSA, Madison, WI.

Tsadilas, C.D., Theodora Matsi, N. Barbayiannis, and D. Dimoyiannis. 1995. Influence of sewage sludge applications on soil properties and on the distribution and availability of heavy metal fractions. Commun. Soil Sci. Plant Anal. 26(15&126): 2603–2619.

Tsadilas, C.D., and V. Samaras, 1999. Influence of sewage sludge application on oil quality: I. Organic matter, pH, phosphorus, potassium, and inorganic nitrogen (in this publication).

USEPA. 1993. Part 503-Standard for the Use or Disposal of Sewage Sludge. Federal Register 58:9387–9404.

Wen, G., J.P. Winter, R.P. Voroney, and T.E. Bates.1997. Potassium availability with application of sewage sludge, and sludge and manure composts in field experiments. Nutrient Cyc. Agroecosyst. 47:233–241.

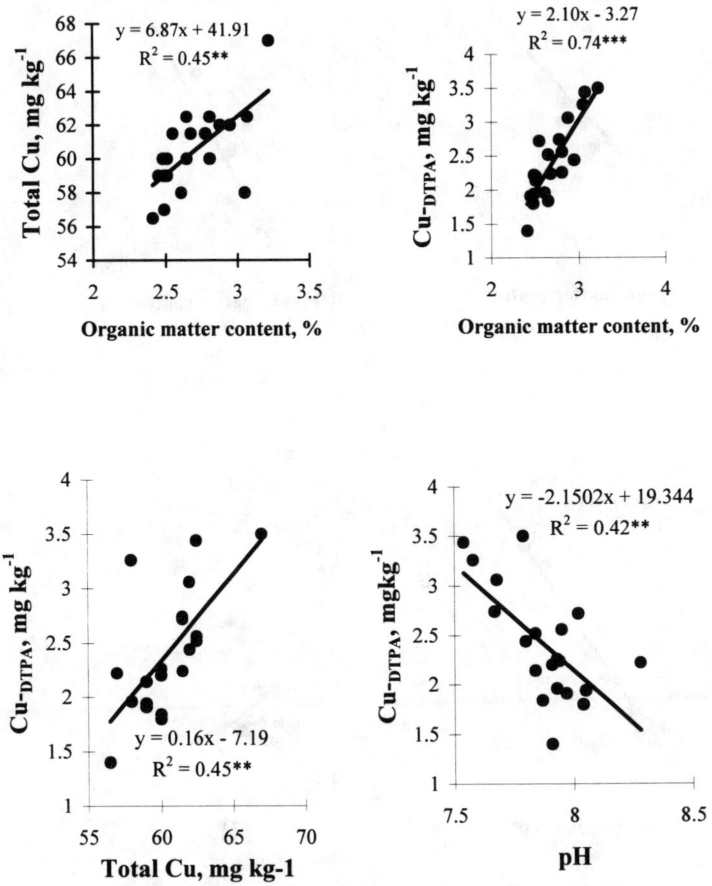

Fig. 1. Relationship between total Cu and Cu-$_{DTPA}$ with matter content and pH.

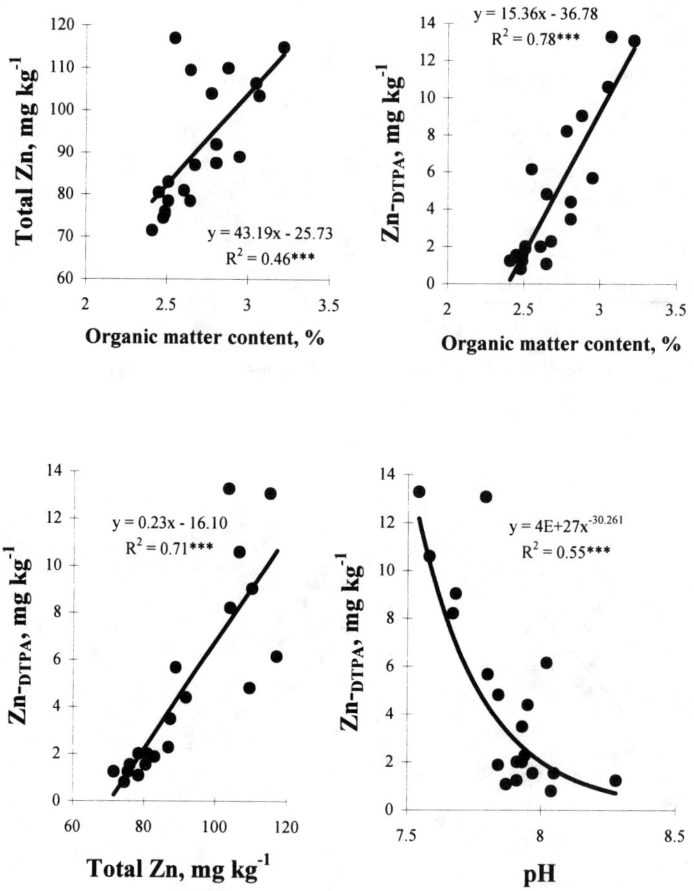

Fig. 2. Relationship between total Zn and Zn-$_{DTPA}$ with organic matter content and pH.

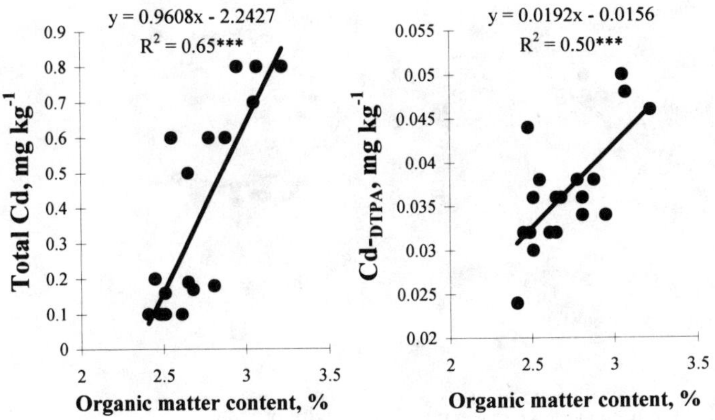

Fig. 3. Relationship between total Cd and Cd-$_{DTPA}$ with organic matter content.

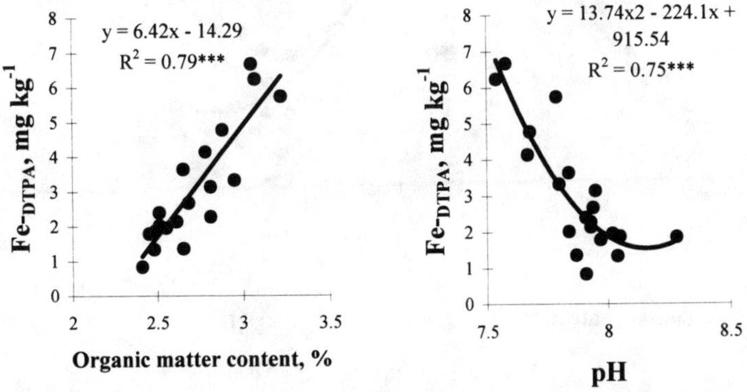

Fig. 4. Relationship between Fe extractable by DTPA with organic matter content and soil pH.

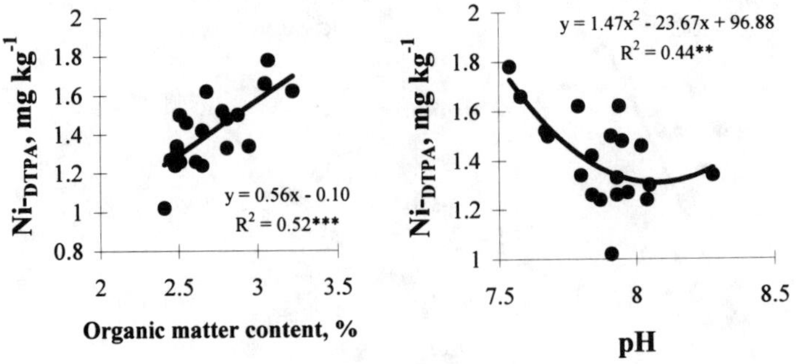

Fig. 5. Relationship between Ni extracted by DTPA with organic matter content and pH.

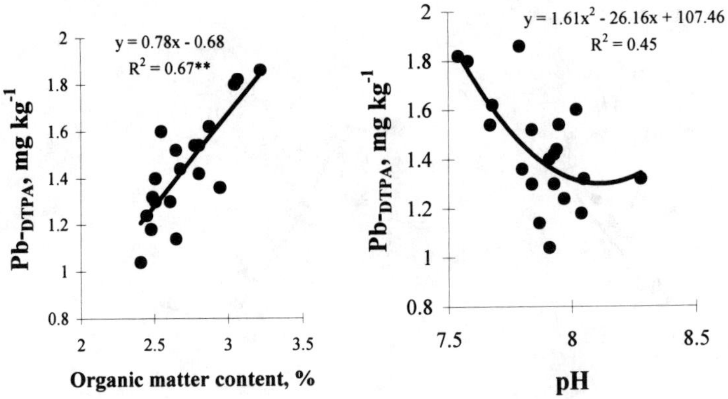

Fig. 6. Relationship between Pb extracted by DTPA with organic matter content and pH.

Heavy Metals Content of Commercial Inorganic Fertilizers Marketed in the Kingdom of Saudi Arabia

A. S. Modaihsh, and M. S. Al-Sewailem
Soil Science Department
King Saud University
Riyadh, Kingdom of Saudi Arabia

ABSTRACT

In recent years much concern has been given to toxic heavy metals that enter the human food chain. Inorganic fertilizers are considered among the potential avenues of such entry. In this work, we report the analyses of 77 samples of commercial fertilizers, marketed in the Kingdom of Saudi Arabia, for their heavy metal concentrations. Fertilizer samples included 20 samples of phosphatic fertilizers (MAP,DAP,TSP), 11 samples of liquid fertilizers, 34 samples of water soluble multiple nutrient fertilizers (WSMF), and 12 samples of solid multiple nutrient fertilizers (SMNF). Concentrations of heavy metals varied according to the type of fertilizer and the tested metal (Cr levels were the highest and Co were the lowest). The data revealed that Cd ranged from 36.8 to <1 $mgkg^{-1}$. Nevertheless, the average Cd content was 32.2 $mgkg^{-1}$ for the phosphatic fertilizers, 13.4 $mgkg^{-1}$for the liquid fertilizers, 18.4 for the (SMNF), and 4.5 $mgkg^{-1}$ for the (WSMF). Concentrations of Pb, Ni, Co, and Cr in the phosphatic fertilizers averaged 17.8, 72.3, 12.9, and 276.8 $mgkg^{-1}$, respectively. However, the corresponding average values of these elements, in the liquid fertilizers, were 13.3, 19.4, 12.5, and 85.1 $mgkg^{-1}$. In the (SMNF) were 14.5, 44.7, 11.7, and 162.0 $mgkg^{-1}$ and in the (WSMF) samples were 10.0, 7.8, 7.4, and 12.5 $mgkg^{-1}$. Data showed that Cd, Co, and Ni concentrations were lower than the tolerance limits for heavy metal addition, and apart from Cr metal, concentrations of the other heavy metals were comparable to those recorded worldwide.

INTRODUCTION

Increased concern about the contamination of soil and water resources with heavy metals has occurred in recent years. Adverse health effects consequent upon consumption of contaminated feeds have also received much attention (Singh, 1991, 1994). Heavy metals naturally occur in all soils in minute quantities, but can accumulate in agricultural soils from various sources, such as fertilizers, organic amendment, atmospheric deposition and urban and industrial activities. These metals are not essential nutrients for plant and animals, however, sufficiently high concentrations can become toxic and constitute serious health problems whenever they enter into the human food chain (Oliver, 1997).

Schroeder and Balassa 1963 were the first to alert that fertilizers are implicated in raising some heavy metals concentrations in food crops; and since then much work has been performed to investigate the impact of impurities in fertilizers on

Copyright © 1999 ASA-CSSA-SSSA, 677 South Segoe Road, Madison, WI 53711, USA. *Proceedings of the Fourth International Conference on Precision Agriculture.*

crop uptake of potentially toxic elements. The main source of fertilizer-derived heavy metals in soils is from phosphatic fertilizers; manufactured from phosphate rocks (PRs), that contain various metals as minor constituents in the ores (Allaway, 1971; Kpomblekou & Tabatabai, 1994)

Recently, Kongshaug et al., (1992) gave a comprehensive account of some heavy metals concentrations found in various phosphate rock deposits. It is believed that these heavy metals , present as impurities in PRs, are transferred to the fertilizers during the production process. Williams and David (1973) found close relationship between concentration of Cd in TSP and their respective PRs sources in Australia. Also, Wakefied (1980) reported that TSP contained 60 to 70% of Cd found in PRs.

Analysis of fertilizers commercially marketed in different parts of the world such as India, Italy, Australia, NewZealand, England and USA showed that all P fertilizers contain significant and varying amounts of heavy metals (Williams & David, 1973; Arora et al. 1975; Pezzarossa et al, 1990). Lately, Charter et al. (1993) showed that TSP, MAP, and DAP marketed in Iowa contain variable concentrations of many trace and heavy metals.

Several studies have shown that heavy metals in phosphatic fertilizers can accumulate in soil and become readily available to plants (Williams & David, 1976; Lee & Keeney 1975). In a more recent study, Mcloughlin et al. (1996) assessed the potential for contamination by phosphate fertilizers and concluded that Cd, F will accumulate at faster rate than As, Pb, or Hg. This issue has been tackled earlier by Sauerbeck, (1992) He compiled a list for the concentration ranges of some elements present in PRs and compared them with their corresponding averages in the earth crust. He concluded that in terms of fertilizer use, elements that can be considered as a potential risk of accumulation in soils are As, Cd, Cr, F, Sr, Th, U, and Zn.

In the Kingdom of Saudi Arabia, inorganic fertilizers are the top chemicals being consumed in the agricultural sector. It is believed that most growers in the Kingdom overdose the applied inorganic fertilizers, substantially due to their lower cost.

Presently, there are no regulations in the Kingdom of Saudi Arabia governing maximum permissible concentrations (MPCs) of heavy metals in fertilizers. Since there is a lack of information, regarding the concentration of heavy metals in fertilizers marketed in the Kingdom, this work was initiated to assess the content of heavy metals in most fertilizers used in the Kingdom and to evaluate the risks of their potential accumulation in soil.

MATERIALS AND METHODS

The fertilizer materials were obtained from the various companies dealing with agrochemicals. These materials represented most of the fertilizers marketed in the Kingdom of Saudi Arabia. The fertilizer samples were grouped into four categories. Namely, phosphatic, liquid, water-soluble multiple nutrients fertilizers (WSMF), and solid multiple nutrients fertilizers (SMNF).

The collected fertilizers were digested using the method of the Association of Official Agricultural Chemists (AoAc), (Williams, 1984). A sample of 5.0 g was dissolved in 10-mL concentration HCL in 100-mL beaker. The beaker was covered with watchglass, and the contents were boiled on a hot plate for approximately 30 min. The contents were then evaporated to near dryness. After cooling 20, mL of 0.1 *M* HCL was added, and the contents were gently boiled. The contents were quantitatively transferred into 100-mL volumetric flask by filtering through Watman no. 2 filter paper. The residue was thoroughly washed with 0.1 *M* HCL and the volume was adjusted with the same solution. The digest obtained was analyzed for Pb, Cd, Ni, Co, and Cr by using a Perkin Elemer (Model 2380) atomic absorption spectrophotometer. Nitrogen was determined in the digest by Kjeldahl method (Bremner, 1965). Phosphorus was determined colorimerically and K was determined using flamephotometery according to standard methods

RESULTS AND DISSCUSSION

Metal concentrations representing 20 phosphatic fertilizers (MAP, DAP,TSP), 11 liquid fertilizers, 34 water soluble fertilizers (WSMF), and 12 solid multiple nutrient fertilizers(SMNF) are given in Tables (1 to 4). The maximum, minimum and average metal concentrations for the studied fertilizers (77 samples) are shown in Table 5. The data indicated that concentration of heavy metals varied considerably with metal and the type of fertilizers. Cr levels were the highest (119.5 mg kg^{-1}) and Co were the lowest (10.3 mg kg^{-1}). Average concentration of Cr was highest in phosphatic fertilizers (276.8 mg kg^{-1}; Table 1) whereas its concentration in the WSMF was as low as 13.8 mg kg^{-1} (Table 4). The averages of the heavy metal content in phosphatic fertilizers were 17.8 mg kg^{-1} Pb, 32.2 mg kg^{-1} Cd, 72.3 mg kg^{-1} Ni, 12.9 mg kg^{-1} Co, and 276.8 mg kg^{-1} Cr in the phosphatic fertilizers. Nevertheless the corresponding value for the liquid fertilizers were 13.3 Pb, 13.4 Cd, 19.4 Ni, 12.5 Co, and 85.1 Cr (Table 2). As for the SMNF they were 14.5, 18.4, 44.7, 11.7, and 162 (Table 4), for the WSMF they were 10.0, 4.5, 7.8, 7.4, and 12.5. mg kg^{-1}, respectively (Table 3).

In general, the concentrations of these heavy metals in the 20 phosphatic fertilizer samples were higher than those in other types of fertilizers. Further more, the data showed that the concentrations of Pb, Co, and Cr were relatively higher in the TSP fertilizer samples (No. 15, 16, 17, and 18) as compared with their concentrations in the other types of phosphatic fertilizers, e.g., MAP and DAP. It is worth mentioning that the concentrations of these metals in the (WSMF) were very small. This may be attributed to the production process used in the manufacturing of WSMF; this process basically gives highly refined products.

In contrast, analysis of 97 solid commercial fertilizers in Iowa (Charter et al, 1993), showed that the mean values were 11, 15, 17, 12, and 133 mg kg^{-1} for Pb, Cd, Ni, Co, and Cr, respectively, in the TSP samples. The corresponding values were 9.1,7.1, 17, 10, and 57 for the MAP samples and 9.8, 10, 19, 8.2, and 71 for the DAP samples. Finck (1992) compiled the average heavy metal concentrations in phosphate rock in different parts of the world. He estimated the concentration of average heavy metals in fertilizer to be 66, 165, and 189 mg kg^{-1} for Pb, Cd, and Ni, respectively.

In the liquid fertilizers, the concentrations of the heavy metals were remarkably variable. Apart from Co, the values of Pb, Cd, Ni, Cr highly correlated with the P content of the fertilizers (Table 5). The SMNF exhibited the same noticeable features of the liquid fertilizers.

It is evident that Cd is the element of most concern and received much of the attention, particularly given its behavior and phytoavailability in soil (Mcloughlin et al. 1996). In this research, the Cd concentration of phosphatic samples analyzed ranged from 22.7 to 36.8mg kg^{-1} with an average of 32.2 mg kg^{-1} (Table 1). In contrast to these values the range of Cd concentration was 1.9 to 27.1 mg kg^{-1} for the liquid fertilizers, 4.4 to 28.2 mg kg^{-1} for the SMNF, and 0.0 to 32.7 mg kg^{-1} for the WSMF.

The obtained values were comparable to Cd concentration in fertilizers analyzed in other parts of the world. For instance, Cd concentration ranged from 6.8 to 47 mg kg^{-1} with a median of 8.1 in fertilizer marketed in Iowa, 1.5 to 9.7 mg kg^{-1} in Wisconsin (Lee & Keeney, 1975). 18 to 91 mg kg^{-1} in Australian fertilizers, (Williams & David, 1973), and >0.1 to 30 mg kg^{-1} in Sweden, (Stenstrom & Vahter, 1974).

The average applied Cd can be estimated from a given P rate. In the Kingdom of Saudi Arabia, the average application rate of phosphatic fertilizers rarely exceeds 400 kg ha^{-1} yr^{-1}. Considering the average Cd concentration of P fertilizers marketed in the Kingdom, which is about 15 mg kg^{-1}, the total expected amount of Cd annually applied would reach about 6 g of Cd ha^{-1}. This value is far less than the tolerance limit in soil; being 100 mg kg^{-1} for German soils (Finck, 1992). Concentrations of Cd in fertilizers obtained from this work demonstrate that it is unlikely that Cd from inorganic fertilizer will have an impact on soils. In contrast fertilizer from Florida phosphate contain <10 mg kg^{-1} Cd, contributed 0.3 to 1.2g Cd ha^{-1} yr^{-1} to soil in long term fertility experiment (Mortvedt, 1987). On the other hand, phosphatic fertilizer manufactured from western deposits containing an average of 174 mg kg^{-1} Cd contributed 100g Cd ha^{-1} yr^{-1}. to the soil in a 36 yr field trial in California. This raised the concentration in soil from 0.07 mg kg^{-1} in the control plot to 1.0 mg kg^{-1} Cd in the fertilized plots (Mulla et al. 1980). Nriagu, (1980) estimated that phosphatic fertilizers with an average Cd content of 7 mg kg^{-1} could contribute an amount of 660t Cd yr^{-1} into the environment on a global scale.

The current data revealed that Cd, Co, and Ni concentrations were generally lower than the tolerance limits for heavy metal addition. And apart from Cr metal, concentrations of the other heavy metals were comparable to those recorded worldwide. Although Cr was relatively high, still it stands much lower than the tolerance limit in soil, which is 100 mg kg^{-1}. The data presented here showed that it would take hundreds years of P applications, at the normal rates, in order to reach the tolerable limit of most of the heavy metals. However, other possible inputs of heavy metals to agricultural soils, such as sewage sludge and aerial deposition should be taken into consideration for a precise evaluation of accumulation of heavy metals. Although the present data showed that continuous application of P fertilizer would unlikely lead to accumulation of heavy metals, however, it should be noted that the quality of ores, e.g., phosphate rocks, has been declining with time and the ores now used in the production of P fertilizers contain more impurities than those used two decades ago (Liekam, 1989). In addition to this, it should also be noted that, due to lower cost of fertilizers, growers in the Kingdom tend to apply high amounts of P fertilizers. Therefore, inorganic fertilizer as well as organic fertilizer, which can contain high levels of heavy metals, should always be monitored and should be considered in environmental assessment.

Table 1. Heavy metals concentration (mg kg^{-1}) in the phosphatic fertilizers.

No.	N%	P%	K%	Pb	Cd	Ni	Co	Cr
1	18.0	46.0	0.0	11.2	34.2	52.9	5.5	215.3
2	18.0	46.0	0.0	12.0	34.0	52.8	5.4	210.0
3	11.0	52.0	0.0	14.3	33.9	84.7	10.4	238.7
4	11.0	52.0	0.0	13.0	33.6	85.2	10.4	236.0
5	11.0	52.0	0.0	14.3	33.5	82.4	11.7	260.0
6	11.0	52.0	0.0	14.4	34.0	82.8	12.0	260.0
7	18.0	46.0	0.0	14.5	31.0	70.5	9.3	216.0
8	18.0	46.0	0.0	16.4	31.0	71.0	9.6	220.0
9	18.0	46.0	0.0	12.8	31.4	67.7	10.9	199.9
10	18.0	46.0	0.0	12.8	32.8	66.6	10.4	200.0
11	18.0	46.0	0.0	14.3	22.7	62.1	12.3	314.7
12	18.0	46.0	0.0	14.2	26.0	73.4	12.0	314.0
13	18.0	46.0	0.0	12.5	28.2	67.3	11.4	236.5
14	18.0	46.0	0.0	12.6	28.2	66.6	14.4	236.0
15	0.0	46.0	0.0	32.0	36.3	67.7	20.6	402.0
16	0.0	46.0	0.0	32.0	36.0	73.2	21.2	396.0
17	0.0	46.0	0.0	32.4	36.0	77.2	20.0	410.0
18	0.0	46.0	0.0	31.6	36.8	79.8	20.6	400.0
19	11.0	52.0	0.0	16.7	31.8	81.1	14.5	280.7
20	11.0	52.0	0.0	16.4	32.0	81.0	14.8	290.0
Maximum				32.4	36.8	85.2	21.2	410.0
Minimum				11.2	22.7	52.8	5.4	199.9
Average				17.8	32.2	72.3	12.9	276.8
$r^{\dagger}$				NS	NS	***	NS	NS

*, **, ***, Significant at the 0.05, 0.01, and 0.001 probability levels, respectively.
† NS, nonsignificant at the 0.05 level.
‡ *r*, coefficient of correlation (P and Heavy metals).

Table 2. Heavy metals concentration (mg kg^{-1}) of the liquid fertilizers.

No.	N%	P%	K%	Pb	Cd	Ni	Co	Cr
1	2.0	52.0	8.0	20.0	27.0	36.1	12.6	64.0
2	7.0	5.0	5.0	4.1	3.6	4.6	3.0	0.0
3	2.0	8.0	28.0	48.5	7.3	14.5	17.7	51.7
4	2.0	8.0	28.0	10.7	7.5	14.7	17.6	50.7
5	13.0	46.0	0.0	10.0	27.1	34.3	7.5	200.0
6	13.0	46.0	0.0	9.9	25.0	31.0	8.9	203.3
7	0.0	0.0	36.0	10.0	2.1	5.6	21.1	0.0
8	0.0	0.0	30.0	9.8	1.9	5.8	21.1	0.0
9	14.0	28.0	3.0	8.1	18.0	25.6	8.6	159.0
10	14.0	28.0	3.0	8.3	18.3	25.5	8.1	142.7
11*	14.0	14.0	14.0	7.1	9.5	15.6	11.2	64.7
Maximum				48.5	27.1	36.1	21.1	203.3
Minimum				4.1	1.9	4.6	3.0	0.0
Average				13.3	13.4	19.4	12.5	85.1
r				NS	***	***	NS	**

*, **, ***, Significant at the 0.05, 0.01, and 0.001 probability levels, respectively.
† NS, nonsignificant at the 0.05 level.
‡ *r*, coefficient of correlation (P and Heavy metals).

Table 3. Heavy metals concentration (mg kg^{-1}) of the (WSMF).

No.	N%	P%	K%	Pb	Cd	Ni	Co	Cr
1	20	20	20	8.9	12.0	3.9	7.1	0.0
2	15	30	15	9.9	3.6	4.3	5.8	0.0
3	15	15	30	8.4	2.9	4.9	8.0	0.0
4	17	6	18	8.7	2.7	4.3	6.7	0.0
5	12	4	24	15.2	3.1	6.5	8.3	0.0
6	15	15	30	12.0	3.4	5.9	8.2	0.0
7	20	20	20	11.5	3.1	6.6	6.1	0.0
8	22	7	7	8.6	2.1	4.2	4.1	0.0
9	12	0	43	13.7	2.5	6.0	9.7	0.0
10	15	30	15	12.5	3.5	5.6	5.8	0.0
11	12	4	24	14.4	2.4	6.6	8.4	0.0
12	28	14	14	8.2	2.6	4.1	4.8	0.0
13	17	6	18	6.5	2.6	4.3	6.2	0.0
14	20	20	20	9.3	3.0	5.6	6.3	0.0
15	28	14	14	6.9	1.6	2.5	4.2	0.0
16	12	4	24	12.9	3.0	6.4	7.6	0.0
17	15	15	30	10.4	2.9	5.2	7.9	0.0
18	15	30	15	11.1	3.6	5.3	5.7	0.0
19	20	20	20	9.8	2.7	5.3	6.5	0.0
20	21	7	7	8.3	1.8	5.6	4.5	0.0
21	15	30	15	9.8	3.1	6.9	8.5	0.0
22	15	30	15	11.8	5.7	11.9	10.8	29.3
25	0	0	50	17.3	5.8	9.9	15.2	0.0
26	13	0	45	8.3	2.5	6.2	10.5	0.0
27	46	0	0	0.0	0.0	0.0	0.0	0.0
28	13	3	43	7.6	4.1	7.3	11.9	0.0
29	16	40	6	11.7	32.7	45.5	8.0	236.0
31	20	20	20	8.7	2.7	5.3	7.9	0.0
32	23	23	0	13.8	14.3	31.1	7.7	164.0
33	25	25	18	4.9	1.3	8.0	5.8	0.0
34	20	5	30	7.9	1.9	5.9	10.7	0.0
Maximum				17.3	32.7	45.5	15.2	236.0
Minimum				0.0	0.0	0.0	0.0	0.0
Average				10.0	4.5	7.8	7.4	12.5
r				NS	**	**	NS	**

*, **, ***, Significant at the 0.05, 0.01, and 0.001 probability levels, respectively.
† NS, nonsignificant at the 0.05 level.
‡ *r*, coefficient of correlation (P and Heavy metals).

Table 4. Metal concentrations (mg kg^{-1}) in solid multiple nutrient fertilizers (SMNF).

No.	N%	P%	K%	Pb	Cd	Ni	Co	Cr
1	28	28	0	12.0	20.0	43.7	5.0	161.3
2	14	38	10	15.8	27.2	60.0	12.3	200.0
3	14	38	10	15.3	28.2	61.5	12.3	213.3
4	11	29	19	15.3	17.9	48.8	14.9	189.3
5	11	29	19	16.3	18.7	49.2	17.3	201.3
6	28	28	0	12.3	20.5	42.2	7.3	126.2
7	28	28	0	11.9	19.3	37.5	6.3	156.7
8	16	20	0	11.5	12.9	40.0	7.5	145.3
9	12	35	8	14.8	21.7	60.2	12.7	200.7
10	18	18	5	16.1	4.4	24.2	15.2	40.0
11	12	27	18	16.2	22.3	37.9	16.1	180.0
12	23	23	0	16.7	7.9	31.0	13.3	130.0
Maximum				16.7	28.2	61.5	17.33	213.3
Minimum				11.5	4.40	24.2	5.00	40.00
Average				14.5	18.4	44.7	11.7	162.0
r				NS	***	***	NS	***

*, **, ***, Significant at the 0.05, 0.01, and 0.001 probability levels, respectively.
† NS, nonsignificant at the 0.05 level.
‡ *r*, coefficient of correlation (P and Heavy metals).

Table 5. Metal concentrations (mg kg^{-1}) of the different types of fertilizers (77samples).

	N%	P%	K%	Pb	Cd	Ni	Co	Cr
Maximum	46	52	50	48.5	36.8	85.2	21.2	410
Minimum	0	26.9	0	0	0	0	0	0
Average	14.9	17.3	12.1	13.2	15.6	32.9	10.3	119.5
r				**	***	***	NS	***

*, **, ***, Significant at the 0.05, 0.01, and 0.001 probability levels, respectively.
† NS, nonsignificant at the 0.05 level.
‡ *r*, coefficient of correlation (P and Heavy metals).

REFERENCES

Allaway, C.H. 1971. Feed and food quality in relation to fertilizer use. p. 533–566. *In* R.A. Olson (ed.) Fertilizer technology and use. SSSA, Madison, WI.

Arora, C.L., V.K Nayyar, and. N.S Randhawa. 1975. Note on secondary and micro-element contents of fertilizers and manures. Indian J. Agric. Sci. 45:80–85

Bremner, J.M. 1965. Inorganic forms of nitrogen. p. 1179–1237. *In* C.A Black (ed.) Methods of soil analysis. Part 2. Agron. Monogr. 9. ASA, CSSA, and SSSA, Madison, WI.

Charter, R.A., M.A. Tabatabai, and J.W. Schafer. 1993. Metal content of fertilizer marketed in Iowa. Commun. Soil Sci. Plant Anal. 24:961–972.

Finck, A., 1992. Dungung der Dungung: Grunlagen und Anleitung zur Dungung der Kulturpflanzer. 2nd ed., VCH Verlagsgesellschaft mbH, Weinheim, Germany

Kongshaug, G., O.C. Bockman, O. Kaarstad, and H. Morka. 1992 Inputs of trace element to soils and plants Proc. Chemical Climatology and Geomedical Problems, Norsk Hydro, Oslo, Norway.

Kpomblekou, A.K., and M.A. Tabatabai. 1994. Metal contents of phosphate rocks. Commun. Soil Sci. Plant Anal. 25:2871–2882.

Lee, K.W., and D.R. Keeney. 1975. Cadmium and zinc additions to Wisconsin soils by commercial fertilizers and waste sludge application. Water Air Soil Pollut. 5:109–112.

Liekam, D.F. 1989. DAP vs. MAP manufacturing/marketing implications. p. 75–82. *In* Proc. of the 19th North Central Extension–Industry Soil Fertility Conf., St. Louis, MO.

Mclaughlin, M.J., K.G. Tiller, R. Nacdu, and D.P Stevens. 1996. Review: The behaviour and environmental impact of contaminants in fertilizers. Aust. J. Soil Res. 34:1–54.

Mortvedt, J.J. 1987. Cadmium levels in soils and plants from some long-term soil fertility experiments in the United States of America. J. Environ. Qual. 16:137–142.

Mulla, D.J., A.L. Page, and Ganje, T.J. 1980. Cadmium accumulations and bioavailability in soils from long-term phosphorus fertilization. J. Environ. Qual. 9:408–412.

Nriagu, J.O. (ed.) 1980. Cadmium in the environment: 1. Ecological cycling. John Wiley, New York.

Oliver, M.A. 1997: Soil and human health: A review. Euro. J. Soil Sci. 48:573–592.

Pezzarossa, B., F. Malorgio, F. Lubrano, L. Tognoni, and G. Petruzzeli 1990. Phosphatic fertilizers as sources of heavy metals in protected cultivation. Commun. Soil Sci. Plant Anal. 21:737–751.

Sauerbeck, D. 1992. Conditions controlling the bioavailability of trace elements and heavy metals derived from phosphate fertilizers in soils. p. 419–48. *In* Proc. Int. IMPHOS Conf. on Phosphorus, Life and Environment. Inst. Mondial du Phosphate, Casablanca.

Schroeder, H.A., and J.J. Balassa. 1963. Cadmium: Uptake by vegetables from superphosphate and soil. Science (Washington, DC) 140:819–820.

Singh, B.R. 1991. Unwanted components of commercial fertilizers and their agricultural effects. The Fertilizer Soc., London. England.

Singh B.R. 1994. Trace element availability to plants in agriultural soils, with special emphasis on fertilizer inputs. Environ. Rev. 2:133–146.

Stenstron, T., and M. Vahter. 1974. Cadmium and lead in Swedish commercial fertilizers. Ambio 3:90–91.

Wakefield, ZT. 1980. Distribution of cadmium and selected heavy metals in phosphate fertilizer processing. Bull. Y-159. Nat. Fertilizer Dev. Ctr., Tennessee Valley Authority, Muscle Shoals, AL

Williams, C.H., and D.J. David. 1976. The accumulation in soil of cadmium residues from phosphate fertilizers and their effect on the cadmium content of plants. Soil Sci. 121:86–93.

Williams C.H., and D.J. David. 1973. The effect of superphosphate on the cadmium content of soils and plants. Aust. J. Soil Res. 11:43–56

Williams, S. (ed.) 1984. Official methods of analysis. Assoc. of Official Analytical Chemists, Arlington, VA.

TECHNOLOGY TRANSFER

The Growth and Development of Precision Agriculture Service Providers

William W. Casady, and Raymond E. Massey
Crops Focus Team, Missouri Commercial Agriculture Program
University of Missouri, Columbia, Missouri

ABSTRACT

A survey of agricultural service providers was conducted to help identify adoption rates and the status of precision farming techniques in Missouri. Results of the survey provide useful information to agribusinesses designing precision farming business strategies and to University Extension educators assisting farmers in making decisions about precision agriculture. The survey revealed that precision farming activities have developed in clusters around agricultural service providers. Established precision farming service providers offer technical support, services, and educational opportunities for farmers in the community. The survey also revealed that agricultural service providers add precision farming services because they predict a growing demand for the services and determine that they must be equipped to meet that demand to remain competitive, especially in fertilizer sales. Guidelines for estimating the costs of precision agriculture services were developed. Although many similarities exist in pricing structures for precision agriculture services, prices were more similar among service providers within geographic locale such as the bootheel or the West Central regions of Missouri. While the return on the investment in precision agriculture tools and services depends on the influence of many variables, the farm level costs can be predicted with confidence.

INTRODUCTION

Farm service businesses in the Corn Belt of the Midwest region of the USA have an important role in establishing successful and sustainable precision agriculture activities in a farming community. Adoption of new management practices often occurs in clusters or within communities of farmers. For example, the growth of residue management practices reported by the National Crop Residue Management Survey (CTIC, 1996) is often connected to areas of already dense conservation activity. Agricultural service industries are often at the nucleus of clusters providing a center for technical support, services, and educational programs offered by institutions such as universities, businesses and farmer associations. Innovative farmers within such a community who demand new services are the catalysts for development of new program (Makowski et al. 1987). Agricultural service providers with confidence in and commitment to supporting new farm practices often become the foundation for a cluster.

Copyright © 1999 ASA-CSSA-SSSA, 677 South Segoe Road, Madison, WI 53711, USA. *Proceedings of the Fourth International Conference on Precision Agriculture.*

Precision farming involves the adaptation to both new technologies and new levels of management. The technologically advanced equipment necessary to implement site-specific fertility and pest control plans can require a significant investment for both farmers and agricultural service providers. The size and scope of a farm service provider significantly affects the philosophy and approach that are adopted as the business begins to offer precision farming services. Small businesses are more likely to select an approach that minimizes their capital investment, such as selling technology. Larger businesses may invest in new equipment and offer a whole range of new services. The choices made by a local agricultural service provider have a large impact on the development of precision agriculture in the community or within a precision-farming cluster.

Most farmers in Missouri and the rest of the corn belt of the US Midwest rely on the services provided by farm service businesses for variable rate application of inputs to make use of the knowledge gained with precision agriculture tools. Farm service businesses must identify or establish a committed group of farmers to be able to provide those services. Farm service businesses and farm communities that have already adopted precision agriculture serve as models and as an information resource for development of precision agriculture programs in other communities.

Yield maps, soil nutrient maps and other field data provide the basis for acquiring new levels of knowledge about fields that can be used to make more specific management decisions. These maps often provide a hands-on learning experience that contributes to improved management habits. Maps can provide clues to pest, drainage, and soil quality problems, other than nutrient deficiencies, that may be limiting production. Although it is difficult to place a value on improved management, certain types of management decisions have the potential to pay for the technology on a single field. The combination of intensive soil management, site-specific production monitoring, scouting and improved management opportunities represents the widest scope of precision farming activities found on most farms.

The tools and expertise required to support a precision farming program can require a substantial investment of time and money for both farmers and dealers. Farmer's costs include the price of owning global positioning system (GPS) receivers with differential corrections (DGPS), yield monitors, computers, variable rate controllers, and the costs of soil testing, as well as contracted services for soil sampling and variable rate applications. An agricultural service provider may purchase several DGPS receivers, all-terrain vehicles, field computers, office computers, extensive software packages, controllers, and variable rate application equipment.

This paper examines the philosophies and service provider–farmer interactions that are involved in precision farming adoption. The survey provides guidelines for estimating the costs of managing fields and farms using precision agriculture tools. While the returns to farmers from precision agriculture tools and services depend on the influence of many variables, the cost of participation in precision agriculture programs and precision agriculture technology can be easily estimated.

METHODS

Agribusinesses that provide precision agriculture services were identified via email contact with regional and state extension personnel and other personal contact. Each business was contacted by telephone to establish a relationship. These 27 businesses then received a questionnaire and a request for the precision farming product and service sales literature that is provided to their customers. A post-survey telephone interview was used to verify the accuracy of the information. More than one-half of the programs had less than one year of activity and did not have an established format or price sheet. Hence, only 14 surveys were completed and information from only 9 was complete enough to be included in the summary.

RESULTS

Responses were obtained from all agribusinesses in Missouri that had performed precision soil testing and variable rate application services for at least one season.

Costs for soil testing, mapping and variable rate application services are summarized in Table 1. The table provides price information for nine organizations. Programs 3, 5, and 7 include variations based on the size of the soil-sampling grid. Field boundary and tract mapping was offered as a separate service by organizations 3, 5, and 6; costs for boundary and tract mapping services were included in all packages when intensive soil sampling was performed to produce nutrient maps and spread maps.

The premium charges for variable rate applications of lime and fertilizer in Table 1 are the additional charges incurred for use of variable rate services and do not include basic spreading charges, which are sometimes included with the purchase of the product. Variable rate applications of lime were typically available with a single product truck; the typical premium was \$1.50, with a range of \$0.00 to \$3.00. Organizations 3 and 5 applied variable rates of lime only in palletized form with a multiple product truck; the additional cost of palletized lime, relative to rock lime, must be considered when estimating the cost of this service. The premium charges for variable rate nutrient applications ranged from \$1.00 to \$5.00 for multiple product trucks and \$1.00 to \$2.50 for single product trucks.

All precision agriculture packages in the survey used the premise of a four-year period for soil analysis. For estimating average annual costs, packages that relied on single product trucks were assumed to be used annually, i.e., four times in a 4-yr period; multiple product trucks were assumed to be used twice in a 4-yr period. The average annual cost (Table 1) based on a 4-year period included the following: (i) boundary, soil test maps, analysis and spread map; (ii) the premium for one variable rate lime application; and (iii) the premium for four single product applications, or two multiple product applications. The estimated annual average costs for packages that measure soil properties on a scale of 2.5 to 3.0 acres averaged \$3.66 per acre and ranged from \$2.63 to \$5.00 per acre.

Table 1. Price list for precision agriculture services offered by farm business services in Missouri.

Precision agriculture program examples	Ex 1	Ex 2	Ex 3a	Ex 3b	Ex 4	Ex 5a	Ex 5b	Ex 5c	Ex 6	Ex 7a	Ex 7b	Ex 8	Ex 9
Grid size (acres)	2.5	2.8	2.8	20.0	2.5	2.5	5	10.0	2.5	2.5	5. 0	3.0	2.5
Boundary–tract maps	---	---	$2.00	$2.00	---	$1.00	$1.00	$1.00	$1.50	---	---	---	---
Soil test and maps	$5.50	$7.50	$6.75	$3.75	$7.00	$10.00	$8.50	$7.50	$8.50	$6.50	$5.50	$11.00	$9.50
Premium charge For variable rate Lime applications †	$1.00	$1.50	†	†	$3.00	†	†	†	$0.00	$1.50	$1.50	$1.50	$1.00
Premium charge For variable rate Nutrient applications ($ acre^{-1}) ††	$1.00 SP (x 4)	$2.00 MP (x 2)	$4.75 MP (x 2)	$2.75 MP (x 2)	$3.50 MP $2.50 SP	$5.00 MP (x 2)	$5.00 MP (x 2)	$5.00 MP (x 2)	$2.50 SP (x 4)	$3.25 MP (x 2)	$3.25 MP (x 2)	$1.00 SP (x 4)	$1.00 MP (x 2)
Yield map from data						NA	NA	NA		NA	NA	Incl.	Incl.
Enrolled in program	$1.00	Incl.	$0.10	$0.10	$0.50				$0.25				
Not enrolled		$0.25	$0.60	$0.60					$0.75				
Average annual cost (4 yr base) §	**$2.63**	**$3.25**	**$4.06**	$2.31	**$4.00**	**$5.00**	$4.63	$4.38	**$3.58**	**$3.63**	$3.38	**$3.63** ¶	**$3.13**
Average annual cost including yield map	$3.63	$3.25	$4.16	$2.41	$4.50	NA	NA	NA	$3.83	NA	NA	$3.63	$3.13

† Palletized lime is the only available method offered for variable rates of lime in these programs.

†† SP, MP: variable rate applications are made with either a single product (SP) or multiple product (MP) truck. Multiple product truck use is estimated to be twice in 4 years; single product trucks are estimated to run four times.

§ Includes soils sampling, 1 VR lime appl., and 2 VR fert appl. with MP truck or 4 VR appl. with SP truck. Boldface entries highlight average costs for packages with soil testing units between 2.5 and 3.0 acres.

¶ A comprehensive premium management package is offered for $2.25 acre^{-1}. Refer to discussion.

DISCUSSION

The surveys revealed various trends not only in pricing structures, but also in philosophies and in the range and types of services offered.

Precision Agriculture Philosophies

Agribusinesses are shaping the evolution of precision farming practices by promoting precision farming programs that complement their vision of how their business and clientele can profit from precision farming while minimizing problems. This vision becomes their philosophy about how precision farming should evolve. Two basic philosophies have evolved, which are generally related to the size of the business.

It is relatively easy for a small agribusiness to get started with a distributorship for yield monitoring and GPS equipment. The initial cost to inventory and the management required to install and service yield monitors is less than investing in variable rate equipment. Customers that have gathered yield data and have knowledge of the variability that occurs within fields are more likely to become committed to using other types of precision data, such as soil sampling, to complement the yield data and improve management of their farms. Hence, small agricultural service providers may advocate first examining variability in productivity with yield monitors. During the first year or two, as customers collect yield data, small agribusinesses can begin to purchase equipment to help meet the larger demand for soil testing and to modify existing single product spreading equipment to provide variable rates.

Agribusinesses that provide custom services for a large number of customers have been more able to purchase some of the equipment required to provide an effective intensive soil testing service and variable rate nutrient applications. The initial cost of the GPS equipment and the variable rate application equipment may be relatively small compared to their total investment in equipment needed to provide traditional services. The GPS equipment increases their efficiency in other more traditional soil testing and application services. Hence, larger agriculture service providers tend to introduce farm level precision management by enrolling some land in an intensive soil-sampling program. This strategy attempts to remove soil nutrient status from the list of factors that cause variability. Yield monitoring data is considered to be valuable feedback, but the focus has been on an intense nutrient management program.

Yield Monitoring

Many small farm budgets will not allow the purchase of much of the new technology. Large farms have the advantage of spreading the cost of the technology over a larger number of land units. Yield monitors are the most likely initial purchase of technology by farmers. The cost of operating a yield monitor or similar equipment is primarily the fixed cost of owning the equipment; a complete system with DGPS may cost US$7000-US$10,000. Hence, the unit cost of using a yield monitor is inversely proportional to the acres of land covered. Farm service businesses sometimes offer a leasing program for customers who are interested in adopting yield monitors but still feel uncertain about the technology. For example, a three-year lease-to-own program provides the customer with installation, service and support for the first three years.

Although our survey revealed only 24% of farms enrolled in a farm service sponsored precision agriculture program have yield monitors, 93% of those farm operators equipped their yield monitors with DGPS to create yield maps. Few farm service businesses plan to use yield data to assist in augmenting fertility decisions based on crop removal; however, some plan to use the maps to assist the farmer in identifying other production limiting factors such as soil compaction and pests such as soybean cyst nematode.

Soil Testing

Most farmers find that they can often purchase soil sampling and mapping services from their local agricultural service provider at very reasonable prices. Soil samples are more efficiently drawn using an all-terrain vehicle with GPS. The farmer does not need to supply labor or management for acquiring samples or developing soil test maps. In all programs surveyed, the farmer pays for the service of acquiring data; hence, the data belongs to the farm owner or farm operator who purchased the services. The cost of acquiring and testing intensive soil samples has a similar unit cost for large and small farms.

Some agribusinesses offer a very basic mapping package that includes only maps of field boundaries and certain other geographical features that may be used to assist in managing fields (Table 1). These maps may be overlaid with soil survey or aerial photography maps to provide information useful for planning tile drainage installations, tillage and cropping practices. The maps also provide precise area calculations for management purposes.

Farmers may soon be able to purchase a variety of other soil property maps using new soil sensors that can estimate soil properties such as organic matter content, CEC and soil moisture (Hummel et al., 1996). Soil conductivity maps may also become a purchased service. Apparent soil conductivity measured using electromagnetic induction (Doolittle et al., 1994) is useful for estimating the depth of topsoil above claypans in Mexico (fine, montmorillonitic, mesic Udollic Ochraqualfs) soils in Missouri. Electromagnetic induction data may be useful for predicting productivity of soils (Kitchen et al., 1995) and for customizing crop nutrient recommendations.

Variable Rate Applications

The most needed service that a farm service business provides to farmers who engage in precision farming is the application of variable rates or formulations of soil nutrients and farm chemicals. The relatively high cost of variable rate application equipment makes owning such equipment prohibitively expensive for most farms.

Most agriculture service providers already offer custom applications of fertilizer and pesticides; however, variable rate technology is most accessible to larger businesses. Large agricultural service providers can offer better services by spreading the cost of the infrastructure necessary to provide those services among many farmers. The degree of sophistication of analysis software and equipment for applying variable rates of inputs is most often related to the size of the customer base.

Several approaches to providing variable rates of soil amendments are currently in use. Multiple product trucks designed to provide variable rates and formulations from among several products can require a capital investment as high as $300,000. Our survey revealed that the premium for variable rate application services with this type of equipment averaged $3.55 in Missouri but ranged from $1.00 to $5.00 acre^{-1}.

Smaller agribusinesses tend to modify existing spreader trucks to spread variable rates of single products. Such modifications provide a basic entry-level unit to support the demand for variable rate services for as little as $15,000 to $20,000. While not as efficient as trucks that can simultaneously spread multiple products, costing as much as $300,000, these units offers a very good solution for correcting pH problems and variability of single nutrients.

Variable rate applications of only one or two products can be offered at smaller costs compared to those for multiple product spreaders. The premium for variable rate applications with a single product spreader averaged $1.50, with a range of $1.00 to $2.50 acre^{-1}. However, two or more treatments may need to be made on some fields. If only one treatment is needed, the cost to the farmer for this service is significantly less than that incurred for the use of multiple product equipment (Table 1). If two treatments are required, the cost is comparable to that for the multiple product equipment. If more than two treatments are required, the single product application costs exceed those charged for the use of multiple product equipment.

Development of Services

Agricultural service providers that have been successful in creating precision farming programs and services have often developed these programs over a period of several years, experimenting with several approaches before making major purchases. New responsibilities that accompany a precision farming program often require creating a new position. A precision farming manager will usually devote a majority of working time to soil sampling, mapping, developing precision agronomic recommendations, maintaining data, and providing learning experiences for farmers.

Successful farm service organizations have developed precision agriculture programs by first building relationships through other programs. Well-organized commercial programs designed to manage soil nutrients using more traditional approaches have been successful in increasing farmer awareness of the value of adequate and accurate soil testing and fertility management. The success of these programs depends partly on providing increased educational opportunities to help farmers understand how soil nutrients affect yield and profitability. Farmers may be offered additional services such as a record keeping system and soil sampling in exchange for a commitment to following a nutrient plan designed by a staff agronomist. A fertilizer budget may be developed to provide the farmer with assurance that the program will be affordable.

Agricultural service providers will typically recommend that a farmer use variable rate applications every two years. A farmer may also elect to buy fertilizer from someone else or to use a straight blanket rate of fertilizer. The farmer is not bound to use the variable rate application services that are available. A farmer may also consider enrolling only a portion of the farm in a soil testing-oriented precision agriculture program. The only long-term contract specifically identified in most programs is for a four-year soil sampling and advising program.

Many farm service businesses are beginning plans to offer a wider array of services to meet the growing demand for site-specific management of crops. Intensive scouting using predetermined routes through a field offers farmers a more thorough approach to scouting. Results of the scouting expedition are recorded as the scout travels through the field and are plotted on a map to develop a treatment plan. Many dealers are beginning to offer spot treatments, variable rates, or variable formulations of pesticides to treat only the affected area. A regular scouting program using mapping procedures can target the problem and eradicate a pest before it becomes economically significant. Premium management programs (Table 1) based on precision agriculture data offer a more comprehensive service that assists in optimizing farm management decisions. Farmer involvement with analysis and interpretation of the data leads to increased management opportunities that often accompany the new range of services available with precision agriculture tools.

Healthy dealer-farmer relationships are extremely important to both parties to minimize risk and to increase the probability of success of precision farming activities. Customers must usually rely on dealers to provide the variable rate application services that will help to complete the objective of increasing productivity and profitability. Dealers rely on their customer base to become committed to using the new equipment and services to recover the costs of providing those services.

SUMMARY

The average annual cost for intensive soil-based precision agriculture programs in Missouri was estimated to be \$3.66 acre^{-1}. This cost was based on a 4-yr cycle that included intensive soil sampling on a scale of 2.5 to 3.0 acres, development of soil nutrient maps, one variable rate lime application, and either two fertilizer applications with a multiple-product truck or four fertilizer applications with a single product truck.

While the range for the estimated costs varied from \$2.63 to \$5.00 per acre, all but two of the estimated costs fell within \$0.41 of the mean or within a range from \$3.25 to \$4.07. Actual costs for precision agriculture services depend on the quality and quantity of services selected.

REFERENCES

Conservation Technology Information Center. 1996. National crop residue management survey. CTIC, West Lafayette, IN.

Doolittle, J.A., K.A. Sudduth, N.R. Kitchen, and S.J. Indorante. 1994. Estimating depths to claypans using electromagnetic induction methods. J. Soil Water Conserv. 49(6):572–575.

Hummel, J.W., L.D. Gaultney, and K.A. Sudduth. 1996. Soil property sensing for site-specific crop management. Computer Electron. Agric. 14:121–136.

Kitchen, N.R., K.A. Sudduth, S.J. Birrell, and S.T. Drummond. 1995a. Spatial prediction of crop productivity using electromagnetic induction. p. 229. *In* Agronomy abstracts. ASA, Madison, WI.

Makowski, T.J., A.J. Sofranko, and J.C. van Es. 1987. Innovators and neighbors: Behavioral influence in the decision to adopt conservation technology. Rural Sociological Society.

REFERENCES

Yield Mapping of On-Farm Cooperative Fields in the Southeast Coastal Plain

John Sadler
Joe Millen
USDA-ARS Coastal Plains Soil, Water, and Plant Research Center
Florence, South Carolina

Patrick Fussell
Josh Spencer
USDA-NRCS
Kenansville, North Carolina

Wilson Spencer
USDA-NRCS
Clinton, North Carolina

ABSTRACT

Long-term research in the SE Coastal Plain shows that soil variability is widespread. Areas of low-yielding soils within fields often significantly reduce yield below that expected for the typical soils within the field. Farmers, though qualitatively aware of both variability and its effect on yield, appear to perceive that purported economic and environmental benefits of variable-rate technology do not justify the initial cost. They need data on economic effects of field-scale variability to allow rational strategic decisions. A multi-agency project was funded to both document existing variability in on-farm yields and to communicate the significance of the problem. Yield monitors were installed on three combines in Duplin and Sampson Counties, NC. These monitors collected data during 1997, totaling 900 ha of wheat and 120 ha of rye, followed by approximately 1500 ha of corn and similar area of soybean. Preliminary data were processed in vendor's yield mapping software. For further analyses and presentation, they were aggregated into ARC/Info GIS. Dramatic variability was documented both within and among fields, operators, and soil types.

INTRODUCTION

Adoption of site-specific farming in the Southeastern Coastal Plain has lagged that in other areas of the USA and world, despite research that indicates soil and concomitant yield variation is a significant problem. Information that would change the wait-and-see attitude farmers have toward the technology would include evidence of either economic or environmental benefits, which have been promised but not conclusively demonstrated. A clear need exists in the Southeast for data documenting the extent of the variability problem in space and time and for data

Copyright © 1999 ASA-CSSA-SSSA, 677 South Segoe Road, Madison, WI 53711, USA. *Proceedings of the Fourth International Conference on Precision Agriculture.*

documenting the economic and environmental effect of site-specific management.

Consequently, funding was obtained to address these issues in a multi-agency, watershed-scale cooperative research and demonstration project within the ASEQ (Agricultural Systems for Environmental Quality) projects. The CSREES-funded project, titled "Management Practices to Reduce Nonpoint Source Pollution on a Watershed Basis", included cooperators from Biological and Agricultural Engineering and from Cooperative Extension Service, both of North Carolina State University; USDA-NRCS at the state, district, and county levels; USDA-ARS at Florence, SC; US Geological Survey; and several local farmer-cooperators. This project followed a 5-yr documentation of the conditions in the Herrings Marsh Run watershed, for which the water quality status was described by Stone et al. (1995). Within the current project, the site-specific management effort was one of three objectives. Paraphrased for this context, this objective was to improve and adopt precision farming as a best management practice, with the following subobjectives: (i) to show existing variation in crop yield with combine monitors, (ii) to use computer models to predict yield and relate precision farming to water quality, and (iii) to improve and encourage site-specific nitrogen management. The purpose of this paper is to present preliminary results for the first subobjective.

MATERIALS AND METHODS

Two farmer cooperators were involved in the project. Most of the data collected were from Duplin and Sampson Counties, North Carolina. Fields in Wayne, Bladen, and Pender Counties were also studied (Fig. 1), but most of these fields were proximal to the Duplin or Sampson County borders.

The cooperator based in Duplin County operated two John Deere 9500[1] (Deere & Co., Moline, IL) combines with 20-ft grain and 4-row corn headers. Both had GreenStar yield monitors installed in March 1997. The DGPS units used satellite-based differential correction. They wrote to 5-MB data cards, which were read into JDMap V2.1.1 software.

The cooperator based in Sampson County operated two Case 2188 combines with 20-ft grain and 4-row corn headers. One machine previously had an AFS yield monitor without DGPS. A DGPS unit (GPS2000, AgLeader Technology, Ames, IA) was installed by project personnel on that machine before wheat harvest in June 1997. This unit used the Ft. Macon Coast Guard beacon for differential correction. The unit wrote to 1-MB cards, which were read into AgLink V5.2.1 (AGRIS Corp., Roswell, GA) software.

---

[1] Mention of trademarks is for information only. No endorsement implied by USDA-ARS or its cooperators.

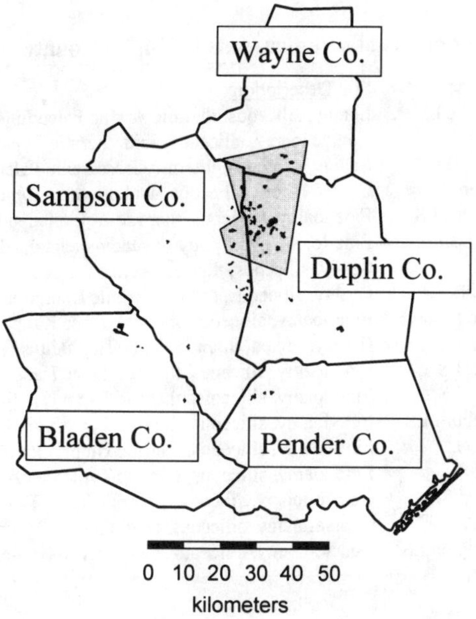

Fig. 1. Location of all corn and wheat on-farm cooperator's fields harvested during the 1997 harvest season. The shaded area encloses fields for Cooperator 1; all other fields are Cooperator 2.

Project personnel set up the monitors, trained the operators in daily monitor-related tasks, and calibrated the monitors using load totals determined with portable truck scales or scale tickets. During harvest, the operators entered field names, crops, and activated the data collection. Project personnel periodically exchanged cards and read data into a personal computer to prepare preliminary yield maps. Data were also transmitted to the server at USDA-ARS Florence via dial-up networking on a regular basis. In Florence, these data were examined for DGPS problems and operator artifacts, such as erroneous crop codes, field names, turns and trips across the field to unload with the header down, etc. Errant passes were straightened, null passes and turns were deleted, and field names and crop codes were corrected as possible. Data from the two combines for the Duplin County cooperator were merged, which required adjacent-pass comparisons to calibrate one of the two combines on some fields. Data from all sources were merged into one comprehensive data set using the AGLink software, then exported to an ASCII format for importing into ARC/Info.

For Duplin County, the soil survey (Goldston et al., 1958)) was available in GIS format. The Duplin County yield data from both cooperators was overlaid in ARC/Info to determine the soil map unit (Table 1) associated with each data point. The resulting ARC/Info table was exported to SAS (SAS Institute, 1990) for summary statistics by soil type. Summary statistics and analyses conducted in SAS include analysis of variance by soil type (for Duplin County data only) and distributions of yield by field, operator, and for the whole data set.

Table 1. Descriptions of soil map units in Duplin County, North Carolina.

Unit	Soil name	Soil Description
AuB	Autryville LS	Loamy, siliceous, thermic Arenic Paleudults
BbA	Bibb SL	Coarse loamy, siliceous, acid, thermic Typic Fluvaquents
BnB	Blanton fS	Loamy, siliceous, thermic Grossarenic Paleudults
FoA	Foreston fS	Coarse loamy, siliceous, acid, thermic Typic Fluvaquents
GoA	Goldsboro LS	Fine loamy, siliceous, thermic Aquic Paleudults
JoA	Johns LS	Fine-loamy over sandy or sandy-skeletal, siliceous, thermic Aquic Hapludults
LnA	Leon S	Sandy, siliceous, thermic Aeric Alaquods
LsB	Lucy LS	Loamy, siliceous, thermic Arenic Kandiudults
McC	Marvyn LS and	Fine-loamy, siliceous, thermic Typic Kanhapludults
	Gritney SL	Clayey, mixed, thermic Aquic Hapludults
NoA	Norfolk LS	Fine-loamy, siliceous, thermic Typic Kandiudults
NoB	Norfolk LS	Fine-loamy, siliceous, thermic Typic Kandiudults
OrB	Orangeburg LS	Fine-loamy, siliceous, thermic Typic Kandiudults
PnA	Pantego L	Fine-loamy, siliceous, thermic Umbric Paleaquults
RaA	Rains LS	Fine-loamy, siliceous, thermic Typic Paleaquults
RuB	Rumford SL	Coarse-loamy, siliceous, thermic Typic Hapludults
ToA	Torhunta fSL	Coarse-loamy, siliceous, acid, thermic Typic Humaquepts
WoA	Woodington LS	Coarse-loamy, siliceous, thermic Typic Paleaquults

RESULTS AND DISCUSSION

Representative Yield Maps

Within the project, 1997 corn yield maps demonstrated wide variation both within and among fields. Three fields for cooperator 1 (Duplin Co.) averaged 5.3 Mg ha^{-1} (Fig. 2). These three fields averaged from 4.3 to 6.2 Mg ha^{-1}, and areas within the fields ranged from <1 to approximately 10 Mg ha^{-1}. Patterns in the yield maps suggest soil type variations consistent with Carolina Bays, including bounding arcs with dramatically reduced yields. In Field 104, the areas south of the Carolina Bays consistently yielded above 7 Mg ha^{-1}, and in field 102, almost all yielded below 7 Mg ha^{-1}. For the same cooperator, six additional fields with essentially the same overall average (Fig. 3) show that the variation is widespread. Wheat yield maps not shown here show similar patterns. In the second year of the project, we hope to find whether corn yield patterns correspond to wheat yield patterns in the same fields.

Representative Distributions of the Yields by Cooperator and Fields

Wheat 97

Cumulative frequency distributions of wheat yield (Fig. 4) showed a difference in variation but no real difference in median yield between cooperators. Frequency distributions by field and cooperator (Fig. 5) show the reason for that result. There are 5 fields for Cooperator 2 in which the median yield is less than the

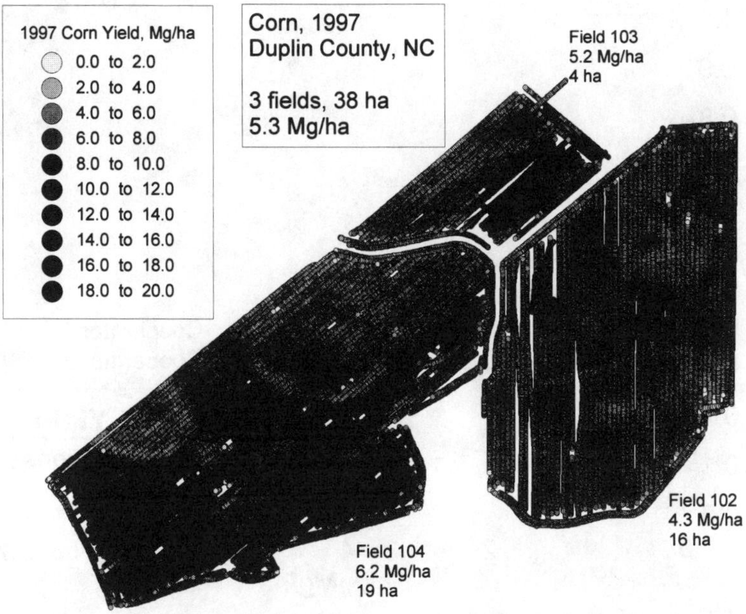

Fig. 2. Yield maps for three representative corn fields in Duplin County, NC.

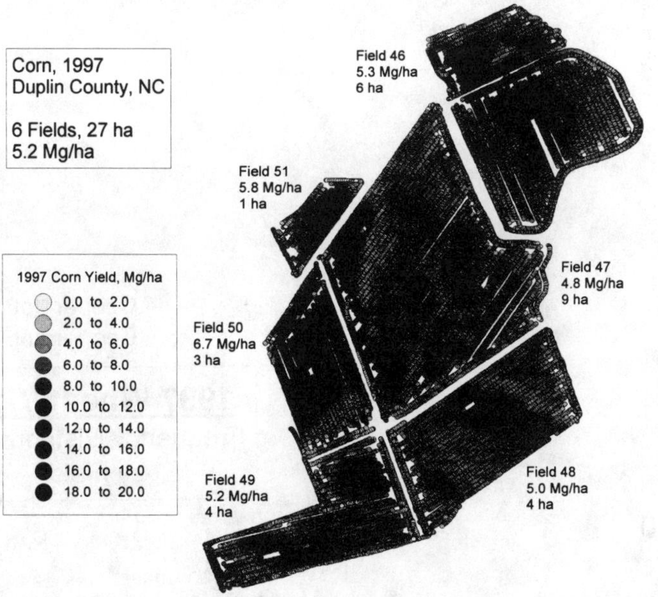

Fig. 3. Yield maps for six representative corn fields in Duplin County, NC.

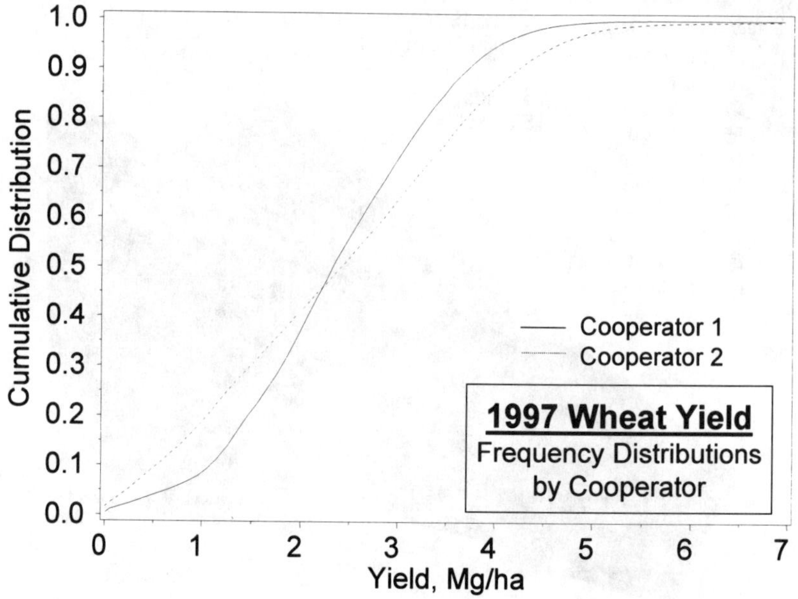

Fig. 4. Cumulative frequency distribution of 1997 wheat yield by cooperator.

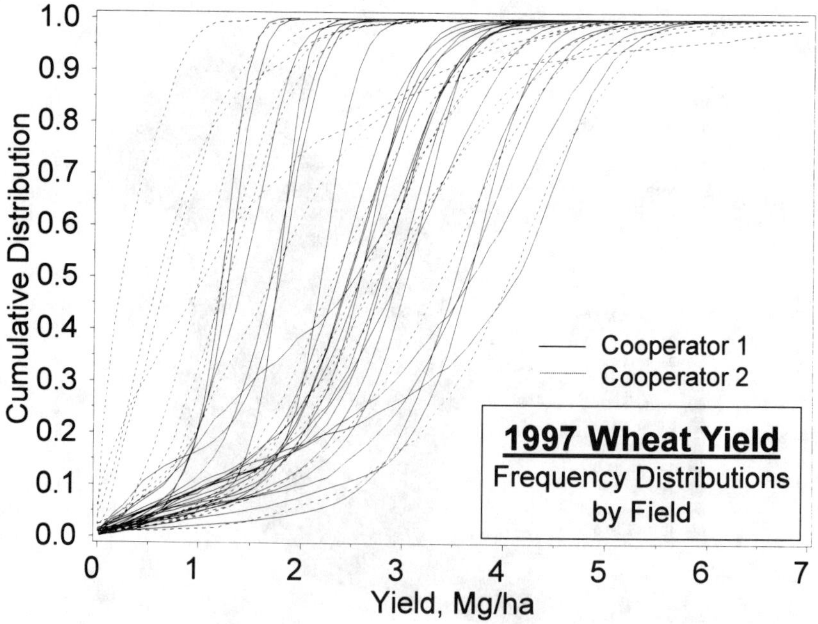

Fig. 5. Cumulative frequency distribution of 1997 wheat yield by field and cooperator.

minimum median yield for all fields of Cooperator 1. Distributions within most fields for Cooperator 2 have longer tails on the upper end, which offsets the five low-yielding fields.

Corn 1997

Similar results occurred for the 1997 corn yield data. The median for Cooperator 1 (Fig. 6) is approximately 0.5 Mg ha^{-1} higher than that for Cooperator 2, but the main difference was in the variation shown. Part of the reason for these differences came from the types of farming operations used by the two cooperators. Cooperator 1 primarily harvested rainfed fields in which he managed the crops, presumably uniformly. On the other hand, Cooperator 2 harvested additional fields for other growers, including several that were irrigated or used as spray fields for swine wastewater. The distributions by fields and cooperators (Fig. 7) supported this observation. The bulk of cooperator 1 fields clustered near the central portion of the graph, while those for Cooperator 2 fell into three groups. There were two fields with median yield less than 1 Mg ha^{-1}, a cluster with medians similar to those for cooperator 1, and five with substantially higher yields, all greater than 8 Mg ha^{-1}.

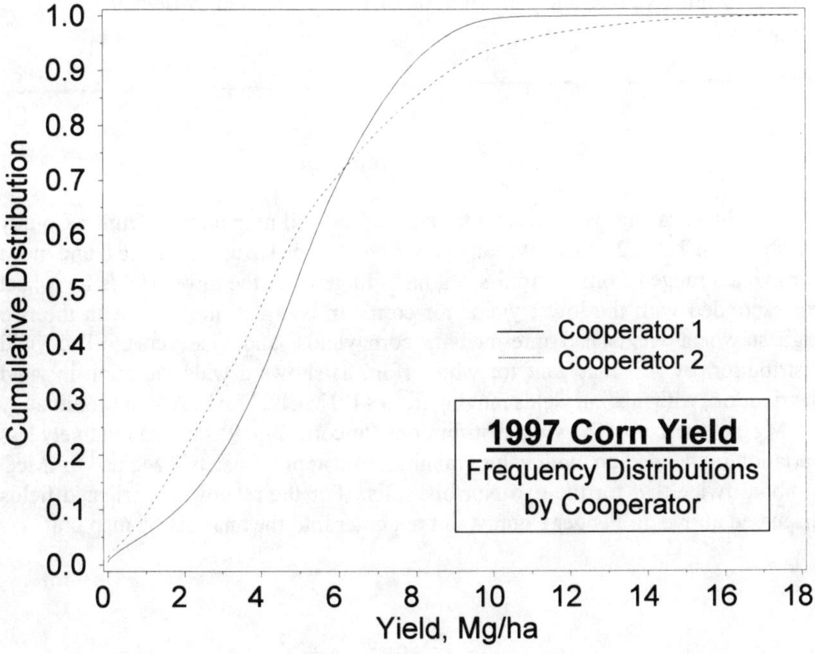

Fig. 6. Cumulative frequency distribution of 1997 corn yield by cooperator.

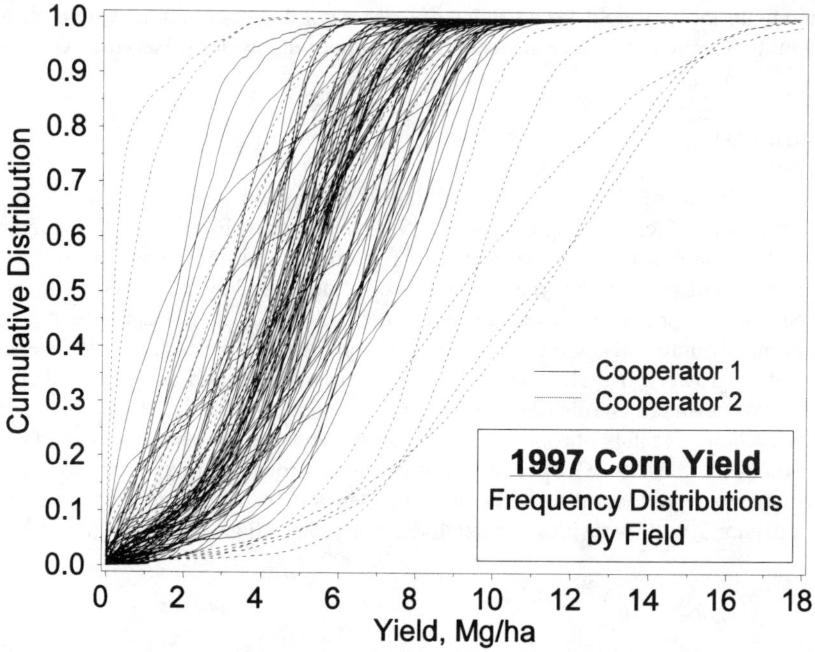

Fig. 7. Cumulative frequency distribution of 1997 corn yield by field and cooperator.

Analysis by Map Unit

The corn and wheat results for each of the soil map units in Duplin County are shown in Table 2. Mean wheat yields ranged from 1.2 to 3.5 Mg ha^{-1}, and mean corn yields ranged from 3.1 to 5.8 Mg ha^{-1}. In general, the lower yields for wheat corresponded with the lower yields for corn. In contrast, the soils with the two highest wheat yields had intermediate corn yields, and vice versa. The yield distribution by soil map unit for wheat (Fig. 8) shows a wide variation in yield distribution, with median yields ranging from <1.0 Mg ha^{-1} for LnA to approximately 3.5 Mg ha^{-1} for ToA. The yield distributions for corn (Fig. 9) showed relatively less variation in the median, with values ranging from approximately 3 Mg ha^{-1} for McC to about twice that for the two Norfolk soils. For the record, the irrigated fields discussed above, in Pender County, did not enter into the analysis by map unit.

Table 2. Means and standard deviations of yields, number of points, and area harvested of soil map units in Duplin County.

	Corn				Wheat			
	Yield	Std	N	Area	Yield	Std	N	Area
Soil	Mg/ha	Mg/ha	-	ha	Mg/ha	Mg/ha	-	ha
AuB	3.88	2.39	8141	9.96	2.16	0.87	17360	17.98
BbA	3.56	2.75	755	0.74	1.60	1.04	152	0.17
FoA	4.72	2.50	20496	23.87	2.57	1.08	20173	25.76
GoA	4.80	1.96	88101	109.36	2.83	1.04	78308	97.38
LnA	none				1.18	1.17	3148	8.39
LsB	3.92	1.80	1222	1.71	1.86	0.67	2487	3.21
McC	3.07	2.23	1209	1.53	1.38	0.62	3442	4.89
NoA	5.79	2.57	58749	74.42	2.18	0.95	44326	54.46
NoB	5.65	2.71	15647	18.96	2.48	0.88	19886	22.22
OrB	none				2.40	0.80	7272	6.95
PnA	3.98	2.31	608	0.65	2.53	0.80	8982	7.01
RaA	4.71	2.01	120560	146.35	2.31	1.02	80770	109.16
RuA	none				2.84	0.92	1887	2.25
RuB	4.42	2.39	727	0.76	none			
ToA	4.30	1.93	7098	17.65	3.53	1.01	3675	14.76
WoA	3.94	2.19	5841	9.65	3.08	1.16	8879	34.65
non-Duplin	4.89	2.82	240588	482.39	2.18	1.32	12308	48.72
Total	4.90	2.51	569742	898.02	2.46	1.06	313055	457.94
LSD(0.05)	0.15				0.07			

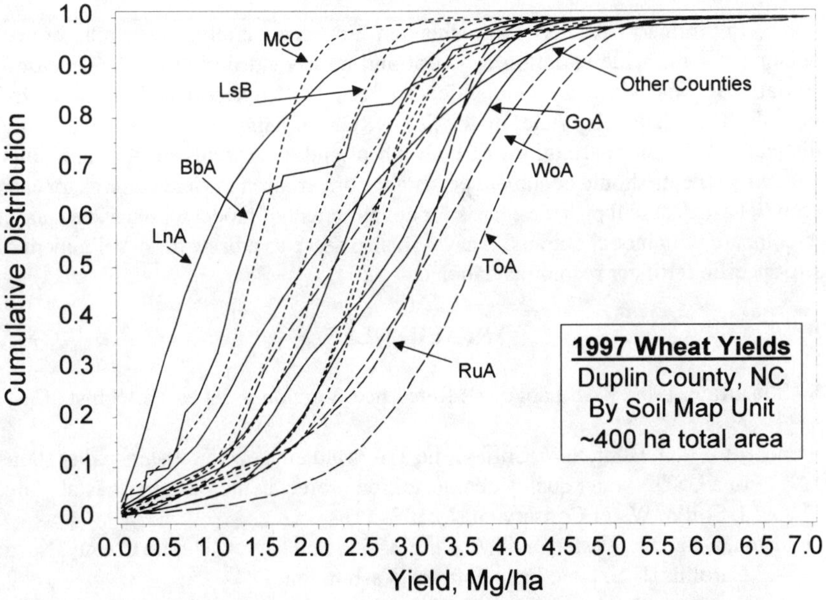

Fig. 8. Cumulative frequency distribution of 1997 wheat yield by soil map unit. The combined curve for all fields outside Duplin County, for which soil information is not available in digital form, is shown for comparison.

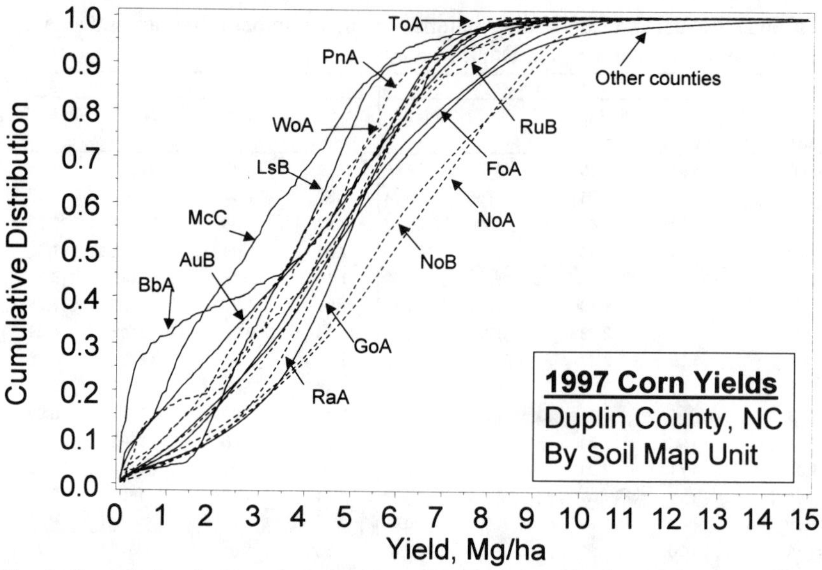

Fig. 9. Cumulative frequency distribution of 1997 corn yield by soil map unit. The combined curve for all fields outside Duplin County, for which soil information is not available in digital form, is shown for comparison.

SUMMARY AND CONCLUSIONS

Preliminary results from this on-farm cooperative research project documented the wide variation in wheat and corn yield during the 1997 season. Variation in yield within and among fields reflected differences in both operators and in soils. Correlation of yields to USDA-NRCS soil map units documented the distinctly different distributions of both wheat and corn yields on different soils. Second-year data should document persistence or variation in yield patterns for each crop. These data will provide a basis for development of model inputs and a dataset to compare with model outputs. They will also serve as a basis for development of site-specific fertilizer recommendations.

REFERENCES

SAS Institute. 1990. SAS Language: Reference. Version 6. 1st ed.. SAS Inst., Cary, NC.

Stone, K.C., P.G. Hunt, S.W. Coffey, and T.A. Matheny. 1995. Water quality status of a USDA water quality demonstration project in the Eastern Coastal Plain. J. Soil & Water Conserv. 50(5):567–571.

Goldston, E.F., D.L. Kaster, and J.A. King. 1958. Soil survey, Duplin County, North Carolina. U.S. Gov. Print. Office, Washington, DC.

Precision Farming: From Technology to Decisions A Case Study

Kim L. Fleming, and Dwayne G. Westfall
Department of Soil and Crop Science
Colorado State University
Fort Collins, Colorado

Dale F. Heermann
USDA-ARS
Fort Collins, Colorado

D. Bruce Bosley
Cooperative Extension
Colorado State University
Fort Collins, Colorado

Frank B. Peairs
Department of Entomology
Colorado State University
Fort Collins, Colorado

Philip Westra
Department of Weed Science
Colorado State University
Fort Collins, Colorado

ABSTRACT

Computers, global positioning systems (GPS), and geographical information system (GIS) technology are enabling farmers and agricultural professionals to enter a new era in farm management, often referred to as precision farming. The objectives for this case are to evaluate the different approaches to precision farming of a farmer and fertilizer consultant and determine how each might improve their utilization of the new technology. The case will follow a fertilizer consultant's decisions on how best to position himself and his business to provide precision farming services. We will examine the interaction of the consultant and the farmer as they work to develop a precision farming program. Through evaluating the management approaches followed by these two businesses, others beginning to adopt precision farming technologies can follow a more effective decision making process.

INTRODUCTION

In the Fall of 1995 Dan Wiens a fertilizer consultant in northeast Colorado was challenged by a customer. Earlier that year Larry Rothe, an area farmer tried a variable rate (VRT) fertilizer program based on grid soil sampling on 2.5 acre grids.

Copyright © 1999 ASA-CSSA-SSSA, 677 South Segoe Road, Madison, WI 53711, USA. *Proceedings of the Fourth International Conference on Precision Agriculture.*

He was disappointed with the application maps because he felt they failed to capture the hills and other areas in the fields where variable rate application was needed. In addition, the program did not provide the flexibility to modify the application map based on criteria beyond grid sampling. Larry needed some way to more accurately identify regions in his fields where variable rates were needed. During this same time much of the research in variable rate application was indicating the cost of grid sampling to the intensity needed to create accurate maps was prohibitive (Hammond, 1992; Franzen & Peck, 1994; Gotway et al., 1996). This was the motivation behind the development of Dan's VRT program.

Dan's Background

Dan grew up on a small cotton, alfalfa, and corn farm near Madera, CA. He graduated from California Poly with a B.S. in Agriculture Business Management in 1970 and earned his MBA from Golden Gate University in 1973, he also is a Certified Crop Advisor. He worked as a sales representative for Uniroyal Chemical from 1970 to 1977. In 1975 he was transferred to Fort Morgan, CO. In 1977 he started his own fertilizer business called Morgan Soil. During this time he purchased a 200 acre irrigated farm that he continues to manage today. He later sold the fertilizer business to Centennial Ag and has worked for them until now as a fertilizer consultant. Dan has always had a love for computers, buying his first one in 1980, an Apple II. Dan comments, "I've always wanted to tell computers what to do and not vice versa."

Larry's Background

Larry was raised on the family farm north of Greeley, CO. He graduated from Colorado State University in 1970 with a B.S. in Farm and Ranch Management. From 1970 to 1972 he worked for his father and uncles on the farm. In 1973, he took a position with First Western Bank in Sacramento, CA. After a year he returned to Colorado and began farming near Wiggins. He now farms 3000 acres of irrigated corn, sugar beets, and onions. He bought his first computer in 1980 and learned how to program, developing a small database program. He saw the potential for field and crop management early on and has constantly been an early adopter of new technologies.

THE CASE

In early January 1996 as Dan discussed their dilemma with developing a viable VRT program with Larry, his eyes came to rest on an aerial photograph of Larry's farm. He had seen the photo many times before but this time saw it in a new context. As he studied the variability in soil color he asked Larry, "why can't we use this to develop our variable rate application maps?"

Later that spring Dan consulted with Dr. Dave Wagner a remote sensing specialist from Colorado State University. Dr. Wagner described the work he had been doing with bare soil aerial imeragy. His work indicated bare soil photos could be effective in identifying areas of high and low soil organic matter (SOM). Other research also indicated the grey tone pattern in black and white aerial photos is often a reflection of properties that impact crop yield (McCann et al., 1996). Bhatti et al. (1991) found SOM can be effectively determined from remote sensing and is significantly correlated with yield. Dan later contracted with Dr. Wagner to photograph three of Larry's fields. Dr. Wagner introduced Dan to several types of image processing software to enhance, georeference, and analyze the aerial photos. Larry felt the photos could be very effective in identifying regions in his fields for variable rate application.

During the winter of 1997, Dan and Larry began examining ASCS photos to determine if they could be effective in identifying variation in soil color. They found that generally they were taken too late in the season to indicate differences in soil color. Photos taken when the fields were in sugar beets, however, tended to reveal the differences in soil color, due to their later canopy cover. After examining several years of photos they were able to identify effective images for most of Larry's farms.

Dan used Adobe Photoshop to enhance and process the images, he then georeferenced the images in a GIS package. Using these images as a template Larry drew vector lines on the maps indicating areas of high, medium, and low productivity based on soil color, topography, and his past management experience on the fields. These early photos and regions represented their first attempt at generating farmer developed management zones for variable rate application (Fig. 1).

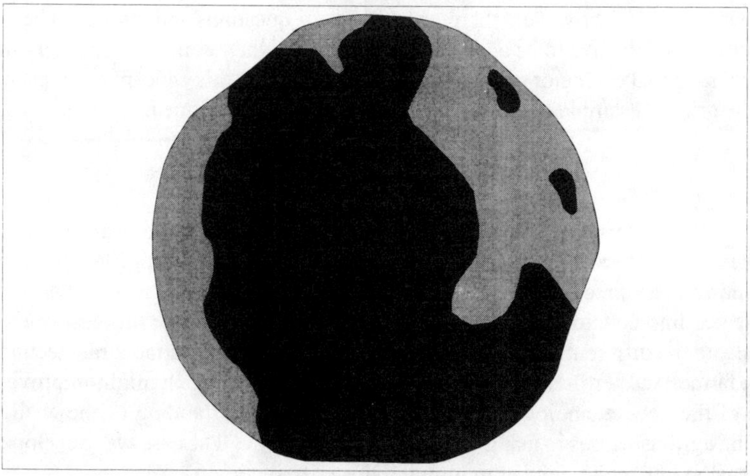

Fig. 1. Farmer developed management zones in center pivot irrigated corn field; (Low productivity = light gray, medium productivity = dark gray, and high productivity = black).

During this same time frame Dan installed variable rate controller equipment on one of his company's Terragator liquid fertilizer applicators and developed the software to vary rates based on the management zones. A single liquid fertilizer mix was formulated to fit each field and was variably applied. As much as 40% more was applied to the lightest color soils and 40% less on the dark color soils. Using this method Dan applied variable rate fertilizer on 2500 acres of corn, sugar beets, and onions on Larry's farm. Larry and Dan estimated an increase in net returns of about $15.00 to $20.00 per acre in some cases. However, since no formal trials or check treatments were applied this was only an estimate.

Beginning in the 1997 growing season Larry also became a cooperator on a field scale precision farming project with Colorado State University and the USDA-ARS (Exhibit 1). Dan is very practical minded businessman and tends to cast a skeptical eye at much of the research being done in academia. He felt as some do, that industry is always one or two steps ahead of university research. Larry feels any advantage he can find will help position himself better for the major changes in agriculture expected during the next few years and having early access to the data being generated out of the project may be one more advantage on his side. Kim Fleming, project manager for the research group was intrigued by their work and asked if he could correlate the project data with the management zones (Fleming et al., 1998). Fleming felt that the application of precision farming technologies has raced ahead of research into its value and saw this as an excellent opportunity to work with industry in determining the value of the technologies. Although skeptical of their work, Dan felt this could give him additional information to validate the management zones. Many of the soil and crop parameters being measured in the project followed the trends indicted by the management zones (Exhibit 2).

Dan needed to decide what his next step was. The management zone concept showed promise, however, there are still many questions remaining. These and other issues will have to be addressed to determine whether it will be prudent for his business to make a major investment to further develop this concept or invest in one of the grid soil sample based VRT programs already developed.

TEACHING NOTE

Computers, GPS, and GIS technology are enabling farmers and agricultural professionals to enter a new era in farm management. The challenge facing everyone adopting these precision farming methods is turning the massive databases being collected into better management decisions. The objectives for this case are to: (i). Evaluate the different approaches to precision farming and variable rate technology of a farmer and fertilizer consultant; (ii). Determine how each might improve their use of the new technology; and (iii). Develop an understanding of the challenges facing agribusinesses in using variable rate technology. The case was developed for extension, university, and agricultural industry trainers.

Notes to Facilitators

Debate over the value of grid soil sampling and the intensity needed to develop accurate maps is expected. The key will be using the synergy produced from differing viewpoints to develop constructive solutions to the complex problem Dan faces. Some key questions I would hope that would be raised are:

1. How effective will other farmers be in developing the management zones based on their aerial photos and experience. Larry felt very comfortable with

the process. Using farmer input helps give him ownership and control of the program; however, Larry is an early adaptor, some producers may feel intimidated by the technology and prefer more of the control to be in the hands of the consultant.

2. How practical is it to get aerial photos of the quality needed to develop the management zones on a wide scale basis and train people to properly enhance, georeference, and analyze the photos?

3. The results from the analysis of the data generated from the research project look positive, however will the concepts apply to other fields and other parts of the country?

4. What has other research in this area indicated? The papers referenced in the case could be provided to the students and in some teaching environments a literature search may be an appropriate exercise.

5. What other types of remote sensed data may be effective in developing the zones?

An important component that is missing that Dan needs to use to make valid decisions are on farm trials comparing his management zones with grid soil sampling approaches to variable rate application. Developing the experimental design and protocol for such a trial could be another valuable outcome from using this case. Depending on the depth of analysis desired the discussion of the case could last from an hour to a full day.

REFERENCES

Bhatti, A.U., D.J. Mulla, and B.E. Frazier. 1991. Estimation of soil properties and wheat yields on complex eroded hills using geostatistics and thematic mapper images. Remote Sens. Environ. 37:181–191.

Fleming, K.L., D.G. Westfall, D.W. Weins, L.E. Rothe, J.E. Cipra, and D.F. Heermann. 1999. Evaluating farmer developed management zone maps for precision farming. Proc. 4th Int. Conf. on Precision Agriculture. ASA Misc. Publ., ASA, CSSA, and SSSA, Madison , WI.

Franzen, D.W., and T.R. Peck. 1994. Sampling for site specific management. p.535–551. *In* Proc. 2nd Int. Conf. on Site-Specific Management for Agricultural Systems. ASA, CSSA, and SSSA, Madison, WI.

Gotway, C.A., R.B. Ferguson, and G.W. Hergert. 1996. The effects of mapping scale on variable-rate fertilizer recommendations for corn. p.321–330. *In* Proc. 3rd Int. Conf. on Site-Specific Management for Agricultural Systems. ASA, CSSA, and SSSA, Madison, WI.

Hammond, M.W. 1992. Cost analysis of variable fertility management of phosphorus and potassium for potato production in central Washington. p. 213–228. *In* P.C. Robert et al., (ed.) Soil Specific Crop Management. ASA, CSSA, and SSSA, Madison WI.

McCann, B.L., D.J. Pennick, C. van Kessel, and F.L. Walley. 1996. The development of management units for site-specific farming. p.295–302. *In* Proc. 3rd Int. Conf. on Site-Specific Management for Agricultural Systems. ASA, CSSA, SSSA, Madison, WI.

APPENDIX

Exhibit One: The USDA ARS/Colorado State University Precision Farming Project

Multi-disciplinary field scale research is needed in precision farming. Colorado State University and the USDA Agricultural Research Service Water Management Unit (ARSWMU) in Fort Collins have assembled a multi-disciplinary team to assess the technical and economic feasibility of precision farming. This project involves 18 scientist, two farmers, three extension specialists, and four graduate students working on two center pivot irrigated fields near Wiggins, CO. Disciplines involved include soil fertility, crop science, weed science, entomology, plant pathology, system engineering, remote sensing, GIS, irrigation engineering, agricultural economics, and statistics.

Our initial goal, before precision farming treatments are applied is to complete two years of intensive and coordinated data collection and analysis. With these data we can begin to identify and quantify the factors contributing to yield variability under sprinkler irrigated conditions. In addition, through intensive data collection we will develop and evaluate various aspects of precision farming sampling methods and strategies, along with analysis techniques. Ultimately we hope to develop models to predict the effect of spatial variability on the profitability of precision farming. The level of spatial variability within a field is in itself variable, thus increased production and savings in input cost from site specific management should also vary. We then hope to incorporate those models into a decision support system we feel is essential in moving precision farming from the early adopter phase into the mainstream farming community.

Exhibit Two: Management Zone Evaluation

Soil organic matter, NO_3–N, K, Zn, and corn yield followed the trends indicated by the management zones with the highest nutrient and yield in the high productivity zones, intermediate levels in the medium zones, and lowest levels in the low productivity zones. Levels of SOM, K, and yield were significantly different in all management zones. Nitrate levels were significantly different between the high–low and high–medium zones while Zn levels were significantly different between the high and low productivity zones. Phosphorus did not follow the trends seen with the other nutrients, with higher levels seen in the lower productivity zones and lower levels in the high productivity zones. Perhaps the immobility of P would account for this. The lower produtivity areas would tend to use less phosphorus resulting in a build up in these areas. These data would indicate the method used to determine the management zones were effective in identifying areas of high, medium, and low productivity.

Results from the texture data show similar trends. Clay and silt levels were significantly higher in the high productivity zone, intermediate in the medium zones, and lowest in the low productivity zones while sand followed the opposite trend. These results also validate the management zones. These soils are quite sandy overall and higher productivity in areas of less sand and higher clay levels would be expected due to the higher water holding and cation exchange capacity. Differences were also significant across all management zones for both the EM38 and Veris 3100 conductivity data.

To determine why corn yields follow the trends indicated by the management zones all parameters were correlated with yield. SOM, % sand, silt and clay, EM 38, and Veris 3100 conductivity data were all significantly correlated with yield. High SOM, % clay, and conductivity would result in higher sources of nutrients such as NO_3 and K, higher cation-exchange capacity, and higher water holding capacity all resulting in higher yields. The % sand was negatively correlated with yield.

SUMMARY

Grid soil sampling to the intensity required to generate accurate variable rate application maps may not be feasible because of the time and expense required. Determining fertility management zones may be a more economical method of developing variable rate application maps. It's well documented that landscape position correlates well with soil parameters and crop yield. It's equally well documented that soil color correlates with SOM and this relationship can be captured in aerial photos. Producers have knowledge of which parts of a field produce good yield and which are low in production. This allows for the identification of different management zones based on past production history. Farmer developed management zones based on these relationships were compared to soil nutrient, texture, conductivity, and crop yield. These soil and crop parameters followed the trends indicated by the management zones. Our initial analysis indicates this method of developing management zones may be effective in developing effective variable rate applications maps.

Decision Making Strategies for Fertilizer Use in Precision Agriculture

S. Haneklaus, D. Schroeder, and E. Schnug

Institute of Plant Nutrition and Soil Science
Federal Agricultural Research Center
Braunschweig, Germany

ABSTRACT

Precision Agriculture will be an integrative part of agricultural farm management practices in future and the exploitation of this tool for fertilizer use will highly depend on the decision making strategies used for the calculation of variable rate fertilizer applications. Fertilization with basic nutrients such as P, K, or Mg traditionally relies on soil analysis. Geocoded grid sampling or directed sampling procedures provide information about spatial variability of nutrient contents in soil. The local fertilizer requirement can easily be determined employing the BOLIDES software which sets geocoded soil analysis data in relation to geocoded yield data. Geocoded yield maps offer the possibility to calculate the nutrient demand on the basis of nutrient removal of each crop and whole crop rotations, respectively and thus follow the concept of balanced fertilization. The optimization of the fertilizer rate for highly mobile nutrients such as nitrogen preferably goes along with the spatial variability of key variables as the available nitrogen content in the soil is extremely variable in space and time. The possible benefits of variable rate fertilization include a locally higher productivity, improvement of yield stability, decreased risk of lodging, improved crop quality, reduced nutrient losses to the environment and economic savings.

INTRODUCTION

Facing a world population of almost 6 billion it is evident that it is an enormous responsibility of agriculture to feed everyone and fertilization plays a key role in this task. At the same time environmental demands on agricultural production are increasing worldwide and are closely intertwined with the quality of foodstuff, soils and drinking water. Thus fertilization not only affects the aspect of food security, but also environmental protection. One major problem fulfilling both requirements equally is the ubiquitous variability of soil fertility features together with uniform treatment of fields causing an inefficient factor utilization and unnecessary environmental impacts. Though the technology for variable rate output of nutrients in order to address the small scale variability of soil parameters is available (Clark & McGuckin, 1996; Schroeder et al., 1997), fertilizer strategies which satisfy economic and ecological demands need to be elaborated and verified (Haneklaus et al., 1996). This implies the withdrawal from traditional

Copyright © 1999 ASA-CSSA-SSSA, 677 South Segoe Road, Madison, WI 53711, USA. *Proceedings of the Fourth International Conference on Precision Agriculture.*

fertilizer customs in favor of new sophisticated approaches. In this paper different decision making strategies for fertilizer use in precision agriculture will be presented and evaluated. Fertilizer strategies based on on-line sensors for the determination of soil features (Lui et al., 1996; Birrell & Hummel, 1997) or different procedures of remote sensing (Panten et al., 1998) are not discussed as they have not become practical options yet.

MATERIALS AND METHODS

On an experimental farm in Mariensee near Hanover, Lower Saxony (52° 33' N; 9° 28' E) new fertilizer strategies have been tested since 1996. Basic soil data and results of geostatistical analysis are given by Haneklaus et al. (1997). Allocation of and navigation to sampling positions were carried out by a differentially corrected Global Positioning System (DGPS).

Soil analysis: Top soil samples (0 – 0.2 m) were taken in a 30∗30 m grid, airdried, and sieved at 2-mm. Available phosphate (P_2O_5) contents were determined in the CAL extract according to Schueller (1969). The plant available $N_{min.}$ content (0–0.9 m) was determined according to Anonymous (1991).

Plant analysis: The whole above ground mass of oats was taken at beginning of stem elongation (growth stage GS 32; Zadoks et al., 1974). The plant material was dried at 85° C in a ventilated oven until constancy of weight and then fine-ground to a particle size of <2 μm using an ultracentrifugal mill. The total N content was determined according to Kjeldahl.

For geostatistical analysis the VARIOWIN software package (Pannatier, 1994) was used and Digital Agro Resource Maps were established by the LORIS software package (Schroeder et al., 1997).

RESULTS AND DISCUSSION

Strategy for Variable Rate Fertilization Based on Soil Analysis

Traditionally, fertilizer recommendations for the major plant nutrients P, K, Mg, and lime rely on soil analysis data. Federal agricultural laboratories still recommend the mixing of 15 to 30 sample cores per hectare (Anonymous 1998) in order to evaluate the nutrient status of the soil. But whatever procedure for collecting these sample cores is selected (Finck, 1979), it interferes with the variability of soil features (Fig. 1) and therefore cannot reflect the local nutrient supply. In the example given in Fig. 1 the available phosphate content varied between 6 and 32 mg $100g^{-1}$ in the single test plots, whereby especially the sampling procedures B-D cover spatial differences. This strengthens the premise that representative samples firstly should be taken within an area of not more than 10 m^2 and secondly have to be geocoded in order to transfer the small-scale heterogeneity of soil features into variable fertilizer rates. Additionally temporal changes in the nutrient content can only be proved when samples are georeferenced.

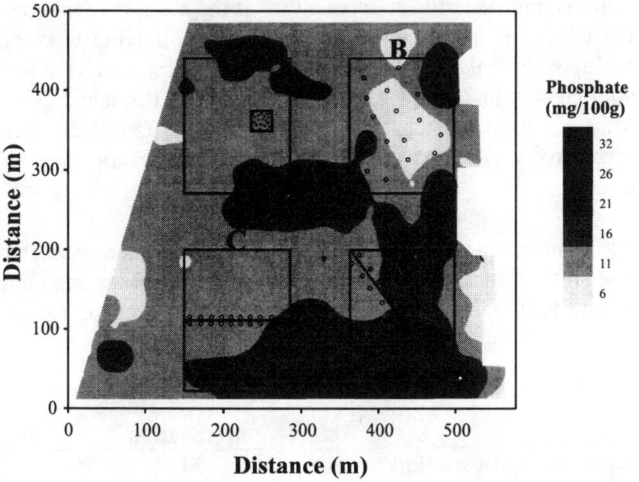

Fig. 1. Influence of the spatial variability of available phosphate contents in soil on different sampling procedures (A-D) proposed by Finck (1979).

Soil samples need to be taken in a density where spatial correlations can be determined in order process them to Digital Agro Resource Maps (DARMs; Haneklaus et al., 1997). Geocoded grid soil sampling and analysis as carried out in Mariensee is, however, not operational on farm level (Haneklaus et al. 1997; Schnug et al. 1998). Sampling efforts can be reduced efficiently employing the equifertile concept. Equifertiles can be derived inductively or deductively (Schnug et al. 1994). Equifertiles are areas in the field with identical or similar productive capacity in space and time. Within equifertiles permanent monitoring points are located which reflect temporal changes in the corresponding equifertile areas and thus provide information about causing factors for differences in productivity. If these points are chosen for directed sampling, they reflect the variability of soil parameters, like available nutrient contents in soil without losing the spatial correlation. A detailed description of directed sampling is given by Schnug et al. (1998).

For the determination of the spatial variability of key variables such as organic matter content, soil texture or geomorphology, the 'self-surveying' procedure was developed in order to acquire this information economically (Haneklaus et al. 1996, 1998). The relevance of this procedure for the development of fertilizer strategies for variable N applications and the possibility of a total relinquishment of soil analysis in favor of fertilization according to nutrient removal is discussed below.

A uniform fertilizer application of major basic nutrients such as P, K and Mg will be associated with an undesired over and under supply with the nutrient in many areas of the field when compared with fertilization based on soil analysis data together with standard fertilizer recommendations. Variable rate fertilization will result in a higher fertilizer expenditure, if the mean nutrient content in the soil is in the range of optimum supply or higher. Following this approach, lower fertilizer input can only be expected, if the mean nutrient content is in the range of severe deficiency. In Table 1 fertilizer expenditure, economic expenses and area related share of fertilizer rates according to phosphate requirement are summarized for variable rate fertilizer application in comparison with a uniform treatment.

Table 1. Comparison of fertilizer expenditure, economic expenses and area related shares of fertilizer rates according to phosphate requirement between variable rate fertilizer application and uniform treatment for a field (11.7 ha) in Lower Saxony.

	Variable rate application	**Uniform application**
Fertilizer expenditure (kg)	748	703
Mean application rate (kg ha^{-1})	64	60
Economic expenses (rel.†)	106	100
Share of fertilizer rates acc. to nutrient requirement (%)	100	70

†100% = uniform application rate.

The phosphate contents in the soil ranged from 11 to 51 mg 100g^{-1} with a mean value of 18.5 mg 100 g^{-1} which is in the range of optimum supply. The results in Table 1 further reveal that over and undersupply with phosphate results in an inefficient use of resources as 30% of the area of the field would receive an application rate above or below the demand.

Fertilizer application rates based on soil analysis regularly go along with recommendation data supplied by the laboratory. The basic fertilizer response curves are, however, restricted to a limited number of field trials under varying soil and climatic conditions. In comparison, the technical tools of Precision Agriculture such as geocoded on-line yield measurements and variable rate application techniques offer the possibility to carry out field experimentation on a large scale basis with a whole field as one large experimental site and with a much higher number of replicates than usual in field application (Schroeder & Schnug 1995). For crop fertilization, the full exploitation of the soil analytical techniques is hindered by the lack of accepted calibration to categorize soils according to the degree to which crop growth is limited by their nutrient status. This can be overcome using the BOLIDES (Boundary Line Development System) software by which field related critical values for soil or plant analysis can be derived (Schnug et al., 1995, 1996). An example for the evaluation of the critical phosphate range of a field in northern Germany is given in Fig. 3.

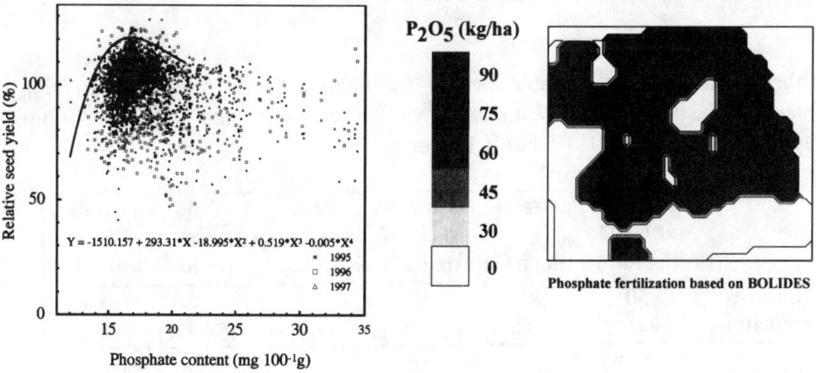

Fig. 3. Evaluation of the optimum range of phosphate contents in the top soil by BOLIDES for a field (11.7 ha) in Lower Saxony.

The results reveal that the optimum range is 15 to 19 mg $100g^{-1}$ P_2O_5 in order to achieve maximum yields. The results have been consistent during 3 yrs of yield mapping. Based on the results of BOLIDES only those zones in the field will be fertilized where the nutrient content is sub-optimum. Thus >65% of the fertilizer expenditure and costs can be saved in comparison with variable rate application based on official fertilizer recommendation and 63% in comparison with a uniform treatment of the field (Table 1). Facing limited natural resources of phosphate which are estimated to satisfy agricultural demands for only another 100 yrs (Anonymous, 1996), decicion making strategies for fertilization should promote a careful use of these resources.

In the case of N fertilization, the avoidance of undesired N losses to the environment is of special interest. In the following example the N fertilization was split into two applications with an uniform rate of 50 kg ha^{-1} at GS23 and 30 kg ha^{-1} at GS32. The first variable N fertilizer rate was calculated dependent on the variation of the $N_{min.}$ contents in the soil directly before application (Table 2). The second variable N rate was omitted as the $N_{min.}$ contents ranged from 45 to 200 kg ha^{-1} N and thus were sufficient to satisfy the demand of the crop. The influence of the spatial and temporal variability of $N_{min.}$ contents in soil on the suitability as a criterion for the N demand will be discussed later on. The efficiency of the variable rate application was proved by analyzing plant tissue for N at the main growth stage and geocoded yield mapping at harvest time (Table 2).

Nitrogen savings ranged from 46 to 71% (Table 2). This approach is therefore supposed to have a strong environmental benefit expressed in reduced N losses to the environment. The strategy yielded no statistically significant differences in the plant N status at stem elongation in comparison with the uniformly treated area of the field. This proves that the variable rate provided an adequate amount of nitrogen. A spatial correlation of $N_{min.}$ contents was verified for the first sampling date at GS23 (Fig. 2), however not for the second one at GS31.

Table 2. Influence of variable rate N fertilization based on the mean $N_{min.}$ contents in the soil on plant N status and yield of oats in comparison with uniform N application on a field (11.5 ha) in Lower Saxony.

	$N_{min.}$ at GS23 (kg ha^{-1})	N-rate (kg ha^{-1})	Plant N at GS 32 (mg g^{-1})	Yield‡ (dt ha^{-1})	Yield§ (dt ha^{-1})	Area (ha)
Uniform application	50 (29–102)	80†	26 (19–36)	62 (42–75)	76 (54–97)	3.1
Variable rate application	55 (28–100)	23–43	28 (15–40)	75 (50–109)	70 (51–98)	8.4

†Calculated N-fertilizer demand = 80 kg ha^{-1} with 50 kg ha^{-1} at GS23 + 30 kg ha^{-1} at GS32. ‡Manual yield determination (1 m^2) at sampling locations. §Geocoded on-line yield monitoring.

Variable rates of N resulted in distinctively higher yields when yield was determined manually and exactly at the locations chosen for soil and plant sampling. In contrast, the range of yield employing a yield monitoring system was higher where a uniform amount of fertilizer was applied. This is probably related to the following factors: the missing spatial correlation of $N_{min.}$ contents at the second sampling date shows that plant available N varies over distances shorter than 30 m. It is well known from sulfate which is as mobile as nitrate in soil, that sampling distances of ≤25 m are required to find a spatial correlation (Schnug & Haneklaus, 1998). Thus insufficiently supplied zones remained hidden and could not verify their yield potential. Another reason that contributed to the yield differences is simply the fact that the uniformly treated area is in the middle of the field so that side effects by surrounding hedges only affected crop growth under variable fertilization.

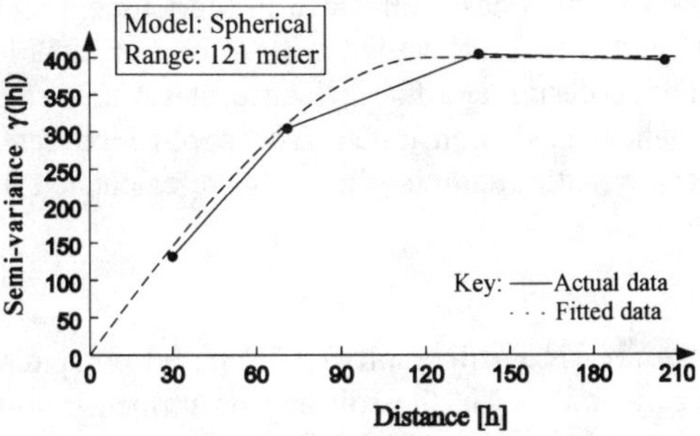

Fig. 2. Variogram for $N_{min.}$ contents in the soil on a field in Lower Saxony.

As already mentioned above there are several problems in using the spatial variability of $N_{min.}$ contents in the soil as a basis for the fertilizer recommendation which will prevent practical use. These are high sampling density, high temporal fluctuance and high analytical costs. Even with sampling distances of 30 m, spatial correlations need not necessarily be found so that the local fertilizer demand cannot be prognosed with sufficient accuracy. The fact that spatial correlations could be verified only at the first sampling date (Fig. 2) stresses that results cannot be transferred to other sampling dates within the vegetative period or years.

Strategy for Variable Rate Fertilization Based on Long-Term Stable Soil Features

Variable rate N fertilization is of utmost interest from an ecological and productive viewpoint, but the prognosis of plant available N is difficult because of its high spatial and temporal variability (see above). Basic long-term variable parameters such as texture, organic matter content, and landscape geomorphology highly influence the productivity of soils (Franko, 1996; Kersebaum, 1996; Richter, 1996; Thompson & Robert 1995). Knowledge about the spatial variability of these key variables offers the opportunity to establish equifertiles (see above) and is essential not only for the prognosis of nutrient dynamics but also for directed soil sampling in order to acquire information about medium term variable (e.g., available P) or short-term variable (e.g., NO_3) factors with reduced sampling effort. At the moment no practicable approach is available to gather information efficiently about time constant soil parameters via external sources. The self-surveying approach, which combines GPS navigation and positioning with human sensory capabilities, efficiently provides geocoded information of time constant soil parameters such as geomorphology, organic matter and clay content in optional density (Haneklaus et al., 1998). A practical approach in using differences in landscape geomorphology for the varibale output of nitrogen is reported by Haneklaus et al. (1996).

Since 1996 variable rate N fertilization based on the spatial variability of the organic matter and clay content has been tested in Mariensee. Fields were divided into two equal parts and the N rate varied with a total of 20% less N input. In Table 3 the experimental design and the influence of variable rate N application on yield of winter wheat is demonstrated. The application rates for all other combinations of organic matter and clay in soil were calculated in the form of a mirror-inverted matrix.

Table 3. Reduced variable N fertilization (-20%) based on spatial variability of organic matter and clay contents in the soil and its influence on yield of winter wheat on a field in northern Germany (11.7 ha)

Organic matter	**Clay content**	**Variable N rate**[†]	**Yield**
	Relative values (%)		(t ha^{-1})
0 - 17	0 - 17	93-100	no data
17 - 34	17 - 34	86-93	11.1
34 -51	34 -51	79-86	11.6
51 - 68	51 - 68	72-79	11.6
68 - 85	68 - 85	65-72	10.7
> 85	> 85	58-65	11.1

[†]100% corresponds with the average uniform application rate of 130 kg ha^{-1} N.

This approach can be used in fields or landscapes with varying contents of both parameters whereby the number of classes should be chosen according to the range of variation. The N rate can be designed individually for each combination of classes with N savings in optional amounts. In the example given in Table 3 the variable rate N application resulted in higher yields with an average of 11.2 t ha^{-1} in comparison with 10.9 t ha^{-1} in the uniformly applied area of the field. During a 2 yrs period yields kept stable following this concept on different fields and with the full environmental benefit of a 20% reduced N burden. It will be the responsibility of each farm manager to elaborate the best model in the sense of sustainable agricultural production.

Strategy for Variable Rate Fertilization Based on Balanced Fertilization

Schumann et al. (1997) summarize the goals and problems of balanced fertilization as follows: nutrient balances should be applied in future on the farm or field level to control and influence nutrient surpluses. Farms with sole cash crop production have lowest (30–70 kg ha^{-1} N; 35 kg ha^{-1} P) nutrient surpluses compared with pure animal production (130–250 kg ha^{-1} N; 90 kg ha^{-1} P). Nutrient surpluses not only depend on the total amount of nutrients applied but in fact on any factor affecting biomass production. Although this explains the high variability of nutrient surpluses, the main summarizing factor determining surpluses is the crop type. The reasons for this are differences between the physiological nutrient demand of the plants and that, which is removed from the field with harvest products. Thus oilseed rape and sugar beet crops (the last excluding use of leaves for animal nutrition) always show higher nutrient surpluses than, for instance, cereal crops. This leads to the conclusion that figures for acceptable nutrient surpluses need to be defined in the first place for individual crops or rotation types rather than for soil quality parameters or the risk of seepage or erosion. The time span over which balances are made should consider the ability of following crops and catch crops to make use of the nutrient surpluses of previous crops.

From this background, variable K application rates were calculated on the basis of the nutrient removal of two winter barley and one *Faba* beans crop during 3 yrs and the nutrient removal of these three crops (Fig. 4). The implementation of new yield sensors for non-grain crops such as sugar beet or potatoes (Walter et al., 1996; Schneider et al., 1997) will enable the farmer to calculate fertilizer rates based on nutrient removal of the whole crop rotation.

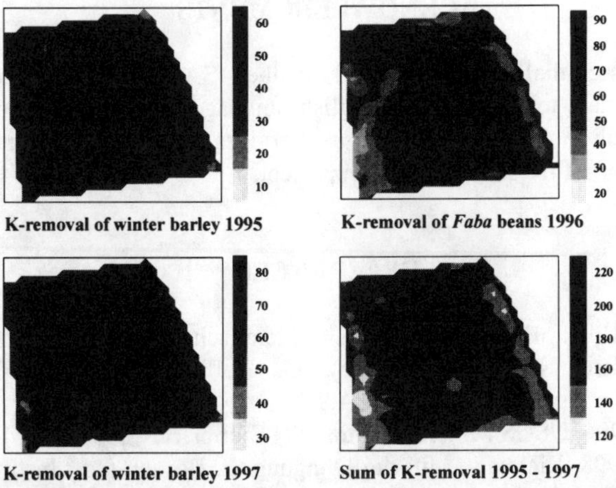

Fig. 4. Variable rate K fertilization based on the spatial variability of nutrient removal of winter barley in 1995 and 1997 and *Faba* beans in 1996 compared with K removal of all three crops on a field (11.7 ha) in Lower Saxony.

The K removal reflects the annual variability of yields (Fig. 4). Consistently over all 3 yrs, the headlands show the lowest yield potential and K removal, respectively. Following this approach it is possible to verify a balanced nutrient management. But it is not possible to evaluate optimum nutrient ranges in order to reduce zones with over-supply and adjust zones with under-supply in the field.

CONCLUSIONS

Decision making strategies for fertilizer use in agriculture traditionally rely on soil analysis. Disadvantages of this approach are that the spatial variability of soil nutrient content is not related to nutrient removal and that general fertilizer recommendations do not reflect local demand. If the strategy is based on the concept of balanced fertilization, an oversupply with nutrients can be avoided, but a yield limiting undersupply may remain hidden without accompanying soil analysis. A field based determination of optimum nutrient ranges by BOLIDES offers the possibility for a variable rate output according to local requirements.

Especially for N, a variable rate application strategy that follows the spatial variability of long term stable key soil features proved suitable in order to adjust rates properly according to the local demand. Using this approach an average of 20% of fertilizers was saved annually for the benefit of both farmer and environment.

ACKNOWLEDGMENTS

The authors cordially thank Dr. K.C. Walker (SAC, Aberdeen) for his never ending willingness to improve the English language of our papers. Furthermore the authors wish to thank KEMIRA Agro Services s.a. (Wavre, Belgium) for the financial support of the LORIS research project.

REFERENCES

Anonymous 1991. Bestimmung von mineralischem (Nitrat-) Stickstoff in Bodenprofilen ($N_{min.}$-Labormethode). G. Hoffman, ed. Methodenbuch I. VDLUFA Verlag, Darmstadt, Kap. A.9.1.4.1

Anonymous 1996. USGS Mineral Commodity Summaries, 1996.

Anonymous 1998. Richtwerte für die Düngung. P. Boysen (ed.) Institutszentrum LUFA/ITL Kiel der Landwirtschaftskammer Schleswig-Holstein

Birrell, S.J., and J.W. Hummel, 1997. Multi-sensor ISFET system for soil analysis. p. 459–468. *In* J.V. Stafford (ed.) 1st European Conf. on Precision Agriculture, Warwick. 1997. Vol. II, BIOS Scientific Publ. Ltd., London.

Clark, R.L., and R.L. McGuckin, 1996. Variable rate application technology: An overview. p. 855–862. *In*. P.C. Robert et al., (ed.) Proc. of the 3rd Int. Conf. on Precision Agriculture, Minneapolis, MN. 23–26 June, 1996

Finck, A. 1979 Dünger und Düngung. Verlag Chemie, Weinheim, New York.

Franko, U. 1996. Simulation der Kohlenstoff-Stickstoff-Dynamik in Agrarlandschaften. Landbauforschung Völkenrode 3/1996:114–120.

Haneklaus, S., D. Schroeder, and E. Schnug, 1996. Strategies for fertilizer recommendations based on digital agro resource maps. p. 361–368. *In* P.C. Robert et al., (ed.) Proc. of the 3rd Int. Conf. on Precision Agriculture, Minneapolis, MI. 23–26 June, 1996.

Haneklaus, S., I. Ruehling, D. Schroeder, and E. Schnug, 1997. Studies on the variability of soil and crop fertility parameters and yields in different landscapes of Northern Germany. p. 821–826. *In* J.V. Stafford (ed.) 1st European Conf. Precision Agriculture, Warwick 1997, Vol. II. BIOS Scientific Publishers Ltd., London.

Haneklaus, S., H. M. Paulsen, D. Schröder, U. Leopold, and E. Schnug, 1998. Self-Surveying - A Strategy for Efficient Mapping of the Spatial Variability of Time Constant Soil Parameters. Commun. Soil Sci. Plant Anal. (in press)

Kersebaum, K.C. 1996. Modellierung der N-Dynamik zur Stickstoff-Düngungsoptimierung auf heterogenen Standorten - Voraussetzungen für den operationellen Einsatz in der Praxis. Landbauforschung Völkenrode 3/1996. 134–140.

Lui, W., S. K. Upadhyaya, T. Kataoka, and S. Shibusawa, 1996. Development of a texture/soil compaction sensor. p. 617–630. *In* P.C. Robert et al., (ed.) Proc. of the 3rd Int. Conf. on Precision Agriculture, Minneapolis, MI. 23-26 June, 1996.

Pannatier, Y. 1994. Variowin. Software for spatial data analysis in 2D. Springer Verlag, New York.

Panten, K., S. Haneklaus, M. Vanoverstraaten, D. Schroeder, and E. Schnug, 1998. Remote sensing as an aid for the spatial management of nutrients. This volume

Richter, G.M. 1996. Variabilität und Schätzgenauigkeit von N-Mineralisation, Ertrag und N-Bedarf - Methoden zur Bewertung der "Ist-Analyse". Landbauforschung Völkenrode 3/1996. 121–126.

Schneider, S.M., R.A. Boydston, S. Han, R.G. Evans, and R.H. Campbell, 1997. Mapping of potato yield and quality. p. 253–261. *In* J.V. Stafford (ed.) 1st European Conf. Precision Agriculture, Warwick 1997, Vol. I, BIOS Scientific Publishers Ltd., London.

Schnug, E., S. Haneklaus, and D.E.P. Murphy, 1994. Equifertiles - an innovative concept for efficient sampling in the local resource management of agricultural soils. Aspects Appl. Biol. 37:63–72.

Schnug, E., J. Heym, D.E.P. Murphy, 1995. Boundary line determination technique (BOLIDES). p. 899–908. *In* Site-Specific Management for Agricultural Systems, ASA, CSSA, and SSSA, Madison, WI.

Schnug, E., J. Heym, and F. Achwan, 1996. Establishing critical values for soil and plant analysis by means of the boundary line development system (Bolides). - Commun. Soil Sci. Plant Anal. 27:(13, 14):2739–2748.

Schnug, E., K. Panten, and S. Haneklaus, 1998. Soil sampling and nutrient recommendations - the future. Commun. Soil Sci. Plant Anal. (in press)

Schnug, E., and S. Haneklaus, 1998. S-book Verlag Kluwer (in press)

Schroeder, D., and E. Schnug, 1995. Application of large scale yield mapping to field experimentation. Aspects Appl. Biol. 43:117–124.

Schroeder, D., S. Haneklaus, and E. Schnug, 1997. Information management in Precision Agriculture with LORIS. p. 821–826. *In* J.V. Stafford (ed.) 1st European Conf. on Precision Agriculture, Warwick. Vol. II. BIOS Scientific Publishers Ltd., London.

Schueller, H. 1969. Die CAL-Methode, eine neue Methode zur Bestimmung des pflanzenverfügbaren Phosphats in Böden. Z. Pflanzenernährung und Bodenkde 123:48–63

Schumann, M., M. Kücke, and E. Schnug, 1997. Fallstudien und Konzeption zur Einführung bilanzorientierter Düngung in der deutschen Landwirtschaft. Landbauforschung Völkenrode, Sonderheft 180

Thompson, W.H., and P.C. Robert, 1995. Evaluation of mapping strategies for variable rate applications. p. 303–323. *In* Site-Specific management for Agricultural Systems. Proc. of the 2nd Int. Conf., 27–30 Mar., 1994. ASA, CSSA, and SSSA, Madison, WI.

Walter, J.D., V.L. Hofman, and L.F. Backer, 1996. Site-specific sugar beet yield monitoring. p. 835–854. *In* P.C. Robert et al., (ed.) Proc. of the 3rd Int. Conf. on Precision Agriculture, Minneapolis, MN. 23–26 June, 1996, ASA, CSSA, and SSSA, Madison, WI.

Zadoks, J.C., T.T. Chang, and C.F. Konzak, 1974. A decimal code for the growth stages of cereals. Weed Res. 14:415–421.

Using the Decision Support System for Agriculture (DSS4AG) for Wheat Fertilization

Reed L. Hoskinson
J. Richard Hess
Lockheed Martin Idaho Technologies Company
Idaho National Engineering and Environmental Laboratory
Idaho Falls, Idaho

ABSTRACT

The DSS4Ag is an expert system under development at the INEEL through the Site-Specific Technologies for Agriculture (SST4Ag) precision farming research project. The system uses artificial intelligence and other computer information technologies to assist in making spatial, site-specific management decisions. Using this decision support system, we generated a variable-rate fertilizer recommendation recipe for a 135 acre wheat (*Triticum aestivum* L.) field with the goal of optimum economic return, not maximum yield. The field was split into blocks, alternately fertilized with the variable-rate recipe and with the uniform application method used by the farmer. The DSS4Ag fertilizer recipe reduced fertilizer costs 39.7% and yields 3.3%, which resulted in a net economic gain of 2.8% as compared with the uniform application used by the farmer.

INTRODUCTION

It is recognized that new and practical management solutions will require system-level integration of multi-faceted relationships, including all the processes (perception, reasoning, intuition, etc.) through which knowledge can be gained about the multi-faceted character of specific sites. Such a cognitive system theory also recognizes that the larger system is composed of cognitive subsystems. Three of those subsystems employed here include the cognitive process of the practitioner's experience, the cognitive process of experts' scientific discoveries, and autonomous computational algorithms that act in a cognitive manner. The practitioner and expert cognitive systems outline the parameters–boundaries and provide generalized knowledge for the autonomous computational cognitive system, which then can produce optimal site-selective knowledge for management recommendations.

THE DECISION SUPPORT SYSTEM PHILOSOPHY

The general methodology of a decision support system is driven off a spatial database of data characterizing the growing environment on a site-specific basis. This methodology brings a new approach to precision agriculture. This notion of a paradigm shift to explain the scientific revolution that is precision agriculture has been discussed by Schefcik and Pete (1996) and in the National

Copyright © 1999 ASA-CSSA-SSSA, 677 South Segoe Road, Madison, WI 53711, USA. *Proceedings of the Fourth International Conference on Precision Agriculture.*

Research Council report on precision agriculture in the 21st Century (1997). Others (Wilson, 1997, Lowenberg-DeBoer, 1997, Dunn, 1997) have also discussed this shift to high technology. Wilson (1997) described farming as one of the hottest frontiers for technology application, and emphasized that the analysis is the most important and most difficult part of precision farming. This was further emphasized at the 3rd International Conference on Precision Agriculture where 15 workgroups ranked the need for decision support systems the most important (Robert, 1996).

We depict this information revolution in agriculture, called precision agriculture or site-specific agriculture, as an historic trend in the agricultural decision-making process. First was the Farmer Knowledge system (the cognitive process of the practitioner's experience) which made the decisions. Then, the Scientific Knowledge System (cognitive process of experts' scientific discoveries) developed in support of the Farmer Knowledge System. Now, the paradigm shift, the information revolution in agriculture, is what we refer to as the Information Knowledge System (autonomous computational algorithms).

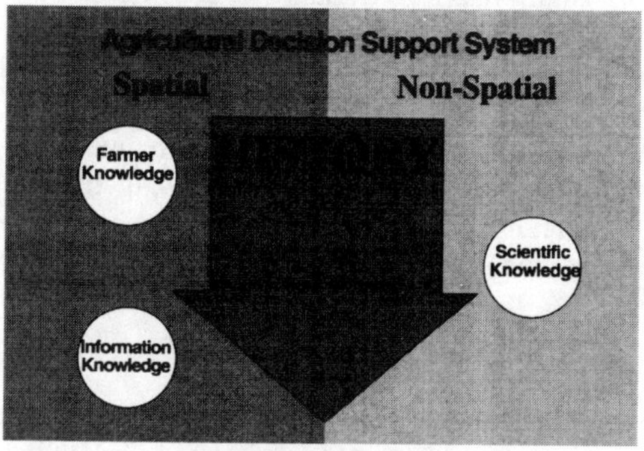

Fig. 1. Historic trend in agricultural decision support.

Cognitive Farmer Knowledge System

Since the earliest times, the agricultural decision support system relied on Farmer Knowledge for decision making. This system was spatial in nature, in that the farmer's knowledge of the growing environment included a spatial attribute, i.e., that the crop did better in the southeast corner of the field, or in the swale, or that the area that used to be an old homestead needed less fertilizer.

There are several positive attributes to this Farmer Knowledge System. This system is, after all, the foundation of agriculture. The farmer knew best the land farmed, the local and regional environmental conditions and patterns, the role of local and regional economics, etc. This system is highly credible to the decision-maker, the farmer, and is directly applicable to their farm. This system is also free, in that it needs no additional capital equipment, such as yield monitors, sensors, or computers. It also takes no additional time to manage administrative overhead.

There are, however, also several negative attributes to the Farmer Knowledge System. First and foremost, it takes a lifetime of experience to acquire. And then, that knowledge is almost impossible to pass on to the next generation. Worse yet, there is no data validation for the knowledge gained. This knowledge is sometimes wrong, and sometimes right for the wrong reason.

Cognitive Scientific Knowledge System

The Scientific Knowledge System developed to support and enhance the agricultural system, through a better scientific understanding of agriculture. This knowledge tends to be nonspatial, in that the knowledge better describes the mechanistics of the system, i.e., the physiology of a plant, the movement of water through a soil, or the interactions of soil nutrients among themselves. Landmarks for this system might be seen as the establishment of the USDA and the Land Grant Universities in the past century.

The knowledge from this Scientific Knowledge System is highly credible, respected, and trusted. It represents much of the fundamental knowledge of agriculture. The theoretical knowledge of this system, however, is often too complex to easily implement into practical applications.

For application, this knowledge is often represented by generalized equations. The complexity of the agriculture system forces generalization of the results, which causes these equations to be limiting in the number of variables they include. Those variables are usually crop, or soil type, or species specific. In some cases, the knowledge represented by these equations is applied to conditions beyond the bounds from which it was developed, since these bounds are often not described for the user. The Scientific Knowledge System also takes time to develop, to use, to apply, and to understand.

Cognitive Information Knowledge System

The Information Knowledge System is the current revolution in agriculture. It brings new methodologies for analyzing and understanding volumes of data. Much of its origin must be owed to the availability of the Global Positioning System (GPS), and the development of geographic information systems (GIS). The GPS and GIS give the capability to collect large volumes of spatial data, and the ability to analyze the data into information that helps to understand the spatial variability that occurs.

It is this ability to begin to understand the information that is locked in the large volumes of spatial data that sets the Information Knowledge System apart. An example of this concept of database mining to develop previously unknown rules and relations is described by Simoudis et al. (1994).

Such tools are often referred to as decision support systems (DSS). A DSS is an expert system that can continue to learn as more data and information becomes available. The system is unbiased since it is driven off the information. This type of system is also very valuable in that it is readily transferable, to the next generation, or the next land manager. In fact, the system itself becomes a valuable, marketable asset. Because the specificity (such as crop or variety) is in the data and information, and not in the DSS, the methodology is usable in new applications.

The negative aspects of the Information Knowledge System are that it is presently seen as a black box, or as a complex system that is not understood or is misunderstood. Many of these systems require complex computing capabilities, and are time consuming. The knowledge in the system is bounded by the extrema of the data from which the information was developed. Also, although the Information Knowledge System is not limited by the number of parameters it can consider, there remains the question of which parameters are needed or not needed.

DECISION SUPPORT SYSTEM FOR AGRICULTURE (DSS4Ag)

The information in the SST4Ag database is used to generate historic and current *statements* used by DSS4Ag. The optimum solution produced (the recipe) is an economic evaluation of cost/benefit, if prior production records are available for comparison, or is a maximum production if no prior data is available. The output options include such things as GIS maps of the recipe, predicted yield, or spatial economics, lists of total inputs and their costs, and/or machine-format instructions to control application of the recipe using variable-rate spreaders or pivots.

The DSS4Ag Uses *Facts* and *Statements*

Statements are individual descriptions of one characterization dataset that describes conditions that occur or occurred at one specific point, or site. An example of a *statement* is the set of soil nutrient parameter values as measured at one geographic point in a farm field at a certain time, along with the crop yield value that was harvested at that point. Historic *statements* are from previously occurring instances, and can be used to generate and/or update the *facts*. Current *statements* describe presently existing conditions, and are the baseline that is modified based on DSS4Ag's analysis of the *facts*.

Facts are descriptors that summarize the limits that bound some phenomena, as determined from the historic *statements*. An example of a *fact* is a descriptor relating an occurrence of a crop yield value to a set of minimum and/or maximum values for multiple soil nutrient parameters that occurred geographically where the crop yield value occurred.

The set of *facts* is generated from the historic *statements* using artificial intelligence technology. The *facts* used can be from the local farm (generated from historic *statements* characterizing the local farm), from a nearby farm, from a regional farm, or from a fundamental set of *facts*. A fundamental set of *facts* can be used as a starting point whenever no other *facts* are available. As the DSS4Ag is initially used, in the early stages of application of precision farming, the only *facts* available to a farmer may be a set produced elsewhere. In most cases, the best *facts* available would be those from as nearby as possible, to take advantage of minimum change in soils, climate, etc. Concerns over this data sharing are a separate issue, which has been pointed out elsewhere (Dunn, 1997; AFBF Task Force, 1995).

As precision agriculture grows, and as the DSS4Ag user accumulates a larger personal database for their farm, the *facts* generated from their farm would be the most applicable. The set of *facts* can be regenerated as additional historic *statements* become available. As the DSS4Ag becomes widely used over several years, this regeneration will produce more and more beneficial *facts*.

USE OF DSS4Ag FOR WHEAT

The DSS4Ag was used in spring, 1997, to develop the optimum recipe for the variable-rate application of multiple fertilizers on a 135 acre field on a wheat–wheat crop rotation, near Ashton, ID. The 1997 recipe was generated from yield and soil nutrient spatial data sets collected from a potato–wheat crop rotation in previous years. These data were input to the geographic information system (GIS) and maps of the spatial variability in the nutrients were developed and used as part of the input to the DSS4Ag.

The DSS4Ag also requires as input a forecast of the market price of the crop at harvest. The cooperating farmer (Hess Farms) forecast a price of \$3.35 bushel^{-1}.

The DSS4Ag research field was divided into 12 blocks, running North-South. Each block was about 210 ft wide, based on 3 passes by the Ag-Chem Equipment Co., 1903 Dry Soilection applicator, which had a 70 ft spray boom. This 210 ft block width also accommodated 8 or 9 passes by the New Holland combines at harvest time. Each block was on the order of 11 acres in size.

The recipe was generated so that the blocks would be fertilized in alternating applications, where one block was the treatment (fertilized at a variable rates determined by the DSS4Ag), and the alternating block was the control (fertilized at the uniform rates specified by the farmer).

Cognitive Modification of the Recipe

Because of the difference in crop rotations, Hess Farms (the Farmer Knowledge System), following university recommendations (the Scientific Knowledge System), adjusted the DSS4Ag recipe. They specified an additional 10 pounds of urea be applied per acre to aid in nitrogen replacement for straw decomposition not accounted for in the DSS4Ag data set. The DSS4Ag recipe was input to the AgChem Equipment Co., Falcon controller on the Soilection machine along with the uniform application recipe, and the fertilizer was applied 12 May, 1997.

HARVEST

The field was swathed into windrows on 2 Sept., 1997. Windrows along the blocks' borders that included wheat from both a treatment and a control block were flagged and thrashed separately, and the straw spread before starting the harvest of the individual blocks.

The New Holland combines used were equipped with GPS and grain yield monitors, and with load cell sensors on the concaves that measured the force against them, which is a measure of the biomass. Both yield and biomass were logged with GPS, allowing for mapping of their spatial variability.

Within each block, the grain was thrashed and the straw dropped back into windrows. For each block the combine was unloaded into an individual truck, which was weighed on the farm's scales as well as at the elevator. After the combining was completed, the windrowed straw was baled by block, loaded onto a truck, and weighed on the farm's scales.

Therefore, for each block, data was collected on the weight of both grain and straw produced. These scale measurements were used to calibrate the yield and biomass sensors.

RESULTS

Fertilization

The following table shows the comparisons of fertilizers used in the DSS4Ag treatment blocks and the Hess Farms control blocks.

This fertilization savings of $13.72 per acre represents a 39.7% cost per acre savings using the DSS4Ag compared with the uniform fertilization recommended by Hess Farms.

Yield

The following map (Fig. 1) shows the spatial variability in yield in all of South Field 3 in 1997. Because most of the west-most treatment block was outside the center pivot irrigated area and unirrigated, and most of the east-most control block that was outside the center pivot irrigated area was irrigated with handlines extending from the adjacent field, the analyses have been done using the area within the center pivot, which was all irrigated by the center pivot.

Table 1. Fertilizers used.

		DSS4Ag Blocks (treatment)		Hess Farms (control)	
Fertilizer	Cost ton^{-1}	Pounds	Cost	Pounds	Cost
Urea	$298	9828	$1464.37	7212	$1074.59
Ammonium phosphate	$305	20	$3.05	3255	$496.39
Ammonium sulfate	$198	199	**$19.70**	8440	$835.56
Boron	$808	65	$26.26		
		Total cost	$1513.38		$2406.54
		Acres	72.73		69.70
		$/acre	$20.81		$34.53

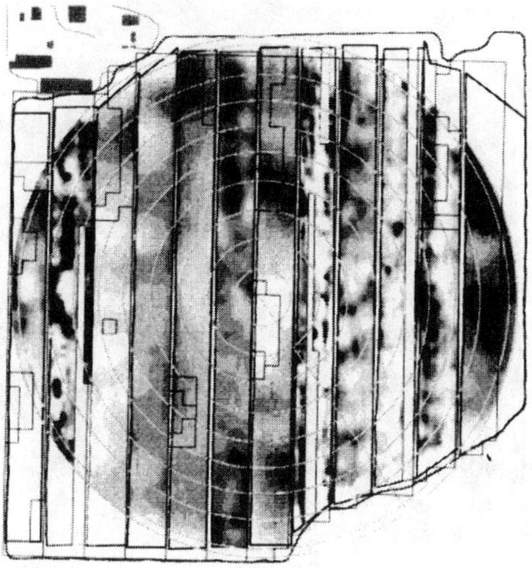

Fig. 1. Yield map, South Field 3, wheat, 1997.

For the area within the pivot, the yield in the treatment averaged 72.7 bu acre^{-1} and in the control the average was 75.2 bu acre^{-1}. This 2.5 bu acre^{-1} difference is a 3.3% reduction in yield in the blocks fertilized with the DSS4Ag recipe.

Biomass and Harvest Index

The straw from each block also was baled and weighed. These data were used to calibrate the load cell system on the concaves of the combines. The load cells measured the force against the concaves, which was calibrated to measure the total biomass passing through the threshing chamber. These data were used to generate the following map in Fig. 2 of the spatial variability in the total biomass produced across the field.

Across the entire field, both within and outside of the pivots, the biomass in the treatment blocks averaged 2375.5 lbs A^{-1}, and averaged 2900.3 lbs A^{-1} in the control blocks. This difference of 524.8 lbs A^{-1} less biomass in the treatment blocks (a reduction of about 18.1% biomass) is advantageous from the standpoint of less straw to decompose in the future, since typically the straw has no market value and is spread back onto the field and plowed under rather than being baled.

These same data represent a difference in Harvest Index (HI) of 0.60 in the control blocks and 0.63 in the treatment blocks. This 0.03 difference could be thought of as a 5% increase in the crop's production efficiency.

Economics

Using the forecast market price of \$3.35 bu^{-1}, the sales loss using DSS4Ag would be \$8.38 acre^{-1}, but the fertilizer savings of \$13.72 gives a net benefit of \$5.34 acre^{-1}.

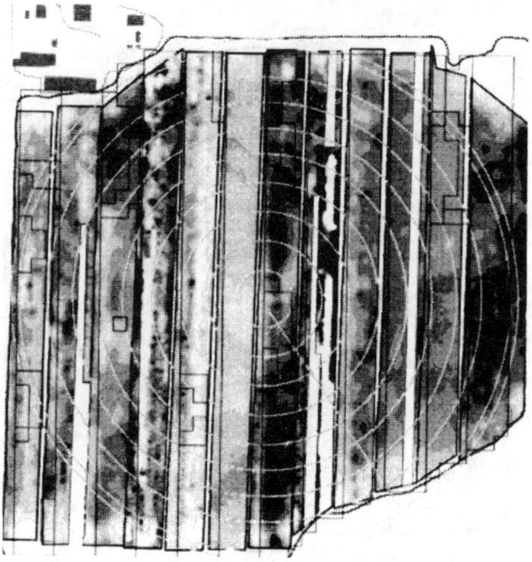

Fig. 2. Biomass map, South Field 3, wheat, 1997.

The actual average market price at which Hess Farms sold their grain from South Field 3 was \$3.17 bu^{-1}. Therefore, the sales loss using DSS4Ag was \$7.93 $acre^{-1}$, and with the fertilizer savings of \$13.72 a net benefit of \$5.79 $acre^{-1}$ was realized.

LIMITATIONS

There are several things about the DSS4Ag test that can be improved. As the SST4Ag research continues, these limitations will be addressed.

1997 DSS4Ag *Facts*

The *facts* used in the DSS4Ag to generate the recipe had several limitations. As discussed earlier, the *facts* can be regenerated as multiple years' data are obtained, making the *facts* more representative as more years and acres of information is used.

Also, the 1997 *facts* were limited to only a few nutrients, since those nutrients were the only ones that could be managed with available fertilizers. Micronutrients that are presently only available through foliar post-emergent sprays were not included.

Crop Rotation

The 1997 *facts* also were generated from a set of historic *statements* representative of a wheat following potatoes rotation, but the recipe was applied in a wheat following wheat rotation.

Typically, the soil fertility following a potato crop remains somewhat higher in sulfates and phosphates than after a wheat crop. As more historic *statements* become available through the crop rotation cycles, the *facts* will better represent the different categories of soil conditions.

CONCLUSIONS

The field test of the DSS4Ag in 1997 successfully demonstrated that the methodology works. A recipe was generated by DSS4Ag, which reduced fertilizer use by 39.7%, with only a 3.3% decrease in yield. Economically, this use of DSS4Ag produced an actual \$5.79 acre^{-1} benefit. Additional economic benefit might be expected from the reduced biomass produced.

The results were positive even though the current system had many limitations as used, giving expectations that even better results might be expected in future years. As the DSS4Ag methodology becomes widely used, the improvements in the *facts* will produce increased economic benefit.

ACKNOWLEDGMENTS

Work supported through the INEEL Laboratory Directed Research & Development (LDRD) Program under DOE Idaho Operations Office Contract DE-AC07-94ID13223. The authors express their thanks to Hess Farms, Inc., and Mr. John Hess, President, Hess Farms, for cooperation and support of this research.

REFERENCES

AFBF GPS/GIS Task Force. 1995. Final report and recommendations of the AFBF GPS/GIS Task Force. Am. Farm Bureau Fed..

Dunn, R.F., Jr. 1997. How far will you go for great information? Precision Ag Illustrated. July/August, 1997.

Lowenberg-DeBoer, J. 1997. Taking a broader view of precision farming benefits. Modern Agriculture, the J. Site-Specific Crop Manage. 1(2):32–33.

National Research Council. 1997. Precision agriculture in the 21st century: Geospatial information technologies in crop management. Committee on Assessing Crop Yield – Site-Specific Farming, Information Systems, and Research Opportunities, Board on Agriculture.

Robert, P. 1996. Summary of workgroups: Precision agriculture research & development needs. p. 1193–1194. *In* P.C. Robert et al. (ed.) 1996 Proc. 3rd Int. Conf. on Precision Agriculture. Minneapolis, MN. 23-26 June, 1996. ASA, CSSA, and SSSA, Madison, WI.

Schefcik, M.E., and J. Pete. 1996. Precision farming, an approach to problem solving, p.14–18. *In* Great Plains Soil Fertility Conf. Proc.. Denver, CO. 5–6 Mar., 1996.

Simoudis, E., B. Livezey, and R. Kerber. 1994. Integrating inductive and deductive reasoning for database mining, *In* Proc. of the AAAI Knowledge Discovery in Databases Workshop, Seattle, WA. August, 1994.

Wilson, J. D. 1997. Precision agriculture: Farming goes high tech. *Modern Agric., J. Site-Specific Crop Manage.* 1(1):21–23.

Analysis, Design, and Implementation of an Information System for Spatially Variable Field Management

J. van Bergeijk, D. Goense, and L. Speelman
Department of Agricultural, Environmental and Systems Technology
Wageningen Agricultural University
Wageningen, the Netherlands

ABSTRACT

A characteristic of precision agriculture research is the cooperation between different research disciplines. Especially when research is conducted at farm scale level, the data, collected by various measurement methods, has to be available in a well described format to the participants. Based on an information model described by Goense et al. (1996), a database structure was implemented to facilitate on farm research. Part of the database incorporates storage of farm management actions with their respective spatial measurements linked to geographical primitives like locations and areas. Functionality has been extended to use crop growth simulation models to optimize prescription rates. The implementation in a relational database management system offers the opportunity to use standard GIS software as a graphical front end. Specific tasks like data exchange with field equipment has been implemented by a structured query language (SQL) interface of the database management system.

INTRODUCTION

The basis for site specific field management is that crop and soil characteristics vary at a scale smaller than the field size. To further optimize crop production, field operations no longer act uniformly at field scale but adapt working conditions to the within field variations. An example is the NO_3 fertilizer application where the application rate depends on the local crop demand, the local NO_3 availability in the soil and the local NO_3 leaching risk of the soil.

A rational approach to obtain the set points for spatially variable field operations requires processing of data from various information sources. In the example of a NO_3 fertilizer application, the following information sources are likely to be used: (i) Yield mapping activities during previous seasons provide an estimate of the target yield. (ii) Remote sensing images indicate and possibly quantify certain crop characteristics. (iii) Analysis of soil samples, taken at different locations provide a spatial measure of the soil characteristics. This information is an input to crop growth simulation models that facilitate the search for an optimal application rate in both the spatial and the temporal dimension (Booltink & Verhagen, 1997). Finally, the actual NO_3 application itself has to be monitored to ensure correct evaluation of NO_3 efficiency of the crop at the end of the growing season.

In precision agriculture research, the data exchange between information sources introduces potential hazards. For instance, when the use of data is limited due

Copyright © 1999 ASA-CSSA-SSSA, 677 South Segoe Road, Madison, WI 53711, USA. *Proceedings of the Fourth International Conference on Precision Agriculture.*

to differences in scale or resolution at which data is collected, these limits should be part of the data. Similarly, when models to support farm management decisions are used, care has to be taken to use the model only within its validated and calibrated ranges. In order to reduce these risks and to facilitate the exchange of data, an information model provides a method to specify units, validity ranges and a structure of the data. Furthermore, the implementation of an information model on a computer system can serve as a platform for information exchange, used by researchers from different disciplines working on a precision farming project.

An information model for spatially variable field operations has been described by Goense et al. (1996). A more generic information model for agricultural systems was build within the ESPRIT III project titled "Computer Integrated Agriculture". An important aspect of these models is the link between geographic entities and field operation entities. The structure of the geographic entities is similar to the basis of a geographical information system (GIS) (Burrough, 1986). Examples of integration of crop growth simulation models with GIS and decision support systems (DSS) are DSSAT (Jones, 1993) and APSIM (McCown et al, 1996). The definition of a general structure for soil data has been initiated by the IBSNAT project (Uehara & Tsuji, 1993) while efforts on standardization of crop data and crop model interfaces are continued within the ICASA program.

In this study, two different information modeling methodologies will be evaluated to construct an information model and implement two levels of an information system for site specific field operations. The scope of the first implementation is the farm management level while the second system is intended to control variable rate field operations. The system at farm management level is open to interact with decision support systems while the system at field operational level is subject to standards for data exchange on agricultural equipment.

INFORMATION MODELING

The purpose of information modeling is to define a structure for the data storage and data handling requirements of an organization. Information modeling itself can be seen as a design process. For this process, several methods are available to analyze the requirements of an organization and to translate these requirements into an information system. In this paper both a design process intended for implementation of a relational database (Structured Analysis) and the use of an object oriented approach for data handling will be discussed.

Structured Analysis and Design

The design process of a relational database using structured analysis and design is depicted in Fig. 1. A computer aided software engineering tool (CASE-tool) was used to maintain a consistent data dictionary (DD) of the information model. The model was graphically constructed by creation of entity relationship diagrams (ERD) and data flow (DFD) diagrams. The entities in the ERD were defined by adding the relevant properties together with their corresponding database variable types.

The ERD was translated into a relational database scheme by the generation of a structured query language (SQL) data definition query. The CASE tool is capable

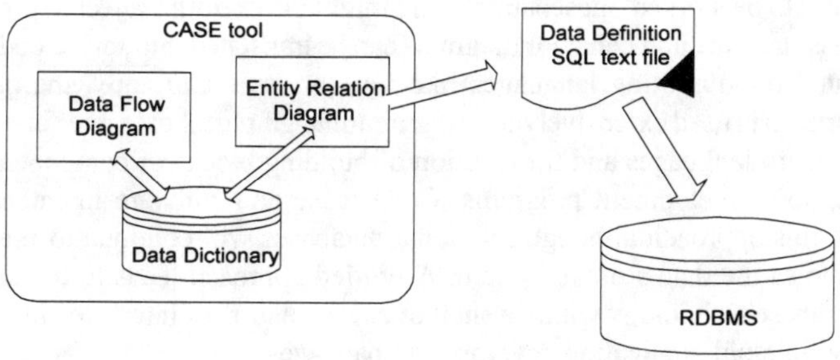

Fig. 1. Structured analysis and design to implement a relational database.

to generate and export these type of queries as an ASCII text file. The text file can be imported by a specific relational database management system (RDBMS) to generate the tables from the entities and to enforce consistency according to the specified constraints on the relations. The strict separation between the information analysis phase and the final implementation in a specific RDBMS, makes it easy to migrate from, for instance, a single user implementation on a personal computer, to a multi user database in a network environment.

After implementation in a RDBMS different users can write routine's to add data to, or retrieve data from the database. Rapid application development programs and form or query builders for the major programming languages are suitable tools to connect to relational databases and write custom analysis routines. For specific tasks like presentation of geographical data, the more specialized GIS packages offer capabilities to connect to remote databases through the use of a SQL interface.

Object Oriented Analysis and Design

A major difference between structured design and object oriented design is that the entities in structured design only contain data while the classes in an object oriented approach contain both data and algorithms. This gives the opportunity to store behavior specific to a certain class together with the data of that class in order to simplify its external interface. Especially when maintenance of consistency for entities with many relations in a relational database becomes difficult, an object oriented approach is a more elegant way to model and implement an information system.

The different CASE-tools for object oriented analysis and design offered a number of graphical notation methods for creation of class structure diagrams and class behavior diagrams. After a decennium of coexistence of these methodologies, recently, the unified modeling language (UML) has been proposed as the common language for information modeling. The UML started as a unification of previous methods formulated by Booch, Rumbaugh, and Jacobson (Fowler & Scott, 1997). Later on, also parts of the more real time systems oriented Shlaer-Mellor method were added. It is important to notice that UML merely defines the modeling language, i.e. the graphical diagrams and their symbols. This language is one part of an object oriented analysis and design method. The other part, the *process* of analysis and design, has to be worked out separately and might differ from project to project.

The class structure and behavior diagrams can be translated into source code for object oriented programming languages like C++ or Java. Currently, the object oriented approach is used extensively for programming graphical user interfaces, for application macro languages and for creation of building blocks or components in rapid application development programs. Code reuse and software maintenance benefit from this approach although relational databases will continue to play an important role as the data storage system. A unified approach leads to the use of relational database technology within a shell or encapsulation of interface objects as demonstrated in rapid application development packages.

FUNCTIONAL REQUIREMENTS

The scope of the information system is limited to precision agriculture research at the farm level. The specification of the requirements will be subdivided into the themes user interaction, data storage requirements and systems integration.

User Interaction

The intended users of the information system are researchers, farm managers and students. Typical use by the researchers is to store crop and soil analysis data and to perform queries to extract data for use in external models or procedures under development. The farm manager is responsible for storage of farm operational data. Preparation of field tasks is a typical farm management task, although interaction with research might be needed regarding the details of experiments on specific fields. The user interface on the farm can be restricted to a number of forms, including geographical presentations, for the farm management tasks. On the research side, the system has to be open to a range of data analysis methods implemented in different programming languages and operating on different hardware platforms.

Data Storage

The information system for spatially variable field operations has to store data on the following items:

- farm structure: e.g., fields, crop production units, equipment and personnel
- farm operations: store field tasks and related data like yield samples.
- soil properties: store soil sample analysis results.
- crop characteristics: store sample characteristics, store remote sensing images.
- climatological data: link to local or regional weather stations.

Systems Integration

Next to a method for storage and retrieval of data, the major components that the information system has to interface to, are:

- geographical presentation tools
- geostatistical algorithms
- external models and algorithms for specific analysis or decision support tasks
- external data sources on for example weather data and market information
- the computer systems for process control and data collection
- when available: existing databases on either the farm or at the research station

STRUCTURED ANALYSIS RESULTS

The results of the structured analysis method have been split up in four thematic diagrams; the basic farm entities (Fig. 2), the farm operations entities (Fig. 3), the specification entities (Fig. 4) and the geometry entities (Fig. 5). Each diagram will be discussed briefly, for more detail and arguments for specific entity types and relations, the reader is referred to Goense et al. 1996.

Farm Structure

Figure 2 illustrates the concepts of both identifying and associative relations. For instance, field and partfield are both topographical entities (tpgobject identifies field or partfield) while one field can be related to more than one partfield through an associative relation. This is the case when field is subdivided by a farmer into several partfields for a relatively short period. The tpgobject has a foreign key to a geometry entity which describes for instance the border polygon of the field or partfield. In the temporal dimension, a partfield can be associated with more than one cultivation. In the spatial dimension a single cultivation covers the entire partfield. The distinction between equipment and implement was introduced to match implement combinations like seedbed preparation and seeding by one piece of equipment. A general approach for administration of products entering or leaving the farm consists of the right hand entities in Fig. 2. Most basic farm entities have a one to one relation with an accountobject for the purpose of identification and ownership

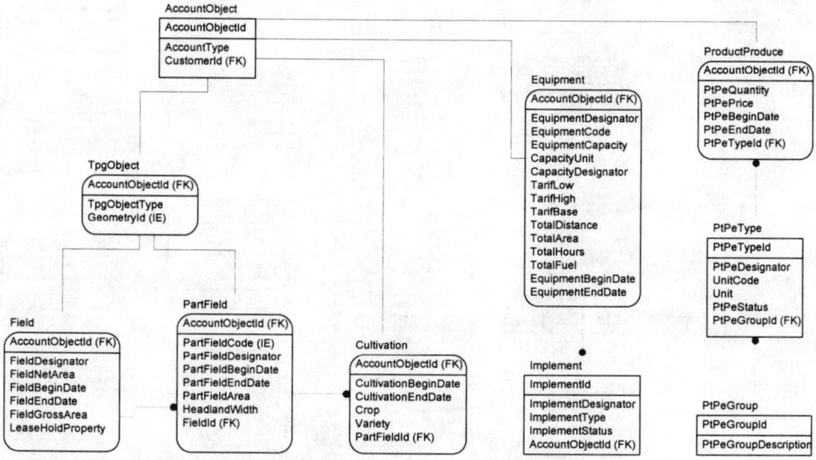

Fig. 2. Entity relationship diagram of the basic farm entities.

as well as to facilitate economic accountability.

Farm Operations

The preparation and administration of farm operations has modeled by the triplet job, task and operation as depicted in the center of Fig. 3. These three entities represent different aggregation levels. A task can consist of several operations and involves allocation of several pieces of equipment. A logical boundary of a task are the physically connected pieces of equipment, e.g a tractor and a fertilizer spreader perform one task that involves two pieces of equipment. For field operations, the task is performed on an accountobject of the type partfield. The multiple features or specifications of an operation are stored as measurement, processdata or consumptionyield tupels. Measurements are used for data with no direct agronomic meaning, e.g., raw sensor readings with in many cases volt as unit. An example of processdata is working depth for soil tillage operations, which is considered to have a direct agronomic meaning. Consumption-yield is used for both setpoint values for consumption of for instance fertilizers and agrochemicals as for yield values of products harvested. The spatial link of these entities will be discussed in the specification and geometry sections.

Specification

The previously described entities measurement, processdata and consumptionyield are subtypes of the supertype specification. Other subtypes of specification are soilfeature, cropfeature and parameter. Soilfeature is associated to partfield to describe the soil characteristics. In a similar way, cropfeature is related

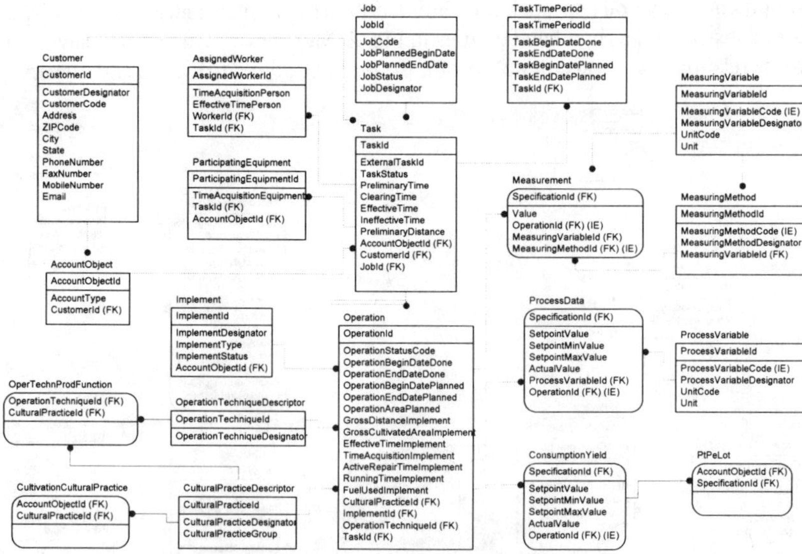

Fig. 3. Entity relationship diagram of the farm operations.

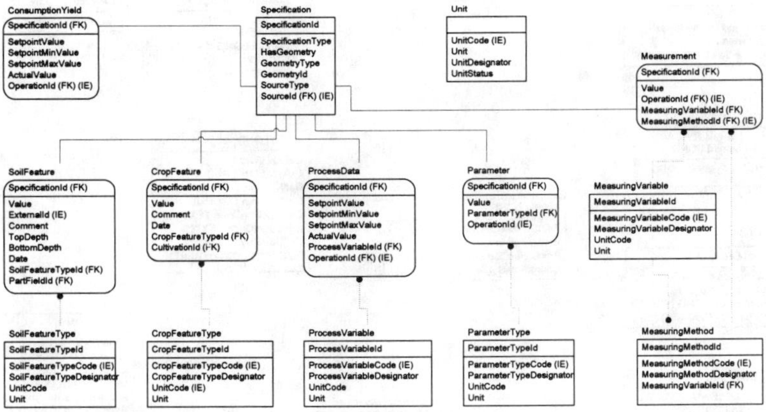

Fig. 4. Entity relationship diagram of the specification model.

to cultivation to specify crop features. The supertype specification has an optional relation to a geometry to locate or bound the specification to a geographic entity. Also the source of a specification has to be specified. When for instance a crop growth model is used to estimate standing yield for a certain location, the result is stored in the database as a cropfeature with a specification linked to the model run and to the geometry representing the location.

Geometry

The basic geometry types described in Fig. 5 are Polygon, PatternPoint, GridPoint and LanePoint. The Polygon and PatternPoint entity types represent vector based GIS data types, the GridPoint is similar to a cell in a raster based GIS and the LanePoint is a specific geographic type to locate field activities in row crops along tramlanes. All geometry types have references down to Point. The Point entity type is bound to CoordinateSystem that describes either globally or locally used coordinate systems.

OBJECT ORIENTED RESULTS

Geometry

The geometry part as described in the structured analysis results will serve as an example to be modeled according to the object oriented paradigm. From this part, the vector based entity types can be remodeled using the multi valued vector map approach (Kufoniyi, 1995). Figure 6 shows an example of a static structure diagram of the classes involved in modeling a multi valued vector map. A class symbol consists of a title box, an attributes box and a methods box. Three different types of relations between the classes are present in this diagram. On the top, *inheritance* relations exists from feature to polygon, from feature to line and from feature to

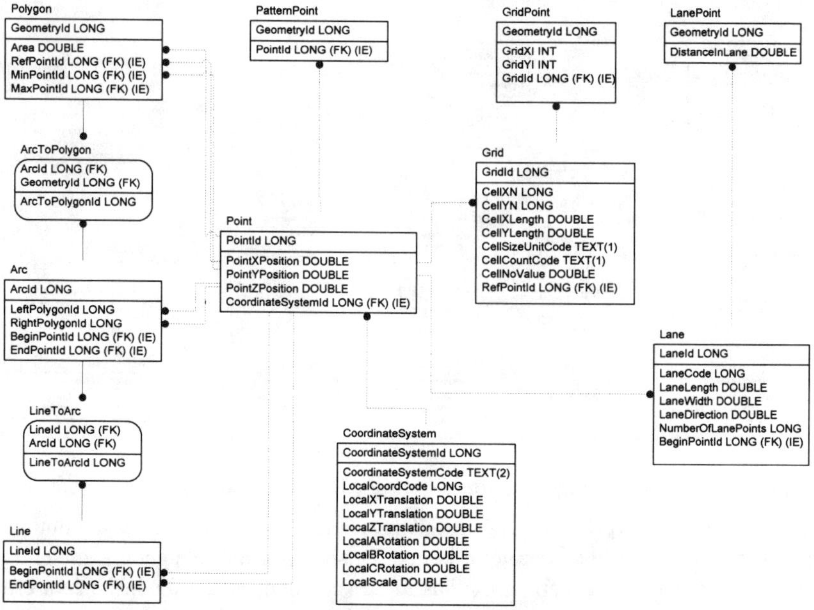

Fig. 5. Entity relationship diagram of the geometry entities.

point. At the bottom, the arc class is an *aggregation* of the node class. The other relations in Fig. 6 are *associative* relations with different cardinality levels, as specified next to the relations.

The feature class represents a value that belongs to a geometry. An instance of a polygon, a line or a point class inherits the value attributes and methods of the feature class. When an object oriented programming environment is used to implement a task controller for precision agriculture field operations, the instances of polygon or point can be used to specify the site specific application rates. An example that has been worked out is to use polygons to specify the different areas in a field that require different fertilizer application rates. Retrieval of the setpoint application rate while driving across the field consisted of querying the set of polygons for a polygon that covered the actual position. For this purpose, the polygon class has a method CoversNode that returns whether the position of the fertilizer spreader, which itself is an instance of the node class, is inside the boundary of a polygon instance. The implementation of a similar procedure with a relational set of geometry entities would separate the functionality to determine whether a node is covered by a polygon from the polygon description. The object oriented approach to combine certain functionality with the data necessary to meet the functionality requirements in one class, makes it easier to reuse for instance the geometry classes in different information systems.

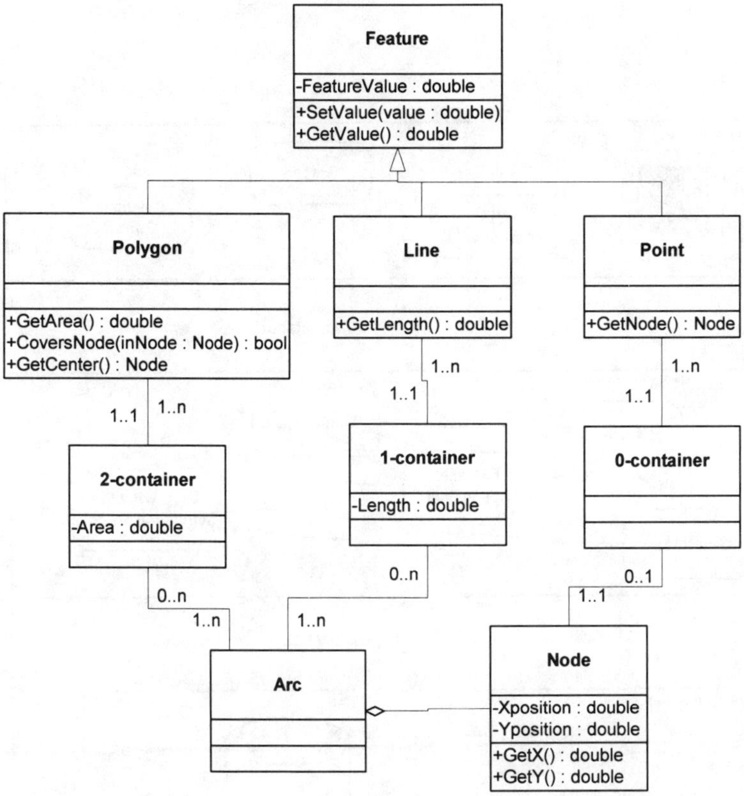

Fig. 6. UML static structure diagram of a vector oriented geometry.

Crop growth modeling

One approach that combines the results of a structured analysis with the advantages of an object oriented model, is to use a multi tier structure. In a multi tier structure, the information system is split up into several levels to separate for instance the data persistence functionality, the application logic and the user interface. While for data persistency a relational database might suit, the application logic can be build up using an object oriented approach and the user interface is build up from classes provided by graphical user interface components.

Figure 7 depicts a system diagram of an information system build to optimize fertilizer rates for site specific fertilizer applications by using crop growth simulations. The bottom layer consists of a number of relational databases whereof the management part and the geometry part were discussed in previous sections. Specific for the crop growth simulation model, databases were constructed for crop and soil parameters and for climatic data. The application logic tier depicts some of the classes responsible for both a consistent link to the persistency tier and an interface to the top level tier. For example the field class contains methods to allow

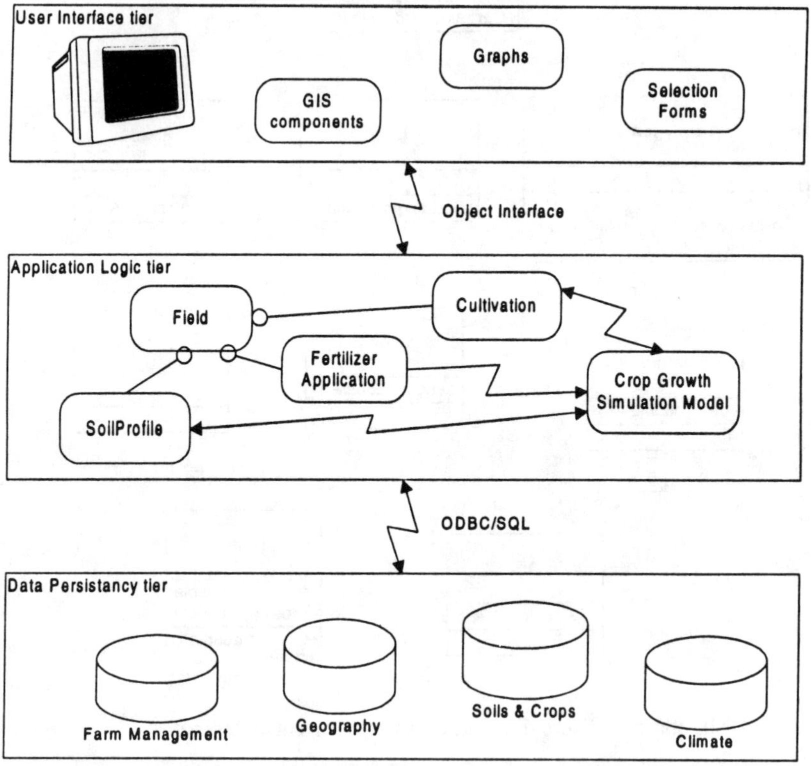

Afbeelding 7 Multi tier system diagram for site specific fertilizer rate optimization

its reconstruction out of the relevant databases from the lower tier while it simultaneous provides an interface to front-end routines of the user interface tier. In this way, the details on persistency of the field attributes are hidden from the user interface components like for instance a map to display an overview of fields. In a similar way, the simulation model class hides the details of the actual implementation of a crop growth simulation model. Instead, the interface of the simulation model class describes the information needed from other classes to be able to calculate the crop yield and fertilizer leaching characteristics of a fertilizer application scenario.

DISCUSSION

Modeling in general is to find a suitable level of abstraction while not throwing away too much detail. In information modeling this is no different. It is therefore arguable which entity types in a structured analysis have to be formulated and which entity types have to be generalized to another level of abstraction. More abstract entity types require additional information before an implementation is possible. On the other hand a too rigid information model with few abstract entity types reduces flexibility of the information system. In an object oriented approach, the levels of abstraction vary with the different inheritance levels in a class diagram. When used correctly, inheritance is a powerful method to model abstraction levels and is better supported by an object oriented environment than the super/subtype entities in a relational implementation.

The possibility to use SQL is an advantage of implementing a structured design in a relational database. Based on SQL, a query to present data in a GIS is easy to implement. Storage of data is more difficult because additional application domain knowledge is required in order to maintain database consistency. Possible methods to enforce consistency are the use of stored procedures in a relational database and the definition of wrapper objects that encapsulate persistency rules. The option to use an object repository instead of a relational database will become more important in near future but currently lacks an industry wide accepted standard interface.

Another important objective of information modeling is to construct a common reference for applications from different manufacturers that have to communicate. In agriculture, equipment from various manufacturers has to communicate with a number of farm management system implementations. Currently, activities to define a standard for communication on farm equipment takes place at international (ISO) level. One part of this standard is a link to farm management by the use of an ADIS protocol. Several entity types that were discussed in the structured analysis results, are part of the proposal to define a data dictionary for communication between farm management and field equipment.

A point of concern is the use of the described information system in on farm research. The implementation in a relational database in combination with a set of SQL queries is sufficient when participants just use the system as an information source. As soon as information requirements become more sophisticated the implementation has to go further by creation of for instance specific input and output components.

CONCLUSION

The combination of the structured analysis method and the object orientated approach to implement an information system is suitable for an environment were different applications have to share information. Data storage is deferred to a relational database while data access is handled by objects that model both application logic and data. Throughout the information analysis the data dictionary was the most stable part. The data dictionary formulated by the structured analysis was reusable for the definition of the attributes of classes in an object oriented design. Nevertheless, switching from a structured analysis method to an object oriented analysis method requires a complete different view on relations between classes. Classes get responsibilities, and their instances, the objects, communicate. Entity types are just data and have to be managed by separately formulated algorithms.

Open, standardized systems are necessary to facilitate communication in a multi vendor, multi user and multi disciplinary environment. Agriculture and agricultural research is such an environment and therefore a strong effort is needed to standardize certain levels of data exchange.

REFERENCES

Booltink, H.W.G., and J. Verhagen. 1997. Using decision support systems to optimize barley management on spatial variable soil. p. 219233. *In* M..J. Kropff et al., (ed.) Applications of systems approaches at the field level, Dordrecht, the Netherlands.

Burrough, P.A. 1986. Principles of geographical information systems for land resource assessment. Oxford.

CAMASE, newsletter, AB-DLO, P.O.Box 14, 6700 AA, Wageningen

Fowler, M., and K. Scott. 1997. UML distilled: Applying the standard object modeling language. Addison Wesley Longman.

Goense, D., J.W. Hofstee, and J. van Bergeijk. 1996. An information model to describe systems for spatially variable field operations. Computers Electron. Agric. 14:197–214.

Jones J.W. 1993. Decision support systems for agricultural development. p. 459–471 *In* F.W.T. Penning de Vries et al., (ed.) Systems approaches for Agricultural Development. Kluwer Academic Publ., Dordrecht, the Netherlands.

Kufoniyi, O. 1995. Spatial coincidence modelling, automated database updating and data consistency in vector GIS. Int. Inst. for Aerospace Surv. and Earth Sci. Publ. 28. IIASES, Enschede, the Netherlands.

Uehara, G., and G.Y. Tsuji, 1993, The IBSNAT project. p. 504–514. *In* F.W.T. Penning de Vries et al., (ed.) Systems approaches for Agricultural Development. Kluwer Academic Publishers, Dordrecht, the Netherlands.

McCown, R.L., G.L. Hammer, J.N.G. Hargreaves, D.P. Holzworth, and D.M. Freebairn. 1996. APSIM: A novel software system for model development, model testing, and simulation in agricultural systems research. Agric. Syst. 50:255–271.

Adoption of Precision Agriculture Technologies by U.S. Corn Producers

Stan G. Daberkow, and William D. McBride
USDA Economic Research Service
Washington, DC.

ABSTRACT

Corn (*zea mays* L.) producers are the largest users of cropland and agri-chemicals in U.S. agriculture and represent a major market for precision agriculture technologies. Based on a USDA survey of corn producing farms in 16 states, about 9% used some aspect of precision agriculture for corn production in 1996, representing nearly one-fifth of 1996 harvested corn acreage. Among precision agriculture adopters, 70% used some form of grid soil-sampling, 32 percent used VRT for lime or fertilizer application, and 54% used a yield monitor. A logit analysis indicated that farmers were more likely to adopt precision technologies if they farmed a large number of corn acres, earned a sizable farm income, and had high expected corn yields. The probability of adoption was also higher for farm operators using a computerized record system, who were <50 yrs of age, and who relied on crop consultants for precision agriculture information.

INTRODUCTION

U.S. corn producers, by virtue of their significant use of the nation's cropland and extensive agri-chemical, seed, and tillage applications, represent a potentially dominant market for precision agriculture technologies. According to the 1992 Census of Agriculture, more than one-half million farms, or more than one-third of all farms with harvested cropland, grew corn for grain or seed. In 1996, the 79.5 million acres planted to corn accounted for nearly one-fourth of all cropland in the USA (USDA-NASS, 1997). Corn production used nearly 60% (224.4 million lb. a.i.) of all agricultural herbicides and >35% (20.8 million lb. a.i.) of all insecticides used on the major U.S. crops in 1992 (Lin et al., 1995). Nearly 44% (4.9 million nutrient tons) of all commercial N fertilizer used in 1992 in the USA, 45% (1.9 million nutrient tons) of phosphate fertilizer, and 46% (2.3 million nutrient tons) of potash fertilizer was applied to corn land (Lin, et al., 1995). Hybrid corn seed accounts for an estimated 30 percent or over $1.5 billion of the expenditures farmers made for seed in the USA in 1994 (USDA-ERS, 1995). Clearly, corn farmers have an economic incentive to manage these inputs efficiently to increase yields or reduce costs of production.

Much of the U.S. corn is grown on or near environmentally sensitive lands that requires more intensive management. For example, in 1995 about 20% of the corn was grown on land designated as HEL and additional corn was produced near wetlands, shallow aquifers, rivers, streams, lakes, karst areas, etc. (USDA-ERS,

Copyright © 1999 ASA-CSSA-SSSA, 677 South Segoe Road, Madison, WI 53711, USA. *Proceedings of the Fourth International Conference on Precision Agriculture.*

1997). Corn production accounts for nearly one-third of the cropland on which conservation tillage is practiced (USDA, ERS, 1997).

Precision agriculture technologies offer a way to manage sub-field variability of soils, pests, landscape, and microclimates by spatially adjusting input use to maximize profits and potentially reducing environmental risks. Hence, the level and rate of adoption of precision agriculture technologies has implications for farm income as well as for the land and water resources associated with production agriculture. Furthermore, the characteristics of the farms and farm operators who have begun to adopt these spatial technologies offer insights to policy-makers concerned with improving farm income and reducing environmental risk.

OBJECTIVE

The objectives of this paper are three-fold: (i) present survey-based data on the extent of adoption of precision agriculture technologies by U.S. corn producers; (ii) offer a socio-economic profile of corn farms and farmers who are early adopters of these technologies; and (iii) use logit analysis to identify the key operator, farm and financial characteristics associated with the decision to adopt a precision agriculture technology.

DATA AND METHODS

Data collected in the 1996 Agricultural Resource Management Study (ARMS) were used to examine the use of precision agriculture technologies in corn production. One version of the 1996 ARMS collected detailed information about corn production practices and costs, and farm financial information. Sampling and data collection for the corn version of the ARMS involved a three phase process (Kott & Fetter, 1997). Phase 1 involved screening a sample of producers to determine whether or not each farm produced corn for grain. For Phase 2, production practice and cost information was collected on a randomly selected corn field for a sample of corn growers. All respondents to the Phase 2 interview were questioned about farm financial, management, and demographic information in Phase 3.

Respondents to all three phases of the 1996 ARMS for corn included 950 farms in 16 States. The states and their regional designation are: North Central— IL, IA, IN, OH, MI, MN, MO, WI; Southeast— KY, NC, SC; Plains— NE, KS, SD, TX; and Northeast—PA. The target population of this sample is farms planting any corn with the intention of harvesting grain. Each sampled farm represents a number of similar farms in the population as indicated by it's expansion factor. The expansion factor, or survey weight, is determined from the selection probability of each farm and expands the ARMS sample to represent the population of corn farms[1].

---

[1]A general farm version of the Phase 3 ARMS collected information about the use of precision agriculture technologies on a broader population of farms, but lacked the detailed information collected in the corn version. Because of the larger sample (1 673 farms), the general farm version was used to estimate population totals for corn farms and acreage in the 16 States. All other estimates presented in this study are based on the corn version (950 farms).

MEASURING ADOPTION

A farm operator is classified as an adopter of precision farming for corn production if, (i) any corn acres were soil grid sampled and mapped, or (ii) any corn acres were fertilized or limed with variable rate technology (VRT), or (iii) any corn acres were harvested using a combine equipped with a yield monitor. Because precision agriculture includes a relatively new set of technologies, this definition is intentionally broad to include the spectrum of farms experimenting with one or more components of precision agriculture.

The estimated number of corn farms and the corn acreage on farms using precision agriculture technologies is shown in Table 1. Using our broad definition of adoption, about 31 000 corn farms (approximately 9% of all corn farms) in the 16 states, with a range of 23 000 to 40 000 farms[2], used one or more precision agriculture technology for corn production during the 1996 season. Among the specific technologies, 7% of farms used grid samples/maps, 4% applied fertilizer or lime with variable rate technology, and 6% used a yield monitor during corn harvest. Only 4% of farms used the yield monitor information to develop yield maps.

The farm adoption estimate of about 9%, with no >7% for any individual technology, suggests that the diffusion of these technologies is very early in the typical adoption process (Rogers, 1983). These early adopters, however, controlled a disproportionally large share of the corn acreage. The 9% of farms using any precision agriculture technology farmed 19% of corn acreage, indicating that adoption has primarily occurred on larger farms.

Corn producers employed the different precision agriculture technologies on different shares of their corn acreage (Table 2). Among the adopters of precision agriculture technologies, soil grid sampling/mapping was the most widely used technology with 70% of farms sampling–mapping 64% of their corn acres. About 60% of adopters reported sampling in 2.5 acre grids, with 43% indicating a sampling frequency of once every 4 yrs. VRT was the least frequently used technology—36% of the adopters used VRT on about 55% of their corn acreage. Yield monitors were used by just over one-half the adopters (54%) on nearly all of their corn acres (94%). Yield monitors were used on an average

---

[2]A 95% confidence interval around the estimate. Because estimated totals are from a sample survey, and not a complete census, interval ranges are included to indicate the extent of sampling error.

Table 1. Estimated number of corn farms and corn acreage on farms using precision agriculture technologies, 1996.†

Technology	Percentage mean	Percentage interval‡	Number mean	Number interval‡
Farms:				
soil grid samples–maps	7.1	5.1–9.2	24,107	17,098–31,116
variable rate technology	3.7	2.7–4.6	12,358	9,176–15,541
yield monitor	5.5	4.4–6.6	18,690	14,966–22,414
yield maps	3.9	2.7–5.2	13,239	8,972–17,505
total§	9.2	6.7–11.7	31,116	22,719–39,513
Acreage (thousands)¶:				
soil grid samples–maps	15.8	12.5–19.0	9,513	7,526–11,501
variable rate technology	8.4	6.2–10.7	5,092	3,733– 6,451
yield monitor	15.6	12.0–19.3	9,441	7,218–11,664
yield maps	12.4	8.2–16.4	7,478	4,977–9,894
total§	18.7	14.4–22.9	11,271	8,710–13,832

† Corn farms are those planting any corn for grain in 16 major corn producing States, including: IL, IA, IN, KS, KY, MI, MN, MO, OH, NE, NC, PA, SC, SD, TX, and WI.
‡ 95% confidence interval.
§ The total is based on a sample of 1 673 farms while the percent distribution among the technologies is based on a sample of 950 farms. The estimated percentage of distribution is used to distribute the estimated total among the technologies.
¶ Total corn acres harvested for grain.

Table 2. The extent to which precision agriculture technologies were used in the corn enterprise on adopting farms†, 1996

Technology	Percentage of farms	Percentage of acreage	Average acreage
Soil grid samples–maps‡	69.9	63.8	267
Variable rate technology	35.8	55.3	242
Yield monitor	54.2	93.7	502
Yield maps	38.3	98.7	593

† Adopters of one or more precision agriculture technology.
‡ Initial grid soil sampling of fields used for 1996 corn production may have occurred prior to 1996.

of >500 acres per farm. Not all farms reporting use of a yield monitor, however, indicated that the data was (or will be) used for field mapping (38 versus the 54%). Under utilization of the yield data may be indicative of the newness of these technologies and suggests that some early adopters may still be learning how to fully utilize the technology.

COMPARING ADOPTERS AND NON-ADOPTERS

The comprehensive nature of the ARMS provided data on a variety of operator, farm structural and financial, and corn enterprise characteristics. Given the sufficiently large number of respondents indicating the use of one or more precision agriculture technologies, a number of traits of adopters and non-adopters could be compared using a difference of means test.

Operator characteristics

The personal characteristics of farm operators who have adopted some form of precision agriculture technology differ in a variety of ways from non-adopters (Table 3). While the average age of operator of the two groups was nearly the same, a significantly larger share of adopters (nearly 70%) were less than age 50 yrs of age compared with less than one-half of the non-adopters. A smaller share of adopters had only a high school or less education compared with the non-adopters, whereas a larger share of adopters had completed college. More than 90% of the adopters gave farming as their major occupation as opposed to only 75% of non-adopters—indicating more management time available for crop production decision-making. Likewise, adopters had gained more experience with computers in the farm business than non-adopters. The sources of information about precision agriculture differed between the two groups. Both groups relied heavily on farm suppliers and dealers, but adopters were more likely to seek out crop consultants and rely less on the Extension Service. In general, adopters were younger, more educated, less likely to work off the farm, more likely to seek out crop consultants, and more extensive computer users relative to non-adopters.

Based on several measures of risk preference, adopters appeared to be less risk averse than non-adopters. Adopters had a significantly higher debt–asset ratio, had less crop and income diversity, and owned a smaller share of the land they farmed relative to non-adopters. A high debt–asset ratio indicates a willingness to accept greater financial risk. While adopters had a larger share of their gross cash income from crops, non-adopters relied more heavily on both crops and livestock. Diversification is one strategy to reduced both production and market risks. Nearly 85% of the acres harvested by adopters consisted of two crops: corn and soybeans, whereas non-adopters were much more diversified. Even though adopters owned a smaller share of their corn acres (thus risking the loss of cropland in future years), they reduced their production and/or financial risk by share renting significantly more of their cropland than did non-adopters.

Table 3. Operator characteristics of precision agriculture adopters and non-adopters: corn producers, 1996

Item	Unit	Adopters	Non-adopters	Total
Operator Characteristics				
Age	years	49	52	52
Age distribution:				
less than 50 years	percentage of farms	69†	48	50
50 years or more	percentage of farms	31†	52	50
Education:				
high school or less	percentage of farms	37†	62	59
attended college	percentage of farms	35	24	26
completed college	percentage of farms	27†	14	15
Major occupation:				
farming	percentage of farms	91†	75	77
other	percentage of farms	9†	25	23
Computer record use	percentage of farms	30†	12	14
Sources of Information about Precision Agriculture				
extension service	percentage of farms	10†	23	22
crop consultants	percentage of farms	35†	14	16
farm supply/dealer	percentage of farms	71	56	57
event or demos	percentage of farms	12†	6	6
newsletter/trade magazine	percentage of farms	40	37	37
Risk Preferences				
Debt-to-assets	ratio	0.23†	0.14	0.15
Income sources:				
livestock	percentage of income	22†	38	34
crops	percentage of income	65†	50	53
government payments	percentage of income	4	4	4
other	percentage of income	9	8	9
Acres harvested by crop:				
corn	percentage of acres	48†	39	40
soybeans	percentage of acres	37†	28	30
wheat	percentage of acres	6	9	8
other	percentage of acres	9†	24	21
Land tenure of corn acres:				
owned	percentage of acreage	33‡	44	41
cash rented	percentage of acreage	29	31	31
share rented	percentage of acreage	38‡	24	28

† Significantly different from non-adopters at the 10% level.
‡ Significantly different from non-adopters at the 5% level.

Farm Structural and Financial Characteristics

By nearly any standard farm size or financial measure, those farms, which have begun to use one or more precision technologies, are bigger and more profitable than other corn farms (Table 4). Acres operated, acres harvested, asset values, return on equity, and net income measures were between 1.5 to >3 times larger for adopting farms relative to non-adopters. Adopting farms reported much higher normal and actual yields which likely reflects higher inherent soil productivity on these farms. The distribution of farms by sales class confirms the correlation of size with adoption of precision agriculture technologies. While over one-half of the adopting farms have sales of $250K or more, <20% of non-adopting farms were of that size. Nevertheless, about 18% of the adopting farms had less than $100,000 in gross sales in 1996. With net cash and farm income over $90,000 in 1996, adopting farms had the financial ability to experiment with this new technology. Even though adopting farms are much larger than non-adopters, the average net worth per farm is not statistically different between the two groups. This may reflect the more risk averse nature of non-adopters as indicated by their relatively modest use of debt.

The location and farm type of precision technology adopters may reflect the availability of vendors as well as demand for such services. The vast majority (70%) of the corn farms (adopting and non-adopting) are located in the North Central states with over one-third of all corn farms located in IN, IL, and IA. However, over one-half of the adopting farms were in the three central Cornbelt states and over 2/3 of the non-adopters were located in the other states. The adopting farms are overwhelmingly specialized in cash grain production rather than in livestock production relative to non-adopters. Most precision farming technologies are focused on crop rather than livestock production.

LOGIT ANALYSIS

The above analysis of traits of adopters and non-adopters utilized a univariate approach where differences between the means of selected characteristics for adopters and non-adopters were compared using a pairwise statistical test. A binary choice model, using the logit specification, was also used to examine the adoption decision in a multi-variate framework. Binary choice models are appropriate when the choice between two alternatives depends on the characteristics of the decision-maker (Pindyck & Rubinfeld, 1981). The logit model is used to describe the relationship between farm–farmer characteristics and the probability of adopting precision agriculture technologies.

Specification of the Logit Model

The dependent variable of the logit model is a binary variable equal to 1 if one or more precision agriculture technology was used in corn production and 0 otherwise. Based on the results from the univariate analysis, we selected the following operator and farm regressors for the logit model: Size: planted corn acres; Income: net cash farm income; Land productivity: corn yield normally expected; Operator age: 1 if the operator is <50 yrs of age, 0 otherwise; Operator

Table 4. Farm size, finances, and the type and location of precision agriculture adopters and non-adopters: corn producers, 1996.

Item	Unit	Adopters	Non-adopters	Total
Farm Size				
Acres harvested	acres	894 †	399	449
Sales class:				
$0-$99,999	percentage of farms	18 †	53	50
$100,000-$249,999	percentage of farms	29	31	31
$250,000-$499,999	percentage of farms	36 †	10	13
$500,000 or more	percentage of farms	17 †	5	6
Corn acreage and yields				
acres planted	acres	434 †	162	189
corn yield (actual)	bushels per acre	147 †	122	128
corn yield (normal)	bushels per acre	148 †	130	134
Farm Finances				
Income statement:				
gross cash income	$1 000 per farm	341 †	142	162
net cash income	$1 000 per farm	100 †	36	43
net farm income	$1 000 per farm	91 †	37	42
Balance sheet:				
assets	$1 000 per farm	972 †	652	686
liabilities	$1 000 per farm	221 †	90	104
equity	$1 000 per farm	751	563	582
Return on equity	percentage	12.1 †	6.6	7.3
Farm Type and Location				
Type:				
cash grain	percentage of farms	79 †	51	54
other crop	percentage of farms	1 †	9	8
livestock	percentage of farms	20 †	40	38
Location:				
north central	percentage of farms	77	69	70
IL, IN, & IA	percentage of farms	55 †	32	34
plains	percentage of farms	19	17	17
southeast	percentage of farms	3 †	8	7
northeast	percentage of farms	- †	6	6

† Significantly different from non-adopters at the 10 percent level.

education: 1 if the operator graduated from college, 0 otherwise; Use of a related technology: 1 if computer records were used for farm income and expense accounting, 0 otherwise; Livestock: percent of gross farm income from livestock sales; Land tenure: 1 if more than 40% of corn acreage was owned, 0 otherwise; Location: 1 if the farm was in either IN, IL, or IA, 0 otherwise; and Information sources: 1 if the information source was one of the main sources of information about precision agriculture technologies reported by the farmer, 0 otherwise. The four sources include the extension service, crop consultants or advisors, farm supply or chemical dealers, and special events or demonstration projects. These sources are those farmers are more likely to actively seek out for themselves.

Parameters of the logit model are estimated using the survey weights in a weighted least squares version of the maximum likelihood method. Due to the complex design of the ARMS sample, standard errors are estimated using a jacknife replication approach.

RESULTS

One-half of the hypothesized operator characteristics included in the logit analysis significantly influenced the technology adoption decision (Table 5). Precision technologies were most likely adopted by operators who were <50 yrs of age, used computers for record-keeping, and used crop consultants for information about precision farming. Education level was not statistically significant. Other sources of information about precision farming, such as input suppliers and farm shows–events, were also not statistically significant.

None of the variables used to assess risk preferences were statistically significant even though the univariate measures presented above indicated that adopters may be less risk averse. The debt–asset ratio did have a positive sign whereas the share of gross cash income from livestock sales, an indication of enterprise diversity, was negatively associated with adoption. Given that precision farming is currently oriented towards crop production, this variable may be an indication of the amount of managerial time available for crop technologies. The share of corn acreage owned by the operator was not related to the adoption decision and in fact had a negative sign. Apparently, even though field level data for precision technology is site specific, farm ownership is not a major consideration in the adoption decision.

Farm size, profitability, productivity, and location were statistically significant and positively influenced the precision technology adoption decision. The probability of adoption increases as the number of corn acres increases—a result which is consistent with most other studies of technology adoption. The ability to pay for new technology, as measured by net cash income, raises the probability of adoption. Increased inherent land productivity also led to enhanced adoption levels—which may be an indication that such technology is more profitable on high yielding fields or perhaps high yielding fields have more variability with respect to yield limiting factors. Finally, location of the farm in the central U.S. Cornbelt states (IL, IA, and IN) also increased the probability of adoption. Most likely the location variable reflects the availability of precision technology vendors, since much of the early precision farming equipment (i.e., yield monitors) was first introduced in these states.

Table 5. Logit regression results for precision agriculture adoption in corn production, 1996.

Variable	Variable description	Parameter estimate	Standard error§
Intercept	-	-5.8160 ‡	1.2086
Size planted corn acres		0.0017 ‡	0.0004
Income	net cash farm income ($1,000)	0.0018 ‡	0.0007
Land productivity	normal corn yield	0.0112 †	0.0061
Operator age	less than 50 yrs	0.6435 ‡	0.2947
Operator education	completed college	0.4998	0.4555
Related technology	computer record use	0.5530 †	0.3032
Livestock	income percentage from livestock	-0.0076	0.0060
Debt-to-asset ratio	>0.15	0.3880	0.3177
Land tenure	>40% owned	-0.1463	0.3443
Location	IN, IL, or IA	0.5994 ‡	0.2744
Information source	extension service	-0.1117	0.4892
Information source	crop consultant	0.8878 ‡	0.3329
Information source	input supplier	0.8627	0.5146
Information source	event or demo	0.7412	0.4459
Overall measures:			
Number of observations	950		
-2 log likelihood function	56 616with 14 df ($P = 0.0001$)		
McFadden R^2	0.24		
Concordant percentage¶	80.6		
Correct percentage	91.0		

† Significant at the 10 percent level.
‡ Significant at the 5 percent level.
§ A jacknife variance estimator was used with 15 replicates. Therefore, the critical t-values are 2.145 at the 5% level and 1.761 and the 10% level.
¶ Percentage of pairs of observations with different responses in which the larger response has a higher probability than the smaller response.

SUMMARY AND CONCLUSIONS

Corn production represents a potentially large market for precision agriculture technology because of significant natural and man-made resources used by corn farmers. By 1996, nearly 10% of all corn farms reported use of some aspect of precision agriculture—grid soil sampling, VRT for lime or fertilizer application, or yield monitors. These early adopters of precision agriculture technologies tended to be <50 yrs of age, have completed college, and be a full-time farmer relative to non-adopters. Early adopters had farms which were significantly larger in size (i.e., assets, acres farmed, corn acres, and gross sales), more leveraged, renting a large share of their acreage, more specialized in cash grain production, producing corn at lower costs, and more profitable than non-adopting farms. Precision farming adopters were more likely to have higher expected and actual corn yields than non-adopters.

A logit analysis indicated that precision agriculture adopters were more likely to farm more corn acres, earn greater cash farm income, be located in the central Cornbelt (i.e., IL, IN, or IA), and have higher expected corn yields. The probability of adoption was also higher for farm operators using a computerized farm record system, who were <50 yrs of age, and rely on crop consultants for information on precision farming. Our measures of risk preference were not significant in the logit analysis, nor was our measure of educational attainment.

From a policy perspective, several implications can be drawn from this analysis of early adopters of precision farming technologies. Large and profitable farms and younger operators, especially in the central Cornbelt, are adopting precision agriculture with little public assistance. If government policy-makers decide to encourage more widespread adoption, such public policies should be targeted toward small farms with older operators who are focused on livestock; especially those operators with little experience with computer related technologies. Additional training and resources for Extension Service personnel might also enhance the precision agriculture adoption rate. Educational programs to improve producer's computer literacy, such as for farm record-keeping, may be another avenue to pursue. Furthermore, policy-makers may want to consider cost-sharing to encourage adoption, especially if precision agriculture is shown to be environmentally beneficial.

REFERENCES

Kott, P.S. and M. Fetter. 1997. A multi-phase sample design to co-ordinate surveys and limit response burden. 1997 Joint Statistical Meetings (ASA, ENAR, WNAR, IMS, SSC).

Lin, B-W, M. Padgitt, L. Bull, H. Delvo, D. Shank, and H. Taylor. 1995. Pesticide and Fertilizer Use and Trends in U.S. Agriculture. USDA Econ. Res. Serv.. U.S. Dep. USDA AER no. 717. U.S. Gov. Print. Office, Madison, WI.

Pindyck, R.S., and D. L. Rubinfeld. 1981. *Econometric Models and Economic Forecasts*. 2nd ed. McGraw-Hill Book Co., New York.

Rogers, E.M. 1983. Diffusion of Innovations. 3rd ed. The Free Press, New York.

USDA-ERS, *Agricultural Resources and Environmental Indicators. 1996–1997.* USDA, Agric. Handb. 712. Washington, DC.

USDA-ERS. 1995. *AREI Updates: 1994 seed use, Costs, Trade*. No. 4.

USDA-NASS. 1997. *Agricultural statistics.* U.S. Gov. Print. Office, Washington, DC.

Information System for Farms Using Precision Agriculture Techniques: Compatibility with EDI Standards

Steffe Jerome, and Grenier Gilbert
Information System Laboratory
ENITA de Bordeaux, France

ABSTRACT

There have been many developments in technology, such as precision agriculture, as well as an evolution in the relationship between the farm and its environment (administration, advisers, suppliers, etc.). All this now makes a review of the Information System of the farm necessary. In order to better understand the new environment of the farm and to better meet the new information demand from end-users, we have tried to define an up-dated global Information System for farming. This global Information System take into account the needs of Precision Farming.

INTRODUCTION

Precision agriculture necessitates both the collection of great deal of data coming from various sources (farm equipment, soil sampling, remote sensing, etc.) and the exchange of information inside the farm (between Management Computer System and Process Computers) as well as between the farm and its commercial or technical environment (advisers, cooperatives, suppliers, etc.).

Hence the necessity to reconsider the Information System of the farm both to integrate the new technical data generated by the development of Precision Agriculture, and to allow data interchange between different Information Systems.

The need for a Global and Integrated Information System

The Farm Information System is too Compartmentalized

In France, we have a considerable amount of specific products for farmers : for example software for accountancy, another for forecasting, yet another for crop management or for herd management, etc. This situation can be explained by the past development of the market.

Over the last 20 yrs, all software developments have been, above all, ad hoc, in response to a specific management problem. Everyone adopted a sectorial approach in meeting demand. Problems were solved sequentially and each program was developed independently. In France, the situation is such that we now have more than 250 agricultural software products on the market. Each company has its own specific software but even within a firm, many specific products are found (for example, the firm ISAGRI, which is the market leader, has more than 24 different software products for managing a farm).

Copyright © 1999 ASA-CSSA-SSSA, 677 South Segoe Road, Madison, WI 53711, USA. *Proceedings of the Fourth International Conference on Precision Agriculture.*

The management field is now well covered but, at the same time, none of those specific software products today satisfies farmers' demands. The more products the farmer has, the more time he will need to spend in learning to use them. Most of the time, the farmer has two or three software products on his farm that treat a specific management problem, all of which often use common data.

The links between these products, however, seldom exist. This means that the farmer is often forced to type the same data twice or thrice ! In this case, there is not only a huge waste of time but also another problem: the coherence of data is not ensured.

The farmer can not use 100% of his time for management : therefore, he wants software offering him an overall response to problems. A study carried out by Taponnier and Desjeux (1994) showed that, nowadays, farmers want software capable of modelling the whole farm. This means multi-function software as users are no longer ready to waste time.

The information system of the farm must be better integrated to offer the farmer the necessary overall response to his problems. This is all the more necessary as the increasing complexity of the environment of the farm makes these problems more acute.

The Farm Environment is More and More Complex

Since the beginning of the 1980s, many changes have appeared in the farm environment making it more and more complex. This increased complexity can be ascribed to :

- the evolution of legislation: because of subsidies, European regulations have a crucial role in the management of a farm. With the evolution of the Common Agricultural Policy and the appearance of new needs (traceability, environmental data, etc.), this role should be reinforced over the next years. This will certainly add more complexity to the information system of the farm.
- More numerous and complex relationships with external actors: the consequences of the increase in commercial exchanges is a rise in the quantity of exchanges between the farm and its environment.
- New organization of the agricultural market : to gain productivity, relationships between firms and farmers have become more and more impersonal, as computing has replaced human contact. The EDI (Electronic Data Interchanges) will soon become the standard for information exchange. Farms which do not adapt to this evolution will be unable to survive.
- New boundaries of the farm: to cope with a drop in prices, many farmers have adopted new alliance strategies. They have created new structures, such as cooperatives, or they have established agreements between their farms. This allows them, for example, to share harvest costs, to obtain reductions on purchases, to exchange crops and fields, and to create a manpower pool, etc. With this approach, it has become more and more difficult to define the boundaries of the farm and consequently, its information system.
- Diversification of activities: the list of farm activities is no longer limited to the mere production of agricultural goods. For small farms, it has become more and more difficult to survive only with agriculture. The solvency of

a farm is today that much more in question as the farm's size is limited. Many small farms are, therefore, diversifying their activities. For example, very huge growth rates for "green tourism" can be observed in France.

Due to these different changes, the environment of the farm has become more and more complex. Information flows have rapidly increased and their nature is rapidly evolving. The management of the farm information system must take into account these new constraints.

If Information Technologies are part of the explanation for the increased complexity, it is also the occasion to offers occasion to discover new opportunities.

Development of Information Technologies Now Offers New Perspectives

In agriculture, as in the other sectors of activity, the growth of information technologies began about 20 yrs ago. This growth concerned two well identified domains :

- on the one hand, management of the farm,
- and on the other hand, process automation and on-board electronic devices.

These two domains, process and management, remained for long time disconnected, in part because the same actors did not participate in both of these domains and, also, because the functional link between process and management was not easy to establish.

The first function of on-board electronic devices on mobile equipment was to control the machine (measure of the strain in order to control the work depth of the plow and / or the tractor slippage rate, measure of the sprayer flow to control application rate of chemical products,...).

This function was therefore a process oriented function, and was considered relevant at the management level because data from sensors were (and remain today) real-time data, and were not used as data for management purposes.

In 1985, the Massey-Ferguson company designed one of the first devices for data interchange between a tractor and the farm (the Mémotronic device, or TRS for Tractor Recording System). Part of the data collected by the on-board computer Datatronic could be transferred, via a transfer device, from this on-board computer to the farm management computer (general data concerning implements, crop, chemical products, workers, etc. could be transferred in the reverse direction).

This device, of which only 10 copies were manufactured, was tested in France jointly by Massey-Ferguson, by software companies (ISAGRI, ENITA de Bordeaux) and by FNCUMA (Federation of Cooperatives for Farm Machinery Utilization).

Unfortunately it was not possible, at that time, to obtain a link between the on-board computer and major software used in France for crop management, for two main reasons :

- it would have been necessary to define new data and/or rules for decision-making concerning some data, such as for example *effective job time* and *stop time* (after what period of stoppage of the machine is the time no longer calculated as part of the effective job time? How can maintenance time and waiting time (which are linked to a task), and repair time (which could be independent of the current task) be differentiated? How can time for road travelling and for in-field working be distinguished?),
- it was necessary for the user to complete the part of the recordings that could not be automated, which considerably increased the work-load in manual data recording (what driver, what implement(s), what crop, what field, what product(s), what rate, etc.). This increase in the work-load was such that, often, recordings were incomplete, therefore unusable by software for crop management.

Starting at this period, much research work was undertaken with the objective to define what data could be collected by on-board computers, and what data could be transferred to (or from) the management computer (Zwanepoel, 1990).

A further step came from the change announced in electronic architecture terms for agricultural mobile equipment, with the replacement of analog electronics to a digital electronics architecture. This change means new capabilities in data recording and interchanging terms, especially for all data from sensors located on the implements. But such architecture necessitates a standardization of data because the data exchange has to be conceived of in the context of a multi brand environment.

In Germany, work on a bus for mobile equipment, based on CAN bus version 2.0A, was started in 1985 by the LAV. This work resulted in DIN standards (DIN 9684), which have now been published.

Since 1990, sub-committee SC19 of ISO/TC23 has been working on agricultural electronics standardization. Work is divided into three working groups:

- WG1, for data interchange between mobile equipment, and also between mobile process computers and management computers (this group is in charge of drawing up the ISO 11783 standard, which is close to DIN 9684 standard, except for the choice of CAN bus version 2.0B),
- WG2, for data interchange between stationary process computers and management computers (in charge of drawing up the ISO 11787 and 11788 standards). This group had defined ADIS (Agricultural Data Interchange Syntax), which is a syntax allowing for the use of an ISO data dictionary, as well as a national data dictionary or a private data dictionary,
- WG3, for electronic animal identification (in charge of drawing up the ISO 11784 and 11785 standards).

The adoption of the CAN bus by many agricultural equipment manufacturers, and the standardization of the linking tractor-implement (standard ISO 11783 or standard DIN 9684) will lead to a standardization of data that are going to be collected and exchanged. Those changes are going to transform the tractor into a technical data recording device, and allow for the use of this data for farm management.

Growth of Precision Agriculture

At least in Europe, precision agriculture is poised to take off. The emergence of this new concept is linked to the tremendous progress in information technologies. It is becoming possible, at a reasonable cost, to measure crop and field variability and, consequently, to optimize crop management. This is due to capabilities that are now offered in terms of measurement, registering and analysis of a very large quantity of information (Stafford, 1996).

These new capabilities, concerning management of in-field variability, will induce a profound evolution both in equipment and management software design. Indeed, by nature, precision agriculture establishes a very strong link between process and management computers :

- data collected is linked to the process itself (yield, soil characteristics, etc.), and is not connected (or less) to equipment control or automation. This data is useful for management and decision-making processes,
- collected data come from multiple sources (on-board sensors, soil sampling, remote sensing, etc.), and the work comparing miscellaneous data will be done, mainly, at the management level. Therefore, there is an obligation to transfer this information from the equipment to the management computer, in order to analyze it and to establish application maps. After that, it is necessary to transfer data again in the reverse direction, to carry out variable rate applications.

Precision Agriculture also necessitates having a reliable means, such as D-GPS, to be able to locate both measurements and actions conducted in the field. As a consequence, this necessity allows for bypassing a difficulty that was found using the TRS systems designed by Massey Ferguson. Indeed, thanks to a location system, it is now possible to determine automatically in which field the tractor or the equipment is working, or if it is travelling on the road, etc.

With this, a quasi-complete automatic data recording could be designed, so the driver work-load, involving manual data recording, could be reduced to near zero.

Additionally, the new demands from consumers concerning product quality and treacability, will incite farmers to buy equipment allowing them to automatically record what actions they have performed on crops. Tools developed for precision agriculture (location devices, mapping software, etc.) will be precious help in meeting these new demands in a simple manner.

These various tools that farmers are going to need, have to be designed taking into account the necessity of EDI (Electronic Data Interchange) at all levels: between mobile equipment (tractor-implement, between mobile equipment and management computer, between the farm and its technical or commercial partners, etc.) in a multi brand environment. The definition of a global information system of the farm is, therefore, more and more difficult to delineate when the boundaries of this area to be covered are far-flung.

Towards a Global Information System

To develop multi-function software, which would allow for the integration of both process and management domains and that would meet the expectations of farmers, necessitates defining a global Information System.

This objective, however, may appear overly-ambitious. A simpler method would consist of connecting, by mean of gateways, the various Information Systems developed for specific problems. Some data could be exchanged between these Information Systems without the farmer having to enter them several times.

But with such a method, the coherence of data could not be ensured: updating data in a given Information System must be carried out in the other Information Systems in which this data appears.

The definition of a global Information System ensures the coherence of data all the time. The Dutch (Goense & Hofstee, 1994), have already proposed such a method with the CIA (for Computer Integrated Agriculture).

The CIA model is set up as one model for the whole primary agricultural sector. This model was built on a high level of abstraction. For example: processes, such as feeding or fertilizing, are described as one process *executing work*. The work is not executed on a field, a crop or an animal group, but on an *account object*. This theoretical approach offer a way to save cost and time in software designing.

Our approach is the same as concerns the need of a global Information System, but the way chosen to design it was different. For us, there exists a strong need to ensure the capability of EDI for technical data from the farm to the outside, especially to the final consumer. To our mind, the farmer will have, in the future, to record in a complete manner all the technical operations carried out in the field. He will have to be able to supply the chain of information for treacability of products and technical operations (the farmer must be able to provide answers to questions like: where, when, what action, what product, what quantity, etc.).

In a multi-actor environment, the use of data that is easily understood by everyone appears to us as mandatory, and as more important than the reduction of costs or time spent in developing software..

Methodology Used

During the initial phase, we inventoried data elements appearing in main management and technical software used by farmers in France, and compared the definition of these data elements (Persiault, 1996). This inventory allowed us to identify basic data found in the relevant software, as well as more marginal data.

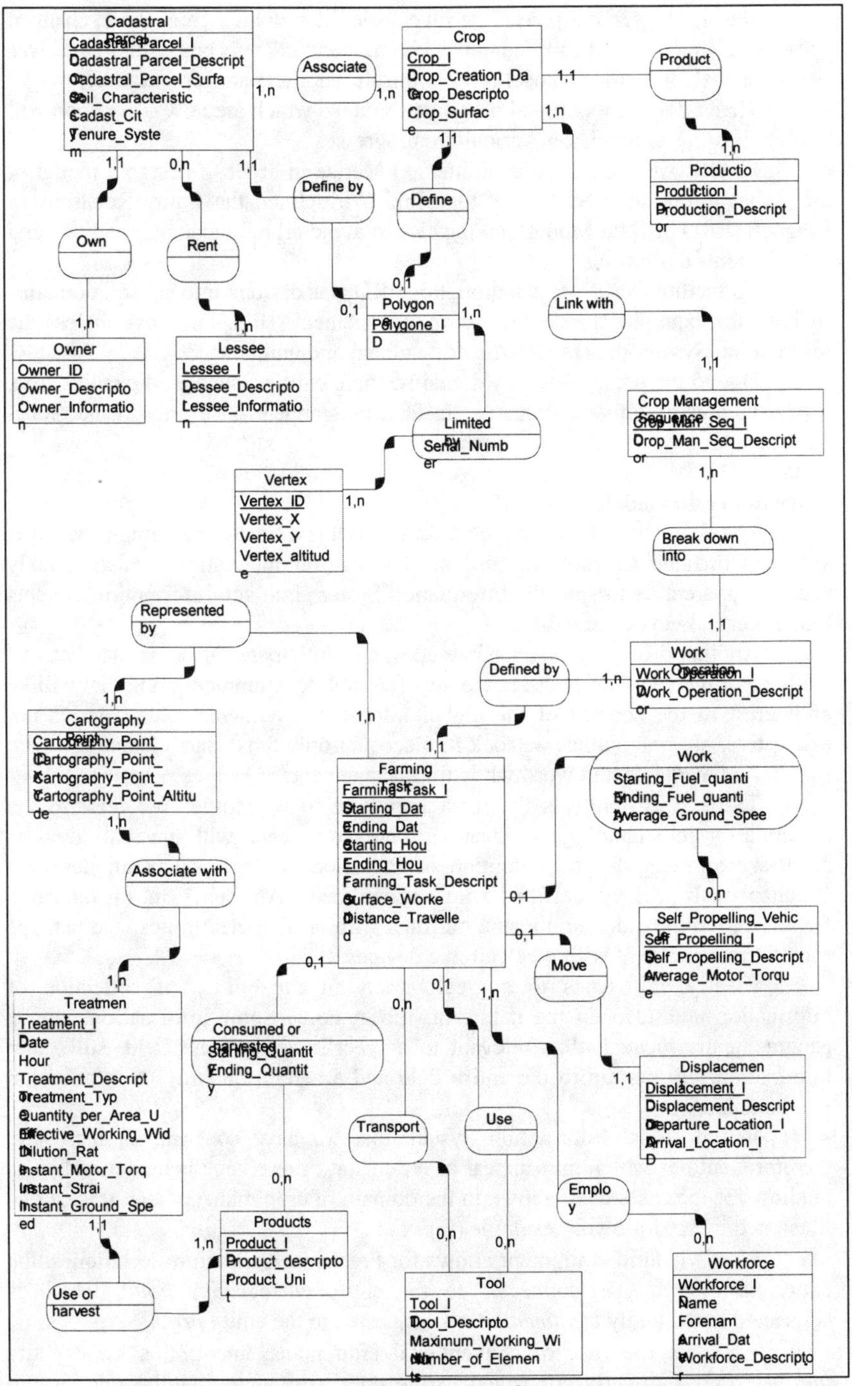

Fig. 1. : Entities relationship diagram for crop management.

During the second phase, we inventoried data which could be exchanged between a tractor and implements, via a standardized bus: this data was simultaneously identified as useful (or not useful) at the management level.

Thereafter, we examined new data elements which are now — or soon will be — appearing as Precision Agriculture progresses.

Finally, we used the French method Merise in order to design a model of the global Information System of the farm. We defined the Entity Relationship Diagram (ERD or Data Model), taking care to avoid all redundancy in entities and data elements definitions.

To facilitate ease presentation, this ERD was divided into several domains, such as, for example, the domain "crop management" (Fig. 1). Nevertheless the Information System thus modeled is truly global and unique.

Due to the presentation by domains, some entities, such as the entity crop, appear in several domains, whereas, in fact, they appear once only in the global data model.

Discussion and Conclusion

The designing of such a global data model is a time-consuming procedure, and it is difficult to carry to fruition. The natural inclination, which quickly became apparent, is to split the Information System into sub-information systems (one for each kind of domain).

Another difficulty we encountered was to anticipate data elements that will really be provided by on-board computers (on mobile equipment), and that will be of interest in the context of the global Information System. Our work is not exhaustive, but pragmatic: we took into account only those data elements we are sure to find available and worthwhile for management.

The last difficulty is the most important to overcome. We refer to the dissimilarity in technological levels: some equipment will have all devices necessary allowing for the adoption of Precision Agriculture techniques and, simultaneously, will be outfitted with standardized CAN bus. But on the same farm there will be older equipment, outfitted with analog electronics, and perhaps some farm equipment with no electronic devices.

Clearly, this leads to a great variety in capabilities of collection of information and, also, in the nature of information. Some information will be geographically located, thus relevant to a specific area of the field, still other information will pertain to the entire field and a certain quantity of information will be unknown.

In the global Information System that we have constructed, there are, therefore, entities which may appear as redundant : these have been defined so as to allow for the cases cited above. In the domain of crop management, this can be illustrated by the following example (Fig. 1) :

- if farm equipment allows for Precision Agriculture techniques, the entity *farming task* is connected to the entity *cartography point*, which is connected to the entity *treatment*, itself connected to the entity *products*,
- in the case of conventional crop management, this same entity *farming task* is directly connected to the entity *products* and the relationship concerns only quantities of products applied (at a constant rate) or harvested.

The circularity that seems to appear in the entity diagram relationships is necessary in order to process, differently, these two possibilities that are linked to differing technological levels.

The current phase of our work encompasses the completion of a multi-function software (accounting, crop management, etc.) based on this global Information System. The first version of this software should be operational before the end of this year.

REFERENCES

Goense D., J.W. Hofstee, 1994. *An open system architecture for application of information technology in crop production.* p. 1356–1363. *In* Proc. of the 12th World Congress on Agricultural Engineering, CIGR, Milano. Aug. 29 – Sept. 1 1994.

Persiault, F. 1996. *Elaboration d'un dictionnaire de données pour l'EDI en agriculture designing of a data dictionary for EDI in agriculture* M.S. thesis. ENITA de Bordeaux.

Stafford, J.V. 1996. *Essential technology for precision agriculture.* p. 595–604. *In* Proc. of the 3rd Int. Conf. on Precision Agriculture, Minneapolis, MN. 23–26 June, ASA, CSSA, and SSSA, Madison, WI.

Taponier S., and Desjeux D. .1994. *Informatique, décision et marché de l'information en agriculture*, L'Harmattan.

Zwaenepoel P. 1990. Standardization of the communication between agricultural equipment and farm computer. *In* Proc. of the Int. Conf. on Agricultural Engineering, AgEng 90, Berlin, 24–26 Oct. 1990.

Technology-Mediated Learning Environment in Precision Agriculture

U. Sunday Tim
Department of Agricultural and Biosystems Engineering
Iowa State University
Ames, Iowa

ABSTRACT

The changing demographics of American higher education are placing new demands on institutions. In 1994, for example, an estimated five million working adults were enrolled part-time in U.S. colleges and universities. That number masks an even larger adult population who want to pursue a college education or take a course but cannot attend a traditional college because of campus inaccessibility, inconvenient class hours, or family responsibility. Colleges and universities are required to respond to these paradigm shifts in education by reengineering courses and curricula to incorporate emerging technologies that break the constraints of time and place of learning. Realizing the potentials and promise of emerging technologies in facilitating change, colleges and universities are navigating toward networked, student-centered learning environments. This paper describes major components of an authentic, Web-based learning environment in precision agriculture that incorporates an interactive multimedia instructional/resource materials, virtual field trips, K-12 activities, glossary and links, and other resource materials that enhance learning.

INTRODUCTION

Recent economic pressure and increasing global competition in the production of food and fiber has caused farmers and producers across the nation to adapt sustainable production practices and keep pace with rapid advances in technology. Moreover, farmers realize that the American public expects agriculture to engage in production practices that are environmentally sensitive. To meet these challenges, farmers are seeking new production systems and technologies that enhance productive efficiency and profitability and at the same time do not deplete the natural resource upon which they depend. A farming system that is gaining widespread acceptance is precision farming, also called precision agriculture, site-specific farming, site-specific crop management, and site-specific resource management.

Precision agriculture benefits from the rapidly evolving spatial information technologies including global positioning systems (GPS), geographic information systems (GIS), remote sensing (RS), yield monitors, and variable rate technology. Precision agriculture offers an unparalleled ability to characterize the nature and extent of variability within a field and to develop sound farm management practices in response to such variability. In general, the potential benefits of precision agriculture include: (i) collection and manipulation of

Copyright © 1999 ASA-CSSA-SSSA, 677 South Segoe Road, Madison, WI 53711, USA. *Proceedings of the Fourth International Conference on Precision Agriculture.*

spatially referenced field data for improved understanding of farming systems; and (ii) precise placement of agricultural inputs to improve net economic return, environmental quality, and global competitiveness. By allowing optimal use of resources on a site-specific basis, precision agriculture has been shown to be economically consistent with and in many cases superior to conventional farming systems. Despite these benefits, however, many in the agribusiness and agriculture education community believe that certain missing links must be bridged in many areas if precision agriculture is to reach its full potential. One such area is education and training to: (i) provide much needed trained and educated workforce, (ii) enable agricultural science and engineering graduates gain increased confidence in the ability to perform in the highly technological area of precision agriculture, and (iii) provide professional training and continuing education to a large constituent of the adult workforce. With the projected growth of the precision agriculture industry in the next millennium, there is a lack of experienced and competent workforce to help design, implement, and manage technological systems such as GIS, GPS, and RS. Anecdotal evidence from agribusiness and the industry indicates a gap between the demand and supply of well-qualified graduates, interns, and professionals to occupy new jobs in precision agriculture.

Higher education in agriculture is changing very rapidly. The way students learn or the way instructors deliver their lessons is being altered by the recent developments in high-performance computing, information systems, and communication technologies. A growing body of evidence indicates that these technologies can expand access to learning resources and improve the education experience regardless of where and when that learning takes place. In particular, information and communication technologies such as the Internet and the World Wide Web (WWW or the Web) can enable learners to access educational materials whenever and wherever they want. In so doing, these technologies have weakened the grip of traditional learning experiences and have made geographic, social, and political boundaries in education less relevant. There are now hundreds of examples of course materials being made available via a central, hyper-linked Web environment and significant number of institutions are delivering distance education courses over the Web. Typical examples include the Open University (http://keats.open.ac.uk/) and the World Lecture Hall (http://www.utexas.edu/world/lecture/index.html). Several consortia of universities and individual universities now offer online courses through the Web.

The explosion of the Web in popularity and accessibility, and as an effective means to disseminate course and curriculum materials, is shaping the way institutions provide distance or distributed educational experiences to their student clients. Hiltz (1993) remarked that a technology-enhanced educational environment facilitates collaborative learning as well as active and independent learning, and exceeds the traditional classroom in its ability to provide student access to course materials on an around-the-clock basis.

A recently released National Research Council (NRC) report "Precision Agriculture in the 21st Century: Geospatial Information Technologies in Crop Management" has emphasized the need for development and implementation of education and training programs in precision agriculture. The report (NRC, 1997) concluded

> *"In the twenty-first century, agricultural professionals using information technologies will play an increasingly important role in crop production and natural resource management. It is imperative that educational institutions modify their curricula and teaching methods to educate and train students and professionals in the interdisciplinary approaches underlying precision agriculture."*

Presently, there is a paucity of curricula, course offerings, and distributed learning environments that provide educational experiences for on- or off-campus learners in the field of precision agriculture.

This paper describes the core components of an authentic learning environment developed to promote on- or off-campus education, professional training, and outreach programs in precision agriculture. The learning environment also supports an undergraduate course in precision agriculture, which is delivered synchronous and asynchronous to a diverse group of learning communities. The remainder of this paper is organized as follows. The first section details the primary components and features of the authentic learning environment in precision agriculture. This is followed by a description of the instructional–resource materials. The final section summarizes a plan to implement and evaluate the benefits and the potential educational impact of the learning environment.

THE LEARNING ENVIRONMENT

During the past decade, development of interactive hypermedia learning environments using high-performance computing, information, and communications technologies has generated new learning experiences for students. Various models of learning environments have been developed, including the use of the Internet and the Web as an information delivery tool. Under this model, content-specific instructional materials for a course are made available to students via the Web, while the instructor and students meet in the traditional classroom. At the other end of the continuum, a growing number of institutions are moving beyond the use of the Web to deliver instructional materials and have developed virtual learning environments that incorporate emerging multimedia and hypermedia technologies.

The authentic learning environment for education and training in precision agriculture articulates the constructivist principle of learning (Rowland, 1995; El-Tigi & Branch, 1997) and promotes, as Dede (1995) succinctly puts it a case-based learning by doing. The learning environment was developed to: (i) provide an alternative, self-directed, and learner-centered environment for knowledge acquisition in precision agriculture; (ii) provide a communications medium for collaborative learning; (iii) provide comprehensive online learning resources that enhance on- or off-campus course offerings in precision agriculture; (iv) provide authentic activities for all levels of education hierarchy; (v) develop basic resources and learning materials that are scalable or adaptable for use in other curricula and courses; and (vi) accommodate diverse community of learners and learning styles.

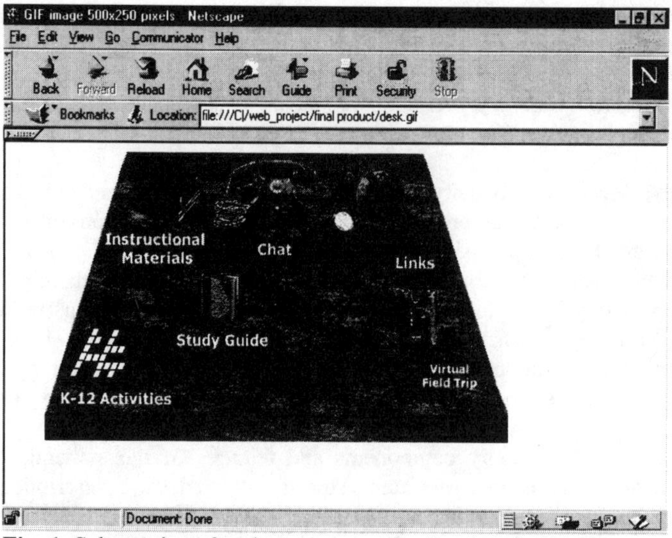

Fig. 1. Schematics of major components of authentic learning environment.

Figure 1 shows the major components of the authentic learning environment in precision agriculture. The major features of the learning environment include: comprehensive, modular instructional–study materials; bulletin board systems; study guide; searchable glossary of terms and concepts in precision agriculture; extensive list of frequently asked questions (FAQs); resource materials such as links to software and equipment vendors, interactive chat facilities; K–12 activities; and virtual field trips. The learning environment also features uniquely designed activities for K–12 students. These activities enforce concepts in agricultural science and technology and include crossword puzzles, interactive word search, and other fun games. Some of the features in the learning environment utilize the multidimensionality of the Web such as hyper-linking and interactive forms, as well as virtual reality modeling language (VRML) and Java. Both Java and VRML are used to create simulations, tutorials, and activities that improve understanding of basic fundamentals of precision agriculture.

High performance computing and emerging platform-independent programming languages (such as Java) have also enabled new tools to be developed to enhance learning. For example, virtual excursions to tourist areas, which are common on the Internet and the Web, can have considerable learning value, particularly if the learner experience them together, share and discuss their experiences, and reinforce them with other learning inputs. Within the learning environment, virtual field trips will be used to provide visual, cognitive, and experiential base, and to motivate further activity and learning in precision agriculture. Here, virtual field trip is used as a metaphoric name for the idea that interactive, networked-accessible sites involving integrated texts, images, videos, and animated data are so uniquely organized that a single or group of remote learners gets the feeling of actually being in the location. A virtual field trip to a museum, for example, can be organized to mimic or even improve upon what one would see in a real working tour of a museum.

The multimedia aspects of most virtual field trip sites also hold significant educational value. Beyond the reality of almost instant and expansive access to vital information lies another benefit of virtual field trip—the opportunity to engage visual materials and simulation tools and techniques for visualizing data. The field of precision agriculture, for example, demands assimilation of a wide range of images, maps, graphics, and data, and generally requires these to be synthesized if it is to enhance understanding of underlying principles. To this end, a number of virtual field trips have been designed for the learning environment in precision agriculture. This includes the combination of images, videos, animations, and virtual reality simulations to mimic crop harvesting and yield monitoring. At the end of the virtual field trip participants are engaged with short multiple choice questions, developed using Java, to demonstrate how much they have learnt about the concepts or topics related to the trip. Simple simulation exercises are provided to further enforce understanding of fundamentals of precision agriculture.

THE INSTRUCTIONAL MATERIALS

A variety of instructional materials design models have been developed over the years to represent different learning styles and educational contexts (Edmonds et al., 1994; Gustafson, 1991; El-Tigi & Branch, 1997). Dick and Carey (1996) proposed a systematic instructional materials design model that consists of five distinct, but interrelated, phases: analysis, design, development, implementation and evaluation. Heinich et al. (1993) present the ASSURE model that incorporate analysis of learners, stating of objectives, selection of media and materials, utilization of media and materials, requirement of learner participation, and evaluation and revision. A combination of these models were used in the design of instructional and resource models for precision agriculture education. The instructional and resource materials created for the learning environment exhibits the following characteristics: (i) provide authentic learning experiences and enhance students understanding of key concepts in and technologies of precision agriculture, (ii) focus on substantive up-to-date information, and (iii) provide opportunity for active involvement of learners in group investigations and experiments that enhance their quantitative reasoning and problem solving skills.

Figure 2 summarizes the structure and contents of the instructional materials for the learning environment. The current prototype instructional and resource materials consist of fourteen lecture modules to support an undergraduate course and experiential learning in precision agriculture. These modules cover important concepts, topics, and technologies of precision agriculture, as well as issues of implementation and adoption. Each module constitutes an instructional framework of cognitive apprenticeship in which the apprentice (i.e., learner) is engaged in learning through interactive study materials, resource materials, assignments and review questions, pre-tests, self-paced quizzes, and the case library consisting of maps, graphics, data, research reports,

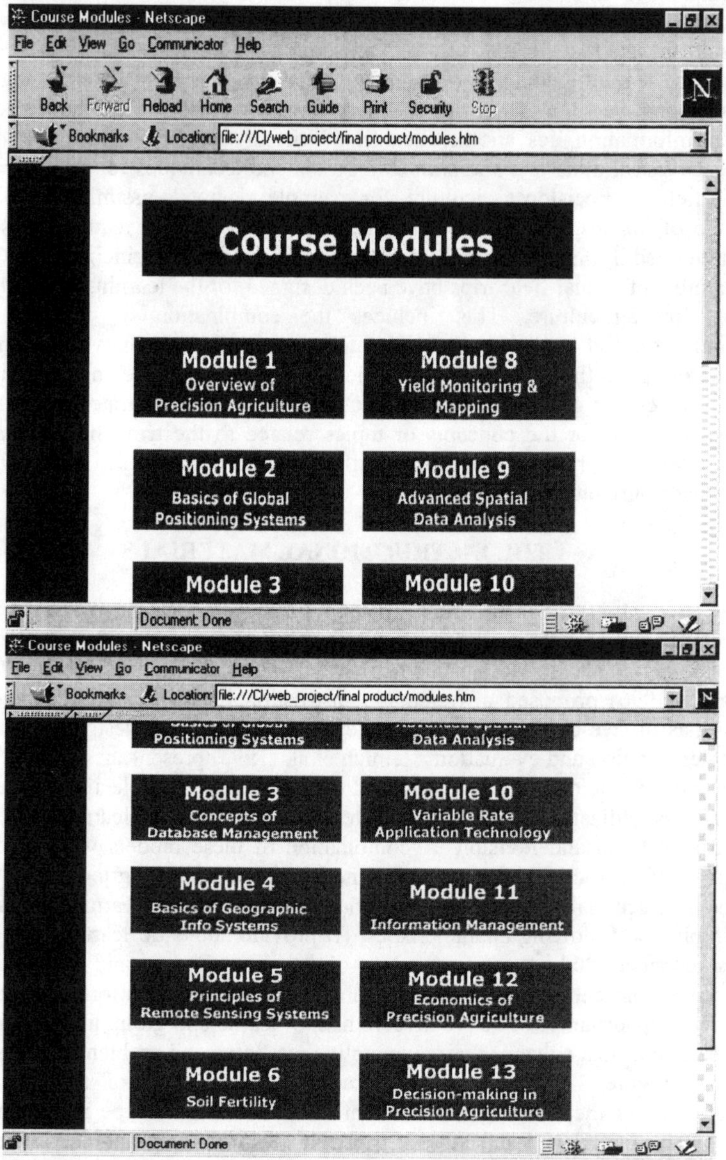

Fig. 2. Partial view of the contents of the instructional modules.

and results from demonstration projects in precision agriculture. Overall, the instructional materials are created such that the user (i.e., learner) is provided with concise, up-to-date, and highly relevant information in precision agriculture and is not overwhelmed with too much detail.

The interactive study materials for each module can be accessed through a table of contents page, which presents an overview of the module, the learning objectives, topics covered in the module, suggested reading lists and reference materials, a set of key terms, and review questions. Each topic is linked to a set of notes that correspond roughly to the set of transparencies about the topic that can be used in a classroom. Underneath these notes are elaboration pages that discuss the specific topic in more depth. Some of these elaboration pages contain links to maps, graphics, video clips, visual images, and interactive activities. In each topic, the learner can select a set of exercises designed to deepen his/her understanding and comprehension, apply knowledge to new situation, and be engaged in the construction of artifacts that are the focus of the topic. The resource materials that accompany instructional materials consists of links to services in precision agriculture, as well as links to background materials, case studies, and Web pages on software, hardware, equipment, and jobs.

IMPLEMENTATION AND EVALUATION PLAN

A debate continues in the curriculum design research literature about whether Web-based learning environments improve learning. This debate stems from the observation that, after many years of research on instructional media, no consistent significant effect in learning outcomes from any medium has been demonstrated. The debate becomes more interesting and complicated when the Web is used as a medium for learning, as opposed to a tool for delivering instructional materials or courseware. There is an obvious interest to contribute to this debate and to determine the potential impact, value, and benefit of our learning environment. The logical approach to this impact assessment is through implementation and evaluation of the learning environment. Our implementation plan involves the use of the learning environment in an undergraduate course titled *Precision Farming Systems (AST 333)* given every fall semester since 1996. The course currently has an enrollment limit of 40 students and this limit has been exceeded since the first year of offering. Participants in the course are derived from several colleges and curricula.

A large body of literature exists related to the evaluation of distributed learning environments and instructional materials (Dick & Carey, 1990; Biner et al., 1995; Reeves, 1992). Bramble and Martin (1995) evaluated a series of military courses taught using different distance learning technologies with the goal of addressing issues of academic performance and degree of student interactivity. Biner et al. (1995) conducted an analysis of factors associated with student satisfaction in a college-level, video-based distance learning course. For the learning environment and instructional materials described in this paper, two phases of evaluation are planned--formative evaluation and summative evaluation. The primary objectives of these evaluations will be to determine if the contents of the learning environment meet the needs of a diverse community of learners and to also assess whether significant changes to the content, resource materials, study materials, and assessment tools (e.g., quizzes and assignments) are warranted. Formative evaluation will use standard instruments and survey techniques to collect information on not only the structure and function of the learning environment, but also on its impact on learning in highly technological area of precision agriculture. Summative evaluation will involve the combined use of focus groups and questionnaires to query students (and learners) about the benefits, consistency, and completeness of the instructional materials including the virtual field trips and interactive activities. Summative evaluation will also collect information on student's reaction to the study notes, pre-tests, and self-paced quizzes. Overall, implementation and evaluation of the learning environment and its contents will provide us the opportunity to test novel approaches to design and delivery of authentic learning environments in precision agriculture.

SUMMARY

The convergence of information, high-performance computing, and communications technologies is having a profound impact on not only the workplace and home, but also on higher education and the way students learn. Colleges and universities across the globe are under increasing pressure to improve the quality of educational programs and at the same time increase the output and reduce cost. Many policy makers see technology as one way of meeting these requirements. Technology such as the Internet and World Wide Web is providing an increasingly rich variety of media through which to present learning materials and to enhance learning. The easy accessibility of the Internet has allowed educational institutions to design, implement, and evaluate authentic learning environments.

In this paper, important components of an authentic learning environment for precision agriculture education and training were described. The learning environment addresses the need for development of instructional materials and learning resources that enable diverse community of learners to attain high level of competence and skills in precision agriculture. The prototype learning environment, as well as the instructional and resource materials can be accessed at the following URL: http://www.ae.iastate.edu/Ast333x/ast333.html). The learning environment is password restricted. Interested users should contact the author by sending e-mail to tim@iastate.edu. When fully implemented, the learning environment will not only meet the increasing demand for flexible, structured, and self-paced virtual learning environments in precision agriculture, but also strengthen outreach and continuing education programs in agriculture. One of the many realities of a changing American economy and revolution in information technology is the fact that life-long learning skills have reached a stage of paramount importance. Americans who have the required technical skills will have a competitive advantage in the marketplace of the next decade.

ACKNOWLEDGMENT

Funds to support development of the learning environment for precision agriculture education were obtained from the College of Agriculture and Iowa State University Extension.

REFERENCES

Biner, P.M., R.S. Dean, and A.E. Mellinger. 1994. Factors underlying distance learning of televised college-level courses. Am. J. Distance Educ. 8:60–67.

Bramble, W.J., and B.L. Martin. 1995. The Florida tele-training project: Military training using compressed video. Am. J. Distance Educ. 8:60–67.

Dede, C.J. 1995. The evolution of constructivist learning environment: Immersion in distributed, virtual worlds. Educ. Tech. 35 46–52.

Dick, W., and L. Carey. 1990. The systematic design of instruction. 3rd ed.. Scott, Foresman & Company, Glenview, IL.

Dick, W., and L. Carey. 1996. The systematic design of instruction. 4th ed.. HarperCollins College Publ., New York.

Edmonds, G., R.C. Branch, and P. Mukherjee. 1994. A conceptual framework for comparing instructional design models. Educ. Tech. Res. Dev. 42:55–62.

El-Tigi, M., and R.B. Branch. 1997. Designing for interaction, learner control, and feedback during Web-based learning. Educ. Tech. 37:23–29.

Gustafson, K.L. 1991. Survey of instructional development models. 2nd ed.. ERIC Clearinghouse on information resources. Syracuse Univ., New York.

Heinich, R., M. Molinda, and J.D. Russell. 1993. Instructional media. Macmillan Publ., New York.

Hiltz, S.R. 1993. Correlates of learning in a virtual classroom. Int. J. Man-machine Stud. 39:71–98.

National Research Council. 1997. Precision agriculture in the 21st Century: Geospatial and information technologies in crop management. Natl. Academy Press, Washington, DC.

Rowland, G. 1995. Instructional design and creativity: A response to the criticized. Educ. Tech. 35:17–22.

Weather Induced Variability in Site-Specific Management Profitability: A Case Study

R. P. Braga
J. W. Jones

Agricultural and Biological Engineering Department
University of Florida
Gainesville, Florida

B. Basso

Crop and Soil Sciences Department
Michigan State University
East Lansing, Michigan

ABSTRACT

Economic profitability is a key issue in the adoption of any new technology that requires high investment. Most of the published work on precision farming profitability is being done assuming extremely well behaved crop response curves for major inputs. Tools like crop simulation modeling and new wave optimum search procedures can be used to overcome these limitations. This paper presents the results of a case study on the profitability of N site specific management in a rainfed corn field. A set of 35 yrs of weather data is used along with CERES-Maize crop model to compare the conventional N application schedule with the averaged optimized site specific one. The benefits of N site-specific management are extremely variable with soil type and weather year.

INTRODUCTION

The spatially-variable crop management (SVCM) market has been characterized by being extremely technology driven. Inevitably, this leads to strong speculation on the returns and cost-effectiveness of farmers' investments. The lack of adequate and long-term research facilitated this situation. Although several economic and agronomic questions are yet to be answered, the agricultural engineers have already a myriad of solutions. Because the technical and the economic optimums rarely coincide, research is needed to get insight into problems like: Does it pay to farm more precisely? If it pays, what level of precision is needed for farmers to get more profit? What is the risk associated with weather variability? Does it pay for most of the years or only in the good ones?

This paper focuses on a case study to answer some of these questions. The theoretical framework of this problem is similar to the one presented in Schmitz and Moss (1997). Its contribution relies on the tools used (adaptive simulated annealing and crop simulation models) and the incorporation of risk associated with weather. Also the inter-relationships between soil type and weather are addressed. Does weather risk vary with soil type?

Copyright © 1999 ASA-CSSA-SSSA, 677 South Segoe Road, Madison, WI 53711, USA. *Proceedings of the Fourth International Conference on Precision Agriculture.*

Reetz and Fixen (1995) point out some desirable characteristics of economic analysis for the agricultural sector, in general, and for site-specific management in particular. Any economic analysis should be as realistic as possible. Only in this way does it provide sufficient detail to evaluate individual components of a technology package and show how each affect the cost and return per unit area as well as per unit costs of production. It should also give useful information for different levels of each technology component and be flexible enough to be easily updated for price changes of inputs and products. Ultimately, a measure of the risk associated with each package–component should be incorporated.

Crop simulation models are easy-to-use tools that describe the complex relationships between crop varieties, soil types, weather, management practices and crop yield. The response curve of any crop yield to major inputs is a never smooth theoretical curve. The agricultural system is too complex too allow that to happen. Most crop models operate on a daily time step and use dry matter production rate and phenology as key processes. Potential biomass production rate is mainly driven by photosinthetically active radiation. The produced biomass on any day is then partitioned between the plant organs (stems, roots, leaves or grains) that are growing at that time. The phenology of the crop is estimated using the accumulation of daily thermal time, basically the daily average temperature minus the appropriate base temperature. So, by dynamically simulating the growth, development and partitioning of a crop, these models are capable of giving yield estimations for combinations of factors and able to do it in a non-continuos manner if that is the case (almost always true for most common yield limiting factors). Thornton and McGregor (1988) have shown the usefulness of this tool in identifying the optimum management regimes for crop enterprises.

The use of crop simulation models to derive the yield response curve limits the type of search algorithm to be used to look for the best management practice. More classical optimization algorithms require yield response curves to be derivable. That is not the case for curves generated from a crop models. So, two distinct paths can be followed at this point: derive regression models from the crop model's output, or, use an alternative optimization algorithm. In this present study we have chosen the second path for three reasons. Deriving regression models results in filtering the crop model output which might lead to information loss. Classical optimization algorithms do not perform acceptably in complex problems (multi-optima). Finally the alternative optimization algorithms are extremely attractive for their simplicity, flexibility and ease of use.

The main weakness of classical optimization algorithms is exactly their inability to escape from local minima. Also, an extreme dependence on the initial estimates and neighborhood size makes these methods inappropriate for complex cases. Of course, if several initial conditions and neighborhood size are tested, the performance can improve significantly, but the computational time will increase as well.

The new generation of optimization algorithms overcomes these disadvantages. Particular reference is made to Simulated Annealing (SA), Tabu Search and Genetic Algorithms (Press et al., 1989). If fact, these algorithms guarantee the universal minima by exploring the objective function's entire surface and trying to optimize it while moving both uphill and downhill. This, of course, is done through the acceptance of a temporary deterioration of the objective function. An implementation of SA developed by Goffe et al. (1994) was applied in this study.

The objective of this study is to try to answer some of the questions related to the economical rationale of adopting site-specific crop management practices. The most important questions are: is it more profitable to farm with higher level of detail? If it is, at which level of detail? and how does that profitability change with the weather year? In other words, is site-specific crop management more or less risky than conventional farming. This study specifically addresses the total amount of nitrogen to apply (split between sowing and side-dressing) to a corn field with four major types of soil. The paper also illustrates how tools like crop models and optimization procedures can be applied in such studies.

MATERIAL AND METHODS

The case-study focuses on a 60 ha non-irrigated corn field planted with a high potential yielding hybrid in Michigan. The field has four major types of soil with different organic matter contents and soil water related properties (water holding capacity). The variable-rate N sprayers are available on a rental basis increasing the total nitrogen application cost to \$9 ha^{-1}. The farmer is mainly concern with increasing net returns. To keep on farming the conventional way there is no need for any additional investment, and the total nitrogen applied is 180 kg ha^{-1} split in two applications (100 + 80).

The soil types are assumed to be perfectly separated and, because of the different properties, have different yield potentials. The soil type will coincide with the management units and will be addressed to as Soil1, Soil2, Soil3 and Soil4. The proportion of each soil in the whole is respectively: 41, 15, 32, and 12% of the whole area.

The conventional N application to be compared with the site-specific management was based on the 180 kg ha^{-1} nitrogen application (100 + 80). The site-specific management was based on the averaged N management for the optimized net return of each soil and year. The comparison between both management practices was then made based on the area weighted field net return average.

The general expression for the net return is:

$$NR = Y \cdot G_p - N \cdot N_p - Y \cdot H_c - F_c - SSC$$

Where NR is the net return in (\$ ha^{-1}); Y is the crop yield in (kg ha^{-1}); Gp is the grain price in (\$ kg^{-1}); Hc is the harvest cost (\$ kg^{-1}); N is the amount of applied nitrogen (kg ha^{-1}); Np is the nitrogen price (\$ kg^{-1}), Fc is the fixed cost (\$ kg^{-1}) and SSC is additional cost for site-specific management (\$ ha^{-1}). For the whole field, NR can be written as:

$$NR = \sum_{i=1}^{n} \frac{a_i}{A} \left(Y_i \cdot G_p - N_i \cdot N_p - Y_i \cdot H_c \right) - F_c - SSC$$

Where n is the number of management units, a_i the area of management unit i and A the total area of the field. Furthermore, it is assumed that G_p is constant year to year and H_c is always dependent on the amount of grain harvest (the fixed part of harvest cost is included in F_c). The values used were: $G_p = 0.13$ \$ kg^{-1}, $H_c = 0.0055$ \$ ha^{-1}, $N_p = 0.3$ \$ kg^{-1}, $F_c = 195$ \$ ha^{-1} and SSC = 8 \$ ha^{-1}. It was assumed that no additional cost arrives from doing two applications rather than one in case the optimal management dictates no nitrogen applied in either sowing or dressing.

The crop model used was Ceres-Maize V3.1 (Hoogenboom et al., 1994). The weather input was a set of 35 yrs (1961 to 1995) of daily values of radiation, maximum and minimum and precipitation temperature for East Lansing, MI. The four soil types were created from existing soil surveys of the field. The corn was sowed on 5 May, 7.2 plant m^{-2}, 56 cm rows (22 in.). First N application was applied on the day of sowing and the side dressing took place 40 days later.

To get the best N management practice for each soil and year, the constraint of N established for both the amount at sowing and at dressing was 0 to 150 kg ha^{-1}. This results in a total of 22 500 possible combinations. The described procedure to find the site-specific crop management for each soil (average for the optimized 35 yrs) was used because it would not be realistic to assume that the farmer would change the N amount on a year by year basis.

RESULTS AND DISCUSSION

Figure 1 shows the yield and net return response to the different weather years for each soil type using conventional N management. Soils 1 and 4 have higher physical and economical potential with average net returns of 931.90 \$ ha^{-1} (standard deviation 336.20) and 937.39 \$ ha^{-1} (standard deviation 335.10), respectively. On the other hand Soils 2 and 3 have a considerably poorer: averages of net return of 567.20 \$ ha^{-1} (standard deviation 378.96) and 675.10 \$ ha^{-1} (standard deviation 369.33), respectively.

Figure 2 shows a subset of the loss function evolution during the simulated annealing process. It is noticeable that it not only evolves to better solutions, but it also searches new areas assuming temporary worse solutions. Figure 3 shows the searched combinations of nitrogen level in the optimization process. It can be seen that the procedure does not search randomly (as often thought) but explores the whole area first and then concentrates on the most promising areas. Of the 22500 combinations, only about 2000 to 3000 are searched.

Figure 4 shows the effects of different weather years 1962, 1988, and 1970, respectively bad, average and good years, on the corn production curve in response to total nitrogen applied for Soil 4. Figure 5 shows the effect of soil on the corn production response to total N applied.

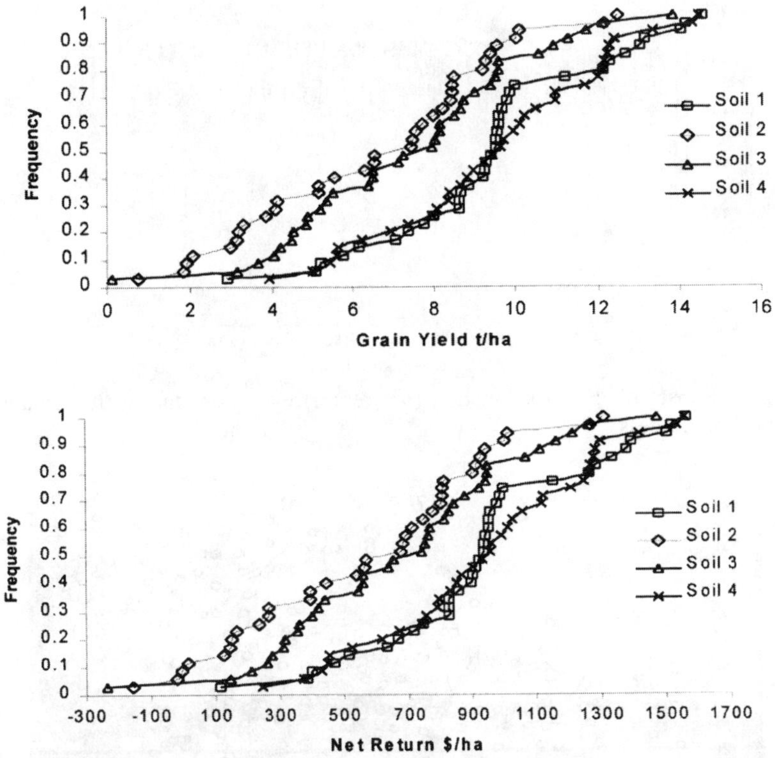

Fig. 1. Yield and net return response to the different weather years for each soil type used in the case study.

As expected, better weather years led to higher grain yields for similar N applications. The N response curve shows that the model is sensitive to the effect of excess application. These response curves are good illustrations of the usefulness of process-based crop models in economic analysis. No theoretical input response curve, intrinsically empirical, could capture the complexity of interactions involved. In this approach, each combination of management practice, soil type and year are unique and taken into account in the economic analysis. Lumping soil types and weather years do not accurately describe the studied scenarios.

Table 1 shows the resultant N application amounts for both sowing (N1) and side-dressing (N2) that optimize net return for each soil. These represent the average optimized management for each combination of soil type and weather year.

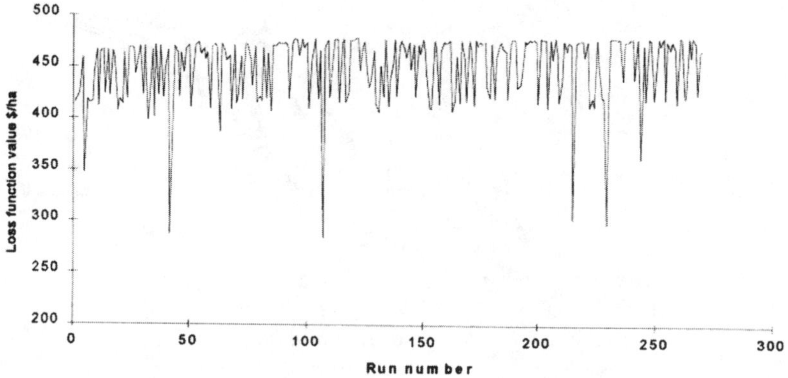

Fig. 2. Subset of the loss function (net return) evolution during the simulated annealing process.

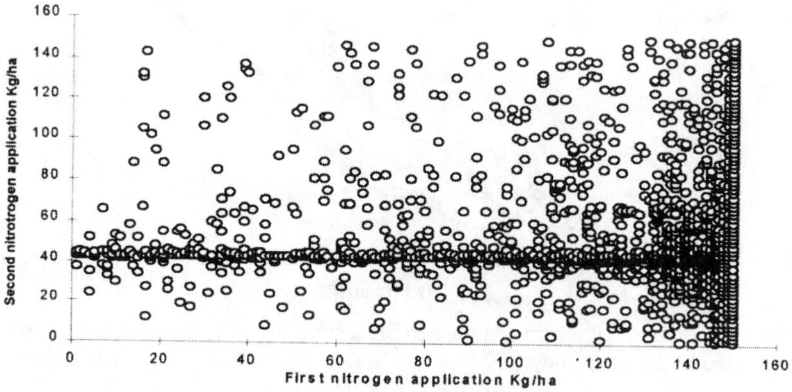

Fig. 3. Searched combinations of N level in the optimization process.

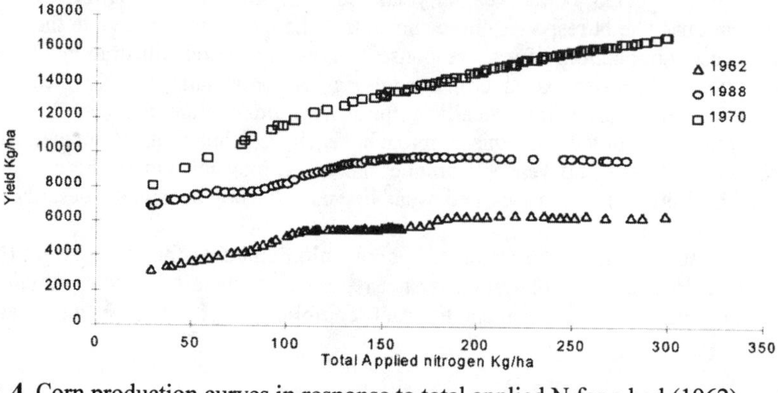

Fig. 4. Corn production curves in response to total applied N for a bad (1962), and average (1988) and a good (1970) weather year.

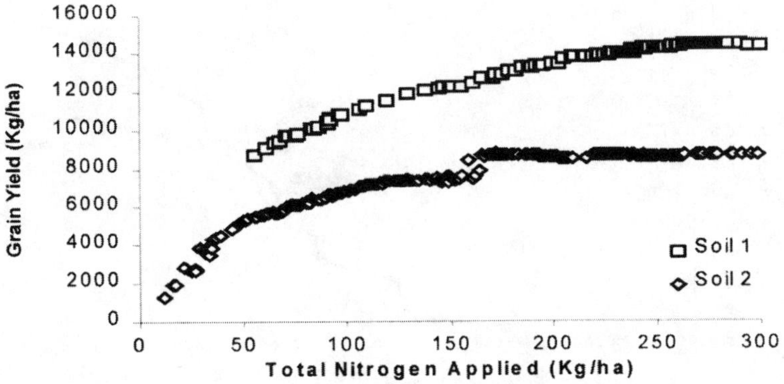

Fig. 5. Corn production curve in response to total applied N for a high potential soil (Soil 1) and a low one (Soil 2).

There was no relevant relation between soil yield potential and N application strategy. The soils with higher potential (1 and 4) did not show any higher or lower total applied N. The same is true for the distribution between sowing and side-dressing applications where no pattern could be associated with the soil potential. This observation also suggests the complexity of the interactions involved and the need for simulation tools.

In Figure 6 the cumulative frequency curve of the whole field net return for both conventional and site-specific management is presented. Figures 7 through 10 show the same data for each soil type. Table 2 presents net return averages and standard deviation for the whole field and each soil as well.

Although, whole field gains for site-specific N management are more frequently higher relative to conventional, the results are highly dependent on soil type and its proportion in the field. Also evident is the tendency for similar results of the two management strategies in extreme conditions (either very good or bad weather years). In Soils 2, 3 and 4 the benefits of the site-specific management were not obvious. So, if Soil 1 were predominant, the whole field results would have been considerably higher. This illustrates the obvious importance of the proportion of areas with different potential soils and its impact on whole field results.

Table 1. Averaged optimized nitrogen amounts for both sowing (N1) and side-dressing (N2) applications (Kg/ha).

Soil	N1	N2	Total
1	139	43	182
2	40	140	180
3	150	41	191
4	122	45	167

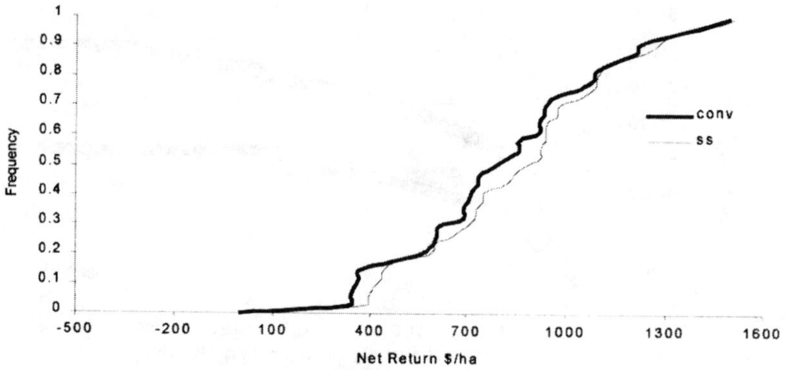

Fig. 6. Cumulative frequency curves for the net return of the whole field with conventional (conv) and site-specific (ss) N management.

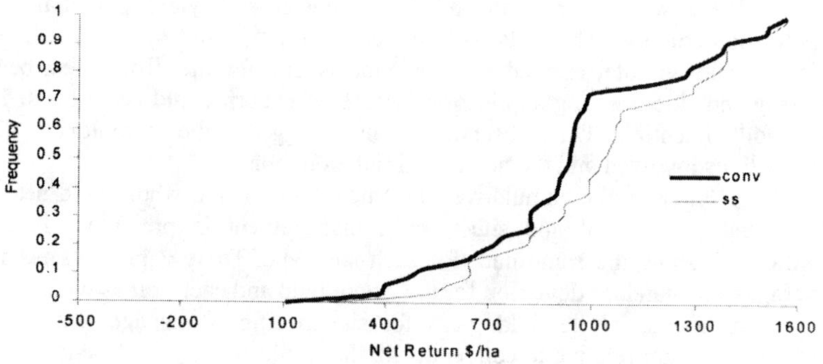

Fig. 7. Cumulative frequency curves for the net return of Soil 1 with conventional (conv) and site-specific (ss) N management.

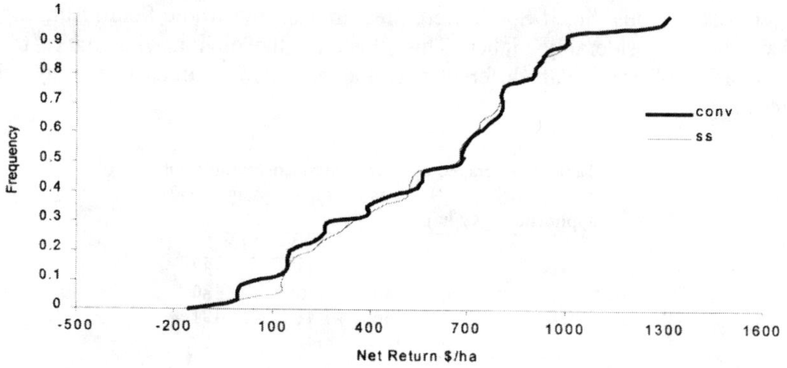

Fig. 8. Cumulative frequency curves for the net return of Soil 2 with conventional (conv) and site-specific (ss) N management.

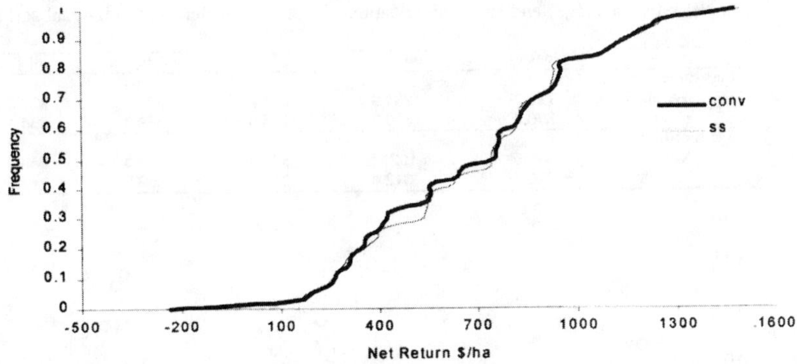

Fig. 9. Cumulative frequency curves for the net return of Soil 3 with conventional (conv) and site-specific (ss) N management.

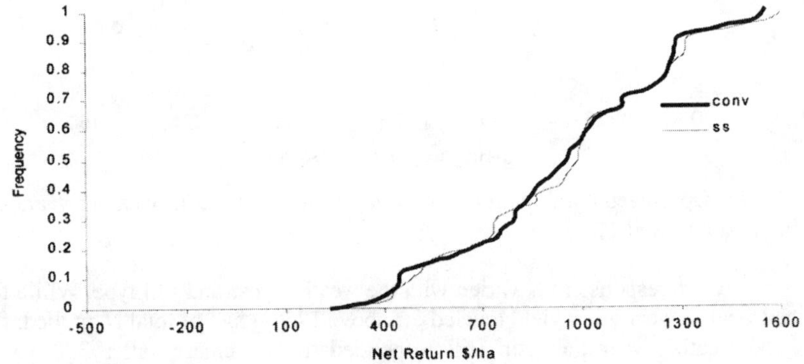

Fig. 10. Cumulative frequency curves for the net return of Soil 1 with conventional (conv) and site-specific (ss) N management.

The fact that the spatially variable N application had a higher per unit N price also had a major impact on the results. This factor shifted the recommendations to lower N amounts, positioning the crop at a lower yield performance. It would have been interesting to see the impact of future price drops in the analysis.

The site-specific management was not able to reduce the risk associated with weather. Both the whole field and the individual field results have similar standard deviations for the set of 35 weather years. These results might reflect a drawback of the approach used to prescribe the site-specific N management. Considering averaged optimized recommendations moves one step behind in the spatial and temporal crop management. Different results are expected if instead of optimizing for 35 individuals years independently, we had optimized for the same 35 yrs in a sequence considering the carry-over effects. Figure 11 illustrates the variability found for optimized N applications for Soil 1 (for the other soils, results are similar).

Table 2. Net return averages and standard deviations for the whole field and individual soil type ($/ha).

	Whole field	Soil 1	Soil 2	Soil 3	Soil 4
Conventionl					
Average	795.7	931.9	567.2	675.1	937.3
Stdv	336.3	336.3	378.9	369.3	335.1
Site-specific					
Average	841.5	1027.5	579.7	685.4	949.9
Stdv	326.7	312.4	361.8	369.1	342.6

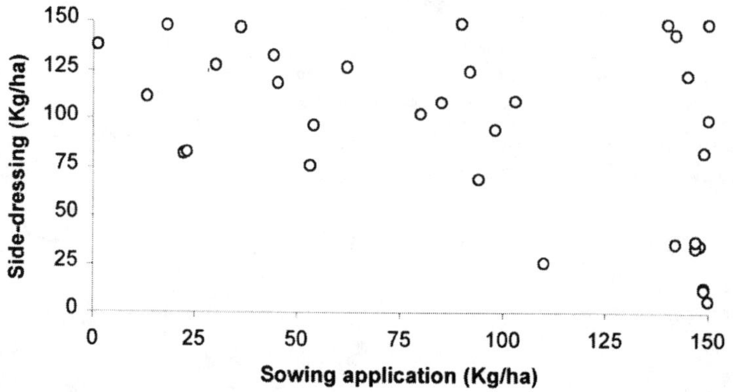

Fig. 11. Optimized combinations for N applications in the 35 weather years used in the study (Soil 1).

Corn response to N varied with the weather year and soil type. While for a bad weather year grain yield leveled off above 175 kg ha^{-1} of total N applied, for a good weather year the crop still responded to N increases after 300 kg ha^{-1} applications. On the other hand, a high potential yielding soil only leveled after 275 kg ha^{-1} total N applied, while a poorer soil this occurred above 175 kg/ha. This demonstrates that it is impossible to capture the complexity of crop response to main inputs by using classic well-behaved models.

The benefits of N site-specific management are extremely variable with weather year, soil types existing in a field and the proportion of each soil within the field. Cumulative frequency curves for the returns of both management options shown than for some years, specifically very good and very bad ones, the benefits of site-specific management is not evident. On the average, the whole field gain was 45.8 $ ha^{-1}. The gains were not consistent for all soils. It varied from 95.1 $ ha^{-1} to 12.5 $ ha^{-1}, suggesting the importance of the proportions of the different soil type in the field.

Site-specific management was not able to significantly reduce weather risk associated with weather. It had a net return standard deviation of 326.7 $ ha^{-1} whereas for conventional management that value was 336.3.

These results stress the importance for developing more detailed economic analysis for major investments that take into account the specific characteristics of each field, weather and soil environment. In other words, an economic analysis to provide valuable information for spatially variable crop management has to very close to reality. In this task, process based crop models and advanced optimization procedures can play a decisive role.

REFERENCES

Goffe, W.L., G. Ferrier, and J. Rogers. 1994. Global optimization of statistical functions with simulated annealing. Journal of Econometrics, 60: 65–99.

Hoogenboom, G., J.W. Jones, P.W. Wilkens, W.D. Batchelor, W.T. Bowen, L.A. Hunt, N.B. Pickering, U. Singh, D.C. Godwin, B. Baer, K.J. Boote, J.T. Ritchie, and J.W. White. 1994. Crop models. p. 95–244. *In* G. Y. Tsuji et al. (ed.). DSSAT Version 3. Vol. 2. Univ. of Hawaii, Honolulu.

Press, W.H., B.P. Flannery, S. Teukolsky, W. Vetterling. 1989. Numerical recipes. The art of scientific computing. Cambridge Univ. Press.

Reetz, H.F., and P.E. Fixen. 1995. Economic analysis of site-specific nutrient management systems. p. 743–752. *In* Proc. of the 2nd intl. Conf. on Site-Specific Management for Agricultural Systems. Minneapolis, MN. 27-30 March 1994. ASA, CSSA, and SSSA, Madison, WI.

Schmitz, T.G., and C.B. Moss 1997. Insvesting in precision agriculture. Submitted to Quarterly Journal of Econ.

Thornton, P.K., and M.J. McGregor 1988. The identification of optimum management enterprises. Outlook Agric., 17: 158–162.

Agroindustrial Sector Global Evaluation System

George Sugai
Department of Agriculture and Animal Science
University of Brasília
Brasília-DF, Brazil

Francisco Assis Nascimento
Department of Electrical Engineering
University of Brasília
Brasília-DF, Brazil

Paulo Roberto Meneses
Department of Remote Sensing
University of Brasília
Brasília-DF, Brazil

Jogi Takechi, and Yoshihiko Sugai
Secretariat for Strategic Management - SEA
EMBRAPA-Brazilian Agricultural Research Corporation
Brasília-DF, Brazil

ABSTRACT

In this paper what has been developed is a mathematical model with the objective of determining stages for developing operational procedures such as harvest, handling, transportation, industrialization, and distribution that, helped by a business management system, gave room for the creation of not only an instrument of making decisions, but also an instrument of administration

Agroindustrial Spatial Decision Support System

The use of the space and time are still the main characteristics of the farming and cattle raising sector. The present state of the agriculture has not been independent from the natural resources, land, climate, precipitation, vegetation as its main factors.

Land is still agriculture's basic resource; the results of its use are yet regarded as a fundamental factor of production. Simultaneously, time also has its share giving characteristics to the farming and cattle raising. It is not possible to harvest corn within 24 hr from planting as well as it is not possible to abate cattle within 24 hr of its birth, reaching 250 lb for the carcass. A lot of time is yet to be spent in order to move further in such direction, inasmuch as the technological progress is still slow.

The development of models in terms of the functioning of the farming and cattle raising sector has advanced. The first experiences in the field refer back to the 1950s and took place in the University of Iowa State.

Copyright © 1999 ASA-CSSA-SSSA, 677 South Segoe Road, Madison, WI 53711, USA. *Proceedings of the Fourth International Conference on Precision Agriculture.*

The university still keeps its group of technicians that has always dedicated to such a study, coming up with more refined results each time. More recently, the U.S. Agricultural Department and the Canadian Government have also invested in this sort of project.

Nowadays, our technical staff has shown results concerning the application of these models, and now we are using such technology in Brazil together with our partners, therefore searching for the continuous development and betterment of a growing number of users of this technology.

The businessman knows that the global economy, away from being just a terminology, became an everyday action. Such a man also knows that the market place is not moved by sudden impulses, and that it takes into account data of great importance. Moreover, the businessman is well aware that information is vital for making the right decisions.

The new companies that work looking forward to the 21th century know that such valuable issue — information — can't be set aside.

The Agroindustrial Sector Global Evaluation System tries to support companies with strong technical basis looking planning decisions.

Planning accounts for the allocation of scare resources in several alternatives. Therefore, a well-structured system can preview and project a demand for a product, evaluate the impact of agrarian policies, and analyze the agroindustrial capacity on scientific basis.

The modern systems must be both practical and operational, providing data, evaluations, projections, simulations and suggestions for those who make decisions dealing with on-line and real-time basis therefore maximizing the productivity.

The use of our system will allow a global analysis of the deveopment of the agricultural sector both at macroeconomic and microeconomic levels, taking into account the aspects related to the allocation of the resources in the productive process, acquisition of products, industrialization, distribution to inward market, as well as to imports and exports. It will also allow a more accurate analysis of ex-ante, and ex-postagricultural policy.

As an initial stage, this system is concerned with characterization, organization, improvement of product information, transportation, storage, refrigerator, slaughter, agroindustries and consumption. The next stage consists of the building of the model, its applications, and its political and economical interpretations.

One of the major difficulties pertinent to such a project is the huge amount of necessary information about its various components. Although such information exists, it can't be promptly used in terms of the analytical project presented.

A big effort will be carried out viewing information improvement, harvest, organization, systematization and manipulation of data in order to make them available, usable, and automatic to the system.

Data will be integrated into the system which will build a matrix to be developed(Linear Programming Models) and later provide a report. Together with SAP/R3, this report will permit a more efficient administration which will, consequently, solve complex issues regarding planning and management of the productive chain. In addition, through a geographical information system(GIS), it will be possible to have a clear view of all the agroindustrial complex (Fig. 1).

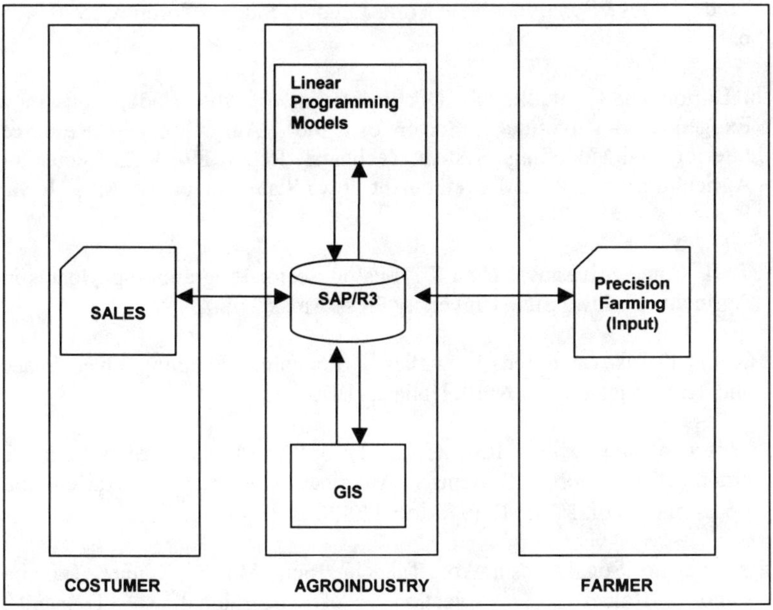

Fig. 1 Agroindustrial decision support system.

CONCLUSION

Indeed, what has been developed is a mathematical model with the objective of determining stages for developing operational procedures such as harvest, handling, transportation, industrialization, and distribution so that helped by a business management system, provides for the creation of not only an instrument of making decisions, but also an instrument of administration and geographical reference.

The real intention is to develop a management tool provided with complex systems – market simulation and investment system – business administration, byophysical and hydrologic models, and geographical information system that make part of a perfect balance.

REFERENCES

Baumes, Harry S. and McCarl, Bruce A . "Linear Programming and Social Welfare: Model Formulation and Objetive Function Alternatives". Canadian Journal of Agricultural Economics vol.26(3). 1978.

Benevenuto, A. Projeção da Demanda de Grãos no Brasil – 1992/95, mimeografado – SEA/EMBRAPA, 1992.

Bressler, Jr. , R. G. and King, Richard A . Market, Prices and Inter regional Trade. Wiley&Sona, Inc. New York, London, Sidney, Toronto, 1970. 426 p.

English, Burton and C. Ettalli. "A Documentation of the Endogenous and Exogenous – Livestock Sector of the Agricultural – Resource Interregional Modelling System. Technical Report 89-TR12. Center for Agricultural and Rural Development, Iowa State University, Ames Iowa, U.S.A

Heady, Earl O.and Srivastava, Uma K. "Spatial Sector Programming Models in Agriculture". Iowa State University Press/Ames. U.S.A 1975.

Judge, G. and T. Takayama (eds.), Studies in Economic Planning Over Space and Time. Amsterdam: North Holland. 1973

Ladd, George W. and Dennis R. Lifferth, 1975, "An Analysis of Alternative Grain Distribution Systems", American Journal of Agricultural Economics, Vol. 57, n° 13 (agosto): 420-30.

Menezes, Luciano, Sugai, Yoshihiko, Takechi, Jogi, Marra, Renner, Teixeira Filho Rafael "Projection of Regional Demand". CEE/SEA/EMBRAPA, Brasília, DF, Brazil. 1997

Peeters, Ludo. "An EC Feed Grain Spatial Equilibrium Model for a Policy Analysis", Center for Agricultural and Rural Development, Iowa State University, Ames Iowa, U.S.A Working Paper 90 – WP6I, June 1990

Sistemas de Produção dos Centros Nacionais de Pesquisa de Arroz e Feijão, Milho e Sorgo, Trigo e Soja da Empresa Brasileira de Pesquisa Agropecuária – EMBRAPA. Mimeografado, 1991.

Sugai, Y. "Global Model of Agricultural Sector – Grains in Brazil"N°. 25 of Agricultural Policy Study Series, IPA – PNUD/UN. Ministry of Planning, Brazil, 1994.

Swarchiter, Jayme L. "Grafos e Algoritmos Computacionais". São Paulo. Editora Campus Ltda. 1989.

Taylor, C.R., Blokland, P.J., Swanson, E.R. and K.K. Frohberg. "Two National – Equilibrium Models of Production: Cost Minimization and Surplus Maximization". Urbana, University of Illinois. 1977.

Thomas D., J. Straus and M.M.T.L. Barboza. Estimating the Impact of Income and Price Changes on Consumption in Brazil. Pesquisa e Planejamento Econômico, 1991.

Willet, Keith. "Single – and Multi-commodity Models of Spatial Equilibrium in a Linear Programming Framework". Canadian Journal of Agricultural Economics, 31/1983.

Management-Oriented Model Guided Within-Season Nitrogen Management

MengBo Li and Russell S. Yost

Department of Agronomy and Soil Science
University of Hawaii at Manoa
Honolulu, Hawaii

ABSTRACT

Within-season management decision aids are important for precision NO_3 management because the final fate of the N in soil–plant systems largely depends on within-season events and management. Uncertainty in future weather challenges N models for within-season management. MOM uses weather forecasts to estimate rainfall in the near future and simulates other components in the soil–plant systems. In addition to its management-oriented optimization, MOM-guided within-season management has the advantages of (i) High efficiency in predicting timely information. Users are advised of the probable N status of soil–plant systems in advance of sensors and soil tests. (ii) Low cost to implement. No within-season soil or tissue sampling and testing are required except an initial soil test. (iii) Transparency of the systems' status. Daily descriptions of the N cycle in soil-plant systems during the cropping season graphically advise users how to control the fate of N. MOM also presents within-season estimates of leachate NO_3 and mineralized N, which are not provided by standard soil tests. Within-season observed data on precipitation and crop growth update MOM-guided management with current events, which improve the precision with which MOM traces the N cycle in soil–plant systems. MOM-guided within-season management was not designed to match future events exactly, but to dynamically adjust probable consequences of management strategies to fit changing conditions within a cropping season. Two scenarios illustrate how MOM can help in precision N management for maximizing profits and yields while minimizing NO_3 leaching by updating management of irrigation and fertilization within-season.

INTRODUCTION

Within-season management decision aids are very important to N management because the final fate of N in soil–plant systems largely depends on the within-season management. Nitrogen is the plant nutrient that is applied in the largest quantity, is the most costly, and is environmentally most often hazardous. The status of inorganic N in soils rapidly changes with crop growth and management practices during a cropping season. In tropic upland field conditions, almost all ammonium was converted to NO_3 in one month (Khan et al., 1986). Nitrate is highly mobile in most soils and can easily move below the root zone, where NO_3 becomes inaccessible to crops and is subject to leaching into groundwater. In high leaching risk areas with sandy soils or heavy rainfall, N fate is very sensitive to rainfall events,

Copyright © 1999 ASA-CSSA-SSSA, 677 South Segoe Road, Madison, WI 53711, USA. *Proceedings of the Fourth International Conference on Precision Agriculture.*

and to the rates and timing of N fertilization and irrigation. Theoretically, N management strategies can be scheduled and planned off-season based on historical weather data, in which the N fertilizer supply can be synchronized with the plant needs. However, the uncertainty in rainfall results in sub-optimal management if based on a fixed schedule. For example, if unexpectedly heavy rainfall occurs during or just after a large application of N, a significant amount of N fertilizer intended for crop uptake may be leached beyond the root zone. The N fertilizer intended for crop growth then becomes a potential N pollutant, which imposes two impacts on the soil–plant system. One is a shortage in N supply for crop growth and another is the risk of damaging the environment. Facing this kind of within-season change, the management schedule should be flexible for within-season adjustment to avoid serious leaching events while supplying sufficient N for crop growth.

Real time control is an approach to within-season management. For example, sensor technologies and soil tests are used to provide feedback of soil–plant systems for adjusting management practices. When a N model is used to guide within-season N management, it must deal with the uncertainty of coming weather. Stochastic models are usually designed to estimate spatial and temporal uncertainties in prediction; however, so far stochastic models of soil–plant systems are difficult to apply to field management (Ling, 1996). No model guided within-season N management has been reported.

Management-Oriented Modeling (MOM) developed by Li and Yost (1998a) is not a stochastic model, but a two-way modeling system that simulates natural processes and human management practices. This two-way modeling style can be used in a real time control system for within-season management. Objectives of this study were (i) examine weather generator models and monitoring N with soil tests or sensors for within-season management, (ii) construct a prototype (concept and method) of model (MOM) guided within-season management to dynamically optimize N management, (iii) test the prototype with two scenarios of N management in tropics. MOM is designed for multiple objectives and multiple users. When MOM focuses on assisting N management within a real cropping season, this MOM working mode is called MOM-guided within-season management, differing from models for general management.

Weather Generator Models

For many agricultural models, one of the greatest uncertainties is future weather. To deal with this unknown, a stochastic weather generator is usually added to N simulation systems. An example of this type of weather generator, WGEN, was developed by Richardson and Wright (1984). WGEN works as a random weather generator that produces daily maximum and minimum temperature, rainfall, and solar radiation based on latitude and longitude for the cotton-heliothis hybrid systems (Stone & Schaub, 1990). Another weather simulation model is NATCOVER, which was developed to improve the accuracy of crop simulation models (Wang & Whisler, 1996). Constructed from historical daily weather data for the GOSSYM/COMAX cotton simulation model, NATCOVER generates several weather patterns based on long-term climatic data under normal conditions, and with six other hypothetical weather scenarios.

Weather generators are useful in agricultural models for general application in research, teaching, planning, and assessment. If models are used for within-season applications, weather generators still face challenges. Stone and Schaub (1990) compared the weather data generated by WGEN with measured field data and historical averages. Patterns of the generated data looked much like actual field data; however, the generated data were less likely to match field conditions than the average data on any particular day. Although the averages provide better estimates on given days, Plant and Stone (1991) concluded that the average season is not a realistic approximation of a particular season's weather. There is no such as a thing as a typical year.

Soil Tests and Sensor Monitoring Within season

Binford et al. (1996) proposed within-season soil testing for monitoring nitrogen management based on sugar beet experiments. In experiments during the 1993, 1994, and 1995 seasons in western Nebraska and Wyoming, 10 rates of N (0 to 304 kg N ha^{-1} in 34 kg N ha^{-1} increments) were applied in four replications before planting. Soil samples, 0 to 30 cm, were collected at 2-wk intervals and analyzed for NO_3 concentration. They found that net returns to N fertilization decreased significantly as the soil NO_3 concentration increased to 40 mg N kg^{-1}. On-site soil NO_3 testing as a method of monitoring N fertilizer management was also suggested by Marx et al. (1996). The on-site monitoring practice was called the pre-sidedress soil nitrate test (PSNT). Nitrate was extracted from 10 mL of field-moisture soil, measured by displacement, and analyzed using a quick-test field kit. The field test results were adjusted for differences caused by soil texture and moisture, based on correction factors calibrated from standard laboratory methods.

Another technology used in nitrogen within-season management is real-time control using sensors. To monitor crop N status, Schepers et al. (1996) mounted N sensors on a high-clearance sprayer and interfaced with the spray control system of the equipment. The sensor readings from the adequately fertilized strip were compared with those from adjacent strips that were likely to develop N stress. If needed, N fertilizer was applied to field strips in the spring. Blackmer and White (1996) reported using remote sensing to identify spatial patterns of N fertilization. Sensor technologies have been successfully applied in many agricultural fields, including monitoring N fertilization for some crops based on tissue N deficiency. Nitrogen sensors monitor crop N status by detecting its color. Sensors can only determine crop N requirements when it can detect the symptom of a crop N stress. For many crops, however, N fertilizer should be applied a few days or a week in advance of when a crop needs N or the crop shows N stress. In other words, the sensor's recommendation for N fertilizer is usually too late to be effective in meeting the N demand of these crops. This situation occurred in most experiments reported by Schepers et al. (1996).

Using MOM to Guide Within-Season Management

To make decisions for within-season N management without a decision-aid, a decision-maker usually first collects the necessary data that include current crop N status, soil N and moisture status, and precipitation (Fig. 1). The data can be collected from soil tests, sensors, and weather forecasts. Then the decision-maker analyzes the data and estimates the amounts and timing of N fertilizer and irrigation needed in the following weeks.

Considering the capability of MOM in predicting N status in soil-plant systems at acceptable accuracy (Li & Yost, 1998; Yost & Li, 1998), MOM was also developed for within-season simulations to optimize nitrogen management, called MOM-guided within-season management. If MOM is calibrated and validated to local conditions, it can simulate the above decision-making process without within-season data from soil tests or sensors, except for initial site conditions. Figure 2. illustrates the concept that MOM navigates within-season N management by mimicking the process above. The primary purposes of MOM-guided within-season N management are (i) use simulated data to substitute for within-season soil and tissue test data, (ii) use the model to dynamically monitor and select appropriate within-season N fertilization and irrigation. Assuming MOM was calibrated and validated for a specific site and a decision-maker made initial soil tests for soil N and moisture just before planting, we describe MOM-guided within-season N management (Fig. 2):

1. Run MOM before planting to schedule seasonal management strategies, based on initial conditions of the soil-plant system and weather data from historic averages (or from weather generators if available). MOM simulates the N cycle of the entire cropping season and predicts the optimum possible crop yields, profits, leached NO_3, and other outputs, and describes the required management strategies of N fertilizer and irrigation.
2. Run MOM weekly (or at shorter intervals if necessary) to update and monitor the current status of N in the soil–plant system.
   - 2.1 Input actual precipitation, irrigation, and N fertilization of the past weeks to update current status of N and water in the soil–plant system of the model. MOM displays the simulated status of the soil–plant system before TODAY in solid lines (Fig. 3).
   - 2.2 Input the forecast amounts of precipitation for the following weeks if they are significantly different from those in MOM databases. Then MOM resimulates and updates the status of the soil–plant system after TODAY, shown in dotted lines (Fig. 3). MOM also reevaluates the management strategies after TODAY and updates the management schedule for the following weeks.
3. <u>Rainfall</u> is always uncertain during within-season management but is important to simulate N movement in soils. Weekly observed precipitation inputs are necessary for MOM-guided within-season management. Updating weekly air temperature and ET is not required by MOM unless the season climate changes fundamentally. Some simple functions of crop uptake N related to the crop <u>growth observations</u>, such as $N_{uptake} = f(LAI, Height_{crop})$, can be established by users. These functions can be used to update growth

curves of the crop demand N by easily observed data. Updating crop demand is helpful when the crop grows abnormally due to unpredicted events such as serious pests, diseases, or crop damage. Within season soil test data are also helpful in tracing soil N status, but not required because MOM uses its simulated data to monitor the N status of the soil–plant system (Fig. 2). If MOM is calibrated and validated to a specific site, the accuracy of the simulated soil data may not be less than those of limited field sample tests (Yost & Li, 1998).

4. MOM-guided within-season management can run in many ways:

4.1. On-site or on-field. Users can run MOM for a specific crop in a field. A single MOM database is needed for a cropping season.

4.2. On-farm or on-watershed. Multiple MOM databases are needed for diverse crops and soils in an area. Farmers, watershed managers, extension agents, and consultants may wish to use MOM this way.

4.3. Soil test reports with MOM. If a sample analysis requires recommendations for N fertilization, MOM may be useful for consultants in (i) demonstrating various N management options to clients, (ii) optimizing the recommendations for within-season management, and (iii) illustrating the effects of recommendations on the soil-plant system in advance.

Within-season management using MOM is a dynamic optimization process. MOM always optimizes the management schedules at weekly or shorter intervals, based on within-season monitoring and updating of the status of the soil-plant system. A perfectly optimal management schedule for the whole cropping season is not guaranteed by MOM-guided within-season management, because of many uncertainties beyond the control of either decision-makers or models. PAST events cannot be revised before TODAY (Fig. 3). However, MOM-guided within-season management can improve management in the near future (e.g., the following week) after TODAY. For example, NO_3 leaching events occurred in days 5 to 7 and days 22 to 26 after planting in Fig. 3, no matter whether these were failures of the rain forecasts or other reasons. MOM just simulates the past situation and focuses on changing the management schedule after TODAY to avoid the coming leaching peak in the near future (30–35 d after planting). In other words, MOM can help the decision-maker reduce the anticipated leaching peaks by selecting an alterative schedule of fertilization and irrigation. MOM uses a weather forecast to predict effects of coming water events on the N cycle in soil–plant systems. This in turn allows MOM to search new management schedules to update recommendations that adjust for the coming effects. MOM does not try to perfectly control all events during the whole cropping season but it dynamically traces crop requirements and updates fertilization and irrigation schedules to fit changing weather in the near future. In comparing Fig. 3 (status at 28 d after planting) with Fig. 4 (status at 91 d after planting), MOM did not exactly predict the patterns of water events and NO_3 leaching between 28 and 91 d after planting. MOM reduced nitrate leaching, however, by updating management of irrigation and fertilization during the period. This situation illustrates a key purpose of MOM-guided within-season management: *not to match future events exactly, but to update model systems within-season and dynamically adjust management strategies to fit changing conditions*. In addition to its management-oriented optimization, using MOM-guided within-season management has the advantages of:

1. **High efficiency**. Users can know the probable status of N and water in their soil-plant systems in a few seconds, without waiting for the results of sample tests. MOM-guided within-season management advises when and how much N that a crop demands in advance of sensors and soil tests.
2. **Low cost**. There is no cost of the N sensor hardware, and no within-season soil or tissue sampling and testing other than an initial soil test.
3. **Transparency**. Daily pictures of the simulated N cycle in soil–plant systems graphically show users predictions of where their N fertilizers would have been, where they will be, and how to control them. MOM also presents estimates of N fate in soil–plant systems such as leachate nitrate and mineralized N, which are not provided by standard soil tests.

A major limitation of MOM-guided within-season management is that all simulation results are based on the assumption that MOM is correctly calibrated and validated to specific sites and crops; however, the discussion above illustrates that MOM-guided within-season management is promising for precision N management.

Scenarios of MOM Optimized Management

Validation of MOM and MOM guided within-season management should include two parts: simulation of natural processes and simulation of management activities, because MOM was designed as a two-way modeling tool (Yost & Li, 1998). Validation of MOM with respect to the natural processes is to compare the agreement (or closeness) of the model predictions with the observations. This is a validation of the MOM simulator, N-SIMULATOR, which was initially evaluated by Yost and Li (1998). The validation of management activities is to compare the differences between the results of existing management practices produced with those that MOM suggests. It is to test whether MOM improved the existing management strategies or not, and the degree of improvement. This validation requires datasets that consist of, at least, two types of observed data: results produced under existing management practices and results produced under MOM-guided management. In addition to analytical data of soil–plant systems, the dataset should include profits, yields and leachate NO_3, which are three sub-goals of MOM. Unfortunately no such dataset was available when MOM was initially developed; however, an approximate test of MOM predictions for N management can be obtained by testing its simulator (Li & Yost, 1998). To evaluate effects that MOM adjusts N management to fit changing conditions during a cropping season, two scenarios of MOM-guided N management for upland crops in the tropics were examined in this study. The information associated with data sources was not released here because the discussion of MOM scenarios may imply that some existing practices might have negative effects on the environment, which was not determined in this study.

To evaluate how much difference that MOM made from the original management practices, all compared data are simulation results based on the original management practices or MOM suggested management. Profit ($ ha^{-1}) in MOM is simply calculated by

$$
\begin{aligned}
Profit = {} & MarketValue \cdot CropYield / (1 + LoanInterest) \\
& - Price_{Fertilizer} \cdot Amount_{Fertilizer} \\
& - Price_{Water} \cdot Amount_{Irrigation} \cdot 10\ t\ ha^{-1}/mm \\
& - Cost_{FertilizerToFarm} \\
& - Cost_{(FertilizerToField + LaborToApply)} \cdot Applications_{Fertilizer} \\
& - Cost_{(WaterToField + LaborToIrrigate)} \cdot Applications_{Irrigation} \\
& - Cost_{FertilizerInIrrigation} \cdot Applications_{FertilizerInIrrigation} \\
& - OtherCost
\end{aligned}
$$

The economic factors and units in scenarios were assumed as follows:

$$MarketValue = 0.50\ \$\ kg^{-1}$$
$$LoanInterest = 12\ \%\ year^{-1}$$
$$Price_{Fertilizer} = 0.80\ \$\ kgN^{-1}$$
$$Price_{Water} = 0.05\ \$\ tonne^{-1}$$
$$Cost_{FertilizerToFarm} = 5.00\ \$\ ha^{-1}$$
$$Cost_{(FertilizerToField + LaborToApply)} = 5.00 + 100.00\ \$\ ha^{-1}$$
$$Cost_{(WaterToField + LaborToIrrigate)} = 0.00 + 0.20\ \$\ ha^{-1}$$
$$Cost_{FertilizerInIrrigation} = 0.00\ \$\ ha^{-1}$$
$$OtherCost = 350.00\ \$\ ha^{-1} \text{ (includes planting and harvest costs)}$$

Note that the profit and economic factors here are simply set to estimate relative costs associated with management activities. They may not be complete nor involve all current marketing factors. MOM may consider these factors in further development with economic assistance. During the MOM optimization search in scenarios, the three MOM sub-goals, high profits, high yields and less nitrate leaching, were assigned the same weights for convenience.

Scenario-1

This is an example of cropping in a tropical wet season (udic soil moisture regime). MOM first examined the native water supply potential during the cropping season and determined the amounts and timing of supplemental irrigation to meet possible shortages in meeting crop requirements (Fig. 5a). When the supply index of soil N or soil water supply potential is equal to 0, the soil supply is equal to the crop requirement (Li & Yost, 1998). A line at the index of zero, called *sufficient line*, was drawn in diagrams of soil N or soil water supply analysis to indicate where the soil supply satisfies the crop requirement. The soil water supply indexes of three soil layers under native conditions were close to *sufficient line*, only slightly lower than it (Fig. 5a). In other words, rainfall would provide most of the water the crop needs, so not much irrigation was needed. The "native" soil N supply potential was evaluated in Fig. 5b. MOM predicted that N supply shortage in the major root zone would occur during the crop's rapid growth stage (about 50–100 d after planting), at which time the native soil N supply index of the surface layer (0–30 cm) was less than the *sufficient line*. So a small amount of N fertilizer was suggested shortly before and during this stage (Fig. 5b). Measured soil NO_3 at the beginning of cropping was 28 mg N kg^{-1} (about 92 kg N ha^{-1}, assuming soil BD = 1.1) in the major root zone (0–30 cm), 26 mg N kg^{-1} (about 85 kg N ha^{-1}) in the minor root zone (30–60 cm, assuming soil BD = 1.1), and 34 mg N kg^{-1} (about 123 kg N ha^{-1}, assuming soil BD = 1.2) in the transition zone (60–90 cm) in the scenario. Given the same goal weights to the three sub-goals of high profits, high yields and less NO_3 leaching, MOM suggested 106 mm irrigation and 42 kg N ha^{-1} for the scenario (Table 1, Fig. 5a and Fig. 5b). Based on the MOM suggested management, soil N supply potential during the cropping was simulated in Fig. 5c. Comparing Fig. 5c with Fig. 5b, the soil N supply index of the major root zone (0–30 cm) rose slightly over *sufficient line* after MOM suggested N fertilizer had been applied. It implies that MOM suggested management nearly matched the crop requirements during the growth. Inorganic N in the soil profile was simulated in Fig. 5d. Inorganic N (mostly in NO_3 form) concentration in the root zone became low at the end of the cropping; however, inorganic N remained high in the transition zone, where only a small fraction of the N was utilized by the crop. Inorganic N in the transition zone was a potential leaching source unless the following crops develop deep roots in this layer.

Table 1. MOM suggested N fertilization and irrigation schedule in scenario 1.

Week	Fertilizer *kgN ha^{-1}*	Irrigation *mm*
1	0.0	0.0
2	0.0	5.4
3	0.0	0.3
4	0.0	0.0
5	0.0	4.2
6	10.0	12.8
7	0.0	8.7
8	0.0	14.7
9	0.0	18.0
10	0.0	10.8
11	32.0	8.7
12	0.0	1.0
13	0.0	12.5
14	0.0	0.0
15	0.0	0.0
16	0.0	9.2
Sum	42.0	106.3

The large rain near the end of the cropping season might leach this inorganic N out of the root zone as occurred in Scenario 1 (Fig. 5e). Comparing simulation results (crop uptake N, percolation, and leachate NO_3) of the original management practice (Fig. 5e) with those the MOM-suggested management (Fig. 5f), suggests that MOM recommendations controlled much of the excessive percolation, which in turn reduced nitrate leaching. Although MOM could not control the severe leaching peak at 100 d after planting, which was caused by a large rain, event, MOM reduced the predicted leached NO_3 from 175 to 67 kg N ha^{-1}. Compared with the original management, MOM also reduced costs by reducing the amounts and number of applications of fertilizer, while maintaining the yield. Finally, MOM suggested management returned \$600 ha^{-1} more profit and 108 kg N ha^{-1} less leached NO_3 than the original management in Scenario-1 (Table 2).

Table 2. Summary of two scenarios of MOM optimizing N management.†

Scenario	Management	Profit *\$ ha⁻¹*	Yield *kg ha⁻¹*	Leachate N *kgN ha⁻¹*	Fertilizer *kgN ha⁻¹*	Irrigation *mm*
1	Original	3314	9092	175	200	323
	MOM	3986	9589	67	42	106
2	Original	3697	9728	40	197	387
	MOM	4439	11001	8	224	271

† All data of profit, yield, and leachate N are results simulated by N-SIMULATOR. Fertilizer and Irrigation data come from original datasets or MOM suggestions.

Scenario-2

Scenario-2 represents a cropping system during a tropical dry season. The native water supply potential of the scenario was simulated in Fig. 6a, with a primary irrigation schedule that MOM suggested to match the crop requirements. Fig. 6a shows that the water supply in the major root zone was insufficient and MOM recommended more irrigation for the crop during this dry season than that in scenario-1 (wet season). Fig. 6b illustrates the native soil N supply potential of the site during cropping and MOM suggested N fertilization during the crop rapid growth period. The soil NO_3 contents at the beginning of cropping were 10 mg N kg^{-1} (about 34 kg N ha^{-1}, assuming soil BD = 1.1) in the major root zone (0–30 cm), 11 mg N kg^{-1} (about 37 kg N ha^{-1}) in the minor root zone (30–60 cm, assuming soil BD = 1.1), and 15 mg N kg^{-1} (about 52 kg N ha^{-1}, assuming soil BD = 1.2) in the transition zone (60–90 cm). Initial soil inorganic N of the scenario was less than that of Scenario 1. More fertilizer and irrigation, 224 kg N ha^{-1} with 271 mm irrigation in total, were suggested for Scenario 2 (timing shown in Fig. 6a and Fig. 6b). Soil N supply potential under MOM management was simulated in Fig. 6c and soil inorganic N was simulated in Fig. 6d. Inorganic N in the soil profile increased during crop growth but after harvest returned to approximate the same levels as before planting. No extra inorganic N was accumulated in the root zone under MOM suggested management. Fig. 6e shows crop uptake N, percolation, and leachate nitrate of original management practice and Fig. 6f shows the same measurements under MOM suggested management. Simulated profit, yield and leachate nitrate of the scenario were listed in Table 2. Except for some savings in irrigation, the original management practice was close to the MOM suggestion. MOM would not have greatly improved profit and yield in this scenario; however, some reduction in NO_3 leaching, from 40 to 8 kg N ha^{-1}, was predicted. Scenario-1 and scenario-2 demonstrate that careful N management is necessary to minimize NO_3 leaching for cropping in a tropical wet season. With uncertain rainfall, that is a difficult management objective even with a decision-aid.

CONCLUSIONS

Advantages of MOM-Guided, within-season N management are:

- Rapid, efficient predictions of crop N status in advance to the actual deficiency – in contrast to diagnoses with sensors, color charts, and soil tests.
- Low cost – no costly sensors or sensor technology needed, reduced soil sampling
- "Transparency" – daily "pictures" of the simulated N cycle graphically show users predictions of both moisture and N status, the need for additional N and irrigation, and options to minimize N leaching.

Disadvantages:

Requires calibration – this appears possible with historical, daily rainfall records, soil moisture retention (drained upper limit and lower limit). Calibration / validation is also somewhat difficult because of the few datasets available to rigorous test the model.

This type of modeling of within-season N and water status should be a useful guide to explore options for N and water management leading to improved management towards the complex, multiple objectives of increased production, high economic benefit, and minimal N leaching below the root zone.

ACKNOWLEDGMENT

The authors acknowledge U.S. Federal Government Hatch Project F93-272-F-531-8-145, which has fully supported this study.

REFERENCES

Binford, G.D., A.D. Blaylock, and D.D. Baltensperger. 1996. In-season soil testing for nitrogen management in sugar beets. p. 313. *In* 1996 Agronomy Abstracts. ASA, Madison, WI.

Blackmer, A.M., and S.E. White. 1996. Remote sensing to identify spatial patterns in optimal rates of nitrogen fertilization. *In* P. C. Robert et al., (ed.) Proc. of the 3rd Int. Conf. on Precision Agriculture. Minneapolis, MN. ASA, CSSA, and SSSA, Madison, WI.

Khan, M.A., R.E. Green, and T. Liang. 1986. Nitrogen transformations in soils: Experimental and mathematical consideration for computer modeling. HITAHR Res. Ser. 045. Univ. of Hawaii, Honolulu.

Li, MengBo, and R.S. Yost. 1998. Management-oriented modeling: Optimizing nitrogen management with artificial intelligence. Manuscript in preparation.

Ling, G. 1996. Assessment of nitrate leaching in the unsaturated zone on Oahu. Ph.D. diss. Dep. of Geology and Geophysics, Univ. of Hawaii, Honolulu.

Marx, E.S., J.M. Hart, and N.W. Christensen. 1996. On-site soil nitrate testing using a quick-test field kit. p. 316. *In* Agronomy Abstracts. ASA, Madison, WI.

Plant, R.E., and N.D. Stone. 1991. Knowledge-based systems in agriculture. McGraw-Hill, New York.

Richardson, C.W., and D.A. Wright. 1984. WGEN: A model for generating daily weather records. USDA, Washington, DC.

Schepers, J.S., T.M. Blackmer, and T. Shah. 1996. Real-time nitrogen management for corn production. p. 257. *In* Agronomy abstracts. ASA, Madison, WI.

Stone, N.D., and L.P. Schaub. 1990. A hybrid expert system/simulation model for the analysis of pest management strategies. *AI Appl. Nat. Resour. Manage.*. 4(2):17–26.

Wang, X.N., and F.D. Whisler. 1996. Development of "NATCOVER" future weather patterns and A service tool for crop models. p. 60. *In* Agronomy Abstracts. ASA, Madison, WI.

Yost, R.S., and MengBo Li. 1998. Nitrogen-simulator predicting nitrate leaching in root zones: II. Applications. Manuscript in preparation.

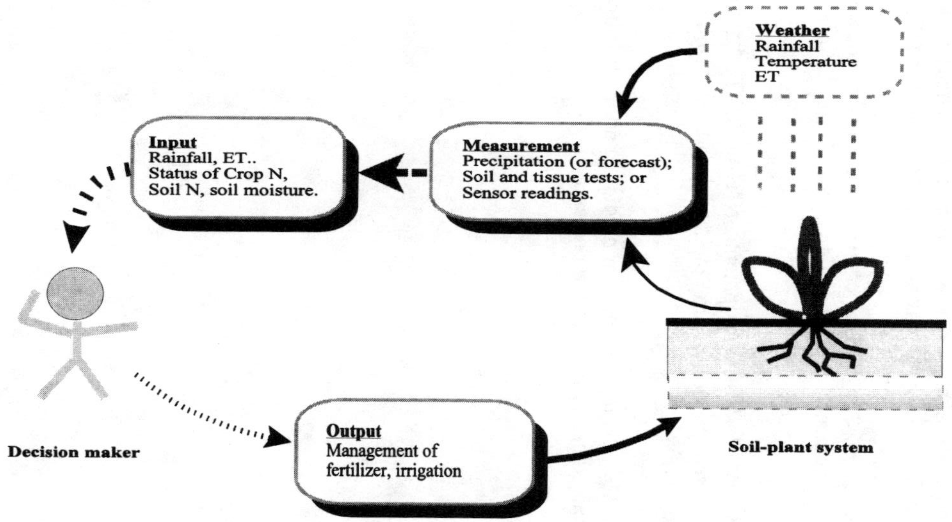

Fig. 1. A diagram of within-season N management without decision-aid. Current information of the soil-plant system is collected from soil tests, sensors, and weather forecasts. A decision-maker analyzes data and estimates the amounts and timing of N fertilizer and irrigation in following weeks.

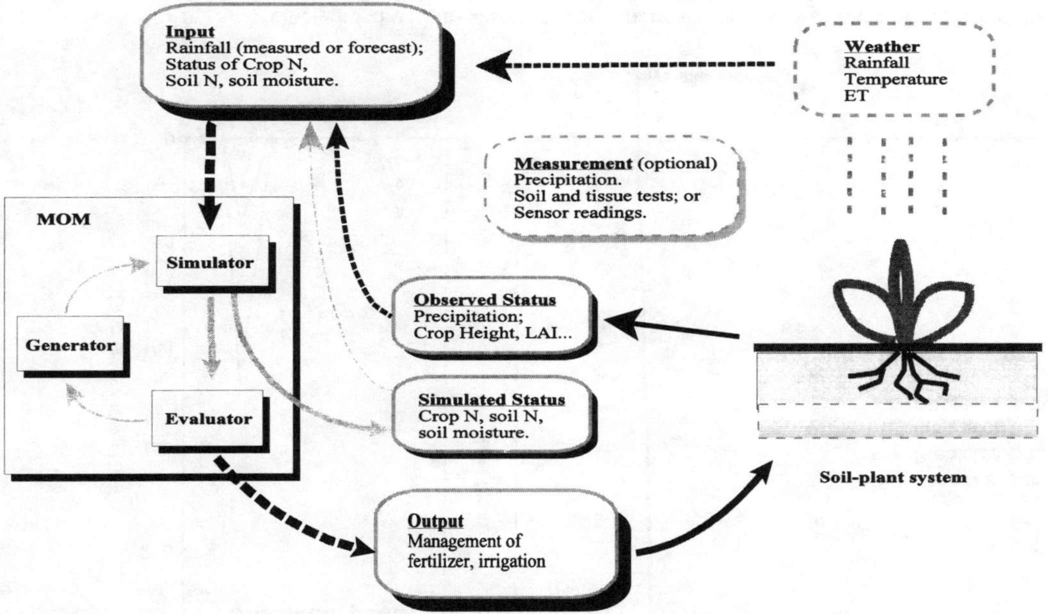

Fig. 2. A diagram of MOM-guided within-season N management. Current information of the soil-plant system comes from simulated data and weather forecasts. MOM's evaluator analyzes data and optimizes the management schedule for the following weeks.

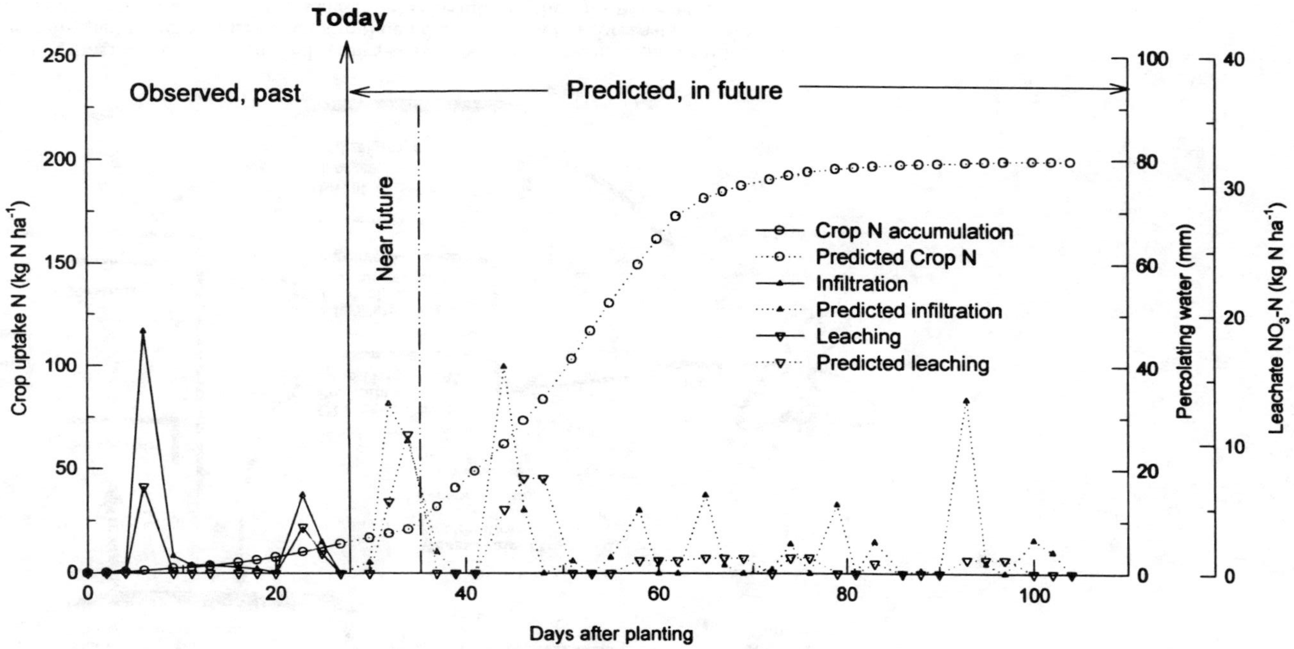

Fig. 3. MOM-guided within-season management shows status of the soil-plant system before TODAY in solid lines, status after TODAY in dotted lines. MOM can do nothing for the past but can change the management in the future.

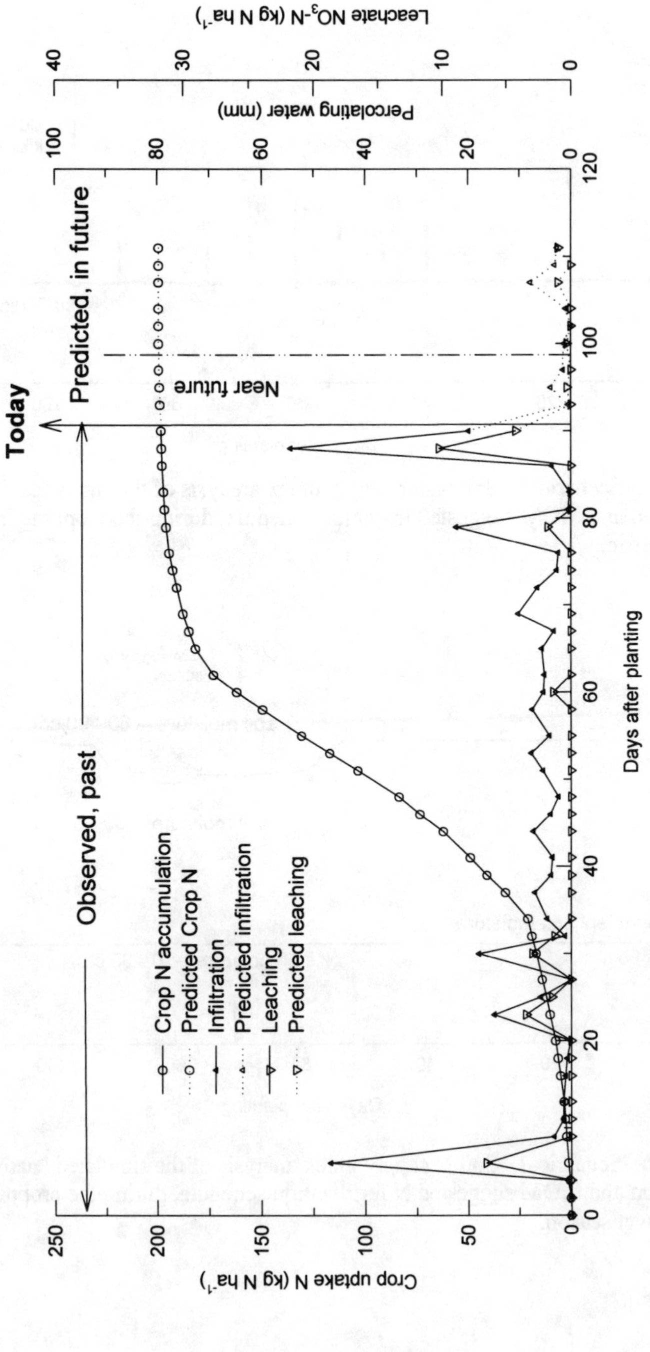

Fig. 4. MOM-guided within-season management may not perfectly update the schedule to fit uncertainty of rainfall. It improved management in near future. Compare with Fig. 3.

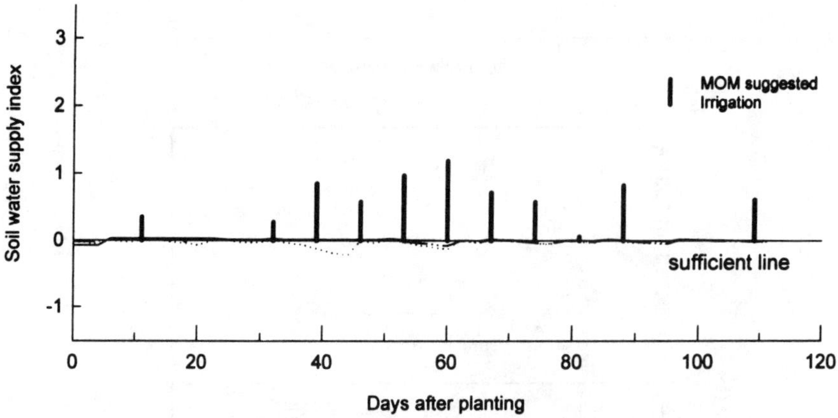

Fig. 5a. Scenario-1. Soil water supply index analysis of the simulated "native" situation and MOM suggested irrigation schedule, during the cropping in the wet season.

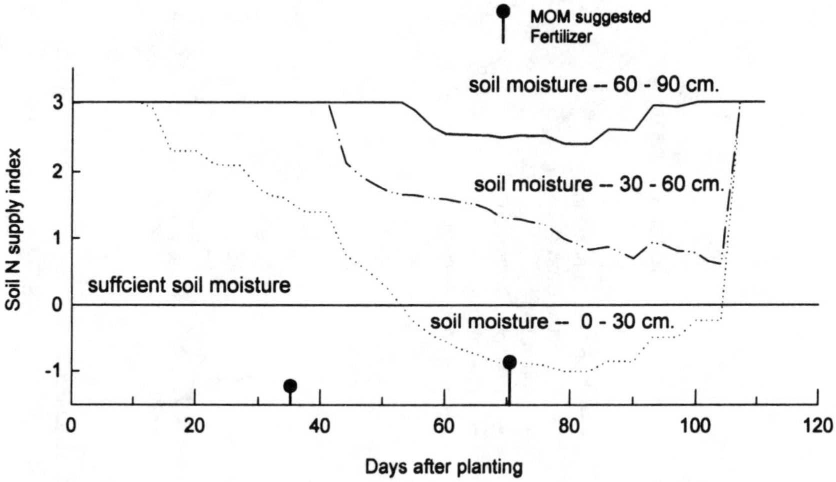

Fig. 5b. Scenario-1. Soil N supply index analysis of the simulated "native" situation and MOM suggested N fertilization schedule, during the cropping in the wet season.

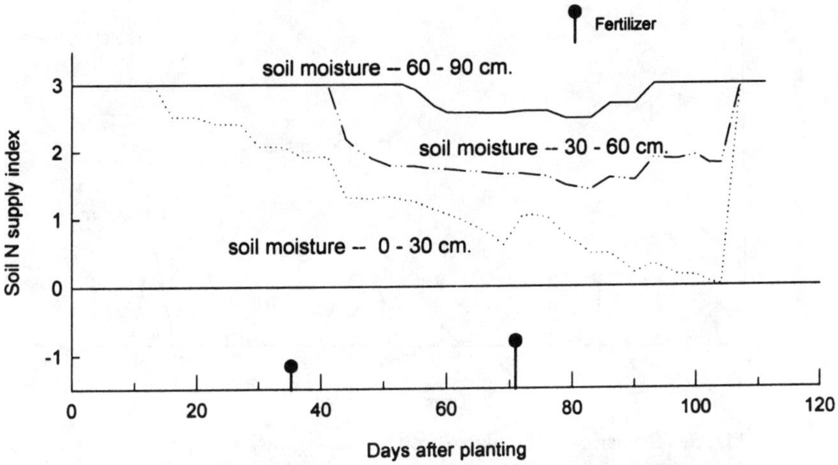

Fig. 5c. Scenario-1. Soil N supply index analysis of the simulated data under MOM-guided management schedule, during cropping in the wet season.

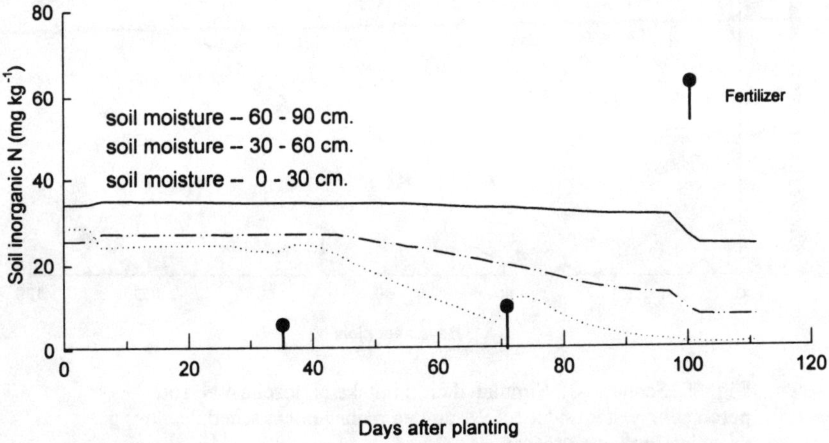

Fig. 5d. Scenario-1. Simulated inorganic N (NO_3-N and NH_4-N) in the soil profile under MOM-guided management schedule, during the cropping in the wet season.

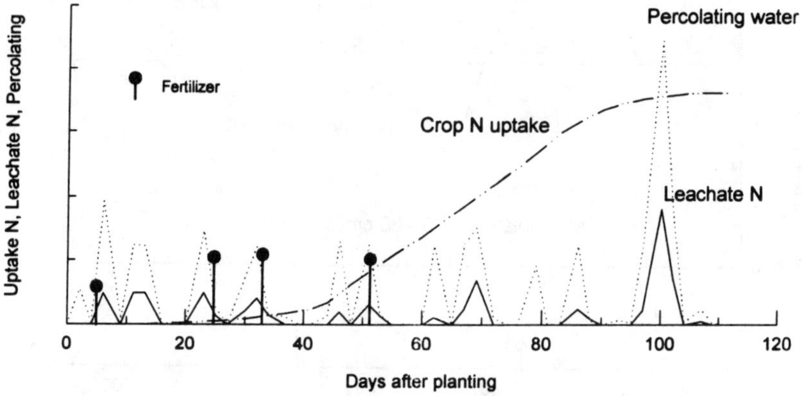

Fig. 5e. Scenario-1. Simulated crop uptake N, leachate N, and percolating water under the original management conditions during cropping in the wet season.

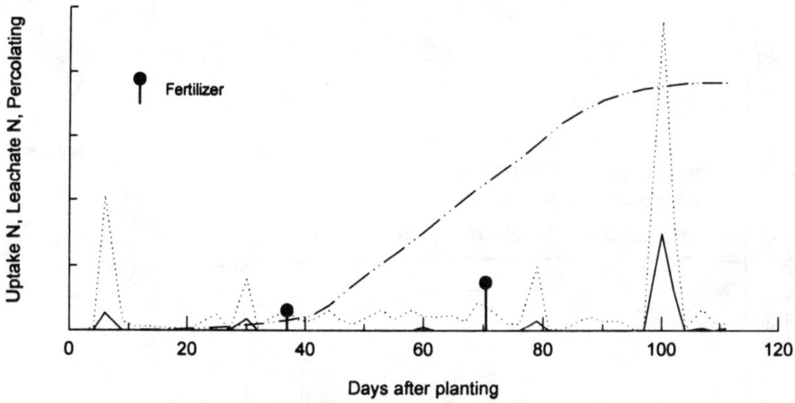

Fig. 5f. Scenario-1. Simulated crop uptake N, leachate N, and percolating water under MOM-guided management schedule, during cropping in the wet season.

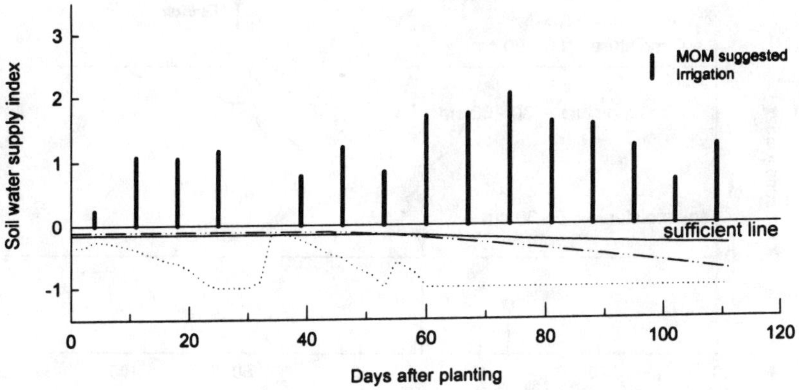

Fig. 6a. Scenario-2. Soil water supply index analysis of the simulated "native" situation and MOM-guided irrigation schedule, during the cropping in the dry season.

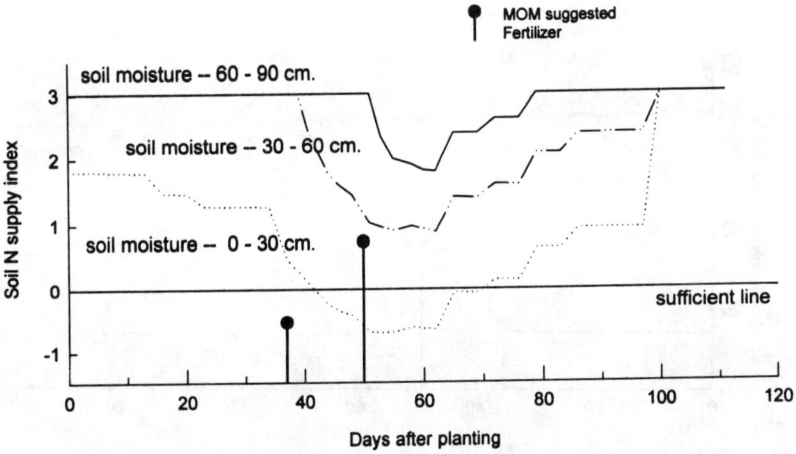

Fig. 6b. Scenario-2. Soil N supply index analysis of the simulated "native" situation and MOM-guided N fertilization schedule, during the cropping in the dry season.

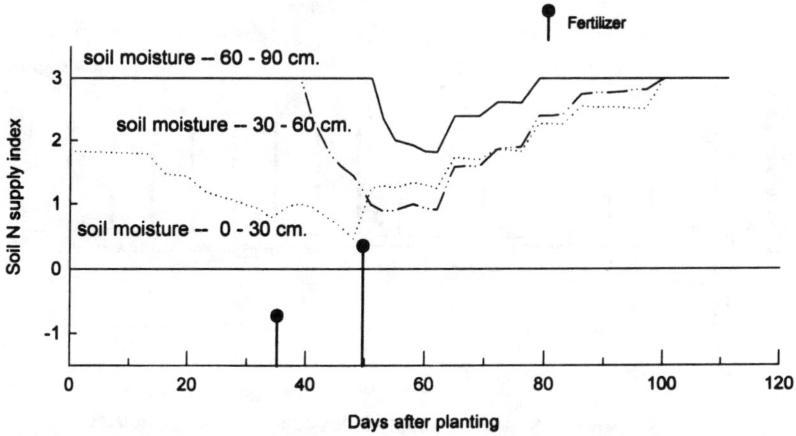

Fig. 6c. Scenario-2. Soil N supply index analysis of the simulated results under MOM-guided management schedule, during cropping in the dry season.

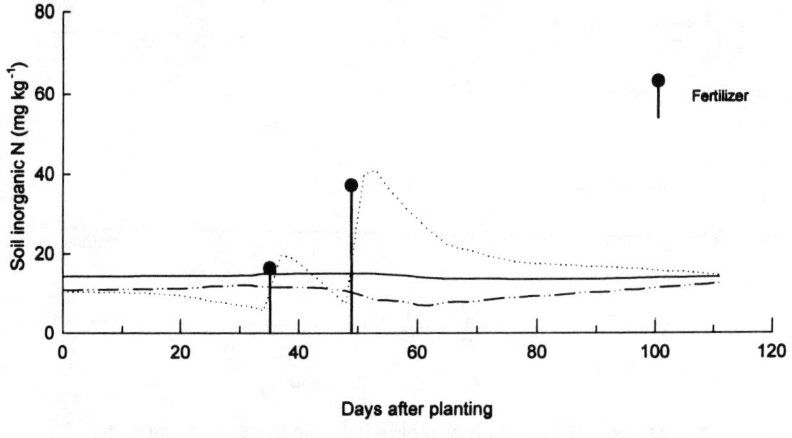

Fig. 6d. Scenario-2. Simulated inorganic N (NO_3-N and NH_4-N) in the soil profile under MOM-guided management schedule, during cropping in the dry season.

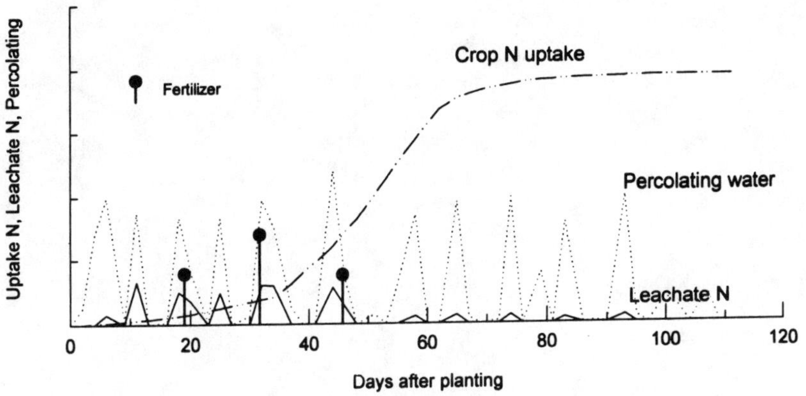

Fig. 6e. Scenario-2. Simulated crop uptake N, leachate N, and percolating under the original management conditions of the dataset, during cropping in the dry season.

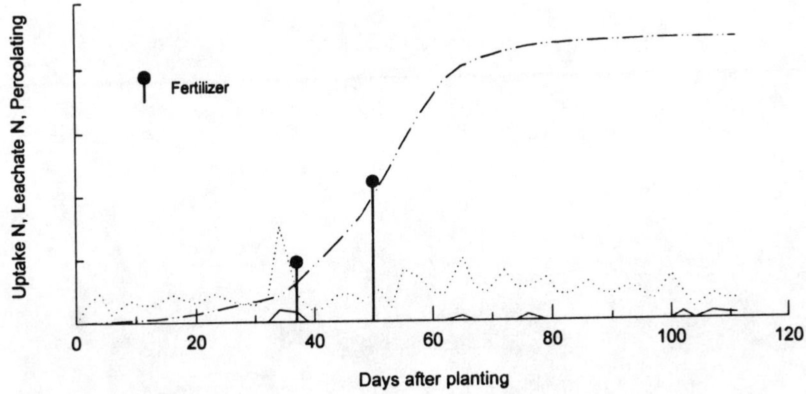

Fig. 6f. Scenario-2. Simulated crop uptake N, leachate N, and percolating under MOM-guided management schedule, during cropping in the dry season.

Precision Agriculture at the University of Minnesota-Crookston: Combining Hands-On With High Tech

Aziz Rahman

Agriculture Management Division
University of Minnesota
Crookston, Minnesota

ABSTRACT

Since the founding of an agricultural experiment station in Crookston more than 100 years ago, the University of Minnesota has recognized the importance of agriculture to the region. Over the years, University services have evolved to meet the needs of farmers and agri-businesses. Today, the University of Minnesota, Crookston, the Northwest Experiment Station, and University of Minnesota Extension Service provide essential support and leadership to this dynamic and increasingly capital intensive, technology-driven industry. The University of Minnesota, Crookston (UMC), has issued laptop computers to all full-time students for 4 years, and according to Microsoft, was one of the first to understand that today's college graduates need highly developed competence in technology. Since about one-half of UMC's students are enrolled in agricultural programs, an institutional commitment was made to serve the emerging precision agriculture industry and to provide students education in new agricultural technologies. In late 1996 and early 1997, UMC faculty and a team of industry advisors crafted the first learner outcomes in precision agriculture. A course was developed, and 26 students took the first class in the fall of 1997. In addition, UMC faculty and its Continuing Education team collaborated with leading precision agriculture vendors and practitioners to offer an intensive day-long training workshop for professionals in agricultural extension, agri-business, and production. This presentation describes UMC's experience in precision agriculture course design and implementation. It reviews the use of technology enhanced teaching in combination with hands-on, experiential learning to achieve desired learner outcomes. UMC's collaborative course management and delivery strategy will be highlighted. The experience gained to date may benefit those interested in developing similar educational initiatives, and will contribute to the advancement of productive dialogue between educators and other professionals with similar and complementary interests.

Copyright © 1999 ASA-CSSA-SSSA, 677 South Segoe Road, Madison, WI 53711, USA. *Proceedings of the Fourth International Conference on Precision Agriculture.*

Site-Specific Agriculture Workshops Assist Farmers to Understand and Manage Soil-Landscape Variability

Ronald L. McNeil
Brenda J. Sawyer
T. W. Goddard

LandWise
Lethbridge, Alberta, Canada

ABSTRACT

Private industry and government agriculture departments have collaborated sine 1994 to provide workshops on precision agriculture in Alberta. The 2-day workshops have been completed by 424 participants, 79% of whom are farmers. The goal of the workshops is to enhance understanding of land resources and precision-farming technology. The participants use enlargements of aerial photography to map field variability. The variability maps are used, along with soil survey information, to develop management zones and soil-sampling strategies. The components of precision farming technology are discussed. The farmers work with agribusiness personnel and consultants to develop a site-specific management plan most suitable for their land base, economic situation, and management scheme. Course evaluations indicate the most valuable features of the workshops include the practical mapping exercises, the explanation of soil and landscape variability, the discussion of strategies for site-specific management, in particular soil sampling, and the building of awareness for implementing precision farming.

Copyright © 1999 ASA-CSSA-SSSA, 677 South Segoe Road, Madison, WI 53711, USA. *Proceedings of the Fourth International Conference on Precision Agriculture.*

Validation and Implementation of a Coordinated Precision Agricultural Curriculum

Terry Brase

Hawkeye Community College
Waterloo, Iowa

ABSTRACT

A Precision Agriculture 2-yr Associate Degree, focusing on the use of emerging technologies in agriculture, has been implemented at Hawkeye Community College as a result of a Phase I NSF/ATE grant. Valication and implementation of this curriculum at other community colleges and its use by other types of educational institutions is the focus of a Phase II NSF/ATE grant. Creating a Midwest regional network of educational institutions and industry will result in a coordinated effort to develop curriculum material that will provide basic math and science skill applied to agriculture, specific technology skills needed by industry and articulation with universities. Specific components of this project include: educational opportunities for current teachers or preservice teachers; instructional–curriculum material for use as a complete course, integrated into current courses or laboratories to give hands-on experience in the technology; and working with other teachers to develop curriculum materials.

Copyright © 1999 ASA-CSSA-SSSA, 677 South Segoe Road, Madison, WI 53711, USA. *Proceedings of the Fourth International Conference on Precision Agriculture.*

APPENDIX I

SUMMARY OF WORKING GROUP DISCUSSIONS ON EDUCATIONAL CURRICULUM

Compiled by Pierre C. Robert.

Sixteen workgroups comprised of a combination of conference participants based on affiliations – industry/agribusiness, producer, researcher, and other, convened in separate meeting rooms on Tuesday from 10:40 AM to 12:00 with the task of recommending curriculum in precision agriculture for technical, college undergraduate and graduate, and professional levels. This is a summary of comments and recommendations made by participants assembled from reports and flip charts.

A. GENERAL COMMENTS made by several work groups

- There is an urgent need for educational programs in Precision Agriculture (PA) at all levels: technical, college undergraduate and graduate, and professional.

- There is a need to encourage K-12 involvement. Summer workshops should be organized in a combined University / Industry program to train the teachers.

- PA education should include practical, hands-on modules similar to methods used at CALPOLY.

- PA does not need a new set of specific courses, but the infusion of PA concepts into the existing coursework with the exception of some new courses relating to spatial techniques. This requires the training of educators in the PA concept.

- Educational methods recommended by several Work Groups (WGs) are:
  - emphasis on thinking and communication skills
  - real life experience
  - internship on PA in agribusiness, industry, and on-farm university research
  - internship on farms for students without farm experience
  - use of system approach in crop management classes
  - some classes should include team work
  - public/private cooperation in some aspects of teaching

B. COLLEGE UNDERGRADUATE EDUCATION

Undergraduate curriculum in precision agriculture is a high priority need. Presently, a major in precision agriculture is not required. A minor based on present coursework modified to include PA concepts and some new classes as listed in B2 should be appropriate.

The most common topics identified by WGs as a base for a curriculum are the following
(there is no ranking since the number of occurrences were very similar):

1. Basic education:
- basic sciences
- statistics
- general soils with emphasis on spatial variability
- general agronomy (plant physiology, fertility, weed, pest, and disease)
- precision agriculture overview in a multidisciplinary approach
- farm / crop management, including risk analysis and management
- communication (written and oral)
- business (basic, accounting), economics

2. Precision agriculture complement:
- GIS/GPS using examples related to agriculture
- spatial statistics, sampling, field experimental design
- record keeping, information systems, information management, data analysis
- computer skills, electronics, software use, Internet
- PA machinery, mechanical issues, sensors
- remote sensing

3. Educational methods:
- scientific thinking
- logic skills
- problem solving
- decision making
- extensive use of case studies
- career seminar (1 WG)
- undergraduate project/thesis (1 WG)

4. Other skills indicated by one or two WGs:
- biology
- ecology
- environmental protection
- meteorology, weather probabilities, hydrology
- modeling
- international experience, foreign language

C. COLLEGE GRADUATE EDUCATION

Comments made by the WGs are fewer and of general interest. The classwork recommended includes all PA topics mentioned for undergraduate student education but at a more advanced level.

Some specific suggestions by individual WGs are:

• Keep today's topics and add advanced PA courses in area of interest

• Academic path should emphasis:
- basic sciences
- spatial analysis
- GIS, survey
- advanced computer skills
- international experience

• Important topics:
- spatial variability
- data acquisition and analysis

• Advanced classes on:
- spatial statistics
- GIS
- distance tools
- communication skills

• Teaching methods should include problem solving and system analysis.

• Exposure to all aspects of PA, plus:
- landscape concept
- information analysis
- system, control theories
- management, integration skills

• PA should be studied as an agricultural system

• A non research curriculum should be offered

• One WG suggested a 18 months program comprising shortcourses at a graduate level

D. TECHNICAL EDUCATION

Several curriculum concepts are suggested by different WGs.

1. Two-year degree

Two different programs are suggested for a two-year degree:

1.1. Technical degrees in 'Agronomy Technology' and in 'Computer Technology'.

A basic program followed by topics related to each degree:

a. Agronomy technology (information acquisition, scouting):
- precision agriculture overview
- surveying

- soil sciences
- plant physiology
- fertilizer
- weeds
- farm management

b. Computer technology (information processing):
- precision agriculture overview
- basics in geostatistics
- GIS
- GPS
- Computers

1.2. Core program followed by five specializations.

a. Core program:
- basic coursework
- precision agriculture overview

b. Specialization in one of the following options:
- surveying, GIS, GPS
- computer, data processing
- machinery, electronics
- remote sensing
- farm management

1.3. Core program followed by three specializations

a. Core program (same as 1.2.a)

b. Specialization in:
- field operations
- precision technologies: installation, use, and maintenance
- information management

c. Methods:
- short courses on specific topics
- skill based courses
- hands-on
- partnership with PA industry

2. One-year degree

A one-year program is recommended by two WGs with specialization in:

1) machine operation, custom application and
2) field surveying, sampling, and scouting.

The recommended coursework is:
- general introduction to the concept of PA
- spatial variability
- practical agricultural training on a regional basis

- basic soil/crop production
- basic equipment/practices
- GIS, GPS, navigation
- machinery, sensor, controller (operation, calibration, maintenance)
- basic computer skills, software
- data acquisition, data processing, data quality
- surveying, sampling, and scouting

Recommended teaching methods are:
- practical use of systems
- hands-on exercises
- trouble shooting (mechanics, electronics)
- team work
- internship
- ability to continue to learn (CCA, CEU)
- program schedule should be adapted to work-time

3. Other recommendations by single WGs:

- Short technical programs in:
  - custom application
  - data management and mapping
- Classes for business skill development (accounting)
- Teaching of written and oral communication skills
- Awareness in statistics.

E. PROFESSIONAL EDUCATION

Some WGs indicated important educational needs for ag-professionals, field agronomists, in PA related topics and skills.

Training needs identified for practitioners are:

- natural resources/agronomy for PA
- spatial statistics, design of sampling procedures, and on-farm experiments
- GIS, GPS, remote sensing
- data analysis and interpretation
- equipment for SSCM: new farm machinery and sensors
- use of computer tools: database management and analytical tools
- business updates: management and marketing
- communication skills
- legal and regulatory aspects

- human resources

Suggested teaching methods are:

- training the trainer format
- problem solving using decision cases
- exposure to end-user perspectives and economic realities
- hands-on experience, student involvement and interaction
- use of equipment with industry
- use of distance education when possible
- sabbatical leaves in selected PA sector (6-18 months)
- 1-3 weeks shortcourses

APPENDIX II – PARTICIPANT LIST

Paul Aakne
University of Minnesota
Crookston, MN
(218) 281-8104
paakre@mail.crk.umn.edu

Ric Abernethy
Farm Chemicals Magazine
37733 Euclid Ave.
Willoughby, OH 44094
(440) 942-2000

Matthew Adams
CSIRO
Private Bag, P.O.
Wembley, WA
Australia, 6014
(61) 8-9333-6381
matthewa@per.bms.csiro.au

John Ahlrichs
Cenex/Land O'Lakes

Miguel Ahumada
Bear Creek Orchards, Inc.
2518 S. Pacific Highway
Medford, OR 97501
(541) 776-2121
ahumada@bco.com

Robert Aiken
USDA-ARS
Box 400
Akron, CO 80720
(970) 345-2259
aiken@gpsr.colostate.edu

Grant Aldridge
Novartis Crop Protection, Inc.
2034 Burbridge Lane
Hillard, OH 43026
(641) 777-1668
grant.aldridge@cp-novartis.com

Per-Anders Algerbo
Swedish Inst. of Ag. Engineer.
Box 7033
Uppsale
Sweden, S-750 07
(46) 1830-3300
per-auders.algerbo@yti.slu.se

Ryan Allen
RDI Technologies, Inc.
300 Hwy. 23
Suite One
Spicer, MN 56288
(320) 796-0019 fx 796-0048

Raymond Allmaras
USDA-ARS
439 Borlaug Hall
St. Paul, MN 55108
(612) 625-1741
allmaras@soils.umn.edu

Abdullah Almodaihsh
University of King Saud
Box 2460
Riyadh
Saudi Arabia
(966) 1-4678445

Jared Alsdorf
Terra Industries
6739 Gulon Rd.
Indianapolis, IN 46268
(219) 967-4229
jalsdorf@terraindustries.com

LaVinia Altebaumer
Resource21
7257 S. Tuscon Way
Suite 200
Englewood, CO 80112

Peter Amadio
Mayo Foundation

C. R. Amerman
USDA/ARS
Room 233, Bldg. 005
BARC-W
Beltsville, MD 20705
(301) 504-6441
cra@ars.usda.gov

Kevin Anderson
Geo Farm, Inc.
Route 2
Box 230
Lake Crystal, MN 56055
(507) 947-3362
geofarm@mnic.net

Noel Anderson
Case Concord, Inc.
3000 7th Ave. N.
Fargo, ND 58102
(701) 277-3618
nanderson@casecorp.com

Gordy Andrews
Andrews Maple Hill Farms
RFD 2
Evansville, WI 53536
(608) 876-6455

Jared Anez
Anez Consulting
723 135th Ave. S.W.
Willmar, MN 56201
(320) 995-6619
anezconsulting@willmar.com

Thomas Arnsperger
The MITRE Corp.
1820 Dolley Madison Blvd.
McLean, VA 22102
(703) 883-6388
tarnsper@mitre.org

Selcuk Arslan
Iowa State University
Davidson Hall
Ames, IA 50010
(515) 572-4740
sarslan@iastate.edu

Stefanie Aschmann
USDA-NRCS
c/o NAC, UNL East Campus
Lincoln, NE 68583-0822
(402) 437-5178 ex 43
saschmann@aol.com

Lanny Ashlock
University of Arkansas
Box 391
Little Rock, AR 72203
(501) 671-2278
lashlock@vaex.edu

Bruce Baier
Ag Partners, LLC
Box 38
Albert City, IA 50510
(712) 843-2291
lbbaier@nwiowa.com

Lyman Baker
Case Corporation
700 State St.
Racine, WI 53404

Luiz Antonio Balastreire
ESALQ/USP
Av. Padua Dias, 11
Piraciaba, Sao Paula
Brazil, 13418-900
(55) 19-4330934
carpa.ciagri.usp.br

Aaron Baldwin
Cargill Ltd.
Box 369
Rayroe, Saskatchewan
Canada, S0A 3J0
(306) 746-4625
aaronbaldwin@cargill.com

Eric Bandy
Novartis Seeds, Inc.
7500 Olson Memorial Hwy
Golden Valley, MN 55340
(612) 593-7107

Yohana Barata
Precision Agriculture Center

Richard Barnhisel
University of Kentucky
Agronomy Department
Lexington, KY 40546-0091
(606) 257-8627
rbarnhis@ca.uky.edu

Jennifer Barrett
Omnistar, Inc.
8200 West Glen
Houston, TX 77063
(713) 785-5850

Glenn Bathke
Novartis Seeds
1501 Eisenhower C6
Northfield, MN 55057
(507) 645-5399
glenn.bathke@seeds.novartis.c
om

Paul Baukol
Centrol, Inc.
2219 30 1/2 Ave. S.
Fargo, ND 58103
(701) 274-8203
pbaukol@fm-net.com

Patricia Bayer
Berlin
Germany
(49) 30-7570-4620
ytbayer@aol.com

York-Th Bayer
Bayer Consulting
Haiserin Augusta Str. 78
Berlin, Germany, 12103
(49) 30-7570-4620
fx (612) 625-2207
ytbayer@aol.com

Kate Beatty
2815 State Rd 225 E
Battle Ground, IN 47920
(765) 567-4840

Marshall Beatty
Emerge
900 Technology Park Dr.
Bldg. 8
Billerica, MA

Kevin Beckard
Cottage Grove Cooperative
203 W. Cottage Grove Rd.
Cottage Grove, WI 53527
(608) 251-9010 ex 283

Rick Behrdmore
mPower 3
419 18th St.
Greely, CO 80631
(907) 356-4400
rbeardmore@mpower3.com

Cindy Bell
Pacific Meridian Resources
421 S. W. 6th Ave. Suite 850
Portland, OR 97204
(503) 228-8708 fx 228-8751
portland@pacificmeridian.com

Thomas Bell
Stanford University
136D Escondido Village
Stanford, CA 94305
(650) 725-6378
tombell@stanford.edu

Dan Bellanger
Precison Ag Illust. Magazine
15444 Clayton Rd.
Suite 314
Ballwin, MO 63011
(314) 527-4001 fx 527-4120
dan@precisionag.com

Gary Belschner
Leica Geosystems, Inc.
3155 Medlock Bridge Rd.
Norcross, GA 30071
(770) 447-6361

Annick Berghman
The Schneider Corporation
3020 N. Post Rd.
Indianapolis, IN 46226-6518
(317) 898-8282
aberghman@theschneidercorp.com

Manuel Bermudez
Iowa State University
3411 Agronomy Hall
Ames, IA 50011

David Bevly
Stanford University
Escondido Village #2E
Stanford, CA 94305
(650) 497-1495
dmbevly@stanford.edu

Charles Biggar
United Agri Products
Box 55
Kasota, MN 56050
(507) 931-6660
biggar@uap.mn-ia.com

Brian Bills
Ag Chem Equipment Co.
5720 Smetana Dr.
Minnetonka, MN 55343
(612) 945-5873
bbills@agchem.com

Mark Bishop
Rhone-Poulenc Ag. Company
Box 12014
Triangle Park, NC

Gunther Bittner
GEO TEC electronics GmbH
Am Soldnermoos 17
Hallbergmoos, Bavaria
Germany, 85399
498116009911

Blaine Blaesing
Resource21
7257 S. Tuscon Way
Suite 200
Englewood, CO 80112

Bruce Blakesley
Watonwan Farm Service
Box 161A
RR1
Amboy, MN 56010
(507) 375-3355
bblake@mctc.net

Paul Blom
Penn. State University
501 ASI Bldg.
University Park, PA 16802-3508
(814) 863-7657
pxb32@psu.edu

Jurg Blumenthal
University of Nebraska
4502 Ave. 1
Scottsbluff, NE 69361
(308) 632-1372
blumenth@Aunlvm.unl.edu

Rick Bobbitt
Red Hen Systems, Inc.
800 Stockton Ave.
Ft. Collins, CO 80524
(970) 493-3952
rick@redhensystems.com

Richard Bohling
DEKALB Genetics Corporation
3100 Syeamore Rd.
Dekalb, IL 60115
(815) 758-9442
dbohling@dekalb.com

Denis Boisgontier
I.T.C.F.
Station Experimentale
Boigneville
France 91720
(33) 16499-2211
dboisgontier@itcf.fr

Charles Bolte
Ag Source
106 N. Cecil St.
Bonduel, WI 54107
(715) 758-2178

Steve Borgelt
University of Missouri
249 Agr. Engineering Bldg.
Columbia, MO 65211
(573) 882-7549
borgelts@missouri.edu

David Bosch
USDA-ARS
Box 946
Tifton, GA 31754
(912) 386-3899
dbosch@tifton.cpes.peachnet.edu

J. Bouma
Wageningen Ag. University
Box 37
Wageningen, 6700AA
The Netherlands

Tyler Bouse
Spectrum Technologies
221 Ottawa Bend Dr.
Morris, IL 60450

N. Dennis Bowman
Unviersity of Illinois
1401D Regency Dr., E.
Champaign, IL 61874
(217) 333-4901
bowmand@mail.aces.uiuc.edu

Broughton Boydell
Australian Ctr. for Prec. Agr.
Romaka
Binigoy, NSW
Australia, 2394
(61) 29351-2947
b.boydell@agec.usyd.edu.au

Judith Bradow
USDA, ARS, SRRC
110 Robert E. Lee Blvd.
New Orleans, LA 70124
(505) 286-4479
jbradow@nola.strc.usda.gov

Terry Bradshaw
Northeast Area Ext. Office
Kansas State University
1515 College Ave.
Manhattan, KS 66502
(785) 532-5833

Ricardo Braga
University of Florida
1 Frazier Rogers Hall
Box 190570
Gainsville, FL
(517) 351-5949
rbraga@agen.ufl.edu

Rob Bramley
CSIRO Land & Water
Davies Lab., PMB Aitkenvale
Townsville, Queensland
Australia, 4814
(61) 7-4753-8591
rob.bramley@tvl.clw.csiro.au

Stewart Brandt
Agric. & AgriFood Canada
Scott Research Farm, Box 10
Scott, Saskatchewan
Canada, S0K 4A0
(306) 247-2011
brandts@agr.ca

Sebastian Braum
University of Minnesota
1991 Upper Buford Circle
St. Paul, MN 55108
(612) 625-1767
sbraum@soils.umn.edu

Christoph Brenk
BASF Aktiengesellschaft
Box 120
Limburgerhoi
Germany, 67114
(49) 621-60-27248
christoph@msm.basf-ag.de

Tom Bride
IGF Ins.-Geo Ag Plus
6407 N. Upland Terrace
Peoria, IL
(309) 239-6536

Roger Brook
Michigan State Univversity
210 Ford Hall
E. Lansing, MI 48840
(517) 353-4456

Bert Broos
Katholieke Universiteit Leuven
Kardinaal Mercierlaan 92
BlokE
Leuven, Vlaams Brabant
Belgium, 3001
(0032) 16-32.016.13

Sandy Browne
Field Worker Products, Limited
H6-1477 Bayview Ave.
Toronto, Ontario
Canada, M4G 3B2
(416) 483-3485 fx 483-7069

Mike Brugger
Brugger Farms
N4390 Hwy. 59
Monroe, WI 53566
(608) 325-1355

John Brumett
University of Missouri
Box 68
Monticello, MO 63457
brumettj@missouri.edu

Roz Buick
Trimble Navigation Limited
9290 Bond St.
Suite 102
Overland Park, KS 66214
(913) 495-2700

Paul Bullock
NoeTix Research, Inc.
403-265 Carling Ave.
Ottawa, Ontario
Canada K1S 2E1
(631) 236-1555
info@noetix.on.ca

Daniel Bunnell
Raven Industries
Box 5107
Sioux Falls, SD 57117

Ken Burgess
Red Hen Systems, Inc.
800 Stockton Ave.
Unit 2
Ft. Collins, CO 80524
(970) 493-3952
kburgess@redhensystems.co

Roger Burkhart
Deere & Company
One John Deere Place
Moline, IL 61265
(309) 765-4365
roger@90.deere.com

D. Tom Burmood
Mycogen Seeds
1340 Corporate Center Curve
Eagan, MN 55121
(612) 405-5908
burmoodt@mycogen.com

Leonard Burton
Farmland Industries
3315 N Oak St
Kansas City, MO 64151
(816) 459-6070

Steve Burtt
Ag & Ag-Food Canada
K.W. Neatby Bldng, 960
Carling
Ottawa, Ontairo
Canada, K1A 0C6
(613) 759-1529

David Buss
Datawise
203 W. Cottage Grove Rd.
Cottage Grove, WI 53527
(608) 251-9010 ex 274
datawise@inxpress.net

John Byrnes
Farm Journal/Top Producer
10705 34th Ave., N.
Plymouth, MN 55441-2450
(612) 513-9880
johnbyrnes@aol.com

Miguel Calmon
University of Florida
Frazier Rogers Hall
Gainesville, FL 32611
(352) 392-9112
mcalmon@agen.ufl.edu

Athyna Cambouris
Agricult. & Agri. Food Canada
2560 baul Hochelago
Ste-Fay, Quebec
Canada, G1V 2J3
(418) 657-7980 ex 255
cambouris@em.agr.ca

Paul Cameron
University of Minnesota

Gabe Camp
Grower Service Corp.
221 W. Lake Lansing Rd.
Suite 102
E. Lansing, MI 48906
(517) 333-8788
grower@growers.net

Ron Campbell
Harvest Master
1740 N. Research Pky.
Logan, UT 84341
(435) 787-4675
rcampbell@harvestmaster.com

James Capron
Cornell Coop. Extension
423 Snell Rd.
Geneva, NY 14456
(315) 331-8415
japron@cce.cornell.edu

Dan Caraylannis
SPOT Image Corporation
1897 Preston White Dr.
Reston, VA 20191
(703) 715-3100

Alan Carlson
Advanced Precision Systems
Box 906
Brandon, SD 57005
(605) 582-3647
adapncsy@worldnet.att.net

Paula Carper
Paula Carper Crop Ins.
Box 461
Holyoke, CO 80734
(970) 854-4596
pcarper@ria.net

Paul Carter
Purdue University
1150 Lilly Hall of Life Sci.
West Layfette, IN 47907-
1150
(765) 494-6247
cart@purdue.edu

Andy Caruso
IGF Ins.-Geo Ag Plus
2330 E.P. True Pkwy. #9
West Des Moines, IA
(515) 633-1180

William Casady
University of Missouri
222 Ag. Eng. Buildnig
Columbia, MO 65211
(573) 882-2731
casadyw@missouri.edu

Jose Casanova
University of Valladolid
Faculty of Sciences
Valladolid
Spain, 47071
(34) 983-423130
jois@cpd.uva.es

Gabriel Casasola
Pasa S.A.
Ruta 9-KM. 79, 4
Campana, Buenos Aires
Argentina, 2804
(54) 589-36493
gcasasol@pecom.com.ar

R. Jason Cathcart
University of Guelph
131 Richards Bldg.
Guelph, ON
Canada, N1G 2W1
(519) 824-4120 ex 4264
jcathcar@lrs.voquelph.ca

Allan Cattanach
American Crystal Sugar
101 N. 3rd St.
Moorhead, MN 56560
(218) 236-4487
a.cattana@crystalsugar.com

Jiyul Chang
South Dakota State
University
Ag Hall #216
Plant Science Dept. SDSU
Brookings, SD 57007
(605) 688-5105

Yona Chen
University of Minnesota
1991 Upper Bufford Circle
St. Paul, MN 55108
(612) 625-4229
yonachen@agri.huji.ac.il

H.H. Cheng
University of Minnesota

Norm Chervany
University of Minnesota

She-Kong Chong
Southern Illinois University
Carbondale, IL 62901-4415
(618) 453-2496
skchong@siu.edu

Corey Christensen
RDI Technologies, Inc.
300 Hwy. 23 Suite One
Spicer, MN 56288
(320) 796-0019

Tom Christensen
Zeneca
6945 Vista Dr.
W. Des Moines, IA 50266
(515) 222-4835

Jerry Christenson
American Crystal Sugar Co.
Box 600
Crookston, MN 56716
(218) 281-0107

Colin Christy
Veris Technologies
601 N. Broadway
Salina, KS 67401
(785) 825-1998

Dick Chronowski
ESRI
380 New York St.
Redlands, CA 92373

Rex Clark
University of Georgia
Driftmier Engineering Center
Athens, GA 30602
(706) 542-0864
rclark@bae.uga.edu

David Clay
South Dakota State Universiy
Plant Science Department
Brookings, SD 57007
(605) 688-5081
clayd@ur.sdstate.edu

Sharon Clay
South Dakota State
University
Plant Science Department
Brookings, SD 57007

Andrew Clock
Terra PCS
5280 W. 3005
Marion, IN
(765) 922-7783
aclock@comteck.com

Michelle Clyde
University of Manitoba
Box 801
Dauphin, Manitoba
Canada, R7N 3B3
638-8656 or 622-2244

Kevin Cobb
Trimble Navigation Limited
9290 Bond St.
Suite 102
Overland Park, KS 66214
(913) 495-2700

Sylvia A. M. Colburn
Crop Technology, Inc.
8318 Hidden Trail Lane
Spring, TX 77379-8722
(713) 973-2767
colburn@soildoctor.com

John Colburn, Jr.
Crop Technology, Inc.
8318 Hidden Trail Lane
Spring, TX 77379-8722
(713) 973-2767
colburn@soildoctor.com

Tom Colvin
USDA/ARS
2150 Pammel Dr.
Ames, IA 50011
(515) 294-8125
colvin@mstl.gov

Greg Cook
DICKEY-John Corporation
5200 DICKEY-john Rd.
Auburn, IL 62615-0010

Simon Cook
CSIRO
Private Bag, P.O.
Wembley, WA
Australia, 6014
(61) 8-9333-6138
simonc@per.dms.csiro.au

A. Ananda Coomaraswamy
Tea Research Institute
Talawakele
Sri Lanka

Kelly Ann Copas
Emerge
900 Technology Park Dr.
Bldg. 8
Billerica, MA 01821
(978) 262-0668 fx 262-0700
karauenzahn@emerge.wsicorp.com

Mark Copenhaver

Bernard Coquil
Matra Marconi Space
31 Rue Des Cosmonautes
Toulouse
France 31402
(33) 5-6219-6650
bernard.coquil@tps.mms.fr

Jose Cora
UNESP/FAPESP
Rod Carlos Tonanni Km 05
Jaboticabal, Sao Paulo
Brazil, 14870-000
(016) 223-2500
cora@fcau.unesp.br

Steve Core
Corn Plus

Paul Cornwell
Can Grow, Inc.
2971 Old Walnut Rd.
Alvinston, Ontarion
Canada, N0N 1AO
(519) 847-5747
pcornwell@cangrow.com

Rob Corry
University of Michigan
430 E. University
Ann Arbor, MI 48109-1115
(734) 763-9893
rcorry@umich.edu

Al Cotter
MCGA
21161 York Rd.
Hutchenson, MN 55350
(320) 578-9146
alcotter@hutchtel.net

David Cox
Davco Farming
Pelican Rd.
AYR, Queensland
Australia, 4807
(61) 7-4782-7575

Graeme Cox
NCEA Australia
USQ, Toowoomba
Australia, 4350
(61) 7-4631-1713
coxg@usq.edu.au

Michael Cox
Mississippi State University
Box 9555
Mississippi St., MS 39762
(601) 325-2311
mcox@onyx.msstate.edu

Max Crandall
ESRI
380 New York St.
Redlands, CA 92373
(909) 793-2853
mcrandall@esri.com

Bill Creath
MAPINFO Corporation
One Global View
Troy, NY 12180

John Creighton
Creighton Farms
14203 Greenvale Rd.
Lena, IL 61048
(815) 369-9051

Ian Crosihwaite
BGA Rural Services
Youngman St., Box 328
Kingaroy, Old
Australia 4610
(07) 416-22311

Jose Cueva
Cadilga, S.A.
3 Ave. S. #120
San Pedro Sula
Honduras
(504) 557-4100
cadelga@simon.intentel.hn

Carlos Eduardo Cugnasca
University of Sao Paulo
R. Des Ferreira Franca
40-AP 1028, Sao Paulo
Brazil 05446-900
(55) 11 818 5366
cecugnas@usp.br

Dan Culp
Terra PCS
14526 W. 300 S.
Francesville, IN 41946
(219) 567-7984
dculp@skyenet.net

Joe Curless
MCS/Case Corp
RR2 Box 79
Princeton, IL 61356
(309) 264-5086
jcurless@casecorp.com

Steve Curley
The Consulting Co.
14075 Foxtail Lane
Apple Valley, MN
(612) 431-7940
stevec@ens.net

Art Curtis
AGRIS Corporation
300 Grimes Bridge Rd.
Roswell, GA 30075

Stan Daberkow
USDA-Economic Research Service
1800 M St., N.W.
Washington, DC 20036-5831
(202) 694-5535
daberkow@econ.ag.gov

Dave Dahgren
Cargill, Inc.
101 3rd AVe., S.W.
Clarion, IA 50525
(515) 532-2834

Howard Dahl
Concord Environmental
Equip.
Route 1
Box 78
Hawley, MN 56549

Ken Dalenberg
Scattered Acres Farm
3014 N 1500 E Rd
Mansfield, IL 61854
(217) 489-9017
kennethd@ncsa.uiuc.edu

Michael Dalgleish
Golden River LTD
Churchill Re.
Bichester, Oxon
United Kingdom 8X6 7XT
011 44 1809 362800
md@goldenriver.com

Jerome Damboise
Eastern Canada Soil & Water
1010 ch. de l'Eglise
Saint-Andre, New Brunswick
Canada E3Y 2X9
(506) 475-4040
jdambois@cuslm.ca

Peter Dampney
AGAS Consulting, Ltd.
Boxworth Research Center
Boxworth, Cambridge
United Kingdom C83 8NN
444 1954 268213
peter_dampney@ adas.co.uk

Gary Dau
IGF Ins.-Geo Ag Plus
2334 N. 42 Rd.
Sheridan, IL 60551
(815) 469-09451

Craig Daughtry
USDA ARS Remote Sensing
Building 007
10300 Baltimore Ave.
Beltsville, MD 21045
(301) 504-5015
cdaughtry@asrr.arsusda.gov

Joan Davenport
Washington State University
24106 N. Bunn Rd.
Prosser, WA 99350
(509) 786-9384
jdavenp@tricity.wsu.edu

Glenn Davis
University of Missouri
241 Agric. Eng. Bldg.
Columbia, MO 65211
(573) 882-9301
DavisJG@missouri.edu

Scott De Jong
Midwest Farmers Co-op
Box 288
Orange City, IA 51041
(712) 737-4944
dejong@nwidt.com

Jeff Dearborn
Ag-Chem Equipment Co.
5720 Smetana Dr.
Minnetonka, MN 55343

Mark DeBower
Novartis Seeds, Inc.
7500 Olson Memorial Hwy.
Golden Valley, MN 55427
(612) 593-7276
mark.debower@seeds.novartis.com

Jean-Claude Deguise
Canada Centre for Remote Sens.
588 Booth St.
Ottawa, ON
Canada K1A 0Y7
(613) 947-1229

Jorge Delgado
USDA-ARS
301 S. Howes St.
Room 407
Ft. Collins, CO 80521
(970) 490-8260
jdelgado@lamer.colostate.edu

Sofia Delin
SLU, Dept. of Ag. Research
Box 234
Skara
Sweden, SE-532 23
46 511 67235
sofia.delin@jvsk.slu.se

Markus Demmel
Tech. Universitaet Muenchen
AM Staudengarten 2
Freising
Germany 85350
(49) 8161-713834
demmel@ban.tec.agrar.tu-

Larry Den Hartog
Midwest Farmers Co-op
4119 300th St.
Boyden, IA 51234

Greg Derksen
Flexi-Coil Ltd.
Box 1928
Saskatoon, Saskatchewan
Canada S7K 3S5
(306) 644-7600 ex 430
gderksen@flexiciol.com

Al Despain
Agri Northwest
Box 2308
Tri-Cities, WA 99302
(509) 735-6461
aldespain@agrinorthwest.com

Alex Desselk
Skyview Technologies
Box 608
Walker, LA

Chester Dickerson
Draper Dickderson Enterprises
11313 Willowbrook Dr.
Potomac, MD 20854-2568
(301) 983-9796
chetdick@aol.com

Sally Dickerson
Draper Dickerson Enterprises
11313 Willowbrook Dr.
Potomac, MD 20854-2568
(301) 983-9796
chetdick@aol.com

Doug Dickman
MCS Consulting Services
1100 S. Countyline Rd.
Maple Park, IL 60151
(815) 827-3108
ddickman@casecorp.com

Kenan Diker
USDA-ARS-NPA
AERC-CSU
Fort Collins, CO 80523

Huseyin Dikici
University of Minnesota
1991 Buford Circle
St. Paul, MN 55108
(612) 625-1767
hdikici@soils.umn.edu

Carl Dillon
University of Arkansas
221 Acriculture Building
Fayetteville, AR 72701
(501) 575-2279
cdillon@comp uark.edu

Lowell Disrud
North Dakota State University
Biosystems & Engineering
1221 Albrecht Blvd., Box 5626
Fargo, ND 58105-5626
(701) 231-7271
disrud@plains.nodak.edu

Jeanette Dockray
MN

John Dockray
Growforce Australia, Ltd.
1808 Ipswich Rd.
Rocklea, Brisban, Queensland
Australia 4106
07 3875 9926

Paul Dodds
Cargill Limited
Route 4
Clinton, Ontario
Canada, N0M 1L0
(519)233-3423

Tom Doerge
Pioneer Hi-Bred International
7100 NW 62nd Ave.
Johnston, Iowa 50131
(515) 334-6999
doergeta@phibred.com

Jerry Dohrman
Ag Tech Marketing
5940 Wyhgate Lane
Minnetonka, MN 55345
(612) 934-0681

Jeff Dolezal
Position Inc.
1609 Roosevelt Dr.
Atlantic, IA 50022
(713) 243-3752

Stephen Donohue
Virginia Tech
Dept. of Crop & Soil Environ.
Blackburg, VA 24061
(540) 231-9740
donohue@vt.edu

John Douglas
Growmark, Inc.
1701 Towanda AVe.
Bloomington, IL 61701
(309) 557-6244

Greg Downing
Rockweel International
267 Red Pine Trail
Hudson, WI
(715) 549-6331

Ben Drake
ERDAS, Inc.
2801 Buford Hwy.
Atlanta, GA 30032
(404) 248-9000 fx 248-9400
www.erdas.com

Darrin Drollinger
Ag Electronics Association
10 S. Riverside Plaza
Suite 1220
Chicago, IL 60606-3700
(312) 321-1470
ddrollinger@emi.org

Sharon Drumm
ARS-NPS
Bldg. 005, Room 223
BARL-West
Beltsville, MD 20705
(301) 504-5416
sdd@ars.usda.gov

Paul Drummond
Veris Technologies
601 N. Broadway
Salina, KS 67401
(785) 825-1998

Scott Drummond
University of Missouri
159 Agricultural Engr. Bldng.
Columbia, MO 65211
(573) 882-1146
DrummondS@missouri.edu

Marek Dudka
Univ. of Wisconsin-Madison
2369 Effirgham Way
Sun Prairie, WI 53590
(668) 936-0706
mdudka@students.wisc.edu

Wayne Dulaney
USDA-ARS-RSML
Bldg. 007
Room 008 Barc W.
Beltsville, ND 20705-2350
(301) 504-6076
wdulaney@asrr.arsusda.gov

Matt Duncan
Crop Production Services
Box 43
Ferris, IL 62336-0043
(217) 746-3861

Dennis Dunivan
RESOURCE21
7257 South Tucson Way
Englewood, CO 80112

Kelly Dupont
Inst. for Technical Develop.
Bldg. 1103 Suite 118
Stennes Sp. Ctr, MS 395509
228-688-2509
jkdupont@iftd.org

Jeffrey Durrence
University of Georgia
Box 748
Tifton, GA 31793-0748
(912) 386-3377
jdurrenc@tifton.cpes.peachnet.edu

Mr. Durrstein
Satcon System GmbH
Obertheres, Bavaria
Germany, 87531
9521 7072 fx 9521 1350
satcon.com@t-online.de

Jason Eagleton
RDI Technologies, Inc.
300 Hwy. 23
Suite One
Spicer, MN 56288
(320) 796-0019

Neal Eash
Southwest Experiment Station
23669 130th St.
Lamberton, MN 56152
(507) 752-7372
eash@ssu.southwest.msus.edu

Stephen Ebelhar
University of Illinois
Dixon Springs Ag. Center
Simpson, IL 62985
(618) 695-2790
sebelhar@uiuc.edu

Charlotte Eberlein
University of Idaho
Twin Falls R & E Center
Box 1827
Twin Falls, ID 83303-1827
(208) 736-3600
ceberl@uidaho.edu

Stanley Ehler
Kansas State University
1211 Houston
Manhattan, KS 66502-4354
(785) 539-9138
momehler@flinthills.com

D. Ehlert
Institute of Ag. Engineering
Max-Eyth-Allee 100
Potsdam
Germany, 14469
49 337 56994701
dehlert@atbuni-potsdam.de

Dan Eklund
Eklund Ag. Consulting
906 Park Cr.
Huxley, IA 50124
(515) 597-2873
ekagcen@pcpartner.net

Nael El-Hout
United States Sugar Corp.
P. O. Drawer 1207
Clewiston, FL 33440
(941) 902-2489

Sharif El-Hout
MN

Craig Elliott
Rockwell Collins, Inc.
350 Collins Rd., N.E.
Cedar Rapids, IA 52498
(319) 255-1179
caelliot@collins.rockwell.com

Charles Ellis
University of Missouri
880 W. College
Troy, MO 63379
(314) 528-4613
ellisc@ext.missouri.edu

Roger Elmore
University fo Nebraska
Box 66
Clay Center, NE 68933
(402) 762-4433
relmore@unl.edu

Michael Elms
Texas Tech Univ.-Farm Oper.
6116 Nashville Ave.
Lubbock, TX 79413
(806) 790-8595

Burton English
University of Tennessee
308 Morgan Hall
Box 1071
Knoxville, TN 37901-1071
(423) 974-3716
benglish@utk.edu

John Enterline
Analytical Spectral Devices
5335 Sterling Dr.
Suite A
Boulder, CO 80301
(303) 444-6522
enterline@asdi.com

Bruce Epler
ERDAS, Inc.
2801 Buford Hwy.
Atlanta, GA 30032
(404) 248-9000

Bruce Erickson
Purdue University
1150 Lilly Hall of Life Sci.
West Lafayette, IN 47907-1150
(765) 494-4788
bjericks@perdue.edu

Steve Ernst
Lockhead Martin
3333 Pilot Knob Rd.
Eagan, MN 55124
(612) 456-2333

Teresa Esser
Kluwer Academic Publishers
101 Philip Dr.
Norwell, MA 02061
(781) 871-6600 fx 871-7507
tesser@wkap.com

Michael Everett
Blondes Farm Supplu of MI
11985 Strait Rd.
Hanover, MI 49241
(517) 629-3546
blondes1@dmci.net

Chris Fagan
University of Nebraska
120 Keim Hall
Lincoln, NE 68583
(402) 472-1594
cfagan@unlgrad1.unl.edu

Dean Fairchild
Cargill
Box 9300
Minneapolis, MN 55440
(612) 742-2061

Steve Faivre
Case Corporation
1100 County Line Rd.
Maple Park, IL 60151
(815) 827-3108
sfaivre@casecorp.com

A/Razag Falatah
King Saud University
Box 2460
Riyadh
Saudi Arabia 11451
966 01 4678461
midress@ksu.ed.sa

Lanny Faleide
Agr Imagis
5174 30th St. N.E.
Maddock, ND 58348
(701) 438-2242 fx 438-2870
agriimagis@stellarnet.com

Jay Fallick
Iowa State University
124 Davidson Hall
Ames, IA 50011
(515) 294-7350
fallick@iastate.edu

Anne Fancher
Case Corporation
7100 S. County Line Rd.
Burr Ridge, IL 60521
(630) 789-7125
afancher@casecorp.com

Antonio Carlos Felix Ribeiro
University of Brasilia
Campus Ddncy Ribeiro ICC
Ald Sul Fac
Agronomia,Brasilia
Brazil 70-910-900

Richard Ferguson
University of Nebraska
Box 66
Clay Center, NE 68933
(402) 762-3535
rferguson@unl.edu

Mark Fering
John Deere Precision Farming
501 River Dr.
Moline, IL 61265
(309) 765-7276
mf26075@deere.com

Filip Feyaerts
K. U. LEUVEN-ESAT-PSI
Kardinaal Mercierlaan 94
Heverlee
Belgium, B 300
32 1632 1708
filip.feyaerts@esat.kufeuren.ac.be

Tim Fiez
Washington State University
Box 646420
Pullman, WA 99164-6420
(509) 335-2997
fiez@wsu.edu

David Fike
TeeJet Technologies
7200 France So.
Suite 128
Edina, MN 55435
(612) 831-1559

Kara Fischer
Concord Environmental Equip.
Route 1
Box 78
Hawley, MN 56549
(218) 937-5100 fx 937-5101
cee@rrnet.com

Judi Fitzsimmons
AGRIS
300 Grimes Bridge Rd.
Roswell, GA 30075

Paul Fixen
Potash & Phosphate Inst.
772 22nd Ave. S.
Brookings, SD 57006
(605) 692-6280
pfixen@ppi-far.org

Shelby Fleischer
Penn State
501 ASI Bldg.
University Park, PA 16802
(814) 863-7788
sjf4@psu.edu

Kim Fleming
Colorado State University
2307 Sceap Ct.
Ft. Collins, CO 80526
451-8663
kfleming@lily.aerc.colostate.edu

Ben Foley
Incitec Fertilizers
6 Spalding St.
Ararat, Victoria
Australia, 3377
0419 756576
ben.foley@incitec.com.au

Randy Follman
Cenex/Land O' Lakes
5600 Cenex Dr.
MS 370
Inv. Gr. Hgts., MN 55077
(612) 451-5047
rlfoll@cnxlol.com

Karl Foord
University of Minnesota
4100 220th St., W.
Farmington, MN 55024-9539
(612) 891-7703
kfoord@extension.umn.edu

Zachary Fore
Cenex/Land O'Lakes
P.O. Box 64089
St. Paul, MN 55164
(612) 451-4626
zfore@cnxlol.com

Clyde Fraisse
USDA-Ag. Research Service
University of Missouri
140 Ag. Engr. Bldng.
Coulmbia, MO 65211
(573) 882-1148
cfraisse@showme.missouri.edu

Dave Franzen
North Dakota State University
Box 5758
Fargo, ND 58105-5758

Keith Frick
Ag-Chem Equipment Co.
5720 Smetana Dr.
Minnetonka, MN 55353

Mary Friedricksen
532 Hayward Ave.
Ames, IA 50014
(515) 296-2495
maryfred@iastate.edu

Robert Fritz
South Dakota State University
NPBL 247
Box 2140C
Brookings, SD 57007
(605) 688-4746
pedology.ur@sdstate.edu

Dennis Fuchs
Stearns County SWCD
110 2nd St. S. Suite 128
Waite Park, MN 56387
(320) 251-6718
dennis.fuchs@co.stearns.mn.us

David Fuhr
Airborne Data Systems
25338 290th St.
Wabasso, MN 56293
(507) 984-5419
airdat@rconnect.com

Max Fuxa
Cenex Land O'Lakes Agro. Co.
1731 31st St. S.
Moorhead, MN 56560
(218) 233-7218
mfuxa@cnxlol.com

Jim Gaebel
The Toro Company
8111 Lyndale Ave., S.
Bloomington, MN 55420
(612) 887-8897
jim.gaebel@toro.com

Dale Gandrud
Gandy Co.
Owatonna, MN 55060
(507) 451-5430
dgandrud@gandy.net

Luis Garcia Torres
Institute for Sustainable Ag.
14080
Cordoba
Spain

Salvador Garibay
ETH-Zurich
CH-8092
Zurich
Switzerland
(41) 1-632 4237
salvador.garibay@ipw.agrl.ethz.ch

Pierre-Yves Gasser
Ag-Knowledge
Box 540, 15B Chemis Dr,
Verger
Chelsea, Quebec
Canada, JOX 1N0
(819) 827-3762

Larry Gaultney
DuPont Agricultural Products
Box 30
Newark, DE 19714
(320) 366-6587
lawrence.d.gaultney@usa.dupont.com

Sharon Gauquie
Rawson Control Systems, Inc.
116 2nd St., S.E.
Oelwein, IA 50662
(319) 283-2225 fx 283-1360
rawson@rawsoncontrol.com

Lee Gholz
Precision Farming Enterprises
1111 Kennedy Place Suite 2
Davis, CA 95616
(916) 785-1946
pfe@davis.com

Stephanie Giard
Resource21
7257 S. Tuscon Way
Suite 200
Englewood, CO 80112

Daniel Ginting
University of Minnesota
1991 Upper Buford Circle
St Paul, MN 55108
dginting@soils.umn.edu

Jean Gleichsher
Fort Hays State University
Agriculture Dept.
Hays, KS 67601

Brad Glenn
IGF Ins.-Geo Ag Plus
RR1 Box 102
Stanford, IL

Daan Goense
Bomenwegh
IMAG-DLO
Wageningen
Netherlands
31 317 476462
d.goense@imag.dlo.nc

Liesbeth Goerse

Todd Golly
Ag Prophets

Shannon Gomes
Cedar Basin Crop Consulting
208 Rainbow Dr.
Waverly, IA 50677
(319) 352-1227
cbcc@sbt.net

Kikmet Gonal
Kansas State University
Throckmorton Hall, Room 2729
Agronomy Dept.
Manhatten, KS 66506
(519) 824-4120 ex 6976
hgoudy@plant.uoguelph.ca

Heather Goudy
University of Guelph/Crop Sci.
Guelph, ON
Canada, N1G 2W1

Dave Goughnour
SST Develpoment Group, Inc.
824 N. Country Club Rd.
Stillwater, MN 74075

Rich Gould
TeeJet Technologies
7200 France Ave. So. Suite 128
Edina, MN 55435
(612) 831-1559

John Grace
Resource21
7257 S. Tuscon Way Suite 200
Englewwod, CO 80112

Carrie Graff
University of Minnesota
1226 Rose Vista Ct #7
Roseville, MN 55113
(561) 917-7198
cgraff@soils.umn.edu

Susan Grecu
NSCSS
20 W. 36th #114
Kansas City, KS 64111
(816) 753-3674
bpecks@hotmail.com

Thomas Green
Gempler's, Inc.
Box 270
Mt. Horeb, WI 53572
(608) 437-4883 fx 437-5383

Mike Greene
Case Corporation
1100 S. Countyline Rd.
Box 364
Maple Park, IL 60151
(815) 827-3108
mgreene@casecorp.com

Richard Greene
Mark II Agronomy
12170 Leech Rd.
Durand, IL 61024
(815) 248-2679

Eric Gregory
Morse Bros. Aerial Applicators
General Delivery
Starbuck, MB
Canada, R0G2P0
(204) 735-2258
morsebro@escape.ca

Gilbert Grenier
Enita de Bordeaux
BP 207
Gradignan
France 33175
33 557 350776
grenier@enital.fr

Timothy Gress
Inst. for Tech. Development
Bldg. 1103
Suite 118
Stennis Sp. Ctr, MS 39509
(228) 688-2509
tgress@iftd.org

Simon Griffin
SOYL Ltd.
Blackdown Farm
Southampton, Hants
United Kingdom 5032 1H5
44 01962 777687
100340.1507@compuserve.com

Steven Griffin
IGF Insurance
6000 Grand Ave.
Des Moines, IA 50312
igfinsurance.com

Terry Griffin
University of Arkansas
221 Agri. Bldg.
Fayetteville, AR 72701
(501) 575-2256
tgriffin@comp.uark.edu

Brant Groen
Ridgewater College
Box 1097
Willmar, MN 56201
(320) 231-7647
bgroen@ridgewater.mascu.edu

Ed Groholshi
Innovative Farmers
1700 11 mzle rd
Burlington, MI 49029
(517) 765-2110

Michelle Guertin

Peter Guertin

Bob Gunzenhauser
Geo-Agra Resource LC
4610 Toronto
Ames, IA 50014
(515) 296-1984
bgunzy@netins.net

John Hadam
KVH Industries, Inc.
50 Enterprise Center
Middletown, RI 02842
(401) 847-3327 fx 845-8190
jhadam@kvh.com

Jens Hald
Agris Corp.
300 Grimes Bridge
Roswell, GA 30075
(770) 552-6522
jenshald@post10.tele.dk

Gene Hall
Starlink Incorporated
6400 Highway 290 E.
Suite 202
Austin, TX 78723

Thomas Hall
North Dakota State University
1221 Albrecht Blvd.
Box 5626
Fargo, ND 58105
(701) 297-3919
hallt7@asme.org

Shufeng Han
Case Corporation
75600 County Line Rd.
Burr Ridge, IL 60521
(630) 887-5470
shan@casecorp.com

Silvia Haneklaus
Inst. of Plant Nutr & Soil Sci
Bundesallee 50
Braunschweig
Germany, D-38116
49 531 596778
haneklaus@pb.fal.de

Glenn Hanson
Midwest Ag Services
1414 5th Ave., S.W.
Jamestown, ND 58401
(701) 252-0580
hansog@midwest-ag.com

John Harapiak
WestCo Marketing
Box 2500
Calgary, Alberta
Canada T2P 2N1
(403) 249-1121
jt.harapiak@westcoag.com

Ian Harley
Farmworks Software
Box 250
Hamilton, IN 46742
(219) 488-3388 fx 488-3737
Farmwork@farmworks.com

Mike Harms
Agri-Tech FS
1037 N Fourth
Platteville, WI 53818
(608) 776-4600

Joe Harroun
Cargill Inc.
2301 Crosby Rd.
Wayzata, MN 55391
(612) 742-6476
joe_harroun@cargill.com

Galen Hart
Resource 21
6457 Eppard St.
Falls Church, VA 22042
(703) 536-1977
galenhart@aol.com

Mike Hassell
Agri Northwest
Box 2308
Tri-Cities, WA 99302
(509) 735-6461
mikehassell@agrinorthwest.c
om

Hirofumi Hayama
Zen-nch Unico America Corp.
245 Park Ave.
32nd Floor
New York, NY 10167
(212) 916-3322
hayama@zen-nohusa.com

Adam Hayes
Ont. Minstry of Ag., Food & Ru
Box 400
Ridgetown, Ontario
Canada, N0P2C0
(519) 674-1624-1
ahayes@omafra.gov.on.ca

John Heard
Manitoba Agriculture
Box 1149
Carman, Manitoba
Canada, R0G 0J0
(204) 745-2040
jheard@agr.gov.mb.ca

Rob Heater
Stahlbush Island Farms, Inc.
3122 Stahlbush Island Rd.
Corvallis, OR 97333
(541) 757-1497

Rob Heckore
Trimble Navigation Limited
9290 Bon St.
Suite 102
Overland Park, KS 66214

Ronnie Heiniger
North Carolina State Univ.
Vernon James Research Center
207 Research Station Rd.
Plymouth, NC 27962

Torben Heisel
Danish Inst. of Agri. Sciences
Research Centre Flakkebjerg
DK-4200 Slagelse
Denmark,
45 58 11 33 92
torbin.heisel@agrsci.dk

Devin Helming
Cargill, Inc.
415 S. Grove
Suite 1
Blue Earth, MN 56013
(507)526-2290

Jeff Hemenway
USDA-NRCS
Federal Building
200 4th St. S.W.
Huron, SD 57350-2475
(605) 352-1239
jhemenway@sd.nrcs.usda.gov

Larry Hendrickson
Case Corporation
7 S 600 County Line Rd.
Burr Ridge, IL 60521
(630) 887-3924
lhendrickson@casecorp.com

Gordon Henrikson
13786 Hanover Way
Apple Valley, MN 55124
(612) 431-3940
gordon_henrikson@planar.com

Stan Herdina
MN

Chad Hertz
Hertz Farm Management
Box 2396
Waterloo, IA 50704
(319) 234-1949
chertz@wat.hfmgt.com

Richard Hess
Lockheed Martin Idaho Tech.
Box 1625
Idaho Falls, ID 83415-3710
(208) 526-0115
jrh@inel.gov

Brian Hicks
NETTJEWYYNNT Farm
19465 Co. Hwy. 8
Tracy, MN
(507) 629-4684
bdhicks@juno.com

Kim Hildebrand
Southern Illinois University
Carbondale, IL 62901-4415
(618) 529-3438
kkh7d@siu.edu

Anna Hill
Farm Chemicals Magazine
37733 Euclid Ave.
Willoughby, Ohio 44094
(440) 942-2000
annahill@meisterpubl.com

Paul Hinz
Iowa State University
Dept. of Statistics
Ames, IA 50011
(515) 294-8948
phinz@iastate.edu

Tom Hirose
Noetix Research, Inc.
#403-265 Carling Ave.
Ottawa, Ontario
Canada K1S 2E1
(613) 236-1555 fx 236-1870
info@noetix.on.ca

Garry Hnatowich
Saskatchewan Wheat Pool
201-407 Downey Rd.
Saskatoon, Saskatchewan
Canada S7N 408
(306) 668-6643
ghnatowich@innovationplace.com

Mark Hockel
Cargill, Inc.
845 1st Ave.
Bingham Lake, MN 56118
(507) 831-1729
cargilbl@rconnect.com

Rich Hodupp
MSU Extension
Lapeer Co. Extension
1575 Suncrest Dr.
Lapeer, MI 48446-1138
(810) 667-0341
hodupp@msuc.msu.edu

Vern Hofman
North Dakota State University
1221 Albrecht Blvd.
Box 5626
Fargo, ND 58105
(701) 231-7240
vhofman@ndsuext.edu

Mark Holoubek
Resource 21
7257 S. Tuscon Way
Englewood, CO 80112

Alberto Honda
Maguines Agricoles
JACTO
Pompeia, Sao Paulo
Brazil 17-580-000
014 4521811
jacto@jacto.com.br

Bryan Hoover
Agri Northwest
Box 2308
Tri-Cities, WA 99320
(509) 735-6461
bryanhoover@agrinorthwest.com

Aishania Hopkins
University of Minnesota
1991 Upper Buford Circle
St. Paul, MN 55108
(612) 625-1767
ahopkins@soils.umn.edu

Jeffrey Hopkins
Ohio State University
2120 Fyffe Rd.
Room 327
Columbus, OH 43210
(314) 292-9329
hopkins.35@osu.edu

Greg Horstmeier
Farm Journal
222 S. Jefferson
Mexico, MO 65265
(573) 581-9643
fjgreg@aol.com

Reed Hoskinson
Lockheed Martin Idaho Tech.
Box 1625
Idaho Falls, ID 83415
(208) 526-1211
hos@inel.gov

George Huber
Trimble Navigation Limited
9290 Bond St.
Suite 102
Overland Park, KS 66214
(913) 495-2700

Brian Huberty
USDA-NRCS
2820 Walton Commons #123
Madison, WI 53718
(608) 224-3014

Berman Hudson
USDA-NRCS
Box 2890
Washington, DC 20013
(202) 720-1809
berman.hudson@usda.gov

Tod Hudson
Missouri Dept. of Natural Res.
Box 176
Jefferson City, MO
(573) 751-8728

Miles Huffaker
Rutgers Cooperative Extension
51 Cheney Rd.
Suite 1
Woodstown, NJ 08098
(609) 769-0090
salem@aesop.rutgers.edu

Jerry Huffman
Dow Agro Sciences
1604 Devonshire Dr.
Champaign, IL 61821
(217) 355-3910
jrhuffman@dowagro.com

David Hughes
MFA Incorporated
201 Ray Young Dr.
Columbia, MO 65201-3599
(573) 876-5461

Marilyn Hughes
Rutgers Cooperative
Extension
Rutgers University
14 College Farm Rd.
New Bruinswick, NJ 08903
(732) 932-1582
mghughes@crssa.rutgers.edu

Daniel Humburg
South Dakota State University
Box 2120
Brookings, SD 57007
(605) 688-5658
humburgd@mg.sdstate.edu

John Hummel
USDA-Agric. Res. Serv.
1304 W. Penn Ave.
Urbana, IL 61801
(217) 333-0808
jhummel@uiuc.edu

Andrew Hunt
Hunt Farms
2355 W. Higgins
Morris, IL 60450
(815) 942-4634
ahunt@uti.com

Todd Hustrulid
University of Minnesota
1390 Eckles Ave.
St. Paul, MN 55108
(612) 625-1708
hust0038@tc.umn.edu

Michihisa Iida
Kyoto University
Sakyo-ku
Kyoto
Japan 606-8502
81 75 753 6168
iida@elam.kais.kyoto-u.ac.jp

Jerry Jackson
AGRIS Corporation
300 Grimes Bridge Rd.
Roswell, GA 30075

Barry Jacobson
Cargill
Box 5699
Minneapolis, MN 55440-5699
(612) 742-3989
Barry_Jacobson@cargill.com

Andreas Jarfe
ZALF, e.v.
Eberswalder Str. 84
Mancheberg
Germany 15374
0049 33432 82257
amueller@zalf.de

Paul Jasa
University of NEBR Coop.
Ext.
202 LWC E. Campus
Lincoln, NE 68583-0726
(402) 472-6715
bsen023@unlum.unl.edu

Dan Jaynes
USDA-ARS-NSTL
2150 Pammel Dr.
Ames, IA 50011
(515) 294-8243
jaynes@nstl.gov

Richard Jenny
AGSCO, Inc.
Box 13458
Grand Forks, ND 58208-3458
(701) 775-5325
richardjenny@hotmail.com

Mary Jetland
Ag-Chem Equipment Company
5720 Semtana Dr.
Minnetonka, MN 55343

Chris Johannsen
Purdue University
1150 Lilly Hall
West Layfette, IN 47907-1150
(465) 494-6248
johan@purdue.edu

Ed John
MapInfo Corporation
One Global View
Troy, NY 12180
(897) 397-9790

Dale Johnson
John Deere Precision Farming
501 River Dr.
Moline, IL 61265
(309) 765-7278
mn22100@deere.com

Darin Johnson
204 5th St NE
Halstad, MN 56548
(701) 799-7463
dajohnson@means.net

Gregg Johnson
University of Minnesota
35838 120th St.
Waseca, MN 56093
(507) 835-3620
john5510@tc.umn.edu

Randy Johnson
University of Minnesota

Richard Johnson
Texas Tech University
USDA-ARS-SRRC
Box 19687
New Orleans, LA 70179
(504) 286-4515
rjohnson@nola.srrc.usda.gov

Kezelee Jones
Michigan State University
14 Chittenden Hall
E. Lansing, MI 48824
(517) 355-6953
jonesk14@pilot.msu.edu

Kah Joo Goh
App. Agr. Research SDN. BHD.
Locked Bag #212, Sg.Guloh P.O.
Sungai Buloh, Selangor
Malaysia 47000
(603) 656-1152
aarsb@po.jaring.my

Sean Jordal
Case Corp./MCS
1100 S. County Line Rd.
Maple Park, IL 60151
(815) 757-8518
sjordal@casecorp.com

Robert Jordan
USDA-ARS-NPA
AERC-CSU
Fort Collins, CO
rob@lily.aerc.colostate.edu

Jason Kahabka
Cornell University
1003 Bladfield Hall
Ithaca, NY 14853
(607) 272-8996
jek15@cornell-edu

Joseph Kaplan
University of Minnesota
1390
Eckles Ave.
St. Paul, MN 55108
(612) 376-5791

Douglas Karlen
USDA-ARS
NSTL 2150 Pammel Dr.
Ames, IA 50011-4012
(515) 294-3336
dkarlen@nstl.gov

Jeff Keiser
Terra Industries
Box 6000
600 4th St.
Souix City, IA 51102
(712) 233-6175
jkeiser@terraindustries.com

Dennis Kemmesat
Trimble Navigation, Limited
9290 Bond St.
Suite 102
Overland Park, KS 66214
(913) 495-2700

Pete Kennedy
Case-Tyler Corporation
E. Hwy. 12
Benson, MN 56215
(320) 843-3333
petek@team-tyler.com

Joe Kennicker
Cenex Land O'Lakes
Box 307
Hazel Green, WI 53811-0307
(608) 854-2802
cenexhgr@hitechag.com

D.J. Kerns
mPower 3
POB 1289
Greenly, CO 80634-1289
(970) 356-8920
djkerns@loveland.uap.com

Greg Kerr
Kerr Agronomics, Inc.
629 Sunset Lane
River Falls, WI 54022
(715) 425-8447
gkerr@iname.com

Mark Kerr
AGSCO, Inc.
Box 13458
Grand Forks, ND 58201
(701) 775-2944
mkerr@mailcity.com

Mark Kessler
Purdue University
9100 N. White Oak Lane #218
Bayside, WI 53217
(414)352-5999
mckessler@worldnet.att.net

B. R. Khakural
University of Minnesota

Nader Khatib
University of Colorado
Campus Box 429
Boulder, CO 80309
(303) 492-2968
khatib@rastro.colorado.edu

Kevin Kilgus
Top Soil Precision Ag
Route 1
Box 9
Cropsey, IL 61731
(309) 377-2702
kkilgus@dave-world.net

Tom Kill
John Deere Precision Farming
501 River Dr.
Moline, IL 61265

Wook Kim
Iowa State University
136 Town
Ames, IA 50011
(515) 294-2672
wook@iastate.edu

Bradley King
University of Idaho
1693 S. 2700 W.
Box AA
Aberdeen, ID 83210-0530
(208) 397-4184
bradk@uidaho.edu

Newell Kitchen
USDA-Ag. Research Service
Univ. of Missouri
240 Ag. Eng. Bldng.
Columbia, MO 65211
(573) 882-1138
kitchenn@missouri.edu

Kelly Klein
Northeast Area Ext. Office
Kansas State University
1515College Ave.
Manhattan, KS 66502
(785) 532-5833

Matt Klein
Centrak
210 E. Kiowa
Fort Morgan, CO 80701

Lee Klossner
University of Minnesota
Box 428
Lamberton, MN 56152
(507) 752-7372
kloss001@tc.umn.edu

Greg Knoblauch
Resource21
7257 south Tucson Way
Englewood, CO 80112

Alan Koehler
Ag-Chem Equipment Co.
5720 Smetana Dr.
Minnetonka, MN 55343

Dwight Koehler
Koehler Search & Placement
5324 Nicholas St.
Omaha, NE 68132
(402) 553-6947
dkoeh73328@aol.com

Connie Kohut
Agrium Inc.
Bag 20
Redwater, Alberta
Canada, T0A 2WO
(403) 998-6141
gkohut@agrium.com

Takashi Kosaki
Kyoto University
Lab of Soils
Kyoto
Japan, 606-8502
81 75 753 6101
kosakit@kais.koyoto-u.ac.jp

Mark Krause
Terra Industries
600 4th St.
Sioux City, IA 51102
(712) 233-6102
mkrause@terraindustries.com

Alexandra Kravchenko
University of Illinois
1102 S. Goodwin Ave.
Urbana, IL 61801
(217) 333-9452
kravchen@uiuc.edu

Charles Krueger
Penn State University
116 Agr Sci And Ind Bldg.
University Park, PA 16802
(814) 863-7624
ckrueger@psu.edu

Michael Krumpelman
University of Missouri
159 Agric. Engr. Bldg.
Columbia, MO 65211
(573) 882-1146
krumpelman@missouri.edu

Matt Krusemark
University of Minnesota

Randy Kutcher
Agriculture Canada
Box 1240
Melfort, Saskatchewan
Canada, SOE 1AO
(306) 752-2776 ex232
kutcherr@em.agr.ca

Joe Kuznia
Novartis Seeds
317 330th St.
Stanton, MN 55018
(507) 663-7633
joseph.kuznia@seeds.novartis.com

Craig Kvien
NESPAL/ University of
Georgia
PO Box 748
Tifton, GA 31793
(912)386-7274
ckvien@tifton.cpes.peachnet.edu

Stephen Ladek
Farmer's Software Assoc.
Box 660
Fort Collins, CO 80522
(970) 493-1722
ladek@farmsoft.com

Mark Lage
Country Side Compt Svc
1708 Nettle Ln
Sheffield, IA 50475
(515) 892-4287
mlage@frontiernet.net

Don Lamker
Cargill
Box 9300 MS #19
Minneapolis, MN 55440
(612) 742-7131
don_a_lamker@cargill.com

Amy Landgraff
Cenex/Land O'Lakes
Box 64089
St. Paul, MN 55164
(612) 451-5268
aland@cnxlol.com

Juan Landivar
Texas A & M University
Ag. Research & Extension Ctr.
Rt. 2, Box 589
Corpus Christi, TX 78406
(512) 265-9201
jlandiva@tamu.edu

Darian Landolt
MCS Consulting Services
1100 S. Countyline Rd.
Box 364
Maple Park, IL 60151
(815) 827-3108
dandolt@casecorp.com

Melissa Landon
Farmers's Software Assoc.
Box 660
Fort Collins, CO 80522
(970) 493-1722
mladon@farmsoft.com

Frank Lang
Flexi Coil
2414 Royl Ave.
Saskatoon, Saskachawean
Canada, S7L 7C5
(306) 664-7684
flang@flexicoil.com

Art Lange
Trimble Navigation Limited
9290 Bond St.
Suite 102
Overland Park, KS 66214

Rob Langford
Dow AgroSciences
RR 3
Strathroy, Ontario
Canada, N7G 3H5
(519) 245-9187

Phil Laprezioso
MAPINFO Corp.
One Glolbal View
Troy, NY 12180
(518) 285-7140
phil.laprezioso@mapinfo.com

Norman Larsen
MCS
17520 Keslinger Rd.
Maple Park, IL 60151
(815) 827-3818
dlarson@casecorp.com

Ed Larson
Larson Acres
4243 Croak Rd.
Evansville, WI 53536
(608) 882-4303

Jim Larson
Haug Implement Co.
Box 1055
Willmar, MN 56201
(320) 235-8115

William Larson
University of Minnesota
Department of Soil, Water &
Climate
St. Paul, MN 55108

Robert Lascano
Texas A & M University
Rt. 3, Box 219
Lubbock, TX 79401
(806) 746-4025
r-lascano@tamu.edu

Laurent Layrol
Geosys
12300 Marion Lane #2317
Minnetonka, MN 55305
(612) 541-5358
layrol@geosys.fr

David Lee
Rutgers Cooperative Extension
51 Cheney Rd. Suite 1
Woodstown, NJ 08098
(609)769-0090
salem@AESOP.rutgers.edu

Jooho Lee
Iowa State University
136 Town
Ames, IA 50011
(515) 294-2672
jlee@iastate.edu

Tamara Lee
Starlink Incorporated
6400 Hwy. 290, E.
Suite 202
Austin, TX 78723

Leslie Legg
Precision Agriculture Center
439 Borlaug Hall
1991 Upper Buford Circle
St. Paul, MN 55108
(612) 624-4224 fx 624-4223
llegg@soils.umn.edu

Damien Lepoutre
Geosys
12 300 Marion Lane #2317
Minnetonka, MN 55305
(612) 541-5358
lepoutre@geosys.fr

Gregory Leth
Terra
302 Byron Ave.
Byron, MN 55920
(507) 775-2900
agronomy@means.net

Vivienne Lethbridge
Queensland Cotton
Box 859
Dalby, Queensland
Australia 4405
fdviv@qcotton.com.au

Tim Lindgren
Cargill Industries
2301 Crosby Rd.
Wayzata, MN 55391
(612) 742-6476
tim_lindgren@cargill.com

Dennis Lindsay
Lindsay Farms/Monsanto

Kevin Little
Resource21
7257 S. Tuscon Way
Englewood, CO 80112

Mark Little
John Deere Precision Farming
501 River Dr.
Moline, IL 61265
(309) 765-7431
mi42207@deere.com

Ke Liu
MN

Shi Shi Liu
MN

Zhoujing Liu
South Dakota State University
Plant Science Department
Ag Hall 206
Brookings, SD 57007
precisf@ur.sdstate.edu

Greg Livingston
IGF Ins.-Geo Ag Plus
2054 Rustic Lane
Winterset, IA 50273
(515) 633-1178

Sara Loechel
University of Maryland
Bldg. 007
Room 008
Beltsville, MD 20705
(301)504-6823
sloechel@asrr.arsusda.gov

Sally Logsdon
NSTL
2150 Pammel Dr.
Ames, IA 50011
(515) 294-8265
logsdon@nstl.gov

Dan Long
MSU Northern Agr. Res. Center
HC 36
Box 43
Havre, MT 59501
(406) 265-6115
dlong@campus.montana.mci.net

Ray Long
North Carolina State Univ.
Box 7620
Raleigh, NC 27695
(919) 515-3520
ray_long@ncsu.edu

Alberto Lopes
R Fradigue Cout/NHO 532 AP 162
Sao Paulo
Brazil, 05416 010
55 11 8145431 or 55 11 8185271
alberto@lopestelecom.com

Allan Lorenc
Rawson Control Systems, Inc.
116 2nd St. S.E.
Olewin, IA 50662
(319) 283-2225 fx 283-1360
rawson@rawsoncontrol.com

Kevin Lowe
Midwest Technologies Inc
2733 E Ash
Springfield, IL 62703
(217) 753-8424
klowe@mid-tech.com

Jess Lowenberg-Deboe
Purdue University
1145 Krannert
W. Lafayette, IN 47907-1145
(765) 494-4230
lowenberg-deboer@agecon.edu

Shen Lu
Cargill Ltd.
300-240 Grahm Ave.
Winnipeg, Manitoba
Canada
(204) 947-6107
shenlu@cargill.com

Carlos Adolfo Luna
CENICANA
Calle 58 N No. 3 BN-110
Cali, Valle
Colombia, 9138
57 2 664 80 25
caluna@cenicana.org

Eric Lund
Veris Technologies
601 N. Broadway
Salina, KS 67401
(785) 825-1998 fx 825-2097

Rose Lundeen
Ntl. Ctr. for Atmospheric Rsh.
Box 3000
Boulder, CO 80302
(303) 497-8390
lundeen@ucar.edu

Lisa Lunz
Haskell Ag Lab-Univ. of NE
57905 866 Rd.
Concord, NE 68728-2828
(402) 584-2846
nerc022@unlvm.unl.edu

Ruth Lutticken
Hous Holzl
Raahood 2
83137 Schoustett
Germany
0049 8075 18839
holzl@gpsct-ouluie.de

Stacey Lyle
Leica Geosystems, Inc.
3155 Medlock Bridge
Morcross, 30071
(770) 447-6361

Takemi Machida
Ibaraki University
Ami, Inashiki
Ibaraki
Japan 3000393
81 298 88 8628
machida@agr.ibaraki.ac.jp

Bob MacMillan
Land Mapper Environmental
7415 118 A St.
Edmonton, Alberta
Canada T6G 1V4
(403) 435-4531
bobmacm@planet.eon.net

Nancy Macy
AGRIS Corporation
300 Grimes Bridge Rd.
Roswell, GA 30075

Ted Macy
AGRIS Corporation
300 Grimes Bridge Rd.
Roswell, GA 30075
(770) 236-5161
daedmoondson@agos.com

Shivakumar Mahajanashetti
University of Tennessee
Box 1071
Knoxville, TN 37901-1071
(423) 974-7658
smahajan@agecon2.ag.utk.edu

Virg Mahlum
Concord Environmental Equip.
Route 1
Boz 78
Hawley, MN 56549
(218) 937-5100 fx 397-5101
cce@rrnet.com

Mike Mailander
Louisana State University
167 Doran LSU
Baton Rouge, LA 70803
(504) 388-1058
mmailand@gumbo.bae.lsu.edu

Antonio Mallarino
Iowa State University
Department of Agronomy
Ames, IA 50011
(515) 294-6200
amallar@iastate.edu

Gary Malzer
University of Minnesota
1991 Upper Buford Circle
St. Paul, MN 55108
(612) 625-6728
gmalzer@soils.umn.edu

Grant Mangold
@g Innovator/@g Online
774 South River Rd.
Linn Grove, IA 51033
(712) 296-3615
innovator@agriculture.com

John Mann
Ag-Chem Equipment Co., Inc.
5720 Smetana Dr.
Suite 100
Minnetonka, MN 55343

Grant Manning
University of Manitoba
RR3
Pilot Mound, MB
Canada, R0G 1P0
(204) 825-2884
wnmanni0@cc.umanitoba.ca

Rosa Marchetti
Istituto Sperimentale Agr.
Vlaje Caduti in Guerra, 134
Modena, MO
Italy, 141100
39 59 230454
lsa-mo@planeta.lt

Antonio Mareos Coelho
University of Nebraska
120 Keim Hall
Lincoln, NE 68583
(402) 472-1594
aeoelho@unlgrad1.unl.edu

Derrel Martin
University of Nebraska
Room 231 L.W. Chase Hall
Lincoln, NE 68583-0726
(402) 472-1586 fx 472-6338
dmartin@unlinfo.unl.edu

Mike Martin
University of Minnesota

Tim Masek
University of Nebraska
120 Keim Hall
Lincoln, NE 68583
(402) 472-1594
tmasekunlgrad1.unl.edu

Jack Masse
I.T.C.F.
Station Experimentale
Boigneville
France, 91720
33 164992250
itcfocp@worldnet.fr

Brian Matuschka
Matuschka Farms
RMB 2090
Horsham, Victoria
Australia, 3401
61 0353844216
brian@comcirc.com

Robert Mayfield
John Deere Precision Farming
Box 8000
Waterloo, IA 50704-8000
(319) 292-8824
re28062@deere.com

Alex McBratney
University of Sydney
Aust. Ctr. for Prec. Ag.
Rose St. Bldg., Sydney NSW
Australia, 2006
61 2 9351 3214
alexmcbartney@acgs.asyd.edu.au

Elaine McCallum
Farmer's Software Association
800 Stockton Ave #2
Ft. Collins, CO 80524

Ed McClenahan
Cornell University
232 Emerson Hall
Ithaca, NY 14853
(607) 255-2275
ejm2@cornell.edu

Darryl McCrae
Iama Horsham
22 Hamilton St.
Horsham, Victoria
Australia, 3400
61 0343825411

Barry McFadden
St. Clair Agri-Services, Ltd.
Route 4
Wallaceburg, Ontario
Canada, N8A 4L1
(519) 627-6089
bmcfaddn@mnsi.net

Mark McGovern
Ag. & Ag-Food Canada
K.W. Neatby Bldng, 960 Carling
Ottawa, Ontairo
Canada, K1A 0C6
(613) 759-1529
mcgovernm@em.agr.ca

Donald McGrath
Iboco
7081 Hwy 71 N.E.
Willmar, MN 56201
(320) 235-9529

Tom McGraw
M.I.S.S.
5220 Co. Rd. 11
Buffalo Lake, MN
(320) 833-2200

Neil McLaughlin
Agric. & Agri-Food Canada
Ecorc, Bldg. 74 CEF
Ottawa, Ontario
Canada, K2K 1W2
(613) 759-1534
mclaughlinn@em.agr.ca

Ed McNamara
Goodhue County Soil & Water
104 E. 3rd Ave.
Box 335
Goodhue, MN 55027
(612) 923-5286

Ron McNeil
Landwise, Inc.
#214-905-1 Ave. S. Suite 407
Lethbridge, Alberta
Canada, T1J 4M7
(403) 320-0407

Jonathan Medway
Charles Sturt University
Box 588
Wagga Wagga, New South Wales
Australia, 2650
61 02 69 332177
jmedway@csu.adu.au

Eric Meek
UAP Richter Area Ag
Box 257
Carlinville, IL 62626
(212) 854-4408
ericrmeek@hotmail.com

Rob Meyer
Red Top Farms

Norman Mieth
Terra Industries
Box 385
709 Centennial Rd.
Wayne, NE 68787
(402) 375-3170
nmieth@kdsi.net

Ron Milby
Growmark, Inc.
1701 Towanda Ave.
Bloomington, IL 61701
(309) 557-6251

Gaines Miles
Purdue University
1146 Ag. & Biological Eng.
West Lafayette, IN 47907-1146
(765) 494-1223

Bruce Miller
Andrews Maple Hill Farms
13804 W. Co. Rd. A
Evansville, WI 53536

Robert Miller
Colorado State University
812 Cambridge Drive
Fort Collins, CO 80525
(970) 493-4382
robm846@aol.com

Montie Milner
MCS Consulting Services
Box 602
Elburn, IL 60119
(815) 827-3108
mmilner@casecorp.com

Jose Paulo Molin
ESALQ/USP
Av Padua Dias 11
Piracicaba, SP
Brazil, 13418-900
(019) 429-4165
jpmolin@carpa.ciagri.usp.br

Steven Moore
Dean Lee Research Station
8105 E. Campus Ave.
Alexandria, LA 71302
(318) 473-6520
smoorealex@aol.com

Susan Moran
USDA-ARS-SWRC
2000 E. Allen Rd.
Tucson, AZ 85719
(520) 670-6380 ex171
moran@tucson.ars.ag.gov

Alan Moritz
Cenex/Land O'Lakes
Agronomy Co
201 E. Main St.
Aspinwall, IA 51432-0008
(712) 653-3441
aspinwal@pionet.net

Russ Morman
Ag Leader Technology
1203A Airport Rd.
Ames, IA 50010

Phil Morris
KVH Industries, Inc.
50 Enterprise Center
Middletown, RI 02842
(401) 847-3327

Dave Mortensen
University of Nebraska

Ty Morton
USDA-ARS-NPA
AERC-CSU
Fort Collins, CO 80523
(970) 491-8511
ty@lily.aerc.colostate.edu

Alan Moulin
Agriculture & Agri-Food Canada
Box 1000A, RR3
Brandon, Manitoba
Canada, R7A 5Y3
(204) 726-7650
moulin@em.agr.ca

David Mulla
Precision Agriculture Center

Phil Mullins
Western Farm Service
509 W. Webster
Suite 201
Stockton, CA 95203
(209) 547-2637

Deena Murphy
South Dakota State University
Box 2207A
Brookings, SD 57007
(605) 688-4755
murphydl@hotmail.com

Brad Murray
Cenex Land O'Lakes
4584 Hwy. 80, S.
Platteville, WI 53818
(608) 348-9500
skidoobrad@aol.com

Tim Murray
Case/Concord, Inc.
3000 7th Ave., N.
Fargo, ND 58102
(701) 277-0383
tmurray@casecorp.com

Lance Murrell
Erny's Fertilizer

T. Scott Murrell
Potash & Phospate Institute
14030 Norway St., N.W.
Andover, MN 55304
(612) 755-3444 fx 755-1119
smurrell@ppi-far.com

Emerson Nafziger
University of Illinois
1102 S. Goodwin
Urbana, IL 61801
(217) 333-4424
ednaf@uiuc.edu

Mustapha Naimi
University of Minnesota

Craig Nelson
Mycogen Seeds
1340 Corporate Center Curve
Eagan, MN 55121-1233
(612) 405-5975
nelson@mycogen.com

Teri Nelson
NEOWA FS, Inc.
1877 110th St.
Hazleton, IA 50641

Dave Nerpal
Ag-Chem Equipment Co.
5720 Smetana Dr.
Minnetonka, MN 55343

Amnon Nevo
Volcani Center
p.o.b.
Israel
nevo@agri.gov.il

Jason Newton
Terra PCS
5050 N. 900 E.
Hope, IN 47246
(812) 546-5481
Newtonj@hsonline.net

Charles Nichols
SPOT Image Corporation
107 S. Pionte
Edwardsville, IL 62025
(618) 656-7460 fx 656-4190

Donald Nielsen
University of California
1004 Pine Lane
Davis, CA 95616
(530) 753-5760

Gerald Nielsen
Montana State University
827 Leon Johnson Hall
Bozeman, MT 59717
(406) 994-5075
nielsen@montana.edu

Norm Nielsen
Trimble Navigation Limited
9390 Bond St.
Suite 102
Overland Park, KS 66214
(913) 495-2700

Kazuya Niigawa
Ag. Coop. Zen-Noh
Higashi Yahata SF1, Kanagawa
Japan
001181 463 22 7703
p8904027@2k.zennoh.or.jp

Kosuke Noborio
Iowa State University
Dept. of Agronomy
Agronomy Hall
Ames, IA 50011-1010
(515) 294-7856
noboriok@iastate.edu

Sheilah Nolan
Alberta Agriculture
#206, 7000 113th St.
Edmonton, Alberta
Canada, T6H 5TH
(403) 427-3719
snolan@ulysses.sis.ualberta.ca

Michel Nolin
Agriculture & Agri Food Canada
350 Franquet St. Enterance 20
Sainte-Foy, Quebec
Canada, G1P 4P3
(418) 648-7730
nolinm@em.agr.ca

Peter Nowak
University of Wisconsin
1450 Linden Dr.
Madison, WI 53706
(608) 265-3581

John Nowatzki
North Dakota State Universsity
1221 Albrecht Blvd.
Box 5626
Fargo, ND 58105
(701) 231-8213
jnowatzt@ndsuext.nodak.edu

Bruce Nowlin
Blue Earth Agronomics
Route 2, Box 230
Lake Crystal, MN 56055
(507) 947-3362
bluesoil@mnic.net

Anthony Nugteren
University of Minnesota

Scott Nusbaum
Farm Work's Software
Box 250
Hamilton, IN 46742

Don O'Neall
Linx Systems
3001 1/2 Gill St
Bloomington, IL 61704

Larry Oldham
Mississippi State University
Box 9555
Ms. State, MS 39762
(601) 325-2311
loldham@onyx.msstate.edu

Alan Olness
USDA-Ag. Research Service
803 Iowa Ave.
Morris, MN 56267
(320) 589-3411 ex131
aolness@mail.mrsars.usda.gov

Shawn Olsen
ESRI
380 New York St.
Redlands, CA 92373

Joan Olson
Joan Olson Communications
8432 Rich Ave.
Bloomington, MN 55437
(612) 896-1937
joanols@ix.netcom.com

Kent Olson
University of Minnesota
Applied Economics
1994 Buford Ave.
St. Paul, MN 55108
(612) 625-7723
kolson@dept.agecon.umn.edu

Anders Olsson
Ostergotlands lans hushallning
Box 275
581 02 Linkoping
Sweeden
4613129520
Anders.olsson@hs-e.hush.se

Ron Ondrejka
Textron Systems
301 E. Carlisle Ave.
Whitefish Bay, WI 53217
(414) 964-7963

John Oolman
Agri-Growth, Inc.
Route 1, Box 180
Hollandale, MN 56045
(507) 889-4371
info@agrigrowth.com

Caleb Oriade
University of Arkansas
221 Agriculture Building
Fayetteville, AR 72701
(501) 575-6839
coriade@comp.uark.edu

Dennis Oser
MO Tech
PO Box 142
Bunceton, MO 65237
(660) 427-5215

Lee Ott
Omnisar, Inc.
8200 Westglen
Houston, TX 77063
(710) 785-5850
lott@omnistar.com

Mark Otto
Agri-Business Consultants
2720 Alpha Access
Lansing, MI 48910
(517) 482-7506
otto.abc@ix.netcom.com

Esther Owubah
Purdue University
1150 Lilly Hall
West Layfette, IN 47907-1150
(765) 494-0796
eowubah@dept.agry.purdue.edu

Emilio Oyarzabal
Monsanto
9739 Colby Ave.
Olive, IA 50325
(515) 331-7009

Doug Page
Profit-Trac Soil Sampeling
Box 519
Dodge Center, MN 55927
(507) 374-2326 fx374-6328

Shauna Page
Resource21
7257 S. Tuscon Way
Suite 200
Englewood, CO 80112

Carlos Paglis
Michigan State University
Crop & Soil Services
East Lansing, MI 48824
(514) 353-4705
paglisca@pilot.msu.edu

Terry Panbecker
Ag-Chem Eqipment Co.
5720 Smetana Dr.
Minnetonka, MN 55343

Bernard Panneton
Agriculture Canada
430 Boul Gouin
St. Jean-Sur-Richelieu, Quebec
Canada, J3B 3E6
(514) 346-4494 ex205
pannetonb@em.agr.ca

Kerstin Panten
Inst. of Plant Nutr. & Soil Sc
Bundesallee 50
Braunschweig
Germany
49 531 596332
panten@pb.fal.de

Karrie Papini
Spectrum Technologies, Inc.
23839 W. Andrew Rd.
Plainfield, IL 60544

Sid Parks
Growmark, Inc.
1701 Towanda AVe.
Bloomington, IL 61701
(309) 557-6250

Mike Pastushak
PCI Geomatics Group
50 W. Wilmot St.
Richmond Hill, Ontairo
Canada, L4B 1M5
(905) 764-0614
pastushate@pcigeomatics.com

Dennis Paulinski
Lockhead Martin
Box 64525
St. Paul, MN 55164
(612) 456-2332
dennis.w.paulinski@lmco.com

Bernie Paulson
McPherson Crop Management

Susan Pautzke
RDI Technologies, Inc.
300 Hwy. 23
Suite One
Spicer, MN 56288

Kyle Payne
UAP/Richter
1208 Cambridge
Rantoul, IL 61866
(217) 892-8593
kpayne@pdnt.com

Joel Paz
Iowa State University
124 Davidson Hall
Ames, IA 50011
(515) 294-7350
jpaz@iastate.edu

Sergio Miranda Paz
University of Sao Paulo
Rua Jesuino Arruda, 187
Apto 52, Sao Paulo
Brazil 04532-918
55 11 8185271
empaz@lag.pcs.usp.br

Dan Pennock
University of Saskatchewan
Department of Soil Science
Saskatoon, Saskatchewan
Canada S7H 5A8
(306) 966-6852
pennock@sisc.usask.ca

Mark Perger
Cargill, Inc.
3030 Lexington Ave.
Suite 700
Eagan, MN 55121-2268
(651) 406-5113
mark.perger@cargill.com

Calvin Perry
University of Georgia
Box 748
Tifton, GA 31793
(912) 386-3377
perrycd@tifton.cpes.peachnet.edu

Carleton Perry
St. Paul Software

Gary Petersen
Penn State University 116
ASI Building
University Park, PA 16802
(814) 865-1540
gwp2@psu.edu

Lynn Petersen
Agrium U.S. Inc.
601 W. Riverside Ave.
Spokane, WA 99220-2540
(509) 838-4600
lpeterse@agrium.com

Perry Petersen
Precision Ag Consultant
522 silver lane
sergeant bluff, IA 51054
(712) 943-3745
perryp@aol.com

Todd Peterson
Pioneer Hi-Bred Int. Inc.
7100 N.W. 62nd Ave.
Box 1150
Johnston, IA 50131-1150
(515) 334-6707
petersonta@phibred.com

Stuart Pettygrove
University of California-Davis
Lawr, One Shields Ave.
Davis, CA 95616-8627
gspettygrove@ucdvis.edu

Betsy Pfister
Red Hen Systems, Inc.
800 Stockton Ave. Unit 2
Ft. Collins, CO 80524
(970) 493-3952
betsy@redhensystems.com

Giovanni Piccinni
Texas Agric. Exp. Station
2301 Experiment Station Rd.
Bushland, TX 79012
(806) 354-5804
g-piccinni@tamu.edu

Francis Pierce
Michigan State University
Room 564 PSSB
E. Lansing, MI 48824
(517) 355-6892
piercef@pilot.msu.edu

M. De Almeida Pierossi
Copersucar Technology Center
Box 162
Piracicaba, SP
Brazil, 13400-970
55 19 429-8114
marcelo@azul.ctc.com

Tom Pistorius
Trimble Naviation Limited
9290 Bond St. Suite 102
Overland Park, KS 66214
(913) 495-2700

Richard Plant
University of California
Dept of Agronomy & Range Sci.
One Shields Ave.
Davin, CA 95616
(530) 752-1705
replant@ucdavis.edu

Stuart Pocknee
NESPAL/University of Georgia
P.O. Box 748
Tifton, GA 31793
(912) 386-7057
spocknee@tifton.cpes.peachnet.edu

John Pointon
Omnistar, Inc.
8200 W. Glen
Houston, TX 77063
(713) 785-5850 fx 785-5164
dgps@omnistar.com

Pascal Pollet
K. U. LEUVEN-ESAT-PSI
Kardinaal Mercierlaan 94
Heverlee
Belgium, B 300
3216321708
pascal.pollet@esat.kuleuven.ac.

Paul Porter
University of Minnesota
Box 428
Lamberton, MN 56152
(507) 752-7372
pporter@tc.umn.edu

Scott Porter
Midwest Technologies, Inc.
2733 E. Ash St.
Springfield, IL 62703
(217) 753-8424 fx 753-8426

John Potts
United States Sugar Corp.
P. O. Drawer 1207
Clewiston, FL 33440
(941) 902-2260
jpottsky@juno.com

J. Power
6120 S. 30th St.
Lincoln, NE 68516
(402) 423-6685
jp32349@navix.net

Larisa Pozdnyakova
University of Wyoming
Box 3354
Laramie, WY 82071
(307) 766-5082
pozd@uwyo.edu

Aziz Rahman
University of MN - Crookston
106 Tudor Ct
Crookston, MN 56716
(218) 281-8103
arahman@mail.crk.umn.edu

Gary Ransom
Agri Northwest
Box 2308
Tri-Cities, WA 99302
(509) 735-6461
garyransm@agrinorthwest.com

John Rathjen
Highway Equipment Company
616 D Ave., N.W.
Cedar Rapids, IA 52405
(319) 363-8281 fx 363-8284

Nancy Read
MMCD
2099 University Ave West
St Paul, MN 55104
(651) 643-8386
nancread@visi.com

Natalie Rector
MSU Extension
315 W. Green St.
Marshall, MI 49068
(616) 781-0785
rector@msue.msu.edu

Dr. Chandra Reddy
Alabama A & M University
Huntsville, AL 35762
(265) 858-4191
aamcxr01@asnaam.aamu.edu

Patrick Reeg
NEOWA FS, Inc.
1877 110th St.
Hazleton, IA 50641
(319) 636-2071
neowafs@sbtek.net

Paul Reep
Milestone Technology
1970 Belmont
Bozeman, MT 83404
(208) 528-8110
miletk@ida.net

Brian Rees
MFA
465 W. Marion
Marshall, MO 65340
(660) 886-3377
mfamarshall@dido.com

Harold Reetz
Potash & Phosphate Institute
111 E. Washington St.
Monticello, IL 61856-1640
(217) 762-2074
hreetz@ppi-far.org

Stewart Reeve
Precision Ag Illustrated
930 Kehrs Mill Rd.
Ballwin, MO 63011
(314) 527-4001
stu@precisionag.com

Corey Reiff
American Prarie, Inc.
422 4th St. S.E.
Medford, MN 55049
(320) 234-6580
aprairie@hhutchtel.net

William Reinert
Precision Farming Enerprises
2850 Spafford
Davis, CA 95616
(530) 758-1946
pfe@davis.com

Andrew Rekow
Stanford University
GP-B Hansen Labs
Stanford, CA 94305-4085
(650) 725-5769
arekow@leland.stanford.edu

Darryl Rester
La Coop. Extension Service
PO Box 25100
Baton Rouge, LA 70894-5100
(504) 788-2229

Bill Reynolds
ESRI
380 New York St.
Redlands, CA 92373

Marty Richards
Highway Equipment Company
616 D Ave., N.W.
Cedar Rapids, IA 52405

Ron Riffey
AGRIS Corporation
300 Grimes Bridge Rd.
Roswell, GA 30075

E. Timesa Rigby-Williams
Iowa State University
PO Box 1168
Ames, IA 50014

Chad Ringenberg
Agridata, Inc.
4300 Dartmouth Dr.
Box 8372
Grand Forks, ND 58202-8372
(701) 777-6547
agridata@theguest.net

Dennis Risinger
Agribank
375 Jackson
St Paul, MN 55101

Pierre Robert
University of Minnesota

Terry Roberts
Potash & Phosphate Institute
Suite 704 cn Tower, Midtown Pl
Saskatoon, Saskachewan
Canada, S7K 1J5
(306) 652-3535
troberts@ppi-ppic.org

Kevin Robertson
Vansco Electronics, Ltd.
1305 Clarence Ave.
Winnipeg, Manitoba
Canada, R3T 1T4
(204) 452-6776
krobertson@vansco.mb.ca

Robert Rogers
ERIM International, Inc.
Box 134008
Ann Arbor, MI 48113-4008
(734) 994-1200 ex3382
rrogers@erim-int.com

Glaucio Roloff
Univ. Federal Do Parana
Ecorc, Central Exper. Farm
Bldg. 74, Ottawa, Ontario
Canada, K1A 0C6
(613) 759-1535
roloffg@em.agr.ca

A. Romanelli
ESRI
380 New York St.
Redlands, CA 92373

A. J. Romanelli
GIS Solutions, Inc.
2612 Farragut Dr.
Springfield, IL 62704
(217) 546-3839
gisspi@gis-solutions.com

Tom Rooney
Trimble Navigation Limited
9290 Bond St.
Suite 102
Overland Park, KS 66214
(913) 495-2700

Martin Rosek
Terra Industries, Inc.
Box 218
Oakley, MI 48649
(517) 845-2200
mrosek@centuryinter.net

Carl Rosen
University of Minnesota
Dept. of Soil, Water & Climate
St. Paul, MN 55108
(612) 625-8114
crosen@soils.umn.edu

Barry Rosof
Agrium, Inc.
10333 Southport Rd. S.W.
Calgary, Alberta
Canada T2T 1N1
(403) 258-8377
brosof@agrium.com

Kenton Ross
Purdue University
Department of Agronomy
West Layfette, IN 47907-1150
(765) 494-6247
kenton@purdue.edu

Kevin Royal
Micro Images, Inc.
201 N. 8th St.
Lincoln, NE 58508
(402) 477-9554 fx 477-9559
info@microimages.com

Eugene Roytburg
CASE Corporation
7 S. 600 County Line Rd.
Burr Ridge, IL 60521
(630) 887-3903
eroytburg@casecorp.com

W William Rudolph
Midwest Technologies Inc
2733 E Ash St
Springfield, IL 62703
(217) 753-9424
wrudolph@mid-tech.com

Jim Ruen
Edge Communcationa
107 Deer Ridge Rd.
Lanesboro, MN 55949
(507) 467-7770
edgecom@means.net

Martin Rund
Tyler Industries, Inc.
801 Westwood Dr.
Catlin, IL 61817
(217) 427-5736
mbenrund@aol.com

Quentin Rund
Potash & Phosphate Institute
111 E. Washington St.
Monticello, IL 61856-1640
(217) 762-2074
grund@ppi-far.org

Ed Runge
Texas A & M University
Soil & Crop Sciences Dept.
College Station, TX 77843-2474
(409) 845-3041
e-runge@tamu.edu

Charlie Rush
Texas Agric. Exp. Station
2301 Eperiment Station Rd.
Bushland, TX 79012
(806) 354-5804
cm-rush@tamu.edu

John Russnogle
Soybean Digest
5797 S. E. Pine Lane
Holt, MO 64048
(816) 320-3198
scfi@aol.com

Richard Rust
University of Minnesota
Dept. of Soil, Water, & Climate
St. Paul, MN 55108

Robert Ruwoldt
Glenvale Farms
RMB 558 Murtoa
Horsham, Victoria
Australia 3390
61 0353852297
rruwoldt@netconnect.com.au

Phil Rzewnicki
Ohio State University
2001 Fyffe Court
Columbis, OH 43210
(614) 292-0117
rzewnichi.1@osu.edu

E. John Sadler
USDA-ARS
2611 W. Lucas St.
Florence, SC 29501
(843) 669-5203
sadler@florence.ars.usda.gov

Keith Sagehorn
Paula Carper Crop Ins.
Box 461
Holyoke, CO 80734
(970) 854-4596
pcarper@ria.net

Antonio Saraiva
University of Sao Paulo
Caira Postal 61548
Sao Paul
Brazil, 05424-970
55 11 818 5366
amsaraiv@usp.br

Philippe Sarrazin
ENESAD
26 Bd Pelit Jean
Dijon
France, 21000
33 380 772754
p.sarrazin@enesad.fr

Sergio Sartori
Maquinas Agricolas Jacto S.A.
Rua Dr. Luiz Miranda
Pompeia, SP
Brasil, 17580-000
(14) 452-1811
andreia jacto.com.br

Brenda Sawyer
Landwise, Inc.
Box 6097
Innisfail, Alberta
Canada, T4G 1S7
(403) 227-1526
bsawyer@telusplanet.net

Mitch Schefcik
Trimble Navigation, Limited
9290 Bond St.
Suite 102
Overland Park, KS 66214

Van Schelhaamer
MN

James Schepers
USDA ARS, Univ. of Nebraska
119 Keim Hall, East Campus
Lincoln, NE 68583-0915
(402) 472-1513
jscheper@unlinfo.unl.edu

Berlie Schmidt
USDA-CSREES
808 Aerospace Center
901 D St., S.W.
Washington, DC 20250-2210
(202) 401-4504
bschmidt@reeusda.gov

Thomas Schmidt
YMI Square
MN

Thomas Schmidt
GPSplus GimGH
Bergstrasses
Bisburg
Germany, D-82239
9 8624 829935

Bill Schoenecker
Omni Advertising, Inc.
640 Gun Club Rd.
Eagan, MN 55123
(612) 423-3422
omniu@spacestar.com

Paul Schrimpf
Farm Chemicals Magazine
37733 Euclid Ave.
Willoughby, Ohio 44094
(440) 942-2000

Chris Schroeder
AEC/Centrec Consulting Group
3 college park court
savoy, MN
rcs@centrec.com

Ronald Schuler
University of Wisconson
460 Henry Mall
Biol. Sys. Engr.
Madison, WI 53706
(608) 262-0613
rschuler@facstaff.wisc.edu

Gene Schultz
Crop Production Services
37733 Euclid Ave.
Willoughby, OH 44094
(440) 942-2000
anne_hill@masterpubl.com

Don Schuster
Natural Resource Conserv. Ser.
601 Business Loop 70 W.
Suite 250
Columbia, MO 65203
(573) 876-0900

Reinhart Schwaiberger
Agri Con
IM Wiesengrund 4
Jahna, Saxonia
Germany, 04749
004987061638
reinhart.schwaiberger@bnla.baynet.de

Doug Schwenk
SST Development Group, Inc.
824 N. Country Club Rd.
Stillwater, OK 74075

Mike Seal
Inst. for Tech. Develpoment
Bldg. 1103
Suite 118
Stennis Sp. Ctr, MS 39509
(228) 688-2509
mseal@iftd.org

Steve Seamon
Top Soil Precision Ag
1537 Margaret Ave., S.E.
Grand Rapids, MI 49507
(616) 475-5519
sseam@worldnet.att.net

Stephen Searcy
Agri. Eng. Texas A&M Univ.
104 AERL Route 4A-West
Campus
College Station, TX 77843
(409) 845-3668
s-searcy@tamu.edu

Clark Seavert
Oregon State University
2990 Experiment Station Dr.
Hood River, OR 97031
(541) 386-3343
clark.seauert@orst.edu

Edwin Seim
Cal Poly State University
292 Charles Dr.
San Luis Obispo, CA 93401
(612) 544-9461

Brian Selle
UAP/Richter
224 Woodland Trail
Hannibal, MO 63401
(217) 248-5588
uapr@nemonet.com

Mark Sellers
Pacific Crest Corp.
990 Richard Ave.
Suite 110
Santa Clara, CA 95050
(408) 653-2070
msellers@paccrst.com

Bruce Senst
Cenex/Land O'Lakes
468 Hillscourte, N.
Roseville, MN 55113
(651) 451-5375
bsens@cnxlol.com

Francis Sevila
Montpellier Ag. UNL/ENSAM
2 Place Viala-UFR GEGR
Montpellier
France, 34060
334996124221
sevila@ensam.inra.fr

John Shanahan
Colorado State University
Dept. of Crop & Soil Sci.
Ft. Collins, CO 80523
(970) 491-1920
jshanaha@lamar.colostate.edu

Kent Shannon
University of Missouri
238 Ag. Eng. Bldng.
Columbai, MO 65211
(573) 884-2267
kshannon@showme.missouri.edu

Ian Shaw
ESRI
380 New York St.
Redlands, CA 92373

Ian Shaw
Linnet Geometrics Int'l Inc.
1600-444 St. Mary Ave.
Winnipeg, Manitoba
Canada, R3C 3T1
(204) 957-7566
ishaw@linnet.ca

Van Shelhamer
Mont. State Univ., Ag Educ.
126 Cheever Hall
Bozeman, MO 59717
(406) 994-3693
uadcs@montana.edu

Jim Shelton
Cenex Land O'Lakes
Box 26
Juda, WI 53550
(608) 934-5215
a36james@aol.com

Hitoshi Shinjo
Kyoto University
Lab of Soils
Kyoto
Japan, 606-8502
81 75 753 6101
shinhit@kais.kyoto-u.ac.jp

David Shoup
ESRI
380 New York St.
Redlands, CA 92373

Karen Shrider
Gempler's, Inc.
Box 270
Mt. Horeb, WI 53572
(608) 437-4883 fx 437-5383

Sid Siefken
Trimble Navigation Limited
9290 Bond St.
Suite 102
Overland Park, KS 66214
(913) 495-2700

Martin Sihombing
Precision Agriculture Center

Arnie Sinclair
Ag-Chem Eqipment Co.
5720 Semtana Dr.
Minnetonka, MN 55343

Raymond Sinclair
USDA-NRCS
Federal Bldg.
Room 152
Lincoln, NE 68508
(402) 437-5699
rsinclair@nssc.nrcs.usda.gov

Cory Sinn
United Ag. Tech
Box 331
Trimont, MN 56176
(507) 639-6441
agtech@frontiernet.net

Cris Skonard
University of Nebraska
153 L.W. Chase Hall
Lincoln, NE 68583
(402) 472-6340
bsen174@unlum.unl.edu

Andrey Skotnikov
Case Technology Center
7 S. 600 County Line Rd.
Burr Ridge, IL 60521-6975
(630) 887-3011
askotnikov@casecorp.com

Glen Slater
University of Nebraska
Box 66
Clay Center, NE 68933
(402) 762-3535
scrc024@unlum.unl.edu

Kory Smith
Space Imaging
12076 Grant St.
Thornton, CO 80241

Oliver Smith
John Deere Precision Farming
501 River Dr.
Moline, IL 61265
(309) 765-7138
ou90a99@deere.com

Roger Smith
Colorado State University
A.E.R.C. Foothills
Fort Collins, CO 80523-1325
(970) 491-8263
roger@lily.saerc.colostate.edu

Wayne Smith
John Deere Precision Farming
501 River Dr.
Moline, IL 61265
dh32618@deere.com

Carol Snyder
Farmer's Software Association
800 Stockton Ave., #2
Ft. Collins, CO 80524
(970) 493-1722 fx 493-3938
info@farmsoft.com

Troy Snyder
Farm Works Software
Box 250
Hamilton, IN 46742
(219) 488-3388 fx 488-3737
farmwork@farmworks.com

Mike Solohub
University of Saskatchewan
Saskatoon, Saskatchewan
Canada, S7N 5A8
(306) 966-4291

Jeff Speckman
Cenex/Land O'Lakes
Box 64089, MS333
St. Paul, MN 55164
(612) 451-5518
jspec@cnxlol.com

Jerry Speir
Oklahoma State University
RT2 Box 265
Stillwater, OK 74074
(405) 624-2706
speir@okstate.edu

Monty Spencer
A.S. Wilcox & Sons Co.
Union Rd. RD3
Pukekohe
New Zeland
006492386010
monty@aswilcox.co.nz

Dick Spiltz
Helena Chemical
6075 Poplar Ave Ste 500
Memphis, TN

Ancha Srinivasan
Regional Science Institute
4-13, Kita 24 Nishi 2, Kita-ku
Sapporo
Japan, 001-0024
81 11 717 6660
ancha@vvt.co.jp

John Stafford
Silsoe Research Institute
Wrest Park, Silsoe
United Kingdom, MK45 4HS
44 1525 860000
john.stafford@bbsrc.ac.uk

Scott Staggenborg
Kansas State University
1515 College Ave.
Manhattan, KS 66502
(785) 532-2277
sstaggen@oz.oznet.ksu.edu

Tim Starick
Iama Horsham
22 Hamilton St.
Horsham, Victoria
Australia, 3400
61 0353825411

Friedrich Sterlemann
GEO TEC electronics GmbH
AmSoldnermoos 17
Hallbergmoos, Bavaria
Germany, 85399
49 811 6009922

Rob Stermitz
Montana State University
334 Leon Johnson Hall
Box 173120
Bozeman, MT 59717-3120
(406) 994-6034
nielsen@montana.edu

Gene Stevens
University of Missouri
Box 160
Portageville, MO 63873
(573) 379-5431
stevensg@ext.missouri.edu

Brian Steward
University of Illinois
Dept. of Ag. Engineering
1304 W. Pennsylvania Ave.
Urbana, IL 61801
(217) 333-9414
bsteward@uiuc.edu

Chris Stickler
Terra Industries
600 4th St.
Souix City, IA 51101
(712) 233-6559
cstickler@terraindustries.com

Dan Stiffler
Resource21
7257 S. Tuscon Way
Suite 200
Englewood, CO 80112

David Stigberg
Ag Electronice Association
10 S. Riverside Plaza
Suit 1220
Chicago, IL 60606-3710
(312) 321-1470
d-stigberg@cecer.army.mil

Dick Stiltz
Helena Chemical
6075 Poplar Ave.
Memphis, TN 38119
(901) 537-7263

Michael Stolp
Cenex/Land O'Lakes
Box 64089
MS 370
St. Paul, MN 55164
(651) 451-4626

Jeff Stotts
Resource21
7257 S. Tuscon Way
Suite 200
Englewood, CO 80112

Bill Stouffer
Route 4
Box 1073
Napton, MO 65340
(660) 886-3185
bstouffer@dido.com

Rob Stouffer
Precison Insights
1408 N. E. Applewood Crt.
Lee's Summit, MO 64086
(816) 246-9056
stouffer@qni.com

Sarah Stratton
Iowa State University
215 Davidson Hall
Ames, IA 50011
(515) 294-3153
strats@aistate.edu

Cesar Strauss
University of Sao Paulo
Rua Ministro Americo Marco
Antoniq 308, Sao Paulo
Brazil, 05442-090
55 11 577 8599
cstrauss@iagusp.usp.br

Jay Stroh
Novartis Seeds
Route 2
Box 168
Underwood, MN
(218) 826-6380
jay.strah@novartis.seeds.com

Haiping Su
Cargill Research
2301 Crosby Rd.
Wayzata, MN 55391
(612) 242-2497
haiping.su@cargill.com

Kenneth Sudduth
University of Missouri
245 Agric. Engr. Bldg.
Columbia, MO 65211
(573) 882-4090
sudduthk@missouri.edu

George Sugai
University of Brasilia
SCLN-402 61B Lj 33 Asa Norte
Brasilia, DF
Brazil, 70-834-520
55 061 2731840
sugai@tecsoft.softex.br

Friedrich Suhr
Canadian Agra Farming, Inc.
Box 616
Kincardine, ON
Canada
(519) 396-5279
fsuhr@cafoods.com

Jim Sulecki
Farm Chemicals/Ag Consultant
37733 Euclid Ave.
Willoughby, OH 44094
(440) 942-2000
jim_sulecki@meisterpubl.com

Emily Swartz
Ohio State University
130 S. #rd St. #57
Cardington, OH 43315
(614) 292-2283
kaeding.2@osu.edu

Matt Swartz
Ohio State University
1495 W. Longiew Ave.
Mansfield, OH
(419) 747-8761
swantz.52@osu.edu

Scott Swinton
Michigan State University
306 Agriculture Hall
East Lansing, MI 48824-1039
(517) 353-7218
swintons@pilot.msu.edu

David Swisher
Univ. of Missouri-Columbia
159 Agriaulture Engr. Bldg.
Columbia, MO 65211
(573) 882-1146
c60918@showme.missouri.edu

Suzi Takis da Costa
University of Bracilia
SCLN-402 61 B LJ 33 Asa
Norte
Brazil, 70834-520
55 016 2731840
takis@goi.sol.com.br

Nadia Tarqulian
Purdue University
1150 Lilly Hall
West Layfette, IN 47907-1150
(765) 494-0796
ntargulian@bigfoot.com

Paul Taylor
Taylor Farms
1419 Baseline Rd.
Esmond, IL 60129
(815) 393-4441
rtaylor1@rochelle.net

Randy Taylor
Kansas State University
Extension Agricultural Eng.
237 Seaton Hall
Manhattan, KS 66506
(785) 532-5813
rtaylor@falcon.bae.ksu.edu

Tim Taylor
AGRIS Corporation
300 Grimes Bridge Rd.
Roswell, GA 30075

TBA TBA
Soybean Digest
7900 International Dr. #300
Minneapolis, MN 55425

Thomas Teuscher
Humboldt Universitatzw-Berlin
Invalideustrabe 4L
Berlin 10115
Germany

Joe Tevis
Ag Chem
5728 Smetana Dr.
Minnetonka, MN
(612) 945-5868
jtevis@agchem.com

Greg Thomason
Modern Agriculture
13741 E. Rice Pl.
Suite 200
Aurora, CO 80015
(303) 690-2242 ex208
gt@modernag.com

Wayne Thompson
Edaphos Limited

Willard Thompson
Rincon Publishing
Box 370
Carpinteria, CA 93014-0370
(805) 684-6581
willard@riconpublishing.com

Arthur Thorp
Howard University
CSTEA, 2216 6th St., N.W.
Suite 103
Washington, DC 20059
(252) 793-4428
methurma@unity.ncsu.edu

Mary Thurman
NC State University
Crop Science Dept.
Box 7620
Raleigh, NC 27695-7620

Mike Thurow
Spectrum Technologies, Inc.
23839 W. Andrew Rd.
Plainfield, IL 60544

Douglas Tiffany
University of Minnesota
316 D COB
1994 Buford Ave.
St. Paul, MN 55108
(612) 625-6715
dtiffany@dept.agecon.umn.edu

Dennis Timlin
USDA ARS
Bldg. 007 Room 008
10300 Baltimore Ave.
Beltsville, MD 20705
(301) 504-6255
dtimlin@asrr.arsusda.gov

Kim Tofin
PCI Geometrics Group
Gurdwara Rd., Suite 220
Nepean, ON
Canada, K2E 1A2

Marcin Topolewski
Lamb Weston Inc.
2005 Saint St.
Richland, WA 99352
(509) 375-5866
mtopolewski@lumb-
weston.com

Andre Torre-Neto
Embrapa
Rua XV De Novembro, 1452
Box 741, Sao Carlos, Sao Paulo
Brazil, 13560-970
55 16 274 2477
andre@cnpdia.embrapa.br

Bill Town
IAMA
Cooper St.
Dalby, Queensland
Australia, 4405
07 4662 1888
iamairri!icc.com.au

Pat Trail
Grower Service Corp.
221 W. Lake Lansing Rd.
Suite 102
E. Lansing, MI 48906
(517) 333-8788
grower@growers.net

Dean Tranel
Iowa State University
Ames, IA 50011
dmtranel@iastate.edu

Paul Trcka
Terra
109 N. Main St.
Box 99
Austin, MN 55912
(507) 433-0387

Christos Tsadilas
N.A.G.R.E.F.
1 Theufrastos St.
Larissa 41335
Greece

Steve Tupa
Ag-Chem Equipment Co.
5720 Semanta Dr.
Minnetonka, MN 55343

Shrini Upadhyaya
Univ. of California-Davis
Bio. & Ag. Eng. Dept.
Davis, CA 95616
(530) 752-8770
skupadhyaya@ucdavis.edu

B. J. Van Alphen
Wageningen Ag. University
Box 37
Wageningen
The Netherlands, 6700 AA
31 317 484410
jecoen.vanalphen@a10. beng.wau.nl

Harold van Es
Cornell University
162 Emerson Hall
Ithaca, NY 14853-1901
(607) 255-5629
hmv1@cornell.edu

Chris Van Kessel
Univer. of Califorina-Davis
One Shield Ave.
Davis, CA 95616
(530) 752-4377
cvankessel@ucdavis.edu

R. P. van Zuydam
IMAG-DLO Wageningen
Box 43
Wageningen
The Netherlands, 6700 AA
31 317 476300
r.p.vanzuydam@imag.dlo.nl

Marc Vanacht
Metz Vanacht Co.
2245 Kehrs Ridge Dr.
St. Louis, MO 63005

Richard Vanden Heuvel
Cenex/Land O'Lakes
805 Lund St., N.
Hudson, WI 54016
(612) 451-4352
rvande@cnxlol.com

Gary Vander Ploeg
Position Inc
6815 E 40 St SE
Calgary, Alberta
Canada, T2C2W7
(403) 720-0277
gvanderplokg@positioninc

Mireille Vanoverstraeten
Kemira S.A.
Av Einstein 11
1300 Wavre
Belgium, 1300
32 10 232824
mireille.vandverstraeten@kemira.com

Jim Veneziano
Trimble Navigation Limited
9090 Bond St.
Suite 102
Overland Park, KS 66214
(913) 495-2700

Sharon Vennix
Grower Service Corp.
221 W. Lake Lansing Rd.
Suite 102
E. Lansing, MI 48906
(517) 333-8788
grower@growers.net

Maurice Vitosh
Michigan State University
Crop & Soi Science Dept.
East Lansing, MI 48824-1325
(517) 355-0212
vitosh@msue.msu.edu

Matthew Volkmann
USDA-Ag Research Service
University of Missouri
158A Ag. Eng. Bldng.
Coulmbia, MO 65211
(573) 882-1138
mvolkmann@showme.missouri.edu

Peter von Bertoldi
University of Guelph
131 Richards Bldg.
Guelph, Ontairo
Canada, N1G 2W1
(519) 824-4120
pvonbert@lrs.voquelph.ca

Kurt Wael
Agrosat Systems Aps
Stubbenoebingvej 41
DK-4840 NR. Alslev
Denmark, DK
45 70203311
jyttesen@posts.tele.dk

David Wagner
Pennsylvania State University
225 Agricultural/Eng. Bldng.
University Park, PA
(814) 865-3722
dgw4@psu.edu

Gary Wagner
A.W.G. Farms, Inc.
Route 1
Crookston, MN 56716
(218) 281-5120
gwagner@mtrade.org

John Wagner
Terra PCS
14526 W. 300 S.
Francesville, IN 47946
(219) 567-2984
csawagner@skyenet.net

Rich Wagoner
Ntl. Ctr. for Atmospheric Rsh.
Box 3000
Boulder, CO 80307
(303) 497-8404
wagoner@ucar.edu

David Waits
SST Development Group, Inc.
824 N. Country Club Rd.
Stillwater, OK 74075

Sharon Walker
Iowa State University
Ames, IA 50011
sqwalker@iastate.edu

Steve Walker
Terra Industries
6739 Guion Rd.
Indianapolis, IN 46268
(317) 328-1000
swalker@terraindustries.com

Ken Walter
University of Minnesota
AES Rosemount
1605 160th St. W.
Rosemount, MN 55608
(612) 423-2455
walte011@maroon.yc.umn.edu

Shawn Walter
Colorado State University
800 Oval Dr.
Fort Collins, CO 80523
(970) 491-5675
smwater@lamar.colostate.edu

Daniel Walters
University of Nebraska-Lincoln
Dept. of Agronomy
261 Plant Science Bldg.
Lincoln, NE 68583-0915
(402) 472-1506
agro084@unlvm.unl.edu

Dennis Walvoort
Dept. of Soil Science & Geol.
Duiendaal 10, 6701 AR
Wageningen, Gelderland
Netherlands, BODENUMMER
313174723g
walvoort@sc.dlo.nl

Xing-guang Wang
Zheng Zhou University-China
Zheng Zhou, Henan Province
PR China, 450052
0371 5999601
wxg@mail.zzu.edu.cn

Bob Wanzel
Precision Ag Illustrated
930 Kehrs Mill Rd.
#314
Ballwin, MO 63011
(314) 527-4001
bob@precisionag.com

Shane Ward
University College of Dublin
Earlsfort Terrace
Dublin
Ireland, 2
353 1 7067351
shane.ward@ucd.ie

Mitt Wardlaw
Mississippi State University
Dept. of Plant & Soil Sciences
Box 4555, 117 Doran Hall
Mississippi St., MS 39762
(601) 325-3601
mcw7@ra.msstate.edu

John Waterer
Cargill
300-240 Graham Ave.
Winnipeg, Manitoba
Canada
(204) 947-6385
john_waterer@cargill.com

Brad Watkins
USDA-ARS-NRI
BARC-West Bldng.
Room 8
Beltsville, MD 20705
(301) 504-5824
bwatkins@asrr.arsusda.gov

Hal Watkins
Heartland Coop.
2829 Westown Pky.
Suite 350
West Des Moines, IA 50266
(515) 225-1334
halwatki@netins.net

Reid Watkins
ESRI
380 New York St.
Redlands, CA 92373

Corey Weddle
Ag Leader Technology
2202 S. Riverside Dr.
Ames, IA 50010
(515) 232-5363
cweddle@agleader.com

Ken Wedig
John Deere Precision Farming
501 River Dr.
Moline, IL 61265
(309) 765-7267
kw67592@deere.com

Jun Wei
Pioneer Hi-Bred Intern'l.
7100 N. W. 62nd St.
Johnson, IA 50131
(515) 334-6704
weij@phibred.com

Randy Weisz
North Carolina State Univ.
Box 7620
Raleigh, NC 27695-7620
(919) 515-5824
raudy_weisz@ncsu.edu

Monte Weller
Case Corp AFS
1100 South County Line Rd.
Maple Park, IL 60151
(815) 827-3108
mweller@casecorp.com

Ole Wendroth
ZALF
Eberswalder Str. 84
Muencheberg
Germany, 15374
49 33432 70327
owendroth@zalf.de

Terry West
Purdue University
Lily Hall
West Lafayette, IN 47907-1150
(765) 494-4799
twest@dept.agry.purdue.edu

Glenn Wheelock
Trimble Navigation Limited
9290 Bond St.
Suite 102
Overland Park, KS 66214

Brett Whelan
University of Sydney
Aust. Ctr. for Prec. Ag.
McMillan Bldg., AOS,Sydney NSW,
Australia, 2006
61 2 9351 2947
b.whelan@acss.usyd.edu.au

David Whetter
University of Manitoba
70 Rice Rd.
Winnipeg, Manitoba
Canada, R2T 4M2
(204) 727-0045
dwhetter@mb.smpatico.ca

Eric White
Trimble Navigation Limited
9290 Bond St.
Suite 102
Overland Park, KS 66214
(913) 495-2700

Jared White
John Deere Precision Farming
501 River Dr.
Moline, IL 61265

Susan White
Iowa State University
2210 Agronomy Hall
Ames, IA 50011
(515) 294-5429
swhite@iastate.edu

Charles Whitlock
Analytical Services & Material
1 Enterprise Pkwy
Suite 3000
Hampton, VA 23666-5845
(757) 827-4882
c.h.whitlock@larc.nasa.gov

Matt Wiebers
Cargill, Inc.
12800 Whitewater Dr.
Minnetonka, MN 55343

Dan Wiens
Centrak
210 E. Kiowa
Fort Morgan, CO 80701
(970) 542-1850
agritrak@henge.com

Richard Wiese
Crosroads Ag Enterprises
8801 N. W. 112th
Malcolm, NE 68402
(402) 796-2159
dw53434@navix.net

JoAnn Wilcox
Successful Farming
1716 Locust St.
Des Moins, IA 50309
(515) 284-3280
jwilcox@map.com

Lori Wiles
USDA-ARS-NPA
AERC-CSU
Fort Collins, CO 80523
(970) 491-8511
lori@lily.aerc.colostate.edu

Brendan Williams
Pathway Precision Farming
2 Natrimak Rd.
Horsham, Victoria
Australia, 3400
613 53811935
lucyalic@comcirc.com.au

Martin Williams
University of Nebraska
Dept. of Agronomy
362 Plant Science
Lincoln, NE 68583-0915
(402) 427-9563
marty@mortsun.unl.edu

Terry Williams
TeeJet Technologies
7200 France S.
Ste. 128
Edina, MN 55435
(612) 831-1559 fx 831-2571
williate@spray.com

Ray Williford
USDA-ARS
Box 36
Stoneville, MS
(601) 686-5352
rwillifo@ag.gov

Patrick Willis
Resource21

Robert Wilson
Nevada Cooperative Extension
995 Campton St
Ely, MN 89301
(702) 289-4459
rwilson@agntl.ag.unr.edu

Tom Winner

Mr. Wirsching
Satcon System GmbH
Overtheres, Bavaria
Germany, 97531
9521 7072 fx 9521 1350
satcon.com@t0online.de

David Wittry
Iowa State University
16224 140th St.
Brede, IA 51436
(515) 294-9865
dwittry@iastate.edu

Maurice Wolcott
LSU Ag. Center
8150 E. Campus
Alexandrea, LA 71302
(318) 473-6520

Nyle Wollenhaupt
Ag-Chem Equipment Co., Inc.
5720 Smetana Dr., Ste 100
Minnetonka, MN 55343

Gavin Wood
Cranfield University
Barton Rd.
Silsoe, Bedforshire
United Kingdom, MK45 4DT
44 0 1525 863063
g.a.wood@cranfield.ac.uk

Matthew Wood
Intermap Technologies
2 Gurdwara Rd., Suite 200
Nepedn, Ontario
Canada, V2E 1A2
(613) 226-5442
mood@intermap.ca

Randy Wood
Case Corporation
700 State. St.
Racine, WI 53404
(414) 636-0188
rwood@casecorp.com

Jim Woosley
Ag Leaders Technology
1203A Airport Rd.
Ames, IA 50010
(515) 232-5363
jwoosley@agleader.com

David Wright
Red Hen Systems, Inc.
800 Stockton Ave. Unit 2
Ft. Collins, CO 80524
(970) 493-3952
david@redhensystems.com

Herbert Wright
Geophyta, Inc.
2685 C.R. 254
Vickery, OH 43464
(419) 547-8538
geophyta@nwonline.net

Nathan Wright
Geophyta, Inc.
2685 C.R. 254
Vickery, OH 43464
(419) 547-8538
geophyta@nwonline.net

Anne Wrona
National Cotton Council
Box 820285
Memphis, TN 38182-0285
(901) 274-9030
awrona@cotton.org

Terry Wyciskalla
Southern Illinois University
Carbondale, IL 62901-4415
(618) 453-2496
wycist@siu.edu

Dawn Wyse-Pester
Colorado State University
Department of Bio Ag. &
Pest management
Fort Collins, CO 80523
(970) 491-5667
dawnwp@lamar.colostate.edu

Ric Yacouby
Emerge
900 Technology Park Dr.
Bldg. 8
Billerica, MA 01821
(978) 262-0671
rsyacouby@emerge.wsicorp.com

Chenghai Yang
USDA - ARS
2413 E. Hwy. 83
Weslaco, TX 78596
(956) 969-4824
yang@pop.tamu.edu

Russell Yost
Unversity of Hawaii
1910 E. West Rd.
Honolulu, HI 96822
(808) 556-7066
rsyost@hawaii.edu

Jan Yttesen
Datalogisk Aps
Stubbekoebingvej 41
Dk-4840 NR. Alslev
Denmark, DK
45 70203311
jyttesen@post3.tele.dk

Richard Zemmelink
Cargill Ltd.
39109 Talbot Line RR7
St. Thomas, Ontairo
Canada,

Minghua Zhang
Zeneca Ag Products
1200 S. 47th St.
Richmond, CA 94804
(510) 231-1481
minghhua.zhang@agna.zeneca.com

Suling Zhao
Resource21
7257 S. Tuscon Way
Suite 200
Englewood, CO 80112

Everett Zillinger
The Fertilizer Institute
501 2nd St. N.E.
Washington, DC 20002

Demetrio Zourarakis
257 Paddock Dr.
Versailles, KY 40383
(502) 564-3080
dpzour01@mis.net

Philippe Zwaenepoel
Cemagref
les Palaquins
Montoldre
France, 03150
33 4 70 457365
philippe.zwaenepoel@cemagref.fr

APPENDIX – III EXHIBITOR LIST

Ag-Chem Equipment Co., Inc.
5720 Smetana Drive, Ste 100
Minnetonka, MN 55343

Nyle Wollenhaupt
(612) 945-5874; f 933-7432

Ag Consultant Magazine
37733 Euclid Ave
Willoughby, OH 44094

Jim Sulecki
(440) 942-2000; f 942-0662
jim_sulecki@meisterpubl.com

Ag Leader Technology
1203A Airport Rd
Ames, IA 50010

Russ Morman
(515) 232-5363; f 232-3595
rmorman@agleader.com

Agri ImaGIS
5174 30th St. NE
Maddock, ND 58348

Lanny Faleide
(701) 438-2242; f 438-2870
agriimagis@stellarnet.com

Agris Corporation
300 Grimes Bridge Rd.
Roswell, GA 30075

Ted Macy
(770) 236-5161; f 643-2239

Concord Environmental Equipment
RR1, Box 78
Hawley, MN 56549

Virg Mahlum
Gary Johnson
(218) 937-5100; f 937-5101
cee@rrnet.com

Dickey-john Corporation
5200 Dickey-john Road
Auburn, IL 62615

George Suaffer
Dennis Luhn
(217) 438-2211; f 438-6012

Emerge
900 Technology Park Dr.
Bldg 8
Billerica, MA 01821

Kim Rauenzahn
(978) 262-0668; f 262-0700
karauenzahn@emerge.wsicorp.com

Erdas, Inc.
2801 Buford Highway
Atlanta, GA 30032

Connie R. Phillips
Bruce Epler
Ben Drake
(404) 248-9000; f 248-9400
www.erdas.com

ESRI
380 New York St.
Redlands, CA 92373

Max Crandall
(909) 793-2853; f 307-3072
mcrandall@esri.com

Farmer's Software Assoc.
800 Stockton Ave. #2
Ft. Collins, CO 80524

Carol Snyder
(970) 493-1722; f 493-3938
info@farmsoft.com

Farm Works Software
P.O. Box 250
Hamilton, IN 46742

Troy Snyder
(219) 488-3388; f 488-3737
farmwork@farmworks.com

FieldWorker Products Ltd.
H6-1477 Bayview Ave.
Toronto, ON M4G 3B2
Canada

Sandy Browne
(416) 483-3485; f 483-7069
sandy@fieldworker.com

Gempler's, Inc.
P.O. Box 270
Mt. Horeb, WI 53572

Thomas Green
Karen Shvider
(608) 437-4883; f 437-5383

Highway Equipment Co
616 D. Ave. NW
Cedar Rapids, IA 52405

John Rathjen
(319) 363-8281; f 363-8284

Kluwer Academic Publishers
101 Philip Drive
Norwell, MA 02061

Teresa Esser
(781) 871-6600; f 871-7507
tesser@wkap.com

KVH Industries, Inc.
50 Enterprise Center
Middletown, RI 02842

John Hadam
(401) 847-3327; f 845-8190
jhadam@kvh.com

Phil Morris
(401) 847-3327; f 845-8190
pmorris@kvh.com

Leica Geosystems Inc.
3155 Medlock Bridge Rd
Norcross, GA 30071

Debbie Kirkland
(770) 447-6361; f 447-0710
debra.kirkland@leica-lsg.com

MapInfo Corporation
One Global View
Troy, NY 12180

Kevin Hickey
(518) 285-7019; f 285-6070
kevin.hickey@mapinfo.com

MicroImages, Inc.
201 North 8th Street
Lincoln, NE 68508

Kevin Royal
(402) 477-9554; f 477-9559
info@microimages.com

Mid-Tech
2733 E. Ash St.
Springfield, IL 62703

Scott Porter
(217) 753-8424; f 753-8426
aheemer@mid-tech.com

Noetix Research, Inc.
#403-265 Carling Ave.
Ottawa, ON K1S 2E1
Canada

Tom Hirose
(613) 236-1555; f 236-1870
info@noetix.on.ca

Omnistar, Inc.
8200 Westglen
Houston, TX 77063

John Pointon
(713) 785-5850; f 785-5164
dgps@omnistar.com

Pacific Meridian Resources
421 SW 6th Ave. Suite 850
Portland, OR 97204

Cindy Bell
(503) 228-8708; f 228-8751
portland@pacificmeridian.com

Potash & Phosphate Institute
14030 Norway St. NW
Andover, MN 55304

T. Scott Murrell
(612) 755-3444; f 755-1119
smurrell@ppi-far.com

Precision Ag Illustrated
15444 Clayton Rd. Ste. 314
Ballwin, MO 63011

Dan Bellanser
(314) 527-4001; f 527-4120
dan@precisionag.com

Precision Agriculture Center
1991 Upper Buford Circle
439 Borlaug Hall
St. Paul, MN 55108

Leslie Legg
(612) 624-4224; f 625-2208
llegg@soils.umn.edu

Raven Industries
P.O. Box 5107
Sioux Falls, SD 57117

Dugan Petersen
(605) 357-0484; f 331-0426
dpetersn@ideasign.com

Rawson Control Systems, Inc.
116 2nd Street S.E.
Oelwein, IA 50662

Sharon Gauquie
(319) 283-2225; f 283-1360
rawson@rawsoncontrol.com

RDI Technologies, Inc.
300 Highway 23, Suite One
Spicer, MN 56288

Ryan Allen
(320) 796-0019; f 796-0048

RESOURCE21
7257 South Tucson Way
Englewood, CO 80112

Dennis Dunivan
(303) 749-3200; f 749-3298
ddunivan@resource21.com

Satcon System GmbH
Obertheres
Bavaria, 97531
Germany

Mr. Dűrrstein
9521-7072; f 9521-1350
satcon.com@t-online.de

Spectrum Technologies, Inc.
23839 W. Andrew Rd.
Plainfield, IL 60544

Mike Thurow
(800) 248-8873; f 436-4460
specmeters@aol.com

SST Development Group, Inc.
824 N. Country Club Rd.
Stillwater, OK 74075

David Waits
(405) 377-5334; f 377-5746
sst@sstdevgroup.com

Starlink Incorporated
6400 Highway 290 East
Suite 202
Austin, TX 78723

Tamara Lee
(512) 454-5511; f 454-5570
info@starlinkdgps.com

TeeJet Technologies
7200 France So. – Suite 128
Edina, MN 55435

Terry Williams
(612) 831-1559; f 831-2571
williate@spray.com

Trimble Navigation Limited
9290 Bond Street Suite 102
Overland Park, KS 66214

Sue Huber
(913) 495-2700; f 495-2750

Tyler Industries, Inc.
P.O. Box 249
Benson, MN 56215

Martin Rund
(217) 427-5736; f 427-5737
mbenrund@aol.com

Veris Technologies
601 N. Broadway
Salina, KS 67401

Eric Lund
(785) 825-1978; f 825-2097